CW01667810

# An Analog Electronics Companion

Intended for electronicists and for engineers and scientists who have to get involved in circuit design. From mature designers who may have forgotten techniques or who trained before the days of circuit simulation, to neophytes seeking to widen their horizon. A series of largely self-contained essays that may be dipped into at any point. Encourages analysis of circuits supported by simulation to confirm and extend understanding. Includes a CD containing the student version of the powerful and fully functional simulation package PSpice, limited only in the size of circuit it will accept. Includes ready to run schematics for all the applications discussed.

The first three parts of the book cover the maths and physics needed to understand circuit function, analysis and design. Part 4 examines some basic circuit components with reference to their physical and simulation properties. The final and largest part examines the design and function of a wide range of analog systems, using simulation to demonstrate the relationship between analysis and performance. Many references to the literature and to the web are provided throughout to allow ready access to further information.

**Dr Scott Hamilton** was senior lecturer at Manchester University in the UK, where, in addition to his research activities, he spent more than 30 years teaching physics and electronic circuit design to undergraduate and graduate students. He is now retired.

# An Analog Electronics Companion

## Basic Circuit Design for Engineers and Scientists and Introduction to SPICE Simulation

Scott Hamilton
Department of Physics and Astronomy
University of Manchester

CAMBRIDGE
UNIVERSITY PRESS

CAMBRIDGE UNIVERSITY PRESS
Cambridge, New York, Melbourne, Madrid, Cape Town, Singapore, São Paulo

Cambridge University Press
The Edinburgh Building, Cambridge CB2 8RU, UK

Published in the United States of America by Cambridge University Press, New York

www.cambridge.org
Information on this title: www.cambridge.org/9780521798389

© Cambridge University Press 2003

This publication is in copyright. Subject to statutory exception
and to the provisions of relevant collective licensing agreements,
no reproduction of any part may take place without
the written permission of Cambridge University Press.

First published 2003
Reprinted with corrections 2007

Printed in the United Kingdom at the University Press, Cambridge

*A catalogue record for this publication is available from the British Library*

*Library of Congress Cataloguing in Publication data*
Hamilton, Scott, 1932–
An analog electronics companion / Scott Hamilton.
p. cm.
Includes bibliographical references and index.
ISBN 0 521 79838 8
1. Electronics. I. Title.
TK7816 .H35 2001
621.381–dc21 2001043081

ISBN-13 978-0-521-79838-9 hardback
ISBN-13 978-0-521-68780-5 paperback

Cambridge University Press has no responsibility for the persistence or accuracy
of URLs for external or third-party Internet websites referred to in this publication,
and does not guarantee that any content on such websites is, or will
remain, accurate or appropriate.

To Laura,

for her encouragement, support and tolerance

The sum which two married people owe to one another defies calculation. It is an infinite debt, which can only be discharged through all eternity.

JOHANNES WOLFGANG GOETHE (1749–1832)

# Contents

## Part 5 SPICE circuit applications

# Preface

This morning my newspaper contained the obituary of Sir Alan Hodgkin, Nobel Laureate together with Andrew Huxley and John Eccles, in physiology and medicine. What has this to do with our interest in electronics? Well, the prize was awarded for the elucidation of the mechanism of the propagation of electrical impulses along neural fibres, the basis of our own internal electronic system. Before the understanding of these mechanisms the position in this field would have been akin to that of Oersted, Faraday, Ohm, Ampère and Maxwell in trying to understand conduction, since at that time the electron was unknown and, for example, they imagined that an electric field somehow created charge to allow for conduction. The intimate interactions between electrical and biochemical activity are nowadays of great interest with the possibility of constructing electronic–biological systems. The consequences of Maxwell's synthesis of electricity, magnetism and light and the prediction of electromagnetic waves have been immense. Almost everything we shall discuss hinges ultimately on his discoveries and they still stand as a pinnacle in the field of physics:

> If you have bought one of those T-shirts with Maxwell's equations on the front, you may have to worry about its going out of style, but not about its becoming false. We will go on teaching Maxwellian electrodynamics as long as there are scientists.
>
> Steven Weinberg, Physicist, Nobel Laureate (New York Review of books)

Why another book on electronics? Twenty years ago I wrote one prompted by the burgeoning production of integrated circuits and the thought that many, like myself, who were not electronic engineers nevertheless needed to be able to develop circuits for our own use. It has been said that the threat of imminent execution concentrates the mind wonderfully. On a very much lower level, having to present a coherent account of all the various topics one thinks important is a very searching test of one's understanding as one finds all the holes in one's knowledge, so there has been a considerable learning process to go through. Age does have some advantages, one of them being the time to think more deeply, to understand more clearly and to fill in the missing bits. As Kierkegaard observed, 'Life can only be understood backwards; but it must be lived forwards'.

The world's first synthesized drug dates back to Hippocrates, who reported that a willow bark extract relieved aches. On August 10th 1897, Felix Hoffman, a chemist for Bayer, created a synthetic version, now called aspirin. This has alleviated many headaches and one may hope that this book may also.

Now, with the centenary of the discovery of the electron by J. J. Thomson, also in 1897, an essential ingredient in this subject, it seemed appropriate to consider an update. However, in the interval one has become older and more experienced even if not wiser, and one's point of view as to what is important has necessarily changed. This is not a textbook; it is not a serial and coherent treatment of electronics topics; it is essentially a prompt and a companion and a reminder of many things and techniques you may not know or have forgotten (at least those which I find useful and have not forgotten). Experienced engineers will possibly find little new of interest, but I aim more, as before, at the many on the margins or who have not had access or time to learn all they would have liked to. The other very significant development in more recent times has been the use of computer techniques for the simulation of electronic circuits. This has so enabled the analysis of systems compared with what before could reasonably be done by hand, as to make non-access to such a facility a severe disadvantage. Since the software can run successfully on PCs, and the cost is not prohibitive, it allows almost all to make use of it. Again, the book is not intended as a manual on how to use SPICE, the generic form of the software, but rather some indication of how it may be used to help in the design process or to test your more extreme 'what if' ideas. There are of course limitations in relating simulation to actual circuits, but it is my experience that with a little thought in making allowance for 'parasitic' effects it is possible to achieve very close correspondence.

It is also my belief that some knowledge of the physical basis and origins of electronics is rather beneficial. The book is divided into five parts. First is a résumé of the general mathematical tools that may be useful in analysing systems. The treatment is on a fairly straightforward level with the emphasis on usability rather than any mathematical rigour – we assume that the mathematicians have sorted out all the difficulties. Second is an introduction to some of the physics underlying the many techniques used. Most electronics books simply state various laws, e.g. Kirchhoff's laws, without any indication as to their origin or validity. With electronics extending now into far-flung areas where applicability may be questioned, it is as well to have some grasp of the underlying physics. Third is a discussion of a number of circuit analysis techniques of general applicability. Fourth is a consideration of some of the most common circuit elements, in particular their deviations from the ideal in so far as this may affect the models that you may use for simulation. Fifth is the use of simulation as an aid to design. I use a particular flavour of SPICE, PSpice, but I hope that most of what is done will be applicable to all the other flavours. There are many, sometimes very large, texts on the format and use of SPICE which should be consulted to learn the techniques. It is slightly unfortunate that most of these date from the time when it was necessary for you to write out the appropriate netlist for the circuit but it is probably useful to know the general techniques and rules involved so you can understand the limitations and sort out some of the difficulties that can arise. Versions of SPICE are now screen based in that you need only draw the circuit schematic and the software will create the required netlist, which saves considerable time and avoids your entry errors. In this part I have chosen a range of circuits many of which have arisen in my own work (and which I hope means that I have had to think much more about and understand better) and which illustrate techniques that could be of use in more general circuits. It is the techniques rather than the applications that are important. Where appropriate I have sought to compare direct analysis, sometimes using Mathcad©, with SPICE results. The aim is also to encourage you to experiment in more unusual ways: modifications are quickly made, signals which in actual circuits may be difficult to measure are readily observed, and if you make a mistake and pass a current of 1000A you do not get a large puff of smoke! Some circuits can take a lot of simulation time so use a fast PC if you can. Nowadays the cost of a high-speed computer is insignificant compared with the time you will save.

Included with the book is a student, or demonstration, copy of the simulation software PSpice on CD-ROM. This is provided by arrangement with Cadence and I must acknowledge their generous assistance and collaboration in this matter. The software includes most of the full version but is limited as to the size of circuits that may be run and the libraries of models that are so essential. The circuits in the book which have been simulated are included on the CD and most, but not all, will run under the demo version of the software. Some additional libraries, made up for the purpose, are also included. The {circuit}.prb files, which determine the form of the simulation to be run and the output display, are also included to assist in the initial running of the circuits.

It will be evident from the book's contents that I do not subscribe to approaches that avoid the use of mathematics at almost any cost. Mathematics is the language of science and you place yourself at a considerable disadvantage if you cannot speak it competently. It provides the path to deeper understanding of how systems behave and, in particular, it allows you to make predictions. Design is in essence prediction since you are expecting the system to meet the requirements.

> Numbers count in every sense. If you know a thing by its quality, you know it only vaguely. If you know it by its quantity, you begin to know it deeply. You have access to power, and the understanding it provides. Being afraid of quantification is tantamount to disenfranchising yourself, giving up one of the most potent prospects for understanding and changing the world.
>
> Carl Sagan, physicist and astronomer

The application of mathematics should not put you off. Like everything else you will make many mistakes but practice is what is required and you can't get that if you never try.

Get it down. Take chances. It may be bad, but that's the only way you can do anything really good.

William Faulkner

Ever tried. Ever failed. No matter. Try again. Fail again. Fail better.

Samuel Becket

In the mathematical approaches, I have generally tried to give a fairly full account of the sums so that they may be more readily followed, and in many cases you can call on the power of SPICE to validate your conclusions. I have tried to relate the mathematics that has been included to the applications considered later but you should be aware that only a small, but significant, portion of the available techniques is included (a recent handbook runs to 2861 pages: Chen 1995).

I have sought to include a substantial number of references for all the topics referred to so that further information may be readily found. Some will be repetitive but this makes it more likely that you will be able to obtain access. The well-known semiconductor manufacturers provide many models for their products and these are generally accessible on websites if not included in your SPICE. The availability of good models is crucial to the process of simulation but it must be remembered that they are mostly functional rather than transistor level models and do not cover every aspect of the device. Some devices are too difficult to model satisfactorily, especially with acceptable simulation times, and some classes of device still appear to be unmodelled, but there is a great deal that can be achieved.

I hope of course that you will find at least something useful in these pages and that they may prompt you to further investigation. As to errors, I would be most grateful if you would bring these to my attention and I would be happy to discuss as far as I am able any matters that may be of mutual interest. My thanks to my present and past colleagues and to all the correspondents from whom I have received such willing help. In deference to market forces and to the entreaties of the publisher I have used analog rather than analogue both in the title and the text. My apologies to any readers affronted by this craven act.

Technical volumes are generally rather dour affairs with little recourse to levity. As the title of the present volume includes the term companion, as in bedside companion, I feel less constrained and have included a range of quotations, some directly relevant and others that I simply liked. The publisher protests that they may confuse the argument but I hope that they will somewhat lighten the approach.

During the writing of this book MicroSim were subsumed by Orcad and shortly afterwards both became part of Cadence. References should therefore be interpreted in the light of this and enquiries directed appropriately. May I acknowledge the considerable help provided by the above companies over the years and more particularly the assistance of Patrick Goss of MicroSim and Dennis Fitzpatrick of Cadence in dealing with my many queries and observations. The development of the PSpice simulations was primarily carried out using Version 8 of the software. To avoid possible additional errors, and to maintain close corespondence, it is this version that is provided on the CD. It should be noted that the latest issue is several versions ahead, which should be borne in mind if you migrate. The new versions are considerably enhanced but for the purposes of the present applications you are not at a disadvantage. The schematics from Version 8 must be 'imported' into the later versions with possibly some minor adjustments required.

**Scott Hamilton**
Department of Physics and Astronomy, University of Manchester
Manchester M13 9PL. 21st September 2000. Scott.Hamilton@manchester.ac.uk

# List of symbols and abbreviations

| | |
|---|---|
| a | prefix atto, $\times 10^{-18}$ |
| A | ampere, unit of current |
| $A$ | gain of amplifier |
| $A_0$ | zero frequency gain of amplifier |
| ABM | analog behavioural model |
| $B$ | susceptance |
| $\mathbf{B}$ | magnetic flux |
| $B, \mathscr{B}$ | bandwidth |
| $C$ | capacitance |
| C | coulomb, unit of charge |
| $c$ | speed of light, $2.998 \times 10^{8}$ m s$^{-1}$ |
| CCCS | current-controlled current source |
| CCVS | current-controlled voltage source |
| $C_{GD}$ | FET gate–drain capacity |
| $C_{GS}$ | FET gate–source capacity |
| $C_{iss}$ | FET common-source input capacity |
| $C_{oss}$ | FET common-source output capacity |
| $C_{rss}$ | FET common-source reverse transfer capacity |
| CMRR | common-mode rejection ratio |
| $D$ | diode |
| dB | decibel |
| d.c. | direct current or zero frequency (z.f.) |
| E | prefix exa, $\times 10^{18}$ |
| $\mathbf{E}$ | electric field |
| F | farad, unit of capacity |
| $f$ | frequency |
| f | prefix femto, $\times 10^{-15}$ |
| $f_c$ | corner frequency |
| $f_T$ | transition frequency |
| G | prefix giga, $\times 10^{9}$ |
| $G$ | conductance |
| $G$ | gain |
| $G$ | FET gate |
| $G(0)$ | gain at zero frequency |
| $G_0$ | gain at zero frequency |
| $G_{fs}$ | transconductance |
| $G_m$ | transconductance |
| $G(s)$ | gain at frequency $s$ |
| $G(\infty)$ | gain at infinite frequency |
| $G_\infty$ | gain at infinite frequency |
| $\mathbf{H}$ | magnetic field |
| H | henry, unit of inductance |
| $h$ | Planck constant, $6.63 \times 10^{-34}$ J s |
| $H(s)$ | transfer function |
| Hz | hertz, unit of frequency |
| $I_B$ | base current |
| $I_C$ | collector current |
| $I_{DSS}$ | zero-bias saturation current, gate tied to source |
| $I_E$ | emitter current |
| $I_{GSS}$ | gate leakage current, source tied to drain |

| | |
|---|---|
| J | joule, unit of energy |
| K | kelvin, unit of absolute temperature |
| $k_B$ | Boltzmann constant, $1.38 \times 10^{-23}$ J K$^{-1}$ |
| k | prefix kilo, $\times 10^3$ |
| $k$ | coupling factor |
| $L$ | Inductance |
| $L$ | load |
| $\mathcal{L}$ | Laplace transform operator |
| LHP | left half-plane (of complex plane) |
| $\boldsymbol{M}$ | magnetization |
| M | prefix mega, $\times 10^6$ |
| m | prefix milli, $\times 10^{-3}$ |
| m | metre, unit of length |
| $m_e$ | electron rest mass, $9.11 \times 10^{-31}$ kg |
| $\boldsymbol{N}$ | Poynting vector |
| $N$, $n$ | turns |
| n | prefix nano, $\times 10^{-9}$ |
| $n$ | refractive index |
| $N_A$ | Avogadro number, $6.02 \times 10^{23}$ mol$^{-1}$ |
| $P$ | power |
| p | prefix pico, $\times 10^{-12}$ |
| P | prefix peta, $\times 10^{15}$ |
| $Q$ | quality factor |
| $Q$ | charge |
| Q | transistor |
| $q_e$ | electronic charge, $1.602 \times 10^{-19}$ C |
| $R$ | resistor |
| $R_{DS(on)}$ | FET on resistance |
| RHP | right half-plane (of complex plane) |
| $S$ | source |
| $S$ | FET source |
| $S$ | signal |
| S | siemen, unit of conductance |
| s | second, unit of time |
| $s$ | complex frequency |
| SRF | self-resonant frequency |
| $T$ | time constant |
| $T$ | time interval or delay |
| T | prefix tera, $\times 10^{12}$ |
| T | degree absolute or kelvin |
| $t_p$ | pulse width |
| $t_r$ | risetime |
| $t_{rr}$ | reverse recovery time |
| V | volt, unit of potential |
| $V_B$ | base voltage |
| $V_{BE}$ | base–emitter voltage |
| $V_C$ | collector voltage |
| $VCC$ | supply voltage |
| VCCS | voltage-controlled current source |
| $V_{CE}$ | collector–emitter voltage |
| VCVS | voltage-controlled voltage source |
| $V_D$ | drain voltage |
| $V_{DS}$ | drain–source voltage |
| $V_E$ | emitter voltage |
| $V_G$ | gate voltage |
| $V_{GS}$ | gate–source voltage |
| $V_J$ | p-n junction voltage |
| $V_{oc}$ | open circuit voltage |
| $V_S$ | source voltage |
| $V_{th}$ | FET threshold voltage |
| W | watt, unit of power |
| $X$ | reactance |

| | |
|---|---|
| $Y$ | admittance |
| Y | prefix yotta, $\times 10^{24}$ |
| $Z$ | impedance |
| Z | prefix zetta, $\times 10^{21}$ |
| z.f. | zero frequency (d.c.) |
| $\alpha$ | coefficient of resistance |
| $\alpha_0$ | attenuation factor |
| $\beta$ | beta, feedback factor |
| $\beta$ | beta, transistor current gain |
| $\beta_0$ | d.c. current gain |
| $\chi_B$ | magnetic susceptibility |
| $\gamma$ | gamma |
| $\delta(t)$ | delta function |
| $\delta$ | skin depth |
| $\partial$ | partial differentiation |
| $\varepsilon$ | epsilon, permittivity |
| $\varepsilon_0$ | permittivity of free space, $8.85 \times 10^{-12}\ \mathrm{F\,m^{-1}}$ |
| $\gamma$ | magnetogyric ratio |
| $\Gamma$ | contour length |
| $\theta$ | theta, angle |
| $\lambda$ | lambda, wavelength |
| $\mu$ | mu, relative permeability |
| $\mu_0$ | permeability of free space, $4\pi \times 10^{-7}\ \mathrm{H\,m^{-1}}$ |
| μ | prefix micro, $\times 10^{-6}$ |
| $\nu$ | nu, frequency |
| $\Phi$ | magnetic flux |
| $\rho$ | rho, density, resistivity |
| $\sigma$ | sigma, conductivity |
| $\tau$ | tau, time constant |
| $\phi$ | phi, angle |
| $\psi$ | psi, angle |
| $\omega$ | omega, angular frequency |
| $\omega_c$ | corner angular frequency |
| $\omega_T$ | transition angular frequency |
| $\Omega$ | ohm, unit of resistance |
| $\bullet$ | vector dot product |
| $\times$ | vector cross product |
| $\propto$ | proportional to |
| $>$ | greater than |
| $\gg$ | much greater than |
| $\geq$ | greater than or equal to |
| $<$ | less than |
| $\ll$ | much less than |
| $\leq$ | less than or equal to |
| $=$ | equals |
| $\equiv$ | identically equal to |
| $\cong$ | approximately or very nearly equals |
| $\approx$ | of the order of |
| * | multiplication in SPICE expressions |
| $*$ | convolution symbol |
| / | division in SPICE expressions |
| $\langle\ \rangle$ | average value of |
| $\lvert\ \rvert$ | modulus or absolute value |
| $\parallel$ | parallel |
| $\Leftrightarrow$ | Fourier pair |
| $\partial$ | partial differential |
| $\delta$ | a small increment |
| $\nabla$ | del |
| $\Delta$ | a small change or increment |
| exp | exponential |
| $j$ | square root of $-1$ |
| $\mathscr{Im}$ | imaginary part of a complex number |
| $\mathscr{Re}$ | real part of a complex number |
| Ln | logarithm to base $e$ |
| Log | logarithm to base 10 |

# Part 1

# Mathematical techniques

Philosophy is written in this grand book – I mean the universe – which stands continually open to our gaze, but it cannot be understood unless one first learns to comprehend the language and interpret the characters in which it is written. It is written in the language of mathematics, and its characters are triangles, circles, and other geometrical figures, without which it is humanly impossible to understand a single word of it; without these one is wandering about in a dark labyrinth.

Galileo Galilei (1564–1642)

As indicated in the preface, this book is substantially about design and hence prediction. The tools that allow us to extrapolate to create a new design are an understanding of the physical characteristics and limitations of components, mathematical techniques that allow us to determine the values of components and responses to input signals, and of course as much experience as one can get. The latter of course includes making as much use as possible of the experience of others either by personal contact or by consulting the literature.

This part covers much of the basic mathematics that is generally found useful in analysing electronic circuits. There is a fairly widely propagated view that you can get by without much mathematical knowledge but I evidently do not subscribe to this. Many do indeed do very well without recourse to mathematics but they could do so much better with some knowledge, and this book is, in part, an attempt to persuade them to make the effort. We do not present a course on these techniques as that would expand the book far beyond an acceptable size, but rather provide an indication and reminder of what we think is important and useful. Much of the reluctance in this direction is possibly caused by the unattractiveness of heavy numerical computation but this is nowadays generally unnecessary since we have the assistance of many mathematical computational packages and, in our case, the enormous power and convenience of electronic simulation software. With the spread of the ubiquitous PC it is now uncommon for an electronicist to be without access to one.

When carrying out algebraic analysis it is all too easy to make mistakes and great care must be taken when writing out equations. It is often of assistance to check your units to see that they are consistent as this can often be of great use in catching errors. You also need to be prepared to make approximations as the equations for even quite simple circuits become more complex than can be analysed. SPICE can be of considerable assistance in that you may use it to determine at least approximate values for parameters that then allow you to determine the relative magnitudes of terms and hence which may be neglected without serious error. You can then check your final result against SPICE which is able to carry out the analysis without significant approximation. The benefit of the algebraic analysis is that it makes the function of each component evident and provides parameterized design formulae.

Though some of the topics may at first sight seem unexpected, I hope that as you progress through later sections you will come to appreciate their relevance. Some are treated in terms of simply a reminder and some are delved into in a little more detail. As far as possible references to further sources of information are provided.

# 1.1 Trigonometry

The power of instruction is seldom of much efficacy except in those happy dispositions where it is almost superfluous.

Edward Gibbon

It may seem unexpected to find a section on trigonometry, but in electronics you cannot get away from sine waves. The standard definitions of sine, cosine and tangent in terms of the ratio of the sides of a right-angled triangle are shown in Fig. 1.1.1 and Eq. (1.1.1).

For angle $\theta$ and referring to the sides of the triangle as opposite ($o$), adjacent ($a$) and hypotenuse ($h$) we have:

$$\sin\theta = \frac{o}{h}, \qquad \cos\theta = \frac{a}{h}, \qquad \tan\theta = \frac{o}{a} \tag{1.1.1}$$

A common way to represent a sinusoidal wave is to rotate the phasor OA around the origin O at the appropriate rate $\omega$ (in radians per second) and take the projection of OA as a function of time as shown in Fig. 1.1.2.

The corresponding projection along the $x$-axis will produce a cosine wave. This allows us to see the values of the functions at particular points, e.g. at $\omega t = \pi/2$, $\pi$, $3\pi/2$ and $2\pi$ as well as the signs in the four quadrants (Q). These are summarized in Table 1.1.1.

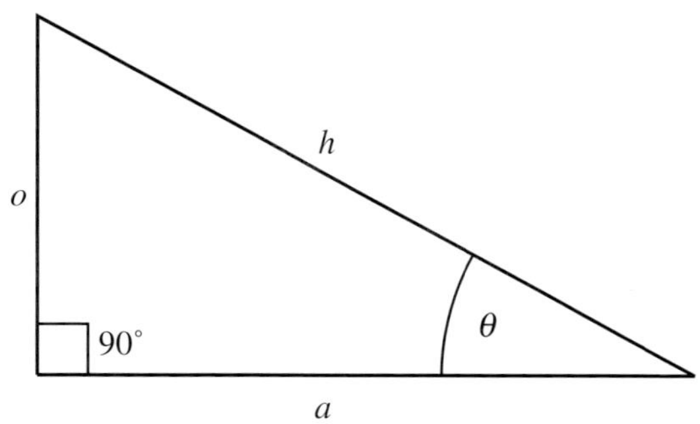


**Fig. 1.1.1** Right-angled triangle.

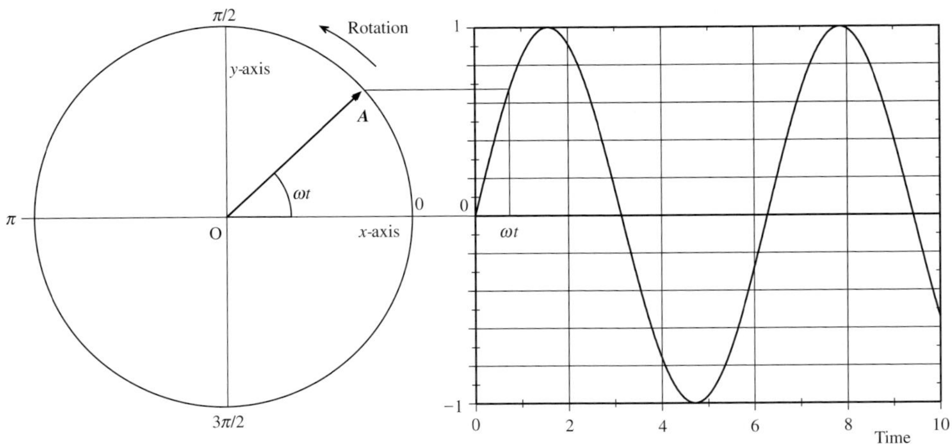


**Fig. 1.1.2** Projection of a rotating vector.

Table 1.1.1 *Values and signs of trigonometrical functions in the four quadrants*

| Angle | 0 | $\pi/2$ | $\pi$ | $3\pi/2$ | $2\pi$ | 1Q | 2Q | 3Q | 4Q |
|---|---|---|---|---|---|---|---|---|---|
| Sin | 0 | 1 | 0 | 1 | 0 | + | + | − | − |
| Cos | 1 | 0 | 1 | 0 | 1 | + | − | − | + |
| Tan | 0 | $\infty$ | 0 | $\infty$ | 0 | + | − | + | − |

Some useful relationships for various trigonometrical expressions are:

(a) $\sin(-\theta) = -\sin\theta$

(b) $\cos(-\theta) = \cos\theta$

(c) $\tan(-\theta) = -\tan\theta$

(d) $\cos(\theta+\phi) = \cos\theta\cdot\cos\phi - \sin\theta\cdot\sin\phi$

(e) $\sin(\theta+\phi) = \sin\theta\cdot\cos\phi + \cos\theta\cdot\sin\phi$

(f) $\sin\theta + \sin\phi = 2\sin\frac{1}{2}(\theta+\phi)\cos\frac{1}{2}(\theta-\phi)$

(g) $\cos\theta + \cos\phi = 2\cos\frac{1}{2}(\theta+\phi)\cos\frac{1}{2}(\theta-\phi)$

(h) $\sin\theta\cdot\cos\phi = \frac{1}{2}[\sin(\theta+\phi) + \sin(\theta-\phi)]$

(i) $\cos\theta\cdot\cos\phi = \frac{1}{2}[\cos(\theta+\phi) + \cos(\theta-\phi)]$

(j) $\sin^2\theta + \cos^2\theta = 1$

(k) $\cos 2\theta = \cos^2\theta - \sin^2\theta = 2\cos^2\theta - 1 = 1 - 2\sin^2\theta$ (1.1.2)

(l) $1 + \cos\theta = 2\cos^2(\theta/2)$

(m) $1 - \cos\theta = 2\sin^2(\theta/2)$

(n) $\sin 2\theta = 2\sin\theta\cos\theta$

(o) $\lim_{\theta \to 0} \frac{\sin \theta}{\theta} = \lim_{\theta \to 0} \frac{\tan \theta}{\theta} = 1$

(p) $\cos(\theta - \phi) = \cos\theta \cdot \cos\phi + \sin\theta \cdot \sin\phi$

(q) $\sin(\theta - \phi) = \sin\theta \cdot \cos\phi - \cos\theta \cdot \sin\phi$

(r) $\sin\theta - \sin\phi = 2\cos\tfrac{1}{2}(\theta + \phi)\,\sin\tfrac{1}{2}(\theta - \phi)$

(s) $\cos\theta - \cos\phi = 2\sin\tfrac{1}{2}(\phi + \theta)\,\sin\tfrac{1}{2}(\phi - \theta)$

(t) $\cos\theta \cdot \sin\phi = \tfrac{1}{2}[\sin(\theta + \phi) - \sin(\theta - \phi)]$

(u) $\sin\theta \cdot \sin\phi = \tfrac{1}{2}[\cos(\theta - \phi) - \cos(\theta + \phi)]$

We will have occasion to refer to some of these in other sections and it should be remembered that the complex exponential expressions (Section 1.5) are often easier to use.

## References and additional sources 1.1

Fink D., Christansen D. (1989): *Electronics Engineer's Handbook*, 3rd Edn, New York: McGraw-Hill. ISBN 0-07-020982-0

Korn G. A., Korn T. M. (1989): Mathematics, formula, definitions, and theorems used in electronics engineering. In Fink D., Christansen D. *Electronics Engineer's Handbook*, 3rd Edn, Section 2, New York: McGraw-Hill.

Lambourne R., Tinker M. (Eds) (2000): *Basic Mathematics for the Physical Sciences*, New York: John Wiley. ISBN 0-471-85207-4.

Langford-Smith F. (1954): *Radio Designer's Handbook*, London: Illife and Sons.

Terman F.E. (1950): *Radio Engineers' Handbook*, New York: McGraw-Hill.

# 1.2 Geometry

For geometry by itself is a rather heavy and clumsy machine. Remember its history, and how it went forward with great bounds when algebra came to its assistance. Later on, the assistant became the master.

Oliver Heaviside (1899): *Electromagnetic Theory*, April 10, Vol. II, p. 124

The relationship between equations and their geometric representation is outlined in Section 1.10. In various places we will need to make use of Cartesian (coordinate) geometry either to draw graphical responses or to determine various parameters from the graphs.

A commonly used representation of the first order, or single pole, response of an operational amplifier in terms of the zero frequency (z.f.) gain $A_0$ and the corner frequency $\omega_c = 1/T$ is given by:

$$A = \frac{A_0}{1 + sT} \tag{1.2.1}$$

which on the log–log scales usually used is as shown in Fig. 1.2.1.

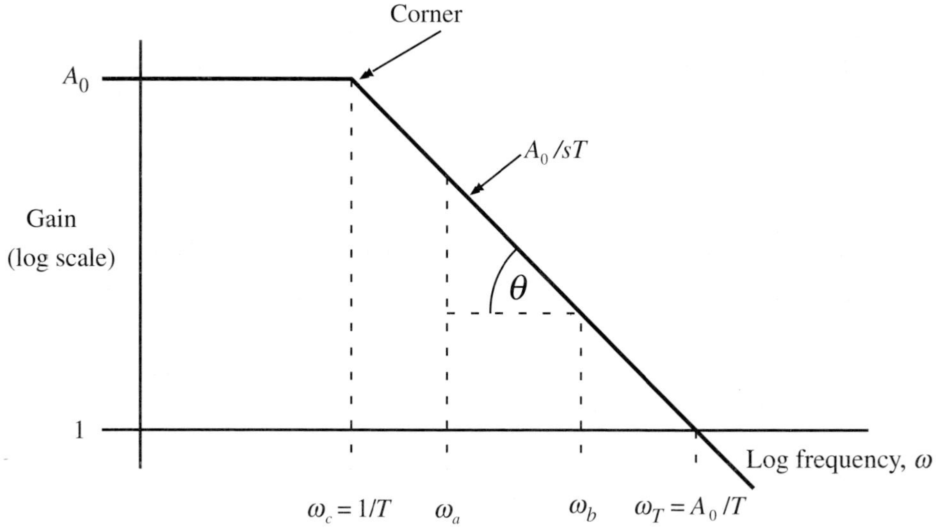


**Fig. 1.2.1** Operational amplifier response.

At low frequency $sT \ll 1$ so the gain is just $A_0$. At high frequency $sT \gg 1$ so the gain is $A_0/sT$ as shown. The graphs are slightly different from those with standard $x$- and $y$-axes, which are of course where $y=0$ and $x=0$, because here we have logarithmic scales. The traditional $x$- and $y$-axes would then be at $-\infty$, which is not useful. We can start by taking $G=1$ and $f=1$ for which the log values are zero ($\log 1=0$). You can of course draw your axes at any value you wish, but decade intervals are preferable. The slope of $A_0/sT$ can be determined as follows. Take two frequencies $\omega_a$ and $\omega_b$ as shown. The slope, which is the tangent of the angle $\theta$, will be given (allowing for the sense of the slope) by:

$$\tan\theta = \frac{\log\left(\dfrac{A_0}{\omega_a T}\right) - \log\left(\dfrac{A_0}{\omega_b T}\right)}{\log(\omega_b) - \log(\omega_a)}$$

$$= \frac{\log\left(\dfrac{\omega_b}{\omega_a}\right)}{\log\left(\dfrac{\omega_b}{\omega_a}\right)} \qquad (1.2.2)$$

$$= 1 \qquad \text{and} \qquad \theta = \tan^{-1}(1) = 45°$$

where we have used the fact that the difference of two logs is the log of the quotient (Section 1.4). Since the slope does not depend on $\omega$ then $A_0/sT$ must be a straight line on the log–log scales.

A point of interest on the amplifier response is the unity-gain (or transition) frequency $\omega_T$ as this defines the region of useful performance and is particularly relevant to considerations of stability. We need to find the value of $\omega$ for which $G=1$. Note that though we use the more general complex frequency $s$ we can simply substitute $\omega$ for $s$ since we are dealing with simple sine waves and are not concerned with phase since this is a graph of amplitude. Thus we have:

$$1 = \frac{A_0}{\omega_T T} \qquad \text{or} \qquad \omega_T = \frac{A_0}{T} \qquad (1.2.3)$$

and remember that $\omega$ is an angular frequency in rad s$^{-1}$ and $f=\omega/2\pi$ Hz. Say we now wish to draw the response for a differentiator which has $G=sR_fC_i$ (Section 5.6). The gain will be 1 when $\omega_D = 1/R_fC_i$ and the slope will be 45°. So fixing point $\omega_D$ and drawing a line at 45° will give the frequency response. To find where this line meets the open-loop response we have:

$$sR_fC_i = \frac{A_0}{sT} \qquad \text{or} \qquad s^2 = \frac{A_0}{R_fC_iT} \qquad \text{so} \qquad \omega_P = \left(\frac{A_0}{R_fC_iT}\right)^{\frac{1}{2}} \qquad (1.2.4)$$

which is shown in Fig. 1.2.2.

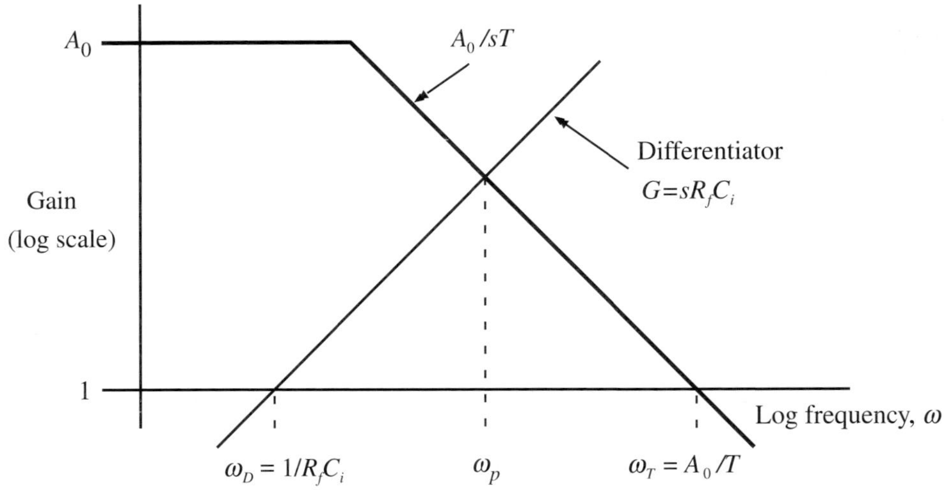


**Fig. 1.2.2** Geometry of differentiator frequency response.

## References and additional sources 1.2


Hambley A. R. (1994): *Electronics. A Top-Down Approach to Computer-Aided Circuit Design*, New York: Macmillan. ISBN 0-02-349335-6. See Appendix B.

Lambourne R., Tinker M. (Eds) (2000): *Basic Mathematics for the Physical Sciences*, New York: John Wiley. ISBN 0-471-85207-4.

Tobey G. E., Graeme J. G., Huelsman L. P. (1971): *Operational Amplifiers. Design and Applications*, New York: McGraw-Hill. Library of Congress Cat. No. 74-163297.

Van Valkenburg M. E. (1982): *Analog Filter Design*, New York: Holt, Rinehart and Winston. ISBN 0-03-059246-1, or 4-8338-0091-3 International Edn. See Chapter 3.

# 1.3 Series expansions

Prof. Klein distinguishes three main classes of mathematicians – the intuitionists, the formalists or algorithmists, and the logicians. Now it is intuition that is most useful in physical mathematics, for that means taking a broad view of a question, apart from the narrowness of special mathematics. For what a physicist wants is a good view of the physics itself in its mathematical relations, and it is quite a secondary matter to have logical demonstrations. The mutual consistency of results is more satisfying, and exceptional peculiarities are ignored. It is more useful than exact mathematics.

But when intuition breaks down, something more rudimentary must take its place. This is groping, and it is experimental work, with of course some induction and deduction going along with it. Now, having started on a physical foundation in the treatment of irrational operators, which was successful, in seeking for explanation of some results, I got beyond the physics altogether, and was left without any guidance save that of untrustworthy intuition in the region of pure quantity. But success may come by the study of failures. So I made a detailed study and close examination of some of the obscurities before alluded to, beginning with numerical groping. The result was to clear up most of the obscurities, correct the errors involved, and by their revision to obtain correct formulae and extend results considerably.

Oliver Heaviside (1899): *Electromagnetic Theory*, April 10, Vol. II, p. 460

Expansion of functions in terms of infinite (usually) series is often a convenient means of obtaining an approximation that is good enough for our purposes. In some cases it also allows us to obtain a relationship between apparently unconnected functions, and one in particular has been of immense importance in our and many other fields. We will list here some of the more useful expansions without derivation (Boas 1966):

$$\begin{aligned}
\sin\theta &= \theta - \frac{\theta^3}{3!} + \frac{\theta^5}{5!} - \frac{\theta^7}{7!} + \cdots \\
\cos\theta &= 1 - \frac{\theta^2}{2!} + \frac{\theta^4}{4!} - \frac{\theta^6}{6!} + \cdots \qquad (1.3.1)\\
e^x &= 1 + x + \frac{x^2}{2!} + \frac{x^3}{3!} + \frac{x^4}{4!} + \cdots
\end{aligned}$$

where

$$n! \equiv n(n-1)(n-2)(n-3)\cdots 1$$

and $n!$ is known as $n$ factorial. Note that, odd though it may seem, $0!=1$. The variable $\theta$ must be in radians.

The binomial series is given by:

$$(1+x)^n=1+nx+\frac{n(n-1)x^2}{2!}+\frac{n(n-1)(n-2)x^3}{3!}+\cdots \tag{1.3.2}$$

valid for $n$ positive or negative and $|x|<1$
which is most frequently used for $x\ll 1$ to give a convenient approximation:

$$(1+x)^n\cong 1+nx \tag{1.3.3}$$

The geometric series in $x$ has a sum $S_n$ to $n$ terms given by:

$$a+ax^2+ax^3+ax^4+\cdots+ax^n+\cdots+, \quad \text{with sum } S_n=\frac{a(1-x^n)}{1-x} \tag{1.3.4}$$

and for $|x|<1$ the sum for an infinite number of terms is:

$$S=\frac{a}{1-x} \tag{1.3.5}$$

In some circumstances, when we wish to find the value of some function as the variable goes to a limit, e.g. zero or infinity, we find that we land up with an indeterminate value such as $0/0$, $\infty/\infty$ or $0\times\infty$. In such circumstances, if there is a proper limit, it may be determined by examining how the function approaches the limit rather than what it appears to do if we just substitute the limiting value of the variable. As an example consider the function (Boas 1966, p. 27):

$$\lim_{x\to 0}\frac{1-e^x}{x}, \text{ which becomes } \frac{0}{0} \text{ for } x=0 \tag{1.3.6}$$

If we expand the exponential using Eq. (1.3.1), then remembering that $x$ is going to become very small:

$$\lim_{x\to 0}\frac{1-e^x}{x}=\lim_{x\to 0}\frac{1-\left(1+x+\dfrac{x^2}{2!}+\cdots\right)}{x}=\lim_{x\to 0}\left(-1-\frac{x}{2!}-\cdots\right)=-1 \tag{1.3.7}$$

Expansion in terms of a series, as in this case, is generally most useful for cases where $x\to 0$, since in the limit the series is reduced to the constant term. There is another approach, known as l'Hôpital's rule (or l'Hospital), which makes use of a Taylor series expansion in terms of derivatives. If the derivative of $f(x)$ is $f'(x)$, then:

$$\lim_{x\to 0}\frac{f(x)}{\phi(x)}=\lim_{x\to 0}\frac{f'(x)}{\phi'(x)} \tag{1.3.8}$$

but the ease of use depends on the complexity of the differentials. The Taylor series referred to allows expansion of a function $f(x)$ around a point $x=a$:

$$f(x)=f(a)+(x-a)f'(a)+\frac{(x-a)^2}{2!}f''(a)+\cdots+\frac{(x-a)^n}{n!}f^n(a) \qquad (1.3.9)$$

where $f(a)$, $f'(a)$, etc. is the value of the quantity for $x=a$. If the expansion is around the origin, $x=0$, the series is sometimes referred to as Maclaurin's series.

## References and additional sources 1.3

Boas M. L. (1966): *Mathematical Methods in the Physical Sciences*, New York: John Wiley. Library of Congress Cat. No. 66-17646.

Courant R. (1937): *Differential and Integral Calculus*, 2nd Edn, Vols 1 and 2. London: Blackie and Son.

James G., Burley D., Clements D., Dyke P., Searl J., Wright J. (1996): *Modern Engineering Mathematics*, 2nd Edn, Wokingham: Addison-Wesley. ISBN 0-201-87761-9.

Pipes L. A. (1958): *Applied Mathematics for Engineers and Physicists*, New York: McGraw-Hill. Library of Congress Cat. No. 57-9434.

Poularikas A. D. (Ed.) (1996): *The Transforms and Applications Handbook*, Boca Raton: CRC Press and IEEE Press. ISBN 0-8493-8342-0.

# 1.4 Logarithms

God created the whole numbers: all the rest is man's work.

Leopold Kronecker (1823–1891)

Logarithms, originally developed to help in complex calculations, are now largely superseded for hand calculation by the ubiquitous calculator or mathematical software on your PC. However, we still need to know about them in a number of circumstances and most particularly in electronics they help us cope with numbers spanning a wide range. In a Bode plot of gain, for example, a linear scale will show only a small portion of the range with any resolution and low gains will be inaccessible. Log scales are very common therefore and the decibel 'unit' is used in many areas from aircraft noise to attenuation in optical fibres.

We list now some of the basic relations for logarithms (Abramowitz and Stegun 1970). Logarithms relate to a base value, say $\beta$, in the following way. For generality we will write **lgm** for reference to an arbitrary base. If:

$$y = \beta^x \qquad \text{then} \qquad \mathrm{lgm}_\beta(y) = x \tag{1.4.1}$$

and on this basis we can write:

$$\begin{aligned}
&\mathrm{lgm}(ab) = \mathrm{lgm}(a) + \mathrm{lgm}(b) \qquad && \mathrm{lgm}\left(\frac{a}{b}\right) = \mathrm{lgm}(a) - \mathrm{lgm}(b) \\
&\mathrm{lgm}(a^n) = n\,\mathrm{lgm}(a) && \mathrm{lgm}\left(a^{\frac{1}{n}}\right) = \frac{1}{n}\mathrm{lgm}(a) \\
&\mathrm{lgm}_\beta(a) = \frac{\mathrm{lgm}_\alpha(a)}{\mathrm{lgm}_\alpha(\beta)}, && \text{for conversion between bases } \beta \text{ and } \alpha \\
&\mathrm{lgm}(1) = 0 && \mathrm{lgm}(0) = -\infty
\end{aligned} \tag{1.4.2}$$

The two common bases are $\beta = 10$, for which is written **log**, and $\beta = e = 2.71828$ . . . , which is written **ln** and which are called natural logarithms. We can then list the additional relations for **ln**:

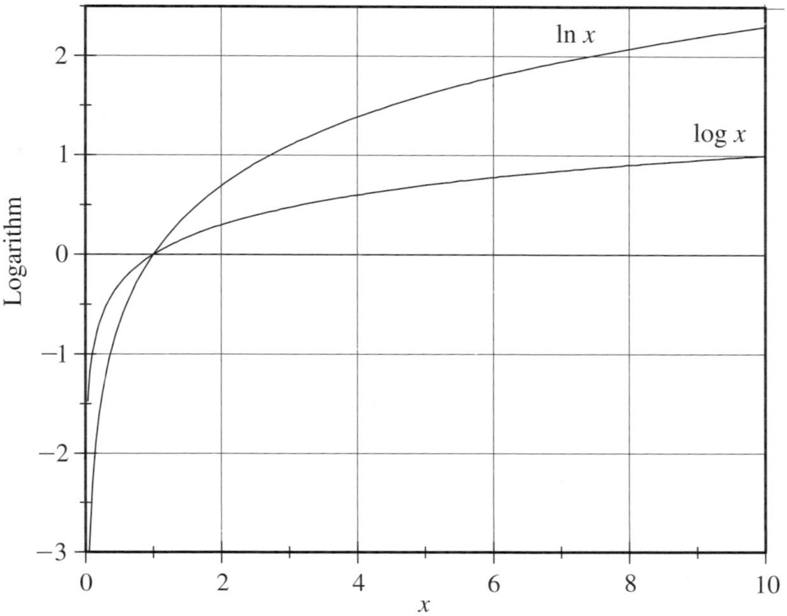


**Fig. 1.4.1** Graph of the logarithmic functions.

$$\ln(1+x)=x-\frac{x^2}{2}+\frac{x^3}{3}-\frac{x^4}{4}+\cdots$$

$$\ln(n+1)=\ln(n)+2\left[\frac{1}{(2n+1)}+\frac{1}{3}\frac{1}{(2n+1)^3}+\frac{1}{5}\frac{1}{(2n+1)^5}+\cdots\right] \quad \text{for } n>0 \quad (1.4.3)$$

$$\frac{d}{dx}[\ln(x)]=\frac{1}{x} \qquad \int\frac{dx}{x}=\ln(x)$$

The value of **ln**(0) $=-\infty$ is the reason *PROBE* in PSpice will refuse to display on a log scale if the data include zero. The use of logarithmic scales is discussed below. The form of the logarithm is shown in Fig. 1.4.1.

It might appear that there are no logarithms for negative numbers, but this is only true if we are restricted to real numbers (Section 1.7). For complex numbers negative arguments are allowed though we will not make use of this possibility. In fact a number, positive or negative will now have an infinite set of logarithms. For example ($n$ is an integer):

$$\ln(-1)=\ln(re^{j\theta})=\ln(r)+\ln(e^{j\theta})=\ln(1)+j(\pi+2n\pi)=j\pi,\ -j\pi,\ 3j\pi,\ \ldots \quad (1.4.4)$$

As mentioned above the very wide range of many of the quantities met in electronics, together with the convenience of simple addition, rather than multiplication, for sequential gains when expressed in a logarithmic scale, prompted the widespread adoption of logarithmic measures. Early measures of attenuation were made in terms of a length of standard cable (Everitt and Anner 1956). To obtain a more generally usable and convenient measure a unit was chosen that closely matched the older version and later it was named the decibel in honour of Alexander Graham Bell. Though the actual unit is the bel, closer agreement with the earlier measures was achieved at one-tenth of this, i.e. the decibel and this has over time become the most commonly used unit. The proper definition is given as a power ratio and could represent both attenuation or gain. If the two powers are $P_1$ and $P_2$ then:

$$\text{Power gain or ratio } G_p = 10\log\left(\frac{P_1}{P_2}\right)\text{dB} \tag{1.4.5}$$

If $P_1$ is greater than $P_2$ then we have gain expressed as +dB whereas if the powers are in the opposite sense then we have attenuation expressed as −dB. If the resistances $R$ at which the powers are measured are the same, then since $P = V^2/R$ we may also write:

$$G_p = 10\log\left(\frac{V_1^2}{V_2^2}\right) = 20\log\left(\frac{V_1}{V_2}\right)\text{dB} \tag{1.4.6}$$

The convenience of logarithmic scales has led to widespread use, sometimes bending the rules, which has caused argument in the literature (Simons 1973; Page 1973), but the improper usages have by now become so established that political correctness has been discarded. When we talk of voltage gains using Eq.(1.4.6) to determine the dB value, the usual difference of impedance levels is ignored. So long as we are all agreed and understand the usage there should be no confusion, but it is as well to be aware of the approximation. It should always be made clear whether power or voltage (or current) is being referred to as the measure will be different. Many derivative units have subsequently been defined such as dBm which refers to a power gain where the reference level (e.g. $P_2$ above) is 1 mW, so that 0 dBm = 1 mW.

An associated reason for using logarithmic scales is that (some of) our senses, e.g. hearing or vision, are logarithmic in sensitivity. This was expressed in the Weber–Fechner law, which holds that 'the minimum change in stimulus necessary to produce a perceptible change in response is proportional to the stimulus already existing' (Everitt and Anner 1956, p. 244). Such responses are illustrated by the audibility sensitivity curves averaged over many subjects in the early 1930s (e.g. Terman 1951); one wonders if measurements at the present time on subjects exposed to the extreme volumes of modern 'music' would reveal the same results.

## References and additional sources 1.4


Abramowitz M., Stegun I. A. (1970): *Handbook of Mathematical Functions with Formulas, Graphs and Mathematical Tables*, Applied Mathematics Series, Washington: National Bureau of Standards.

Everitt W. L., Anner G. E. (1956): *Communication Engineering*, 3rd Edn, New York: McGraw-Hill. Library of Congress Cat. No. 55-12099. See pp. 241–247.

Lambourne R., Tinker M. (Eds) (2000): *Basic Mathematics for the Physical Sciences*, New York: John Wiley. ISBN 0-471-85207-4.

Page C. (1973): Logarithmic quantities and units. *Proc. IEEE* **61**, 1516–1518.

Pipes L. A. (1958): *Applied Mathematics for Engineers and Physicists*, 2nd International Student Edn, New York: McGraw-Hill. Library of Congress Cat. No. 57-9434. See p. 29.

Simons K. (1973): The dB anything. *Proc. IEEE* **61**, 495–496.

Terman F. E. (1951): *Radio Engineering*, 3rd Edn, London: McGraw-Hill. See p. 864.

# 1.5 Exponentials

There must be an ideal world, a sort of mathematician's paradise where everything happens as it does in textbooks.

Bertrand Russell

The exponential function occurs frequently in electronics. It represents phenomena where the rate of change of a variable is proportional to the value of the variable. In a more abstract form we will meet it in Section 1.7 where we will find a most useful relation between it and the trigonometrical functions. Let us consider a more practical circumstance, the charging of a capacitor (Fig. 1.5.1).

We assume that the capacitor is uncharged (this is not essential) and at time $t=0$ the switch is closed. At this instant $V_C$ is zero so the current $i=V_{in}/R=i_0$. We require to find the variation of $V_C$ with time. At any time when the current is $i$ and the charge on $C$ is $Q$, $V_C$ will be given by:

$$V_C=\frac{Q}{C} \quad \text{so that} \quad dV_C=\frac{dQ}{C}=\frac{i\,dt}{C}, \quad \text{where} \quad i=\frac{V_{in}-V_C}{R}$$

$$\text{so}\ \frac{dV_C}{dt}=\frac{i}{C}=\frac{V_{in}-V_C}{RC} \tag{1.5.1}$$

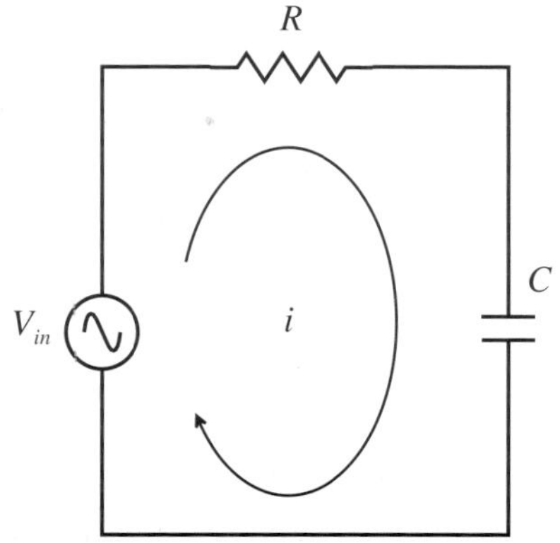


**Fig. 1.5.1** Current flow in a capacitor.

This fits with the statement above about the exponential function except that the rate of change $dV_C/dt$ is proportional to $V_{in} - V_C$, so that as $V_C$ increases towards its final value $V_{in}$ the rate of change will decrease. There is a formal method of solving this differential equation for $V_C$ but we will take the easier path by guessing (knowing) the answer and showing that it agrees with Eq. (1.5.1). We try:

$$V_C = V_{in}(1 - e^{-t/RC}) \tag{1.5.2}$$

To see if this is in agreement with (1.5.1) we differentiate to give:

$$\begin{aligned} \frac{dV_C}{dt} &= 0 + \frac{V_{in}}{RC}e^{-t/RC} \\ &= \frac{V_{in}}{RC}\left(\frac{V_{in} - V_C}{V_{in}}\right) = \frac{V_{in} - V_C}{RC} \end{aligned} \tag{1.5.3}$$

where we have substituted for $e^{-t/RC}$ from Eq. 1.5.2, and see that (1.5.2) is a solution of Eq. (1.5.1). We also have:

$$i = \frac{V_{in} - V_C}{R} = \frac{V_{in}}{R}e^{-t/RC} = i_0\, e^{-t/RC}, \qquad \text{where} \qquad i_0 = \frac{V_{in}}{R} \tag{1.5.4}$$

The initial slope of $V_C$ is given by the value of $dV_C/dt$ at $t=0$. From Eq. (1.5.3):

$$\left(\frac{dV_C}{dt}\right)_{t=0} = \frac{V_{in}}{RC} \tag{1.5.5}$$

The quantity $RC = \tau$ is called the time constant. The initial slope tangent will reach $V_{in}$ at time $\tau$. The exponent $-t/RC$ must be dimensionless so that the units of $RC$ must be time. This can be checked:

$$\begin{aligned} &C = \frac{Q}{V} = \frac{\text{Coulomb}}{\text{Volt}} = \frac{\text{Amp sec}}{\text{Volt}} \qquad \text{and} \qquad R = \frac{\text{Volt}}{\text{Amp}} \\ &\text{so} \qquad RC = \frac{\text{Volt}}{\text{Amp}}\,\frac{\text{Amp sec}}{\text{Volt}} = \text{sec} \end{aligned} \tag{1.5.6}$$

It is often a useful check when doing some complex algebra to examine the consistency of the units of all the terms to see if they are compatible. Any inconsistency can alert you to errors in your algebra. Units are discussed in Section 2.12.

The voltage $V_R$ across $R$ is just the difference between $V_{in}$ and $V_C$. Thus we have from Eq. (1.5.2):

$$V_R = V_{in} - V_C = V_{in}\, e^{-t/RC} \tag{1.5.7}$$

The form of the various functions are shown in Fig. 1.5.2.

Theoretically $V_C$, for example, never reaches $V_{in}$. The time to reach within a given percentage of $V_{in}$ can be calculated and must be allowed for when making more

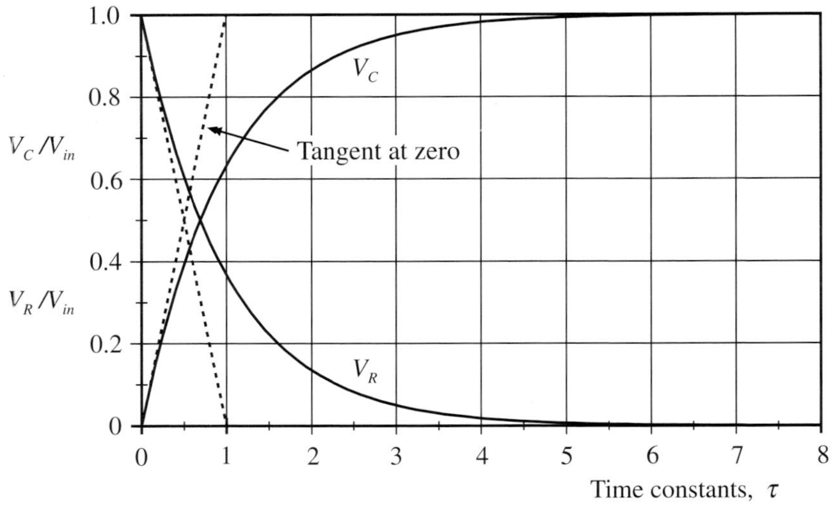


**Fig. 1.5.2** Exponential responses and initial slopes.

Table 1.5.1 *Approach of an exponential function to the final value*

| Time $\tau$ | 0 | 1 | 2 | 3 | 4 | 5 | 6 | 7 |
|---|---|---|---|---|---|---|---|---|
| $V_C/V_{in}$ | 0 | 0.632 | 0.865 | 0.950 | 0.982 | 0.993 | 0.998 | 0.999 |
| $V_R/V_{in}$ | 1 | 0.368 | 0.135 | 0.050 | 0.018 | 0.007 | 0.002 | 0.001 |

accurate measurements. Table 1.5.1 shows the difference as a function of multiples of the time constant $\tau$. For example at time $t=\tau=RC$ the value of $V_C$ is $\cong$63% of $V_{in}$ and you must wait for seven time constants to be within 0.1% of $V_{in}$.

If we consider a similar circuit to Fig. 1.5.1 with an inductor replacing the capacitor then a similar analysis leads to the result:

$$i_L = i_0(1 - e^{-tR/L}), \qquad \text{with} \qquad i_0 = V_{in}/R \tag{1.5.8}$$

and in this case the time constant is $\tau = L/R$.

## References and additional sources 1.5

Courant R. (1937): *Differential and Integral Calculus*, 2nd Edn, Vols 1 and 2, London: Blackie and Son. See Chapter III, Section 6.

Lambourne R., Tinker M. (Eds) (2000): *Basic Mathematics for the Physical Sciences*, New York: John Wiley. ISBN 0-471-85207-4.

Walton A. K. (1987): *Network Analysis and Practice*, Cambridge: Cambridge University Press. ISBN 0-521-31903-X. See Appendix 2.

# 1.6 Vectors

Also, he should remember that unfamiliarity with notation and processes may give an appearance of difficulty that is entirely fictitious, even to an intrinsically easy matter; so that it is necessary to thoroughly master the notation and ideas involved. The best plan is to sit down and work; all that books can do is to show the way.

Oliver Heaviside (1891): *Electromagnetic Theory*, Nov. 13, Vol. I, p. 139

In discussing electromagnetic topics it is necessary to make use of vectors since many of the quantities involved have both magnitude and direction. The algebra of vectors is a little messy but if we can understand the vector relationships things become much neater and easier to write. There are a number of very useful theorems which allow us to transform relations to suit our purposes: we will state and describe how they work. Vectors are written in bold italic type, e.g. $\boldsymbol{A}$.

The addition and subtraction of vectors follows the simple parallelogram geometry as discussed in Section 1.7 and the possible circumstance of three dimensions simply requires two successive operations. Subtraction also follows the same technique. Vectors may be resolved along any suitable set of coordinates, such as Cartesian or polar, but for our purposes we can restrict our choice to Cartesian.

Multiplication presents us with two different possibilities. Any vector $\boldsymbol{A}$, say, may be resolved into components along a chosen set of rectangular coordinates $x$, $y$ and $z$, with components $A_x$, $A_y$ and $A_z$ (Fig. 1.6.1).

The magnitude of the vector is then given by:

$$\text{Magnitude} = (A_x^2 + A_y^2 + A_z^2)^{\frac{1}{2}} \tag{1.6.1}$$

We can define the *scalar product*, shown by $\boldsymbol{A} \bullet \boldsymbol{A}$ (sometimes also called the dot product) by:

$$\boldsymbol{A} \bullet \boldsymbol{A} = A_x^2 + A_y^2 + A_z^2 \tag{1.6.2}$$

which is a scalar, i.e. it has no direction, only magnitude. It is just the square of the length of the vector $\boldsymbol{A}$ and, though the components will change, $\boldsymbol{A} \bullet \boldsymbol{A}$ is independent of the axes chosen. The scalar product of two different vectors $\boldsymbol{A}$ and $\boldsymbol{B}$ is defined in a similar way as:

$$\boldsymbol{A} \bullet \boldsymbol{B} = A_x B_x + A_y B_y + A_z B_z \tag{1.6.3}$$

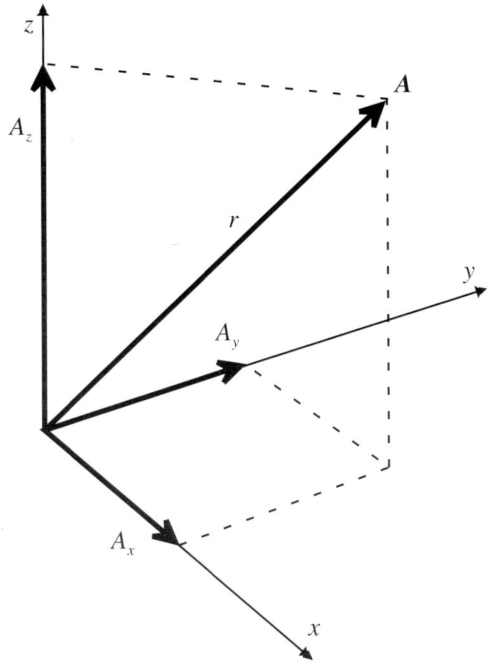


Fig. 1.6.1 Vector components.

and an alternative form can be expressed in terms of the lengths, $A$ and $B$, of the two vectors and the angle $\theta$ between them (Fig. 1.6.2):

$$\boldsymbol{A} \cdot \boldsymbol{B} = AB\cos\theta \qquad (1.6.4)$$

and it is evident that the order of the vectors is immaterial, i.e. $\boldsymbol{A} \cdot \boldsymbol{B} = \boldsymbol{B} \cdot \boldsymbol{A}$.

It is often convenient to make use of *unit vectors* along each of the three axes $x$, $y$ and $z$. These are usually represented by the vector symbols $\boldsymbol{i}, \boldsymbol{j}$ and $\boldsymbol{k}$, respectively (it is perhaps unfortunate that $i$ is also used in complex numbers, and that we use $j$ instead in electronics, but the context should make the meaning clear; there are just not enough symbols to go round for everything to have its own). The scalar products of these vectors can be readily deduced from Eq. (1.6.4) to give:

$$\begin{array}{lll} \boldsymbol{i} \cdot \boldsymbol{i} = 1 & \boldsymbol{j} \cdot \boldsymbol{j} = 1 & \boldsymbol{k} \cdot \boldsymbol{k} = 1 \\ \boldsymbol{i} \cdot \boldsymbol{j} = 0 & \boldsymbol{j} \cdot \boldsymbol{k} = 0 & \boldsymbol{k} \cdot \boldsymbol{i} = 0 \end{array} \qquad (1.6.5)$$

and we can use the unit vectors to express any vector in the form:

$$\boldsymbol{A} = \boldsymbol{i}A_x + \boldsymbol{j}A_y + \boldsymbol{k}A_z \qquad (1.6.6)$$

Some quantities multiply in a quite different way. You may be aware that the force acting on a charge moving in a magnetic field is proportional to the product of velocity and field but acts in a direction normal to both velocity and field. This

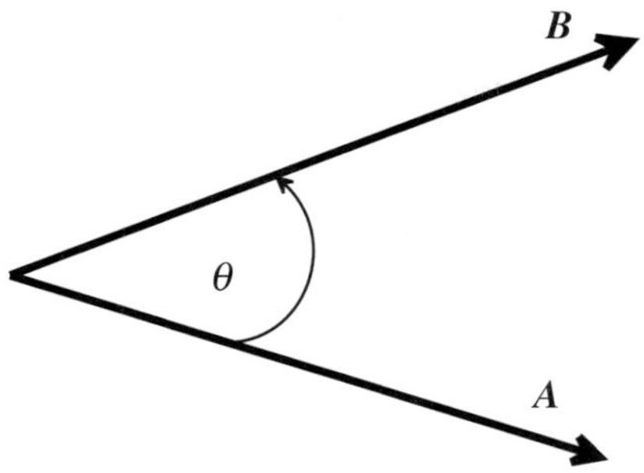


Fig. 1.6.2 Vector scalar or dot product.

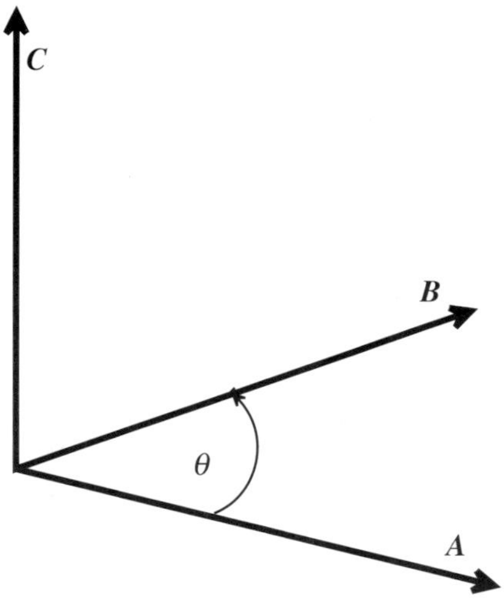


Fig. 1.6.3 Vector cross product.

requires the definition of another form of multiplication known as the *vector or cross product* shown by $\boldsymbol{A} \times \boldsymbol{B}$ which is itself a vector (some books use the symbol $\wedge$ instead of $\times$). The magnitude is most directly defined by:

$$\boldsymbol{A} \times \boldsymbol{B} = AB \sin \theta \tag{1.6.7}$$

and the direction of the vector is normal to the plane containing $\boldsymbol{A}$ and $\boldsymbol{B}$ and in the sense of the advancement of a right-handed screw rotated from $\boldsymbol{A}$ to $\boldsymbol{B}$, and shown as $\boldsymbol{C}$ in Fig. 1.6.3. It is evident that reversing the order of $\boldsymbol{A}$ and $\boldsymbol{B}$ gives the same magnitude but the opposite direction, i.e. $\boldsymbol{B} \times \boldsymbol{A} = -\boldsymbol{C}$.

The consequences for unit vectors are:

$$\begin{aligned} &\boldsymbol{i} \times \boldsymbol{i} = \boldsymbol{j} \times \boldsymbol{j} = \boldsymbol{k} \times \boldsymbol{k} = 0 \\ &\boldsymbol{i} \times \boldsymbol{j} = \boldsymbol{k}, \qquad \boldsymbol{j} \times \boldsymbol{k} = \boldsymbol{i}, \qquad \boldsymbol{k} \times \boldsymbol{i} = \boldsymbol{j} \end{aligned} \tag{1.6.8}$$

If we write the vectors in the form of Eq. (1.6.6) and carry out the multiplications using (1.6.8), then we have:

$$\begin{aligned} \boldsymbol{A}\times\boldsymbol{B} &= (\boldsymbol{i}A_x+\boldsymbol{j}A_y+\boldsymbol{k}A_z)\times(\boldsymbol{i}B_x+\boldsymbol{j}B_y+\boldsymbol{k}B_z) \\ &= \boldsymbol{i}(A_yB_z-A_zB_y)+\boldsymbol{j}(A_zB_x-A_xB_z)+\boldsymbol{k}(A_xB_y-A_yB_x) \\ &= \begin{vmatrix} \boldsymbol{i} & \boldsymbol{j} & \boldsymbol{k} \\ A_x & A_y & A_z \\ B_x & B_y & B_z \end{vmatrix} \end{aligned} \tag{1.6.9}$$

where the last line expresses the relation in the form of a determinant (Section 1.10) and is much easier to recall.

We will be making use of vector algebra for dealing with electromagnetic field quantities, which vary both with respect to position and with time. We therefore need to examine how we can differentiate and integrate vectors. Taking a vector in the form of Eq. (1.6.6), then if say the vector is a function of $t$ we have:

$$\frac{d\boldsymbol{A}}{dt}=\boldsymbol{i}\frac{dA_x}{dt}+\boldsymbol{j}\frac{dA_y}{dt}+\boldsymbol{k}\frac{dA_z}{dt} \tag{1.6.10}$$

which is a vector whose components are the derivatives of the components of $\boldsymbol{A}$. If we have a field representing a simple scalar quantity $\phi(x, y, z)$, (temperature is a good example), then we can ask what is the steepest slope or gradient at any point and this will evidently depend on the slope in the direction of each of the axes. The result is found to be the gradient of $\phi$:

$$\text{grad } \phi=\boldsymbol{i}\frac{\partial\phi}{\partial x}+\boldsymbol{j}\frac{\partial\phi}{\partial y}+\boldsymbol{k}\frac{\partial\phi}{\partial z} \tag{1.6.11}$$

and is evidently also a vector. The use of $\partial$ signifies that when differentiating only quantities depending on what you are differentiating with respect to are relevant, e.g. in the first term those depending on $x$ are relevant while those depending on $y$ or $z$ are considered constants. The form found here arises frequently and it is convenient to define a symbol to represent this in the form of an operator, for example just like say $d/dt$:

$$\boldsymbol{\nabla}=\boldsymbol{i}\frac{\partial}{\partial x}+\boldsymbol{j}\frac{\partial}{\partial y}+\boldsymbol{k}\frac{\partial}{\partial z} \tag{1.6.12}$$

The vector operator $\boldsymbol{\nabla}$ is called *del* and only has meaning when operating on something. Thus we have that $\boldsymbol{\nabla}\phi\equiv\text{grad }\phi$ as in Eq. (1.6.11). $\boldsymbol{\nabla}$ can also operate on a vector. If we have a field described by a vector function $\boldsymbol{V}(x, y, z)$ where the components $V_x$, $V_y$ and $V_z$ of $\boldsymbol{V}$ are functions of $x$, $y$ and $z$:

$$\boldsymbol{V}(x, y, z)=\boldsymbol{i}V_x(x, y, z)+\boldsymbol{j}V_y(x, y, z)+\boldsymbol{k}V_z(x, y, z) \tag{1.6.13}$$

then we can define the *divergence* of $\boldsymbol{V}$ by:

$$\operatorname{div} \boldsymbol{V} \equiv \boldsymbol{\nabla} \cdot \boldsymbol{V} = \frac{\partial V_x}{\partial x} + \frac{\partial V_y}{\partial y} + \frac{\partial V_z}{\partial z} \tag{1.6.14}$$

which is a scalar. We can also define the *curl* of $\boldsymbol{V}$ by:

$$\operatorname{curl} \boldsymbol{V} \equiv \boldsymbol{\nabla} \times \boldsymbol{V} = \boldsymbol{i}\left(\frac{\partial V_z}{\partial y} - \frac{\partial V_y}{\partial z}\right) + \boldsymbol{j}\left(\frac{\partial V_x}{\partial z} - \frac{\partial V_z}{\partial x}\right) + \boldsymbol{k}\left(\frac{\partial V_y}{\partial x} - \frac{\partial V_x}{\partial y}\right) \tag{1.6.15}$$

and this is a vector. Since the gradient is a vector function we can define a further useful relation by taking the divergence of it:

$$\begin{aligned} \operatorname{div} \operatorname{grad} \phi \equiv \boldsymbol{\nabla} \cdot \boldsymbol{\nabla}\phi &= \frac{\partial}{\partial x}\frac{\partial \phi}{\partial x} + \frac{\partial}{\partial y}\frac{\partial \phi}{\partial y} + \frac{\partial}{\partial z}\frac{\partial \phi}{\partial z} \\ &= \frac{\partial^2 \phi}{\partial x^2} + \frac{\partial^2 \phi}{\partial y^2} + \frac{\partial^2 \phi}{\partial z^2} \equiv \nabla^2 \phi \end{aligned} \tag{1.6.16}$$

and this is a very important expression. The operator $\nabla^2$ is called the Laplacian and is a scalar operator. For example an equation of the form:

$$\nabla^2 \phi = \frac{1}{a^2}\frac{\partial^2 \phi}{\partial t^2} \tag{1.6.17}$$

is a wave equation as we will come across in examining the consequences of Maxwell's equations. If we reverse the order of div and grad and apply this to a vector, then:

$$\operatorname{grad} \operatorname{div} \boldsymbol{V} \equiv \boldsymbol{\nabla}(\boldsymbol{\nabla} \cdot \boldsymbol{V})$$

$$= \boldsymbol{i}\left(\frac{\partial^2 V_x}{\partial x^2} + \frac{\partial^2 V_y}{\partial x \partial y} + \frac{\partial^2 V_z}{\partial x \partial z}\right) + \boldsymbol{j}\left(\frac{\partial^2 V_x}{\partial x \partial y} + \frac{\partial V_y}{\partial y^2} + \frac{\partial^2 V_z}{\partial y \partial z}\right) + \boldsymbol{k}\left(\frac{\partial^2 V_x}{\partial x \partial z} + \frac{\partial^2 V_y}{\partial y \partial z} + \frac{\partial^2 V_z}{\partial z^2}\right) \tag{1.6.18}$$

and now it becomes more evident that the symbolic forms can save a lot of writing. Since $\nabla^2$ is a scalar, the operation on a vector is simply a vector with components:

$$\nabla^2 \boldsymbol{V} \equiv \boldsymbol{\nabla} \cdot \boldsymbol{\nabla} \boldsymbol{V} = (\nabla^2 V_x, \nabla^2 V_y, \nabla^2 V_z) \tag{1.6.19}$$

One further relation will be needed which is defined by:

$$\begin{aligned} \boldsymbol{\nabla} \times (\boldsymbol{\nabla} \times \boldsymbol{V}) &= \boldsymbol{\nabla}(\boldsymbol{\nabla} \cdot \boldsymbol{V}) - (\boldsymbol{\nabla} \cdot \boldsymbol{\nabla})\boldsymbol{V} \\ &= \boldsymbol{\nabla}(\boldsymbol{\nabla} \cdot \boldsymbol{V}) - \nabla^2 \boldsymbol{V} \end{aligned}$$

$$\text{or} \quad \operatorname{curl} \operatorname{curl} \boldsymbol{V} = \operatorname{grad} \operatorname{div} \boldsymbol{V} - \operatorname{del}^2 \boldsymbol{V} \tag{1.6.20}$$

The divergence of a cross product will be required in Section 2.2:

$$\boldsymbol{\nabla} \cdot (\boldsymbol{A} \times \boldsymbol{B}) = \boldsymbol{B} \cdot (\boldsymbol{\nabla} \times \boldsymbol{A}) - \boldsymbol{A} \cdot (\boldsymbol{\nabla} \times \boldsymbol{B}) \tag{1.6.21}$$

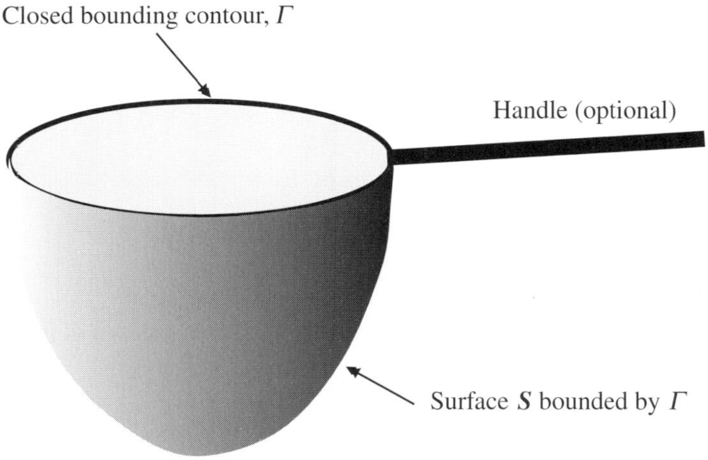


**Fig. 1.6.4** Surface defined by bounding contour.

In manipulating some vector functions it will be necessary to change between line integrals and surface integrals. A very useful relation between these is referred to as Stokes' theorem. Boas (1966) gives as an example a butterfly net with the rim providing the closed bounding contour $\Gamma$ and the net being the surface $S$ bounded by the contour (Fig. 1.6.4).

If we have a vector quantity $\boldsymbol{V}$, with $\boldsymbol{n}$ the normal to the surface at any particular point on it and $d\boldsymbol{l}$ a length on the contour $\Gamma$, then Stokes' theorem tells us that the line integral around the contour is equal to the surface integral of $(\nabla \times \boldsymbol{V}) \cdot \boldsymbol{n}$ over any surface bounded by $\Gamma$:

$$\oint_{\Gamma} \boldsymbol{V} \cdot d\boldsymbol{l} = \int_{S} (\nabla \times \boldsymbol{V}) \cdot \boldsymbol{n}\, ds \tag{1.6.22}$$

and it does not matter what shape the surface $S$ has as long as it is bounded by $\Gamma$.

The divergence theorem is another very useful relation, this time connecting volume and surface integrals. The divergence of a vector $\boldsymbol{E}$ from a volume $V$ bounded by a surface $S$ with $\boldsymbol{n}$ the surface normal as before is:

$$\int_{V} \nabla \cdot \boldsymbol{E}\, dV = \int_{S} \boldsymbol{E} \cdot \boldsymbol{n}\, dS \tag{1.6.23}$$

Some books may write the left-hand side as a triple integral (one for each dimension) and the right-hand side as a double integral, but the meaning is identical.

As an example of the convenience of vector operations we will consider the motion of magnetic moments in a magnetic field. In Section 2.11 we will discuss

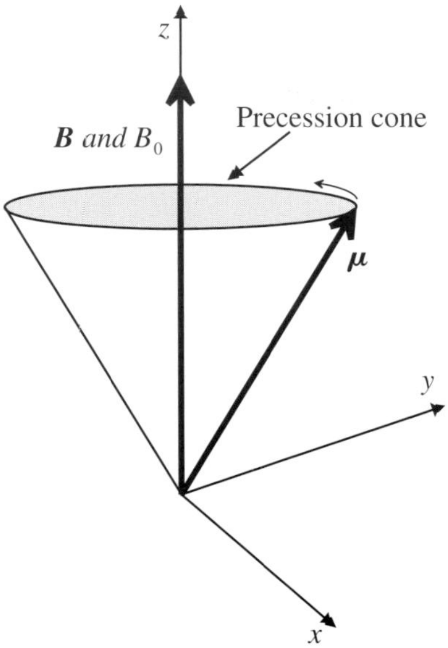


**Fig. 1.6.5** Precession of a magnetic moment $\boldsymbol{\mu}$ around $\boldsymbol{B}$.

magnetic materials. Magnetic effects in materials arise from the atomic-scale magnetic moments of atoms and electrons, and the behaviour of these in both static and varying magnetic fields leads to the magnetic properties of such materials. The essential difference between a classical view of the atomic magnetic moments as like little bar magnets and the quantum view is that the moments possess angular momentum. There is also of course the matter of quantization, but we will be dealing with the macroscopic effects where we can rely on a semi-classical view. A spinning gyroscope, subject to a torque due to gravity, will precess around the vertical as shown in Fig. 1.6.5. In the same way an atomic magnetic moment $\mu$, which also has angular momentum $\boldsymbol{J}$, will precess around the direction of the magnetic field $\boldsymbol{B}$ (Slichter 1964).

The torque acting on the moment is given by $\boldsymbol{\mu} \times \boldsymbol{B}$ (a vector product) and this is equal to the rate of change of angular momentum. Since for atomic moments $\boldsymbol{\mu} = \gamma \boldsymbol{J}$, where $\gamma$ is called the magnetogyric ratio, then:

$$\frac{d\boldsymbol{J}}{dt} = \boldsymbol{\mu} \times \boldsymbol{B} \qquad \text{or} \qquad \frac{d\mu}{dt} = \boldsymbol{\mu} \times \gamma \boldsymbol{B} \tag{1.6.24}$$

This equation, which holds whether or not $\boldsymbol{B}$ is time dependent, tells us that at any instant changes in $\boldsymbol{\mu}$ are perpendicular to both $\boldsymbol{\mu}$ and $\boldsymbol{B}$. An instructive method for solving for the motion of $\boldsymbol{\mu}$ is to transform to a rotating coordinate frame. For a vector function of time $\boldsymbol{F}(t)$ in the frame $(x, y, z)$:

$$\boldsymbol{F}(t) = \boldsymbol{i}F(x) + \boldsymbol{j}F(y) + \boldsymbol{k}F(z) \tag{1.6.25}$$

then in a frame rotating with instantaneous velocity $\boldsymbol{\Omega}$:

$$\frac{d\boldsymbol{i}}{dt} = \boldsymbol{\Omega} \times \boldsymbol{i} \qquad \text{and similarly for the other coordinates} \tag{1.6.26}$$

so that we have:

$$\begin{aligned}\frac{d\boldsymbol{F}(t)}{dt} &= \boldsymbol{i}\frac{dF(x)}{dt} + F(x)\frac{d\boldsymbol{i}}{dt} + \boldsymbol{j}\frac{dF(y)}{dt} + F(y)\frac{d\boldsymbol{j}}{dt} + \boldsymbol{k}\frac{dF(z)}{dt} + F(z)\frac{d\boldsymbol{k}}{dt} \\ &= \boldsymbol{i}\frac{dF(x)}{dt} + \boldsymbol{j}\frac{dF(y)}{dt} + \boldsymbol{k}\frac{dF(z)}{dt} + \boldsymbol{\Omega} \times [(\boldsymbol{i}F(x) + \boldsymbol{j}F(y) + \boldsymbol{k}F(z)] \\ &= \frac{\delta \boldsymbol{F}}{\delta t} + \boldsymbol{\Omega} \times \boldsymbol{F}\end{aligned} \tag{1.6.27}$$

where $\delta \boldsymbol{F}/\delta t$ is the rate of change of $\boldsymbol{F}$ with respect to the frame $(x, y, z)$. Thus if $\delta \boldsymbol{F}/\delta t = 0$, the components of $\boldsymbol{F}$ along $\boldsymbol{i}, \boldsymbol{j}, \boldsymbol{k}$ do not change with time. Using (1.6.27) we can rewrite Eq. (1.6.24) in terms of a coordinate system rotating with an as yet arbitrary angular velocity $\boldsymbol{\Omega}$:

$$\begin{aligned}&\frac{\delta \boldsymbol{\mu}}{\delta t} + \boldsymbol{\Omega} \times \boldsymbol{\mu} = \boldsymbol{\mu} \times \gamma \boldsymbol{B} \qquad \text{or} \\ &\frac{\delta \boldsymbol{\mu}}{\delta t} = \boldsymbol{\mu} \times (\gamma \boldsymbol{B} + \boldsymbol{\Omega})\end{aligned} \tag{1.6.28}$$

where the changed order of the cross product changes the sign (see just below Eq. (1.6.7)). This tells us that the motion of $\boldsymbol{\mu}$ in the rotating frame is the same as in the fixed frame provided we replace $\boldsymbol{B}$ with an effective field $\boldsymbol{B}_e$:

$$\boldsymbol{B}_e = \boldsymbol{B} + \frac{\boldsymbol{\Omega}}{\gamma} \tag{1.6.29}$$

We can now solve for the motion of $\boldsymbol{\mu}$ in a static field $\boldsymbol{B} = \boldsymbol{k}B_0$ (i.e. a field along the $z$-axis) by choosing such that $\boldsymbol{B}_e = 0$:

$$\boldsymbol{\Omega} = -\gamma B_0 \boldsymbol{k} \tag{1.6.30}$$

Since in this reference frame $\delta \boldsymbol{\mu}/\delta t = 0$, $\boldsymbol{\mu}$ remains fixed with respect to $\boldsymbol{i}, \boldsymbol{j}, \boldsymbol{k}$ and the motion with respect to the laboratory is just that of the frame $(x, y, z)$, i.e. it precesses about $\boldsymbol{B}_0$ with angular velocity given by Eq. (1.6.30). This is called the Larmor precession frequency and is that which is detected in magnetic resonance applications like magnetic resonance imaging. To the initial justification for this analysis, that it demonstrates the convenience and effectiveness of vector analysis, we can add the result given by Eq. (1.6.30) that moving in this fashion makes the magnetic field vanish. The quotations from Feynman in Sections 2.6 and 2.7 should be read to appreciate the relevance of this example.

Though the technique used to determine the motion of the magnetic moments may seem somewhat more complex than necessary, when in nuclear magnetic resonance applications an additional oscillating magnetic field is applied normal to the static field, the method becomes most helpful in determining and visualizing the additional complex motions of the nuclear spin magnetic moments.

## References and additional sources 1.6

Boas M. L. (1966): *Mathematical Methods in the Physical Sciences*, New York: John Wiley. Library of Congress Cat. No. 66-17646.

Corson D. R., Lorrain P. (1962): *Introduction to Electromagnetic Fields and Waves*, San Francisco: W. H. Freeman. Library of Congress Cat. No. 62-14193.

Grant I., Philips W. R. (1975): *Electromagnetism*; London: John Wiley. ISBN 0-471-32246-6. 2nd Edn, 1990, ISBN 0-471-92712-0.

James G., Burley D., Clements D., Dyke P., Searl J., Wright J. (1996): *Modern Engineering Mathematics*, 2nd Edn, Wokingham: Addison-Wesley. ISBN 0-201-87761-9.

Lambourne R., Tinker M. (Eds)(2000): *Basic Mathematics for the Physical Sciences*, New York: John Wiley. ISBN 0-471-85207-4.

Pipes L. A. (1958): *Applied Mathematics for Engineers and Physicists*, New York: McGraw-Hill. Library of Congress Cat. No. 57-9434.

Poularikas A. D. (Ed.) (1996): *The Transforms and Applications Handbook*, Boca Raton: CRC Press and IEEE Press. ISBN 0-8493-8342-0.

Slichter C. P. (1964): *Principles of Magnetic Resonance*, New York: Harper and Row. Library of Congress Cat. No. 63-11293. See p. 10.

# 1.7 Complex numbers

When algebra reached a certain stage of development, the imaginary turned up. It was exceptional, however, and unintelligible, and therefore to be evaded, if possible. But it would not submit to be ignored. It demanded consideration, and has since received it. The algebra of real quantity is now a specialisation of the algebra of the complex quantity, say a + bi, and great extensions of mathematical knowledge have arisen out of the investigation of this once impossible and non-existent quantity. It may be questioned whether it is entitled to be called a quantity, but there is no question as to its usefulness, and algebra of real quantity would be imperfect without it.

Oliver Heaviside (1899): *Electromagnetic Theory*, April 10, Vol. II, p. 457

Some people have difficulties with the idea of complex numbers. This possibly arises from the use of the word imaginary and from a one-dimensional view of the world of numbers. As the square root of $-1$ is one of the most important numbers in mathematics it has, like $e$ and $\pi$, its own special symbol $i$. Euler's equation:

$$e^{i\pi} + 1 = 0 \tag{1.7.1}$$

is often quoted as containing five of the most important numbers in mathematics. We shall see later how to evaluate it. In electronics the symbol $i$ is usually used for current so in this subject the symbol $j$ is used instead, but they are identical and can be exchanged anywhere as you wish.

In a one-dimensional world there is indeed no meaning to ask what the square root of $-1$ is. However, we can consider a simple argument to show what it does mean. Think of all the real numbers plotted along an axis as shown in Fig. 1.7.1.

We have marked on the axis some representative numbers and also shown negative numbers to the left of zero. Now take any real number you wish, say 3 for example, which we can represent by the vector $\mathbf{0}\boldsymbol{A}$, and ask what general operator we can think of, i.e. an operator that will work for any real number, which will change 3 to $-3$. The simplest operator is just $-1$, since:

$$-1 \times 3 = -3$$

and this will work for any number. What the operator $-1$ does is to rotate the vector $\mathbf{0}\boldsymbol{A}$ by 180° to $\mathbf{0}\boldsymbol{B}$. This of course means that we must allow a two-dimensional space for our numbers. We can now ask, what operator can we

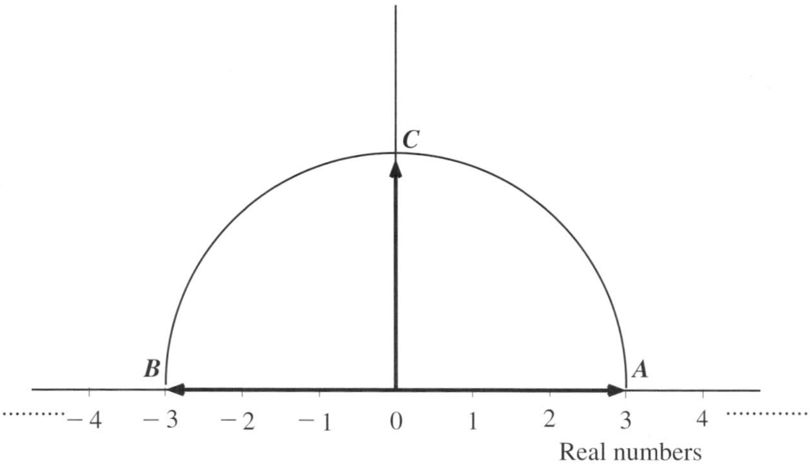


**Fig. 1.7.1** Complex number as a rotation.

imagine that will change our vector from **0***A* to **0***C*? This is not immediately evident but we can say that if we apply this operator twice to rotate from **0***A* to **0***C* and then from **0***C* to **0***B*, then the result will be the same as using −1. If we call this operator Op, say, then we have:

$$\mathrm{Op}\times\mathrm{Op}\times 3=-3 \text{ or } \mathrm{Op}^2(3)=-3 \tag{1.7.2}$$

so we can now write:

$$\mathrm{Op}^2=-1 \qquad \text{or} \qquad \mathrm{Op}=\sqrt{-1}\equiv j \tag{1.7.3}$$

Thus $j$ is an operator that *rotates* a vector by 90°. The '$x$-axis' is called the *real axis* and the '$y$-axis' is called the *imaginary axis*: the number $j3=$ **0C**. A number may have some real part ($\mathcal{R}e$) together with some imaginary ($\mathcal{I}m$) part, and in general we refer to *complex numbers*. A plot of a complex number in the real/imaginary plane is known as an Argand diagram. If we have a complex number $Z=x+jy$ then this is the vector **0***D* shown in Fig. 1.7.2 where the coordinates of $D$ are simply $x$ and $y$.

If the length of the vector **0***D* is $R$ and the angle it makes with the real axis is $\theta$, then we can also write:

$$\begin{aligned}&\text{Real part of } Z=\mathcal{R}e(Z)=x=R\cos\theta \quad \text{and}\\ &\text{Imaginary part of } Z=\mathcal{I}m(Z)=y=R\sin\theta\end{aligned} \tag{1.7.4}$$

so that our complex number can also be expressed as:

$$\begin{aligned}&Z=x+jy=R\,(\cos\theta+j\sin\theta)\\ &\text{where } R=(x^2+y^2)^{\frac{1}{2}} \quad \text{and} \quad \theta=\tan^{-1}(y/x)\end{aligned} \tag{1.7.5}$$

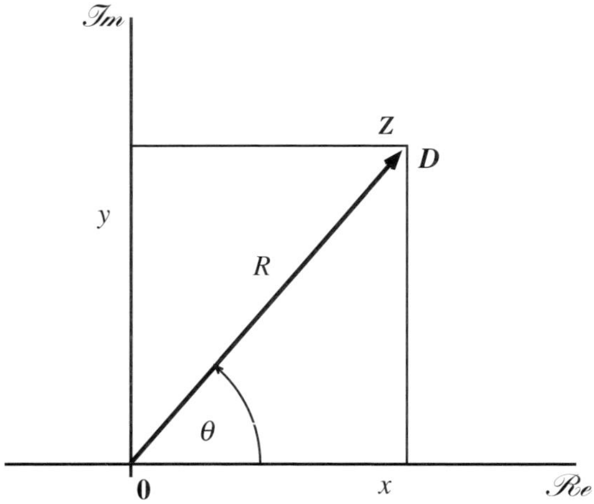


**Fig. 1.7.2** Complex number in polar format.

The quantity $R$ is called the *modulus* or *absolute value* of $Z$ and is written $|Z|$. The absolute value is always positive. The angle $\theta$ is known as the *argument* or phase of $Z$. Thus:

$$Z = \text{mod}\textit{ulus}\,[\cos(\text{arg}\textit{ument}) + j\sin(\text{arg}\textit{ument})] \tag{1.7.6}$$

Most functions can be expressed in the form of an infinite series (Boas 1966). Such series provide a means of computing the value of the function to any desired precision or to find approximations. The series expansions for sin, cos and the exponential (Section 1.3) are:

$$\text{Sin } \theta = \theta - \frac{\theta^3}{3!} + \frac{\theta^5}{5!} - \frac{\theta^7}{7!} + \cdots$$

$$\text{Cos } \theta = 1 - \frac{\theta^2}{2!} + \frac{\theta^4}{4!} - \frac{\theta^6}{6!} + \cdots$$

$$e^{j\theta} = 1 + j\theta + \frac{(j\theta)^2}{2!} + \frac{(j\theta)^3}{3!} + \frac{(j\theta)^4}{4!} + \cdots \tag{1.7.7}$$

$$= \left(1 - \frac{\theta^2}{2!} + \frac{\theta^4}{4!} + \frac{\theta^6}{6!} + \cdots\right) + j\left(\theta - \frac{\theta^3}{3!} + \frac{\theta^5}{5!} + \frac{\theta^7}{7!} + \cdots\right)$$

$$= \cos\theta + j\sin\theta$$

This remarkable relationship between the trigonometrical functions cos and sin and the complex exponential is known as Euler's formula. This is one of the reasons that complex numbers are of such particular use in circuit analysis. Conversely we can write:

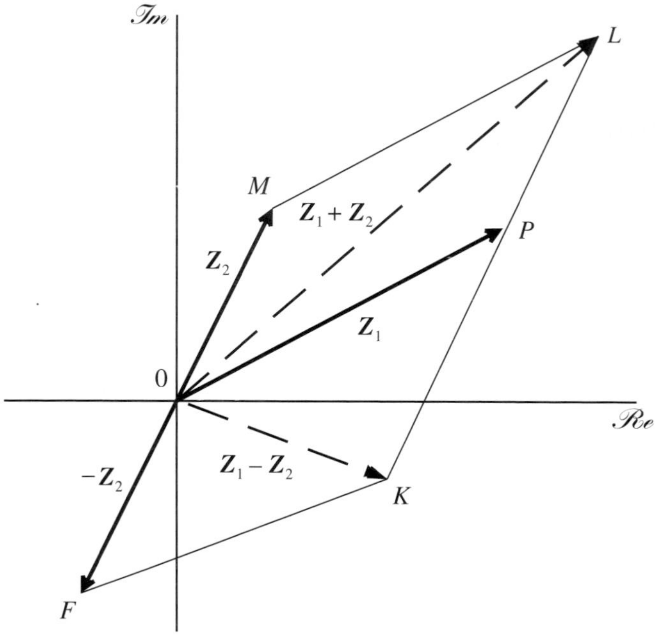


**Fig. 1.7.3** Complex number addition and subtraction.

$$\text{Cos } \theta = \frac{e^{j\theta} + e^{-j\theta}}{2} \qquad \text{and} \qquad \sin \theta = \frac{e^{j\theta} - e^{-j\theta}}{2j}$$

$$\text{Cos } j\,\theta \equiv \cosh \theta = \frac{e^{\theta} + e^{-\theta}}{2}, \qquad \text{the hyperbolic cosine} \tag{1.7.8}$$

$$\text{Sin } j\,\theta \equiv j \sinh \theta = j\frac{e^{\theta} - e^{-\theta}}{2}, \qquad \text{the hyperbolic sine}$$

We now need to consider the algebra of complex numbers, i.e. how you add, subtract, multiply and divide them. Adding is straightforward; if we have two numbers $Z_1$ and $Z_2$:

$$Z_1 = p + jr \qquad \text{and} \qquad Z_2 = s + jt$$

$$\text{then } Z = Z_1 + Z_2 = (p+s) + j(r+t) \tag{1.7.9}$$

which operation is shown in Fig. 1.7.3 for the numbers indicated. This is the same as adding two vectors. Subtraction works in the same way – you may think of finding $-Z_2$ and adding this to $Z_1$:

$$Z = Z_1 - Z_2 = (p-s) + j(r-t) \tag{1.7.10}$$

which is also shown in Fig. 1.7.3 for the same example numerical numbers.

Similarly we may say that if two complex numbers are equal then:

$$\text{if } p+jr=s+jt, \quad \text{then} \quad p=s \quad \text{and} \quad r=t \tag{1.7.11}$$

Using (1.7.7) we can write any complex number in exponential rather than trigonometrical terms (mathematically exponentials are usually much easier to handle than trigonometrical terms):

$$Z=R(\cos\theta+j\sin\theta)=Re^{j\theta} \tag{1.7.12}$$

which allows us to write the $n$th power of a complex number as:

$$Z^n=Re^{jn\theta}=R^n(\cos n\theta+j\sin n\theta) \tag{1.7.13}$$

which is known as De Moivre's theorem. It should be noted that $n$ need not be an integer and can be a fraction, e.g. if $n=1/3$ we get the cube root. The exponential form allows us to carry out multiplication and division much more easily than with the original Cartesian form. Taking:

$$\begin{gathered} Z_1=p+jr=R_1e^{j\theta_1}, \quad \text{and} \quad Z_2=s+jt=R_2e^{j\theta_2} \\ Z=Z_1\cdot Z_2=(p+jr)(s+jt)=R_1R_2e^{j(\theta_1+\theta_2)} \end{gathered} \tag{1.7.14}$$

Thus the rules are that the modulus of a product is the *product* of the moduli:

$$Z=|Z_1\cdot Z_2|=|Z_1|\cdot|Z_2| \tag{1.7.15}$$

and the argument of a product is the *sum* of the arguments:

$$\text{Arg } Z=\text{Arg }(Z_1\cdot Z_2)=\text{Arg } Z_1+\text{Arg } Z_2 \tag{1.7.16}$$

For division we have similarly:

$$Z=\frac{Z_1}{Z_2}=\frac{p+jr}{s+jt}=\frac{R_1e^{j\theta_1}}{R_2e^{j\theta_2}}=\frac{R_1}{R_2}e^{j(\theta_1-\theta_2)} \tag{1.7.17}$$

so the rules are that the modulus of a quotient is the quotient of the moduli:

$$|Z|=\left|\frac{Z_1}{Z_2}\right|=\frac{|Z_1|}{|Z_2|} \tag{1.7.18}$$

and that the argument of a quotient is the difference of the arguments:

$$\text{Arg } Z=\text{Arg}\left(\frac{Z_1}{Z_2}\right)=\text{Arg } Z_1-\text{Arg } Z_2 \tag{1.7.19}$$

The complex conjugate $Z^*$ of a complex number $Z$ is formed by taking the negative of the imaginary part. Thus if:

$$Z=p+jr=R(\cos\theta+j\sin\theta)=Re^{j\theta} \quad \text{then} \quad Z^*=p-jr=R(\cos\theta-j\sin\theta)=Re^{-j\theta} \tag{1.7.20}$$

so that if we take the product of $Z$ and $Z^*$ it is readily seen that:

$$|Z| = R = (Z \bullet Z^*)^{\frac{1}{2}} \tag{1.7.21}$$

which provides a ready way of finding the modulus. The complex conjugate also provides a convenient way of rationalizing a complex quotient. For example, if we have (multiplying the numerator and the denominator by the complex conjugate of the denominator):

$$\begin{aligned} Z &= \frac{p+jr}{s+jt} = \frac{(p+jr)(s-jt)}{(s+jt)(s-jt)} \\ &= \frac{ps+jsr-jpt+rt}{s^2+jts-jts+t^2} = \frac{(ps+rt)+j(sr-pt)}{s^2+t^2} \\ &= \left(\frac{ps+rt}{s^2+t^2}\right) + j\left(\frac{sr-pt}{s^2+t^2}\right) \end{aligned} \tag{1.7.22}$$

which is now in the standard *real* plus *imaginary* form.

In conventional a.c. circuit analysis (Section 3.2) we make great use of the idea of complex numbers, and with the Laplace transform we meet the initially strange idea of a complex frequency. In describing the behaviour of dielectrics and magnetic materials it is also convenient to use these ideas.

And as for Euler's equation, given above as (1.7.1), we may now evaluate it:

$$\begin{aligned} &e^{j\pi} = \cos(\pi) + j\sin(\pi) = -1 + 0 \quad \text{from Eq. (1.7.7) and Table 1.1.1} \\ &\text{so } e^{j\pi} + 1 = -1 + 1 = 0 \end{aligned} \tag{1.7.23}$$

## References and additional sources 1.7

Boas M. L. (1966): *Mathematical Methods in the Physical Sciences*, New York: John Wiley. Library of Congress Cat. No. 66-17646.

James G., Burley D., Clements D., Dyke P., Searl J., Wright J. (1996): *Modern Engineering Mathematics*, 2nd Edn, Wokingham: Addison-Wesley. ISBN 0-201-87761-9.

Pipes L. A. (1958): *Applied Mathematics for Engineers and Physicists*, 2nd Edn, New York: McGraw-Hill. Library of Congress Cat. No. 57-9434.

# 1.8 Differentiation

Every one who has gone seriously into the mathematical theory of a physical subject (though it may be professedly only an ideal theory) knows how important it is not to look upon the symbols as standing for mere quantities (which might have any meaning), but to bear in mind the physics in a broad way, and obtain the important assistance of physical guidance in the actual work of getting solutions. This being the case generally, when the mathematics is well known, it is clear that when one is led to ideas and processes which are not understood, and when one has to find ways of attack, the physical guidance becomes more important still. If it be wanting, we are left nearly in the dark. The Euclidean logical way of development is out of the question. That would mean to stand still. First get on, in any way possible, and let the logic be left for later work.

Oliver Heaviside (1899): *Electromagnetic Theory*, April 10, Vol. II, p. 460

Differentiation is a mathematical process that tells us about rates of change, a factor very much of interest in electronics. If some quantity is changing, say as a function of time, e.g. the voltage across a capacitor, then the differential of the function with respect to time evaluated at a point will tell us the slope of the function at the point, i.e. the slope of the tangent there. We will list here some of the differentials commonly encountered or that we will use. For compactness, a differential of a function $f$ is often written $f'$.

$$\begin{aligned}
&f(x)=ax^n && \frac{df(x)}{dx}=anx^{n-1}, && \text{for } a \text{ and } n \text{ constants}\\
&f(x)=ae^{nx} && \frac{df(x)}{dx}=ane^{nx} && \\
&f(x)=ae^{u(x)} && \frac{df(x)}{dx}=ae^{u(x)}f'u(x) && \\
&f(x)=\ln(x) && \frac{df(x)}{dx}=\frac{1}{x}, && \ln \text{ is to base } e \qquad (1.8.1)\\
&f(x)=\sin(x) && \frac{df(x)}{dx}=\cos(x) && \\
&f(x)=\cos(x) && \frac{df(x)}{dx}=-\sin(x) &&
\end{aligned}$$

$$f(x) = a^x \qquad \frac{df(x)}{dx} = a^x \ln(a), \quad a \neq 1$$

$$f(x) = \tan^{-1} x \qquad \frac{df(x)}{dx} = \frac{1}{1+x^2}$$

Some rules for the differentiation of more complex forms are:

Differentiation of a sum of two functions:

$$f(x) = u + v \qquad \frac{df}{dx} = \frac{du}{dx} + \frac{dv}{dx}$$

Differentiation of a product of two functions:

$$f(x) = u \times v \qquad \frac{df}{dx} = u\frac{dv}{dx} + v\frac{du}{dx} \tag{1.8.2}$$

Differentiation of a quotient of two functions:

$$f(x) = \frac{u}{v} \qquad \frac{df}{dx} = \frac{v\dfrac{du}{dx} - u\dfrac{dv}{dx}}{v^2}$$

Differentiation of a function of a function:

$$f(x) = u[v(x)] \qquad \frac{df}{dx} = \frac{du}{dv} \times \frac{dv}{dx}$$

If $n$ successive differentiations are carried out then the differential is written:

$$\frac{d^n f(x)}{dx^n} \equiv \frac{d^{n-1}}{dx^{n-1}}\left[\frac{df(x)}{dx}\right] \tag{1.8.3}$$

and so on depending on $n$.

The idea of the differential as the slope of the function allows us to find the turning points of the function, i.e. where it is a maximum or a minimum, since at these points the slope will be zero. The slope is positive if the tangent runs from lower left to upper right for our normal Cartesian coordinate system, and negative from lower right to upper left. Thus as the point of interest moves through an extremum it must pass through zero slope. The second derivative $f''$ gives the curvature, i.e. 1/radius, of the curve. The value of $f''$ at the extremum will indicate a minimum if it is positive, and a maximum if it is negative. In the special case that it is zero then the point will be one of inflection. A second form of symbol for indicating differentiation compactly is to place a dot, or dots, above the variable and is used particularly when the differentiation is with respect to time:

$$\frac{dx}{dt} \equiv \dot{x} \quad \text{or} \quad \frac{d^2x}{dt^2} \equiv \ddot{x}, \qquad \text{and so on} \tag{1.8.4}$$

In circumstances where there is more than one variable in the function and we are only interested in the variation with respect to one of them, then we write the differential as a partial differential to indicate this:

For $f(x, t)$ the differential is either $(\partial f/\partial x)_t$ keeping $t$ fixed or $(\partial f/\partial t)_x$ keeping $x$ fixed (1.8.5)

though the subscripts are not always used when the meaning is evident.

## References and additional sources 1.8

Courant R. (1937): *Differential and Integral Calculus*, 2nd Edn, Vols 1 and 2, London: Blackie and Son.

James G., Burley D., Clements D., Dyke P., Searl J., Wright J. (1996): *Modern Engineering Mathematics*, 2nd Edn, Wokingham: Reading: Addison-Wesley. ISBN 0-201-87761-9.

Lambourne R., Tinker M. (Eds) (2000): *Basic Mathematics for the Physical Sciences*, New York: John Wiley. ISBN 0-471-85207-4.

# 1.9 Integration

> We avail ourselves of the labours of the mathematicians, and retranslate their results from the language of the calculus into the language of dynamics, so that our words may call up the mental image, not of some algebraical process, but of some property of moving bodies.
>
> James Clerk Maxwell (1873) *A Treatise on Electricity and Magnetism*, Article 554

From our point of view, integration is the inverse of differentiation, i.e. if we differentiate a function and then integrate the differential we should arrive back at the original function. This is in general not quite true since if we differentiate a constant we get zero so integrating again will leave us with the necessity of adding a constant which we will have to determine separately. This inverse relationship is probably one of the commonest ways of finding integrals, so looking at Section 1.8 is often a good place to start. A definite integral specifies the limits as subscript and superscript and will give a specific value when these limits are inserted. An indefinite integral does not specify the limits and will therefore carry the additional constant to be determined, usually by reference to some known initial conditions. In the standard integrals listed in Eq. (1.9.1) the constant is shown as $C$. There are of course very many common integrals but here we list only a few, primarily those that we will encounter in other sections.

$$\begin{aligned}
&\text{(a)} \quad \int ax^n dx = \frac{ax^{n+1}}{n+1} + C, \qquad \text{where } a \text{ and } n \text{ are constants and } n \neq -1 \\
&\text{(b)} \quad \int e^{ax} dx = \frac{e^{ax}}{a} + C \\
&\text{(c)} \quad \int \frac{dx}{x} = \ln(x) + C \\
&\text{(d)} \quad \int \sin(x) dx = -\cos(x) + C \\
&\text{(e)} \quad \int \cos(x) dx = \sin(x) + C
\end{aligned} \tag{1.9.1}$$

(f) $\int \frac{dx}{(a^2 - x^2)^{\frac{1}{2}}} = \sin^{-1}\left(\frac{x}{a}\right) + C, \qquad \text{for } \left|\frac{x}{a}\right| < 1$

(g) $\int \frac{a\,dx}{a^2 + x^2} = \tan^{-1}\left(\frac{x}{a}\right) + C$

(h) $\int \sin^2(x)dx = \frac{1}{2}x - \frac{1}{4}\sin(2x) + C$

(i) $\int \cos^2(x)dx = \frac{1}{2}x + \frac{1}{4}\sin(2x) + C$ (1.9.1 cont.)

(j) $\int xe^{ax}\,dx = \frac{1}{a^2}(ax - 1)e^{ax} + C$

(k) $\int e^{ax}\sin(bx)dx = \frac{e^{ax}}{a^2 + b^2}[a\sin(bx) - b\cos(bx)] + C$

(l) $\int e^{ax}\cos(bx)dx = \frac{e^{ax}}{a^2 + b^2}[a\cos(bx) + b\sin(bx)] + C$

(m) $\int x\sin(bx)dx = \frac{1}{b^2}\sin(bx) - \frac{x}{b}\cos(bx) + C$

(n) $\int x\cos(bx)dx = \frac{1}{b^2}\cos(bx) + \frac{x}{b}\sin(bx) + C$

If limits are specified then we have:

$$\int_a^b f'(x)dx = [f(x)]_a^b = f(b) - f(a), \qquad \text{where } f'(x) \text{ is the differential of } f(x)$$

$$\int_a^b f(x)dx + \int_b^c f(x)]dx = \int_a^c f(x)dx \tag{1.9.2}$$

$$\int_a^b f(x)dx = \int_a^b \phi(x)dx + \int_a^b \psi(x)dx, \qquad \text{where } f(x) = \phi(x) + \psi(x)$$

and a useful general formula for integrating, referred to as integrating by parts:

$$\int f(x)g'(x)dx = f(x)g(x) - \int g(x)f'(x)dx \tag{1.9.3}$$

In some cases it is convenient to change the variable in an integral as for example in Eq. (3.6.7) where we have the integral:

$$a(t) = \int_{-\infty}^{t} \frac{A\omega_m}{\pi}\operatorname{sinc}[\omega_m(t - \tau)]dt \tag{1.9.4}$$

which is simplified if we make the substitution $x = \omega_m(t - \tau)$. To do this we must also replace $dt$ and evaluate the corresponding limits for the new variable.

Since $x = \omega_m(t - \tau)$,

then $dx = \omega_m dt - 0$ or $dt = dx/\omega_m$

For $t = t$, then $x = \omega_m(t - \tau)$ and for $t = -\infty$, then $x = -\infty$ (1.9.5)

$$\text{so now} \quad a(t) = \int_{-\infty}^{\omega_m(t-\tau)} \frac{A\omega_m}{\pi} \operatorname{sinc}(x) \frac{dx}{\omega_m}$$

A commonly encountered integral arises in cases where we have exponential rises or decays. An example is the decay of oscillations in a resonant circuit as discussed in Section 3.5, where we have the differential equation:

$$\frac{dU}{dt} = \frac{-UR}{L} = \frac{-U\omega_0}{Q} \tag{1.9.6}$$

and we need to integrate this to find how the energy $U$ varies with time $t$. The variables need to be separated, i.e. all the $U$'s on one side and all the $t$'s on the other, which gives:

$$\frac{dU}{U} = \frac{-\omega_0\, dt}{Q}, \quad \text{with the limits for } t = 0,\ U = U_0 \text{ and } t = t,\ U = U$$

$$\text{so} \int_{U_0}^{U} \frac{dU}{U} = \int_{0}^{t} \frac{-\omega_0\, dt}{Q}, \quad \text{and hence} \quad [\ln(U)]_{U_0}^{U} = \left(\frac{-\omega_0\, t}{Q}\right)_0^t \tag{1.9.7}$$

$$\text{or } \ln(U) - \ln(U_0) = \ln\left(\frac{U}{U_0}\right) = \frac{-\omega_0}{Q}(t - 0)$$

and thus $U = U_0 \exp(-\omega_0 t/Q)$

In Section 3.6 we have to evaluate the integral:

$$\begin{aligned} h(t) &= \frac{1}{2\pi} \int_{-\omega_m}^{+\omega_m} (Ae^{-j\omega\tau}) e^{j\omega\tau} d\omega \\ &= \frac{A}{2\pi} \int_{-\omega_m}^{+\omega_m} e^{j\omega(t-\tau)} d\omega \end{aligned} \tag{1.9.8}$$

$$=\frac{A}{2\pi}\left[\frac{e^{j\omega(t-\tau)}}{j(t-\tau)}\right]_{-\omega_m}^{\omega_m} \tag{1.9.8 cont.}$$

$$=\frac{A}{2\pi}\left[\frac{e^{j\omega_m(t-\tau)}-e^{-j\omega_m(t-\tau)}}{j(t-\tau)}\right]$$

$$=\frac{A}{\pi}\frac{\sin[\omega_m(t-\tau)]}{(t-\tau)}, \quad \text{since } \sin(x)=\frac{e^x-e^{-x}}{2j}$$

$$=\left(\frac{A\omega_m}{\pi}\right)\text{sinc}[\omega_m(t-t)]$$

Some other integrals with given limits are:

$$\int_0^{\infty}\frac{\sin x}{x}\,dx=\frac{\pi}{2}$$

$$\int_{-\infty}^{+\infty}\exp(-x^2)\,dx=\sqrt{\pi}$$

$$\int_{-\infty}^{+\infty}\exp(-\pi x^2)\,dx=1$$

$$\int_0^{+\infty}\exp(-ax)\,dx=\frac{1}{a}$$

$$\int_0^{+\infty}x\exp(-ax)\,dx=\frac{1}{a^2}$$

$$\int_0^{+\infty}x^2\exp(-ax)\,dx=\frac{2}{a^3} \tag{1.9.9}$$

$$\int_0^{+\infty}x^{\frac{1}{2}}\exp(-ax)\,dx=\frac{\sqrt{\pi}}{2a\sqrt{a}}$$

$$\int_0^{+\infty}\frac{\exp(-ax)-\exp(-bx)}{x}\,dx=\ln\left(\frac{b}{a}\right)$$

$$\int_0^{+\infty}\exp(-ax)\sin(mx)\,dx=\frac{m}{a^2+m^2}$$

$$\int_0^{+\infty} \exp(-ax)\cos(mx)\,dx = \frac{a}{a^2+m^2}$$
$$\int_0^{\pi} \sin^2(mx)\,dx = \int_0^{\pi} \cos^2(mx)\,dx = \frac{\pi}{2} \qquad (1.9.9\ \text{cont.})$$

## References and additional sources 1.9

Abramowitz M., Stegun I. A. (1970): *Handbook of Mathematical Functions with Formulas, Graphs and Mathematical Tables*, Applied Mathematics Series, Washington: National Bureau of Standards.

Boas M. L. (1966): *Mathematical Methods in the Physical Sciences*, New York: John Wiley. Library of Congress Cat. No. 66-17646.

Courant R. (1937): *Differential and Integral Calculus*, 2nd Edn, Vols 1 and 2, London: Blackie and Son.

Covington M. S. (1954): Two new mathematical results. *Proc. IRE* **42**, 1703. This describes a special integral:

$$\int \frac{dc}{\text{cabin}} = \log \text{cabin} + C = \text{houseboat}$$

James G., Burley D., Clements D., Dyke P., Searl J., Wright J. (1996): *Modern Engineering Mathematics*, 2nd Edn, Wokingham: Addison-Wesley. ISBN 0-201-87761-9.

Lambourne R., Tinker M. (Eds) (2000): *Basic Mathematics for the Physical Sciences*, New York: John Wiley. ISBN 0-471-85207-4.

Poularikas A. D. (Ed.) (1996): *The Transforms and Applications Handbook*, Boca Raton: CRC Press and IEEE Press. ISBN 0-8493-8342-0.

# 1.10 Equations and determinants

What is of greater importance is that the anti-mathematicians sometimes do a deal of mischief. For there are many of a neutral frame of mind, little acquainted themselves with mathematical methods, who are sufficiently impressible to be easily taken in by the gibers and to be prejudiced thereby; and, should they possess some mathematical bent, they may be hindered by their prejudice from giving it a fair development. We cannot all be Newtons or Laplaces, but there is an immense amount of moderate mathematical talent lying latent in the average man I regard as fact; and even the moderate development implied in a working knowledge of simple algebraic equations can, with common sense to assist, be not only the means of valuable mental discipline, but even be of commercial importance (which goes a long way with some people), should one's occupation be a branch of engineering for example.

Oliver Heaviside (1891): *Electromagnetic Theory*, January 16, Vol. I, p. 7

To follow some of the analyses that will be presented it is necessary to have a knowledge of the techniques of dealing with and solving equations. It is also useful to understand the geometrical form of equations. A very simple equation relating a quantity $y$ to a variable $x$ is:

$$y = mx + c \tag{1.10.1}$$

where $m$ and $c$ are constants. We say that $y$ is a function of $x$: $y = f(x)$, where $x$ is the independent variable and $y$ is the dependent variable. For any value of $x$ there will be a corresponding value of $y$. If we plot the relationship in Cartesian coordinates we get Fig. 1.10.1 (line I).

The equation represents a straight line which intercepts the $y$-axis at $c$ and which has a slope of $m$. For any right-angled triangle as shown the slope is the ratio of side $PQ$ to $QR$. If the line slopes to the right, as shown, then the slope is positive. If we simply change the sign of $m$ then the line will be as shown at II. Differentiation is the usual method of determining the slope of a function. In this case we have:

$$\frac{dy}{dx} = m + 0 \tag{1.10.2}$$

so the slope is everywhere the same, i.e. it is independent of $x$ as is evident from the graph. The differential at a particular point tells us the rate-of-change of the func-

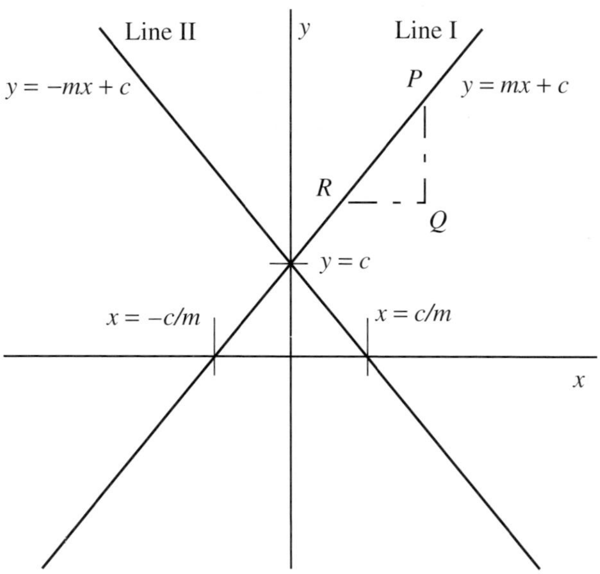


**Fig. 1.10.1** Straight line graphs.

tion at that point. One other point of interest is the value of $x$ for which $y=0$. Putting $y=0$ in (1.10.1) gives for $x$:

$$x=\frac{-c}{m} \tag{1.10.3}$$

and this value is called the root of the equation. Consider now a second order equation (this means that the highest power of the independent variable is 2). Such an equation is also known as a quadratic equation:

$$y=ax^2+bx+c \tag{1.10.4}$$

If this is plotted as before, the result will look like Fig. 1.10.2.

If $a$ is positive the curve will be oriented as shown (I), while if $a$ is negative the curve will be inverted (II). The curve is known as a parabola and is the shape of the face of a cone when it is sliced parallel to the side. The value of $c$ determines where the parabola cuts the $y$-axis, i.e. when $x=0$. The slope is found as above by differentiating to give:

$$\frac{dy}{dx}=2ax+b \tag{1.10.5}$$

so the slope now depends on the particular point chosen. At the maximum (or minimum) value of $y$ the slope will be zero, so (1.10.5) allows us to determine the corresponding value of $x$. We have:

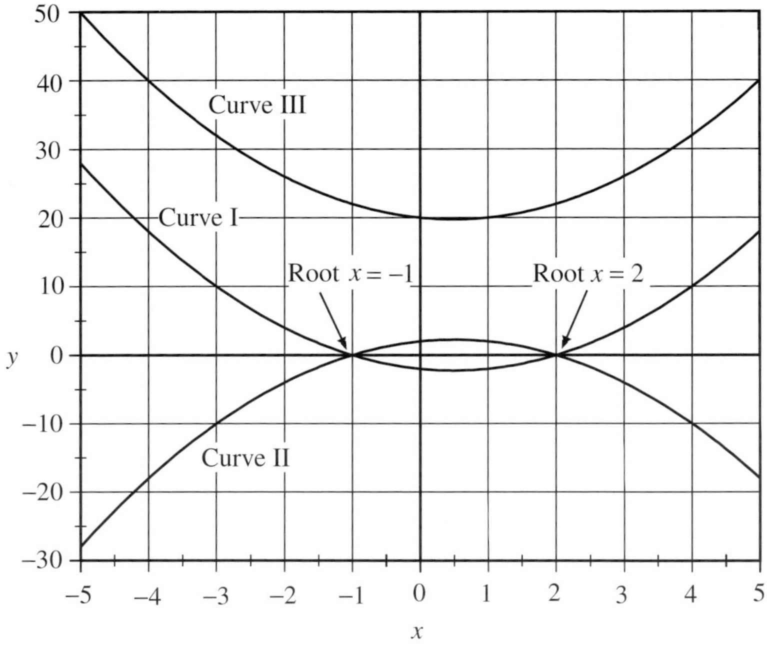


**Fig. 1.10.2** Graphical form of Eq. (1.10.4). The form is known as parabolic.

$$2ax + b = 0 \quad \text{or} \quad x = \frac{-b}{2a} \tag{1.10.6}$$

We are also interested in the roots of the quadratic function, i.e. where $y = 0$. From (1.10.4) with some algebraic manipulation we find the classic result:

$$x_{1,2} = \frac{-b \pm \sqrt{b^2 - 4ac}}{2a} \tag{1.10.7}$$

where we find two roots given by taking either the plus sign before the square-root or the minus sign. An equivalent way of writing our original equation that makes the matter of roots more evident is:

$$\begin{aligned} y &= (x - x_1)(x - x_2) \\ &= x^2 - xx_1 - xx_2 + x_1x_2 \\ &= x^2 - x(x_1 + x_2) + x_1x_2 \end{aligned} \tag{1.10.8}$$

The first line shows the form using the two roots $x_1$ and $x_2$: if $x = x_1$ or $x_2$ then $y = 0$. Multiplying out the brackets gives the third line, which demonstrates from comparison with (1.10.4) that:

$$-b = (x_1 + x_2) = \text{sum of the roots, and } c = x_1x_2 = \text{product of the roots} \tag{1.10.9}$$

It is one of the fundamental theorems of algebra that a function of order $n$ has $n$ roots. Here $n=2$ from the $ax^2$ term, so we must have two roots. Figure 1.10.2 was drawn with $a=1$, $b=-1$and $c=-2$, so you can easily check the values for the minimum and the roots. There are general solutions for some higher order equations (Poularikas 1996) but they are not convenient in the parametric sense so we will not use them. Unfortunately the equations for even quite simple circuits are of higher order than quadratic so we will at times try to approximate them to be able to make use of Eq. (1.10.7) or, where this is not readily done, we can turn to mathematical packages such as Mathcad. Fortunately SPICE is very good at solving high order equations so we can rely on such solutions or see if our approximations were valid, but it is useful to try and understand what is happening by way of direct analysis as far as possible.

For some values of the constants in Eq. (1.10.7) we find that the parabola does not cross the $x$-axis so there would appear to be no roots (Curve III, say). In these cases Eq. (1.10.7) will turn out to have $(b^2-4ac)$ negative, so that the square-root will be imaginary rather than real and we will land up with complex roots (Section 1.7). It should be noted that complex roots must *always* occur in complex conjugate pairs of the form:

$$x_1 = g+jh \quad \text{and} \quad x_2 = g-jh \tag{1.10.10}$$

A third order (cubic) equation will have three roots, at least one of which must be real: the other two can be either real or complex. The general shape of a third order equation is shown in Fig. 1.10.3.

For $x$ large, the value will be dominated by the $x^3$ term so the two 'ends' must go off to infinity as shown ( if we had $-x^3$ then the curve would be reflected in the $x$-axis). The curve must 'intersect' the $x$-axis in three places (i.e. there must be three roots) so the middle part of the curve has to be of the form shown, e.g. Curve I. However, like the second order form there may be complex roots, as for example shown by Curve II. There must therefore always be at least one real root and the other two may be real or a complex conjugate pair. Factoring high order polynomials is difficult but nowadays mathematical computer packages (e.g. Mathcad and others) make this very easy. An example of this is given in Section 3.5. In considering the stability of feedback systems (Section 3.10) it is shown that all the poles of the transfer function (Section 1.12) must lie in the left half-plane. This requires all the roots of the denominator polynomial to be negative. A general test for such a condition is provided by the Routh–Hurwitz techniques, though these do not tell one what the roots actually are. Some convenient relations between the coefficients of the polynomial are available for third and fourth orders, and are given in Section 1.12. An illustration of a much higher order equation is shown in Fig. 1.13.5 (see p. 83) though there is an overriding exponential decay as well which somewhat alters the form.

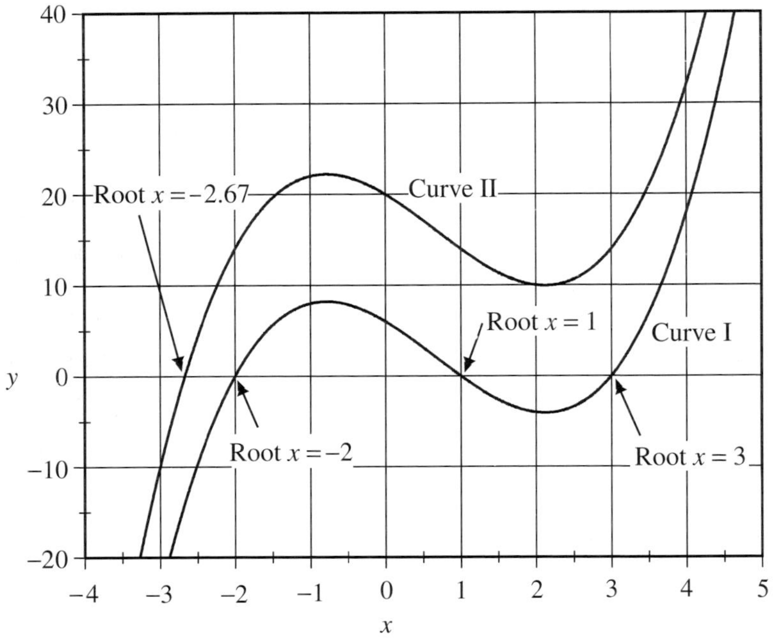


**Fig. 1.10.3** Graph of a cubic equation.

In solving for the voltages and currents in a circuit we use Kirchhoff's laws and arrive at a set of simultaneous equations. The direct approach is to use the formal solution as can be written in the form of *determinants* (Boas 1966). These entities are a shorthand way of writing complex expressions and have a number of rules which allow us to manipulate and evaluate them. A second order determinant, which is written as an ordered array of elements between vertical bars, is for example the equivalent of:

$$a_1b_2 - a_2b_1 \equiv \begin{vmatrix} a_1 & b_1 \\ a_2 & b_2 \end{vmatrix} \tag{1.10.11}$$

and the rule is to cross multiply the elements and add together with the signs alternating as shown:

$$\begin{vmatrix} + & - & + & - \\ - & + & - & + \\ + & - & + & - \\ - & + & - & + \end{vmatrix} \tag{1.10.12}$$

Determinants may have as many rows and columns as you need. Equation (1.10.11) shows how to evaluate a two by two, or second order, determinant. For third or higher order determinants it is necessary to expand successively until a second order is reached. For example, if we have a third order determinant we can expand it from *any* row or *any* column, but here we show the expansion from the first column:

$$\begin{vmatrix} a_1 & b_1 & c_1 \\ a_2 & b_2 & c_2 \\ a_3 & b_3 & c_3 \end{vmatrix} = a_1 \begin{vmatrix} b_2 & c_2 \\ b_3 & c_3 \end{vmatrix} - a_2 \begin{vmatrix} b_1 & c_1 \\ b_3 & c_3 \end{vmatrix} + a_3 \begin{vmatrix} b_1 & c_1 \\ b_2 & c_2 \end{vmatrix} \tag{1.10.13}$$

where to get the second order determinants you cross out the row and column of the expansion element ($a_1$ say for the first) and make allowance for the sign from Eq. (1.10.12).

Let us consider a set of three simultaneous equations for three unknowns $x$, $y$ and $z$:

$$\begin{aligned} a_1x + b_1y + c_1z + d_1 &= 0 \\ a_2x + b_2y + c_2z + d_2 &= 0 \\ a_3x + b_3y + c_3z + d_3 &= 0 \end{aligned} \tag{1.10.14}$$

then the solutions are given by:

$$\frac{x}{\begin{vmatrix} b_1 & c_1 & d_1 \\ b_2 & c_2 & d_2 \\ b_3 & c_3 & d_3 \end{vmatrix}} = \frac{-y}{\begin{vmatrix} a_1 & c_1 & d_1 \\ a_2 & c_2 & d_2 \\ a_3 & c_3 & d_3 \end{vmatrix}} = \frac{z}{\begin{vmatrix} a_1 & b_1 & d_1 \\ a_2 & b_2 & d_2 \\ a_3 & b_3 & d_3 \end{vmatrix}} = \frac{-1}{\begin{vmatrix} a_1 & b_1 & c_1 \\ a_2 & b_2 & c_2 \\ a_3 & b_3 & c_3 \end{vmatrix}} \tag{1.10.15}$$

where the sequence of indices is cyclic and, for example, for $x$ the 'column' of $a$'s is missed out. This procedure is called Cramer's rule. The procedure soon becomes extensive but it is direct and hence readily automated for computer solution. An example of the use of this technique is given in Section 5.21.

In manipulating an equation the simple rules of proportion are often most helpful. Consider the simple form shown:

$$\frac{A}{B} = \frac{C}{D} \tag{1.10.16}$$

where the symbols can represent more complex functions. As long as we carry out the same operation on both sides the result is still true. For example, adding 1 to each side gives:

$$\frac{A}{B} + 1 = \frac{C}{D} + 1 \quad \text{so} \quad \frac{A+B}{B} = \frac{C+D}{D} \quad \text{or stated simply}$$
$$\frac{\text{LTop} + \text{LBottom}}{\text{LBottom}} = \frac{\text{RTop} + \text{RBottom}}{\text{RBottom}} \tag{1.10.17}$$

and many other variants. For example, in Section 4.10 we wished to use Eq. (4.10.6):

$$i = i_0\left[1 - \exp\left(\frac{-\omega_0\, t}{2Q}\right)\right] \quad \text{or} \quad \frac{i}{i_0} = \frac{\left[1 - \exp\left(\frac{-\omega_0\, t}{2Q}\right)\right]}{1} \tag{1.10.18}$$

to obtain an expression for $(i_0 - i)/i_0$. Following the form of Eq. (1.10.17) we can most simply say (not forgetting that '*D*' in this case is 1):

$$\frac{\text{Bottom} - \text{Top}}{\text{Bottom}} \quad \text{to give} \quad \frac{i_0 - i}{i_0} = \frac{1 - \left[1 - \exp\left(\frac{-\omega_0\, t}{2Q}\right)\right]}{1} \tag{1.10.19}$$

$$\text{or} \quad \frac{i_0 - i}{i_0} = \exp\left(\frac{-\omega_0\, t}{2Q}\right)$$

Another form of equation frequently encountered is the differential equation, which is considered in Section 1.13.

## References and additional sources 1.10

Boas M. L. (1966): *Mathematical Methods in the Physical Sciences*, New York: John Wiley. Library of Congress Cat. No. 66-17646.

James G., Burley D., Clements D., Dyke P., Searl J., Wright J. (1996): *Modern Engineering Mathematics*, 2nd Edn, Wokingham: Reading: Addison-Wesley. ISBN 0-201-87761-9.

Jeffrey A. (1996): *Mathematics for Engineers and Scientists*, 5th Edn, London: Chapman and Hall. ISBN 0-412-62150-9.

Korn G. A., Korn T. M. (1989): Mathematics, formula, definitions, and theorems used in electronics engineering. In Fink D., Christansen D. *Electronics Engineer's Handbook*, 3rd Edn, New York: McGraw-Hill, ISBN 0-07-020982-0. Section 2.

Lambourne R., Tinker M. (Eds) (2000): *Basic Mathematics for the Physical Sciences*, New York: John Wiley. ISBN 0-471-85207-4.

Pipes L. A. (1958): *Applied Mathematics for Engineers and Physicists*, New York: McGraw-Hill. Library of Congress Cat. No. 57-9434.

Poularikas A. D. (Ed.) (1996): *The Transforms and Applications Handbook*, Boca Raton: CRC Press and IEEE Press. ISBN 0-8493-8342-0.

Senior T. B. A. (1986): *Mathematical Methods in Electrical Engineering*, Cambridge: Cambridge University Press. ISBN 0-521-30661-1.

Tinker M., Lambourne R. (2000): *Further Mathematics for the Physical Sciences*, New York: John Wiley. ISBN 0-471-86273-3.

# 1.11 Fourier transforms

A great mathematical poem.

Lord Kelvin on Fourier transforms

In investigating the flow of heat Fourier invented the technique of describing a function in terms of some more tractable basis functions, in his case sine waves. He showed how to represent any repetitive function in terms of sine (or cosine) waves. This has proved to be of enormous benefit in many fields and electronics is no exception. His proposition was that any continuous repetitive wave could be represented by an infinite sum of harmonically related sine waves (you can do a lot with an infinite number of pieces!). For a function $f(t)$ of angular frequency $\omega_1$ we can write:

$$f(t) = \frac{a_0}{2} + \sum_{n=1}^{n=\infty} [a_n \cos(n\omega_1 t) + b_n \sin(n\omega_1 t)] \tag{1.11.1}$$

where the first term represents any z.f. offset (the form $a_0/2$ is arbitrary and other forms are used) and $n$ is an integer. The general idea is readily accepted but the problem is to determine the magnitudes of the $a_n$ and $b_n$ coefficients. In terms of the period of the wave $T_1 = 2\pi/\omega_1$, these are found from:

$$a_n = \frac{2}{T_1} \int_{-T_1/2}^{T_1/2} f(t) \cos(n\omega_1 t)\, dt, \quad \text{for} \quad n = 0, 1, 2, 3, \ldots$$

$$b_n = \frac{2}{T_1} \int_{-T_1/2}^{T_1/2} f(t) \sin(n\omega_1 t)\, dt, \quad \text{for} \quad n = 1, 2, 3, \ldots \tag{1.11.2}$$

In terms of the exponential forms for cos and sin (Section 1.3):

$$\cos(n\omega t) = \frac{e^{jn\omega t} + e^{-jn\omega t}}{2} \quad \text{and} \quad \sin(n\omega t) = \frac{e^{jn\omega t} - e^{-jn\omega t}}{2j} \tag{1.11.3}$$

Table 1.11.1 *Harmonic amplitudes for a symmetrical square wave of* $\pm 1$ V *amplitude and given by Eq.* (1.11.5)

| Harmonic, $n$ | Amplitude | Harmonic, $n$ | Amplitude |
|---|---|---|---|
| 1 | 1.2732 | 11 | 0.1157 |
| 3 | 0.4244 | 13 | 0.0979 |
| 5 | 0.2546 | 15 | 0.0849 |
| 7 | 0.1819 | 17 | 0.0749 |
| 9 | 0.1415 | 19 | 0.0670 |

we can write Eq. (1.11.1) in the equivalent form:

$$f(t)=\sum_{n=-\infty}^{n=\infty} F(n)e^{jn\omega_1 t}$$

$$\text{with} \quad F(n)=\tfrac{1}{2}(a_n - jb_n), \quad \text{for} \quad n=0, \pm 1, \pm 2, \pm 3, \ldots \tag{1.11.4}$$

$$=\frac{1}{T_1}\int_{-T_1/2}^{T_1/2} f(t)e^{-jn\omega_1 t}\,dt$$

This gives discrete frequency components at the harmonically related frequencies $n\omega_1$. SPICE provides a very convenient means of demonstrating the correspondence between the original waveform and the harmonic representation. All we need to do is connect the appropriate number of voltage generators in series and add a resistive load (we could alternatively connect current generators in parallel). For a symmetric square wave of amplitude $A$ and frequency $\omega_1 = 2\pi f_1$ and time zero at an edge, the expansion is given by:

$$f(t)=\frac{2A}{\pi}\left[\sin(\omega_1 t)+\frac{1}{3}\sin(3\omega_1 t)+\frac{1}{5}\sin(5\omega_1 t)+\frac{1}{7}\sin(7\omega_1 t)+\cdots\right] \tag{1.11.5}$$

only the odd harmonics having the appropriate symmetry to match that of the square wave with time zero at an edge. If, however, you take the time origin at the centre of a 'square' (as for the pulse diagram in Fig. 1.11.6, see p. 55) then we will find cosine components (put $b = T/2$ in Eq. (1.11.6)). As an example of Eq. (1.11.5) we may take a square wave of amplitude $\pm 1$ V and calculate the amplitude of the harmonics (see Table 1.11.1).

Figure 1.11.1 shows the fit of the series to a 1 kHz square wave for the first, the sum up to the fifth and up to the nineteenth ($v_n$ = sum up to and including harmonic $n$). Figure 1.11.2 shows a corner of the square wave with all the progressive sums to illustrate the curious fact that the overshoot does not appear to get progressively smaller, only narrower, as the number of harmonics included increases

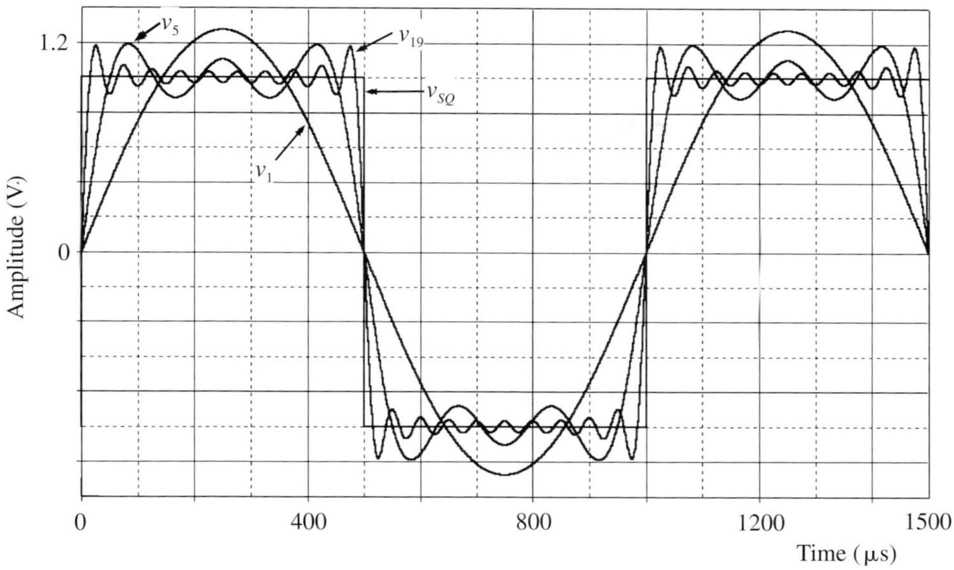


**Fig. 1.11.1** Fourier series fit to a square waveform.

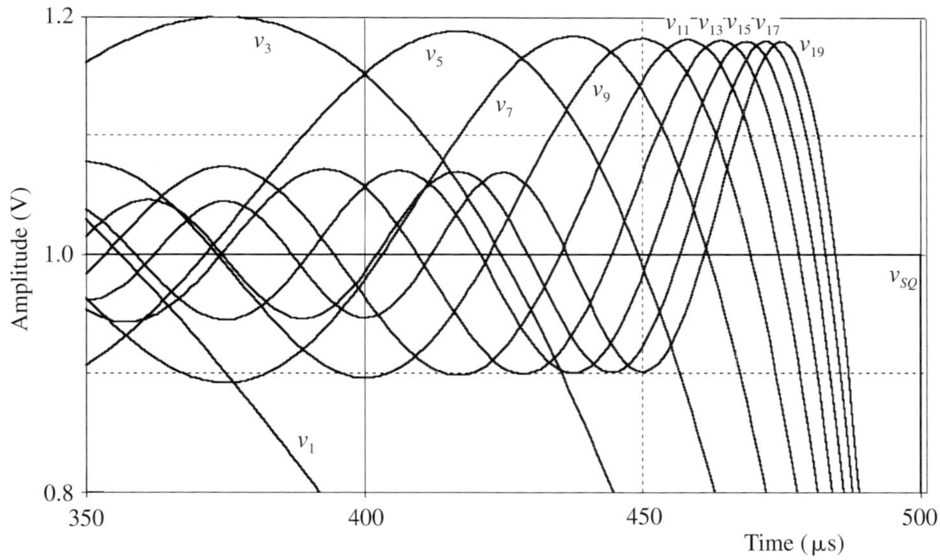


**Fig. 1.11.2** Fourier fit at a corner of the square wave.

(see e.g. Siebert 1986; Prigozy 1993). This phenomenon is called the Gibbs effect after the eminent physicist J. Willard Gibbs, creator of the powerful physical science of statistical mechanics, who first described this in 1899 (Gibbs 1899). The same overshoot is found in Section 3.6 for the 'brick wall' filter. It is also evident that to reproduce the sharp transitions, high frequency harmonics are necessary. Thus low-pass filtering will round off a square wave.

Table 1.11.2 *Harmonic amplitudes for a pulse waveform of $A = +1$V amplitude and width b a quarter of the period $T = 1/f_1$. The constant term is* 0.25 V.

| Harmonic, $n$ | Amplitude | Harmonic, $n$ | Amplitude |
|---|---|---|---|
| 1 | 0.4502 | 6 | −0.1061 |
| 2 | 0.3183 | 7 | −0.0643 |
| 3 | 0.1501 | 8 | 0.0000 |
| 4 | 0.0000 | 9 | 0.0500 |
| 5 | −0.0900 | 10 | 0.0637 |

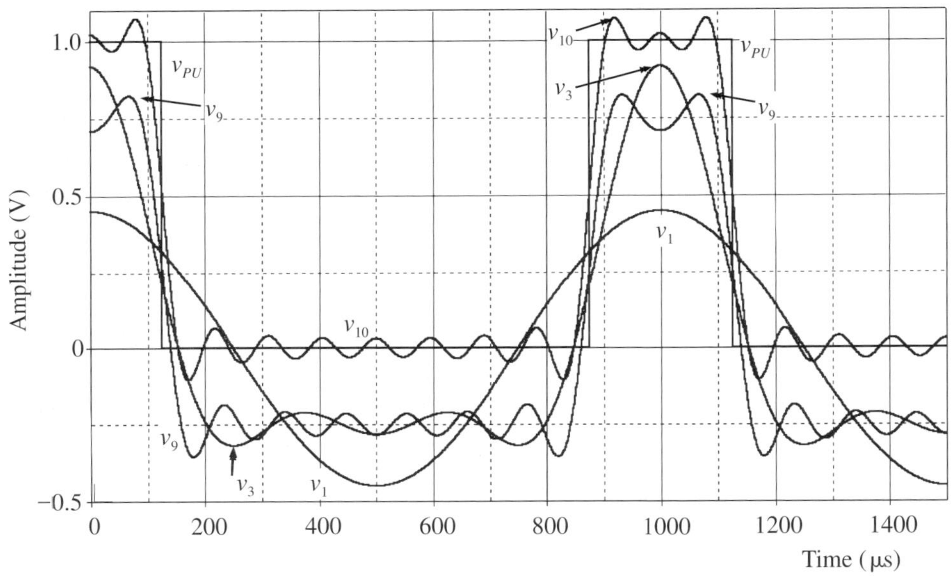


**Fig. 1.11.3** Fourier series fit to a pulse waveform. The labels indicate the sum up to the indicated harmonic. To make the traces more clear the sums, other than the final $v_{10}$, exclude the constant term of 0.25 V and so are offset.

If the wave mark-to-space ratio is not 1:1 then for an amplitude $A$, width $b$, frequency $f_1$ and we take the origin in the centre of $b$, then the series becomes:

$$f(t) = Abf_1 + 2Abf_1 \sum_{n=1}^{\infty} \left[ \frac{\sin(n\pi b f_1)}{n\pi b f_1} \right] \cos(2\pi n f_1 t) \qquad (1.11.6)$$

and now we find that the coefficients can vary in sign. Even and odd values of the harmonic $n$ are present with calculated values as shown in Table 1.11.2 (remember that the angles for the sin function are in radian).

The correspondence between the series and the pulse waveform is shown in Fig. 1.11.3. Harmonics $n = 4$ and 8 are identically zero in this case. To obtain the

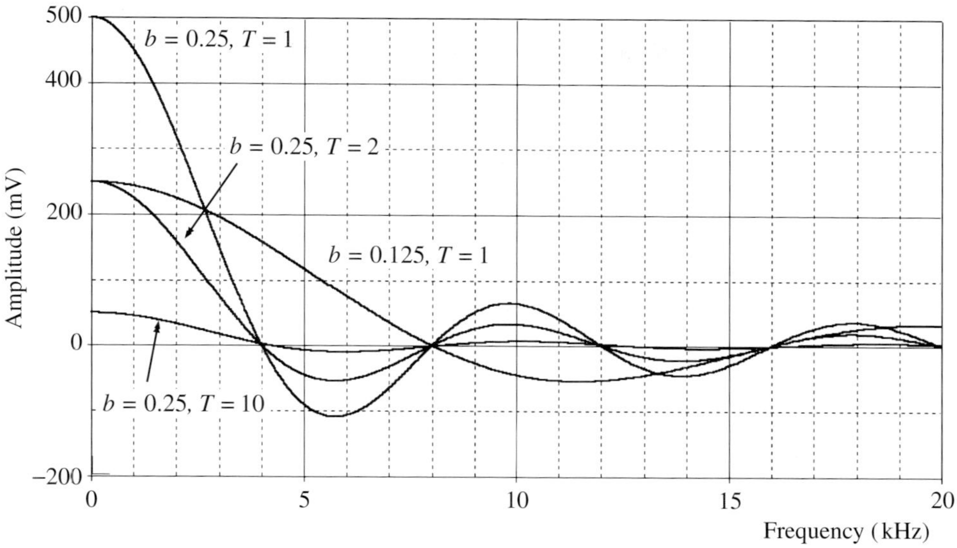


**Fig. 1.11.4** Sinc function for the pulse train Fourier amplitudes. *b* and *T* are given in ms. It may be made one-sided by plotting the *ABS* (absolute) value if desired. The zero amplitude harmonics are evident.

cos generators simply set the *PHASE* attribute of a *VSIN* generator to 90 (degrees).

The distribution of amplitudes in Eq. (1.11.6) is a sinc function (Section 3.6), which we can plot as an alternate means of obtaining the harmonic amplitudes. PSpice does not allow one to plot this directly on the Fourier display so it must be done separately. Simply run an *AC SWEEP* and ask for the function to be plotted. Now the frequency becomes a continuous function so we put $nf_1 = f$, and in this case we have $b = T/m$ with $T = 10^{-3}$ and $m = 4$ (*FREQUENCY* is the SPICE variable):

$$0.5\left[\frac{\sin(n\pi b f_1)}{n\pi b f_1}\right] \Rightarrow 0.5\left[\frac{\sin\left(\dfrac{\pi}{4}\dfrac{f}{10^3}\right)}{\dfrac{\pi}{4}\dfrac{f}{10^3}}\right] = \left[\frac{\sin(7.854\times 10^{-4}\times FREQUENCY)}{(15.708\times 10^{-4}\times FREQUENCY)}\right] \tag{1.11.7}$$

and the result is shown in Fig. 1.11.4. At the appropriate harmonic frequencies the amplitudes match the calculated values and signs in Table 1.11.2.

The inverse process of determining the Fourier content of a waveform is provided by PSpice by running a transient simulation and under *ANALYSIS/TRANSIENT* checking *ENABLE FOURIER*. To get good resolution the *FINAL TIME* should be set to an integer number of cycles and the more cycles the better. The *PRINT STEP* should be set to a small fraction of a cycle as this time is used

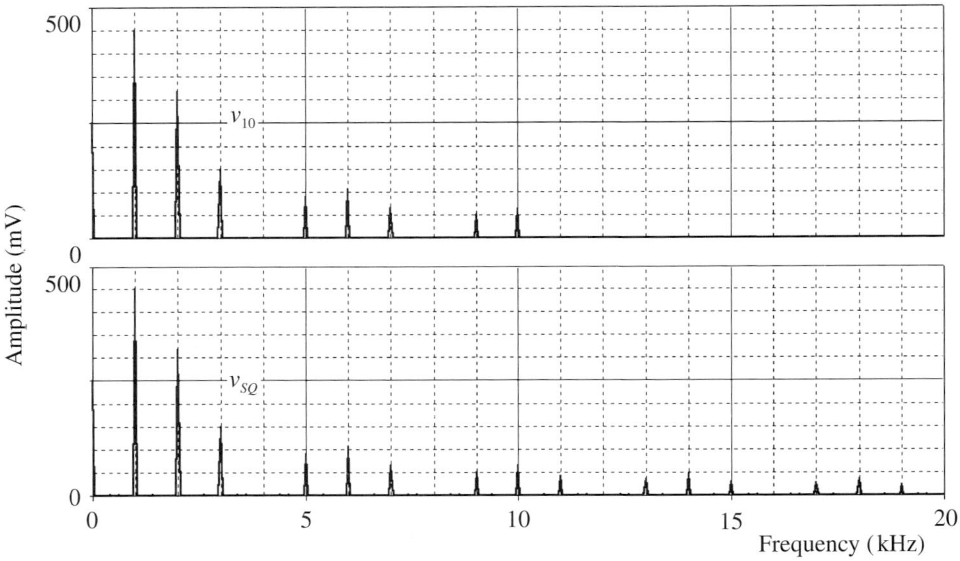


**Fig. 1.11.5** Fourier spectrum of pulse train from Fig. 1.11.3 together with spectrum of $v_{10}$ showing just the ten terms that constituted it. The amplitude of the components matches those in Table 1.11.2 (the SPICE analysis does not indicate the sign of the components).

for the sampling interval of the Fourier computation. OrCAD/MicroSim (1997) and Tuinenga (1988) give advice on the settings. As an example the pulse waveform used in Fig. 1.11.3 was used with a run of 20 ms (=20 cycles) and the results are shown in Fig. 1.11.5.

For convenient reference, we list here the Fourier expressions for a number of common waveforms as shown in Fig. 1.11.6.

$$\begin{aligned}
&\text{Triangular wave} \quad && F(t)=\frac{4A}{\pi^2}\sum_{m=1}^{\infty}\frac{1}{m^2}\cos(m\omega_1 t), \quad \text{with } m \text{ odd} \\
&\text{Saw-tooth wave} \quad && F(t)=\frac{A}{\pi}\sum_{m=1}^{\infty}\frac{1}{m}\sin(m\omega_1 t), \quad \text{with all } m \\
&\text{Half-rectified sine} \quad && F(t)=\frac{A}{\pi}\left[1+\frac{\pi}{2}\cos(\omega_1 t)+\sum_{m=1}^{\infty}(-1)^{m+1}\frac{2\cos(2m\omega_1 t)}{(4m^2-1)}\right] \\
&\text{Full-rectified sine} \quad && F(t)=\frac{A}{\pi}\left[2+\sum_{m=1}^{\infty}(-1)^{m+1}\frac{4\cos(2m\omega_1 t)}{(4m^2-1)}\right]
\end{aligned} \tag{1.11.8}$$

The Fourier theory considered so far applies to periodic waveforms that have been running for effectively an infinite time. If we have, say, an isolated pulse then we have to consider the effect on the transform. The commonly used approach to reach an isolated pulse is to let the period increase while keeping the pulse width fixed. Then

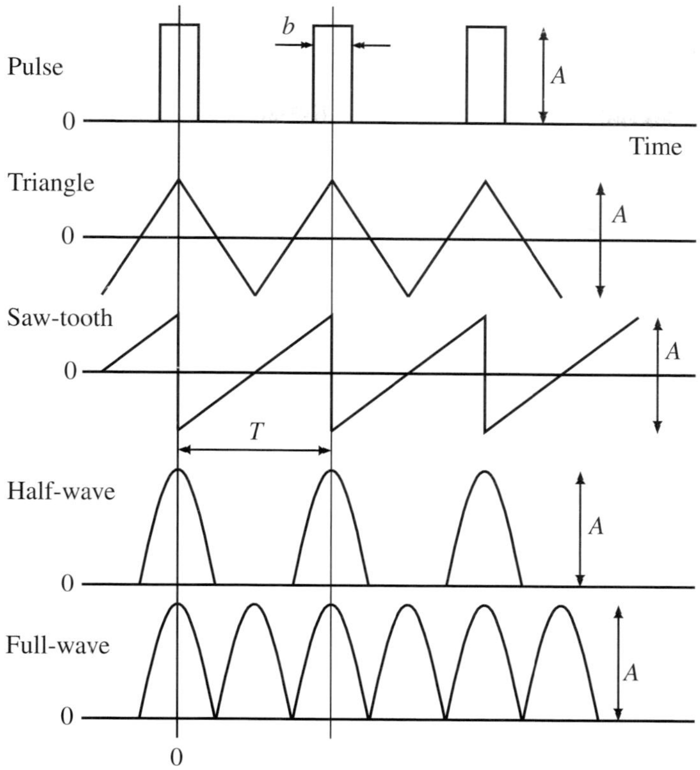


**Fig. 1.11.6** Periodic waveforms giving the Fourier series of Eq. (1.11.8).

we can see what happens as the period $T$ tends to infinity in the limit. PSpice can demonstrate the trend. Figure 1.11.4 shows the appropriate sinc functions where we have kept the pulse width $b$ fixed at a quarter of the original 1 ms period (i.e. 0.25 ms) and then set the period to 2 and 10 ms. The case of $T=1$ ms and $b=0.125$ ms is also shown. The discrete harmonic components are shown in Fig. 1.11.7.

It is evident that as the period increases the number of harmonics in the main lobe increases, and the overall amplitude and the spacing decrease. As the period tends to infinity the spacing becomes infinitesimal and we then have in effect a continuous distribution of frequencies rather than discrete harmonics. The coefficients $F(n)$ become vanishingly small but the product $F(n)T_1$ does not so we can use this as a new variable $F(\omega)$. The frequencies $n\omega_1$ similarly become continuous and we can now write them as just $\omega$, and the frequency $\omega_1$ becomes the infinitesimal $d\omega$. Thus we now have in the limit:

$$f(n)T_1 \Rightarrow F(\omega) = \int_{-\infty}^{+\infty} f(t)\exp(-j\omega t) \tag{1.11.9}$$

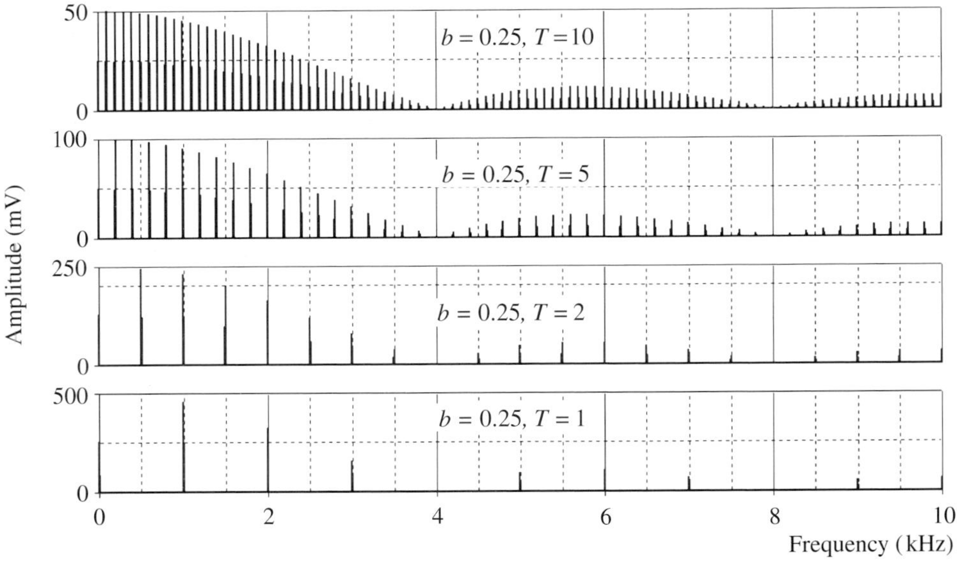


**Fig. 1.11.7** Harmonic amplitudes and spacing for the values of $b$ and $T$ indicated. $b$ and $T$ values are in ms. The sinc function for $T=5$ was omitted from Fig. 1.11.5 for clarity. Run time was 200 ms and print step was 5 μs.

$$\text{and} \quad f(t)=\sum_{n=-\infty}^{+\infty}\frac{F(\omega)}{T_1}\exp(jn\omega_1 t)=\sum_{n=-\infty}^{+\infty}F(\omega)\frac{\omega_1}{2\pi}\exp(jn\omega_1 t) \qquad (1.11.9\text{ cont.})$$

$$\Rightarrow\frac{1}{2\pi}\int_{-\infty}^{+\infty}F(\omega)\exp(-j\omega t)\,d\omega \quad \text{in the limit}$$

These relations for $F(\omega)$ and $f(t)$ define the Fourier transform and are said to be a transform pair, and like the closely related Laplace transforms (Section 1.12), many have been worked out and tabulated (Lighthill 1955; Stuart 1961; Pain 1976; Lynn 1986; Siebert 1986; Champeney 1987; James 1995). If, as in the recording of Fig. 1.11.7 you set the period at 200 ms to match the run time used there, so that there is but one pulse to be analysed, you will find a ‘continuum’ distribution.

Comparing the $b=0.25$, $T=1$ graph with the $b=0.125$, $T=1$ response in Fig. 1.11.4 illustrates the inverse relationship between time and frequency, as is also found in Section 3.6. The shorter pulse has the wider frequency spread. Taken to the limit of a $\delta(t)$ function the spectrum will cover all frequencies with equal amplitude. An infinite frequency range of cosine waves will all be in-phase at $t=0$ and so add, whereas everywhere else they will cancel. Since the $\delta$ function contains all frequencies then any circuit stimulated by one will produce its full range of responses. The use of the $\delta$ function is crucial to the analysis of circuits and PSpice makes use of it to evaluate Laplace expressions (Section 1.14).

The use of impulse functions in science and engineering was popularized by the English physicist P. A. M. Dirac and by Oliver Heaviside long before impulses became 'respectable' mathematically. Indeed we continue to use Dirac's notation, $\delta(t)$, and the unit impulse is often called *Dirac's δ-function*. Both Dirac and Heaviside stressed the idea that $\delta(t)$ was defined in terms of what it 'did'. Thus Dirac said, 'Whenever an improper function [e.g. impulse] appears it will be something which is to be used ultimately in an integrand – the use of improper functions thus does not involve any lack of rigour in the theory, but is merely a convenient notation, enabling us to express in a concise form certain relations which we could, if necessary, rewrite in a form not involving improper functions, but only in a cumbersome way which would tend to obscure the argument.'

W. McC. Siebert 1986, p. 319

Lighthill (1955) dedicated his book to:

Paul Dirac, who saw it must be true, Laurent Schwartz, who proved it, and George Temple, who showed how simple it could be made.

but for a more accessible treatment of the properties of such generalized functions see Kuo (1966).

A Gaussian pulse (but see Stigler 1999) has the property that its transform is also Gaussian though the widths are, as we have seen above, inverse. Expressions for a Gaussian pulse and its corresponding Fourier transform are:

$$f(t) = A\exp\left(\frac{-t^2}{2\sigma^2}\right), \text{ with half-width at } e^{-\frac{1}{2}} \text{ of the peak of } \sigma, \text{ and} \tag{1.11.10}$$

$$F(\nu) = A\sigma(2\pi)^{\frac{1}{2}}\exp(-2\pi^2\nu^2\sigma^2), \text{ with half-width at } e^{-\frac{1}{2}} \text{ of the peak of } (2\pi\sigma)^{-1}$$

To try a well isolated pulse (similar in width to that of Fig. 1.11.3) we use an analog behavioural model (ABM) and define the pulse for $A=1$ and $\sigma^2 = 1$ E$-7$, by:

$$\exp[-(TIME-0.05)*(TIME-0.05)/(2*1\mathrm{E}-7)] \tag{1.11.11}$$

where we have offset the pulse by 50 ms and will use a run of 100 ms to give a frequency resolution of 10 Hz. Figure 1.11.8 shows an expanded view of the pulse and the Fourier spectrum. Measurement of the spectral width at $e^{-0.5} = 0.6065$ of the peak agrees with the calculated value $(2\pi\sigma)^{-1} = (2\pi \times 3.162 \times 10^{-4})^{-1} = 503$ Hz.

It should be noted that PROBE uses a discrete Cooley–Tukey FFT (see Tuinenga 1988) so the overall amplitude does not agree with Eq. (1.11.10).

The Fourier transform thus allows us to transfer readily between time and frequency descriptions of a system. It also signifies that you cannot change one without the other being affected.

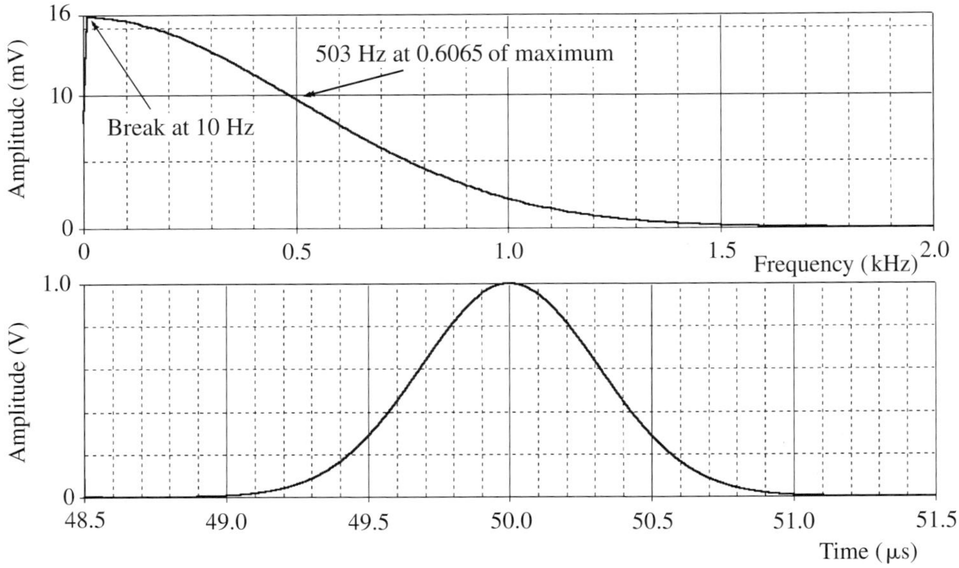


**Fig. 1.11.8** Gaussian pulse and Fourier spectrum for 100 ms run. The lowest frequency is thus at 10 Hz as indicated. (*STEP CEILING*= *PRINT STEP*= 1 μs.)

## SPICE simulation circuits

| | |
|---|---|
| Fig. 1.11.1 | Fourier 1.SCH |
| Fig. 1.11.2 | Fourier 1.SCH |
| Fig. 1.11.3 | Fourier 2.SCH |
| Fig. 1.11.4 | Fourier 5.SCH |
| Fig. 1.11.5 | Fourier 3.SCH |
| Fig. 1.11.7 | Fourier 6.SCH |
| Fig. 1.11.8 | Fourier 7.SCH |

## References and additional sources 1.11

Boas M. L. (1966): *Mathematical Methods in the Physical Sciences*, New York: John Wiley. Library of Congress Cat. No. 66-17646.

Champeney D. C. (1987): *A Handbook of Fourier Theorems*, Cambridge: Cambridge University Press. ISBN 0-521-26503-7.

Fourier J. B. J. (1822): *Théorie Analytique de la Chaleur*, New York: Dover (English version).

Gibbs J. W. (1899): Fourier's series. *Nature* **59**, 606. The adjacent article entitled 'Wireless telegraphy.' makes interesting reading.

James J. F. (1995): *A Student's Guide to Fourier Transforms*, Cambridge: Cambridge University Press. ISBN 0-521-46829-9.

Kuo F. F. (1966): *Network Analysis and Synthesis*, New York: John Wiley. ISBN 0-471-51118-8. See Appendix B.

Lighthill M. J. (1955): *Fourier Analysis and Generalized Functions*, Cambridge: Cambridge University Press.

Lynn P. A. (1986): *Electronic Signals and Systems*, Basingstoke: Macmillan Education. ISBN 0-333-39164-0.

OrCAD/MicroSim (1997): MicroSim Application Hints – How can I get more accurate Fourier analysis of transient results from MicroSim PSpice? www.microsim.com/faq.html

Pain H. J. (1976): *The Physics of Vibrations and Waves*, 2nd Edn, New York: John Wiley. ISBN 0-471-99408-1. The expression for the spectrum of a Gaussian pulse (p. 261) and the width are in error.

Pipes L. A. (1958): *Applied Mathematics for Engineers and Physicists*, New York: McGraw-Hill. Library of Congress Cat. No. 57-9434.

Poularikas A. D. (Ed.) (1996): *The Transforms and Applications Handbook*, Boca Raton: CRC Press and IEEE Press. ISBN 0-8493-8342-0.

Prigozy S. (1993): Teaching the Gibbs phenomenon with PSPICE. *Computers in Education J.* **III** No. 1, January–March, 78–84.

Senior T. B. A. (1986): *Mathematical Methods in Electrical Engineering*, Cambridge: Cambridge University Press. ISBN 0-521-30661-1.

Shoucair F. S. (1989): Joseph Fourier's analytical theory of heat: a legacy to science and engineering. *IEEE Trans*. **EDUC-32**, 359–366.

Siebert W. McC. (1986): *Circuits, Signals, and Systems*, Cambridge Massachusetts: MIT Press/McGraw-Hill. ISBN 0-07-057290-9.

Stigler S. (1999): *Statistics on the Table: the History of Statistical Concepts and Methods*, Cambridge: Harvard University Press. ISBN 0-674-83601-4. See Section 14: Stigler's Law of Eponymy – 'No scientific discovery is named after its original discoverer'. The law itself may be construed to contradict Stigler's eponym. The Gaussian distribution was probably originated by de Moivre. It may also be noted that Laplace used 'Fourier transforms' before Fourier's publication, and that Lagrange used 'Laplace transforms' before Laplace began his career. See p. 278. As Stigler quotes one historian's view (from an unknown source) 'Every scientific discovery is named after the last individual too ungenerous to give due credit to his predecessors'.

Stuart R. D. (1961): *An Introduction to Fourier Analysis*, London: Chapman and Hall, Science Paperback SP21.

Tuinenga P. W. (1988): *SPICE: A Guide to Circuit Simulation and Analysis Using PSPICE*, Prentice Hall. 3rd Edition 1995. ISBN 0-13-158775-7.

# 1.12 Laplace transforms

. . . Marquis Pierre Simon de Laplace (1749–1827), who pointed out the biunique relationship between the two functions and applied the results to the solution of differential equations in a paper published in 1779 with the rather cryptic title 'On what follows'. The real value of the Laplace transform seems not to have been appreciated, however, for over a century, until it was essentially rediscovered and popularized by the eccentric British engineer Oliver Heaviside (1850–1925), whose studies had a major impact on many aspects of modern electrical engineering.

W. McC. Siebert (1986): *Circuit, Signals and Systems*, Cambridge, Mass: MIT Press and McGraw-Hill, p. 44

We can analyse circuits with sinusoidal waveforms using the techniques discussed in Section 3.2, leading to the ideas of complex impedances and phase shifts. But how can we deal with square waves or pulses, or other non-sinusoidal shapes? You could do a Fourier transform to find the equivalent set of sinusoids, use the $j\omega$ approach on each, and then add up the results – not a pleasing prospect.

Many problems are more readily solved by transferring them into a different representation, e.g. the use of logarithms. In the present circumstance we have the problem of an electronic network containing active or passive elements to which is applied an input or *excitation function* (Fig. 1.12.1). The network modifies the input according to its *transfer function* to give an output or *response function,* which can be written algebraically:

$$\text{Response function} = \text{Transfer function} \times \text{Excitation function} \tag{1.12.1}$$

Approaching this in the normal way results in an equation containing a combination of integrals, derivatives, trigonometric terms, etc., which is awkward to deal with even with sinusoidal let alone other waveforms. The attraction of the Laplace transform is that *all* these functions are changed to algebraic forms which can be easily manipulated. Once this is done the problem then arises of transforming back from the Laplace to the original representation as we would do with logarithms by looking up the antilogarithms. To enable this approach to be used with the same facility a very large number of Laplace transforms have been worked out so there is no need to worry about the mathematics (see the references at the end of this section).

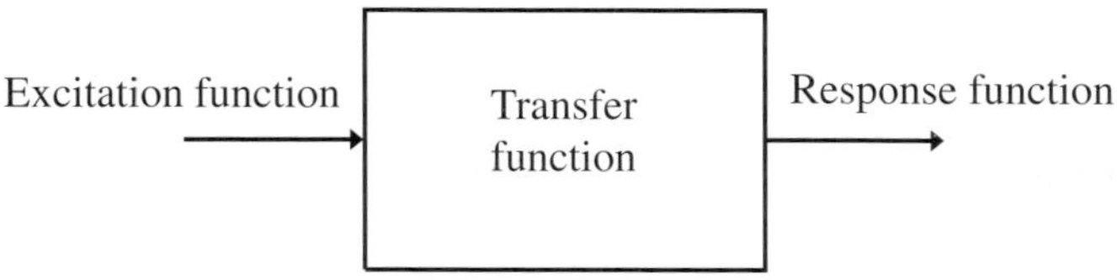


**Fig. 1.12.1** Role of the transfer function.

The technique for writing down the impedance $Z$ of components of a circuit for sinusoidal waveforms is already familiar (see Section 3.2):

$$Z_R = R; \qquad Z_C = \frac{1}{j\omega C} = \frac{-j}{\omega C}; \qquad Z_L = j\omega L \tag{1.12.2}$$

We may look upon $j\omega$ as a sort of operator that operates on the magnitude of the component to give its impedance in both magnitude and phase. This particular form arises from the property of sinusoids that differentiation and integration do not change the *form* of the function. The amplitude and phase may be altered but the shape remains sinusoidal. However, in many cases we are not dealing with the simple sinusoidal form, but with more complex forms such as square, triangular, saw-tooth, pulses, etc. For these the $j\omega$ operator is only applicable in the Fourier sense.

What is needed is a more general operator that is applicable in all cases (even sinusoidal waves have to be switched on). It turns out that the operator to use includes $j\omega$ as we may have expected, but is symmetrical (in the complex number sense) in that it has a real ($\sigma$) as well as an imaginary ($\omega$) part:

$$s = \sigma + j\omega \tag{1.12.3}$$

The $j\omega$ operator involves a transformation between the time ($t$) and the frequency ($\omega$) domains, the connection between the two being the Fourier transform. For the $s$ operator the transform is between the time ($t$) and the complex frequency ($s$) domains, the connection being the Laplace transform. The idea of a complex frequency may seem strange at first but with a little experience one soon gets used to it. The use of $s$ also saves some writing even in sinusoidal only cases, and here we will commonly do that.

The Laplace transform $F(s)$ of some function $f(t)$ is defined (note that it is one sided, i.e. the integration is from 0 to $\infty$; see Section 1.9) by:

$$\mathcal{L}f(t) = F(s) \int_0^\infty f(t) e^{-st}\, dt \tag{1.12.4}$$

and the inverse transform $\mathcal{L}^{-1}$ by:

$$\mathcal{L}^{-1}F(s) = f(t) = \frac{1}{2\pi j} \oint F(s) e^{ts}\, ds \tag{1.12.5}$$

Table 1.12.1 *Some Laplace transforms*

| $f(t)$ | $F(s)$ |
|---|---|
| 1. $\delta(t)$ | 1 (impulse function) |
| 2. $\delta(t-a)$ | $\exp(-as)$ (delayed impulse) |
| 3. $Au(t)$ | $\frac{A}{s}$ (step function, $A=$ constant) |
| 4. $Au(t-a)$ | $\frac{A\exp(-as)}{s}$ (delayed step function) |
| 5. $t$ | $\frac{1}{s^2}$ (ramp function) |
| 6. $\exp(-\alpha t)$ | $\frac{1}{(s+a)}$ |
| 7. $\exp(\alpha t)$ | $\frac{1}{(s-a)}$ |
| 8. $\cos(\omega t)$ | $\frac{s}{(s^2+\omega^2)}$ |
| 9. $\sin(\omega t)$ | $\frac{\omega}{(s^2+\omega^2)}$ |
| 10. $\exp(-\alpha t)\sin(\omega t)$ | $\frac{\omega}{(s+\alpha)^2+\omega^2}$ |
| 11. $\exp(-\alpha t)\cos(\omega t)$ | $\frac{(s+\alpha)}{(s+\alpha)^2+\omega^2}$ |
| 12. $\exp[(\alpha+j\omega)t]$ | $\frac{1}{(s-\alpha-j\omega)}$ |
| 13. $\exp[(\alpha-j\omega)t]$ | $\frac{1}{(s-\alpha+j\omega)}$ |
| 14. $(1-\alpha t)\exp(-\alpha t)$ | $\frac{s}{(s+\alpha)^2}$ |
| 15. $\frac{t^{n-1}\exp(-\alpha t)}{(n-1)!}$ | $\frac{1}{(s+\alpha)^n}$ |
| 16. $\frac{[\exp(-\alpha t)-\exp(-\beta t)]}{(\beta-\alpha)}$ | $\frac{1}{(s+\alpha)(s+\beta)}$ |
| 17. $\frac{[\alpha\exp(-\alpha t)-\beta\exp(-\beta t)]}{(\alpha-\beta)}$ | $\frac{s}{(s-\alpha)(s-\beta)}$ |
| 18. $\frac{[(a_0-\alpha)\exp(-\alpha t)+(\beta-a_0)\exp(-\beta t)]}{(\beta-\alpha)}$ | $\frac{(s+a_0)}{(s+\alpha)(s+\beta)}$ for $\alpha\neq\beta$; see No. 14 for $\alpha=\beta$ |

Table 1.12.1 (*cont.*)

| $f(t)$ | $F(s)$ |
|---|---|
| 19. $\dfrac{(a_0-\alpha)e^{-\alpha t}}{(\alpha^2+\beta^2)}+\left[\dfrac{a_0^2+\beta^2}{\alpha^2\beta^2+\beta^4}\right]^{\frac{1}{2}}\sin(\beta t+\phi)$ | $\dfrac{(s+a_0)}{(s+\alpha)(s^2+\beta^2)}$ |
| $\phi=\tan^{-1}\left(\dfrac{\alpha}{\beta}\right)-\tan^{-1}\left(\dfrac{a_0}{\beta}\right)$ | |
| 20. $\dfrac{(a_0^2+\alpha^2)^{\frac{1}{2}}\sin(\alpha t+\phi)}{\alpha}$ | $\dfrac{(s+a_0)}{(s^2+\alpha^2)}$ |
| $\phi=\tan^{-1}\left(\dfrac{\alpha}{a_0}\right)$ | |
| 21. $\left[\dfrac{1-e^{-\alpha t}-\alpha te^{-\alpha t}}{\alpha^2}\right]$ | $\dfrac{1}{s(s+\alpha)^2}$ |
| 22. $1-\operatorname{erf}\left(\dfrac{a}{2\sqrt{t}}\right)$ | $\dfrac{1}{s}\exp(-a\sqrt{s})$ |
| 23. $\dfrac{1}{\sqrt{\pi t}}\exp\left(\dfrac{-a^2}{4t}\right)$ | $\dfrac{1}{\sqrt{s}}\exp(-a\sqrt{s})$ |

the integral being the line integral around all the poles of $F(s)$ (we will come to poles later). These look rather formidable but are only included here for reference. Common transform pairs and some operational theorems are appended in Tables 1.12.1 and 1.12.2. The unit step function $u(t)$ and unit impulse (or delta) function $\delta(t)$ are central to the use of the transform.

Consider simple $R$, $L$ and $C$ circuit elements. The relations between voltage and current as functions of time $t$ are:

$$v_R(t)=Ri(t);\quad v_L(t)=L\frac{di(t)}{dt};\quad v_C(t)=\frac{1}{C}\int i(t)dt \tag{1.12.6}$$

which give when Laplace transformed (see Table 1.12.1; we use upper case for transformed quantities, e.g. $\mathscr{L}v(t)=V(s)$):

$$\begin{aligned}&V_R(s)=RI(s);\quad V_L(s)=sLI(s);\quad V_C(s)=\frac{I(s)}{sC}\\ &\text{or } Z_R(s)=\frac{V_R(s)}{I(s)}=R;\quad Z_L(s)=\frac{V_L(s)}{I(s)}=sL;\quad Z_C(s)=\frac{V_C(s)}{I(s)}=\frac{1}{sC}\end{aligned} \tag{1.12.7}$$

i.e. just like the sinusoidal forms in Eq. (1.12.2) – in fact those are just for the special case $\sigma=0$, so there is nothing really new to accommodate so far.

Table 1.12.2 *Operational theorems*[a]

| $f(t)$ | $F(s)$ | Theorem |
|---|---|---|
| 1. $af(t)$ | $aF(s)$ | |
| 2. $\exp(at)f(t)$ | $F(s-a)$ | Shifting |
| 3. $f(t/a)$ | $aF(s-a)$ | |
| 4. $\int_0^t f(t)dt$ | $\frac{F(s)}{s}$ | |
| 5. $\lim_{t\to 0} f(t)$ | $\lim_{s\to\infty} sF(s)$ | Initial value |
| 6. $\lim_{t\to\infty} f(t)$ | $\lim_{s\to 0} sF(s)$ | Final value |
| 7. $\int_0^t f_1(t-\tau)f_2(\tau)d\tau$ | $F_1(s)\bullet F_2(s)$ | Convolution |
| 8. $\frac{df(t)}{dt}$ | $sF(s)-f(0)$ | |

*Note:*
[a] For further tables of transforms see the references.

For networks that we will be concerned with the relationship between input $x$ and output $y$ can be expressed in the form of a linear differential equation with constant coefficients:

$$a_0 y + a_1\frac{dy}{dt} + a_2\frac{d^2y}{dt^2} + \cdots = b_0 x + b_1\frac{dx}{dt} + b_2\frac{d^2x}{dt^2} + \cdots \tag{1.12.8}$$

In these cases the Laplace domain *transfer function*, $H(s)$, can be written as a rational function of $s$ with coefficients obtained directly from the differential equation:

$$H(s) = \frac{\text{Output}}{\text{Input}} = \frac{Y}{X} = \frac{a_0 + a_1 s + a_2 s^2 + \cdots + a_m s^m}{b_0 + b_1 s + b_2 s^2 + \cdots + a_n s^n} \tag{1.12.9}$$

where $n \geq m$ since the output usually falls to zero as the frequency $s\to\infty$. As an example consider the simple *LCR* circuit, Fig. 1.12.2. Here the input, $x$, is now the voltage $v_{in}$ and the output, $y$, is $v_{out}$.
Thus:

$$v_{in}(t) = L\frac{di(t)}{dt} + \frac{1}{C}\int i(t)dt + Ri(t)$$

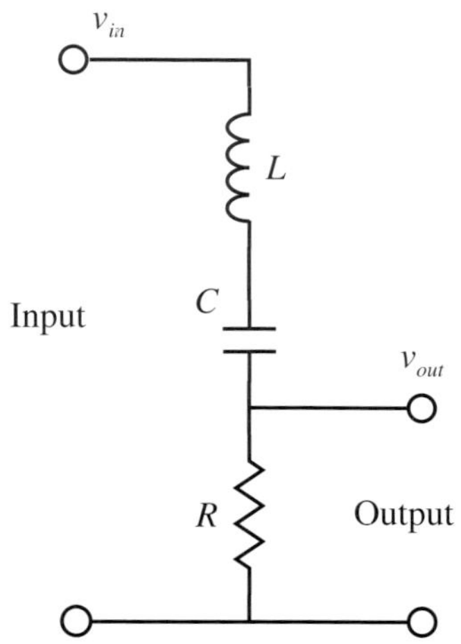


**Fig. 1.12.2** *LCR* circuit for determination of its transfer function.

$$\text{or}\quad V_{in}(s) = sLI(s) + \frac{I(s)}{sC} + RI(s)$$

$$\text{and}\quad v_{out}(t) = Ri(t);\quad V_{out}(s) = RI(s) \tag{1.12.10}$$

$$\text{so}\quad H(s) = \frac{V_{\text{out}}(s)}{V_{\text{in}}(s)} = \frac{R}{sL + 1/sC + R} = \left(\frac{R}{L}\right)\frac{s}{(s^2 + sR/L + 1/LC)}$$

i.e. of the form of Eq. (1.12.9) as indicated. This result could just as easily have been written down using the impedances of Eq. (1.12.7). We can now choose an input form for $v_{in}$ – say a delta function impulse input $\delta(t)$:

$$v_{in}(t) = \delta(t);\quad V_{in}(s) = 1$$

$$\text{so}\quad V_{out}(s) = \left(\frac{R}{L}\right)\frac{s}{(s^2 + sR/L + 1/LC)}$$

$$\text{and}\quad v_{out}(t) = \left(\frac{R}{L}\right)\left[\frac{\beta e^{-\beta t} - \alpha e^{-\alpha t}}{\beta - \alpha}\right],\quad (\alpha \neq \beta,\ \text{Table 1.12.1, No. 17}) \tag{1.12.11}$$

where $\alpha + \beta = R/L$ and $\alpha\beta = 1/LC$ (from roots of denominator)

Solving for $\alpha$ and $\beta$ gives (see Eq. (1.10.7)):

$$\alpha, \beta = \frac{R}{2L} \pm \frac{1}{2}\left(\frac{R^2}{L^2} - \frac{4}{LC}\right)^{\frac{1}{2}} \tag{1.12.12}$$

The form of the response depends on the values of $\alpha$ and $\beta$. If *real* we get simple exponential decays, while if *complex* we get an oscillatory response:

$$\exp[-(a + jb)t] = \exp(-at)[\cos(bt) - j\sin(bt)] \tag{1.12.13}$$

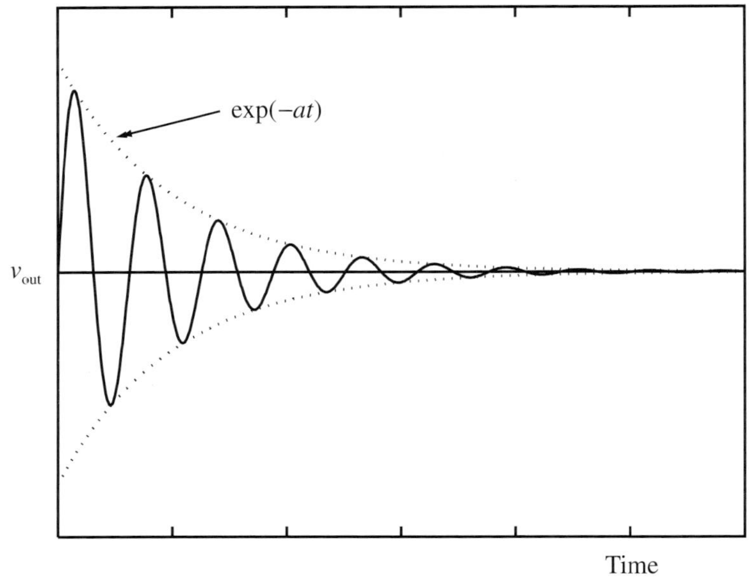


**Fig. 1.12.3** Sinusoidal response with exponential decay.

Thus from Eq. (1.12.12) if:

(i) $4/LC < R^2/L^2$, $\alpha$ and $\beta$ are real giving exponential response

(ii) $4/LC > R^2/L^2$, $\alpha$ and $\beta$ are complex giving oscillatory response

(iii) $4/LC = R^2/L^2$, $\alpha = \beta$, a special case of critical damping requiring use of transform No. 14

In the term $e^{-at}$ in (1.12.13), $a$ must be positive, giving a decaying amplitude, since we have a *passive* network with no source of energy (Fig. 1.12.3).

The response found is the *natural or free response*, i.e. free from any driving function except for the initial impulse $\delta(t)$. If excitation continues after $t = 0$ then the response will be a combination of the free and continuing forced response. This will be considered later under convolution (Section 1.14).

## Poles and zeros

In some applications, such as the consideration of stability in feedback systems, it is the *form* of the transfer function that is of interest rather than the response to a particular input. Since a polynomial of order $n$ has $n$ roots (Section 1.10), Eq. (1.12.9) becomes:

$$H(s) = \text{Const.} \frac{(s - z_1)(s - z_2) \cdots (s - z_m)}{(s - p_1)(s - p_2) \cdots (s - p_n)} \tag{1.12.14}$$

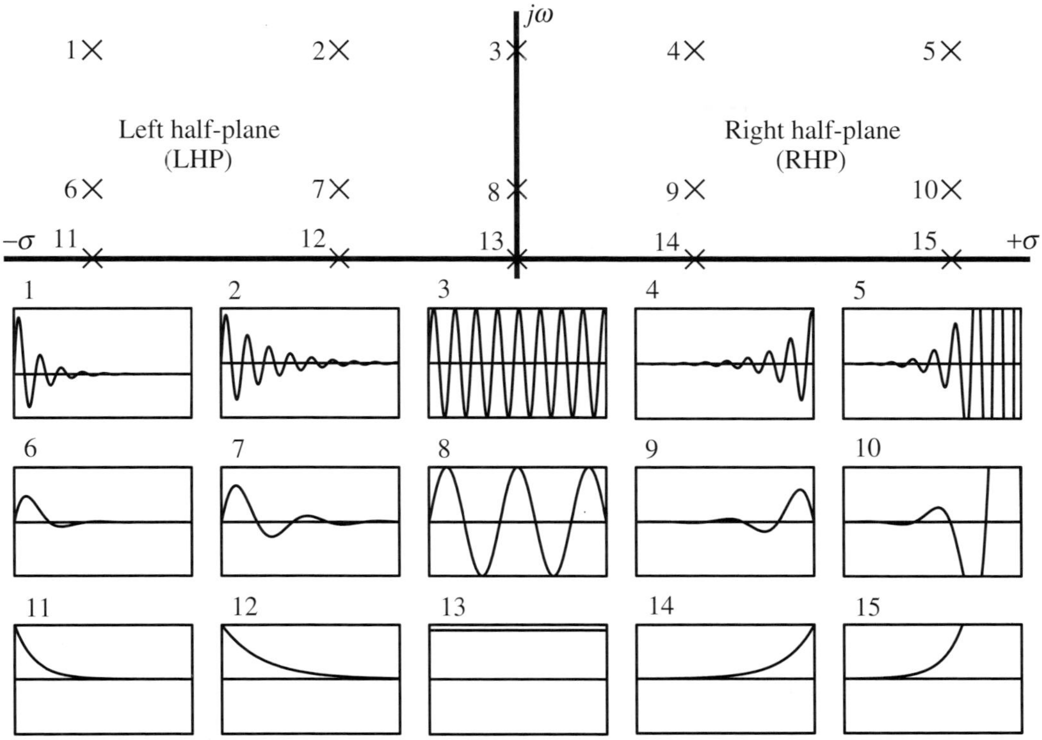


**Fig. 1.12.4** Poles and zeros plotted in the upper half of the $s$-plane, with corresponding response functions.

The roots $z$ of the numerator are called the *zeros* of the transfer function since $H(s)=0$ when $s=z$. The roots $p$ of the denominator are called the *poles* since $H(s)=\infty$ for $s=p$. Since we have assumed $n>m$ then we can write (1.12.14) in terms of partial fractions:

$$H(s)=\frac{K_1}{(s-p_1)}+\frac{K_2}{(s-p_2)}+\cdots+\frac{K_n}{(s-p_n)} \tag{1.12.15}$$

The response of this system to a $\delta(t)$ stimulus is then the sum of $n$ exponentials since:

$$\mathscr{L}^{-1}\frac{K_n}{(s-p_n)}=K_n\exp(p_n t), \quad \text{(Table 1.12.1, No. 7)} \tag{1.12.16}$$

The poles may be real or complex. For the former we get a simple monotonic response, while for the latter we get an oscillatory response as shown in the *LCR* example above. It is usual to plot the poles (shown by a small ×) and zeros (small o) in the complex plane once the roots are known (Fig. 1.12.4).

The waveforms show the impulse response for the correspondingly numbered pole. Conversely, if we have a pole-zero diagram the transfer function can be written down directly except for any multiplicative constants (e.g. $R/L$ in Eq. (1.12.11)). Note that though complex roots always occur in conjugate pairs the complementary root does not give any extra information and so is usually omitted and only the upper half-plane is used. If the poles of a system are all in the left half-plane (LHP) then the system will be *stable* though it might not have a desirable response. Any poles in the right half-plane (RHP) indicate an *unstable* system. Consider a pole with coordinates $p=\sigma+j\omega$, which will lead to a transfer function:

$$H(s)=\frac{1}{s-p}=\frac{1}{s-\sigma-j\omega} \quad \text{which has an inverse transform (Table 1.12.1, No. 12)}$$

$$\exp[(\sigma+j\omega)t]=\exp(\sigma t)\exp(j\omega t)=e^{\sigma t}[\cos(\omega t)+j\sin(\omega t)] \tag{1.12.17}$$

which represents a sinusoidal oscillation with an amplitude dependent on time according to $\exp(\sigma t)$. If $\sigma$ is positive, i.e. the pole $p$ is in the RHP, then the amplitude will increase with time and the system is said to be unstable. If $\sigma$ is negative, i.e. the pole is in the LHP, then the amplitude will decay with time and the system is stable and any transients will die out with time. A passive network must have all its poles in the LHP since there is no source of energy to keep the response increasing with time as there will always be losses. In a system with a pole in the RHP, then even if there is not overt input to cause any response there will always be a switch-on transient or one arising from the inescapable noise in the system (Section 2.13) which will cause a growing response. To sustain the growth the system must have an energy source to make up for the inescapable losses. Thermodynamics ensures that you cannot get something for nothing. The intellectual might of Bell Labs took many years to persuade the Patent Office that Black's proposals for negative feedback were not for some sort of perpetual motion machine.

## Active networks

So far we have only shown examples of passive networks – what happens when active elements (i.e. with gain) are included? We will consider the case of the Wien-bridge oscillator commonly used for 'audio' oscillators (see also Section 5.8). The schematic circuit is shown in Fig. 1.12.5(*a*).

Positive feedback via the $RC$ arm of the bridge determines the *frequency* of oscillation while negative feedback via the resistive arms $R_1$ and $R_2$ (one of these resistors is a thermistor, i.e. a temperature dependent resistor) serves to stabilize the

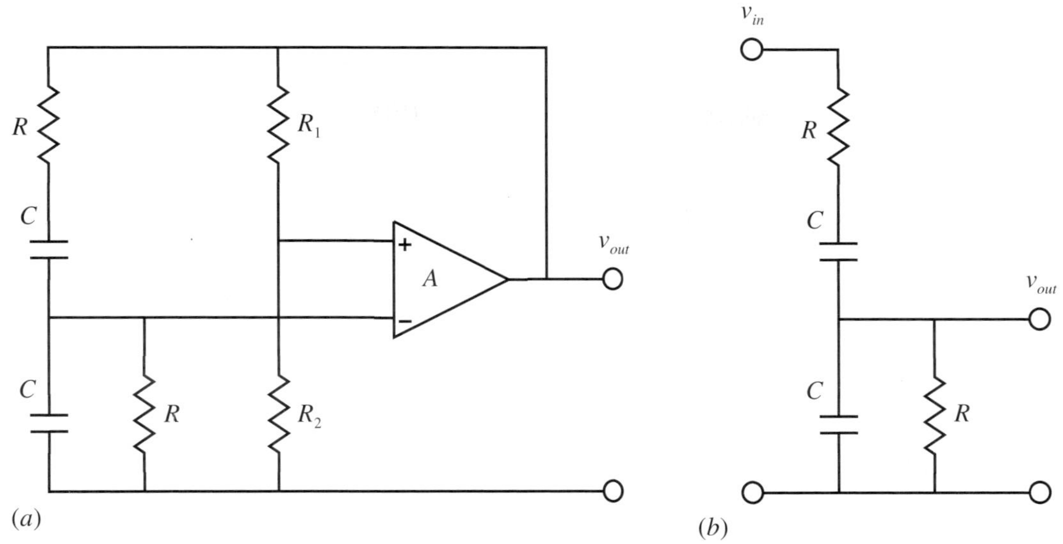


**Fig. 1.12.5** (*a*) Wien oscillator circuit. (*b*) Wien network.

*amplitude* of oscillation. Consider first the *RC* network only (Fig. 1.12.5(*b*)). Viewing it as a potential divider we can write for the two sections:

$$Z_{par}=\frac{R}{1+sCR};\quad Z_{ser}=\frac{1+sCR}{sC}$$
$$\text{or}\quad H(s)=\frac{V_{out}(s)}{V_{in}(s)}=\frac{Z_{par}}{Z_{par}+Z_{ser}}=\frac{sCR}{s^2C^2R^2+3sCR+1} \tag{1.12.18}$$

The poles are given by the roots of the denominator:

$$s_p=\frac{-3RC\pm(9R^2C^2-4R^2C^2)^{\frac{1}{2}}}{2R^2C^2}$$
$$=\frac{-3\pm\sqrt{5}}{2RC}=\frac{-2.62}{RC}\quad\text{or}\quad\frac{-0.38}{RC} \tag{1.12.19}$$

i.e. the poles are real and negative giving exponentially decreasing time functions, e.g. poles 11 or 12 in Fig. 1.12.4. Now introduce the amplifier of gain $A$ such that (the gain will be determined by the negative feedback arm $R_1$ and $R_2$):

$$H(s)\,A=1 \tag{1.12.20}$$

and the network losses will just be made up by the amplifier (a source of energy). Thus from (1.12.18) and (1.12.20):

$$AsCR=s^2C^2R^2+3sCR+1$$
$$\text{or}\quad s^2C^2R^2+sCR(3-A)+1=0 \tag{1.12.21}$$

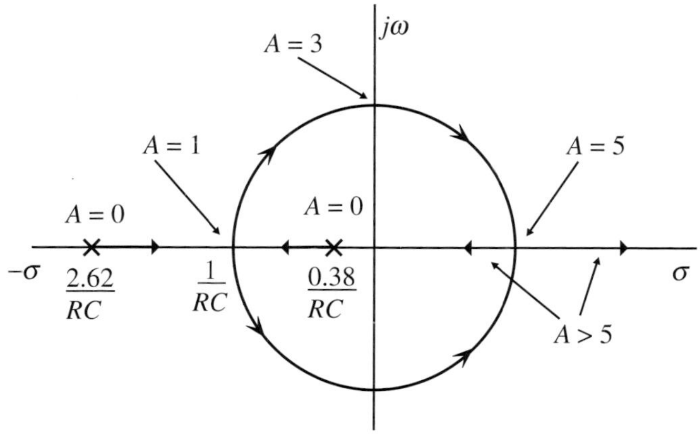


**Fig. 1.12.6** Locus of the poles as a function of gain $A$.

Using the operational theorems (Table 1.12.2) you will recognize this as the equation for simple harmonic motion with damping (the term $sCR(3-A)$). To achieve continuous oscillation of constant amplitude we require zero damping so we want $A=3$. Then:

$$s^2C^2R^2=-1 \quad \text{or} \quad s_p=\frac{\pm j}{RC}=\sigma\pm j\omega \tag{1.12.22}$$

The poles are now on the $j\omega$ axis with $\sigma=0$ so there is no growth or decay of the oscillations (Fig. 1.12.4, pole 3 or 8). The 'real' frequency of oscillation is then:

$$f=\frac{\omega}{2\pi}=\frac{1}{2\pi RC} \tag{1.12.23}$$

The example demonstrates the use of active elements, i.e. gain, to move the poles and zeros around the complex plane to obtain the response we require. If we plot the locus of the poles from Eq. (1.12.21) as a function of $A$ you will obtain Fig. 1.12.6. Once the poles have passed into the right half-plane ($A>3$) the oscillations will grow until limited by the capabilities of the amplifier.

The pole-zero approach is very effective and will be used in many other sections. In operational amplifier circuits the response you will get will depend on the *closed-loop* location of the poles. For a feedback system the closed-loop gain $G$ is given by (note that $A(s)$ may be positive or negative):

$$G(s)=\frac{A(s)}{1-A(s)\,\beta(s)}=\frac{A(s)}{1-L(s)} \tag{1.12.24}$$

where $A(s)$ is the open loop and $L(s)$ the loop gain (see Section 5.3). The poles of $A(s)$ are no longer effective: when $A(s)=\infty$, $L(s)=A(s)\beta(s)=\infty$ as $\beta(s)\neq 0$. The effective poles are those of $G(s)$ and are determined by the roots of:

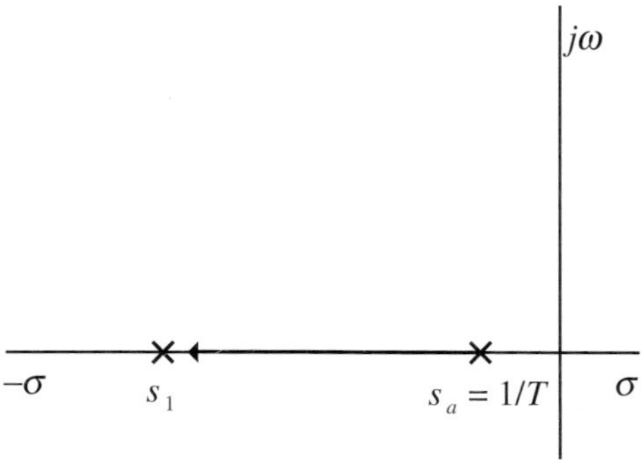


**Fig. 1.12.7** Location of pole for single pole system.

$$1 - L(s) = 0 \tag{1.12.25}$$

Consider a single lag (pole) system with time constant $T$, and say $\beta(s) = \beta_0$, i.e. not a function of frequency (Fig. 1.12.7).

The open-loop gain is then:

$$A(s) = \frac{-A_0}{1 + sT}, \quad \text{i.e. pole at } s_a = \frac{-1}{T} \tag{1.12.26}$$

For the closed loop, the pole for $G(s)$ is found from:

$$1 + sT + A_0\beta_0 = 0, \quad \text{i.e. a pole at } s_1 = \frac{-(A_0\beta_0 + 1)}{T} \tag{1.12.27}$$

Thus as $A_0$ increases, the pole, which is always real, moves along the $-\sigma$ axis further into the left half-plane, i.e. stability increases (Fig. 1.12.7). This is the aim in making operational amplifiers with a single dominant pole so that it is stable under most circumstances.

Now consider what happens if we have a two-pole amplifier with time constants $T_1$ and $T_2$ (say $T_1 > T_2$). Then:

$$A(s) = \frac{-A_0}{(1 + sT_1)(1 + sT_2)}, \quad \text{i.e. pole at } s_a = \frac{-1}{T_1} \quad \text{and} \quad s_b = \frac{-1}{T_2} \tag{1.12.28}$$

and taking $\beta(s) = \beta_0$ as before, the closed-loop poles are given by:

$$1 + s(T_1 + T_2) + s^2 T_1 T_2 + A_0\beta_0 = 0$$
$$\text{or } s_{1,2} = \frac{-(T_1 + T_2) \pm [(T_1 + T_2)^2 - 4(A_0\beta_0 + 1)T_1T_2]^{\frac{1}{2}}}{2T_1T_2} \tag{1.12.29}$$

As $A_0\beta_0$ increases the poles move from the original positions $s_a$, $s_b$ to $s_1$, $s_2$ as shown in Fig. 1.12.8.

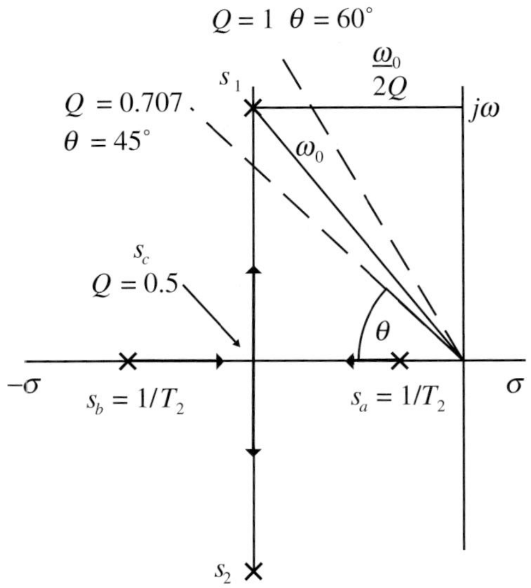


**Fig. 1.12.8** Movement of the poles as a function of gain $A_0$ and relation of pole position to frequency $\omega_0$ and $Q$.

The roots are equal when:

$$(T_1 + T_2)^2 = 4(A_0\beta_0 + 1)T_1T_2 \quad \text{or} \quad s_{1,2} = \frac{-(T_1 + T_2)}{2T_1T_2} = s_c \tag{1.12.30}$$

and for higher values of loop gain the poles become complex and move along a locus parallel to the $j\omega$ axis, i.e. they can never cross into the RHP and hence the system is unconditionally stable (in practice, of course, other poles may become significant at high frequencies which will cause the locus to eventually cross into the RHP). It is helpful to relate the position of the poles to the frequency response of the circuit, which is what you usually measure. We can write Eq. (1.12.29) in terms of the $Q$ of the circuit:

$$1 + \frac{s}{Q\omega_0} + \left(\frac{s}{\omega_0}\right)^2 = 0$$

where $Q^2 = \dfrac{(A_0\beta_0 + 1)T_1T_2}{(T_1 + T_2)^2}$ and $\omega_0^2 = \dfrac{(A_0\beta_0 + 1)}{T_1T_2}$

$$\text{so} \quad s_{1,2} = \frac{-\omega_0}{2Q} \pm \left[\left(\frac{\omega_0}{2Q}\right)^2 - \omega_0^2\right] \tag{1.12.31}$$

Plots of frequency normalized gain $(\omega/\omega_0)$ and phase as a function of $Q$ are shown in Fig. 1.12.9. The geometrical relation of $Q$ and $\omega_0$ to the pole position is shown in Fig. 1.12.8.

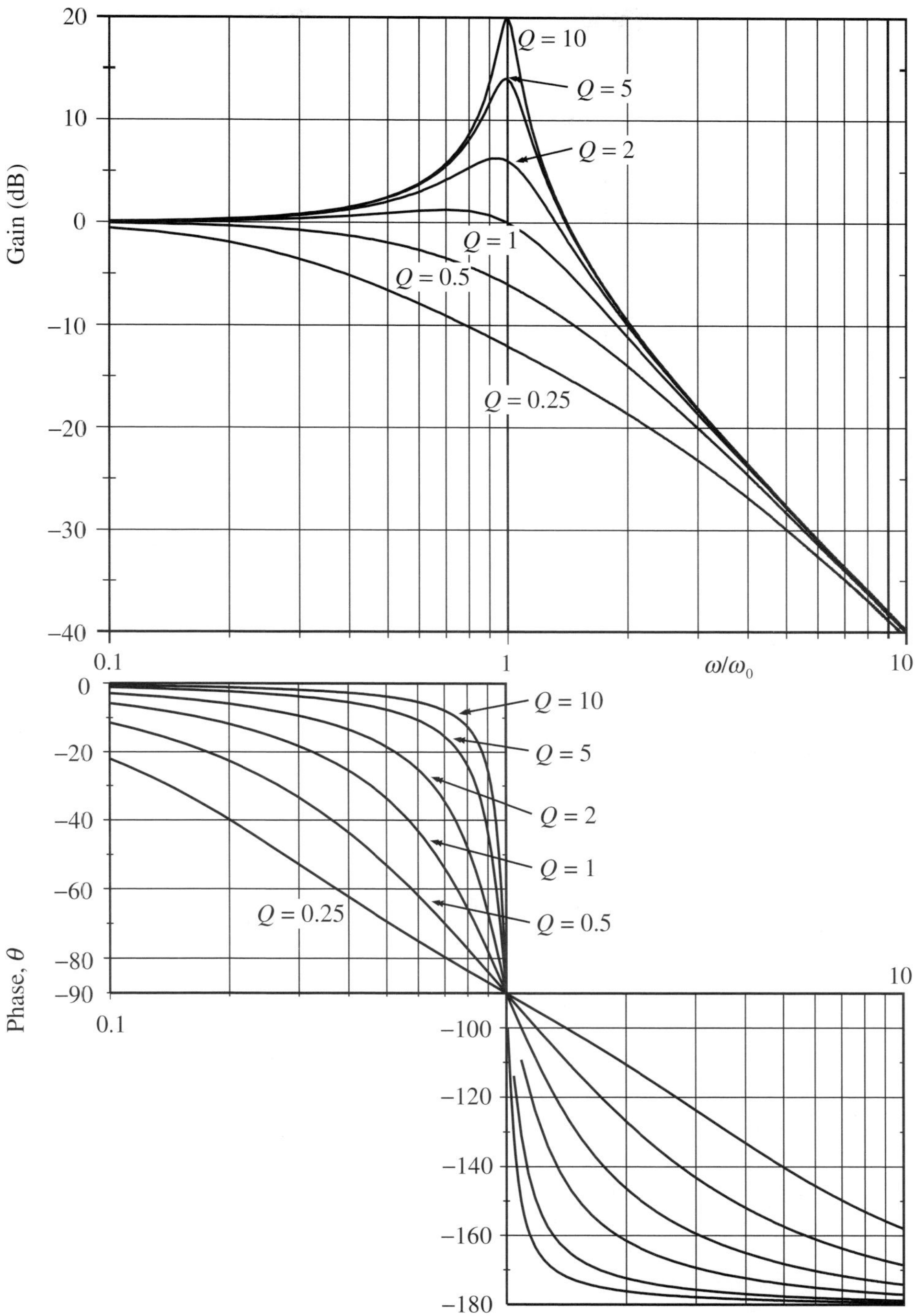


**Fig. 1.12.9** Plots of frequency normalized gain and phase as a function of $Q$.

For second order denominator polynomials it is easy to determine whether the system will be stable. For higher order polynomials we must either determine the pole positions by computation or seek some more general test. The Routh–Hurwitz techniques are applicable for this purpose and though they are not too readily used, some general guidance may be derived. Four useful rules (Siebert 1986, p. 175; Pipes 1958, p. 242) are:

For a polynomial to have all its roots in the left half-plane it is necessary that:

(a) all terms must have the same sign

(b) all powers of $s$ from highest to lowest must have non-zero coefficients, unless all even or all odd powers are absent.

Conditions (a) and (b) are also sufficient for a quadratic polynomial.

(c) For a cubic polynomial, e.g. $s^3+\alpha s^2+\beta s+\gamma$, necessary and sufficient conditions are that $\alpha$, $\beta$, $\gamma>0$ and $\beta>\gamma/\alpha$.

(d) For a quartic polynomial, e.g. $s^4+\alpha s^3+\beta s^2+\gamma s+\delta$, necessary and sufficient conditions are that $\alpha$, $\beta$, $\gamma,\delta>0$ and $\alpha\beta\gamma>\alpha^2\delta+\gamma^2$.

An example of application of the conditions for a cubic is given in Section 5.13.

## References and additional sources 1.12


Abramowitz M., Stegun I. A. (Eds) (1970): *Handbook of Mathematical Functions with Formulas, Graphs and Mathematical Tables*, Applied Mathematics Series, Washington: National Bureau of Standards.

Boas M. L. (1966): *Mathematical Methods in the Physical Sciences*, New York: John Wiley. Library of Congress Cat. No. 66-17646.

Gillespie C. C. (1997): *Pierre-Simon Laplace 1749–1827*, Princeton: Princeton University Press. ISBN 0-891-51185-0.

Holbrook J. G. (1966): *Laplace Transforms for the Electronic Engineer*, 2nd (revised) Edn, Oxford: Pergamon Press. Library of Congress Cat. No. 59-12607.

McCollum P. A., Brown B. F. (1965): *Laplace Tables and Theorems*, New York: Holt, Reinhart, Winston.

Nixon F. E. (1965): *Handbook of Laplace Transforms*, 2nd Edn, Englewood Cliffs: Prentice-Hall. Library of Congress Cat. No. 65-14937.

Oberhettinger F., Badii L. (1970): *Tables of Laplace Transforms*, Berlin: Springer-Verlag. ISBN 3-540-06350-1.

Pipes L. A. (1958): *Applied Mathematics for Engineers and Physicists*, 2nd International Student Edn, New York: McGraw-Hill. Library of Congress Cat. No. 57-9434. Note that this work uses a slightly different definition of the Laplace transform so that for example the table on his page 152 should have all the $g(p)$ functions divided by '$p$' ($p\equiv s$) to agree with the transforms in our Tables 1.12.1 and 1.12.2. See also his pp. 643 *et seq.*

Poularikas A. D. (Ed.) (1996): *The Transforms and Applications Handbook*, Boca Raton: CRC Press and IEEE Press. ISBN 0-8493-8342-0.

Savant C. J. (1965): *Fundamentals of the Laplace Transformation*, New York: McGraw-Hill.
Siebert W. McC. (1986): *Circuits, Signals, and Systems*, Cambridge Mass: MIT Press and McGraw-Hill. ISBN 0-07-057290-9.
Spiegel M. R. (1965): *Theory and Problems of Laplace Transforms*, Schaum's Outline Series. New York: McGraw-Hill.

# 1.13 Differential equations

But the reader may object, Surely the author has got to know the go of it already, and can therefore eliminate the preliminary irregularity and make it logical, not experimental? So he has in a great measure, but he knows better. It is not the proper way under the circumstances, being an unnatural way. It is ever so much easier to the reader to find the go of it first, and it is the natural way. The reader may then be able a little later to see the inner meaning of it himself, with a little assistance. To this extent, however, the historical method can be departed from to the reader's profit. There is no occasion whatever (nor would there be space) to describe the failures which make up the bulk of experimental work. He can be led into successful grooves at once. Of course, I do not write for rigourists (although their attention would be delightful) but for a wider circle of readers who have fewer prejudices, although their mathematical knowledge may be to that of the rigourists as straw to a haystack. It is possible to carry wagon-loads of mathematics under your hat, and yet know nothing whatever about the operational solution of physical differential equations.

Oliver Heaviside (1895): *Electromagnetic Theory*, January 11, Vol. II, p. 33

Much of physics is concerned with deducing the appropriate differential equation that describes the particular phenomena and then trying to find solutions to this equation. The sections on Maxwell's equations provide an example of this and there are many more from Schrödinger's wave equation in quantum mechanics, to that for a simple pendulum as shown in Fig. 1.13.1.

The restoring force $F$, which always acts towards the centre equilibrium position, will be equal to the mass $m$ of the bob times its acceleration according to Newton's law. Resolving the gravitational force $mg$ along the string, which equals the tension in the string, and normal to it which is the force $F$ accelerating the bob we have:

$$F = -mg\sin(\theta) = m\frac{d^2s}{dt^2}, \quad \text{where } s = l\theta \text{ is the distance along the arc}$$

$$\text{so} \quad -g\sin(\theta) = l\frac{d^2\theta}{dt^2}, \qquad \text{and for small } \theta, \quad \sin(\theta) \cong \theta \quad \text{giving} \tag{1.13.1}$$

$$\frac{d^2\theta}{dt^2} = -\frac{g}{l}\theta$$

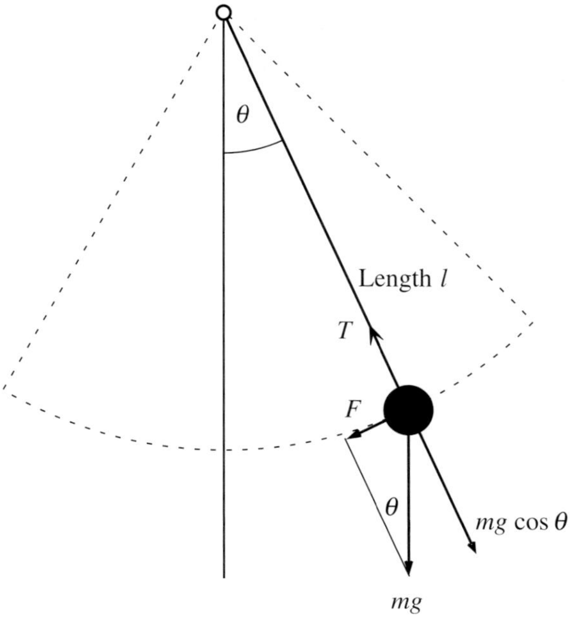


**Fig. 1.13.1** Simple pendulum.

where we have used the approximation for sin($\theta$) when $\theta$ is small (Section 1.1). We now have a differential equation for $\theta$ for which we must seek a solution. The form with sin($\theta$) is equally a differential equation but the solution is more complex and not harmonic. There are direct techniques for deriving solutions but we will use the sometimes simpler approach of guessing (knowing) the form of the solution and showing that it is acceptable. Thus we guess at a sinusoidal form for $\theta$:

$$\theta = A\sin(\omega t) \quad \text{so} \quad \frac{d\theta}{dt} = A\omega\cos(\omega t) \quad \text{and} \quad \frac{d\theta}{dt} = -A\omega^2\sin(\omega t) = -\omega^2\theta$$

$$\text{so our choice is a solution if } \omega^2 = \frac{g}{l} \quad \text{or} \quad \omega = \left(\frac{g}{l}\right)^{\frac{1}{2}} \tag{1.13.2}$$

The motion is therefore what is generally known as simple harmonic with angular frequency $\omega$. Also we can now say that whenever we have an equation of the form of (1.13.1) that the solution will be of the same form. This form of equation represents free or natural vibration or oscillation in the sense that it is unaffected by any outside influence. In using a pendulum in a clock we would of course have some damping present so that energy would have to be supplied to keep the pendulum swinging continuously. The escapement mechanism that enables this is controlled by the swing of the pendulum so that there is no conflict between the swing and the feed of energy via the escapement. If, however, a resonant system is driven by a force

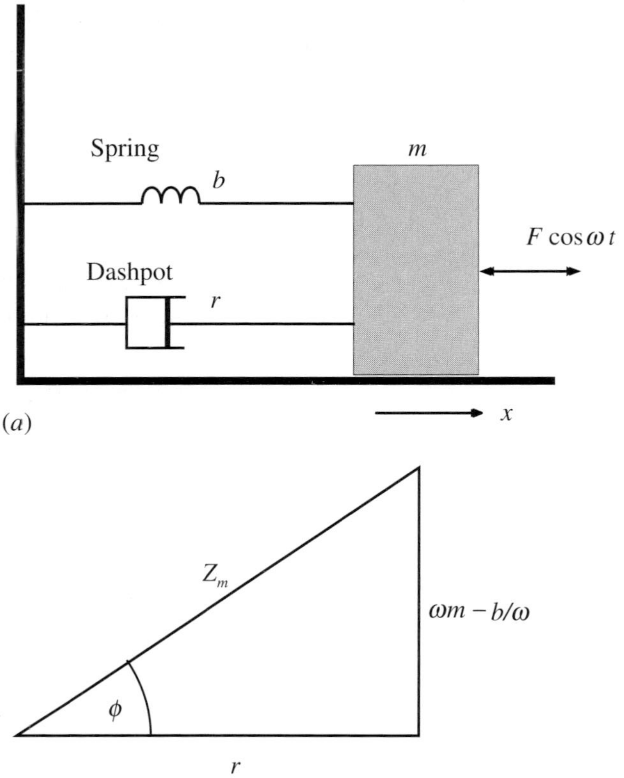


Fig. 1.13.2 (*a*) Mechanical resonant system. (*b*) Vector relations.

of frequency different from the natural frequency then the motion will be different and the differential equation describing the system will have a different solution. When the force is first applied there will be a transient effect but after some time this will have died away and we will be left with the steady-state response. For the mechanical system shown in Fig. 1.13.2(*a*) the equation of motion is:

$$m\ddot{x}+r\dot{x}+bx=F\cos(\omega t)=\mathcal{Re}(Fe^{j\omega t}) \tag{1.13.3}$$

where we use the more convenient exponential notation but should remember that at the end we must extract the real part of the response as the answer (we will drop the $\mathcal{Re}$ during the algebra).

We again guess the form of the (steady-state) solution, differentiate and substitute into Eq. (1.13.3), noting that $A$ may be a complex vector:

$$\boldsymbol{x}=\boldsymbol{A}e^{j\omega t} \quad \text{so} \quad \dot{\boldsymbol{x}}=j\omega\boldsymbol{A}e^{j\omega t}=j\omega\boldsymbol{x} \quad \text{and} \quad \ddot{\boldsymbol{x}}=-\omega^2\boldsymbol{A}e^{j\omega t}=-\omega^2\boldsymbol{x}$$

then (1.13.4)

$$-m\omega^2\boldsymbol{x}+rj\omega\boldsymbol{x}+b\boldsymbol{x}=Fe^{j\omega t} \quad \text{or} \quad (-m\omega^2+rj\omega+b)\boldsymbol{A}e^{j\omega t}=Fe^{j\omega t}$$

This must be true for all $t$ so that we have:

$$A = \frac{F}{-m\omega^2 + rj\omega + b} = \frac{-jF}{\omega r - j(b - m\omega^2)} = \frac{-jF}{\omega \boldsymbol{Z}_m}$$

where (1.13.5)

$$\boldsymbol{Z}_m = r + j\left(\omega m - \frac{b}{\omega}\right) = |\boldsymbol{Z}_m| e^{j\phi} = Z_m e^{j\phi}$$

so $A$ is complex. The mechanical impedance, defined by the ratio of force to velocity, will be seen from (1.13.8) to be $\boldsymbol{Z}_m$. The components may be derived from the vector addition diagram in Fig. 1.13.2(*b*):

$$|\boldsymbol{Z}_m| = Z_m = \left[r^2 + \left(\omega m - \frac{b}{\omega}\right)^2\right]^{\frac{1}{2}} \quad \text{and} \quad \tan^{-1}\phi = \frac{[\omega m - (b/\omega)]}{r} \tag{1.13.6}$$

We can now write the solution for $x$ and extract the real part:

$$x = \frac{-jF}{\omega Z_m} e^{j\omega t} = \frac{-jFe^{j\omega t}}{\omega Z_m e^{j\phi}} = \frac{-jF}{\omega Z_m} \exp(\omega t - \phi)$$

$$= \frac{-jF}{\omega Z_m}[\cos(\omega t - \phi) + j\sin(\omega t - \phi)] \tag{1.13.7}$$

so the real part of $x$ is

$$x = \frac{F}{\omega Z_m} \sin(\omega t - \phi)$$

i.e. it is sinusoidal at the driving frequency with a phase lag of $(90° + \phi)$ though $\phi$ can range between $\pm 90°$ (or $\pm \pi/2$). The velocity is given by:

$$v = \dot{x} = \frac{F}{Z_m} \cos(\omega t - \phi) \tag{1.13.8}$$

which confirms our definition of mechanical impedance above. The phase relationships may best be shown as in Fig. 1.13.3.

If $r$ is not too large then the velocity response as a function of frequency will be a resonance with the peak where $Z_m$ is a minimum, i.e. at frequency:

$$\omega m = \frac{b}{\omega} \quad \text{or} \quad \omega_0 = \left(\frac{b}{m}\right)^{\frac{1}{2}} \tag{1.13.9}$$

For the amplitude $x$ the resonant peak will occur at a slightly different frequency given by the minimum of $\omega Z_m$ rather than $Z_m$. Differentiating and putting equal to zero we get:

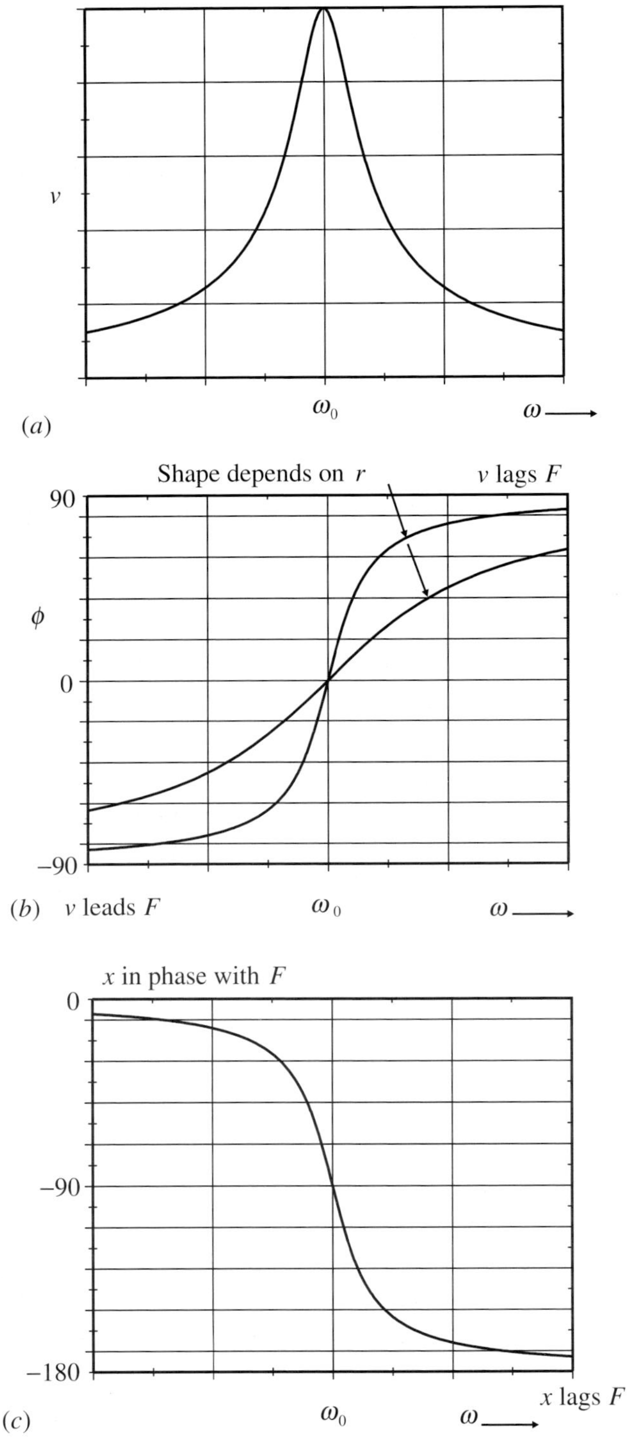


**Fig. 1.13.3** Relationships for forced oscillator. (*a*) Velocity resonant response. (*b*) Force–velocity phase response. (*c*) Force–displacement phase response.

$$\frac{d(\omega Z_m)}{d\omega} = \frac{d}{d\omega}\omega\left[r^2 + \left(\omega m - \frac{b}{\omega}\right)^2\right]^{\frac{1}{2}} = \frac{d}{d\omega}\left[\omega^2 r^2 + \omega^2\left(\omega m - \frac{b}{\omega}\right)^2\right]^{\frac{1}{2}}$$

$$= \frac{d}{d\omega}\left[\omega^2 r^2 + \omega^2\left(\omega^2 m^2 - 2bm + \frac{b^2}{\omega^2}\right)\right]^{\frac{1}{2}} = \frac{d}{d\omega}(\omega^2 r^2 + \omega^4 m^2 - 2bm\omega^2 + b^2)^{\frac{1}{2}}$$

$$= \tfrac{1}{2}(\omega^2 r^2 + \omega^4 m^2 - 2bm\omega^2 + b^2)^{-\frac{1}{2}}(2\omega r^2 + 4\omega^3 m^2 - 4bm\omega + 0)$$
$$= 0 \quad \text{for minimum}$$

thus

$$2\omega r^2 + 4\omega^3 m^2 - 4bm\omega = 0, \quad \text{i.e.} \quad \omega = 0 \quad \text{or} \quad \omega^2 = \frac{b}{m} - \frac{r^2}{2m^2} = \omega_0^2 - \frac{r^2}{2m^2} \qquad (1.13.10)$$

The reason for working through this treatment of a mechanical system is to demonstrate that it does not matter what the physical system is – if the same differential equation describes it, it will have the same solutions. An equivalent electrical resonator would be obtained if we replaced $m$ by $L$, $r$ by $R$ and $b$ by $1/C$ and then $v$ would represent current, $F$ represent voltage and $x$ represent charge (Pipes 1958, pp. 163 and 195). The electrical equivalent is examined in Section 3.5.

In Section 2.9 we discuss some quantum ideas and the harmonic oscillator provides an opportunity to illustrate the differences between the quantum and the classical view. If we observe an oscillator, such as the pendulum considered above, many times and record the position we will obtain a probability distribution showing the likelihood of finding it at any position. Since the velocity passes through zero at the extremes of the motion and is greatest at the centre, we would expect the probability to be greatest at the ends and least at the centre.

For a quantum mechanical harmonic oscillator with one quantum of energy the wavefunction, which gives the probability of finding the system in any position, has the form shown in Fig. 1.13.4.

It is clearly very different from that of the classical oscillator and indicates that the oscillator can be found outside the equivalent limits of oscillation – a consequence of the uncertainty principle. However, if we plot the wavefunction for a quantum oscillator with rather more energy (in this case 10 units), then the wavefunction is as shown in Fig. 1.13.5. This shows how the quantum picture changes to approach the classical distribution at high quantum numbers, an example of Bohr's correspondence principle (Powell and Craseman 1961), but we should not take these comparisons too seriously as we are comparing somewhat different averages.

SPICE, in the context of this book, provides a means of solving differential equations if we can formulate a circuit that has the same equation (Prigozy 1989; Wilson 1996). Consider the simple circuit shown in Fig. 1.13.6(*a*).

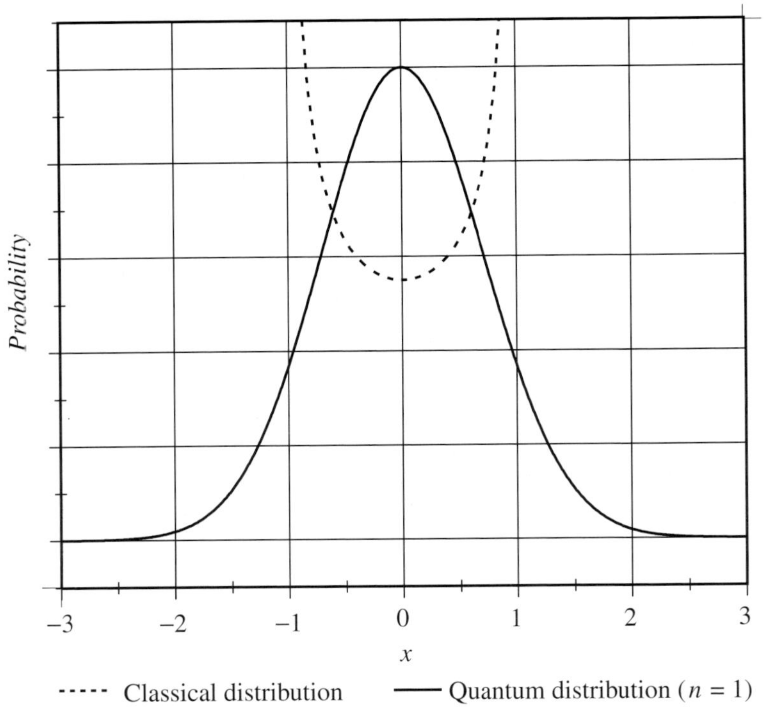


**Fig. 1.13.4** Quantum and classical probability distribution for simple harmonic motion ($n = 1$).

From Eq. (1.5.1), and since here we do not have a voltage source but only the charged capacitor, we have the simple differential equation:

$$\frac{dv}{dt} = \frac{-v}{RC} \tag{1.13.11}$$

which we need to integrate to find the variation of $v$ as a function of $t$. We can set up the circuit as shown in Fig. 1.13.6($b$) using the ABM integrator, multiplier and constant. If $v_{out}$ is the required solution then the input to the integrator must be $dv_{out}/dt$ and the feedback via the multiplier constrains this to be $-1/RC$ times $v_{out}$, where $-1/RC$ is the value of the constant. The initial condition (IC) for the integrator will set the value of $v_{out}$ at time $t = 0$ and the *GAIN* of the integrator will be left at its default value of 1. If you run a transient analysis, for say a time interval of several time constants $RC$ you will obtain the expected exponential decay.

A more interesting example is the solution of the van der Pol equation (van der Pol 1934; Pipes 1958 pp. 691, 701):

$$\frac{d^2y}{dt^2} - \mu(1 - y^2)\frac{dy}{dt} + y = 0 \tag{1.13.12}$$

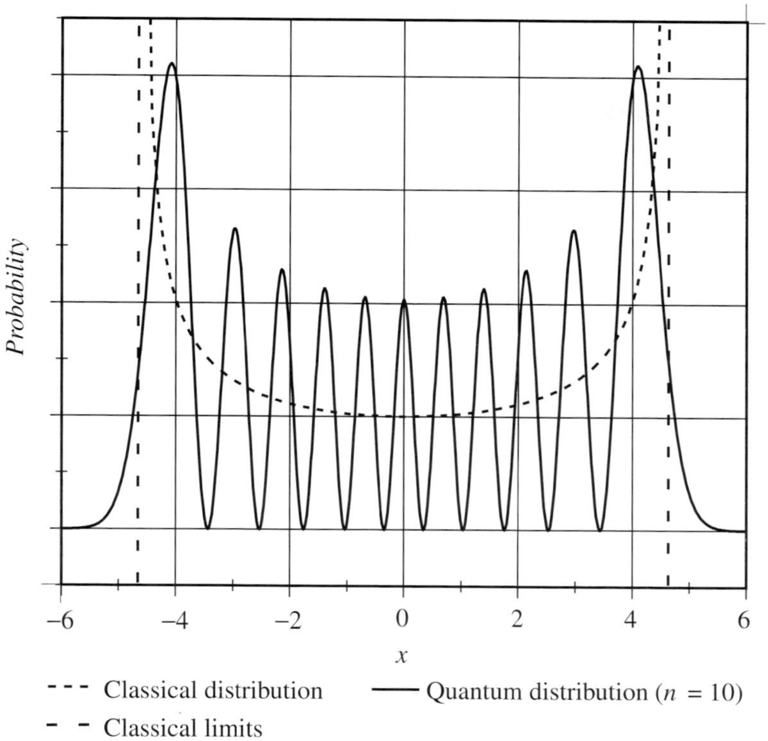


**Fig. 1.13.5** Quantum and classical probability distribution for simple harmonic motion ($n = 10$).

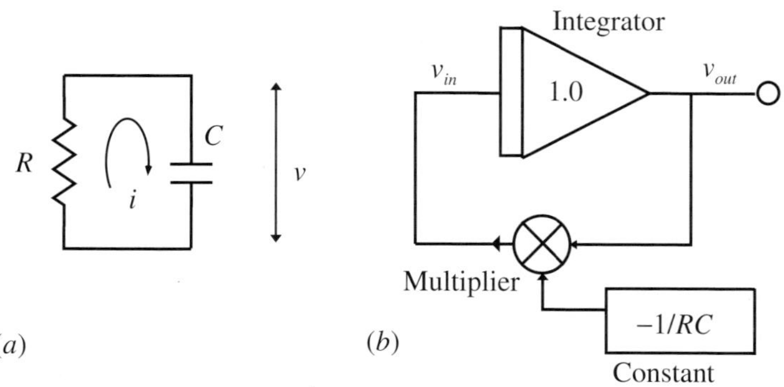


**Fig. 1.13.6** (*a*) Simple RC circuit. (*b*) PSpice differential equation solver.

As shown by Prigozy (1989) this non-linear equation can be simulated using a simple circuit as shown in Fig. 1.13.7, except that here we have replaced his original current controlled voltage source (CCVS) with an equivalent ABM.

In this circuit the current is the analog of $y$ and the voltage across the inductor is the analog of $dy/dt$ (or $\dot{y}$ in Prigozy's paper). The schematic looks incomplete but

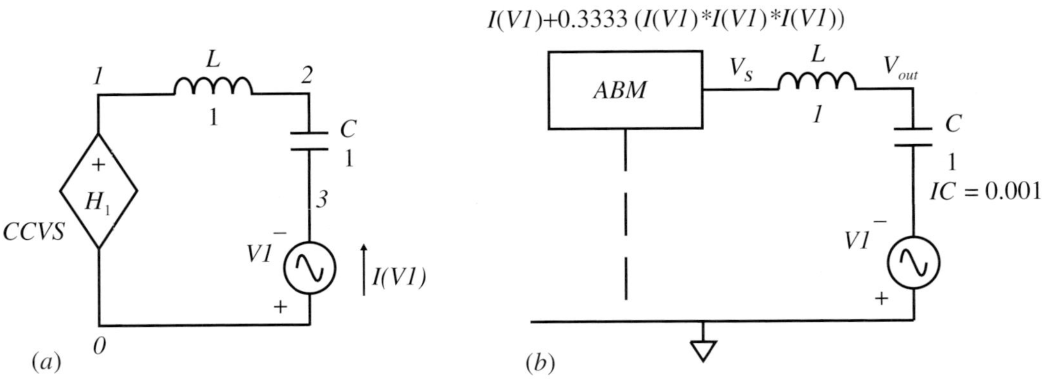


**Fig. 1.13.7** (*a*) Simulation of van der Pol's differential equation. (*b*) Equivalent ABM configuration.

it should be recalled that the ABM has a hidden connection to 0 V (shown dashed). The current that controls the voltage output of the ABM is referenced as that through the null voltage source $V_1$, i.e. $I(V1)$. Prigozy (his Fig. 6) defines a current controlled voltage source $H$ by a polynomial:

$$V=C_0+C_1I+C_2I^2+C_3I^3, \quad \text{with} \quad C_0=0,\ C_1=-1,\ C_2=0,\ C_3=\tfrac{1}{3} \tag{1.13.13}$$

We may then apply Kirchhoff's voltage law to give:

$$V+L\frac{dI}{dt}+\frac{1}{C}\int I\,dt=0, \quad \text{or} \quad -I+\tfrac{1}{3}I^3+L\frac{dI}{dt}+\frac{1}{C}\int I\,dt=0$$

and differentiating gives (1.13.14)

$$\frac{-dI}{dt}+\frac{3}{3}I^2\frac{dI}{dt}+L\frac{d^2I}{dt^2}+\frac{I}{C}=0, \quad \text{or} \quad LC\frac{d^2I}{dt^2}-C(1-I_2)\frac{dI}{dt}+I=0$$

and to match Eq. (1.13.12) we must put $LC=1$, $C=\mu$ and hence $L=1/\mu$. The specification of the ABM is:

$$EXP1=-I(V1)+0.3333*(I(V1)*I(V1)*I(V1)) \tag{1.13.15}$$

so the effective 'resistance' of the system has a linear and a cubic term. Writing the latter as shown rather than as a power avoided convergence problems. It is important that the null voltage source has the polarity orientation shown; $I(V1)$ flows from + to − as indicated. For the component values in the figure (i.e. $\mu=1$), and with the initial condition $IC$ for $C$ set to 0.001, a transient run of 40 s (for $\mu$ high) to 200 s (for $\mu$ low) is appropriate and a plot of $I(V1)$ will show the solution for van der Pol's equation as given by Prigozy. If, in *PROBE*, you 'add a plot' and 'unsync the $x$-axes', you can plot simultaneously the $I(V1)$ *versus* time graph and the phase portrait of $V(V(OUT))-V(V(S))$ as a function of $I(V1)$. If you change

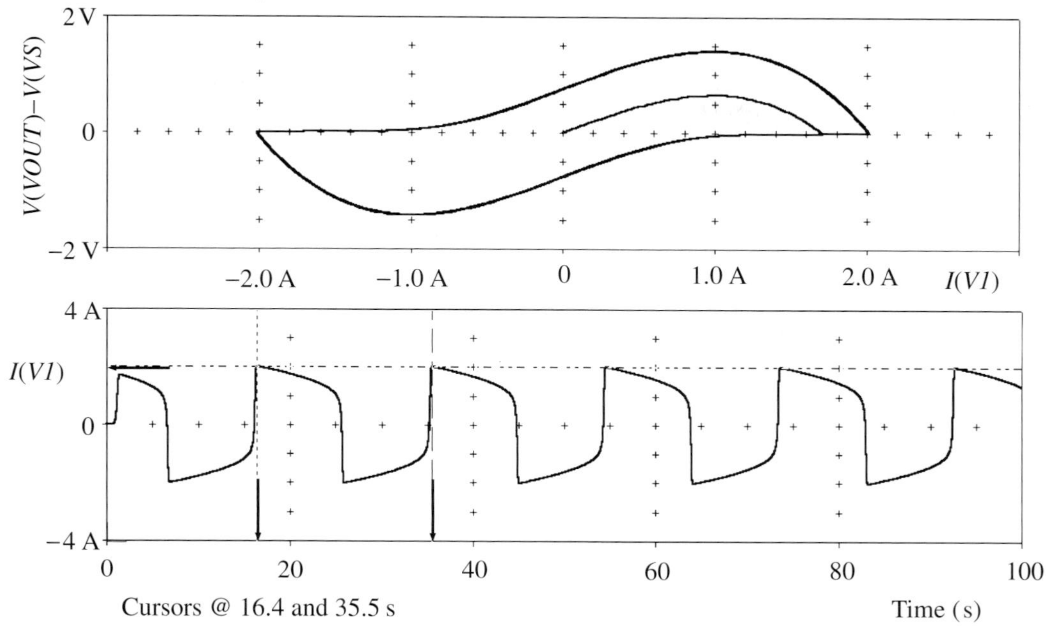


**Fig. 1.13.8** Output of circuit of Fig. 1.13.7 for $\mu=10$ ($C=10$, $IC=0.001$, $L=0.1$).

$\mu$ to 0.1 or 10 you will get the results shown by Pipes on p. 706, Fig. 14.4. The low value of $\mu$ gives near sinusoidal oscillations while the high $\mu$ value gives relaxation type oscillations. The results for $\mu=10$ are shown in Fig. 1.13.8. What van der Pol must have taken days to plot you can see in seconds.

If $IC$ is made high, say 2, then the start point will be outside the limit circle and the phase trajectory will spiral in. Try also $\mu=0.1$ with $IC=0.707$. Shohat (1944) has examined the calculation of the period of the oscillations, particularly for high values of $\mu$. His complex formula (see also Pipes p. 708) will give a close approximation to your simulated result (about 20 s).

## References and additional sources 1.13

James G., Burley D., Clements D., Dyke P., Searl J., Wright J. (1996): *Modern Engineering Mathematics*, 2nd Edn, Wokingham: Addison-Wesley. ISBN 0-201-87761-9.

Kennedy M. P., Chua L. O. (1986): Van der Pol and chaos. *IEEE Trans.* **CAS-33**, 974–980.

Kuo F. F. (1966): *Network Analysis and Synthesis*, New York: John Wiley. See Chapter 4. ISBN 0-471-51118-8.

Pipes L. A. (1958): *Applied Mathematics for Engineers and Physicists*, New York: McGraw-Hill. Library of Congress Cat. No. 57-9434.

Powell, J. L., Crasemann B. (1961): *Quantum Mechanics*, Reading: Addison Wesley. Library of Congress Cat. No. 61-5308.

Prigozy S. (1989): Novel applications of SPICE in engineering education. *IEEE Trans.* **EDUC-32**, 35–38. I am indebted to the author for discussions.
Shohat J. (1944): On van der Pol's and related nonlinear differential equations. *J. Applied Phys.* **15**, 568–574.
van der Pol B. (1926): Relaxation oscillations. *Philosophical Magazine* **2**, 978–992.
van der Pol B. (1927): Forced oscillations in a circuit with non-linear resistance. *Philosophical Magazine* **3**, 65–80.
van der Pol B. (1934): Nonlinear theory of electric oscillations. *Proc. IRE* **22**, 1051–1086.
Wilson I. (1996): *Solving Differential Equations with MicroSim PSPICE*, MicroSim Corporation Application Notes Manual, April, pp. 235–241.

# 1.14 Convolution

'We might have one mathematical fellow, in case we have to calculate something out.' Thomas Alva Edison giving advice to the U.S. President on setting up the Naval Consulting Board 1915.

Robert Watson-Watt (1962): Electrons and elections. *Proc. IRE* **50**, 646–652

In Section 1.12 it was seen that the transfer function of a linear system was given by the impulse response. If we know the impulse response how can we determine the response to some arbitrary input? If we divide up the input into a sequence of impulses of magnitude equal to the input at that particular time then the output at some chosen time will be the sum of the impulse responses arising from all previous impulses. To illustrate this let us choose a case with a simple impulse response $h(t)$ as shown in Fig. 1.14.1. The only effect of a more complex response would be to make the diagram more difficult to follow.

If we consider an impulse at time $\tau$ then the response at our chosen time $t$ (with $t > \tau$) will be given by the magnitude of $h(t)$ at that time, i.e. $h(t - \tau)$.

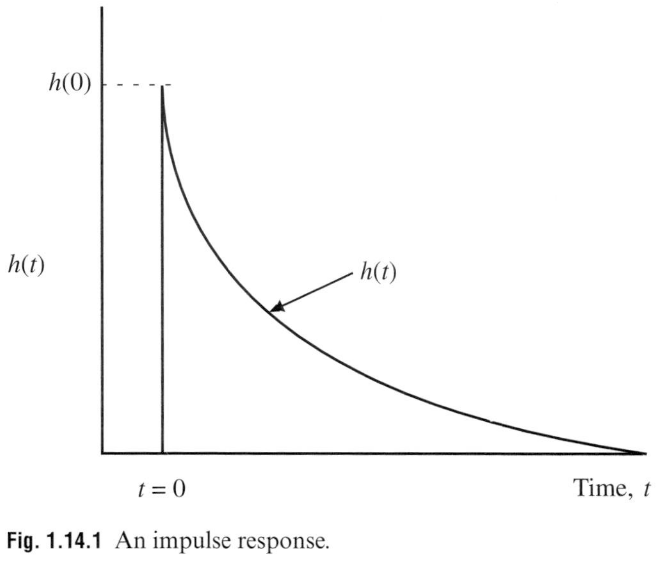


**Fig. 1.14.1** An impulse response.

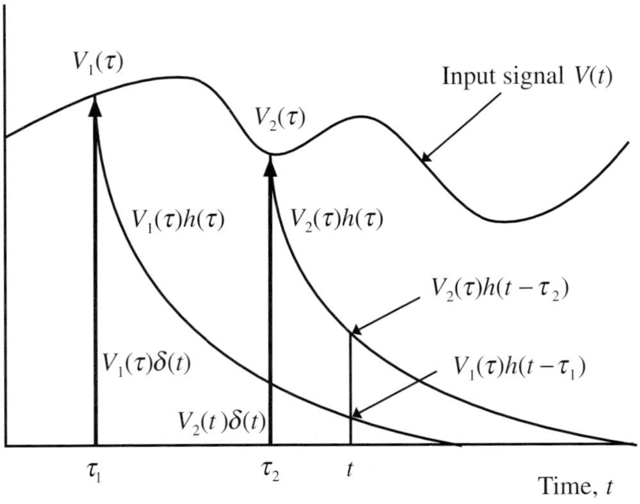


**Fig. 1.14.2** Convolution.

The system output $y_\tau(t)$ arising from a $\delta$ function input $x(\tau)$ at time $\tau$ will be given by:

$$y_\tau(t) = x(\tau)\ \delta(\tau)\ h(t-\tau) \tag{1.14.1}$$

and the net output, arising from all the input previous to $t$, will then be the sum, or in the limit the integral, of all the individual contributions:

$$y(t) = \int_{-\infty}^{t} y_\tau(t) = \int_{-\infty}^{t} x(\tau)\ h(t-\tau)d\tau, \quad \text{or symbolically}$$

$$y(t) = x(\tau)^*\ h(t-\tau) \tag{1.14.2}$$

where the * is the commonly used symbol to indicate convolution.

Convolution is the technique used by PSpice to treat Laplace transform expressions. An inverse Fourier transform is applied to the Laplace expression to find the impulse response. This is then convolved with the input source to find the output (MicroSim 1996). As the Laplace transform is one-sided care is required where the expression results in an impulse response before zero time which leads to causality problems (Section 3.6).

There is a somewhat astonishing, but often useful, relationship between Fourier transforms and convolution. From Eq. (1.14.2) with infinite limits and taking the Fourier transform of both sides:

$$Y(\nu) = \int_{-\infty}^{\infty} y(t)\exp(2\pi j\nu t)dt$$

$$= \int_{-\infty}^{\infty}\int_{-\infty}^{\infty} [x(\tau)h(t-\tau)d\tau]\exp(2\pi j\nu t)dz, \quad \text{and putting } (t-\tau)=z, \quad \text{so } dt=dz$$

$$= \int_{-\infty}^{\infty}\int_{-\infty}^{\infty} [x(\tau)h(z)d\tau]\exp[2\pi j\nu(z+\tau)]dt$$

$$= \int_{-\infty}^{\infty} x(\tau)\exp(2\pi j\nu t)d\tau \bullet \int_{-\infty}^{\infty} h(z)\exp(2\pi j\nu t)dz$$

$$= F[x(\tau)] \bullet F[h(z)] \qquad (1.14.3)$$

and thus $Y(\nu)$, the Fourier transform of the convolution of $x(\tau)$ and $h(t-\tau)$, is equal to the product of the Fourier transforms of $x(\tau)$ and of $h(t-\tau)$, i.e. we have a transform pair (often shown by a symbol like $\Leftrightarrow$):

$$x(t)^{*}\, h(t-\tau) \Leftrightarrow F[x(t)] \bullet F[h(t-\tau)] \qquad (1.14.4)$$

This relation can be useful in determining some more difficult functions.

## References and additional sources 1.14


Boas M. L. (1966): *Mathematical Methods in the Physical Sciences*, New York: John Wiley. Library of Congress Cat. No. 66-17646.

Faulkner E. A. (1969): *Introduction to the Theory of Linear Systems*, London: Chapman and Hall. ISBN 412-09400-2.

James J. F. (1995): *A Student's Guide to Fourier Transforms*, Cambridge: Cambridge University Press. ISBN 0-521-46829-9.

Lynn P. A. (1986): *Electronic Signals and Systems*, Basingstoke: Macmillan. ISBN 0-333-39164-0.

MicroSim (1996): *PSpice User's Guide*, Ver.7.1, October, pp. 6–45.

Pippard A. B. (1985): *Response and Stability*, Cambridge: Cambridge University Press. ISBN 0-521-31994-3. See Chapter 3.

Siebert W. McC. (1986): *Circuits, Signals, and Systems*, Cambridge, Mass: MIT Press/McGraw-Hill. ISBN 0-07-057290-9.

# Part 2

# Physics

> In reflecting on these things one cannot help but be amazed at how short are the memories of many engineers and engineering faculties, and how often the battle to achieve these objectives needs to be refought. At the beginning of World War II, when engineers were presented the problem of developing radar, they were (except in very few cases) found woefully lacking in an ability to cope with such an unconventional situation; and physicists, both theoretical and experimental, had to be called in to do what was essentially an engineering job.
>
> E. A. Guillemin (1962): *Proc. IRE* **50**, 872–878

I hope things are now better than the quotation suggests, but it was penned not long ago by an eminent electronicist of the old school. As a physicist by training, I have found the knowledge that that has provided to be of immense help in the electronics field. This part is a brief introduction to some of the topics that seem relevant and which to some extent are missed out in the training of electronic engineers. It must of course be a rather brief foray into the realm of the physicist but it may give some encouragement to find out more, and will provide some insight into the roots from which the subject of electronics has grown. The move to much higher frequencies, the great advances in electro-optics and the development of many new materials, including the manipulation of single atoms, means that some knowledge of physics is highly beneficial. The topics considered have been confined to those closely related to circuits and their behaviour rather than the more general field.

The synthesis of electricity, magnetism and light by Maxwell is one of the great triumphs of physics. Together with the discovery of the predicted electromagnetic waves by Hertz and the revolution of understanding arising from Einstein's theory of relativity, this edifice stands as the core of our subject. The realization that such waves can propagate in free space without benefit of any medium leaves us with considerable philosophical problems to which mathematics is the only presently satisfactory answer. It is rather like the Indian rope trick, but without even the rope. It is with these thoughts in mind that some of the following sections have been included to try and provide some background. But the quotations, included in some of the sections, from one of the most eminent physicists of our time, Richard Feynman, should be borne in mind when trying to grasp the subtleties of electromagnetism. You can still be a good electronicist without this knowledge, but I think you could be a better one with it.

# 2.1 Current flow

I am inclined to think that an electric current circulating in a closed conductor *is heat*, and becomes capable of producing thermometric effects by being frittered down into smaller local circuits or 'molecular vortices'.

Letter from Professor William Thomson to J. P. Joule, March 31, 1852. See Joule J. P. (1855): On the œconomical production of mechanical effects from chemical forces. *Philosophical Magazine* **5**, 1–5

## Motion of electrons in a conductor

The current in a conductor consists of electrons that are essentially free from their original atoms while the positively charged atomic ions are held in place by the forces which create the solid. As a result of being at some finite temperature $T$ the electrons will be moving with high velocity and the ions will be vibrating about their equilibrium positions. The net charge will of course be zero since there are the same number of electrons as there are positive ions. It is instructive to calculate the effective velocity of the electrons when a current flows, as the result is rather unexpected.

Let us first calculate the density of conduction electrons in copper. One *mole* of copper has a mass of $63.5\times 10^{-3}$ kg and, by definition, this contains $N_A = 6.02\times 10^{23}$ atoms (Avogadro's number), and the density of copper is $\rho = 8920\,\mathrm{kg\,m^{-3}}$. The volume $V$ of one mole of copper is:

$$V = \frac{63.5\times 10^{-3}}{8920} = 7.12\times 10^{-6}\,\mathrm{m^3} \qquad (2.1.1)$$

For copper, each atom contributes one free electron so the density $n$ of conduction electrons is:

$$n = \frac{N_A}{V} = \frac{6.02\times 10^{23}}{7.12\times 10^{-6}} = 8.46\times 10^{28}\,\mathrm{m^{-3}} \qquad (2.1.2)$$

If we consider a current of say 1 A in a wire of cross-section 1 mm$^2$ ($10^{-6}$ m$^2$) then since the current density $j_c$ is related to the drift velocity $v_d$ by ($q_e$ is the electronic charge $1.6\times10^{-19}$ C):

$$j_c = nq_e v_d$$

$$\text{so} \quad v_d = \frac{j_c}{nq_e} = \frac{1}{1\times10^{-6}\times8.46\times10^{28}\times1.6\times10^{-19}} = 74\times10^{-6}\ \text{m s}^{-1} \tag{2.1.3}$$

a very small velocity indeed (about 3 h to travel 1 m!). So currents, or charges, do not flow at the 'speed of light' ($c\approx3\times10^8$ m s$^{-1}$ in free space) but the electromagnetic field does, so the charges all the way round the circuit move 'simultaneously'. A Mexican wave travels around a stadium very much more rapidly than a person could.

The individual electrons do move at high velocity but since they are randomly scattered very frequently it is only the average drift velocity which matters. The average frequency of collisions $\nu_{coll}$ can be found from (Feynman et al. 1964):

$$j_c = \sigma E = \frac{nq_e^2 E}{m_e \nu_{coll}} \tag{2.1.4}$$

where $\sigma$ is the conductivity ($5.8\times10^7$ Sm$^{-1}$) and $m_e$ the mass of the electron ($9.1\times10^{-31}$ kg). Thus:

$$\nu_{coll} = \frac{nq_e^2}{m_e\sigma} = \frac{8.46\times10^{28}\times(1.6\times10^{-19})^2}{9.1\times10^{-31}\times5.8\times10^7} = 4.1\times10^{13}\ \text{s}^{-1} \tag{2.1.5}$$

The electrons are only accelerated by the electric field $E$ for on average about $2.4\times10^{-14}$ s before suffering a collision, so it is not surprising that the drift velocity is so low.

Charge is conserved. If we consider a volume $v$ containing a net charge of density $\rho$, then the outward flux of current of density $\boldsymbol{j}$ through the surface $\boldsymbol{S}$ of volume $v$ must equal the rate of decrease of charge within the volume (Corson and Lorrain 1962):

$$\int_S \boldsymbol{j}\cdot d\boldsymbol{S} = \int_v \frac{\partial\rho}{\partial t}dv \tag{2.1.6}$$

which is the equation of continuity. If we make use of the divergence theorem (Eq. (1.6.23)) then the surface integral can be changed to a volume integral:

$$\int_v \boldsymbol{\nabla}\cdot\boldsymbol{j} = -\int_v \frac{\partial\rho}{\partial t}dv \tag{2.1.7}$$

and since the equation must be true for all $v$ we can equate the integrands to get:

$$\nabla \cdot j = \frac{-\partial \rho}{\partial t} \tag{2.1.8}$$

which is the law of conservation of charge in differential form. This equation allows us to examine the charge density $\rho$ in a conductor. We have from Gauss' law, Eq. (2.7.1(I)):

$$\nabla \cdot E = \frac{\rho}{\varepsilon \varepsilon_0} \tag{2.1.9}$$

and using Ohm's law in general form:

$$j = \sigma E \quad \text{so} \quad \nabla \cdot j = \sigma \nabla \cdot E = \frac{\sigma \rho}{\varepsilon \varepsilon_0} = \frac{-\partial \rho}{\partial t} \tag{2.1.10}$$

using (2.1.8). This equation for $\rho$ can be integrated to give (see Eq. (1.9.6) and following):

$$\rho = \rho_0 \exp\left(\frac{-\sigma t}{\varepsilon \varepsilon_0}\right) = \rho_0 \exp\left(\frac{-t}{\tau}\right) \tag{2.1.11}$$

showing a time constant or relaxation time $\tau$ of:

$$\tau = \frac{\varepsilon \varepsilon_0}{\sigma} \approx 10^{-19}\ \text{s} \tag{2.1.12}$$

where we have taken $\varepsilon \cong 1$ and $\sigma = 5.8 \times 10^7\ \text{Sm}^{-1}$ for copper (the dielectric constant of a conductor presents some difficulty). This shows that if free charges are introduced inside a good conductor the density $\rho$ decays in an extremely short time so that any net charge must reside on the surface only. Though this result is often given in many books it implies a field propagating with a velocity many orders of magnitude greater than the velocity of light. This simple approach to the relaxation is thus rather optimistic and the time is considerably longer (Mott et al. 1972). A different approach by Gruber (1973), who disagrees with Mott et al., suggests that the time is more like the inverse of the collision frequency $\nu_{coll}$, i.e. of the order of $10^{-13}$ s – about $10^6$ times longer. At the atomic level we must use quantum rather than classical theory.

## Current in capacitors

The introduction of displacement current by Maxwell was the key to the formulation of his famous equations and the realization that light was an electromagnetic wave. The displacement current is the current equivalent to the displacement of charges in a dielectric when polarized by an electric field. However, it is more than this in that it occurs even *in vacuo* and the current density is given by:

$$j_D = \varepsilon_0 \frac{dE}{dt} \tag{2.1.13}$$

where $\varepsilon_0$ is the permittivity of free space and $E$ is the electric field.

If a conduction current $I$ flows to charge a capacitor then the rate of change of charge $Q$ is given by:

$$\frac{dQ}{dt} = I \tag{2.1.14}$$

It is indicated in Section 2.3 that current flows *through* a capacitor (see Fig. 2.3.1, p. 103). Here we show that the *displacement* current (in the dielectric or free space) is equal to the *conduction* current in the connecting wire. Consider a circular parallel plate capacitor of radius $a$ and separation $d$. If the area of a plate is $A$ then the capacity is given by:

$$C = \frac{\varepsilon_0 A}{d} = \frac{\varepsilon_0 \pi a^2}{d} \tag{2.1.15}$$

and if $Q$ is the charge on the plates then by Gauss' law we get a uniform field $E$ between the plates:

$$E = \frac{Q}{A\varepsilon_0} = \frac{Q}{\pi a^2 \varepsilon_0} = \frac{V}{d} \tag{2.1.16}$$

where $V$ is the potential across the capacitor. Hence the displacement current $I_D$ is:

$$I_D = \int j_D \, dA = \pi a^2 \varepsilon_0 \frac{dE}{dt} = \frac{\pi a^2 \varepsilon_0}{\pi a^2 \varepsilon_0} \frac{dQ}{dt} = \frac{dQ}{dt} = I \tag{2.1.17}$$

so the displacement current $I_D$ is equal to the conduction charging current $I$.

## References and additional sources 2.1


Corson D. R., Lorrain P. (1962): *Introduction to Electromagnetic Fields and Waves*, San Francisco: W. H. Freeman. Library of Congress Cat. No. 62-14193.

Feynman R. P., Leighton R. B., Sands M. (1964): *The Feynman Lectures on Physics*, Vols I, II, III, Reading, Massachusetts: Addison-Wesley. See Vol. II, p. 32–10, Section 32–6 and Vol. I, Section 43.

Gruber S. (1973): On charge relaxation in good conductors. *Proc. IEEE* **61**, 237–238.

Mott H., Raeburn W. D., Webb W. E. (1972): Charge density in a conducting medium. *Proc. IEEE* **60**, 899–900. See also correction **61**, 238,1973.

# 2.2 Energies

Electrical engineers

A scientist of even quite modest attainments will find from time to time that he receives unsolicited contributions from the general public proposing solutions to the riddle of the physical universe. His correspondent may need a little help with the mathematics or a testimonial to facilitate the publication of their ideas, but they are confident that they have made an important advance. I am sorry to have to say that such items of this character that have come my way have, without exception, proved valueless. Many have not been sufficiently articulate even to attain the status of being wrong. Nor does some kinship with science prove a help in this matter. Some of my most persistent and wrong-headed correspondents have been electrical engineers.

This thought crosses my mind as I, a theoretical physicist by profession, take up my pen to write on matters theological.

John Polkinghorne (1998): Science and creation – the search for understanding (Society for Promoting Christian Knowledge, 1998, p. 1) *Amer. J. Phys.* **66**, 835

A charged capacitor stores energy as does an inductor carrying a current. This is an important concept to understand as it has significant consequences. For a capacitor $C$ or an inductor $L$ the energy stored is given (in joule) by (see Eqs. (4.2.3) and (4.3.2)):

$$\tfrac{1}{2}CV^2 \quad \text{and} \quad \tfrac{1}{2}LI^2 \tag{2.2.1}$$

This storage of energy is what makes, for example, an $LC$ tuned circuit resonate. In this case the energy is alternately stored in the electric field of the capacitor or in the magnetic field of the inductor. We will consider resonance later (Section 3.5). High voltage capacitors should be handled with care and stored with terminals shorted to avoid possibly lethal shocks. The high voltages developed across an inductor when the magnetic field is collapsing provide the spark for your motor car, destroy the transistor driving a relay or provide the rising arcs so beloved in Frankenstein movies!

The availability of two energy storage elements allows the construction of a resonator. If we consider a classic case, the simple pendulum, we can see that there are two forms in which the energy appears. When the pendulum is at one extreme or the other, the velocity is zero and all the energy is potential. When the pendulum is at

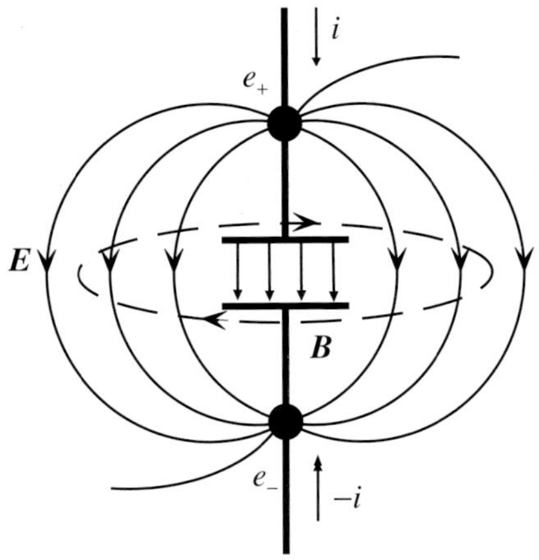


**Fig. 2.2.1** Energy flow into a capacitor.

the midpoint of its motion, the velocity is a maximum and all the energy is kinetic. The energy thus passes back and forth between potential and kinetic so the motion can continue in an oscillatory manner. There is of course some energy loss during each cycle due to air friction, for example, so the amplitude of oscillation will decay with time. To maintain continuous oscillation this loss must be replaced from some source, e.g. the wound spring in a clock. In the case of an electronic pendulum we need also to have the two forms of energy storage and in a resonant circuit this is provided by the static energy stored in the capacitor and the dynamic energy stored in the inductor. In the capacitor the energy is electrostatic and is stored in the electric field and in the inductor it is stored in the magnetic field. As for the pendulum there are losses due to finite resistance, particularly in the inductor, and this loss must again be made up, each cycle, from another source. Any resonant system will have a natural frequency of oscillation determined by its energy storage elements.

The connection between the energy dissipation in a resistive conductor and the surrounding electromagnetic field is examined in Section 2.5. The energy stored in a capacitor is located in the field between the plates and it is of interest as to how it gets there. Consider a current flowing to charge a capacitor as illustrated in Fig. 2.2.1.

With current in the conventional direction there will be an electron approaching from below and a positive charge from above. The electric field will be of the form shown where we overlook any local distortions caused by the capacitor plates. The magnetic field will be concentric with the connecting wires. As the two charges approach the capacitor the field will shrink and the $\boldsymbol{E}$ and $\boldsymbol{B}$ fields, and hence energy, will enter the space between the plates from the sides rather than flowing along the wires, which may not be what you might have expected. For the capacitor we assume the energy will be concentrated between the plates, and using Eq. (4.2.2) for a parallel plate capacitor of area $S$ and separation $d$ we have:

$$\text{Energy} = \tfrac{1}{2}CV^2 = \tfrac{1}{2}\frac{S\varepsilon_0}{d}d^2E^2, \quad \text{since } V = Ed, \quad \text{where } E \text{ is the electric field} \tag{2.2.2}$$

$$\text{So energy density} = \tfrac{1}{2}\varepsilon_0 E^2, \quad \text{since the volume is } Sd$$

Similarly, for a solenoid we may assume that the energy is concentrated in the volume of the solenoid and hence derive an expression for the energy density of the magnetic field in terms of the field. Using Eqs. (2.5.14), (2.5.12) and (2.11.7) we have for a solenoid of length $l$, radius $r$ with $N$ turns per unit length and carrying a current $I$:

$$\text{Energy} = \tfrac{1}{2}LI^2 = \tfrac{1}{2}\frac{\mu_0 N^2 l^2 S}{l}\frac{H^2}{N^2}, \quad \text{since } n = Nl, \; I = \frac{B}{\mu_0 N}, \; B = \mu_0 H, \; S \equiv \pi r^2 \tag{2.2.3}$$

$$\text{So energy density} = \tfrac{1}{2}\mu_0 H^2, \quad \text{since the volume is } Sl$$

Though we have used somewhat idealized conditions a more precise approach leads to the same results.

An electromagnetic wave, which consists of complementary electric and magnetic fields, also carries energy. However, at some distance from the radiator or antenna we have to consider a different approach to describing this energy as we cannot so readily associate it with capacitors and inductors. The fields surrounding an antenna are complex and depend on whether we are close to the antenna (the near field) or at a considerable distance (the far field). There is of course an intermediate region where the two regimes merge and hence cause greater complexity. The scale of ‘far’ in this context is determined by the wavelength of the radiation, i.e. far means many wavelengths. The near field falls off as the square of the distance and is the one involved in, for example, transformer coupling. The far field falls off more slowly and is the one that is responsible for radio transmission over longer distances. At some distance from the antenna the field distribution over a limited area will be essentially planar so we can consider the energy passing through a small area normal to the direction of propagation. We make the assumption that the results of Eqs. (2.2.2) and (2.2.3) are valid for time-varying fields, and a proper investigation confirms this. To determine the flow of energy across unit area normal to the direction of our plane wave we consider the divergence of $\boldsymbol{E}\times\boldsymbol{H}$ (remembering that in free space $\boldsymbol{j}=0$):

$$\begin{aligned}\boldsymbol{\nabla}\cdot(\boldsymbol{E}\times\boldsymbol{H}) &= -\boldsymbol{E}\cdot(\boldsymbol{\nabla}\times\boldsymbol{H}) + \boldsymbol{H}\cdot(\boldsymbol{\nabla}\times\boldsymbol{E}), \quad \text{using Eq. (1.6.21)}\\ &= -\left(\boldsymbol{E}\cdot\varepsilon_0\frac{\partial\boldsymbol{E}}{\partial t}\right) - \left(\boldsymbol{H}\cdot\mu_0\frac{\partial\boldsymbol{H}}{\partial t}\right), \quad \text{using Eqs. (2.7.2) and (2.7.1(II))}\\ &= -\frac{\partial}{\partial t}\left(\frac{\varepsilon_0 E^2}{2} + \frac{\mu_0 H^2}{2}\right)\end{aligned} \tag{2.2.4}$$

Now if we consider a volume $V$ with a bounding surface $S$ with normal $\boldsymbol{n}$, and integrate over the volume:

$$\int_V \boldsymbol{\nabla} \cdot (\boldsymbol{E} \times \boldsymbol{H})\, dv = -\frac{\partial}{\partial t}\int_V \left(\frac{\varepsilon_0 E^2}{2} + \frac{\mu_0 H^2}{2}\right) dv$$

$$\int_S (\boldsymbol{E} \times \boldsymbol{H}) \cdot \boldsymbol{n}\, ds = -\frac{\partial}{\partial t}\int_V \left(\frac{\varepsilon_0 E^2}{2} + \frac{\mu_0 H^2}{2}\right) dv, \quad \text{using Eq. (1.6.23)} \qquad (2.2.5)$$

The integral on the right-hand side is just the sum of the electric and magnetic energy densities over the volume $V$, so the left-hand side must be the outward flux over the surface $S$. In this case the flux is in the direction of the vector $\boldsymbol{E} \times \boldsymbol{H}$, i.e. in the direction of propagation. This result was found by Poynting in 1884 (Buchwald 1988) and the vector $\boldsymbol{N}$ is known as the Poynting vector:

$$\boldsymbol{N} = \boldsymbol{E} \times \boldsymbol{H} \qquad (2.2.6)$$

Some care is required in using the Poynting vector since, for example, a statically charged bar magnet has $\boldsymbol{E}$ and $\boldsymbol{H}$ fields but there is no energy flow. In dynamic conditions we may interpret $\boldsymbol{N}$ as the energy crossing unit area per second in an electromagnetic wave and experiments so far agree with this interpretation. An application is examined in Section 2.5.

## References and additional sources 2.2


Buchwald J. Z. (1988): *From Maxwell to Microphysics*, University of Chicago Press. ISBN 0-226-07883-3. See Chapter 4.

Corson D. R., Lorrain P. (1962): *Introduction to Electromagnetic Fields and Waves*, San Francisco: W. H. Freeman. Library of Congress Cat. No. 62-14193. See Chapter 9.

Feynman R. P., Leighton R. B., Sands M. (1964): *The Feynman Lectures on Physics*, Vols I, II, III, Reading, Mass: Addison-Wesley. Library of Congress Cat. No. 63-20717. See Vol. II, Section 27.

Grant I., Philips W. R. (1990): *Electromagnetism*, 2nd Edn, London: John Wiley. ISBN 0-471-32246-6. ISBN 0-471-92712-0. See Section 11.3.

Haus H. A., Melcher T. R. (1989): *Electromagnetic Fields and Energy*, Englewood Cliffs: Prentice-Hall. ISBN 0-13-249277-6.

Poynting J. H. (1884): On the transfer of energy in the electromagnetic field. *Phil. Trans. Roy. Soc.* **175**, 343–361.

Poynting J. H. (1885): On the connexion between electric current and the electric and magnetic inductions in the surrounding field. *Phil. Trans. Roy. Soc.* **176**, 277–306.

# 2.3 Kirchhoff's laws

Volta's invention of multiplying the Galvanic action repeating its prerequisites arbitrarily and indefinitely is the greatest gift to Galvanism since Galvani. Even the simple action enabled us to penetrate into the system of its effects down to a considerable profoundness; if we are allowed to extend the effects, which seemed to be much too small for a lot of people to be of their interest until that day, up to 60, 80 or 100 and manifold, we will immediately focus all the attention on it, as indeed has owed. But not only for manifolded representation of the already known, Volta's invention is as well qualified excellently for the discovery of new effects of Galvanism, only possible to be registered by the restricted senses because of the enlargement of the corresponding causes.

J. W. Ritter (1800): *Volta's Galvanische Versuche*, Weimar

Kirchhoff's laws are central to our ability to analyse circuits. When they were first enunciated the electron had not yet been discovered and frequencies greater than zero were almost unknown (Kirchhoff 1847). Most books on electronics simply state them with maybe a brief indication of their origin. There is rather more to them than these passing references would indicate, and their central role makes it necessary that we examine their validity carefully (Fano et al. 1960). They arise from conservation laws and must of course be consistent with the overarching Maxwell laws. The first law is based on the more general law of conservation of energy, one of the most fundamental of physics. The proposition is that the energy supplied by the source, e.g. a battery, in producing the current flow in the circuit must be equal to the energy dissipated in the various elements in the circuit. The dissipation is usually as heat at low frequencies but can be radiated as electromagnetic energy at high frequencies. Thus we may state Kirchhoff 's voltage law (KVL) as:

The algebraic sum of the voltage drops around any circuit mesh is zero. (2.3.1)

Algebraic means that the sense of the voltage drops must be allowed for, i.e. the sense of the source voltage will be opposite to that across a passive element, and that phase must be taken into account for alternating voltages. This law also follows from Maxwell's equation that says that the line integral of $E$ around a closed loop is zero (Section 2.7).

The second law arises from the law of conservation of charge which means that charges cannot just disappear or appear at some point in a circuit. This law is thought to be one of the fundamental laws of the universe and all experimental evidence so far supports it. Charge may in effect 'disappear' by being neutralized by a charge of the opposite sign as, for example, in an atom where the negative charges on the electrons are exactly matched by the positive charge on the nucleus. The sign of a charge is of course quite arbitrary; all we must maintain is consistency. The oppositely directed flow of electrons and conventional current does cause some difficulty when first learning the subject. We also consider that elements in our circuits do not support unbalanced charge which would in that case result in additional fields and change the experimentally observed behaviour. Capacitors do illustrate the separation of charge but there is always exact equality of positive and negative charge and so neutrality. If there were evidently charges at some point in our circuit then we expect there will be somewhere a balancing charge, i.e. there will be an additional capacity which we have not allowed for. The precision involved is extreme as illustrated by the discussion of the relativistic origins of the magnetic field in Section 2.6. The consequence of this physical circumstance is that at any node the algebraic sum of the currents at the node must be zero, i.e. current *in* must equal current *out* otherwise there would be an accumulation of charge. This constitutes Kirchhoff's current law (KCL):

The algebraic sum of the currents entering any circuit node must be zero. (2.3.2)

Some authors define this law in terms of charges as 'The charge on a node or in an element is identically zero at all instants of time' (Chen 1995, p. 421), but the significance is the same.

These rules allow us to compute the currents and voltages throughout the circuit for any given conditions. Applying them to a circuit will result in a set of simultaneous equations which will allow the determination of all the voltages and currents in the circuit. A technique for solving such a set of equations is outlined in Section 1.10 and an example is shown in action in Section 5.21. It is evident in the latter that the calculation very soon becomes too complex to carry out readily by hand, but mathematical software packages can now come to our aid. SPICE is geared to the efficient solution of such sets of equations and can also cope with non-linearity which would not be feasible to do by hand.

As mentioned, currents are often forgotten about but they are what carries the signal from one part of the circuit to another. There is an important consequence of Kirchhoff's laws in considering currents. The second law states that the current must be everywhere continuous, i.e. it must flow in a complete loop. So if it 'starts' at some point then it must eventually get back to that point. The route by which it can achieve this may not always be evident from the circuit but you should not overlook the power supplies. These are commonly omitted from circuit diagrams but are the common route for currents to complete their loop.

This requirement for currents sometimes causes some convoluted thinking where capacitors are concerned. It is often said that a current cannot pass 'through'

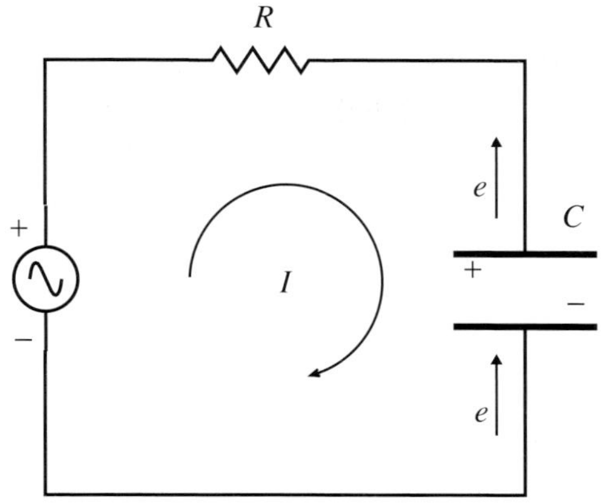


**Fig. 2.3.1** Current flow through a capacitor.

a capacitor, which is not true and which will contradict our acceptance of Kirchhoff's law. Consider a simple circuit as shown in Fig. 2.3.1. If you were asked to analyse this circuit the first thing you would probably do is to draw a loop arrow as shown and label it $I$ – you have immediately shown the current, which must be everywhere continuous, flowing 'through' $C$! When confronted people fall back on the idea of current flowing 'into' the capacitor rather like a bucket of water.

Consider an electron leaving the negative terminal of the battery and reaching the lower plate to produce a negative charge. This electric field causes or requires a positive charge on the upper plate which can only come about by an electron having left the upper plate, i.e. the current flows *through* the capacitor. In a zero frequency case, after the capacitor is charged the current will have dropped to zero and no current then flows *through* the capacitor or in the rest of the circuit. Section 2.1 shows that the displacement current equals the charging current. So do not worry about capacitors: just treat them like any other component. If you have a clip-on current probe for an oscilloscope available try examining the a.c. current around the circuit including passing the capacitor through the probe.

## References and additional sources 2.3


Chen, Wai-Kai (Ed.) (1995): *The Circuits and Filters Handbook*, Boca Raton: CRC Press and IEEE Press. ISBN 0-8493-8341-2.

Fano R. M., Chu L. J., Adler R. B. (1960): *Electromagnetic Fields, Energy, and Forces*, New York: John Wiley.

Kirchhoff G. (1847): On the solution of the equations obtained from the investigations of the linear distribution of galvanic currents. IEEE Trans. **CT-5**, 4–7, 1958. A translation by J. B. O'Toole of the original paper 'Ueber die Auflösung der Gleichungen, auf welche man bei der Untersuching der linearen Vertheilung Galvanisher Ströme geführt wird.' *Ann. Phys. Chem.* **72**, 497–508, December 1847.

# 2.4 Faraday's law and Lenz's law

We are dwarfs mounted on the shoulders of giants, so that we can see more and further than they: yet not by virtue of the keenness of our eyesight, nor through the tallness of our stature, but because we are raised and borne aloft upon that giant mass.

Bernard of Chartres, French scholar of the 12th century. Reused by Isaac Newton in a letter to Robert Hooke, possibly unknowingly

The interaction of magnetic flux and circuits sometimes leads to considerable confusion as to the consequences. The 'simple' idea that a change of flux linked with a circuit generates an e.m.f. can lead you to contradictory results (Cathode Ray 1955). The great discovery by Faraday was that it did not matter whether the flux was changing or whether the circuit was moving, the induced e.m.f. was the same. The law which bears his name was formulated, in the mathematical form we now know, by Maxwell, but the apparently simple equation hides the underlying physics. Faraday's law is usually stated in the form given in Eq. (2.4.1):

$$e = -\frac{d\Phi}{dt} \tag{2.4.1}$$

where $\Phi$ is the flux linked with the circuit of area $A$, e.g. $\Phi = BA$, if the field $B$ is uniform. The negative sign is a consequence of Lenz's law, which states that the sense of the voltage induced in the circuit would, if the circuit is closed, cause a current to flow in a direction such as to itself produce a magnetic field that opposes the original field. This opposition is required for the conservation of energy, as otherwise the induced current would produce an enhancement of the original which would produce more current and so on *ad infinitum*.

The flux linking the circuit may change either due to time-dependent changes in the field $B$ or because the circuit is moving relative to the field. We need to examine the two possibilities separately to see how the e.m.f. arises. First we examine what happens when the circuit moves (or changes its aspect) relative to the field, and this can be demonstrated with a circuit of convenient geometry as shown in Fig. 2.4.1.

The square U-shaped conductor is held stationary while the crossbar, making electrical contact with the U-section, moves to the right with velocity $v$. There is a

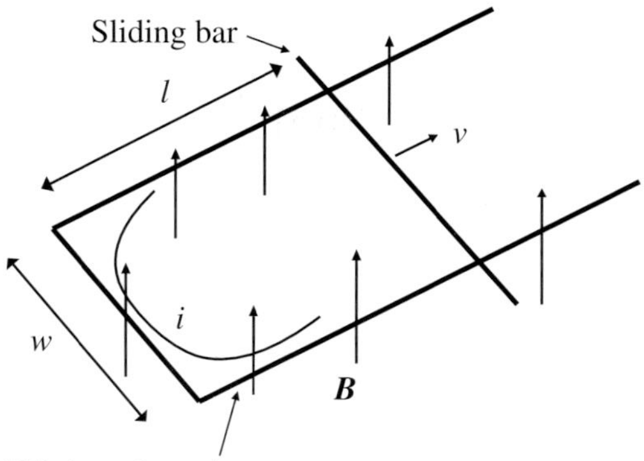


**Fig. 2.4.1** A 'moving' circuit in a steady magnetic field.

constant field $B$ normal to the plane of the circuit. The flux $\Phi$ linked with the circuit will be $B$ times the area $l \times w$, so we expect from Eq. (2.4.1) that the e.m.f. generated will be:

$$e = -\frac{d\Phi}{dt} = \frac{d}{dt}(Blw) = Bw\frac{dl}{dt} = Bwv \qquad (2.4.2)$$

The e.m.f. must be a consequence of the force acting on the free electrons in the conductor arising from the relative motion of conductor and magnetic field. Here only the crossbar is moving so it is in this section of the circuit that the electrons will feel the force. Motion is required, but that in itself is insufficient. If instead of the crossbar moving the whole circuit is moved in a constant field, then there will be no change in the flux linked with it and hence no e.m.f. developed. The general expression for the force $\boldsymbol{F}$ acting on an electric charge $q$ in an electric field $\boldsymbol{E}$ and a magnetic field $\boldsymbol{B}$ is given by:

$$\boldsymbol{F} = q(\boldsymbol{E} + \boldsymbol{v} \times \boldsymbol{B}) \qquad (2.4.3)$$

There is no external electric field $\boldsymbol{E}$ so the force in our example must arise from the $\boldsymbol{v} \times \boldsymbol{B}$ term, which acts in a direction normal to both $\boldsymbol{v}$ and $\boldsymbol{B}$, i.e. along the crossbar. The e.m.f. in a circuit is the integral of the force around the loop. Since the force is constant along the crossbar and zero elsewhere the integral is simply $Bwv$ as found in Eq. (2.4.2).

If we now consider a circumstance where the field $\boldsymbol{B}$ is changing as a function of time then experiment showed that the e.m.f. generated was still given by Eq. (2.4.1). Maxwell's equation (2.7.2 (II)) tells us the relation between the rate of change of the $\boldsymbol{B}$ field and the consequential $\boldsymbol{E}$ field:

$$\nabla \times \boldsymbol{E} = -\frac{\partial \boldsymbol{B}}{\partial t} \qquad (2.4.4)$$

and this is more properly called Faraday's law. This equation can be changed to the integral form by use of Stokes' theorem (Eq. (1.6.22)), where $\Gamma$ is a closed contour and $S$ any surface bounded by $\Gamma \bullet \boldsymbol{n}$ is the unit normal to an element $da$ of the surface and $\boldsymbol{dl}$ is an element of $\Gamma$:

$$\oint_{\Gamma} \boldsymbol{E} \bullet \boldsymbol{dl} = \int_{S} (\boldsymbol{\nabla} \times \boldsymbol{E}) \bullet \boldsymbol{n}\, da = -\int_{S} \frac{d\boldsymbol{B}}{dt} \boldsymbol{n} \bullet da = -\frac{\partial}{\partial t} \int_{S} \boldsymbol{B} \bullet \boldsymbol{n}\, da \qquad (2.4.5)$$

If $\Gamma$ represents the circuit, then the integral of $\boldsymbol{E}$ around the contour is just the e.m.f., and the right-hand side is just the rate of change of the flux through the surface $S$, i.e. the circuit. It is not even necessary that there be an actual conductive circuit: an electric field is generated in the space surrounding the varying $\boldsymbol{B}$ field so that, for example, electrons in the region would be accelerated by the electric field. This is just the case in the particle accelerator known as the betatron (Kerst 1941). The two cases, changing circuit and changing field, both give the same induced e.m.f. even though two quite different phenomena are involved (Feynman et al. 1964).

It is this effect of the reaction of the induced current which prevents electromagnetic fields penetrating far into a good conductor, that allows us to use a conductor to screen sensitive systems. Since electromagnetic interference is nowadays an important matter which can prohibit use of systems that do not comply with regulations, it is necessary to understand both the origins and the techniques of prevention.

## References and additional sources 2.4


'Cathode Ray' (1955): Flux-cutting or flux-linking? *Wireless World*, February, 74–78.

Corson D. R., Lorrain P. (1962): *Introduction to Electromagnetic Fields and Waves*, San Francisco: W. H. Freeman. Library of Congress Cat. No. 62-14193. See Chapter 6.

Feynman R. P., Leighton R. B., Sands M. (1964): *The Feynman Lectures on Physics*, Vols I, II, III, Reading, Mass: Addison-Wesley. Library of Congress Cat. No. 63-20717. See Vol. II, Section 17.

Grant I., Philips W. R. (1975): *Electromagnetism*, 2nd Edn 1990, London: John Wiley. ISBN 0-471-32246-6. ISBN 0-471-92712-0. See Chapter 6.

Kerst D. W. (1941): Acceleration of electrons by magnetic induction. *Phys. Rev*. **60**, 47–53.

Skeldon K. D., Grant A. I., Scott S. A. (1997): A high potential tesla coil impulse generator for lecture demonstrations and science exhibitions. *Amer. J. Phys*. **65**, 744–754.

# 2.5 Currents and fields

In the neighbourhood of a wire carrying a current, the electric tubes may in general be taken as parallel to the wire while the magnetic tubes encircle it. The hypothesis I propose is that the tubes move in upon the wire, their places being supplied by fresh tubes sent out from the seat of the so-called electromotive force. The change in the point of view involved in this hypothesis consists chiefly in this, that induction is regarded as being propagated sideways rather than along the tubes or lines of induction. This seems natural if we are correct in supposing that the energy is so propagated, and if we therefore cease to look upon current as merely something travelling along the conductor carrying it, and in its passage affecting the surrounding medium. As we have no means of examining the medium, to observe what goes on there, but have to be content with studying what takes place in conductors bounded by the medium, the hypothesis is at present incapable of verification. Its use, then, can only be justified if it accounts for known facts better than any other hypothesis.

Poynting J. H. (1885): On the connection between electric current and magnetic induction in the surrounding field. *Philosophical Transactions* **176**, 277–306

## Electromagnetic fields

An improved understanding of some aspects of circuit behaviour is provided by a knowledge of the related electromagnetic fields. The introduction in recent times of mandatory regulations regarding interference and electromagnetic compatibility make this even more necessary.

Though in the frequency range in which we are concerned, i.e. where the wavelength is much longer than the dimensions of our circuit, this means that we do not have to analyse circuits by solving Maxwell's equations: it is most desirable that we have a good picture of what the electric and magnetic fields look like and to understand the inherent relationship between the field view and that in terms of currents and voltages. It is somewhat of a chicken-and-egg situation: do the fields cause the currents or do the currents cause the fields? The answer is that they are just two views of the same thing, i.e. you cannot have one without the other. You use whichever one is appropriate to the particular circumstance. The thought of

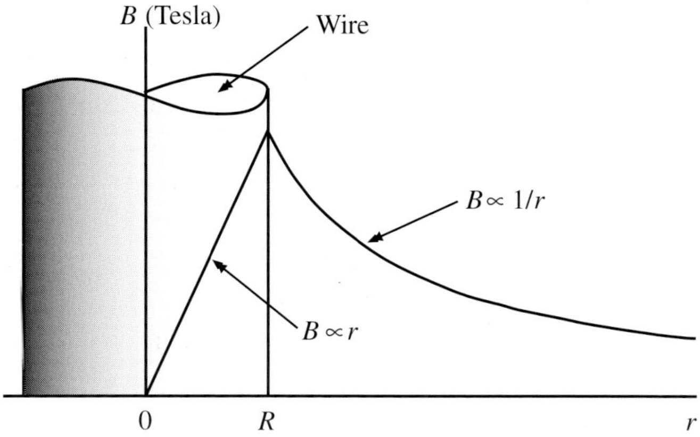


**Fig. 2.5.1** Magnitude of the magnetic induction $B$ as a function of radial distance for a wire carrying a low frequency current. The orientation of the wire relative to the graph is only symbolic.

describing the operation of a transistor say, let alone of a microprocessor, in terms of solutions to Maxwell's equations is too difficult to contemplate. To appreciate the connection between field and circuit descriptions we may consider the simple case of a long (to avoid end effects) straight wire carrying a current. The necessary potential difference between the ends of the wire means that the electric field $\boldsymbol{E}$ must be along the length of the wire and it is this that acts on the electronic charges in the wire to cause them to flow along it.

Ampère's law, the non-existence of isolated magnetic poles and symmetry mean that the magnetic induction $\boldsymbol{B}$ is in the form of closed circular loops around the wire. Ampère's circuital law relates the line integral of the magnetic induction $\boldsymbol{B}$ around a closed path $\boldsymbol{l}$ to the current crossing any surface $\boldsymbol{S}$ bounded by the path. It may be expressed as:

$$\oint \boldsymbol{B} \bullet d\boldsymbol{l} = \mu_0 \int_S \boldsymbol{J} \bullet d\boldsymbol{S} \tag{2.5.1}$$

where $\boldsymbol{J}$ is the current density, and $\mu_0$ is the permeability of free space. We can use this to determine $B$ around a cylindrical conductor like a wire carrying a current $I$ (Fig. 2.5.1). If the current density $J$ is uniform across the cross-section then $J = I/\pi R^2$ and taking a circular path of radius $r > R$ we get:

$$B = \frac{\mu_0 I}{2\pi r} \tag{2.5.2}$$

Inside the wire where $r < R$ we get:

$$B = \frac{\mu_0 I r}{2\pi R^2} \tag{2.5.3}$$

so $B$ varies as shown in Fig. 2.5.1.

The direction of the fields fulfils the requirement of Maxwell's equations that ***E*** and ***B*** must be orthogonal both to one another and to the direction of propagation of the wave $z$ – one can imagine the wavefront moving out from the wire as the current increases with the propagation normal to the wire. If the current is alternating then the fields will increase with the sense in one direction, then decrease to zero and increase in the opposite sense.

The relationship between the two views can be seen clearly if we consider energy or more particularly power losses. Let us examine the relationship between currents and fields in more detail. For a conduction current density $j_c$ we have ($\sigma$ is the conductivity of the conductor):

$$j_c = \sigma E \quad \text{or} \quad E = \frac{j_c}{\sigma} = \frac{I}{\sigma \pi r^2} \tag{2.5.4}$$

For an electromagnetic wave the Poynting vector ***N*** (Section 2.2) gives the flow of energy through an area normal to the direction of propagation: you would, for example, feel the warming from this energy flow as you pass from the shade into full sunlight. The Poynting vector ***N*** is the vector cross product of ***E*** and ***B*** and hence lies along the direction of propagation, but since by symmetry we know the direction we can just take the ordinary product. The magnitude of the Poynting vector (into the surface of the wire) is, from Eq. (2.2.6):

$$N = \frac{EB}{\mu_0} = \frac{I}{\sigma \pi r^2} \frac{1}{\mu_0} \frac{\mu_0 I}{2\pi r} = \frac{I^2}{\sigma 2\pi^2 r^3} \tag{2.5.5}$$

so that the energy per second $P_1$ flowing into a wire segment of length $l$ is:

$$P_1 = N 2\pi r l = \frac{I^2 2\pi r l}{\sigma 2\pi^2 r^3} = \frac{I^2 l}{\sigma \pi r^2} \tag{2.5.6}$$

The resistance of this length $l$ of wire is:

$$R = \frac{1}{\sigma} \frac{l}{\pi r^2} \tag{2.5.7}$$

so that the power dissipation $P_2$ for current $I$ is:

$$P_2 = I^2 R = \frac{I^2 l}{\sigma \pi r^2} = P_1 \tag{2.5.8}$$

This confirms the equivalence of the current and the field views.

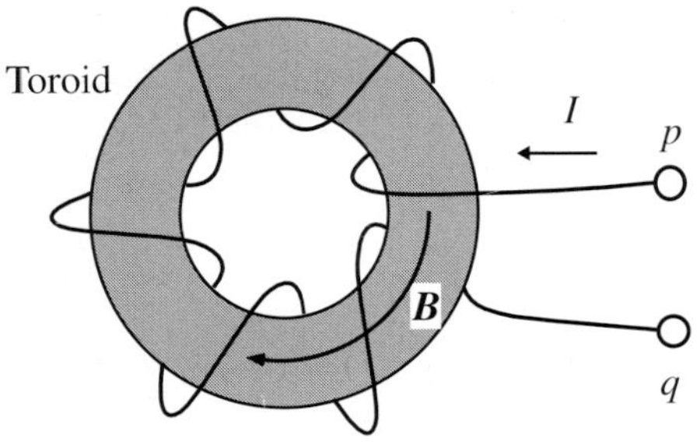


Fig. 2.5.2 Inductor with confined magnetic field.

## Inductance

The creation of the magnetic field around a current means that the source of the current must do work equal to that stored in the field (Section 2.2). This conductor therefore presents an impedance to the flow of current. This impedance is called inductance and we will now examine this from the point of view of the field equations. Consider an inductance where the magnetic field is closely contained, say by winding a number of turns of ideal conductor on a toroid, so that the field does not affect the rest of the circuit (Fig. 2.5.2).

Thus there is no magnetic field near the terminals $p$ and $q$, and we assume that the resistance is negligible and that there are no charges appearing on the wires as the field develops (Section 2.3). We have defined an ideal inductor. When the current $I$ flows a magnetic field $\boldsymbol{B}$, proportional to $I$, will appear. How is this described from the field point of view? The second of Maxwell's equations (Section 2.7) tells us that the line integral of $\boldsymbol{E}$ around a closed loop is equal to the negative of the rate of change of the flux $\Phi$ of $\boldsymbol{B}$ through the loop. If we apply this to our inductor circuit starting at $p$ and travelling inside the wire to $q$, then through the region where there is no field back to $p$, the line integral can be written as:

$$\oint \boldsymbol{E} \cdot d\boldsymbol{l} = \int_p^q \boldsymbol{E} \cdot d\boldsymbol{l}\{\text{coil}\} + \int_q^p \boldsymbol{E} \cdot d\boldsymbol{l}\{\text{air}\} \tag{2.5.9}$$

The electric field in the wire must be zero since we have a perfect conductor, so that the first integral is zero. Since the second line integral is in a region where $\boldsymbol{B}=0$ the integral is independent of the path taken. The potential difference, or voltage $V$, between the terminals $p$ and $q$ is thus just the second integral and from Maxwell's equation (or Faraday's law) we then have:

$$V = -\int_q^p \boldsymbol{E} \cdot d\boldsymbol{l} = -\oint \boldsymbol{E} \cdot d\boldsymbol{l} = \frac{d\Phi}{dt} = L\frac{dI}{dt} \tag{2.5.10}$$

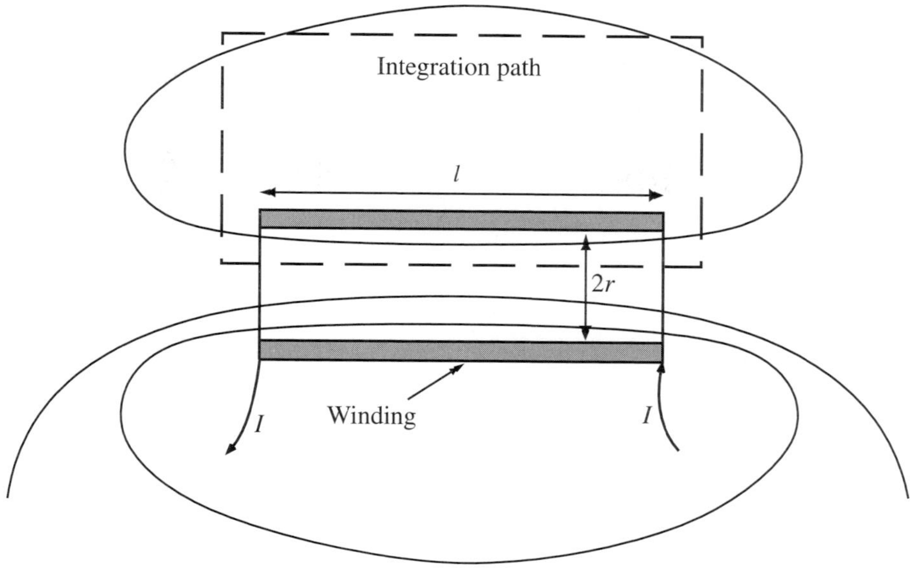


**Fig. 2.5.3** Integration path for a long thin solenoid.

where $L$ is the inductance of the coil. To relate the inductance to the geometry of the coil we first need to determine the field $\boldsymbol{B}$ arising from the current $I$. As an example we consider a long solenoid (long means $l \gg 2r$) as shown in Fig. 2.5.3. The solenoid has $N$ turns per unit length and carries a current $I$.

Ampère's law, (Eq. (2.7.1(IV)), requires that the line integral of $\boldsymbol{B}$ around a closed path as shown is:

$$\oint \boldsymbol{B} \cdot d\boldsymbol{l} = \mu_0 NlI \qquad (2.5.11)$$

since the total current through the path is $Nl \times I$. To a first approximation we consider the field within the solenoid to be parallel to the axis and to be independent of $r$ and, since the space outside the solenoid is very large compared with that within, that the field outside is effectively zero (you should recall that $\boldsymbol{B}$ represents a flux *density*). We then have:

$$Bl = NI\mu_0 l \quad \text{or} \quad B = NI\mu_0 \qquad (2.5.12)$$

The total flux $\Phi$ linked with the coil is then given by:

$$\begin{aligned} \Phi &= B \times \text{cross-sectional area} \times \text{number of turns} \\ &= NI\mu_0 \times \pi r^2 \times Nl \\ &= \mu_0 N^2 \pi r^2 l\, I \\ &= LI \end{aligned} \qquad (2.5.13)$$

where the inductance $L$ is defined by:

$$L = \mu_0 N^2 \pi r^2 l = \frac{\mu_0 n^2 \pi r^2}{l}, \quad \text{where total number of turns } n = Nl \qquad (2.5.14)$$

Practical expressions for the inductance of coils are given in Section 4.3.

## References and additional sources 2.5

Corson D. R., Lorrain P. (1962): *Introduction to Electromagnetic Fields and Waves*, San Francisco: W. H. Freeman. Library of Congress Cat. No. 62-14193.

Fano R. M., Chu L. J., Adler R. B. (1960): *Electromagnetic Fields, Energy, and Forces*, New York: John Wiley.

Feynman R. P., Leighton R. B., Sands M. (1964): *The Feynman Lectures on Physics*, Vols I, II, III, Reading, Mass: Addison-Wesley. Library of Congress Cat. No. 63-20717.

Grant I., Philips W. R. (1975): *Electromagnetism*, London: John Wiley. ISBN 0-471-32246-6. 2nd Edn. 1990. ISBN 0-471-92712-0.

Hofman J. R. (1995): *André-Marie Ampère*, Cambridge: Cambridge University Press. ISBN 0-521-56220-1.

# 2.6 Magnetism and relativity

Everything should be made as simple as possible, but not simpler.

A. Einstein

The discovery by Maxwell of the equations describing the electromagnetic field revealed the intimate connection between electric and magnetic fields, but it was not then realized just how intimate they were. It is reported that Einstein said that with Maxwell one era ended and a new era began. Einstein was subsequently led to his theory of relativity by a 'conviction that the electromotive force acting on a body in motion in a magnetic field was nothing else but an electric field' (Shankland 1973). It turns out that magnetism is a relativistic effect, which also meant that Maxwell's equations were already relativistically correct. We will outline here the relativistic argument that demonstrates the origin of magnetic effects (Rosser 1959, 1961a, b; Feynman 1964; Gibson 1969).

To support the argument we need to state two results from the theory of relativity. If we consider say a particle moving with velocity $v$ relative to the frame of an observer, then a time interval in the frame of the particle will appear to the observer to be longer by the factor $\gamma$:

$$\gamma=\left(1-\frac{v^2}{c^2}\right)^{-\frac{1}{2}} \tag{2.6.1}$$

where $c$ is the velocity of light. $\gamma$ is necessarily $\geq 1$. Second, observers in uniform relative motion agree as to the magnitude of momentum normal to their relative motion.

Consider a particle moving with constant velocity and that is acted on by a force that gives it some transverse momentum $\delta p$ in a time interval $\delta t_0$. Then the force $F_0$, given by the rate of change of momentum, is:

$$F_0=\frac{\delta p}{\delta t_0} \tag{2.6.2}$$

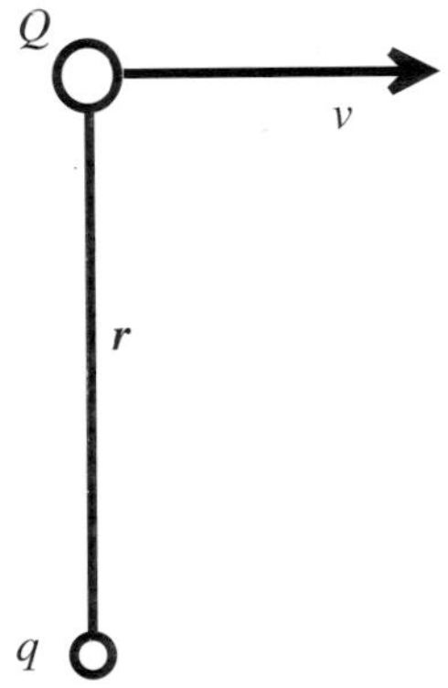


**Fig. 2.6.1** Relativity and magnetism. The interaction of a moving charge $Q$ and a fixed charge $q$.

An observer in the laboratory frame will agree about the magnitude of the momentum change $\delta p$ but will think that the time interval $\delta t$ was longer according to Eq. (2.6.1) and hence that the force acting was:

$$F = \frac{\delta p}{\delta t} = \frac{\delta p}{\gamma \delta t_0} = \frac{1}{\gamma} F_0 \tag{2.6.3}$$

Now let us use these results to examine the interaction between two electric charges $Q$ and $q$, with $Q$ moving with velocity $v$ normal to the line joining the two as shown in Fig. 2.6.1.

If $v=0$ then the field at $q$ due to $Q$ is given by $\boldsymbol{E}$ and the force between them by $\boldsymbol{F}$:

$$\boldsymbol{E} = \frac{1}{4\pi\varepsilon_0} \frac{Q}{r^2} \frac{\boldsymbol{r}}{r} \quad \text{and} \quad \boldsymbol{F} = q\boldsymbol{E} \tag{2.6.4}$$

where the third part of $\boldsymbol{E}$, $\boldsymbol{r}/r$ is just equivalent to a unit vector in the direction $r$. If $v \neq 0$ then the field at $q$ is still $\boldsymbol{E}$ and the force still $q\boldsymbol{E}$ independent of $v$ since charge is relativistically invariant. For an observer in the rest frame of $q$, however, the force on $q$ will appear to be larger according to Eq. (2.6.3). This force $\boldsymbol{F}_0$ is thus:

$$\boldsymbol{F}_0 = \gamma \boldsymbol{F} = \gamma q \boldsymbol{E} = q\boldsymbol{E}_0 \tag{2.6.5}$$

and this observer will say that the field at $q$ is $\boldsymbol{E}_0 = \gamma \boldsymbol{E}$ rather than $\boldsymbol{E}$, i.e. the field transforms as does the force. If now both charges move with velocity $v$ (if the velocities are different the sums are more difficult but the results are the same), the force on $q$ measured by an observer moving with them will still be $q\boldsymbol{E}$ (Fig. 2.6.2).

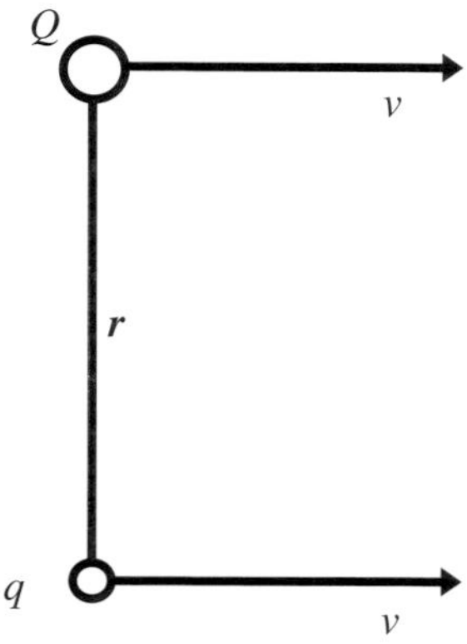


**Fig. 2.6.2** Both charges moving with velocity $v$.

An observer at rest in the laboratory will see the force $\boldsymbol{F}'$ on $q$ as smaller according to Eq. (2.6.3):

$$\boldsymbol{F}' = \frac{1}{\gamma} q\boldsymbol{E} \tag{2.6.6}$$

The force $\boldsymbol{F}'$ may be considered as the sum of a force which is independent of its motion, i.e. given by (2.6.5) together with a force $\boldsymbol{F}_m$ associated with its movement then:

$$\boldsymbol{F}' = \boldsymbol{F}_0 + \boldsymbol{F}_m$$

$$\text{or} \quad \boldsymbol{F}_m = \boldsymbol{F}' - \boldsymbol{F}_0 = q\boldsymbol{E}\left(\frac{1}{\gamma} - \gamma\right) = q\boldsymbol{E}\left[\left(1 - \frac{v^2}{c^2}\right)^{\frac{1}{2}} - \left(1 - \frac{v^2}{c^2}\right)^{-\frac{1}{2}}\right] \tag{2.6.7}$$

and for values of $v \ll c$ we can expand using the binomial theorem (Section 1.3) to the order of the first power in $(v^2/c^2)$ to give:

$$\boldsymbol{F}_m = q\boldsymbol{E}\left[\left(1 - \frac{v^2}{2c^2} - \cdots\right) - \left(1 + \frac{v^2}{2c^2} + \cdots\right)\right] = -q\boldsymbol{E}\left(\frac{v^2}{c^2}\right) \tag{2.6.8}$$

The negative sign indicates that $\boldsymbol{F}_m$ is in the opposite sense to $\boldsymbol{F}_0$, i.e. when $Q$ and $q$ are of the same sign so that $\boldsymbol{F}_0$ is a repulsion, then $\boldsymbol{F}_m$ is an attraction, though the latter can never lead to a net attraction. The force $\boldsymbol{F}_m$ is attributed to magnetic effects. The moving source charge $Q$ is said to produce a magnetic field of flux density $\boldsymbol{B}$ with the characteristic that a test charge $q$ moving through it with a velocity $v$ experiences a force:

$$F_m = qvB = qE\left(\frac{v^2}{c^2}\right) \quad \text{or} \quad B = \frac{Ev}{c^2} \tag{2.6.9}$$

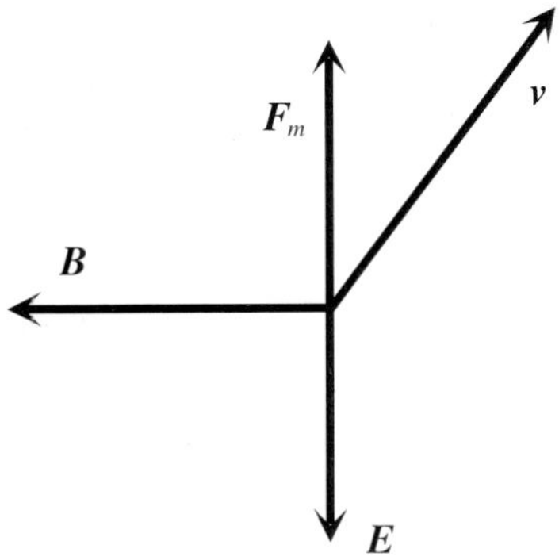


**Fig. 2.6.3** Lorentz force acting on a moving charge.

The result for $B$ gives the magnitude but not the direction, though a more proper derivation does do so. However, we know from experiment that $\boldsymbol{B}$ is normal to both $\boldsymbol{E}$ and $\boldsymbol{v}$ so that we should write (Section 1.6):

$$\boldsymbol{B} = \frac{1}{c^2}\boldsymbol{E} \times \boldsymbol{v} \tag{2.6.10}$$

and the force is normal to both $\boldsymbol{B}$ and $\boldsymbol{v}$, giving:

$$\boldsymbol{F}_m = q\boldsymbol{v} \times \boldsymbol{B} \tag{2.6.11}$$

The total force, including the electrostatic contribution, is called the Lorentz force, and is then:

$$\boldsymbol{F}_m = q(\boldsymbol{E} + \boldsymbol{v} \times \boldsymbol{B}) \tag{2.6.12}$$

We can now examine the force between two parallel wires carrying electric currents. Consider one electron and one positive ion in wire $M$ and the same in wire $N$ on a line normal to the two wires. The positive ions are fixed in position as they represent the charge on the atom of which the wire is made up. The electrons are moving in the same direction and at the instant we consider they are effectively coincident with the ions. In practice, of course, there will be a very large number of ions and electrons so this condition is normally fulfilled. The observer is in the rest frame of the wires and hence also of both ions. The effect of the ion in $N$ on the ion and electron in $M$ will cancel out since they are effectively at the same location. The effect of the electron in $N$ will be different for the ion and the electron in $M$. The force on the ion will be given by $F_0$, i.e. Eq. (2.6.5) with the sign changed since we now have opposite charges, and the force on the electron by $F'$ (Eq. (2.6.6)). The net attractive force is given by:

$$\boldsymbol{F}_0 - \boldsymbol{F}' = \boldsymbol{F}_m = q\boldsymbol{E}\left(\frac{v^2}{c^2}\right) \tag{2.6.13}$$

and the force is just that arising from the magnetic field. The magnetic force is smaller than the electrostatic force by the factor ($v^2/c^2$), which is about $10^{-25}$! The reason we detect any magnetic effect at all is that the wires are electrically neutral to at least this degree so that this minute magnetic field is all that is left and it becomes the dominant term.

The changing face of the fields depending on where you are and how you are moving presents great difficulty in providing a picture of them. I can do no better than to quote one of the outstanding physicists of our time, Richard Feynman (1964, Vol. II p. 1–9).

> The only sensible question is what is *the most convenient* way to look at electrical effects. Some people prefer to represent them as the interaction at a distance of charges, and to use a complicated law. Others love the field lines. They draw field lines all the time, and feel that writing E's and B's is too abstract. The field lines, however, are only a crude way of describing a field, and it is very difficult to give the correct, quantitative laws directly in terms of field lines. Also, the ideas of the field lines do not contain the deepest principle of electrodynamics, which is the superposition principle. Even though we know how the field lines look for one set of charges and what the field lines look like for another set of charges, we don't get any idea about what the field line patterns will look like when both sets are present together. From the mathematical standpoint, on the other hand, superposition is easy – we simply add the two vectors. The field lines have some advantage in giving a vivid picture, but they also have some disadvantages. The direct interaction way of thinking has great advantages when thinking of electrical charges at rest, but has great disadvantages when dealing with charges in rapid motion.
>
> The best way is to use the abstract field idea. That it is abstract is unfortunate, but necessary. The attempts to try to represent the electric field as the motion of some kind of gear wheels, or in terms of lines, or of stresses in some kind of material have used up more effort of physicists than it would have taken simply to get the right answers about electrodynamics. …
>
> In the case of the magnetic field we can make the following point: Suppose that you finally succeeded in making up a picture of the magnetic field in terms of some kind of lines or of gear wheels running through space. Then you try to explain what happens to two charges moving in space, both at the same speed and parallel to each other. Because they are moving, they will behave like two currents and will have a magnetic field associated with them … An observer who was riding along with the two charges, however, would see both charges as stationary, and would say that there is *no* magnetic field. The 'gear wheels' or 'lines' disappear when you ride along with the object! How can gear wheels disappear?! The people who draw field lines are in a similar difficulty. Not only is it not possible to say whether the field lines move or do not move with the charges – they may disappear completely in certain coordinate frames.

For the practising electronicist it is necessary to have some picture of what the fields look like in space. It would be much more difficult to contemplate the form of a transformer, the inductance of a coil, or the effect a ferrite rod may have on the performance of an antenna if we did not have the image of lines of the $\boldsymbol{B}$ field, but be aware of the limitations.

## References and additional sources 2.6

Feynman R. P., Leighton R. B., Sands M. (1964): *The Feynman Lectures on Physics*, Vols I, II, III, Reading, Mass: Addison-Wesley. See Vol. II, p. 1–9. Library of Congress Cat. No. 63-20717.

Gibson W. M. (1969): *Basic Electricity*, Harmondsworth: Penguin Books.

Rosser W. G. V. (1959): Electromagnetism as a second order effect. I. The definition of the ampere. *Contemporary Physics* **1**, 134–141.

Rosser W. G. V. (1961a): Electromagnetism as a second order effect. II. The electric and magnetic fields of a charge moving with uniform velocity. *Contemporary Physics* **1**, 453–466.

Rosser W. G. V. (1961b): Electromagnetism as a second order effect. III. The Biot–Savart law. *Contemporary Physics* **3**, 28–44.

Shankland R. S. (1973): Conversations with Albert Einstein, Part II. *Amer. J. Phys.* **41**, 895–901.

# 2.7 Maxwell's equations

From a long view of the history of mankind – seen from, say ten thousand years from now – there can be little doubt that the most significant event of the 19th century will be judged as Maxwell's discovery of the laws of electrodynamics.

R. P. Feynman, Nobel Laureate, 1964, *Lectures in Physics* Vol. II, pp. 1–11

The experiments of Coulomb (in 1785) led to his law that the force between electric charges was inversely proportional to the square of the distance between them. Much of what the universe is depends on the precision of the square power. There has been no indication that the power is not exactly two and all experiments to measure this have confirmed it; if the power is represented by $2+q$ then experiment shows that $q$ must be less than $10^{-16}$ (e.g. Williams et al. 1971). Experiments by Oersted, Ampère and Faraday (in the early part of the 19th century) led to several equations describing the electric and magnetic fields. Though not then written in the form we now use, we can most effectively use the modern nomenclature. The equations were:

I. $\boldsymbol{\nabla}\cdot\boldsymbol{E}=\rho/\varepsilon_0$ or flux of $\boldsymbol{E}$ through a closed surface = charge inside/$\varepsilon_0$

II. $\boldsymbol{\nabla}\times\boldsymbol{E}=\dfrac{-\partial\boldsymbol{B}}{\partial t}$ or line integral of $\boldsymbol{E}$ around a loop $=\dfrac{-\partial}{\partial t}$ (flux of $\boldsymbol{B}$ through the loop)

III. $\boldsymbol{\nabla}\cdot\boldsymbol{B}=0$ or flux of $\boldsymbol{B}$ through a close surface = 0 (2.7.1)

IV. $\boldsymbol{\nabla}\times\boldsymbol{B}=\mu_0\boldsymbol{j}$ or integral of $\boldsymbol{B}$ around a loop = $\mu_0\times$ current through the loop

V. $\boldsymbol{\nabla}\cdot\boldsymbol{j}=\dfrac{-\partial\rho}{\partial t}$ or flux of current through a closed surface $=\dfrac{-\partial}{\partial t}$ (charge inside)

Equation I is Gauss' law, II is Faraday's law and III follows from the absence of magnetic poles. Equation IV is Ampère's law and V is the equation of continuity or conservation of charge. It is necessary to be very careful in interpreting these relations in terms of how they were understood or viewed at the time (1850–1860). An illuminating account of the development of electromagnetism from this period to the end of the century when the present views became established is given by Buchwald (1988).

James Clerk Maxwell on considering these equations was concerned that there was something evidently wrong with IV, which had been derived from the experiments of Ampère, which had the limitation of only using steady currents; there were no handy signal generators on the shelf in those days. Maxwell took the divergence of both sides of the equation, and noting that the divergence of a curl is identically zero was confronted by the requirement that the divergence of $\boldsymbol{j}$ (i.e. $\boldsymbol{\nabla}\cdot\boldsymbol{j}$) was zero. This cannot generally be zero as shown by the fundamental requirement of Eq. V. This is where Maxwell made his leap into the unknown, since there was no experimental evidence for it, of proposing that Ampère's law should be modified by the addition of a deceptively simple term to be:

$$\boldsymbol{\nabla}\times\boldsymbol{B}=\mu_0\boldsymbol{j}+\mu_0\varepsilon_0\frac{\partial\boldsymbol{E}}{\partial t} \tag{2.7.2}$$

where it should be remembered that at that time $\mu_0$ and $\varepsilon_0$ were just some constants to be determined experimentally. Now if we take the divergence of both sides we get:

$$0=\mu_0\boldsymbol{\nabla}\cdot\boldsymbol{j}+\mu_0\varepsilon_0\boldsymbol{\nabla}\cdot\left(\frac{\partial\boldsymbol{E}}{\partial t}\right)$$

$$\text{so}\quad \boldsymbol{\nabla}\cdot\boldsymbol{j}+\varepsilon_0\frac{\partial}{\partial t}(\boldsymbol{\nabla}\cdot\boldsymbol{E})=\boldsymbol{\nabla}\cdot\boldsymbol{j}+\mu_0\frac{\partial}{\partial t}\left(\frac{\rho}{\varepsilon_0}\right)=0$$

$$\text{and}\quad \boldsymbol{\nabla}\cdot\boldsymbol{j}=\frac{-\partial\rho}{\partial t} \tag{2.7.3}$$

which is just the equation of conservation of charge as in (2.7.1(V)) above. This apparently small adjustment leads to astonishing consequences, and one of the greatest achievements of physics. The new term $\partial\boldsymbol{E}/\partial t$ is called the displacement current in contrast to the conduction current $\boldsymbol{j}$. Section 2.1 considers the application of displacement current to the flow of current in a capacitor.

Let us consider what Maxwell's equations lead to. To take the simplest circumstance we consider free space without currents or charges, so that we may set $\boldsymbol{j}=0$ in Eq. (2.7.2). Then taking the curl of both sides and using (2.7.1(II)) we have:

$$\boldsymbol{\nabla}\times\boldsymbol{\nabla}\times\boldsymbol{B}=\boldsymbol{\nabla}\times\mu_0\varepsilon_0\frac{\partial\boldsymbol{E}}{\partial t}=\mu_0\varepsilon_0\frac{\partial}{\partial t}(\boldsymbol{\nabla}\times\boldsymbol{E})=\mu_0\varepsilon_0\frac{\partial}{\partial t}\left(\frac{-\boldsymbol{B}}{\partial t}\right)=-\mu_0\varepsilon_0\frac{\partial^2\boldsymbol{B}}{\partial t^2} \tag{2.7.4}$$

Using the vector identity (1.6.20) and (2.7.1(III)) we then have:

$$(\boldsymbol{\nabla}\cdot\boldsymbol{B})-\nabla^2\boldsymbol{B}=-\mu_0\varepsilon_0\frac{\partial^2\boldsymbol{B}}{\partial t^2}\qquad\text{or}\qquad \nabla^2\boldsymbol{B}=\mu_0\varepsilon_0\frac{\partial^2\boldsymbol{B}}{\partial t^2} \tag{2.7.5}$$

and following the same procedure with curl $\boldsymbol{E}$ we find also:

$$\nabla^2 \boldsymbol{E} = \mu_0 \varepsilon_0 \frac{\partial^2 \boldsymbol{E}}{\partial t^2} \tag{2.7.6}$$

with both Eqs. (2.7.5) and (2.7.6) in the standard wave equation form, i.e. Maxwell's equations predict electromagnetic waves with velocity $v = 1/(\varepsilon_0 \mu_0)^{1/2}$. The values of $\varepsilon_0$ and $\mu_0$ were not very well known at the time (though well enough for the purpose), but if we use our present values we obtain:

$$v = (8.854 \times 10^{-12} \times 4\pi \times 10^{-7})^{-\frac{1}{2}} = 2.998 \times 10^8 \text{ m s}^{-1} \tag{2.7.7}$$

which was very close to the then known velocity $c$ of light. As Maxwell wrote: 'We can scarcely avoid the inference that light consists in the transverse undulations of the same medium which is the cause of electric and magnetic phenomena.' This was one of the great unifications of physics; as Einstein is reported to have remarked: 'with Maxwell one era ended and a new era began'.

To examine a solution of the wave-equation we will consider a plane wave propagating in the $z$-direction, so that $\boldsymbol{E}$ only varies in the $z$-direction:

$$\boldsymbol{E} = \boldsymbol{E}_0 \exp[j(\omega t - kz)], \quad \text{with} \quad c = \omega/k \tag{2.7.8}$$

and since the solution must also satisfy (2.7.1(I)) with $\rho = 0$, then since:

$$\frac{\partial E_x}{\partial x} = \frac{\partial E_y}{\partial y} = 0, \quad \text{we have} \quad \boldsymbol{\nabla} \cdot \boldsymbol{E} = \frac{\partial E_z}{\partial z} = 0 \tag{2.7.9}$$

and there is no component of $\boldsymbol{E}$, other than a static field, along the direction of propagation. We can set $E_z = 0$ since we are not concerned with any static field. A similar result is found for $\boldsymbol{B}$. Now assuming a wave with the electric field along the $x$-axis (which does not lose us any generality) and differentiating (2.7.8) with respect to $x$ and $t$ (using $\boldsymbol{u}_x$ as a unit vector along the $x$-axis) we have:

$$\begin{aligned} \frac{\partial \boldsymbol{E}}{\partial t} &= \boldsymbol{u}_x E_0 (j\omega) \exp[j(\omega t - kz)] & \frac{\partial \boldsymbol{E}}{\partial z} &= \boldsymbol{u}_x E_0 (-jk) \exp[j(\omega t - kz)] \\ \frac{\partial^2 \boldsymbol{E}}{\partial t^2} &= \boldsymbol{u}_x E_0 (-\omega^2) \exp[j(\omega t - kz)] & \frac{\partial^2 \boldsymbol{E}}{\partial z^2} &= \boldsymbol{u}_x E_0 (-k^2) \exp[j(\omega t - kz)] \end{aligned} \tag{2.7.10}$$

If these are substituted in (2.7.6) then the two sides are equal and hence (2.7.8) is an acceptable solution. To find the corresponding solution for $\boldsymbol{B}$ we use (2.7.1(II)) and the general form for curl from (1.6.9), but using $\boldsymbol{u}_x \equiv \boldsymbol{i}$, etc. to avoid confusion, to get:

$$\frac{-\partial \boldsymbol{B}}{\partial t} = \boldsymbol{u}_y \left( \frac{\partial E_x}{\partial z} \right) = \boldsymbol{u}_y E_0 (-jk) \exp[j(\omega t - kz)] \tag{2.7.11}$$

and integrating this with respect to time gives:

$$B = \int u_y jKE_0 \exp[j(\omega t - kz)]dt$$

$$= u_y jKE_0 \left[ \frac{\exp[j(\omega t - kz)]}{j\omega} \right]$$

$$= u_y \left( \frac{k}{\omega} \right) E = u_y \frac{E}{c} \tag{2.7.12}$$

so $\boldsymbol{B}$ (along $\boldsymbol{u}_y$) is normal to $\boldsymbol{E}$ (along $\boldsymbol{u}_x$) and both are normal to the direction of propagation $z$. $B$ and $E$ are also in phase (there are no $j$ components). If we make use of the relationship $B = \mu_0 H$ (Section 2.11) then we have (if you check the units from Section 2.12):

$$\frac{E}{H} = \mu_0 c = 4\pi \times 10^{-7} \times 3 \times 10^8 = 377\ \Omega \tag{2.7.13}$$

which gives a value for the impedance of free space.

We have jumped forward rather in consideration of Maxwell's discovery. At the time (about 1862) there were great difficulties associated with the understanding of how electromagnetism worked. For some forty years there ensued attempts to relate electromagnetic waves to some sort of medium, or 'luminiferous aether', which, as for other types of wave, was thought necessary to support them. The effort and ingenuity that went into this endeavour was immense and ingenious, but in the end failed since it turned out that no medium was in fact required. The experiments of Michelson and Morley (1887) to detect the motion of the Earth through the aether gave a null result. In the same year Hertz finally managed to generate and detect electromagnetic waves and showed that they had the same properties as light waves, and incidentally discovered the photoelectric effect. Hertz's great discoveries and his interactions with the other researchers in the field are engagingly described by Susskind (1995) and O'Hara and Pricha (1987). In 1897, J. J. Thomson discovered the electron, which revolutionized the understanding of electromagnetic effects and began what we would now recognize as electronics (see e.g. Feffer 1989; Mulligan 1997). From about 1880 to 1900, Hendrik Lorentz developed his ideas of the relationship of macroscopic fields to their microscopic or atomic origins. In 1901, Planck shattered the classical edifice by showing that quantization was necessary to explain the spectrum of black-body radiation. In 1905, Einstein published his theory of relativity, which from our point of view said that electromagnetic waves travel at velocity $c$ irrespective of the velocity of the source (unlike other types of wave) and that there was no need for any sort of medium to support them. In the same year Einstein extended Planck's ideas of quantization to an explanation of the photoelectric effect in a way that made it necessary now to consider that electromagnetic waves were made up of quanta called photons rather than a classical continuum, and that they also interacted with matter as quanta. This view, and a reconciliation of the two, is discussed in Section 2.9.

What do electromagnetic fields look like? Is it possible to visualize them? It is of great assistance to have a picture of the directions and intensities of $\boldsymbol{E}$ and $\boldsymbol{H}$ to consider how we may make use of them or how they interact with our circuits, and the use of field lines (or tubes) as introduced by Faraday are often instructive. However, you should be wary of these representations as they may change depending on how you look at them, and this leads to conceptual difficulties. The following quotation may make you feel that you are not alone in this difficulty:

I have asked you to imagine these electric and magnetic fields. What do you do? Do you know how? How do I imagine the electric and magnetic field? What do I actually see? What are the demands of scientific imagination? Is it any different from trying to imagine that the room is full of invisible angels? No, it is not like imagining invisible angels. It requires a much higher degree of imagination to understand the electromagnetic field than to understand invisible angels. Why? Because to make invisible angels understandable, all I have to do is to alter their properties *a little bit* – I make them slightly visible, and then I can see the shapes of their wings, and bodies, and halos. Once I succeed in imagining a visible angel, the abstraction required – which is to take almost invisible angels and imagine them completely invisible – is relatively easy. So you say, 'professor, please give me an approximate description of the electromagnetic waves, even though it may be slightly inaccurate, so that I too can see them as well as I can see almost invisible angels. Then I will modify the picture to the necessary abstraction.'

I'm sorry I can't do that for you. I don't know how. I have no picture of this electromagnetic field that is in any sense accurate. I have known about the electromagnetic field for a long time – I was in the same position 25 years ago that you are now, and I have had 25 years more of experience thinking about these wiggling waves. When I start describing the magnetic field moving through space, I speak of the $E$- and $B$-fields and wave my arms and you may imagine that I can see them. I'll tell you what I see. I see some kind of vague shadowy, wiggling lines – here and there is an $E$ and $B$ written on them somehow, and perhaps some of the lines have arrows on them – an arrow here or there which disappears when I look too closely at it. When I talk about the fields swishing through space, I have a terrible confusion between the symbols I use to describe the objects and the objects themselves. I cannot really make a picture that is even nearly like the true waves. So if you have some difficulty in making such a picture, you should not be worried that your difficulty is unusual.

Richard Feynman (Feynman et al. (1964): Vol. II, pp. 20–29)

## References and additional sources 2.7


Buchwald J. Z. (1988): *From Maxwell to Microphysics*; Chicago. University of Chicago Press. ISBN 0-226-07883-3.

Farley F. J. M., Bailey J., Picasso E. (1968): Is the special theory right or wrong? *Nature* **217**, 17–18.

Feffer S. M. (1989): Arthur Schuster, J. J. Thomson, and the discovery of the electron. *Historical Studies in the Physical and Biological Sciences* **20**, **(1)**, 33–61.

Feynman R. P., Leighton R. B., Sands M. (1964): *The Feynman Lectures on Physics*, Vols I, II, III, Reading, Mass: Addison-Wesley. Library of Congress Cat. No. 63-20717.

Michelson A. A., Morley E. W. (1887): On the relative motion of the Earth and the luminiferous æther; *Philosophical Magazine*, **24**, 449–463.

Mulligan J. F. (1997): The personal and professional interaction of J. J. Thomson and Arthur Schuster. *Amer. J. Phys.* **65**, 954–963.

O'Hara J. G., Pricha W. (1987): *Hertz and the Maxwellians*, London: Peter Peregrinus. ISBN 0-86341-101-0.

Shankland R. S. (1964): The Michelson–Morley experiment. *Scientific American* **211**, 107–114.

Susskind C. (1995): *Heinrich Hertz: A Short Life*, San Francisco: San Francisco Press. ISBN 0-911302-74-3.

Thomson J. J. (1897): Cathode rays; *Philosophical Magazine*, **44**, 293–316.

Williams E. R., Faller J. E., Hill H. (1971): New experimental test of Coulomb's law: a laboratory upper limit on the photon rest mass. *Phys. Rev. Letters* **26**, 721–724.

# 2.8 Conductivity and the skin effect

> When you can measure what you are speaking about, and express it in numbers, you know something about it; but when you cannot measure it, when you cannot express it in numbers, your knowledge is of a meagre and unsatisfactory kind: it may be the beginning of knowledge, but you have scarcely, in your thoughts, advanced to the stage of science.
>
> William Thomson, Lord Kelvin (1824–1907)

Electric and magnetic fields have to satisfy certain boundary conditions where free space meets a conductor, a magnetic material or a dielectric. For example, you will be familiar with the deviation in the path of a light ray when passing through a lens or a prism (what would Galileo have been without his telescope?). There are fundamental physical requirements that must be fulfilled at a boundary between two media with regard to both the electric and the magnetic fields. We will consider some of these in so far as they affect our electronic circuits. To start with we consider what happens at the surface of a perfect conductor:

(i) At the surface, the electric field $\boldsymbol{E}$ must be normal to it; if it were not then there would be a component of the field parallel to the surface which would cause the charges in the conductor to move until the field was normal to the surface. The electric field inside the conductor must be zero by the same argument.

(ii) A static magnetic field $\boldsymbol{B}$ will not be affected; the direction is not proscribed.

(iii) For a time-varying magnetic field $\boldsymbol{B}$ the normal component at the surface must be zero. This arises because by Faraday's law (Section 2.4) there will be an induced current in the conductor that will be of such a magnitude and sense as to cancel out the applied field (Lenz's law). The component of $\boldsymbol{B}$ parallel to the surface will not be affected.

These boundary conditions mean that an electromagnetic field cannot penetrate a perfect conductor. A closed conducting box will exclude an external field or totally contain an internal field. A wave is in effect perfectly reflected by the conductor. Thus the first resort in trying to screen a system is to enclose it in a conducting box. In a more practical situation we have to make do with conductors that are not perfect and in which, therefore, the strict boundary conditions will be modified to a greater or lesser extent.

To investigate the penetration of an electromagnetic wave into a conductor we will consider a plane wave moving in the $z$-direction incident normally on a conductor. Inside the conductor there will be no net free charges so we may put $\rho=0$, but there will be conductivity so $\sigma\neq 0$, and $\varepsilon$ and $\mu$ may not also be unity. Following the development from (2.7.2) but not putting $\boldsymbol{j}=0$, the field equations can now be written:

$$\frac{\partial^2 E}{\partial z^2}=\varepsilon\varepsilon_0\mu\mu_0\frac{\partial^2 E}{\partial t^2}+\sigma\mu\mu_0\frac{\partial E}{\partial t} \quad \text{and} \quad \frac{\partial^2 H}{\partial z^2}=\varepsilon\varepsilon_0\mu\mu_0\frac{\partial^2 H}{\partial t^2}+\sigma\mu\mu_0\frac{\partial H}{\partial t} \tag{2.8.1}$$

For a good conductor the conduction current, which gives rise to the second term in each equation, will be much larger than the displacement current, which gives rise to the first term. Using the ratio of the first to the second derivative of $\boldsymbol{E}$ from (2.7.11), i.e. $\omega$, we can write the ratio of conduction current $j_C$ to displacement current $j_D$ from (2.8.1) as:

$$\frac{j_C}{j_D}=\frac{\sigma}{\omega\varepsilon\varepsilon_0}=\frac{5.8\times 10^7}{2\pi\times 10^6\times 1\times 8.85\times 10^{-12}}\approx 10^{12} \tag{2.8.2}$$

where the numerical values for 1 MHz are used as an example. So neglecting the displacement current and assuming a solution for $E$ of the form of a damped exponential:

$$E=E_0\exp(-pz)\exp[j(\omega t-kz)]$$

$$\text{then} \quad \frac{\partial E}{\partial z}=-(jk+p)E_0\exp(-pz)\exp[j(\omega t-kz)]=-(jk+p)E$$

$$\text{and} \quad \frac{\partial^2 E}{\partial z^2}=(p+jk)^2E, \quad \text{with} \quad \frac{\partial E}{\partial t}=j\omega E \tag{2.8.3}$$

$$\text{so} \quad (p+jk)^2=p^2+2jpk-k^2=j\omega\sigma\mu\mu_0, \quad \text{from (2.8.1)}$$

Equating the real parts and the imaginary parts yields:

$$p^2-k^2=0 \quad \text{and} \quad 2pk=\omega\sigma\mu\mu_0, \quad \text{so} \quad p=k=\left(\frac{\omega\sigma\mu\mu_0}{2}\right)^{\frac{1}{2}} \tag{2.8.4}$$

so the amplitude of the wave will decay to $1/e$ of the incident amplitude in a distance $\delta=1/p$:

$$\delta=\frac{1}{p}=\left(\frac{2}{\omega\sigma\mu\mu_0}\right)^{\frac{1}{2}} \tag{2.8.5}$$

The velocity $v$ of the wave in the conductor may be found from the wavenumber $k$:

$$k=\frac{\omega}{v}, \quad \text{so} \quad v=\frac{\omega}{k}=\left(\frac{2\omega}{\sigma\mu\mu_0}\right)^{\frac{1}{2}} \tag{2.8.6}$$

Table 2.8.1 *Parameters for plane waves in a good conductor*[a]

| $f$ (Hz) | $\omega$ ($s^{-1}$) | $\delta$ (m) | $v$ ($ms^{-1}$) | $\lambda_C$ (m) | $\lambda_{free}$ (m) |
|---|---|---|---|---|---|
| 50 | 314 | $9.3\times10^{-3}$ | 2.94 | $5.8\times10^{-2}$ | $6\times10^{6}$ |
| 60 | 377 | $8.5\times10^{-3}$ | 3.22 | $5.4\times10^{-2}$ | $5\times10^{6}$ |
| $1\times10^{6}$ | $6.28\times10^{6}$ | $6.6\times10^{-5}$ | $4.15\times10^{2}$ | $4.2\times10^{-4}$ | $3\times10^{2}$ |
| $1\times10^{9}$ | $6.28\times10^{9}$ | $2.1\times10^{-6}$ | $1.3\times10^{4}$ | $1.3\times10^{-5}$ | 0.3 |
| $6\times10^{14}$ | $3.77\times10^{15}$ | $2.7\times10^{-9}$ | $1.02\times10^{7}$ | $1.7\times10^{-8}$ | $5\times10^{-7}$ |

[a] For copper $\sigma=5.8\times10^7\ \mathrm{Sm}^{-1}$, $\mu=1$, $\mu\mu_0\sigma=72.88$. $\lambda_{free}$ is the free-space wavelength.

and hence for frequency $f$ the wavelength in the conductor, $\lambda_C$, is given by:

$$\lambda_C=\frac{v}{f}=\frac{2\pi v}{\omega}=\frac{2\pi}{p}=2\pi\delta \tag{2.8.7}$$

The quantity $\delta$ is, as we shall see below, the effective depth in the conductor in which the currents related to the field will flow, and is known as the skin depth. We may calculate the various quantities for a good conductor like copper (Table 2.8.1):

The wavelength in the conductor is very small and the wave amplitude drops very rapidly with distance. If we take say an amplitude decrease to 1% then we have:

$$\exp\left(\frac{-z}{\delta}\right)=\exp\left(\frac{-2\pi z}{\lambda_C}\right)=0.01,$$

$$\text{so}\quad \frac{-2\pi z}{\lambda_C}=-4.605 \quad\text{or}\quad z=\frac{4.605\lambda_C}{2\pi}=0.73\lambda_C \tag{2.8.8}$$

so the wave penetrates only about one conductor wavelength (Fig. 2.8.1).

It is evident that the skin effect is significant even at mains frequencies and at high frequencies $\delta$ can be very small. Thus even at 50 Hz the dimensions of a conductor are significant. For large currents, e.g. in a power station, it is more effective to have a bussbar of flat rectangular cross-section than an equivalent-area round conductor. The frequency $6\times10^{14}$ Hz corresponds to the green region of the visible spectrum (500 nm) and the results should not be taken too seriously since atomic interactions are now active. We can find the relationship between the electric and the magnetic amplitudes from (2.7.13), with $B=\mu\mu_0 H$:

$$\frac{E}{H}=\frac{\omega\mu\mu_0}{k}=v\mu\mu_0\left(\frac{2\omega\mu\mu_0}{\sigma}\right)^{\frac{1}{2}}=\left(\frac{2\times6.28\times10^6\times4\pi\times10^{-7}}{5.8\times10^7}\right)^{\frac{1}{2}}=5.22\times10^{-4} \tag{2.8.9}$$

so that the wave, and energy, is primarily magnetic. This ratio $E/H$ is also the impedance and may be compared with the free space value of 377 Ω (Eq. (2.7.13)). Though we have not determined it here, the phase difference between $E$ and $H$ is

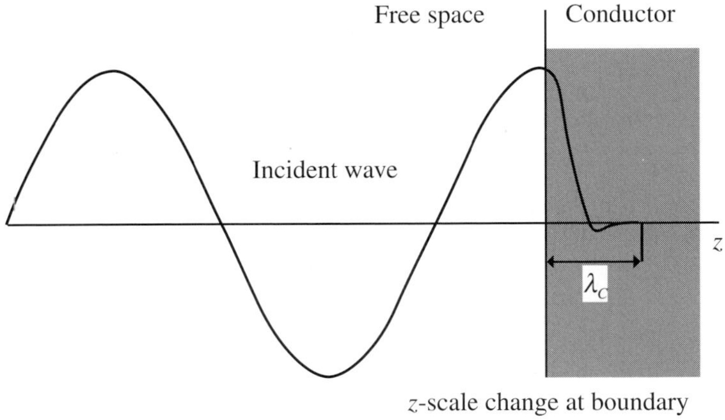


**Fig. 2.8.1** Attenuation of a wave penetrating a conductor. The illustration is rather misleading but it is difficult to show both waves properly. At 1 MHz say, the wavelength in the conductor is about $10^{-6}$ of the free-space wavelength, so the $z$-scale as shown within the conductor is about $10^6$ times that outside.

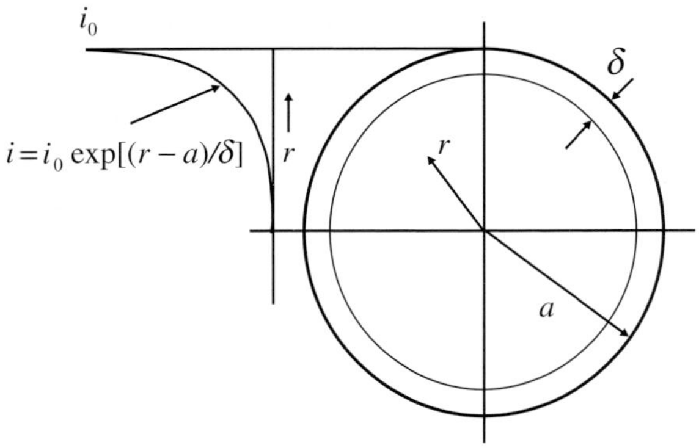


**Fig. 2.8.2** Exponential decay of current density from the surface towards the centre.

now 45°. From a transmission line point of view the large ratio of the impedances also indicates that the incident wave will be almost completely reflected.

Let us return to consideration of the fields outside the long straight wire considered previously (Section 2.5). We tacitly assumed that the fields were zero inside the wire but even for a good conductor like copper the fields as we have seen can penetrate into the wire. The amplitude of the field is attenuated exponentially at a rate given by (2.8.5). This means that the current density in the wire is not uniform across the cross-section – the density will be greater at the surface than towards the centre as illustrated in Fig. 2.8.2.

The effective resistance of the wire is therefore greater since the inner areas make less contribution to the conductivity. A direct approach to calculation of the effective resistance of a wire of circular cross-section leads to Bessel function integrals,

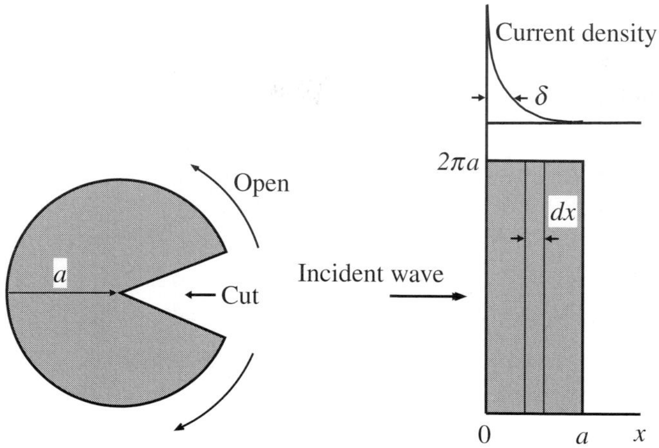


**Fig. 2.8.3** Equivalent plane sheet for a round conductor.

but if the skin depth is very much less than the radius we may simplify the geometry by cutting along a radius and unrolling (and stretching) it to form a flat strip as illustrated in Fig. 2.8.3.

Though this somewhat exaggerates the cross-sectional area of the wire the current density in the expanded region is so low as to have little effect. Since the current will fall exponentially from some surface value $i_0$ the total current $i_T$ flowing will be given by:

$$
\begin{aligned}
i_T &= \int_0^a i_0 \exp(-x/\delta) 2\pi a dx \\
&= 2\pi a i_0 \int_0^a \exp(-x/\delta) dx \\
&= 2\pi a i_0 [-\delta \exp(-x/\delta)]_0^a \\
&= 2\pi a i_0 \delta, \qquad \text{since} \qquad \exp(-a/\delta) \ll 1, \quad \text{for} \quad a \gg \delta
\end{aligned}
\tag{2.8.10}
$$

A depth $\delta$ within which the current density is assumed uniform at $i_0$ will carry an effective current:

$$i = 2\pi a i_0 \delta \tag{2.8.11}$$

i.e. the same as the actual distribution. Hence we can for convenience think of a uniform current density in the skin depth rather than the exponential distribution. The ratio of high frequency $R_{hf}$ to z.f. resistance $R_0$ will then be:

$$\frac{R_{hf}}{R_0} = \frac{a}{2\delta} \tag{2.8.12}$$

The increase in resistance can therefore have a significant effect. If, for example, you measure the characteristics of an inductor at 1 kHz (a typical frequency for a modest *RLC* bridge) then you will get significantly different values for resistance, inductance and *Q* at say 1 MHz. The inductance changes because the current distribution in the wire changes and hence the geometry of the coupling between turns.

In all the above we have made an assumption regarding $\sigma$, that it is constant and hence we may also use the z.f. value in calculations. For the frequency range which concerns us here this is satisfactory, but as noted in Section 2.1 at high enough frequency the conductivity must be examined more closely. When obtaining an expression for the drift velocity and the collision frequency, Eqs. (2.1.3) and (2.1.5), it was implicit that there was no viscous or damping phenomena acting on the electrons. However, at high enough frequency damping becomes significant and we must introduce a term in the equation of motion to represent this. This leads to an expression for the conductivity of the form:

$$\sigma(\omega) = \frac{ne^2\tau}{m_e(1 + j\omega\tau)} \tag{2.8.13}$$

However, further consideration of the propagation of electromagnetic waves in various forms of media will take us too far from our mainly circuit interests. A readable account at the present level is, for example, given by Klein (1970) or Garbuny (1965).

## References and additional sources 2.8


Corson D. R., Lorrain P. (1962): *Introduction to Electromagnetic Fields and Waves*, San Francisco: W. H. Freeman. Library of Congress Cat. No. 62-14193.

Feynman R. P., Leighton R. B., Sands M. (1964): *The Feynman Lectures on Physics*, Vols I, II, III, Reading, Mass: Addison-Wesley. Library of Congress Cat. No. 63-20717. See Vol. II, p. 32-11.

Garbuny M. (1965): *Optical Physics*, New York: Academic Press. Library of Congress Cat. No. 65-19999.

Grivet P. (1970): *The Physics of Transmission Lines at High and Very High Frequencies*, Vols 1 and 2, New York: Academic Press. ISBN 12-303601-1.

Klein M. V. (1970): *Optics*, New York: John Wiley. See Chapter 11. ISBN 471-49080-6.

Matick R. E. (1969): *Transmission Lines for Digital and Communication Networks*, McGraw-Hill. Library of Congress Cat. No. 68-30561.

Ramo S., Whinnery J. R. (1944): *Fields and Waves in Modern Radio*, New York: John Wiley. 2nd Edition 1953. Library of Congress Cat. No. 53-6615.

Ramo S., Whinnery J. R., van Duzer T. (1965): *Fields and Waves in Communication Electronics*, New York: John Wiley.

Spectrum Software (1997): Modeling skin effect. *Spectrum News* Fall Issue, 9–13. www.spectrum-soft.com

Wheeler H. A. (1942): Formulas for the skin effect. *Proc. IRE* **30**, 412–424.

## 2.9 Quantization

So far as the laws of mathematics refer to reality, they are not certain. And so far as they are certain, they do not refer to reality.

Albert Einstein

About the 1890s it became evident that there was a significant difficulty in explaining the spectral distribution of electromagnetic radiation from a black body. The classical theory at that time, due to Rayleigh and Jeans, predicted what became known as the ultraviolet catastrophe – the emission from a black body would continue increasing in intensity as wavelength decreased and so eventually become infinite. This was clearly wrong but there seemed to be no way out of the dilemma on the basis of existing electromagnetic theory. Max Planck sought a way round the problem by imposing certain limits on the modes of the system and arrived at the idea of discrete quanta of energy rather than a continuum, and thus was ushered in a new revolutionary era in physics. Though Planck himself was said to have great difficulty in accepting this new departure, and for some years many others were also unwilling to do so, the success in predicting precisely the spectrum of black-body radiation made it necessary to re-examine the whole existing basis of electromagnetism. Many other experimental phenomena, such as the very well defined emission and absorption spectra of atoms and the effect of particularly magnetic fields on these spectra (the Zeeman effect; Zeeman 1897), and the discovery of the electron by J. J. Thomson in 1897, also strongly indicated the need for quantization. The publication in 1905 by Einstein of the quantum theory of the photoelectric effect and the theory of relativity provided a new and successful basis for the understanding of the electromagnetic field.

Planck had shown that the electromagnetic field had to be quantized and gave the relation between the energy $E$ (in joule) of a quantum and its frequency $\nu$ (in Hz) of the radiation as:

$$E = h\nu, \quad \text{with} \quad h = 6.626 \times 10^{-34}\ \text{Js} \tag{2.9.1}$$

where $h$ is called Planck's constant. Einstein postulated that the velocity $c$ of electromagnetic waves *in vacuo* was fixed and did not depend on the velocity of the source. The measured value is:

$$c = (2.9979250 \pm 0.0000010) \times 10^8 \text{ m s}^{-1} \tag{2.9.2}$$

and this constancy has been tested up to a source velocity of $0.999\,75c$ (Farley et al. 1968). The theory of relativity removed the need for a medium to support the waves and the 'luminiferous aether' that had bedevilled electromagnetism for so many years was swept away.

The photoelectric effect, the emission of electrons when light is absorbed by a material, was another phenomena that could not be satisfactorily explained by the classical theory. Hertz had incidentally discovered the effect though he did not of course know about electrons. Classical waves do carry energy and should therefore be able to transfer enough of this energy, in time, to an electron in a material to enable it to escape the material. Experimentally it was observed that however long you waited, unless the *frequency* of the radiation was above a certain value (depending on the material) no emission would occur. Einstein argued that if the radiation was quantized, then if the energy of a quantum, given by (2.9.1), was insufficient then this is just what you would expect, since the energy is proportional to frequency. Thus it was evident that not only was the electromagnetic field quantized but that interactions with matter were also. From our point of view these ideas raise the question of how are we, when dealing with electronic circuits, to view the electromagnetic field. Some years later when quantum (or wave) mechanics was developed, Heisenberg proposed his uncertainty relation. He showed that in the quantum domain it was impossible to measure certain conjugate quantities simultaneously with precision, i.e. if you tried to measure one with greater and greater precision then the other would become correspondingly less precise (see e.g. Eisberg 1961). The means of improving the precision disturbed the system to make the other quantity less precise. In terms of the imprecision of momentum $\Delta p$ and location $\Delta x$ of a particle the relation is:

$$\Delta p \cdot \Delta x \approx h/2\pi \tag{2.9.3}$$

where $\approx$ means of the order of; the exact equality is dependent on how the measurements are made. The meaning of this expression is that if you try to, say, determine the momentum $p$ more and more precisely then the position $x$ of the particle is correspondingly uncertain. We can derive a similar relation between the total relativistic energy $E$ of the particle and the time $\Delta t$ within which the measurement is made. The de Broglie relation between momentum $p$ and wavelength $\lambda$ is $p = h/\lambda$, and since the photon travels at velocity $c$ we have:

$$p = \frac{h}{\lambda} = \frac{h\nu}{c} = \frac{E}{c}, \quad \text{since } \nu\lambda = c \text{ and using (2.9.1)}$$

$$\text{so} \quad \Delta p = \frac{\Delta E}{c} \quad \text{and} \quad \Delta x = c\Delta t \tag{2.9.4}$$

and thus $\Delta p \cdot \Delta x = \frac{\Delta E}{c} c\Delta t = \Delta E \cdot \Delta t \approx h/2\pi$

We can take this one step further to enable us to comment on the relation between the quantum and the classical Maxwellian picture of an electromagnetic wave. A classical wave is defined by its amplitude and its rate of change of phase, i.e. by its frequency. If there are $n$ photons then the total energy $E = nh\nu$ and the phase $\phi$ of the wave is related to the frequency $\nu$ by $\phi = \omega t = 2\pi\nu t$ so that:

$\Delta E = \Delta n \cdot h\nu$ and $\Delta\phi = 2\pi\nu\Delta t$

$$\text{and} \quad \Delta E \cdot \Delta t = \Delta n \cdot h\nu \frac{\Delta\phi}{2\pi\nu} \approx \frac{h}{2\pi} \tag{2.9.5}$$

or $\Delta n \cdot \Delta\phi \approx 1$

which tells us that we cannot know the amplitude and the phase of the wave with certainty simultaneously. If we treat the photons as if they obey Poissonian statistics (they properly obey Bose–Einstein statistics, which has important consequences but in this context the difference is inconsequential, e.g. Eisberg 1961) then the fluctuation $\Delta n$ in $n$ is given by $n^{\frac{1}{2}}$. Thus as the mean number of photons is increased, i.e. as the intensity is increased, then $\Delta n$ increases as $n^{\frac{1}{2}}$, and hence the uncertainty in $\phi$ is reduced, but the fractional fluctuation $n^{\frac{1}{2}}/n$ decreases so that the wave beam becomes correspondingly better defined both in amplitude and in phase angle (Loudon 1973, p. 152). Thus if we have lots of photons we can correctly treat the wave by classical macroscopic field equations; if we have few photons then we must treat the situation quantum mechanically. It is readily seen that at the frequencies we are concerned with we have a very great numbers of photons so our circuit ideas are valid. For say a power of 1 mW at a frequency of 100 MHz we have:

$$n = \frac{\text{power}}{\text{energy per photon}} = \frac{10^{-3}}{6.626 \times 10^{-34} \times 10^{8}} \frac{\text{Js}^{-1}}{\text{J}} = 1.5 \times 10^{22} \text{ photons s}^{-1} \tag{2.9.6}$$

so there is no need to be concerned with quantum matters. In the optical region of the electromagnetic spectrum where the photon energy is much greater and single photons are readily detected, then a quantum mechanical approach may be necessary. A proper description of electromagnetic fields and their interaction with matter, quantum electrodynamics, is a continuing and difficult process.

## References and additional sources 2.9

Eisberg R. M. (1961): *Fundamentals of Modern Physics*, New York: John Wiley. ISBN 0-471-23463-X.

Farley F. J. M., Bailey J., Picasso E. (1968): Is the special theory right or wrong? *Nature* **217**, 17–18.

Loudon R. (1973): *The Quantum Theory of Light*, Oxford: Oxford University Press. ISBN 0-19-851130-2.

Thomson J. J. (1897): Cathode rays. *Philosophical Magazine*, **44**, 293–316

Zeeman P. (1897): On the influence of magnetism and the nature of the light emitted by a substance. *Philosophical Magazine* **43**, 226–239.

# 2.10 Dielectrics and permittivity

The more things we know, the better equipped we are to understand any one thing and it is a burning pity that our lives are not long enough and not sufficiently free of annoying obstacles, to study all things with the same care and depth.

Vladimir Nabakov author and lepidopterist

Dielectrics are almost by definition insulators. They play two important roles in electronics. First, they serve simply as insulators to isolate the various conductors in our circuit. There are no perfect insulators since in practice all available materials conduct to some degree. Table 2.10.1 lists a number of the common insulators together with their bulk resistivities. Some conductors are included for comparison. For the very good insulators it should be noted that the bulk values are often not the controlling factor since contamination of the surfaces, e.g. by water vapour condensation with the wide range of airborne pollutants now present, can lead to surface conduction. In these extreme cases it is necessary to clean the surface with appropriate solvent, enclose them and provide an effective desiccant to keep them 'dry'.

The second application for dielectrics is to provide increased capacity in a given volume. This arises from the polarizability of the material which is measured by the dielectric constant, or permittivity, $\varepsilon$ also shown in the table. Typical polymer dielectrics have modest values of $\varepsilon$ but some ceramic materials have very high values so allowing very large capacities in a small volume. Capacitors are discussed in Section 4.2. In practice the permittivity is not just a real number but must be given by a complex value:

$$\varepsilon = \varepsilon' - j\varepsilon'' \tag{2.10.1}$$

where $\varepsilon'$ is the value we have been referring to and $\varepsilon''$ indicates the deviation from perfection, i.e. the conductivity. For a capacitor, the current is in quadrature with the voltage so the power dissipation is zero. If the dielectric is not perfect there will be a component of the current in phase with the voltage and hence there will be a corresponding power loss dependent on $\varepsilon''$. If the (usually) small phase angle between 90° and the actual angle is $\theta$ then $\tan\theta$ is called the loss tangent: a small

Table 2.10.1 *Rough guide to the resistivity and permittivity of some insulators, with good conductors for comparison*[a]

| Material | Resistivity ($\Omega$ cm$^{-1}$) | Permittivity, $\varepsilon$ |
|---|---|---|
| Copper | $1.7 \times 10^{-10}$ | – |
| Aluminium | $2.8 \times 10^{-10}$ | – |
| Polyester | $10^{15}$–$10^{16}$ | ~3.2–4 |
| PTFE | $10^{17}$–$10^{18}$ | ~2.1 |
| Glass epoxyPCB | $10^{10}$–$10^{13}$ | ~5 |
| Polyethylene | $10^{16}$–$10^{18}$ | ~2.3 |
| Polystyrene | $10^{16}$–$10^{18}$ | ~2.5 |
| Ceramic | $10^{14}$–$10^{15}$ | ~10–300 000 |

*Note*:
[a] $\varepsilon$ values are for low frequency and somewhat variable depending on manufacture. High resistivity materials are much affected by surface contamination. Ceramic high $\varepsilon$ value materials are substantially temperature and voltage dependent.

value indicates a good dielectric. The power loss is more or less frequency dependent but all dielectrics eventually become quite lossy at high enough frequency and capacitors of inappropriate type can become quite warm in power circuits at even moderate radio frequencies. It is of course just this loss that is responsible for the heating effects in microwave ovens where the dielectric is water and the frequency about 3 GHz. When Maxwell derived his wave-equation for electromagnetic waves he found that the velocity of the predicted waves was given (in our present nomenclature) by:

$$v = (\mu\mu_0\varepsilon\varepsilon_0)^{\frac{1}{2}} \tag{2.10.2}$$

where $\mu_0$ and $\varepsilon_0$ are the permeability and permittivity of free space, and $\mu$ and $\varepsilon$ are the values for the medium (for vacuum or air $\mu = \varepsilon = 1$). If the values are inserted then it is found that $v = c = 3 \times 10^8$ ms$^{-1}$ ($c$ is the usual symbol for the velocity of light in free space). Of course Maxwell did not have very accurate values of the constants but the value he obtained was close enough to the then measured velocity of light to suggest very strongly that light was just an electromagnetic wave. The matter in general was bedevilled for many years by arguments and speculation about the medium, the aether, which supported the waves and this question was not resolved until Einstein proposed his theory of relativity (Buchwald 1988).

If the dielectric constant of a material is measured over the range from zero frequency to optical frequencies (say $6 \times 10^{14}$ Hz in the visible region) the value is found to vary over a wide range. For example, at low frequencies water has a dielectric constant of about 80. What is more readily measured in the visible region is refractive index. Electromagnetic theory shows that there is a relationship between dielectric constant $\varepsilon$ and refractive index $n$ ($n$ is defined as the ratio of velocity of the waves in free space to that in the medium):

$$\varepsilon = n^2 \tag{2.10.3}$$

and for water $n \approx 1.5$ giving $\varepsilon = 2.2$, quite different from the low frequency value. For most electronics this variation is not significant in that the frequencies involved are very much less than optical frequencies. However, with the common use of optical communications nowadays it is important to be aware of the variations. It is sometimes said or implied that optical signals travel at the 'speed of light' while electrical signals travel at some lesser speed. In an optical fibre the velocity will be given by $v = c/n = c/1.5$ for a typical glass. For a coaxial cable, where the wave travels in the dielectric between the inner conductor and the outer screen, the velocity will be $v = c/\varepsilon^{\frac{1}{2}}$ and for polyethylene $\varepsilon = 2.3$ so $\varepsilon^{\frac{1}{2}} \approx 1.5$. Thus the velocity will be the same at about $0.7c$. The bandwidth of the optical fibre is, however, very much greater than the coaxial cable so you can send much more *information* in a given time – but the wave does not get there any sooner.

The mechanism responsible for the dielectric constant is the polarization of the charge distribution of the atoms or molecules (Grant and Philips 1990). This shift of charge causes the faces of the dielectric to become charged and so produces an electric field internal to the dielectric in opposition to the external applied field. With the dielectric say filling the capacitor then for the given quantity of charge on the plates the effective field is lower than it would be without the dielectric so that the potential will be lower and hence the capacity will be higher. Even for the highest electric fields that can be applied before breakdown, say $\sim 10^7$ V m$^{-1}$, the displacement of the charges amounts to only about $1.5 \times 10^{-15}$ m, which is about $10^{-5}$ of the diameter of an atom! The charges do not move far, but there are a very great number of them. This very small movement means that the polarization is closely proportional to the applied field and we do not have to be concerned about non-linear effects in materials with this type of dielectric effect. Materials with ionic type binding of the constituents are also substantially linear, but those with natural polarization like the high permittivity ceramics are substantially voltage sensitive. When the applied field oscillates then the molecular dipoles must respond by changing their orientation to match that of the field direction. At higher frequencies they become less able to keep in phase and the lag means that the current in the capacitor will no longer be in quadrature with the applied voltage. This means that there will be increasing power loss in the capacitor and eventually an inability to perform its intended function.

## References and additional sources 2.10

Bleaney B. I., Bleaney B. (1963): *Electricity and Magnetism*, Oxford: Clarendon Press. See Chapter XVIII 'Theory of the dielectric constant'.

Buchwald J. Z. (1988): *From Maxwell to Microphysics*, Chicago: University of Chicago Press. ISBN 0-226-07883-3.

Corson D. R., Lorrain P. (1962): *Introduction to Electromagnetic Fields and Waves*, San Francisco: W. H. Freeman. Library of Congress Cat. No. 62-14193. See Chapter 3.

Grant I., Philips W. R. (1990): *Electromagnetism*, 2nd Edn, London: John Wiley. ISBN 0-471-92712-0.

Keithley (1992): *Low Level Measurements*, Keithley Instruments handbook.

Matick R. E. (1969): *Transmission Lines for Digital and Communication Networks*, New York: McGraw-Hill. Library of Congress Cat. No. 68-30561.

# 2.11 Magnetic materials

One of the principal objects of theoretical research in any department of knowledge is to find the point of view from which the subject appears in its greatest simplicity.

Josiah Willard Gibbs

The magnetic fields previously discussed have been in free space, or in air as this is effectively the same from our point of view. The introduction of magnetic material can greatly change the fields both in their geometric distribution and their magnitude. There is, however, a difficulty in determining the resulting fields. Consider, say, a solenoid with a core of magnetic material (Fig. 2.11.1). Current in the solenoid produces some field that results in the magnetization $\boldsymbol{M}$ of the core which in itself contributes a further field. But $\boldsymbol{M}$ depends on the total field, which is just the quantity we wish to determine. This difficulty is most readily treated by introducing an additional quantity, called the magnetic intensity $\boldsymbol{H}$, which depends only on free currents due to the flow of charges in a conductor and is independent of magnetization currents. The latter are the equivalent currents that we postulate to produce the magnetization of the magnetic material. These currents represent the effect of the atomic magnetic dipoles in the core material, which will give a macroscopic effect if the material is magnetized.

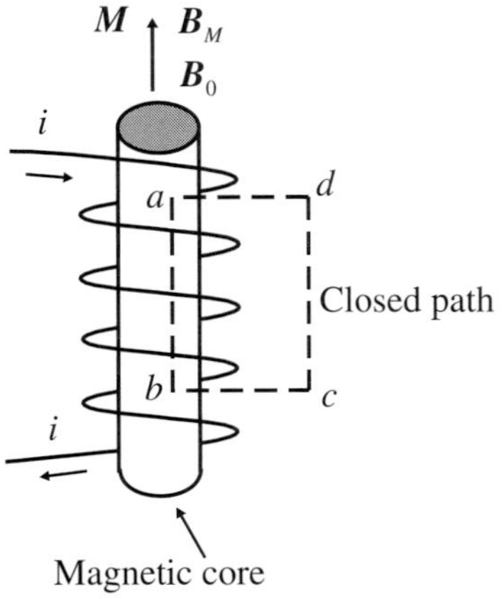


**Fig. 2.11.1** Solenoid with magnetic core.

The line integral of $\boldsymbol{B}$ around a closed path is equal to the line integral of the sum of the field $\boldsymbol{B}_0$ in the absence of the magnetic material and $\boldsymbol{B}_M$ from the magnetization The relation between the $\boldsymbol{M}$, which must be zero outside the core, and the consequential field $\boldsymbol{B}_M$ is:

$$\boldsymbol{B}_M = \mu_0 \boldsymbol{M} \tag{2.11.1}$$

The line integral around the rectangular path as shown in Fig. 2.11.1 will be:

$$\oint \boldsymbol{B} \cdot d\boldsymbol{l} = \oint (\boldsymbol{B}_0 + \boldsymbol{B}_M) \cdot d\boldsymbol{l} = \oint \boldsymbol{B}_0 \cdot d\boldsymbol{l} + \mu_0 \oint \boldsymbol{M} \cdot d\boldsymbol{l}$$

$$\text{or} \quad \oint (\boldsymbol{B} - \mu_0 \boldsymbol{M}) \cdot d\boldsymbol{l} = \oint \boldsymbol{B}_0 \cdot d\boldsymbol{l} \tag{2.11.2}$$

The value of $\boldsymbol{B}_0$ is given by (2.5.12) and depends on the free current $I_f$. Thus:

$$\oint (\boldsymbol{B} - \mu_0 \boldsymbol{M}) \cdot d\boldsymbol{l} = \mu_0 I_f \tag{2.11.3}$$

and if we now define the new quantity $\boldsymbol{H}$:

$$\mu_0 \boldsymbol{H} = \boldsymbol{B} - \mu_0 \boldsymbol{M} \quad \text{or} \quad \boldsymbol{H} = \frac{\boldsymbol{B}}{\mu_0} - \boldsymbol{M} \tag{2.11.4}$$

then (2.11.3) becomes:

$$\oint \boldsymbol{H} \cdot d\boldsymbol{l} = I_f \tag{2.11.5}$$

which is Ampère's law for $\boldsymbol{H}$. The units of $\boldsymbol{H}$ are current over length and hence ampere per metre (sometimes ampere-turns per metre). The line integral of $\boldsymbol{H}$ is sometimes called the magnetomotive force (or magnetomotance) and is similar to the voltage in ordinary circuits. The flux $\Phi = \boldsymbol{B} \times$ area corresponds to the current (Corson and Lorrain 1962, p. 296). We can also write an expression for $\boldsymbol{M}$ in terms of $\boldsymbol{B}$ (rather than $\boldsymbol{B}_M$ as in Eq. (2.11.1)):

$$\boldsymbol{M} = \chi_B \frac{\boldsymbol{B}}{\mu_0} \tag{2.11.6}$$

where $\chi_B$ is the magnetic susceptibility of the material. Using this in Eq. (2.11.4) gives:

$$\boldsymbol{H} = \frac{\boldsymbol{B}}{\mu_0} - \frac{\chi_B \boldsymbol{B}}{\mu_0} \quad \text{or} \quad \boldsymbol{B} = \mu \mu_0 \boldsymbol{H}, \quad \text{where} \quad \mu = (1 - \chi_B)^{-1} \tag{2.11.7}$$

and the dimensionless quantity $\mu$ is called the relative permeability of the material. For iron and many of its alloys, and for ferrite materials, $\mu$ can range from unity

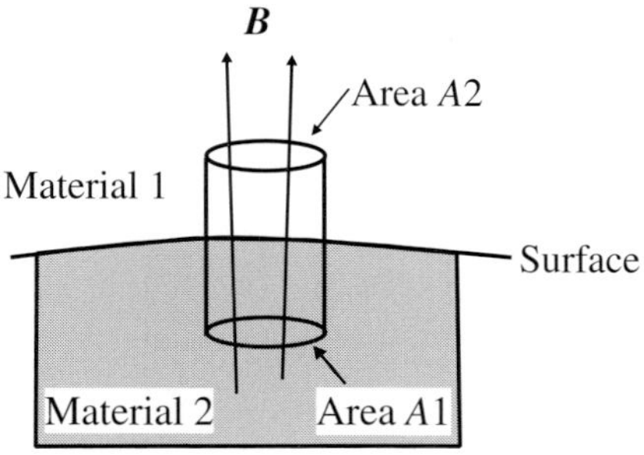


**Fig. 2.11.2** Boundary conditions for ***B***.

to say 1000 or more, so that considerable enhancement of ***B*** may be obtained from the same current. There are limitations, however, in terms of non-linearity as a function of ***B*** and of frequency, and temperature dependence. The field ***H*** differs from ***B*** since it has sources and so, like an electric field, the lines of ***H*** are not continuous. ***B*** has no sources so that the lines must be continuous. To solve for the distribution of ***H*** and ***B*** in a given circumstance involves finding the solution to Eq. (2.11.5). At the boundaries between different media we will have some difficulty since ***H*** has sources there. Though ***B*** and ***H*** are related by (2.11.7), the fields must also satisfy boundary conditions that determine how the fields are related just on each side of the boundary.

Consider a small pillbox shape crossing the boundary as shown in Fig. 2.11.2.

If the length of the pillbox is decreased so that the two ends are infinitesimally close to the surface then since ***B*** has no sources div$\boldsymbol{B}=0$, and the flux out of one end must equal the flux in at the other, i.e:

$$\int_{A1} \boldsymbol{B}\cdot dS + \int_{A2} \boldsymbol{B}\cdot dS = 0 \tag{2.11.8}$$

and since it is the component normal to the surface and the flux enters through one area and exits through the other, then:

$$\int_{A1} \boldsymbol{B}_n\cdot dS = \int_{A2} \boldsymbol{B}_n\cdot dS \tag{2.11.9}$$

where $\boldsymbol{B}_n$ is the component normal to the surface. Since the areas are equal then the normal components must be continuous across the surface, and any solution to the fields must satisfy this condition.

If we consider a closed path *abcd* through the surface as shown in Fig. 2.11.3 then using (2.11.5) we can derive the boundary condition for ***H***.

As we bring the two parts of the path *ab* and *cd* closer to the surface the area

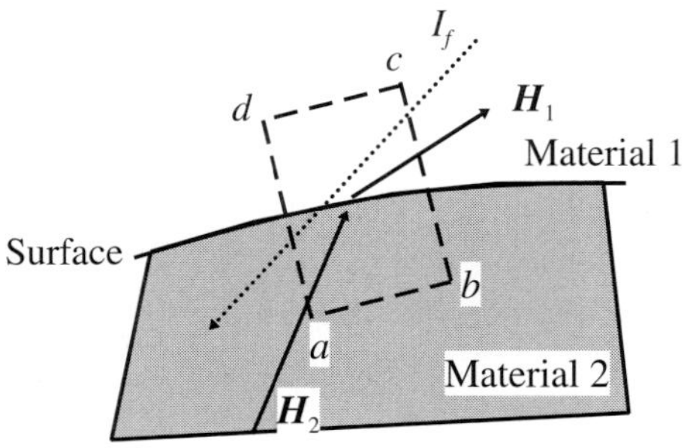


**Fig. 2.11.3** Boundary conditions for $\boldsymbol{H}$.

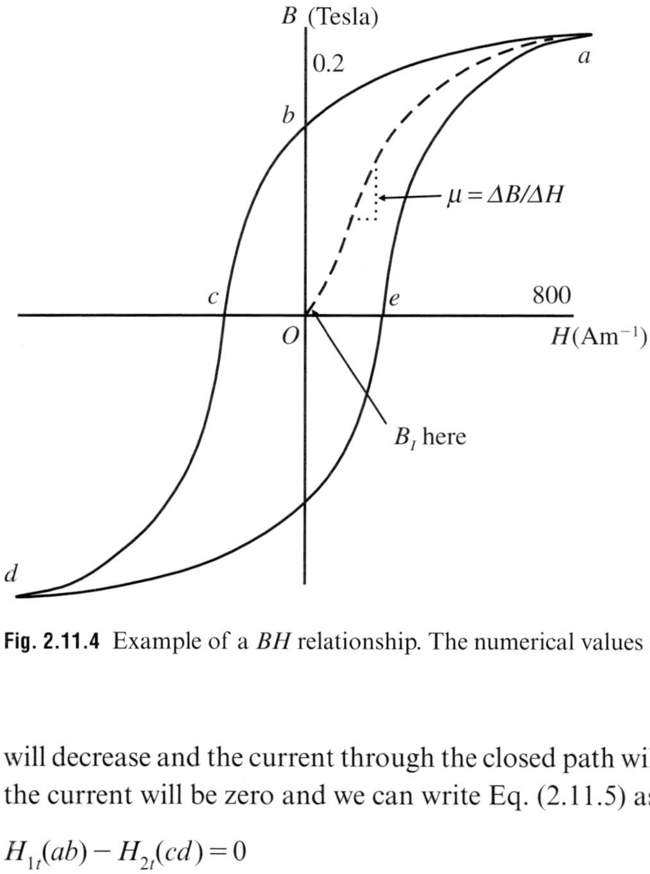


**Fig. 2.11.4** Example of a *BH* relationship. The numerical values are only indicative.

will decrease and the current through the closed path will also decrease. In the limit the current will be zero and we can write Eq. (2.11.5) as:

$$H_{1t}(ab) - H_{2t}(cd) = 0 \tag{2.11.10}$$

where $\boldsymbol{H}_t$ is the component tangential to the surface. Thus we have the condition that the tangential component of $\boldsymbol{H}$ must be continuous across the interface.

Calculation of the distribution of $\boldsymbol{B}$ and $\boldsymbol{H}$, other than for rather simple configurations, presents considerable difficulty. It is made significantly more difficult if the fields are large and the relative permeability $\mu$ becomes non-linear. Measurements of the fields are generally shown as *BH* curves, as illustrated by

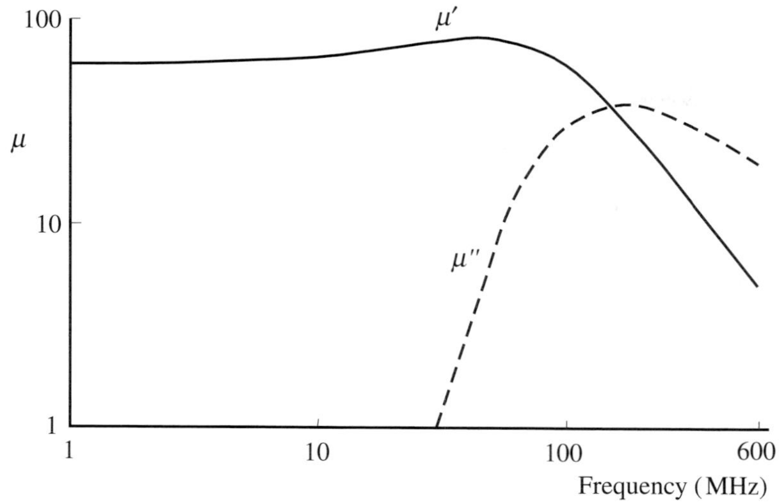


**Fig. 2.11.5** Variation of $\mu$ with frequency.

Fig. 2.11.4 for a ferrite material with an initial permeability of ≈60. Figure 2.11.5 shows the variation of $\mu$ with frequency, and it should be noted that it is a complex quantity. The *BH* relation shows that the permeability is also a function of its history, i.e. the value of $B$ for a given $H$ depends on what has happened before.

Starting with an unmagnetized specimen the initial variation of $B$ with $H$ will follow the path from the origin $O$ to $a$. At $a$ all the microscopic domains in the material are aligned and the material becomes saturated. Further increase of $H$ now results in $B$ increasing as if there were only air. If $H$ is decreased towards zero, $B$ will follow the path $ab$, and the value of $B$ for $H=0$ is called the remanance, i.e. it is magnetized. The value of $H$ in the reverse sense that makes $B=0$ at $c$ is called the coercive force. Further increase of $H$ in this sense duplicates the curve to $d$, and reversing $H$ again brings the system back via $e$ to $a$. This dependence of $B$ on the sense and history of $H$ is called hysteresis and the area inside the curve is a measure of the energy losses in taking the system around the loop. Such losses are of particular importance, for example, in the inductors used in switching power supplies. SPICE simulation of *BH* curves is outlined in Section 3.12.

Knowledge of the detailed field distributions in, for example, a power station generator, a stepper motor, a magnetic head for a disc drive, a magnetic resonance imaging system, a mass spectrometer for breath analysis to see if you suffer from *Helicobactor pylori*, or the deflection circuits in your TV or computer monitor is vital to their proper design. Software packages are nowadays available to help solve for the fields and are the equivalents for magnetic systems of the electronic packages like SPICE. Figure 2.11.6 shows the output of such a package (Vector Fields 1997) for a short coil wound on a long ferrite rod. To enable the calculations, the volumes

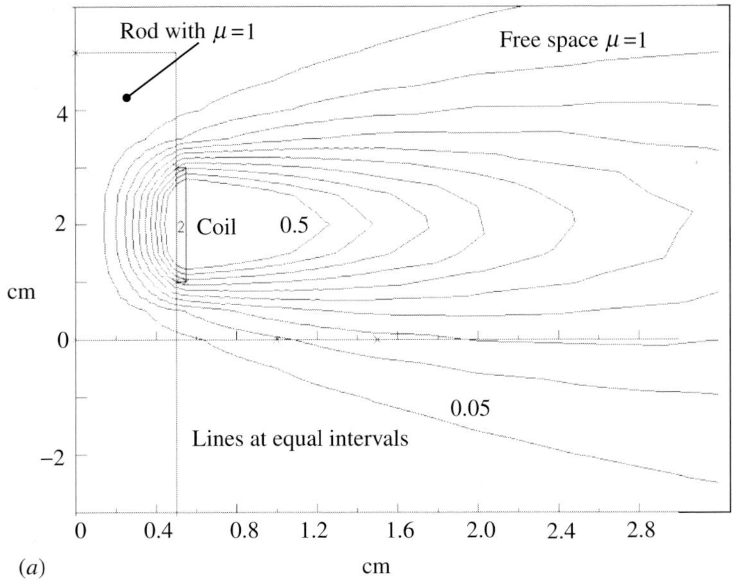


(*a*)

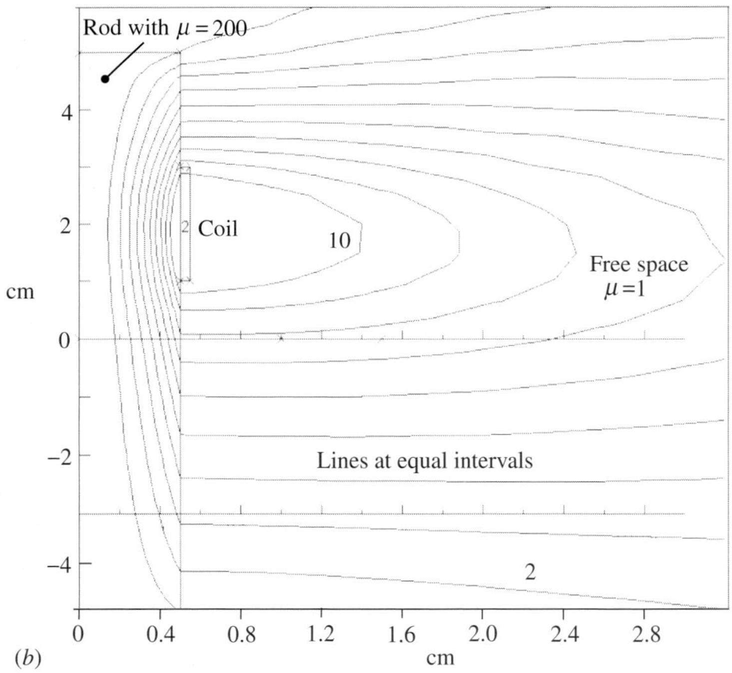


(*b*)

**Fig. 2.11.6** (*a*) Field surrounding a solenoid. (*b*) Field surrounding the same solenoid but with a ferrite rod of $\mu = 200$ as shown. The rod is 10 cm long by 1 cm in diameter and the coil is offset. The coil current is the same in each case and the values on the contours indicate relative intensities. As the system is cylindrically symmetric, only one-half of the field is computed. The irregularities in the far field arise from the coarse finite element division in that region. A total of about 9000 elements were used. Though the different $x$- and $y$-scales distort the diagram, the expanded $x$-axis is necessary to show the field in the ferrite clearly.

of material and free space are divided up into small finite elements and the calculations are followed from element to element until one returns to the start point. If the values at start and end do not agree, an adjustment is made and the loop repeated until they do, i.e. a self-consistent result is obtained. This sort of procedure requires considerable computation, but now PCs are able to handle quite complex systems in acceptable times. Software is also available for plotting the trajectory of a charged particle through electric and magnetic fields, but there is some difficulty in relating interactions in fields with the electronic circuits that produce them.

The status of $\boldsymbol{H}$ is, astonishingly, still subject to heated debate by physicists (Crangle and Gibbs 1994). As one of the older generation, I find $\boldsymbol{H}$ useful so it has been included here, especially as most of the literature, and data on materials, you will encounter makes use of it. But be aware that there may be other views and approaches.

Magnetic materials are available in many types and forms. Those appropriate for use at low frequency, as for example in mains power transformers, will be in the form of iron laminations that are stacked to form the magnetic core. The laminations are insulated from one another to prevent eddy currents which would lead to large power losses. As frequency rises it becomes impossible to make the laminations thin enough and other materials must be used instead. Ferrite materials can have both high permeability and high resistivity which allows their use up to frequencies of say tens or hundreds of megahertz, though there are also special applications in the microwave region. Ferrites are hard, brittle ceramic materials and they are of the form $MeFe_2O_4$, where 'Me' in the most popular types consists of combinations of manganese and zinc (MnZn) or nickel and zinc (NiZn). The former have higher permeabilities and saturation levels but lower resistivities than the latter. The frequency boundary between the two types is around 3 MHz. As the ferrite materials are originally produced in powder form they can be fashioned into many shapes before being sintered at high temperature to produce the final hard material. The common reference to these materials as 'soft magnetic materials' refers to their magnetic properties rather than their mechanical. Table 2.11.1 provides some illustrative figures for the two families.

In addition to the matter of resistivity that may affect the choice of material there is consideration of the permittivity or dielectric constant. This is particularly relevant for the MnZn types where the permittivity may reach $\approx 10^6$ at low frequency, not often referred to in the commercial literature. Such high values, together with high permeability, mean that the velocity of electromagnetic waves in the material becomes very low and hence the wavelength may be reduced to the order of the dimensions of the piece which results in dimensional resonance and additional losses. The permittivity falls off at high frequencies but even in the microwave region it can still be high enough to allow the construction of directional antenna. The high dielectric constant also means that the self-capacity of a coil wound directly on the ferrite will be increased.

Table 2.11.1 *Guide properties of ferrites*

| Material | MnZn | NiZn |
|---|---|---|
| Initial permeability, $\mu_i$ | 2000 | 100 |
| Resistivity, $\rho$ ($\Omega$m) 0 °C | 7 | $5\times10^7$ |
| 20 °C | 4 | $10^7$ |
| 100 °C | 1 | $10^5$ |
| 100 kHz | 2 | $10^5$ |
| 1 MHz | 0.5 | $5\times10^4$ |
| 10 MHz | 0.1 | $10^4$ |
| Permittivity, $\varepsilon_r$ 100 kHz | $2\times10^5$ | 50 |
| 1 MHz | $10^5$ | 25 |
| 10 MHz | $5\times10^4$ | 15 |

## References and additional sources 2.11


Coleman J. E. (1973): Conventions for magnetic quantities in SI. *Amer. J. Phys.* **41**, 221–223.

Corson D. R., Lorrain P. (1962): *Introduction to Electromagnetic Fields and Waves*, San Francisco: W. H. Freeman. Library of Congress Cat. No. 62-14193.

Crangle J., Gibbs M. (1994): Units and unity in magnetism. *Physics World*, November, 31–32. See also *Physics World*, January1995, 19–21; February 1995, 22–23; March, 1995, 23; June 1995, 21–22.

Dobbs E. R. (1993): *Basic Electromagnetism*, London: Chapman and Hall. ISBN 0-412-55570-0.

Feynman R. P., Leighton R. B., Sands M. (1964): *The Feynman Lectures on Physics*, Vols I, II, III, Reading, Mass: Addison-Wesley. Library of Congress Cat. No. 63-20717.

Fish G. E. (1990): Soft magnetic materials. *Proc. IEEE* **78**, 947–972.

Grant I., Philips W. R. (1975): *Electromagnetism*, London: John Wiley. ISBN 0-471-32246-6. 2nd Edn. 1990. ISBN 0-471-92712-0.

Hess J. (1999): An end to leakage. *Siemens Components* **2/99**, 32–34. (Ferrite polymer composite film for gap filling in ferrite assemblies.)

Jiles D. (1991): *Introduction to Magnetism and Magnetic Materials*, London; Chapman and Hall. ISBN 0-412-38630-2.

Jiles D. C., Atherton D. L. (1986): Theory of ferromagnetic hysteresis. *J. Mag. and Magnetic Materials* **61**, 48–60.

Lenz J. E. (1990): A review of magnetic sensors. *Proc. IEEE* **78**, 973–989.

MMG-Neosid: *Soft Ferrite Components and Accessories*, www.mmg-neosid.com

Neosid Pemetzreider Gm2.11: *Soft Ferrite Components*, www.neosid.de

Rose M. J. (1969): Magnetic units in SI. *Mullard Tech. Comm.* January, 223–225.

Snelling E.C. (1969): *Soft Ferrites. Properties and Applications*, London: Iliffe Books. 2nd Edition, Butterworths 1988.

Stopes-Roe H. V. (1969): MKSA, Giorgi and SI. *Nature* **222**, 500–502.

Vector Fields (1997): *PC Opera-2D*, Vector Fields Ltd, Oxford. www.vectorfields.com

Whitworth R., Stopes-Roe H. V. (1971): Experimental demonstration that the couple on a bar magnet depends on ***H***, not ***B***. *Nature* **234**, 31–33.

# 2.12 Units of electromagnetism

. . . save the one on electromagnetic units. There is some progress to report. Of the three stages to Salvation, two have been safely passed through, namely the Awakening and the Repentance. I am not alone in thinking that the third stage, Reformation, is bound to come.

Oliver Heaviside (1899): *Electromagnetic Theory*, April 10, Vol. II, p. iv

Various systems of units have been used over the years for electromagnetic quantities and have caused considerable anguish to many students of the subject. The system first proposed by Giorgi (1901), (Tunbridge 1992), which later became known as the MKS (for metre, kilogram, second) was a long time in gaining acceptance, but eventually sense prevailed. Now that the SI (Système International) system of units has been well established and accepted, at least in the scientific community if not yet completely in the commercial engineering community, it is much easier to work with and be sure that information from different sources is equivalent. All equations must be consistent in the units of the various terms and this is sometimes a useful check on your algebra. The electromagnetic quantities we are concerned with can be defined in terms of four fundamental quantities mass ($M$), length ($L$), time ($T$) and charge ($Q$). There are also three constants, the velocity of light $c$, the permittivity of free space $\varepsilon_0$ and the permeability of free space $\mu_0$ that are related and are to be determined by experiment. Since they are related it is possible to make an arbitrary choice of one of them to make some equations somewhat simpler. The choice of $\mu_0 = 4\pi \times 10^{-7}$ Hm$^{-1}$ has been made. Maxwell's equations lead to the relationship (Section 2.7):

$$c = (\varepsilon_0 \mu_0)^{-\frac{1}{2}} \tag{2.12.1}$$

and since the measured value for the velocity of light is:

$$c = (2.9979250 \pm 0.0000010) \times 10^8 \text{ m s}^{-1} \tag{2.12.2}$$

the derived value of $\varepsilon_0$ is:

$$\varepsilon_0 = (8.854185 \pm 0.000006) \times 10^{-12} \text{ F m}^{-1} \tag{2.12.3}$$

It should be noted that all units are singular; plurals can cause confusion with s (seconds). The various quantities are given in Table 2.12.1 which shows the

Table 2.12.1 *Electromagnetic units, symbols and dimensions*

| Quantity | SI unit | Symbol | Dimensions |
|---|---|---|---|
| Capacitance | Farad | F | $[M^{-1}L^{-2}T^2Q^2]$ |
| Charge | Coulomb | C | $[Q]$ |
| Current | Ampere | A | $[T^{-1}Q]$ |
| Energy | Joule | J | $[ML^2T^{-2}]$ |
| Force | Newton | N | $[MLT^{-2}]$ |
| Inductance | Henry | H | $[ML^2Q^{-2}]$ |
| Potential | Volt | V | $[ML^2T^{-2}Q^{-1}]$ |
| Power | Watt | W | $[ML^2T^{-3}]$ |
| Resistance | Ohm | Ω | $[ML^2T^{-1}Q^{-2}]$ |
| Electric displacement | Coulomb metre$^{-2}$ | D | $[L^{-2}Q]$ |
| Electric field | Volt metre$^{-1}$ | E | $[MLT^{-2}Q^{-1}]$ |
| Magnetic field | Tesla | B | $[MT^{-1}Q^{-1}]$ |
| Magnetic intensity | Ampere metre$^{-1}$ | H | $[L^{-1}T^{-1}Q]$ |

*Note:*
$M$ = mass, kilogram; $L$ = length, metre; $T$ = time, second; $Q$ = charge, coulomb.

dimensions in terms of the four fundamental quantities as given above. Note also that the preferred form for writing units is, for example, as shown above: i.e. F m$^{-1}$ rather than F/m.

The reciprocal of resistance is conductance and has the unit siemen (S).

For the magnetic quantities B and H the older Gaussian units were gauss (B) and oersted (H) and older sources of data (and sometimes even recent) will be in terms of these. The conversions between the units are given by:

1 tesla = $10^4$ gauss and 1 oersted = 79.4 ampere metre$^{-1}$

As an example we can check the consistency of Eqs. (2.1.15) and (2.1.16). In terms of dimensions we have:

$$j_c = nev_d = L^{-3} \times Q \times LT^{-1} = QL^{-2}T^{-1} \quad \text{and}$$

$$j_c = \frac{ne^2E}{m_e v_{\text{coll}}} = \frac{L^{-3} \times Q^2 \times MLT^{-2}Q^{-1}}{M \times T^{-1}} = QL^{-2}T^{-1}$$

There are also a number of constants that will be required:

Boltzmann constant: $k_B = 1.381 \times 10^{-23}$ J K$^{-1}$

Electronic charge: $q_e = 1.602 \times 10^{-19}$ C

Planck constant: $h = 6.626 \times 10^{-34}$ J s

Avogadro number: $N_A = 6.022 \times 10^{23}$ mol$^{-1}$

Electron mass: $m_e = 9.107 \times 10^{-31}$ kg

## References 2.12

Coleman J. E. (1973): Conventions for magnetic quantities in SI. *Amer. J. Phys.* **41**, 221–223.

Giorgi (1901): Unita razionali di electromagnetismo. *Atti dell' A.E.I*; and *Elettricita* (Milan) **20**, 787–788, Dec. Also, Proposals concerning electrical and physical units. *Proceedings of the International Electrical Congress*. St Louis, 1904, Vol. 1, p. 136.

Rose M. J. (1969): Magnetic units in SI. *Mullard Tech. Comm.* January, 223–225.

Tunbridge P. (1992): *Lord Kelvin. His Influence on Electrical Measurements and Units*, London: Peter Peregrinus. ISBN 0-86341-237-8. Includes the story of how the electrical units we now use came to be fixed behind the scenes in the Hôtel Chatham and the Restaurant Chibest, Paris, in 1881.

Zimmerman N. M. (1998): A primer on electrical units in the système international. *Amer. J. Phys.* **66**, 324–331.

# 2.13 Noise

Electromagnetics has been said to be too complicated. This probably came from a simple-minded man.

Oliver Heaviside

Noise is, as they say of the poor, always with us. This does not refer to such external sources such as lightning, emanations from sunspots or from the ubiquitous mobile telephone, but to sources within the electronic devices of our circuits. The noise with which we are concerned is fundamental and inescapable, but if we can understand its origins and characteristics it will allow us to minimize the consequences. The first source arises from thermal fluctuations in the variables or parameters which specify the state of the system, such as voltage, current or energy. Thermodynamically all such quantities are subject to fluctuation and from our point of view it is possible to derive the magnitude of the fluctuations from purely thermodynamic arguments without any reference to the nature of electricity, such as that it is due to discrete electrons. It would not matter if it were a continuum fluid. The original investigation of the form of this fluctuation was carried out by Nyquist (1928) (see also Kittel 1967) and the experimental proof of the theory was provided by Johnson (1928) and it is nowadays commonly referred to as Johnson noise. Nyquist showed that the mean squared voltage $\langle v_{NR}^2 \rangle$ appearing across a resistor $R$ in any frequency interval $\Delta f$ is given by:

$$\langle v_{NR}^2 \rangle = 4k_B TR\,\Delta f \quad \text{volt}^2 \tag{2.13.1}$$

where $T$ is the absolute temperature and $k_B$ is Boltzmann's constant $1.38\times10^{-23}$ J K$^{-1}$

If $R$ is in kΩ and $\Delta f$ in kHz then applying the subscript $k$ we have:

$$\langle v_{NR} \rangle = 0.13(R_k\,\Delta f_k)^{\frac{1}{2}} \quad \mu\text{V r.m.s. at 300 K} \tag{2.13.2}$$

It should be noted that (2.13.1) is derived on the basis of the Rayleigh–Jeans approximation to the distribution of black-body radiation, which is applicable up to millimetre wavelengths at ordinary temperatures. At higher frequencies, or at very low temperatures, it is necessary to use the Planck distribution which gives:

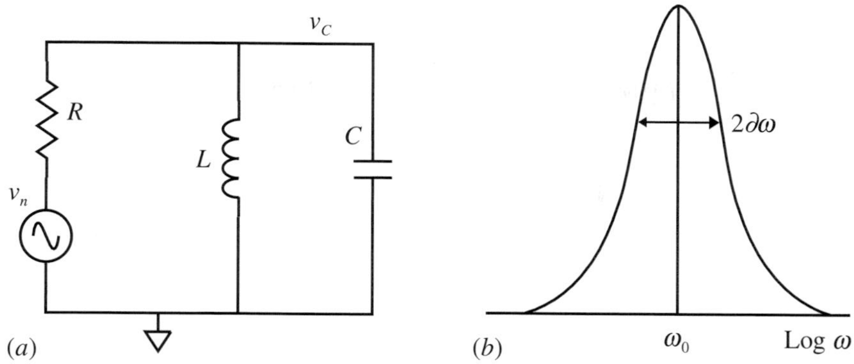


**Fig. 2.13.1** (*a*) Noise source with bandwidth limiting circuit. (*b*) Bandpass characteristic.

$$\langle v_{NR}^2 \rangle = \frac{4hfR\Delta f}{\exp(hf/k_B T) - 1} \tag{2.13.3}$$

where $h$ is Planck's constant $6.625 \times 10^{-34}$ J s. For our present purposes this is not of consequence but electronics now reaches into these very high frequencies so it is as well to be aware. An important aspect of Eq. (2.13.1) is that the noise voltage depends only on the frequency interval $\Delta f$ and not on the frequency, i.e. the distribution is white, a term arising from the optical region where white light contains 'all' frequencies (at least those that we can see). This means that there is no advantage in working at one frequency rather than another, though this will later need to be modified in the light of other forms of noise.

Nyquist derived his result from consideration of thermal equilibrium between two resistors connected to the ends of a transmission line. As an alternative approach to show the relation of Eq. (2.13.1) to classical thermodynamics we can consider the circuit of Fig. 2.13.1:

The generator $v_n$ represents the noise voltage arising from the resistor $R$ at absolute temperature $T$, and $L$ and $C$ define a bandpass for the circuit. From Section 3.5 we have for the centre frequency $\omega_0$ and bandwidth $2\partial\omega$:

$$\omega_0 = (LC)^{-\frac{1}{2}}, \quad \text{or} \quad \omega_0^2 LC = 1$$

$$2\partial\omega = \frac{Q}{\omega_0} = \frac{\omega_0 CR}{\omega_0} = CR, \quad \text{since from (3.5.18)} \quad Q = \omega_0 CR \tag{2.13.4}$$

If $Z_p$ is the impedance of $L$ and $C$ in parallel then:

$$v_C = \frac{v_n Z_p}{R + Z_p} \quad \text{and with} \quad \frac{1}{Z_p} = sC + \frac{1}{sL} \tag{2.13.5}$$

$$v_C = \frac{v_n}{1 + R\left(sC + \frac{1}{sL}\right)} = \frac{v_n}{1 + \omega_0 RC\left(\frac{s}{\omega_0} + \frac{\omega_0}{s\omega_0^2 LC}\right)} = \frac{v_n}{1 + Q\left(\frac{s}{\omega_0} + \frac{\omega_0}{s}\right)} \quad \text{using (2.13.4)}$$

and for the modulus of $v_C$ we then have (using the complex conjugate $v_C^*$):

$$|v_C|^2 = |v_C \cdot v_C^*| = \frac{|v_n|^2}{1 - Q^2\left(\frac{\omega}{\omega_0} - \frac{\omega_0}{\omega}\right)^2} \quad \text{using} \quad s = j\omega$$

$$= \frac{4k_B TR\Delta f}{1 - Q^2\left(\frac{\omega}{\omega_0} - \frac{\omega_0}{\omega}\right)^2} \tag{2.13.6}$$

We now have to integrate over the whole bandwidth. To do this we may make use of the approximations of Eq. (3.5.7) and the Lorentzian function $F(\omega)$ to give:

$$\langle v_C^2 \rangle = \int_{\omega=-\infty}^{\omega=\infty} \frac{4k_B TR\, df}{1 + 4Q^2\left(\frac{\delta_\omega}{\omega_0}\right)^2}, \quad \text{where} \quad \delta_\omega = \omega - \omega_0$$

$$= \frac{4k_B TR}{2\pi} = \int_{\omega=-\infty}^{\omega=\infty} \frac{d\delta_\omega}{1 + 4Q^2\left(\frac{\delta_\omega}{\omega_0}\right)^2}, \quad \text{where } \omega = 2\pi f \text{ so } d\omega = 2\pi f\, df \text{ and } d\omega = d\delta_\omega \tag{2.13.7}$$

$$= \frac{4k_B TR}{2\pi}\frac{\omega_0}{2Q}\int_{-\infty}^{\infty} \frac{dx}{1 + x^2} \quad \text{substituting} \quad \delta_\omega = \frac{\omega_0 x}{2Q} \quad \text{so} \quad d\delta_\omega = \frac{\omega_0}{2\pi}dx$$

$$= \frac{k_B T}{C} \quad \text{using integral (1.9.1(g)) which} = \pi \text{ for these limits, and Eq. (2.13.4).}$$

Now if we find the energy in the capacitor we have using Eq. (4.2.3):

$$\text{Energy} = \tfrac{1}{2}C\langle v_C^2 \rangle = \tfrac{1}{2}C\frac{k_B T}{C} = \tfrac{1}{2}k_B T \tag{2.13.8}$$

which is exactly what classical thermodynamics predicts. The equipartition theorem requires each quadratic term in the system energy to be associated with a thermal energy of $\frac{1}{2}k_B T$. Similarly, for the inductor we have using Eq. (4.3.2):

$$\langle i_L^2 \rangle = \frac{\langle v_C^2 \rangle}{(\omega_0 L)^2} = \frac{\langle v_C^2 \rangle C}{L} = \frac{k_B T}{L} \quad \text{or energy} \quad \tfrac{1}{2}L\langle i_L^2 \rangle = \tfrac{1}{2}L\frac{k_B T}{L} = \tfrac{1}{2}k_B T \tag{2.13.9}$$

so we could have started at this end and worked back to Eq. (2.13.1).

The fact that an electrical current does consist of individual electrons gives rise to a second form of random fluctuation called shot noise, first identified by the nearly eponymous Schottky (1918). The charges are assumed to pass or arrive quite randomly so that Poissonian statistics apply (e.g. Boas 1966). Then if an average number $\langle n\rangle$ are counted in a given interval of time $t$, the variance $\langle \Delta n^2\rangle$ in this number will be given by $\langle n\rangle$. The average current $\langle I\rangle$ and the variance $\langle i_{NS}^2\rangle$ are given by ($q_e$ is the electronic charge $1.602\times 10^{-19}$ C):

$$\langle I\rangle=\langle n\rangle\frac{q_e}{t} \quad \text{and} \quad \langle i_{NS}^2\rangle=\langle \Delta n^2\rangle\frac{q_e^2}{t^2}=\langle n\rangle\frac{q_e^2}{t^2}$$
$$\text{so} \quad \langle i_{NS}^2\rangle=\langle I\rangle\frac{q_e}{t}=2q_e\langle I\rangle\Delta f \quad \text{amp}^2 \tag{2.13.10}$$

The final step in relating the time interval $t$ to frequency interval $\Delta f$ requires some mathematical consideration (Robinson 1974). The arrival of the individual charges is equivalent to a sharp impulse and a measure of the effect requires integration over a convolution of random pulses with the circuit response function (Section 1.14). The result is in effect the same as saying that the bandwidth associated with a time interval of $t$ is $\Delta f=1/2t$, which is compatible with Eq. (3.6.8). The spectrum here is also white, depending only on frequency interval $\Delta f$ as for Johnson noise. The name 'shot' is thought to derive from the old style of manufacture of lead shot. Molten lead, at the top of a tall shot tower was poured out to free fall and break up into large numbers of spherical shot, and the noise of the collection in a metal container at the base of the tower sounded rather like electrical shot noise when made audible. For $I$ in mA and $\Delta f$ in kHz we have:

$$i_{NS}=5.7\times 10^{-4}(I\Delta f)^{\frac{1}{2}} \quad \mu\text{A r.m.s.} \tag{2.13.11}$$

Equations (2.13.1) and (2.13.10) make it evident why noise units given for devices are in terms such as μV per root Hz. Since noise sources will in general be uncorrelated, it is necessary to add them by taking the root of the sum of the squares of the individual contributions.

There is a further form of noise that arises rather more for reasons of imperfection than from fundamental sources and is found to vary roughly as $1/f^{\alpha}$, where $\alpha$ is of the order of unity. This cannot continue down to zero frequency otherwise we would have infinite noise there. The increase as frequency decreases has led to this being termed pink noise, again with reference to visible light and an increase towards the red end of the spectrum. In this case there is a considerable advantage in operating towards higher frequencies if there is a choice. Typically one may expect that pink noise would become negligible compared with the other noise sources in the frequency range 1–10 kHz. For active devices the effective sources of noise are affected by their gain as a function of frequency which results in a rise at higher frequencies (Gray and Meyer 1977).

The meaning of bandwidth in the context of noise calculations is only obvious if we have a square cut-off, whereas practical responses have a finite rate. A simple *RC* low-pass filter has a transfer function:

$$G(f) = \frac{1}{1 + j\omega RC}$$

$$\text{so} \quad |G(f)^2| = G(f)G(f)^* = \frac{1}{1 + j\omega RC}\frac{1}{1 - j\omega RC} = \frac{1}{1 + (\omega RC)^2} \tag{2.13.12}$$

since we have to determine the integrated value of $v^2$. As $G(0) = 1$, the equivalent noise bandwidth is obtained by integrating over the frequency range from zero to infinity:

$$\mathcal{B}_N = \int_0^\infty \frac{df}{[1 + (2\pi fRC)^2]} = \frac{1}{2\pi RC}\int_0^\infty \frac{dx}{1 + x^2}, \quad \text{where } x \equiv 2\pi fRC \tag{2.13.13}$$

$$= \frac{1}{2\pi RC}[\tan^{-1} x]_0^\infty = \frac{1}{4RC}$$ using Eq. (1.9.1(g)), which has value $\pi/2$ for these limits.

which should be compared with the usual $-3$ dB bandwidth for the *RC* circuit of $1/(2\pi RC) = 1/(6.3RC)$.

## References and additional sources 2.13


Barkhausen H. (1919): Two phenomena discovered with the aid of the new amplifier. *Phys. Zeit.* **20**, 401–403. (In German.)

Bell D. A. (1960): *Electrical Noise*, New York: Van Nostrand. Library of Congress Cat. No. 59-11055.

Bennett W. R. (1960): *Electrical Noise*, New York: McGraw-Hill.

Boas M. L. (1966): *Mathematical Methods in the Physical Sciences*, New York: John Wiley. Library of Congress Cat. No. 66-17646. See p. 712.

Bryant J., Counts L. (1990): Op-amp issues – Noise. *Analog Dialogue* **24**(2), 20–21 and **24**(2), 24–25.

Buckingham M. J., Faulkner E. A. (1974): The theory of inherent noise in p-n junction diodes and bipolar transistors. *Radio Electronic Engng* **44**(3), 125–140.

Faulkner E. A. (1968): The design of low-noise audio-frequency amplifiers. *Radio Electronic Engng* **36**, 17–30.

Faulkner E. A. (1975): The principles of impedance optimization and noise matching. *J. Phys. E Sci. Instrum.* **8**, 533–540.

Gray P. R., Meyer R. G. (1977): *Analysis and Design of Analog Integrated Circuits*, New York: John Wiley. IS2.13. 0-471-01367-6. See Chapter 11, 'Noise in integrated circuits', p. 607.

Gupta M. S. (1977): *Electrical Noise: Fundamentals and Sources*, New York: IEEE Press. Reprints of important papers on noise.

Hamilton T. D. S. (1977): *Handbook of Linear Integrated Electronics for Research*, London: McGraw-Hill. ISBN 0-07-084483-6. Many references to noise papers.
Johnson J. B. (1928): Thermal agitation of electricity in conductors. *Physical Review* **32**, 97–109.
Kittel C. (1967): *Elementary Statistical Physics*, New York: John Wiley, Library of Congress Cat. No. 58-12495.
LlacerJ. (1975): Optimal filtering in the presence of dominant 1/*f* noise. *Nuc. Instrum. Meth.* **130**, 565–570.
Morrison R. (1967): *Grounding and Shielding Techniques in Instrumentation*, New York: John Wiley.
Motchenbacher C. D., Fitchen F. C. (1973): *Low-Noise Electronic Design*, New York: John Wiley.
Netzer Y. (1981): The design of low-noise amplifiers. *Proc. IEEE* **69**, 728–741.
Nyquist H. (1928): Thermal agitation of electrical charges in conductors. *Physical Review* **32**, 110–113.
Radeka V. (1988): Low-noise techniques in detectors. *Ann. Rev. Nuclear and Particle Physics* **38**, 217–277.
Robinson F. N. H. (1962): *Noise in Electrical Circuits*, Oxford: Clarendon Press.
Robinson F. N. H. (1974): *Noise and Fluctuations in Electronic Devices and Circuits*, Oxford: Clarendon Press. ISBN 0-19-859319-8.
Ryan A., Scranton T. (1969): D-C amplifier noise revisited. *Analog Dialogue* **3**(1), 3–10.
Schottky W. (1918): Spontaneous current fluctuations in various conductors. *Ann. der Phys.* **57**, 541–567. Also renowned for his work on metal–semiconductor junctions and now, for example, enshrined in Schottky diodes.
Schottky W. (1926): Small shot effect and flicker effect. *Phys. Rev.* **28**, 75–103.
Smith L., Sheingold D. H. (1969): Noise and operational amplifier circuits. *Analog Dialogue* **3**(1), 1–16.
Soderquist D. (1975): *Minimization of Noise in Operational Amplifier Applications*, Precision Monolithics Application Note AN-15, April.
Usher M. J. (1974): Noise and bandwidth. *J. Phys. E Sci. Instrum.* **7**, 957–961.
Van der Ziel A. (1954): *Noise*, Englewood Cliffs.

# Part 3

# Introduction to circuit mathematics

Facts which at first seem improbable will, even on scant explanation, drop the cloak which has hidden them and stand forth in naked and simple beauty.

Galileo Galilei (1564–1642): Dialogues Concerning Two New Sciences, 1638

The basis of our ability to analyse circuits lies in the various circuit laws arising from the requirements of the underlying physics, together with a number of higher level techniques which we can derive from these. Of course there is no dividing line between these, usually mathematical, techniques and any level of more complex systems, but there are a number of more generally useful ideas that practitioners have over time come to agree as necessary for all to know, or at least to know of. We will all have some variation as to what is desirable and what is essential, but I can only write about those that have been important to me. Books which cover 'all' techniques generally tend to be very large and require a considerable number of authors, for example Chen (1995) with 2861 pages and 114 authors! But if you can manage to lift it you will find many things there.

The present book is only a companion whose aim is to provide refreshment or a reminder, or an encouragement to experiment, so the various topics will be approached in this vein rather than going through development from first principles. References are provided which can direct you to further information should you so require.

Division of material between the parts is always to some extent arbitrary or idiosyncratic, so do explore other sections for techniques, making use of the index as an additional guide. Analysis is provided in some detail as far as space allows, together with frequent encouragement to test the outcome against SPICE simulation. Some topics have been examined in greater detail as examples of how SPICE can be used and the insight it can provide. Again the caveat that it can only analyse the circuit you provide using the models it has available so that you still have to try the circuit in the flesh, but there may hopefully be a little less blood about. Having said that I have found that with the inclusion of some realistic allowance for the strays of reality the prognosis is generally very good.

## References

Chen, Wai-Kai (Ed.) (1995): *The Circuits and Filters Handbook*, Boca Raton: CRC Press and IEEE Press.

# 3.1 Circuit laws

The law is the true embodiment
Of everything that's excellent
It has no kind of fault nor flaw
And I my lords embody the law.

Lord Chancellor's song, from *Iolanthe* by W. S. Gilbert and A. Sullivan

One of the main circuit laws, Kirchhoff's law(s), is discussed in Section 2.3 since it refers particularly to the relation between the physics of the system and the circuit view. We now consider several other laws or techniques that assist in analysing circuits.

Ohm's law is probably the earliest 'law', though in fact it is essentially an experimental observation. George Simon Ohm found as a result of experiments on a number of conductors that the relationship of the voltage applied to the current that flowed was linear, and the constant of proportionality was called the resistance:

$$\frac{\text{Voltage}}{\text{Current}} = \text{Resistance} \tag{3.1.1}$$

The reason for the wide range of currents over which the law holds is that the proportion of the free electrons in the conductor that are in effect involved in the sort of currents we normally deal with is tiny (Section. 2.1). This linearity is of course not always so, as for example in a p–n junction diode.

Ohm only had d.c. voltages available and nowadays we work with systems in which the changes occur in very short times indeed. If we try to extend the basic law to cover alternating voltages and currents then the laws of physics indicate that we have to consider additional 'resistances' that arise due to the rate of change, i.e. the consequences of capacity and inductance. Since these also consist of similar linear conductors then we may expect that the relationships would similarly be linear. The form of the 'resistance' in these devices differs from the simple resistor in that they possess storage properties. The effects are examined in Section 3.2.

There are several circuit theorems that can often be used to allow us to examine circuits more easily. In a linear system with more than one source, the voltage or

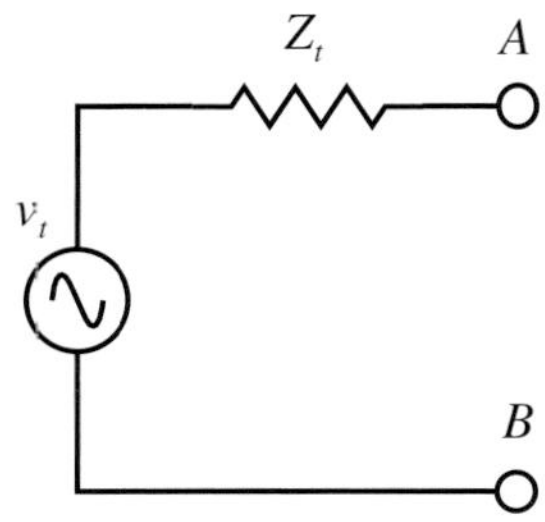


**Fig. 3.1.1** Thévenin equivalent circuit.

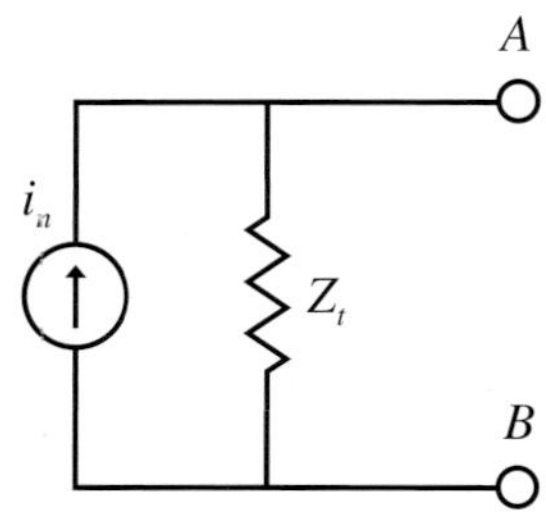


**Fig. 3.1.2** Norton equivalent circuit.

current at any point will be the sum arising from each of the sources separately with the other sources inactive – voltage sources shorted and current sources open circuit. This proposition is known as the superposition theorem, and is only applicable to linear systems. It is examined below.

If we have a complex system it is often helpful to divide it up into several parts to help understanding. If we then wish to consider one part we need to represent the part driving this as simply as possible. Thévenin's theorem (Brittain 1990) provides this facility, and may be written:

Any portion of a linear system between two terminals $A$ and $B$ can be replaced with an equivalent circuit consisting of a voltage generator $v_t$ in series with an impedance $Z_t$. $v_t$ is given by the open-circuit voltage at the terminals $AB$ and $Z_t$ is given by the impedance at the terminals with all the internal independent sources deactivated.

A simple example is a signal generator which from the external point of view may be considered as a voltage source in series with an output resistance, Fig. 3.1.1.

A corollary of Thévenin's theorem is that due to Norton. Instead of using an equivalent circuit consisting of a voltage source and a series resistor one may equally well use a current source in parallel with a resistor. The equivalent is shown in Fig. 3.1.2.

The value of the current generator $i_n$ is the current that would flow if $A$ and $B$ were short circuited. From Thévenin's theorem this current is $i_n = v_t/Z_t$. The par-

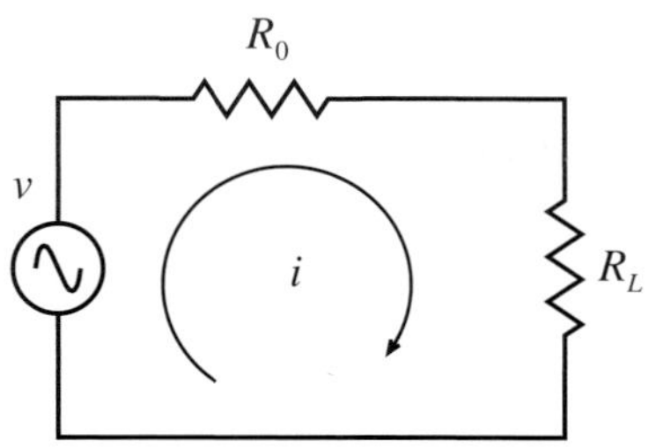


**Fig. 3.1.3** Power transfer circuit.

allel Norton impedance is $Z_t$ since this leads to an open-circuit voltage at $AB$ of $v = i_n Z_t = v_t$.

In interconnecting systems it is often important to obtain the maximum transfer of power from one to the other, e.g. between a power amplifier and a loudspeaker. Consider first a source that can be represented by a Thévenin equivalent voltage $v$ and output resistance $R_0$ connected to a load resistor $R_L$ (Fig. 3.1.3).

The power $P_L$ dissipated in the load will be given by:

$$P_L = i^2 R_L = \frac{v^2 R_L}{(R_0 + R_L)}, \quad \text{since} \quad i = \frac{v}{R_0 + R_L}$$

$$\text{and} \quad \frac{\partial P_L}{\partial R_L} = \frac{(R_0 + R_L)^2 v^2 - 2R_L v^2 (R_0 + R_L)}{(R_0 + R_L)^4} = 0, \quad \text{for maximum} \tag{3.1.2}$$

$$\text{so} \quad (R_0 + R_L)^2 - 2R_L(R_0 + R_L) = 0$$

$$\text{giving} \quad R_0 = R_L \quad \text{and} \quad P_{L\max} = \frac{v^2}{4R_L}$$

Thus for maximum power transfer the load resistance must equal the source resistance and half the power will be dissipated in each. When the load impedance is complex then the conditions must be extended. The power dissipation in the load is now (Section 3.2):

$$P_L = \tfrac{1}{2} ii^* \mathcal{Re}(Z), \quad \text{with} \quad i = \frac{v}{(R_0 + R_L) + j(X_0 + X_L)} \equiv \frac{v}{R + jX} = \frac{v^2 R_L}{2(R^2 + X^2)}$$

$$\text{then} \quad \frac{\partial P_L}{\partial X} = \frac{v^2 R_L (0 - 2X)}{2(R^2 + X^2)^2} = 0, \quad \text{for maximum} \tag{3.1.3}$$

$$\text{so} \quad X = 0, \quad \text{i.e.} \quad X_0 + X_L = 0 \quad \text{or} \quad X_L = -X_0$$

The procedure is repeated for $R$ as in Eq. (3.1.2) to find the same result. Thus the load to match an output impedance $(R + jX)$ is $(R - jX)$, the complex conjugate. Matching is an important concept and will be considered further in Section 3.17. These conditions for maximum energy transfer can be demonstrated with

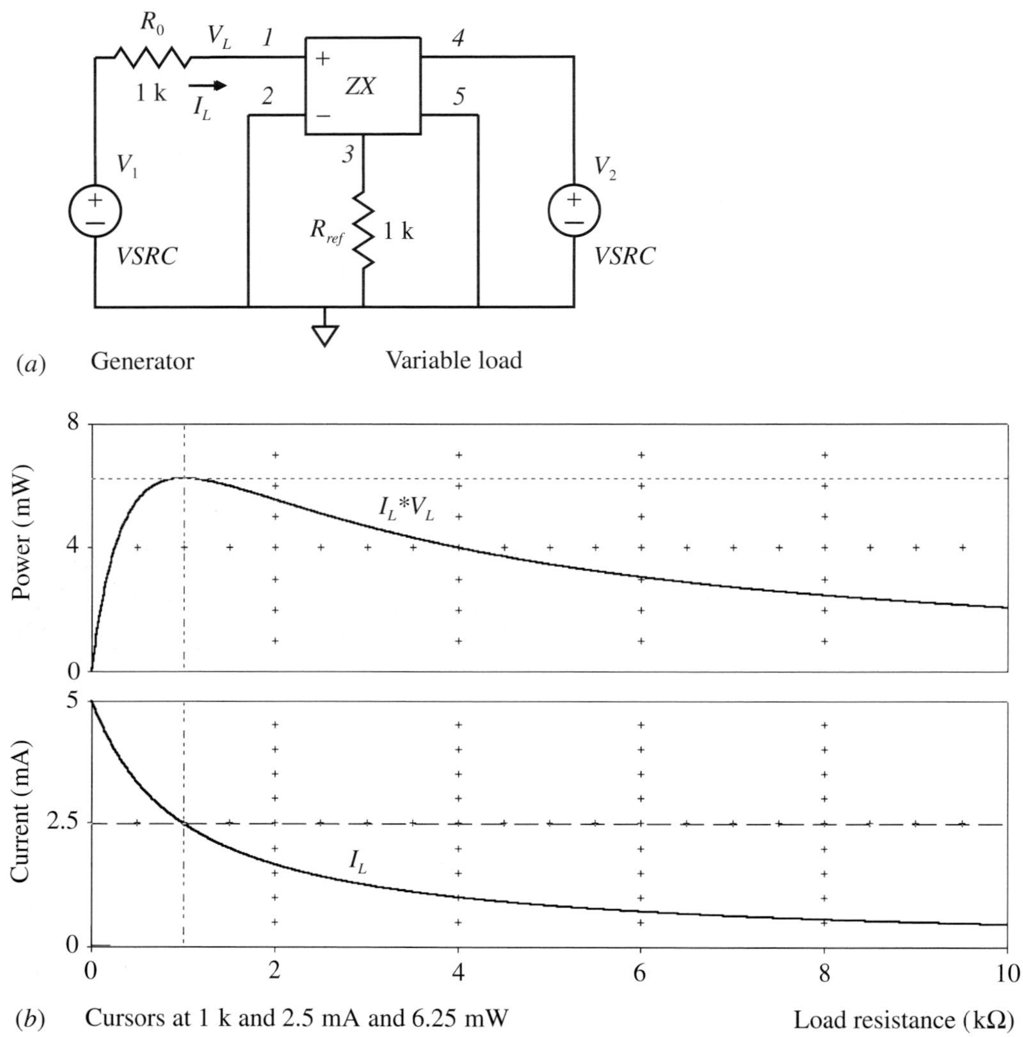


**Fig. 3.1.4** (*a*) Simulation circuit for variation of power transfer with load. (*b*) Simulation response.

PSpice. To vary the load impedance we may use the *ZX* device (Tuinenga 1988) as shown in Fig. 3.1.4.

The control voltage $V_2$ is swept over the range 0–10 V to vary the effective value of resistance between terminals *1* and *2* from zero to 10 kΩ. If you then plot the load power *I*(*R0*)* *V*(*VL*) you will find a maximum at $V_2 = 1$ V, i.e. a load resistance of 1 kΩ, equal to the source resistance. You can if you wish change the *x*-axis to *V*(*VL*)/*I*(*RL*) to have it in ohms. For a reactive source impedance the circuit is shown in Fig. 3.1.5, where we have assumed a capacitive source reactance.

The load and source resistances are made equal and we vary the conjugate reac-

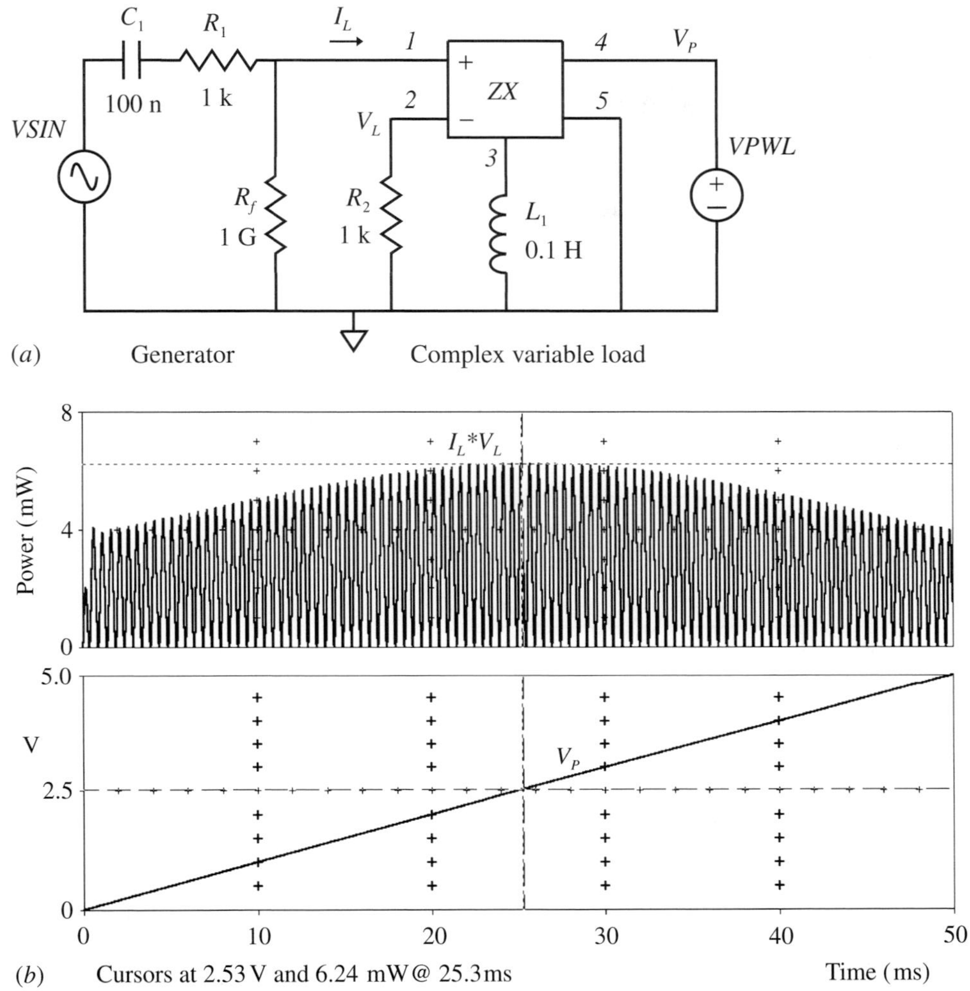


**Fig. 3.1.5** (*a*) Simulation of power transfer for reactive impedances. (*b*) Simulation response.

tance $L$. The source frequency is 1 kHz and *VPWL* is set to vary over the range 0–5 V in 50 ms, i.e. a slow change relative to the sine wave period. If you plot *I(R2)*I(R2)*1000* or *I(R2)*V(VL)* you will get a response (at twice the frequency) with a maximum at 2.53 V, i.e. $L = 0.253$ H. This can be confirmed from the matching requirement in this case of $\omega L = 1/\omega C$.

## Superposition

The superposition theorem is sometimes useful for analysing linear circuits. The theorem states that in a linear bilateral network the current at any point is the

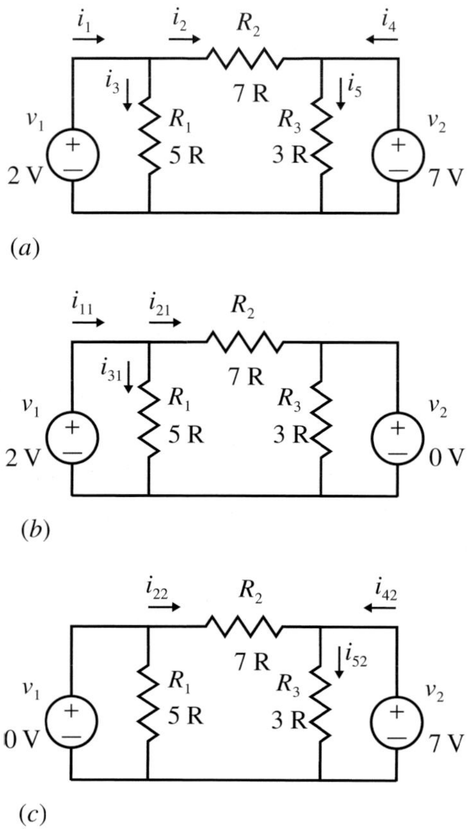


**Fig. 3.1.6** (*a*) Circuit to demonstrate superposition. (*b*) Circuit with $v_2=0$. (*c*) Circuit with $v_1=0$.

algebraic sum of the currents due to each generator separately, with all other independent generators set to their internal impedance. For a voltage source this will be zero and for a current source it will be infinity. We use this technique in Section 3.8 to construct equivalent circuits. The principle may be demonstrated by considering the simple circuit shown in Fig. 3.1.6(*a*) with two voltage generators.

In the first instance we make $v_2=0$ as shown in (*b*) and represent the corresponding currents produced by $v_1$ by appending a 1 to the subscript. The currents are then:

$$i_{31}=\frac{2}{5}=0.400\ \text{A};\quad i_{21}=\frac{2}{7}=0.286\ \text{A};\quad i_{11}=i_{21}+i_{31}=0.686\ \text{A} \tag{3.1.4}$$

With $v_1=0$ and $v_2=7$ V, as shown in (*c*) we have:

$$i_{52}=\frac{7}{3}=2.333\ \text{A};\quad i_{22}=\frac{7}{7}=1.000\ \text{A};\quad i_{42}=i_{22}+i_{52}=3.333\ \text{A} \tag{3.1.5}$$

so, for example, the combined currents with both voltage sources active should be (making due allowance for the senses of the currents):

$$i_2 = i_{21} - i_{22} = 0.286 - 1.000 = -0.714 \text{ A}$$
$$i_1 = i_{11} - i_{22} = 0.686 - 1.000 = -0.314 \text{ A} \qquad (3.1.6)$$
$$i_4 = i_{42} - i_{21} = 3.333 - 0.286 = 3.047 \text{ A}$$

Instead of solving the equations for the case of both sources active to see if the superposition is correct you can set up the circuit in SPICE and let it do the work. Run the *BIAS POINT DETAIL* under *ANALYSIS* and use the *ENABLE BIAS CURRENT DISPLAY* buttons to show the currents. The currents revealed will agree with those of Eq. (3.1.6) and if you click on each current flag a small arrow will show on the component to indicate the sense of the current. Currents from the voltage sources normally flow from the +ve terminal through the circuit and back to the −ve terminal, i.e. it flows internally from − to +. So if the current is negative in Eq. (3.1.6), it should show on the display as flowing in the opposite direction.

## SPICE simulation circuits

| | |
|---|---|
| Fig. 3.1.4(*b*) | Pwrtrfr 1.SCH |
| Fig. 3.1.5(*b*) | Pwrtrfr 2.SCH |
| Fig. 3.1.6(*a*) | Suprpsn1.SCH |

## References and additional sources 3.1

Brittain J. E. (1990): Thévenin's theorem. *IEEE Spectrum*, **27** March, 42.

'Cathode Ray' (1953): Ohm's law. Do we know what it really means? *Wireless World* August, 383–386.

Faulkner E. A. (1966): *Principles of Linear Circuits*, London, Chapman and Hall.

Irwin J. D. (1996): *Basic Engineering Circuit Analysis*, Englewood Cliffs: Prentice Hall. ISBN 0-13-397274-7.

Mancini R. (1999): *Understanding Basic Analog – Circuit Equations Application Report*, Texas Instruments Literature Number SLOA025, July.

Tse C. K. (1998): *Linear Circuit Analysis*, Harlow: Addison Wesley. ISBN 0-201-34296-0. (SPICE based.)

Tuinenga P. W. (1988): *SPICE: A Guide to Circuit Simulation and Analysis Using SPICE,* Prentice Hall. 3rd Edition 1995. ISBN 0-13-158775-7.

Walton A. K. (1987): *Network Analysis and Practice*, Cambridge: Cambridge University Press. ISBN 0-521-31903-X. See Section 3.6.

# 3.2 A.C. theory

January 1889: Columbia College, New York, has decided to have a special course in electrical science, and not a moment too soon, for this has long been seen as a department by itself, and, while allied to other branches of natural philosophy, requires, at least from those who would adopt it as a profession, an undivided attention. In a practical age like this, the most valuable college instruction would seem to be the one that best resembles what its recipients are expected to accomplish outside of it.

Quote from the past, reprinted in *Scientific American* **260** (1), 6, 1989

The analysis of a.c. circuits is the foundation of circuit analysis. The nature of the sine function, at least in linear circuits, is that whatever you do to it the shape of the function is preserved. Adding, subtracting, multiplying, dividing, raising to powers, differentiating and integrating all produce consequences that are still sinusoidal. For more complex waveforms we can operate on the basis of the Fourier representation and preserve this property, though this will result in considerable mathematical effort. Trigonometrical functions are not easy to handle and the equivalents in terms of exponential notation have become predominant. The revolution in circuit analysis in terms of complex quantities and the introduction of the $j\omega$ notation is commonly ascribed to Charles Steinmetz (1897), possibly more commonly remembered for his law of hysteresis (Steinmetz 1892), and we all owe a great debt to him. As Belevitch (1962) observes, 'Steinmetz vulgarized the use of complex quantities'.

Sinusoidally varying quantities may be expressed in trigonometrical or complex exponential form:

$$v = v_0 \cos(\omega t) + jv_0 \sin(\omega t) \equiv v_0 \exp(j\omega t) \tag{3.2.1}$$

The corresponding current is generally not in phase with the voltage, and if the phase difference is $\theta$ then the current $i$ may be written:

$$i = i_0 \cos(\omega t - \theta) + ji_0 \sin(\omega t - \theta) \equiv i_0 \exp(j\omega t - \theta) \tag{3.2.2}$$

and the impedance $Z$ of the circuit is now given by:

$$Z = \frac{v}{i} = \frac{v_0 \exp(j\omega t)}{i_0 \exp(j\omega t - \theta)} = \frac{v_0}{i_0}\exp(j\theta) = \frac{v_0}{i_0}(\cos\theta + j\sin\theta) \tag{3.2.3}$$

and we can derive the modulus and the argument of $Z$ as:

$$\text{Modulus } Z \equiv |Z| = \left[\frac{v_0^2}{i_0^2}(\cos^2\theta + \sin^2\theta)\right]^{\frac{1}{2}} = \frac{v_0}{i_0}, \quad \text{using Eq. (1.1.2(h))}$$

$$\text{Argument } Z \equiv \phi = \tan^{-1}\left(\frac{\sin\theta}{\cos\theta}\right) = \tan^{-1}(\tan\theta) = \theta \tag{3.2.4}$$

The real part of $Z$ is called the resistance $R$ and the imaginary part the reactance $X$:

$$Z = R + jX, \quad \text{where} \quad R = \frac{v_0\cos\theta}{i_0} \quad \text{and} \quad X = \frac{v_0\sin\theta}{i_0} \tag{3.2.5}$$

The magnitude of the impedance $|Z|$ is given by:

$$|Z| = [(R + jX)(R - jX)]^{\frac{1}{2}} = (R^2 + X^2)^{\frac{1}{2}} \tag{3.2.6}$$

and the phase difference between $v$ and $i$ by:

$$\phi = \tan^{-1}\left(\frac{X}{R}\right) \tag{3.2.7}$$

Measurements for a.c. circuits are commonly given in terms of root-mean-square (r.m.s.) values since the ratios and powers vary over a wide range during a cycle. Taking voltage as an example, the magnitude is squared to counteract the change of sign, the mean over a cycle is determined and then the square-root is taken to make up for the squaring in the first place. Thus (angle brackets $\langle\rangle$ indicate an average, which is taken over the period $T = 2\pi/\omega$):

$$v_{\text{r.m.s.}} = \langle v^2\rangle^{\frac{1}{2}} = \left[\frac{1}{T}\int_0^T v_0^2\cos^2(\omega t)dt\right]^{\frac{1}{2}}$$

$$= \left\{\frac{v_0^2}{T}\int_0^T \tfrac{1}{2}[1 + \cos(2\omega t)]dt\right\}, \quad \text{using Eq. (1.1.2(j))} \tag{3.2.8}$$

$$= \left\{\frac{v_0^2}{2T}[t + \sin(2\omega t)]_0^T\right\}^{\frac{1}{2}} = \left(\frac{v_0^2}{2}\right)^{\frac{1}{2}} = \frac{v_0}{\sqrt{2}} = 0.707 v_0$$

with an identical relation for $i_{\text{r.m.s}}$. It should be emphasized that this is only applicable to a simple sine wave; if significant harmonics are present, or for other waveforms, the numerical value will be different. For example, the crest factor, or ratio of peak to r.m.s., which is as seen above $\sqrt{2} = 1.414$ for sine waves, is 1 for a symmetrical square wave and 1.73 for a triangle wave (see e.g. Kitchen and Counts 1986). The power dissipated in the circuit varies through the cycle so we need to

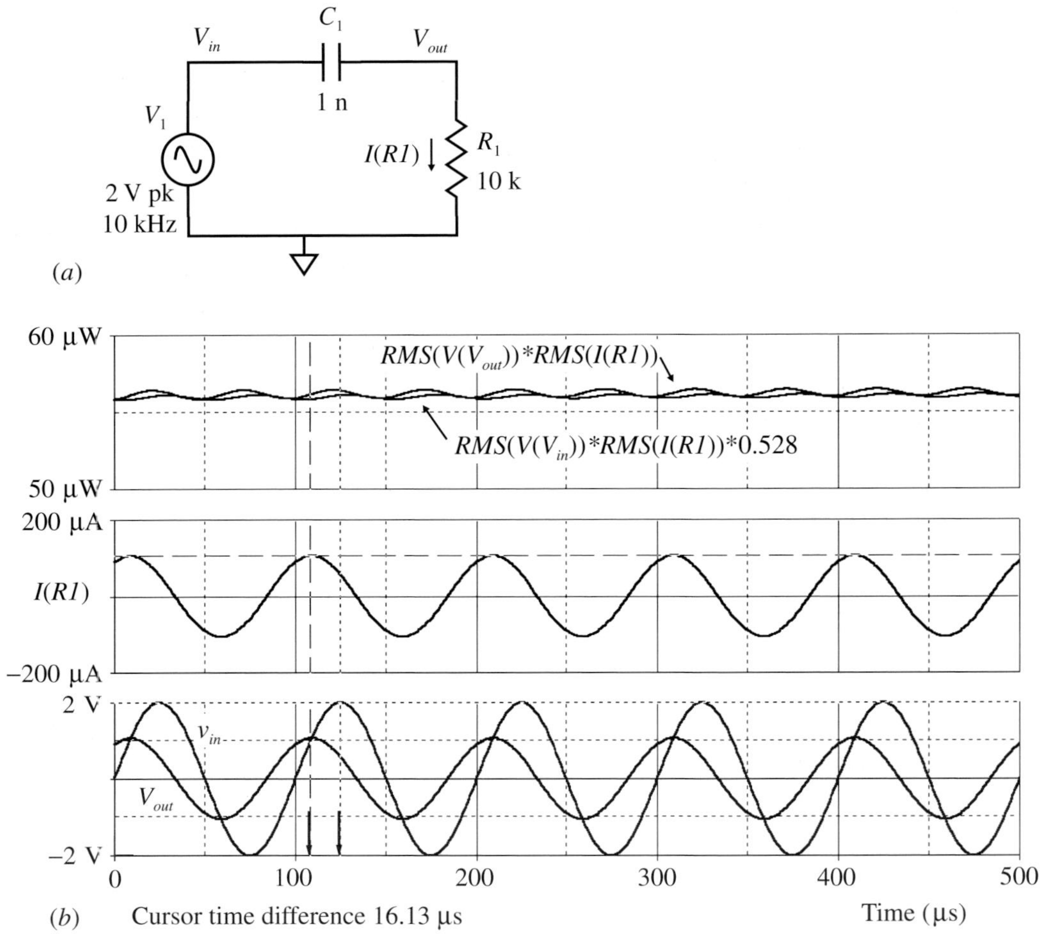


**Fig. 3.2.1** (*a*) Simple *CR* circuit. (*b*) Simulation traces.

find the average. Assuming a phase difference $\phi$ between $v$ and $i$, the instantaneous power $P_i$ is given by:

$$\begin{aligned} P_i &= v_0 \cos(\omega t) \times i_0 \cos(\omega t - \phi) \\ &= \tfrac{1}{2} v_0 i_0 [\cos(2\omega t - \phi) + \cos \phi], \quad \text{using Eq. (1.1.2(g))} \end{aligned} \tag{3.2.9}$$

The first term is oscillatory and over a cycle will average to zero, so the average power dissipation is given by:

$$P = \tfrac{1}{2} v_0 i_0 \cos \phi = v_{r.m.s.} i_{r.m.s.} \cos \phi \tag{3.2.10}$$

and $\cos \phi$ is known as the power factor. These relations may be illustrated by considering the simple circuit shown in Fig. 3.2.1. The simulations shown in (*b*) give the voltages, currents and phase shifts. *PROBE* provides an *RMS* operator and the two results illustrated show the power dissipation given by Eq. (3.2.10). The time delay between $V_{in}$ and $V_{out}$ is 16.13 μs, and since the period in this case is $T=100$ μs this gives a phase shift of $\phi=(16.13\times 360/100)=58°$ and hence $\cos\phi=0.528$. The total power dissipation is seen to be equal to that in $R$ alone, as it should be. The residual ripple in the traces is a consequence of the continuous averaging process in *PROBE*.

For an ideal resistor $R$ there is no dependence on frequency so the relationship between voltage and current is simple and the impedance is just $R$. The effective impedance of a capacitor $C$ for sinusoidal waveforms may be derived from the relationship of current to voltage given by:

$$i=C\frac{dv}{dt} \qquad \text{so for} \qquad v=v_0\cos(\omega t)$$

then $\quad i=-\omega C v_0\sin(\omega t)=-i_0\sin(\omega t), \quad$ where $\quad i_0=\omega C v_0$

Alternatively for

$$v=\mathcal{R}e\, v_0\, e^{-j\omega t}, \qquad \text{then} \qquad i=\mathcal{R}e\, j\omega C v_0\, e^{j\omega t}=\mathcal{R}e\, j i_0\, e^{j\omega t}$$

and the impedence

$$Z_C=\frac{v}{i}=\frac{v_0}{j i_0}=\frac{1}{j\omega C}=\frac{-j}{\omega C} \quad \text{ohm} \tag{3.2.11}$$

For an inductor we have in a similar manner:

$$v=L\frac{di}{dt} \qquad \text{so for} \qquad i=i_0\cos(\omega t)$$

then $\quad v=-\omega L i_0\sin(\omega t)=-v_0\sin(\omega t)$, where $\quad v_0=\omega L i_0$

Alternatively for

$$i=\mathcal{R}e\, i_0\, e^{j\omega t}, \quad \text{then} \quad v=\mathcal{R}e\, j\omega L i_0\, e^{j\omega t}=\mathcal{R}e\, j v_0\, e^{j\omega t}$$

and the impedence

$$Z_L=\frac{v}{i}=\frac{j v_0}{i_0}=j\omega L \quad \text{ohm} \tag{3.2.12}$$

and we can use these impedances to analyse circuits for sinusoidal excitation. The presence of imaginary as well as real components will mean that currents are generally not in phase with applied voltages.

In some circumstances, for example in adding up parallel impedances, it is more convenient to work with reciprocals where the admittance $Y$ is written:

$$Y=1/Z=G+jB \tag{3.2.13}$$

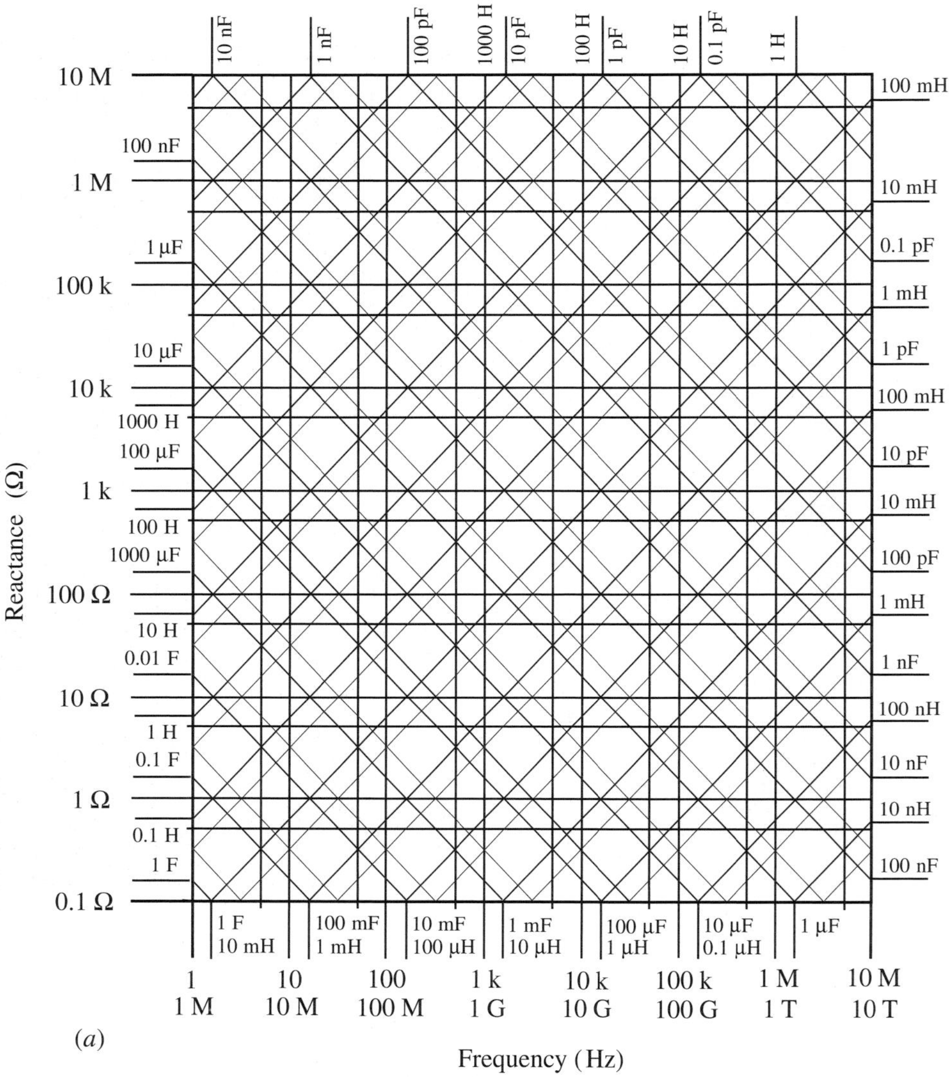


**Fig. 3.2.2** (*a*) Chart to determine reactance as a function of frequency. Always use corresponding scales for *C*, *L* or frequency.

where $G$ = conductance (in siemen) and $B$ = susceptance (in siemen). Note that the unit of admittance was known as mho (reciprocal of ohm) in older references. Conversions between admittance and impedance may be derived as follows. From (3.2.13) and (3.2.5):

$$G + jB = \frac{1}{R + jX} = \frac{R - jX}{(R + jX)\,(R - jX)} = \frac{R - jX}{R^2 + X^2}$$

$$\text{so} \quad G = \frac{R}{R^2 + X^2} \quad \text{and} \quad B = \frac{-X}{R^2 + X^2} \tag{3.2.14}$$

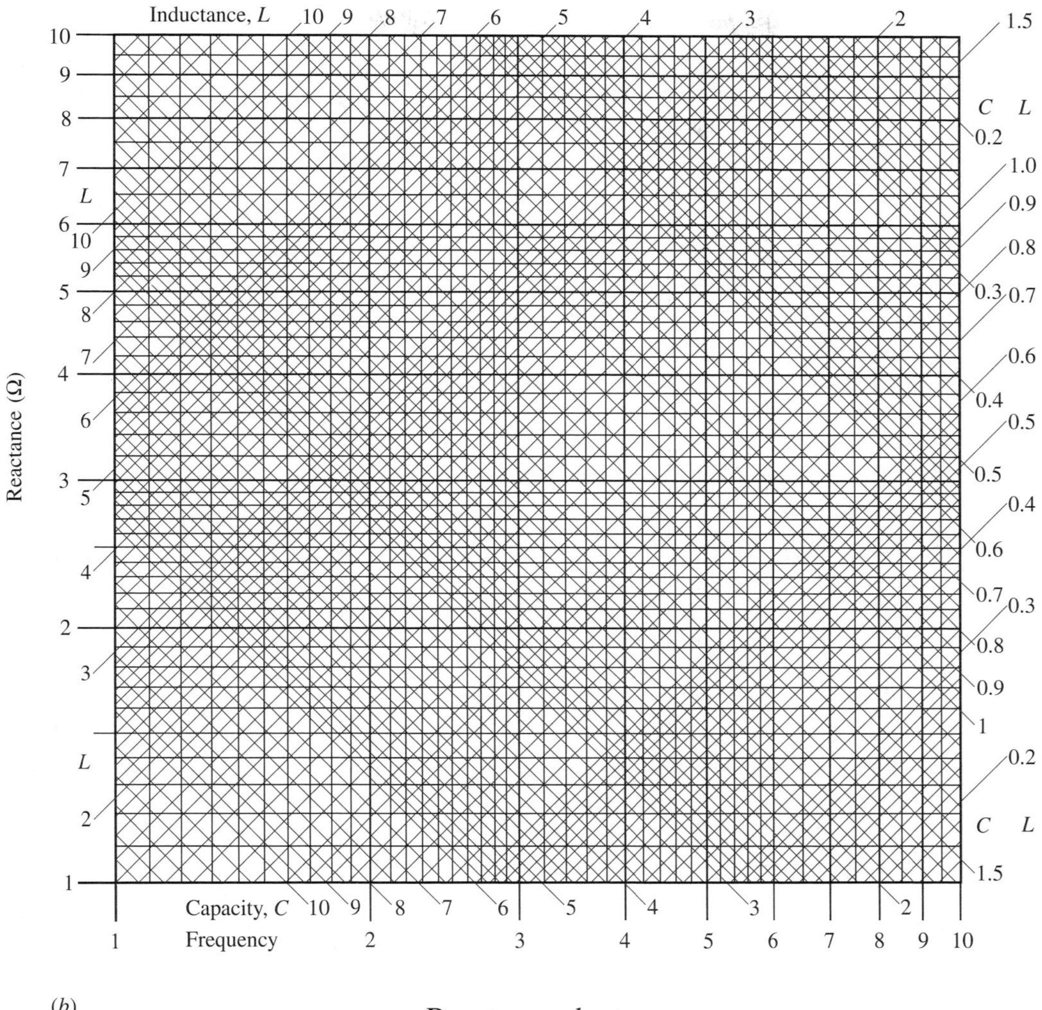


**Fig. 3.2.2** (*b*) One decade frequency–reactance chart for greater resolution.

Similarly $R = \dfrac{G}{G^2 + B^2}$ and $X = \dfrac{-B}{G^2 + B^2}$

To enable quick estimates of the reactance of capacitors or inductors reference may be made to Fig. 3.2.2. The approximate value is obtained from part (*a*) and then an improved value from (*b*). The charts may also be used to determine the resonant frequency of *LC* circuits.

## SPICE simulation circuits

| | |
|---|---|
| Fig. 3.2.1(*b*) | Actheory.SCH |

## References and additional sources 3.2


Belevitch V. (1962): Summary of the history of circuit theory. *Proc. IRE* **50**, 848–855. This is one of the papers in a wide ranging set of reviews on communications and electronics 1912–1962, covering many areas and filling pages 657–1420.

Carter G. W., Richardson A. (1972): *Techniques of Circuit Analysis*, Cambridge: Cambridge University Press.

Clark (Bob), Fazio M., Scott D. (1991): *RMS-to-DC Converters Ease Measurement Tasks*, Analog Devices Application Note.

Dorf R. C., Svoboda J. A. (1998): *Introduction to Electric Circuits*, 4th Edn, New York: John Wiley. ISBN 0-471-19246-5.

Faulkner E. A. (1966): *Principles of Linear Circuits*, London: Chapman and Hall.

Johnson D. E., Johnson J. R., Hilburn J. L. (1989): *Electric Circuit Analysis*, New York: Prentice Hall, and 1992. ISBN 0-13-249509-0.

Kitchen C., Counts L. (1986): *RMS-to-DC Conversion Application Guide*, 2nd Edn, Analog Devices.

Murphy M., Smith J. (1990): Extend rms-to-dc converter's range. *Electronic Design* 12 July. See also 5 September 2000, pp. 177–178, and update by Gilbert B., Whitlow D. pp. 178, 180.

Steinmetz C. P. (1892): On the law of hysteresis. *Amer. Inst. Elec. Engnrs. Trans.* **9**, 3–64. See also *Proc. IEEE* **72**(2), 196-221, 1982.

Steinmetz C. P. (1897): *Theory and Calculation of Alternating Current Phenomena*, New York: Electrical World.

Walton A. K. (1987): *Network Analysis and Practice*, Cambridge: Cambridge University Press. ISBN 0-521-31903-X.

# 3.3 Phasors

No man but a blockhead writes, except for money.

Dr Samuel Johnson

The picture of rotating phasors is very useful for representing phase shifts in circuits by means of phasor diagrams. The diagram represents an instantaneous picture of the magnitude and relative phase of the various voltages and currents. By tradition, increasing time is represented by rotation in a counterclockwise direction, i.e. of increasing angle. The initial difficulty in drawing a phasor diagram is to know which phasor to start with – the wrong choice leads you into difficulty. Take as an example a simple *RC* low-pass filter as shown in Fig. 3.3.1(*a*).

The quantity that is common to *R* and *C* is the current *i*, so we start with this since we know the relationship between this and the voltages $v_R$ and $v_C$. Thus $v_R$ must be in phase with *i* and $v_C$ must lag by 90°, so we can draw the two phasors as shown. The vector sum of $v_R$ and $v_C$ must equal $v_{in}$ and since $v_{out} = v_C$ we see that $v_{out}$ lags $v_{in}$ by the angle $\theta$. If you had started with the phasor $v_{in}$ you would immediately be stuck as you know nothing about the phase relationships between this and any of the other phasors.

If you have a parallel *RC* circuit (Fig. 3.3.2(*a*)) the common factor is now $v_{in}$ so we start with this. You know the relation of the currents $i_R$ and $i_C$ to $v_{in}$ so these can be drawn as shown. The sum of $i_R$ and $i_C$ gives $i_{in}$ so we see that $i_{in}$ leads $v_{in}$ by the angle $\phi$.

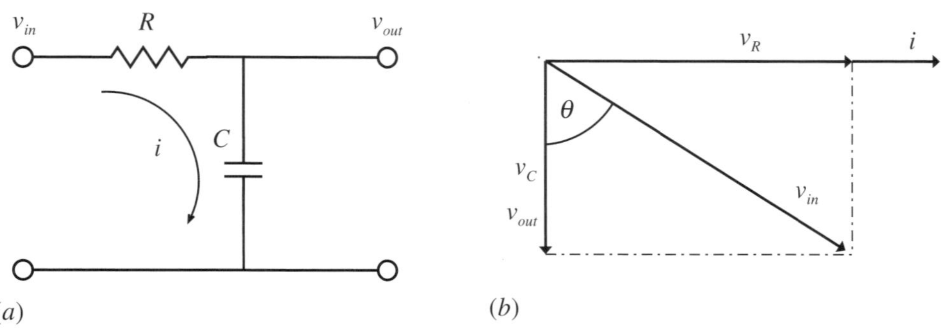


**Fig. 3.3.1** (*a*) Series *RC* circuit. (*b*) Phasor diagram.

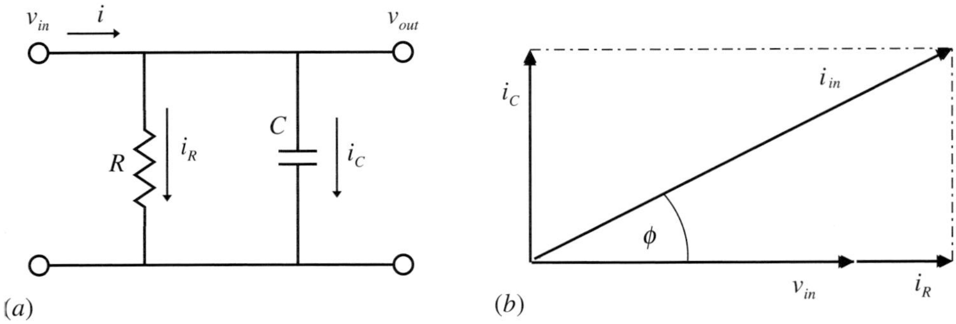


**Fig. 3.3.2** (*a*) Parallel *RC* circuit. (*b*) Phasor diagram.

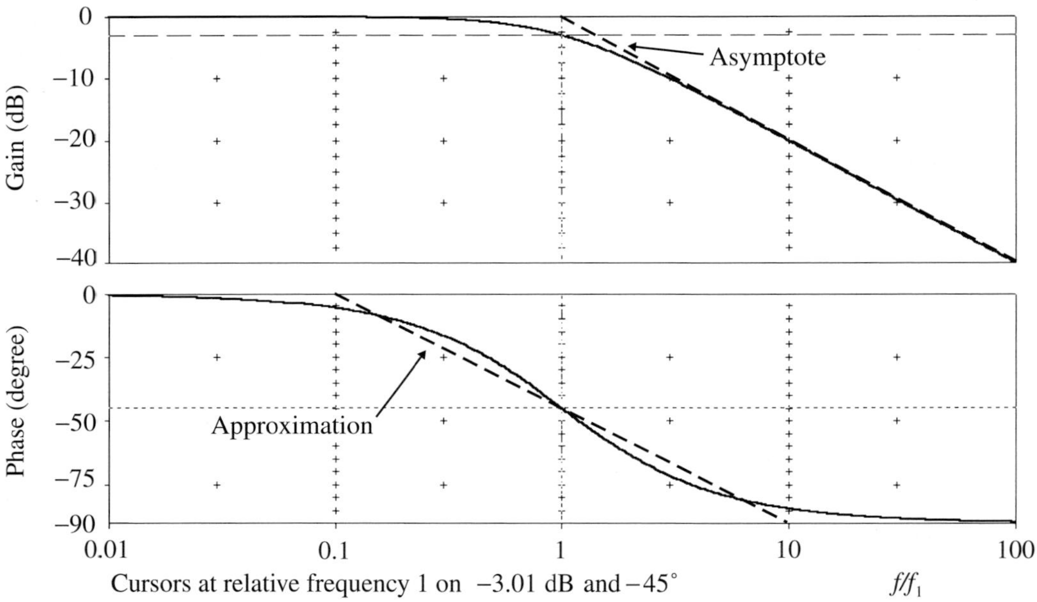


**Fig. 3.3.3** Gain and phase for the series *RC* circuit as a function of relative frequency $f_1 = 1/2\pi RC$.

For the series *RC* circuit of Fig. 3.3.1(*a*) the gain *G* is given by:

$$G = \frac{v_{out}}{v_{in}} = \frac{1/j\omega C}{R + 1/j\omega C} = \frac{1}{1 + j\omega RC} = \frac{1}{1 + j\omega/\omega_1} \qquad (3.3.1)$$

where $\omega_1 = 1/RC$ (you will recall that the unit of $RC$ is seconds and of $\omega$ is $s^{-1}$). The *magnitude* of the gain is:

$$|G| = [1 - (\omega/\omega_1)^2]^{-\frac{1}{2}} \qquad (3.3.2)$$

and the phase shift:

$$\phi = \tan^{-1}(-\omega/\omega_1) = -\tan^{-1}(\omega/\omega_1) \tag{3.3.3}$$

At the corner frequency $\omega_1$ we have:

$$|G| = (1+1)^{-\frac{1}{2}} = 1/(2)^{\frac{1}{2}} = 0.707 \quad \text{or} \quad |G| = 20\log(0.707) = -3.01 \text{ dB}$$
$$\text{and} \quad \phi = -\tan^{-1}(1) = -45° \tag{3.3.4}$$

and taking the asymptotic slope using two frequencies a factor 2 (an octave) or a factor 10 (a decade) apart we have:

$$\begin{aligned} |G_{2\omega}| - |G_{\omega}| &= -20[\log(2\omega/\omega_1) - \log(\omega/\omega_1)] \\ &= -20\log[(2\omega/\omega_1)(\omega_1/\omega)] \\ &= -6.021 \text{ dB/octave} \quad \text{or} \quad -20 \text{ dB/decade} \end{aligned} \tag{3.3.5}$$

A plot of $|G|$ and $\phi$ versus frequency on appropriate scales gives the graphs shown in Fig. 3.3.3 and this form is known as a Bode diagram. The straight line approximations shown are commonly used and are quite accurate enough for most uses. For gain the asymptotic tangents at zero and at high frequency intersect at the corner ($-3$ dB) frequency, and in the phase diagram the maximum error is only 6°, which occurs at frequencies of $0.1f_c$ and $10f_c$, i.e. the phase change is effectively confined to the two decades around the corner frequency.

The phase advance network shown in Fig. 3.3.4 is useful in adjusting the relation of phase to gain as a function of frequency.

A phase retard network is shown in Fig. 3.3.5(*a*).

An instructive example of a phasor diagram is the consideration of the phase shifts in a feedback circuit such as the differentiator as shown in Fig. 5.6.3 (see p. 439). Assuming an input signal $v_P$ at $P$, then at low frequency the output signal will be represented by say $v_{out}$ (Fig. 3.3.6), lagging by 180° due to the inversion in the amplifier. For the defined direction of rotation as shown, lags will move the phasors clockwise.

At higher frequency there will be additional phase lag $\phi$ due to the internal delays in the amplifier to give say an output $v_{oH}$. Now we add the lag and attenuation arising from the feedback components $R_f$ and $C_i$ to give say a phasor $v_Q$ as shown, which now has a component in-phase with the original input $v_P$. All systems will of course reach this sort of position eventually, the critical consideration being whether the gain around the loop is $\geq 1$ with a phase lag of 360°.

## SPICE simulation circuits

| | |
|---|---|
| Fig. 3.3.3 | Pha-ret.SCH |
| Fig. 3.3.4(*b*) | Pha-adv.SCH |
| Fig. 3.3.5(*b*) | Pha-ret.SCH |

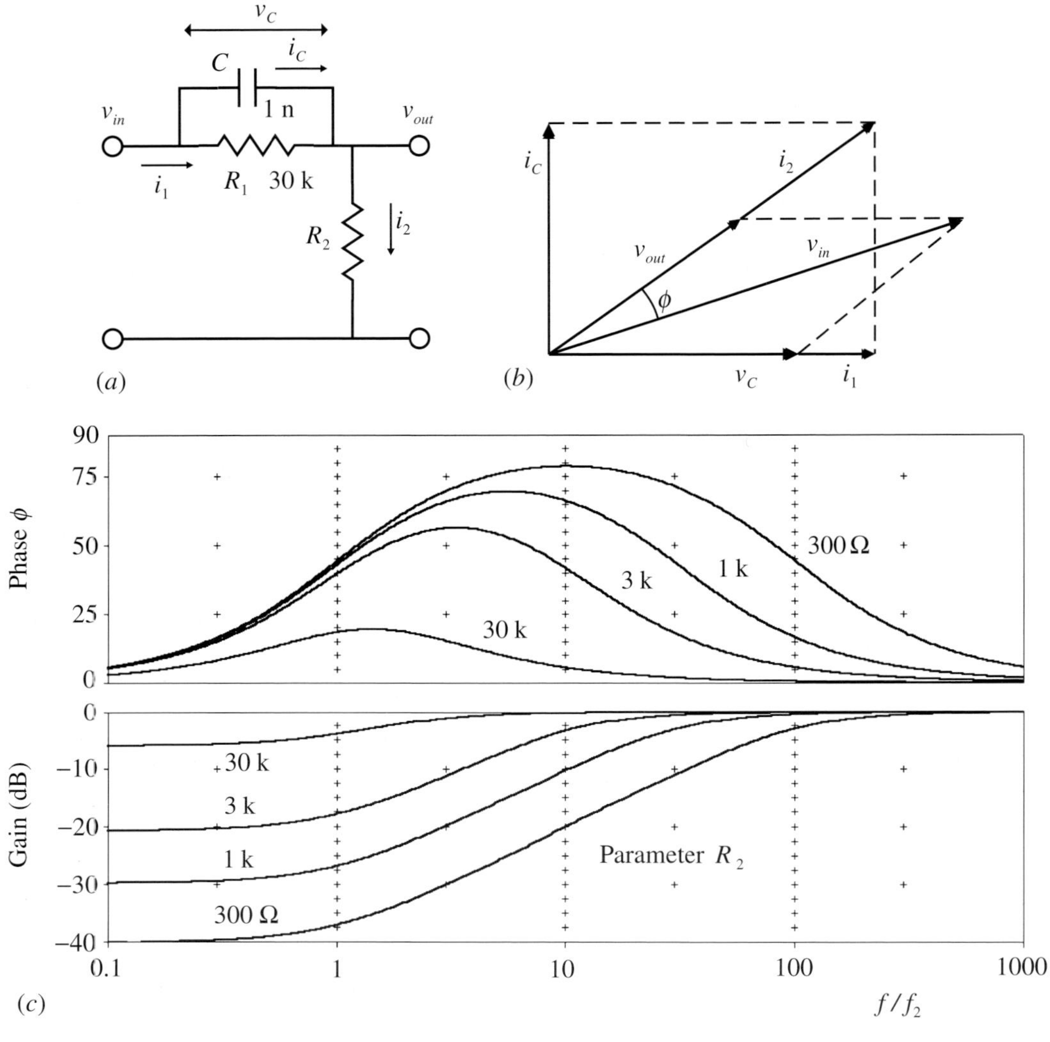


**Fig. 3.3.4** (*a*) Phase advance circuit. (*b*) Phasor diagram. (*c*) Gain and phase as a function of relative frequency $f_2 = 1/2\pi R_1 C$, with the values of $R_2$ shown in the figure.

## References and additional sources 3.3


Budak A. (1991): *Passive and Active Network Analysis and Synthesis*, Boston: Houghton Mifflin (1974). Reprint Waveland Press 1991 (Prospect Heights, Illinois). ISBN 0-88133-625-4.

Irwin J. D. (1996): *Basic Engineering Circuit Analysis*, Englewood Cliffs: Prentice Hall. ISBN 0-13-397274-7.

Walton A. K. (1987): *Network Analysis and Practice*, Cambridge: Cambridge University Press. ISBN 0-521-31903-X.

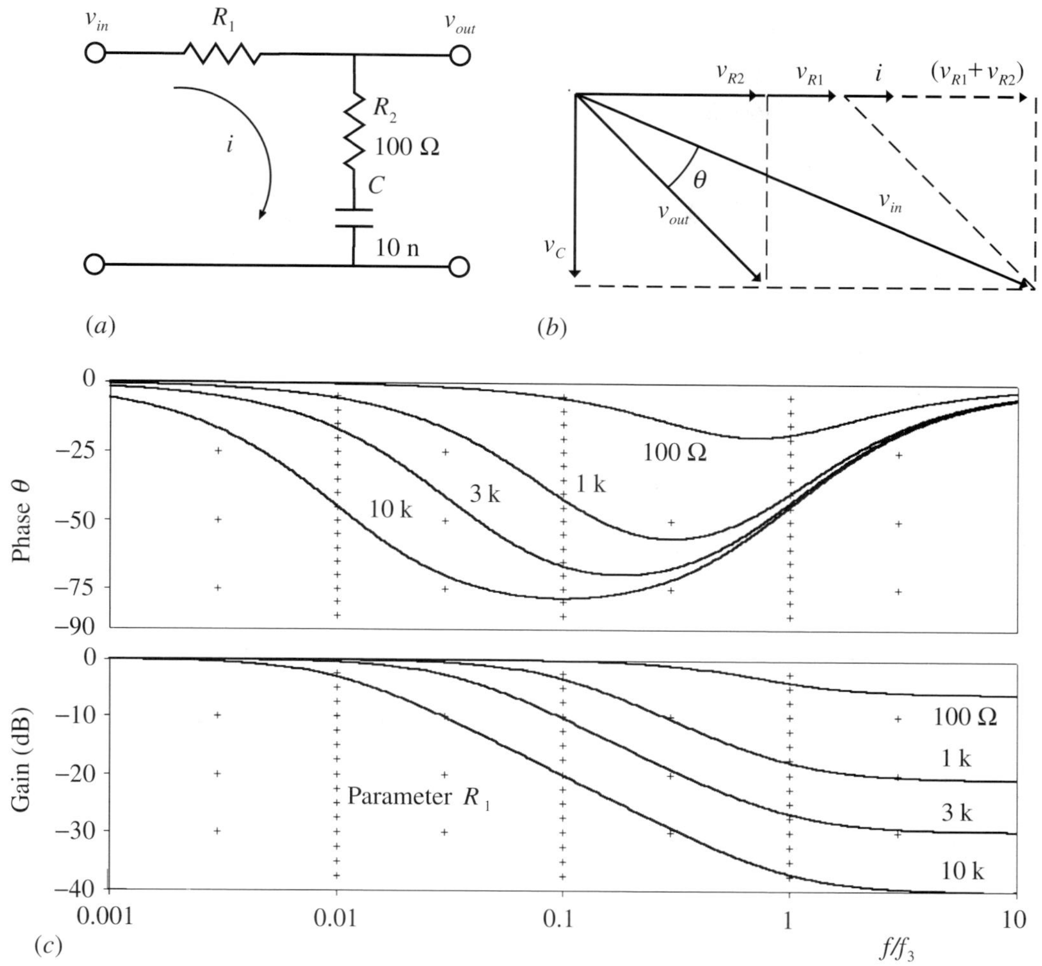


**Fig. 3.3.5** (*a*) Phase retard circuit. (*b*) Phasor diagram. (*c*) Gain and phase as a function of relative frequency $f_3 = 1/2\pi R_2 C$, with the values of $R_1$ shown in the figure.

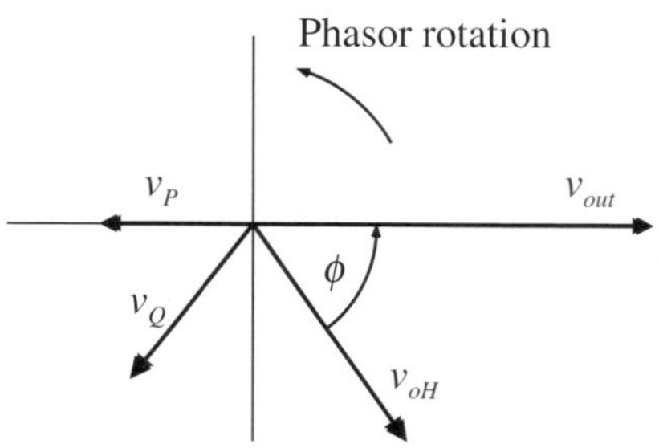


**Fig. 3.3.6** Phasor diagram for a feedback circuit.

# 3.4 Phase and amplitude

Education is the brief interval between ignorance and arrogance.

As we make much use of the representations of the response functions of systems in terms of poles and zeros it is necessary that we also have a means of determining the gain and phase responses as functions of frequency (Kuo 1966). In older books there is much discussion of how to sketch the responses, but now of course we can more rapidly and accurately obtain them via SPICE. However, it is still useful to examine the basic techniques so that we can understand the general effects of poles and zeros in particular positions in the $s$-plane.

Since we are here concerned with the frequency response rather than the consequences of some more complex input function, then taking a simple example transfer function $H(s)$ we may replace $s$ by $j\omega$ to give:

$$H(j\omega)=\frac{K(j\omega-z_0)}{(j\omega-p_0)(j\omega-p_1)} \tag{3.4.1}$$

where $K$ is a constant, and $p_0$ and $p_1$ are conjugate poles. We can express each factor in polar form:

$$(j\omega-z_0)=Z_0\exp(j\phi_0),\quad (j\omega-p_0)=P_0\exp(j\psi_0),\quad (j\omega-p_1)=P_1\exp(j\psi_1) \tag{3.4.2}$$

to give from Eq. (3.4.1):

$$\begin{aligned} H(j\omega) &= \frac{KZ_0\exp(j\phi_0)}{P_0\exp(j\psi_0)\,P_1\exp(j\psi_1)} \\ &= \frac{KZ_0}{P_0P_1}\exp[j(\phi_0-\psi_0-\psi_1)] \end{aligned} \tag{3.4.3}$$

At some particular frequency $\omega_A$, the amplitude $A$ will be given by:

$$A(\omega_A)=K\,\frac{\text{Product of vector magnitudes from zeros to } \omega_A \text{ on } j\omega \text{ axis}}{\text{Product of vector magnitudes from poles to } \omega_A \text{ on } j\omega \text{ axis}} \tag{3.4.4}$$

(where product is often written as the Greek $\Pi$). The phase response at $\omega_A$ is given by:

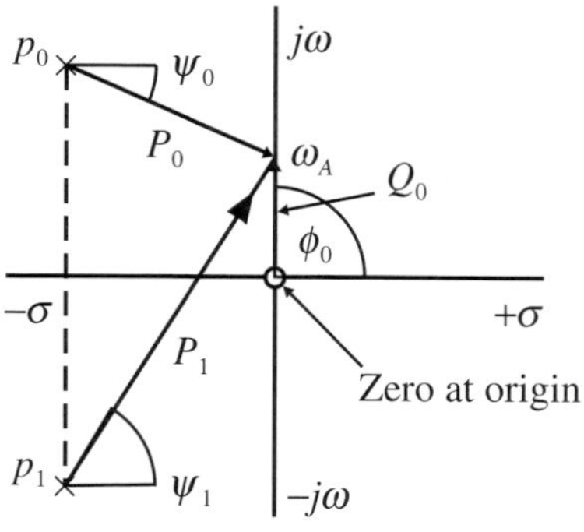


**Fig. 3.4.1** Phase-shift from pole-zero diagram.

$$\theta(\omega_A) = (\text{Sum of angles of magnitude vectors from zeros to } \omega_A \text{ on } j\omega \text{ axis}) - (\text{Sum of angles of magnitude vectors from poles to } \omega_A \text{ on } j\omega \text{ axis}) \quad (3.4.5)$$

The angles are positive measured in the counterclockwise direction from the direction of the positive $\sigma$-axis. If $K$ is negative an extra $-180°$ must be added. To avoid complications with angles greater than $180°$ the smaller angle should be taken and considered as negative, as at $\psi_0$. These relations are illustrated in Fig. 3.4.1, showing $\phi_0$ and $\psi_1$ as positive and $\psi_0$ as negative. The zero $z_0$ is here chosen to be at the origin.

These measures only give the response at the particular frequency $\omega_A$ and must be repeated for all the other frequencies of interest, so it is not an efficient or attractive approach. For a frequency $\omega_A$, as shown, the phase will be given by:

$$\theta(\omega_A) = \phi_0 - \psi_1 - (-\psi_2) = \phi_0 - \psi_1 + \psi_2 \quad (3.4.6)$$

so that at zero frequency and at very high frequency we have:

$$\begin{aligned} \theta_0 &= 90° - \psi_1 + \psi_2 = 90°, \quad \text{since} \quad |\psi_1| = |\psi_2| \\ \theta_\infty &= 90° - 90° + 90° = -90° \end{aligned} \quad (3.4.7)$$

The amplitude will be zero at zero frequency since $Z_0$ is zero, and will again be zero at very high frequency since the magnitude vectors become very large and essentially equal so we then have $A \approx K/\infty = 0$. Somewhere in the region of $\omega$ for $p_0$ there will be a maximum since $P_0$ will be near its minimum. The responses will look something like Fig. 3.4.2, where the relation between the slope of $A$ and the phase should be noted.

On the basis of these ideas we may make the following general observations:

(i) A zero near the $j\omega$ axis at say $\omega_1$ will produce a dip in $A$ and a rapid change of phase from small to large with increase in $\omega$.

(ii) A pole near the $j\omega$ axis at say $\omega_2$ will produce a peak in $A$ and a rapid change of phase from large to small with increase in $\omega$.

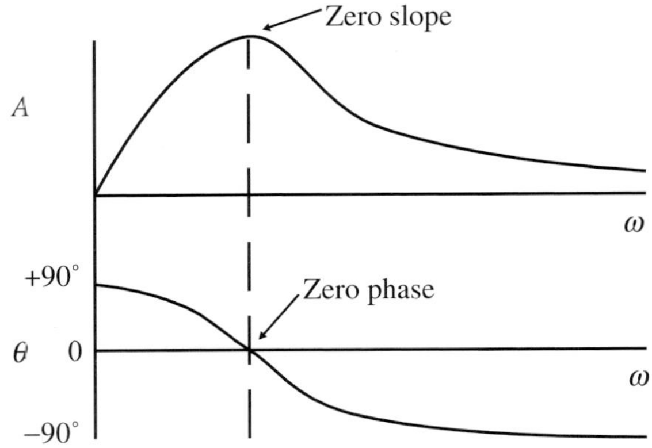


**Fig. 3.4.2** Phase and amplitude response as a function of frequency.

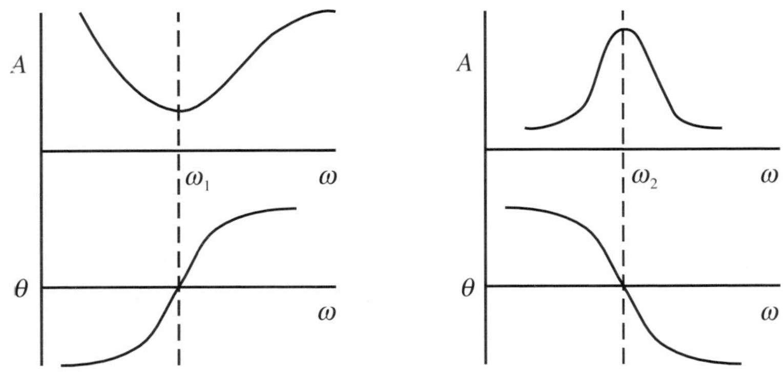


(*a*) (*b*)

**Fig. 3.4.3** (*a*) Phase change for dip in gain response. (*b*) Phase change for peak in gain response.

(iii) Poles and zeros far from the $j\omega$ axis (large $\sigma$) will have little effect on the shape of the response, only effectively scaling the response.

which are illustrated in Fig. 3.4.3.

There can be no poles in the RHP for a stable system but we can have zeros. Consider two configurations with mirror image zeros as shown in Fig. 3.4.4.

It is evident that the phase contributions by the zeros in (*a*) will be substantially less than those in (*b*). A system with the configuration like (*a*) is called a minimum phase system, while one with one or more zeros in the RHP is a non-minimum phase system. This differentiation is evidently significant in, for example, feedback systems. The existence of a zero in the RHP indicates that at some frequency with negative '*a*' in Eq. (1.12.13) the gain falls to zero. The type of network that exhibits such a zero possesses two paths between input and output, such as a bridge or

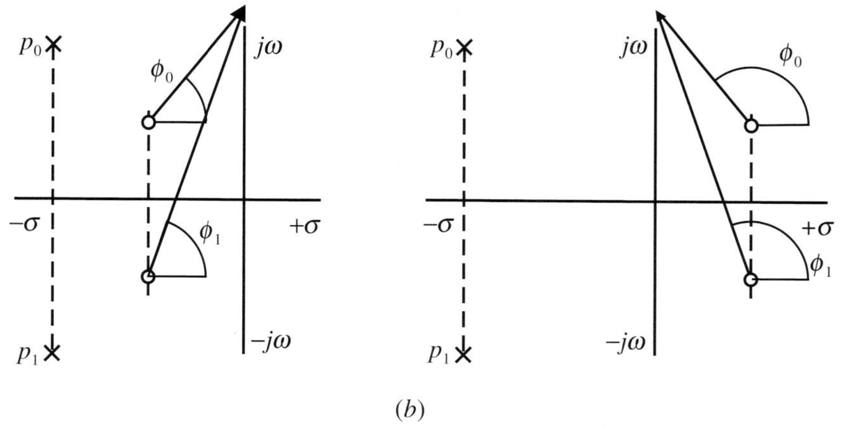


**Fig. 3.4.4** (*a*) Phase for left half-plane zeros. (*b*) Phase for right half-plane zeros.

a twin-tee network, though almost any network will at high enough frequency show such behaviour, but if high enough it will not be of concern (Thomason 1955).

A particular case arises when the poles in the LHP are in mirror image locations to the zeros in the RHP (Fig. 3.4.5(*a*)). In this case the magnitude vectors from the poles exactly match those from the zeros so that there is no variation in $A$ as a function of $\omega$. This corresponds to the so called all-pass response (Section 3.14). The phase will, however, vary substantially as shown in Fig. 3.4.5(*b*).

There is a difficulty in determining the phase angle contribution in Eq. (3.4.5) for zeros like $Z_0$ in Fig. 3.4.5(*a*). For $\omega$ varying from zero frequency up to the same value as for the zero itself ($\omega_0$) we would reckon its contribution to be the smaller angle and for this to be negative. However, when we reach $\omega_0$ the phase contribution could be reckoned as either plus or minus 180°. A correct result will be obtained by continuing with the negative angle even though it will now be greater than 180°. This procedure can be verified using PSpice to compute the phase response for a simple example. With the following locations we also have the corresponding transfer function:

Poles at $-1 \pm j$ and zeros at $1 \pm j$ give a transfer function

$$H(s) = \frac{(s-1+j)(s-1-j)}{(s+1+j)(s+1-j)} = \frac{s^2-2s+2}{s^2+2s+2} \tag{3.4.8}$$

which can be simulated using the PSpice Laplace model with a sin input and resistive load to produce a phase response that agrees with the manual method (remember that the latter is in terms of $\omega$ while PSpice will show results in terms of frequency $= \omega/2\pi$). The PSpice display for amplitude will look terrible until you examine the scale! A check using the 3-pole 3-zero all-pass configuration shown in Kuo (1966, p. 221) may be made using the transfer function:

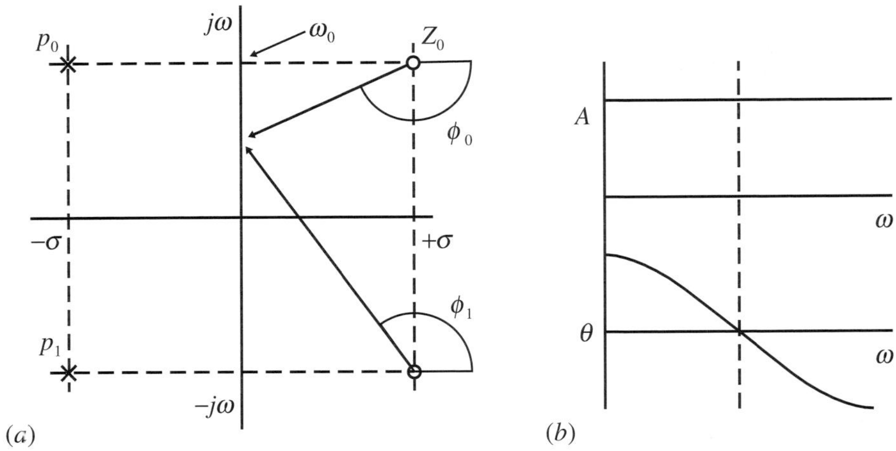


**Fig. 3.4.5** (*a*) Mirror image left-half and right-half zeros. (*b*) Phase and amplitude response.

$$H(s) = \frac{(s-1+j)(s-1-j)(s-2)}{(s+1+j)(s+1-j)(s+2)} = \frac{(s^2-2s+2)(s-2)}{(s^2+2s+2)(s+2)} \qquad (3.4.9)$$

and you will find agreement.

## SPICE simulation circuits

None

## References and additional sources 3.4


Bode H. W. (1940): Relation between attenuation and phase in feedback amplifier design. *Bell Syst. Tech. J.* **19**, 412–454.

Bode H. W. (1947): *Network Analysis and Feedback Amplifier Design*, New York: Van Nostrand.

Irwin J. D. (1996): *Basic Engineering Circuit Analysis*, Englewood Cliffs: Prentice Hall. ISBN 0-13-397274-7.

Kuo F. F. (1966): *Network Analysis and Synthesis*, New York: John Wiley.

Tanner J. A. (1951): Minimum phase networks. *Electronic Engineering* **23**, 418–423.

Thomason J. G. (1955): *Linear Feedback Analysis*, Oxford: Pergamon Press.

# 3.5 Resonance

The test of science is its ability to predict.

Richard P. Feynman (*Lectures in Physics*, Vol. II, p. 41–12).

Though mechanical resonance has been known for a very long time (the walls of Jericho?) the idea of electrical resonance is thought to be due to William Thompson, later better known as Lord Kelvin (the river Kelvin fronts the University of Glasgow) and to Gustav Kirchhoff (Grimsehl 1933, p. 567; see also Blanchard 1941, who credits Maxwell with the first proper understanding and Hertz with producing the first resonance curve). The year 1997 was the centenary of the patent granted to Oliver Lodge for the resonant tuning for wireless telegraphy. He even sued Marconi for breach of this patent (and was pacified by award of a consultancy), but we are now free to use it as we please.

A well known example of a mechanical resonator is the simple pendulum. In this the energy of the system is exchanged between potential and kinetic, i.e. there are two forms in which the energy can be stored. Similarly, for an electrical resonator we must have two forms of storage and these are provided by connecting together an inductor and a capacitor. In the former the energy is stored in the magnetic field while in the latter it is stored in the electric field (Section 2.2). As in the pendulum where air resistance causes loss of energy, in an electrical resonator energy is lost due to resistance, so the amplitude of the oscillation dies away with time.

The inductor and capacitor can be connected in series or in parallel as shown in Figs. 3.5.1 and 3.5.3 (see pp. 184 and 188). We may assume that the capacitor is near ideal for practical purposes but the inductor will have some self-resistance shown as $R$. The series connection Fig. 3.5.1($a$) is somewhat simpler to analyse so we will start with this.

The impedance $Z$ is given by:

$$Z = R + j\omega L + \frac{1}{j\omega C} = R + j\left(\omega L - \frac{1}{\omega C}\right) \tag{3.5.1}$$

so $|Z| = [R^2 + (\omega L - 1/\omega C)^2]^{\frac{1}{2}}$ and $\tan\theta = (\omega L - 1/\omega C)/R$

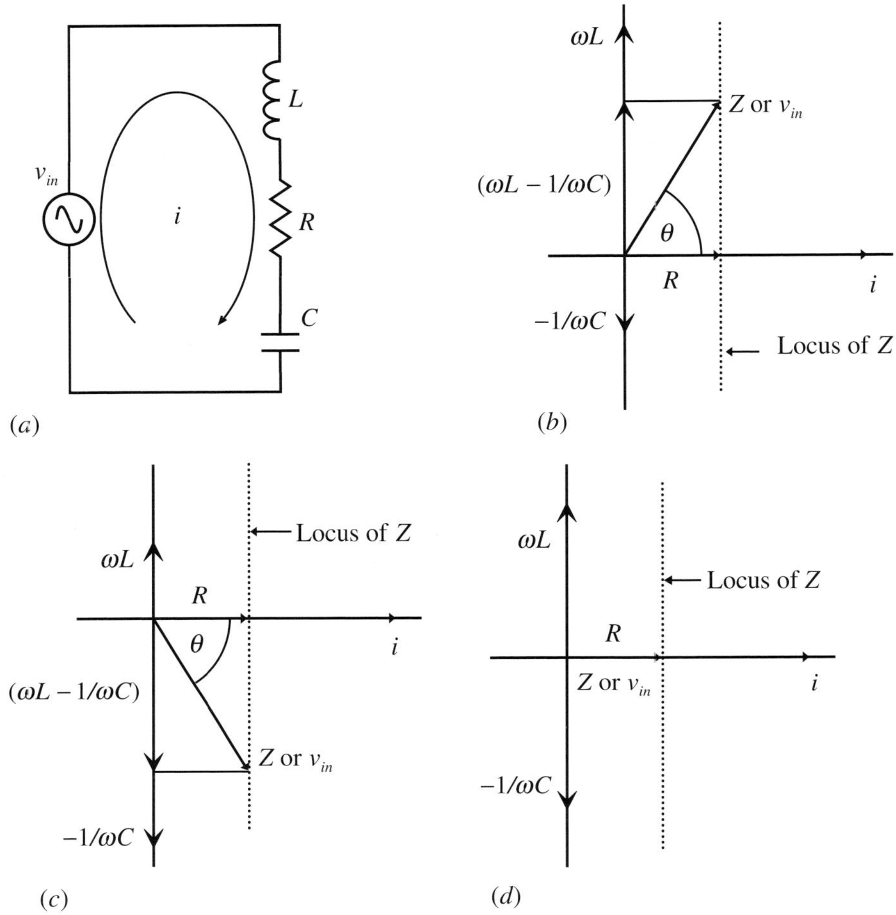


**Fig. 3.5.1** (*a*) Series resonance circuit. (*b*) Phasor diagram: $f > f_{res}$. (*c*) Phasor diagram: $f < f_{res}$. (*d*) Phasor diagram: $f = f_{res}$.

At a particular frequency $\omega_0$ the reactive component will be zero, i.e. when $\omega_0 L = 1/\omega_0 C$:

$$\omega_0^2 = 1/LC \quad \text{or} \quad \omega_0 = (LC)^{-\frac{1}{2}} \tag{3.5.2}$$

and this frequency is called the resonant frequency.

Phasor diagrams can be drawn to show the variation of $Z$ and $\theta$ with frequency (Fig. 3.5.1). The common factor is $i$ so, as explained in Section 3.3, we start with this. The three diagrams show the results for $f \ll f_0$, $f = f_0$ and $f \gg f_0$. The dotted line is the locus of $Z$, and if $R$ is not too large the magnitude of $Z$ varies over a large range with a minimum equal to $R$ at $\omega_0$. At low frequency $i$ leads $v_{in}$ by ~90° and at high frequency $i$ lags $v_{in}$ by ~90°. At resonance $v_{in}$ is in phase with $i$ and $Z$ is purely resistive so the current is given by $i = v_{in}/R$. This current flows through $L$ and $C$ so the voltages across these are:

$$v_L = \frac{v_{in}}{R}(j\omega_0 L) \quad \text{and} \quad v_C = \frac{v_{in}}{R}\left(\frac{-j}{\omega_0 C}\right) = \frac{-v_{in}}{R}(j\omega_0 L) \tag{3.5.3}$$

so that $v_L$ and $v_C$ are equal in magnitude and 180° out of phase to cancel exactly. The factor relating these voltages to the input voltage $v_{in}$ is called the $Q$ (for quality factor) of the circuit and is given by:

$$Q = \frac{\omega_0 L}{R} = \frac{1}{\omega_0 CR} = \frac{1}{R}\left(\frac{L}{C}\right)^{\frac{1}{2}} \tag{3.5.4}$$

In practice the maximum value of $Q$ that can be achieved in the kHz to MHz range is about 100 to 200 (at microwave frequencies, where resonant circuits are in the form of resonant cavities rather than separate $L$ and $C$, $Q$'s of up to $10^4$ are possible; for Mössbauer nuclear resonance absorption $Q$ is about $10^{10}$!). It should be noted that both $Q$ and $L$ are frequency dependent since the distribution of current in the wire is affected by the skin effect (Section 2.8). At resonance the voltages $v_L$ and $v_C$ are increased substantially since:

$$v_L = v_C = v_{in} Q \tag{3.5.5}$$

The $Q$ of the circuit also determines the sharpness of the resonance, a high $Q$ corresponding to a narrow response (see, for example, Section 4.10). We may rewrite Eq. (3.5.1) using (3.5.4) and (3.5.2) to give:

$$\begin{aligned} Z &= R\left[1 + \frac{j}{R}\left(\omega L - \frac{1}{\omega C}\right)\right] \\ &= R\left[1 + \frac{jQ}{\omega_0 L}\left(\omega L - \frac{1}{\omega C}\right)\right], \quad \text{using Eq. (3.5.4)} \\ &= R\left[1 + jQ\left(\frac{\omega L}{\omega_0 L} - \frac{1}{\omega_0 L \omega C}\right)\right] \\ &= R\left[1 + jQ\left(\frac{\omega}{\omega_0} - \frac{\omega_0}{\omega}\right)\right], \quad \text{using Eq. (3.5.2)} \end{aligned} \tag{3.5.6}$$

In examining the region close to resonance there is the difficulty of dealing with the small difference between two large quantities. To cope with this we write $\omega = \omega_0(1+\delta)$ and to ensure $\delta$ is small we restrict ourselves to values of $Q > 10$. The inner bracket can then be approximated:

$$\left(\frac{\omega}{\omega_0} - \frac{\omega_0}{\omega}\right) = (1+\delta) - \left(\frac{1}{1+\delta}\right) = \frac{1 + 2\delta + \delta^2 - 1}{1+\delta} = \delta\left(\frac{2+\delta}{1+\delta}\right) \approx 2\delta, \quad \text{if } \delta \text{ is small}$$

let $F(\omega) \equiv (1 + 2jQ\delta)^{-1} \cong (1 - 2jQ\delta)$, using binomial expansion (Section 1.3)

so $|F(\omega)| = [(1 - 2jQ\delta)(1 + 2jQ\delta)]^{-\frac{1}{2}} = (1 + 4Q^2\delta^2)^{-\frac{1}{2}}$ (3.5.7)

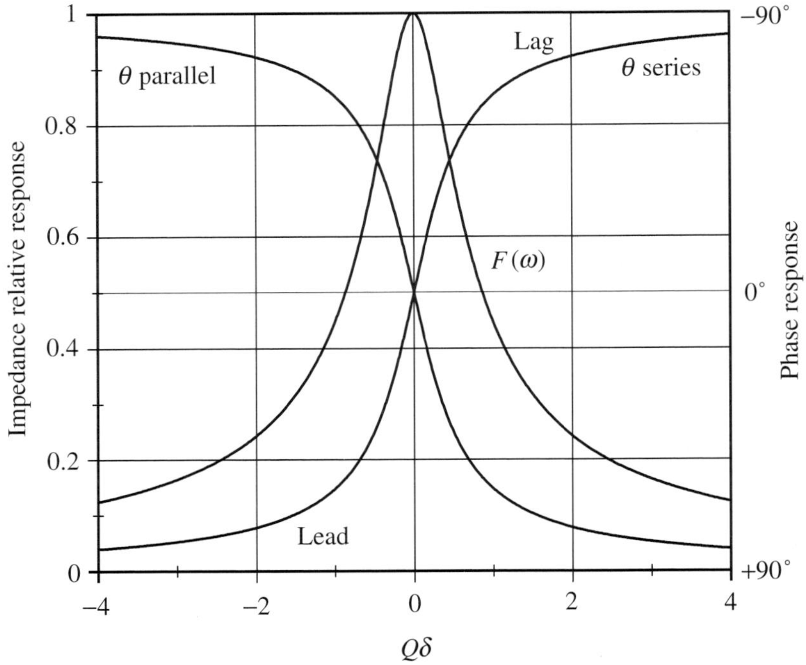


**Fig. 3.5.2** Lorentzian curves.

with $\arg F(\omega) = \tan^{-1}(-2Q\delta)$

then $Z = R(1 + 2jQ\delta) = R/F(\omega)$, with $\delta = \dfrac{\omega - \omega_0}{\omega_0}$

and $|Z| = R(1 + 4Q^2\delta^2)^{\frac{1}{2}} = R/|F(\omega)|$

The function $F(\omega)$ is known as a Lorentzian function and is shown in Fig. 3.5.2.

At frequency $\omega = \omega_0\,(1 + 1/2Q)$ or frequency deviation $(\omega - \omega_0) = \omega_0\,/2Q = \partial\omega$, then from Eq. (3.5.7) we can see that the magnitude $|F(\omega)|$ will have decreased to $1/\sqrt{2} = 0.707$ (or –3 dB) of its peak value at $\omega = \omega_0$. The phase shift at these two points will be ±45°. Thus the $Q$ may be determined for a resonance curve by measuring the frequency difference $\Delta\omega = 2\partial\omega$ between the −3 dB points. Then:

$$Q = \omega_0/\Delta\omega \tag{3.5.8}$$

Useful 'universal' resonance curves are given by Terman (1950).

If the circuit is at resonance and the driving source is switched off then the oscillations will decay exponentially as shown in Fig. 1.12.3. The decay rate is related to the $Q$ of the circuit; if the $Q$ is high the decay will be slow and vice versa. This provides an alternative way of determining the $Q$ of a circuit. If the amplitude of the resonant current is $I_0$ then the energy stored in the circuit will at some instant be all magnetic and hence the energy $U$ will be given by (see Eq. (2.2.1)):

$$U = \tfrac{1}{2}LI_0^2 \tag{3.5.9}$$

and the energy loss per cycle $\Delta U$ is given by the power dissipation in $R$ times the time of one cycle, i.e. the period $T=2\pi/\omega_0$. Thus:

$$\Delta U=\tfrac{1}{2}RI_0^2T=\frac{RI_0^2\pi}{\omega_0} \tag{3.5.10}$$

and so we have:

$$\frac{\text{Stored energy at resonance}}{\text{Energy loss per cycle}}=\frac{U}{\Delta U}=\frac{\omega_0 L}{2\pi R}=\frac{Q}{2\pi} \tag{3.5.11}$$

The amplitude of the oscillation will decay as energy is being dissipated. The form of the decay may be obtained by integrating Eq. (3.5.11). Since $\Delta U$ is the loss per cycle this may be equated to the rate of energy loss $dU/dt$ times the period $T=2\pi/\omega_0$ to give (the negative sign indicates the decrease or loss):

$$\frac{dU}{dt}\frac{2\pi}{\omega_0}=-\Delta U=\frac{-2\pi U}{Q}, \quad \text{so} \quad \int\frac{dU}{U}=\frac{-\omega_0}{Q}dt$$

and thus

$$U=U_0\exp(-\omega_0 t/Q), \quad \text{see Eq. (1.9.7)} \tag{3.5.12}$$

This shows that the energy in the circuit decays with a time constant $\tau_u=Q/\omega_0$. Since energy $\propto$ amplitude squared, then the time constant for the voltage (amplitude) is $\tau_a=2Q/\omega_0$. The same results will hold for the growth of the amplitudes when the drive is first applied except that the form will be $[1-\exp(-\omega_0 t/2Q)]$. In time $\tau_a$ the amplitude will rise to $(1-1/e)=63\%$ of the final amplitude. If $T$ is the period then the number of cycles $N_{63}$ to reach this level will be related to $Q$ by:

$$Q=\frac{\omega_0\tau_a}{2}=\frac{2\pi\, N_{63}T}{2T}=\pi N_{63}, \quad \text{since} \quad \omega_0=\frac{2\pi}{T} \quad \text{and} \quad \tau_a=N_{63}T \tag{3.5.13}$$

providing a simple way to determine $Q$. The ratio of successive amplitudes in a decaying oscillation is given by (the same result would be found for a growing response):

$$\begin{aligned}\frac{v_2}{v_1}&=\frac{v_0\exp(-\omega_0 t_2/2Q)}{v_0\exp(-\omega_0 t_1/2Q)}\\ &=\exp\left[\frac{-\omega_0 t_2}{2Q}+\frac{\omega_0 t_1}{2Q}\right]\\ &=\exp\left[\frac{-\omega_0(t_2-t_1)}{2Q}\right]\\ &=\exp\left(\frac{-\pi}{Q}\right), \quad \text{since} \quad (t_2-t_1)=T=\frac{2\pi}{\omega_0}\end{aligned} \tag{3.5.14}$$

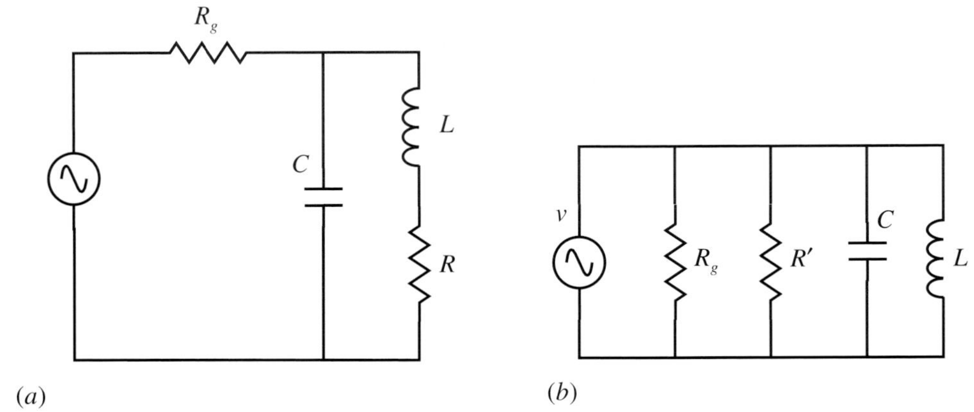


**Fig. 3.5.3** (*a*) Parallel resonance circuit. (*b*) Equivalent parallel resonance circuit.

so $\ln\left(\frac{v_2}{v_1}\right) = \frac{-\pi}{Q}$

and $-\pi/Q$ is called the logarithmic decrement of the circuit.

Figure 3.5.3(*a*) shows a general parallel resonant circuit. In practice it is more appropriate to include some resistance in series with the inductor as shown, but we can transform this to the configuration as shown in Fig. 3.5.3(*b*).

For the parallel circuit it is more convenient to work in terms of admittance $Y$ and consider a current source. We will for the moment allow the current source to have a finite $R_g$ which can be set to infinite if required (Everitt and Anner 1956). The admittance $Y_{LR}$ of the inductor $L$ and resistor $R$ will be:

$$Y_{LR} = \frac{1}{(sL+R)} = \frac{(R - j\omega L)}{R^2 + \omega^2 L^2} \cong \frac{R}{(\omega L)^2} - \frac{j}{\omega L}, \quad \text{for} \quad (\omega L)^2 \gg R^2, \quad \text{i.e. high } Q_L = \frac{\omega L}{R} \tag{3.5.15}$$

which is equivalent to a resistor $R' = (\omega L)^2/R$ in parallel with an inductor $L$. The admittance $Y_e$ of the effective parallel resistance of $R_g$ and $R'$ is then:

$$Y_e = \frac{1}{R_e} = \frac{R}{(\omega L)^2} + \frac{1}{R_g} = \frac{R + \dfrac{(\omega L)^2}{R_g}}{(\omega L)^2} \tag{3.5.16}$$

and if we have a perfect current source then $R_g \to \infty$. In terms of $Q_L$ for the series connection of $L$ and $R$, the equivalent $Q_p$ for the parallel connection of $L$ and $R'$ is given by:

$$Q_L = \frac{\omega L}{R} = \frac{\omega L R'}{(\omega L)^2} = \frac{R'}{\omega L} = Q_p \tag{3.5.17}$$

In terms of the admittance $Y_p$ of the equivalent parallel resonator and the generator current $i$ the voltage $v$ across the circuit is then given by:

$$
\begin{aligned}
v &= \frac{i}{Y_p} = \frac{i}{Y_e + j\left(\omega C - \dfrac{1}{\omega L}\right)} \\
&= \frac{iR_e}{1 + j\left(\dfrac{R_e}{\omega_p L}\right)\left(\omega\omega_p LC - \dfrac{\omega_p}{\omega}\right)}, \quad \text{with} \quad LC = \frac{1}{\omega_p^2}
\end{aligned}
\tag{3.5.18}
$$

$$
\text{or} \quad \frac{v}{v_p} = \frac{Z}{Z_p} = \frac{1}{1 + jQ_e\left(\dfrac{\omega}{\omega_p} - \dfrac{\omega_p}{\omega}\right)}, \quad \text{since } i \text{ is constant with } v_p = iR_e \text{ at } \omega_p
$$

$$
\text{and} \quad Q_e = \frac{R_e}{\omega_p L} = \frac{R_e \omega_p}{\omega_p^2 L} = \omega_p R_e C, \quad \text{is the effective parallel } Q
$$

which is of the same form as for $i\,(\propto 1/Z)$ for the series circuit as given in Eq. (3.5.6). We have implied in (3.5.18) that the resonant frequency $\omega_p = (LC)^{-\frac{1}{2}}$ as for the series circuit, which needs to be justified. We need to write out the full expression for the impedance of Fig. 3.5.3($a$) and rationalize it.

$$
\begin{aligned}
Y &= \frac{1}{R + j\omega L} + j\omega C \\
&= \frac{1 + j\omega C(R + j\omega L)}{R + j\omega L} \\
&= \frac{[1 + j\omega C(R + j\omega L)](R - j\omega L)}{R^2 + \omega^2 L^2} \\
&= \frac{(R - j\omega L) + j\omega C(R^2 + \omega^2 L^2)}{R^2 + \omega^2 L^2} \\
&= \frac{R - j\omega L + j\omega CR^2 + j\omega^3 CL^2}{R^2 + \omega^2 L^2}
\end{aligned}
\tag{3.5.19}
$$

We may make a choice of definition for the resonant frequency: either that which results in maximum impedance or that which results in unity power factor, i.e. current and voltage in phase. For our 'high $Q$' conditions the difference is slight and we choose the latter. This means that the coefficient of $j$ in (3.5.19) should be set to zero:

$$
-j\omega L + j\omega CR^2 + j\omega^3 CL^2 = 0, \quad \text{so} \quad \omega_p^2 = \frac{1}{LC} - \frac{R^2}{L^2} = \frac{1}{LC}\left(1 - \frac{1}{Q_L^2}\right)
\tag{3.5.20}
$$

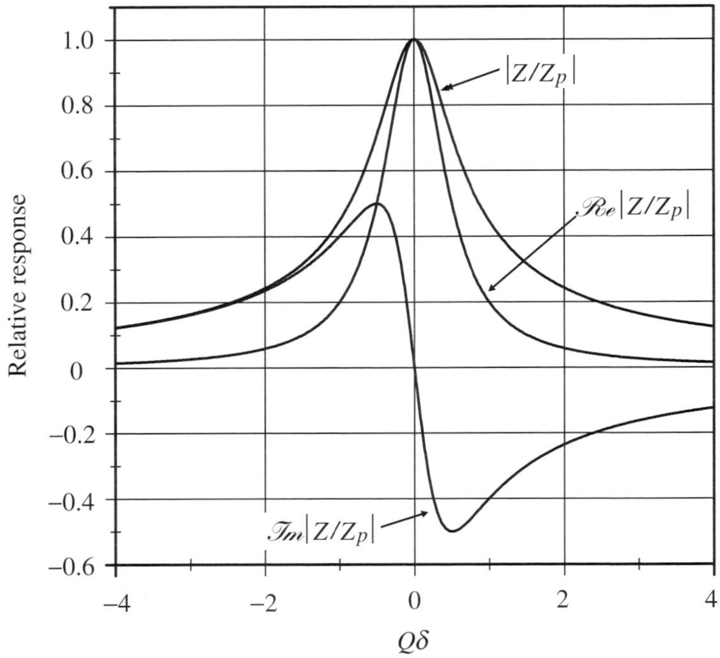


**Fig. 3.5.4** Universal resonance curves. The curves shown are directly calculated but an example of a SPICE simulation may be seen in Resnant1.sch.

so for $Q_L > 10$ the error is less than 1%. At resonance we have from either (3.5.18) or (3.5.19):

$$Z_p = \frac{L}{CR} = \frac{\omega_p^2 L^2}{R} = Q_L \omega_p L = RQ_L^2 \tag{3.5.21}$$

so $Z_p$ will be high and the current $i_p$ from the generator will be low:

$$i_p = \frac{v_{in}}{Z_p} = \frac{v_{in}}{RQ_L^2} \tag{3.5.22}$$

Using the first part of Eq. (3.5.7) we can rewrite (3.5.18) as:

$$\frac{Z}{Z_p} = \frac{1}{1+j2Q\delta} = \frac{1-j2Q\delta}{(1+j2Q\delta)(1-j2Q\delta)} = \frac{1-j2Q\delta}{1+4Q^2\delta^2} = \frac{1}{1+4Q^2\delta^2} - \frac{j2Q\delta}{1+4Q^2\delta^2} \tag{3.5.23}$$

and we can plot the real and the imaginary parts as functions of $Q\delta$ as shown in Fig. 3.5.4, together with the modulus $|Z/Z_p|$.

The form and origin of these curves is of relevance for consideration of absorption and dispersion in dielectrics and ferrites (Sections 4.2 and 2.11).

We consider now some examples of slightly more complex resonant circuits

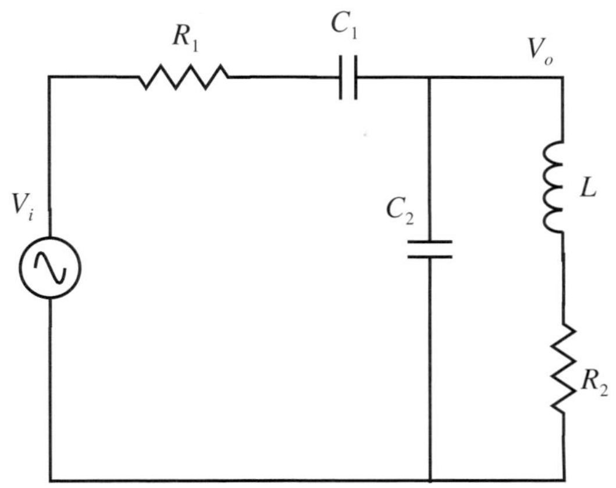


**Fig. 3.5.5** Matching drive to parallel circuit.

which require us to make approximations in the analysis to obtain solutions that we can handle more easily. As will be seen circuits with only a few components generally lead to equations of higher than second order for which it is difficult to find the roots (Section 1.10). In making approximations we are concerned that the simplifications are justified so it is desirable to have some means of assuring ourselves that they are. This is one of the benefits of SPICE in that it solves the full equations without approximation so that we can compare results for given circuit values. In deciding which terms in our analysis we can neglect it is usually necessary to have some idea of the actual circuit values, which of course we probably do not know until we have completed the analysis – a sort of catch-22. Again SPICE comes to our aid by allowing us to experiment to find suitable values to allow the approximations to be considered. The reason for doing the analysis remains since SPICE does not tell us anything parametric about the circuit. The first circuit is shown in Fig. 3.5.5.

The intention is to obtain a large current in $L$ to produce a r.f. magnetic field. The algebra is a bit messy but you just have to work through it. The impedance of the series $R_1$ and $C_1$ is given by:

$$Z_s = R_1 + \frac{1}{sC_1} = \frac{1 + sC_1R_1}{sC_1} \tag{3.5.24}$$

and for the parallel section of $C_2$ and $L$:

$$\frac{1}{Z_p} = sC_2 + \frac{1}{R_2 + sL} = \frac{sC_2(R_2 + sL) + 1}{(R_2 + sL)}$$
$$\text{so} \quad Z_p = \frac{(R_2 + sL)}{1 + sC(R_2 + sL)} \tag{3.5.25}$$

Then considering it as a potential divider we have:

$$\frac{V_o}{V_i}=\frac{Z_p}{Z_p+Z_s}$$

$$=\frac{\dfrac{(R_2+sL)}{1+sC_2(R_2+sL)}}{\left[\dfrac{(R_2+sL)}{1+sC_2(R_2+sL)}\right]+\left[\dfrac{(1+sC_1R_1)}{sC_1}\right]}$$

$$=\frac{sC_1(R_2+sL)}{sC_1(R_2+sL)+(1+sC_1R_1)+sC_2(R_2+sL)(1+sC_1R_1)} \tag{3.5.26}$$

$$=\frac{sC_1(R_2+sL)}{s^2(C_1L+C_1R_1C_2R_2+C_2L)+s(C_1R_2+C_1R_1+C_2R_2)+1+s^3C_2LC_1R_1}$$

$$\cong\frac{s^2C_1L}{s^2L(C_1+C_2)+s(C_1R_2+C_1R_1+C_2R_2)+1}$$

The approximations made to obtain the last line are decided as follows. The circuit is intended for operation at a frequency of 13.56 MHz with $L=1$ μH, $R_1=20$ Ω and $R_2=2$ Ω. PSpice simulation leads to values of $C_1=26$ p and $C_2=110$ p. A high value of $C_2$ was desired to minimize stray capacitive effects. With the frequency $s=8.52\times10^7$ it is found that $C_1R_1C_2R_2$ is small relative to the other terms in $s^2$, and that the term in $s^3$ is also smallish, but for the moment we want to be left with a second order polynomial since we can more readily understand the response. We find the poles of the transfer function by equating the denominator to zero. Comparing this to the general form shown in Eq. (1.12.30) we can put:

$$\frac{\omega_0}{Q}=\frac{(C_1R_2+C_1R_1+C_2R_2)}{L(C_1+C_2)} \quad \text{and} \quad \omega_0^2=\frac{1}{L(C_1+C_2)} \tag{3.5.27}$$

and inserting the actual values above gives:

$$C_1+C_2=137.7\text{ p} \quad \text{and} \quad Q=14.8 \tag{3.5.28}$$

The values for $C_1$ and $C_2$ fit well but if the $Q$ of the simulation peak is measured it is found to be about 31.6, which does not. The calculated value of capacity to resonate at 13.56 MHz with $L=1$ μH is 137.8 p. This demonstrates that $C_1$ and $C_2$ are in effect in parallel across $L$, which we can see from the circuit if we remember that the voltage generator (by definition) has zero impedance. If you short out the generator and stimulate the circuit with a current generator (with an infinite source impedance), e.g. between common and say the junction of $R_1$ and $C_1$, you will get the same response.

The difference in the $Q$ values indicates that our approximations were not good in this respect. This can be made more evident if we compute the amplitude of the voltage $V_o$ across the parallel segment at resonance, but now retaining the term in $s^3$. From Eq. (3.5.26) the term in $s^2$ becomes $-1$ at resonance (putting $s=j\omega$) which cancels out with the $+1$. We now have:

$$\frac{V_o}{V_i}=\frac{j\omega C_1R_2-\omega^2C_1L}{j\omega[(C_1R_2+C_1R_1+C_2R_2)-\omega^2C_2LC_1R_1]}$$

$$=\frac{C_1R_2+j\omega\, C_1L}{[(C_1R_2+C_1R_1+C_2R_2)-\omega^2C_2LC_1R_1]} \tag{3.5.29}$$

$$=\frac{(5.2+j221.5)\times 10^{-11}}{(7.92-4.15)\times 10^{-10}}$$

$$=0.134+j5.875$$

$$\cong j5.88$$

So $V_o$ is 90° out of phase with $V_i$ and the resonant amplification is a factor 5.88. This result indicates that the influence of the $s^3$ term (which contributes the value $-4.15$ in the denominator) on $Q$ is substantial – in this case by a factor $7.92/(7.92-4.15)=2.10$, so that the $Q$ of 14.8 calculated above should be multiplied by 2.10 which gives 31.1 in good agreement. Again we can see the benefit of recourse to SPICE to guide us in our analysis.

It is instructive to see if we can obtain the roots of the third order denominator in Eq. (3.5.26). Filling in the appropriate values for the components leads to the equation:

$$s^3\times 5.72\times 10^{-26}+s^2\times 1.36\times 10^{-16}+s\times 7.92\times 10^{-10}+1=0 \tag{3.5.30}$$

and not the sort of equation to be solved by hand. Mathcad, using Symbolic Solve for Variable, readily gives the three roots (note that $E9\equiv\times 10^9$):

$$p_1=-2.3749\text{ E9},\quad p_2=-1.3653\text{ E6}\pm j8.5788\text{ E7} \tag{3.5.31}$$

We now know the location of the three poles. The real pole $p_1$ is on the real axis, and is so far to the left that it does not concern us. Plotting the complex pole(s) $p_2$ and using Fig. 1.12.8, p. 72, we get Fig. 3.5.6 (plot scaled by E7) and obtain the relations:

$$\omega_0=[(1.365\text{ E6})^2+(8.579\text{ E7})^2]^{\frac{1}{2}}=8.58\text{ E7},\quad \text{so}\quad f_0=13655\text{ kHz}$$

$$\frac{\omega_0}{2Q}=1.365\text{ E6},\quad \text{so}\quad Q=\frac{\omega_0}{2\times 1.365\text{ E6}}=31.4 \tag{3.5.32}$$

in good agreement.

So far we have not considered how the total capacity should be apportioned

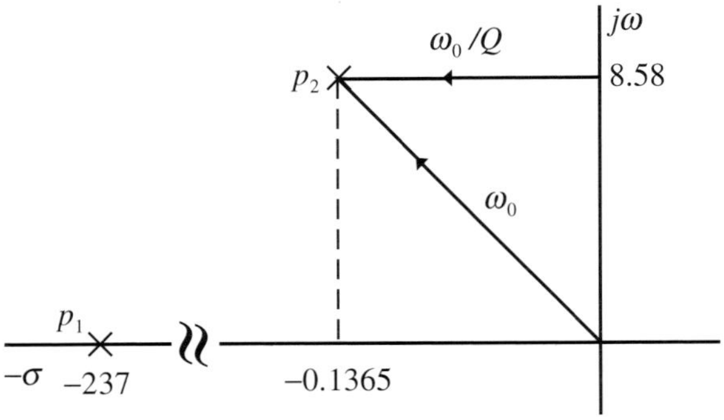


**Fig. 3.5.6** Relation between pole position and $Q$. The plot is scaled by E7.

between $C_1$ and $C_2$, the values used above being derived from some experimentation with PSpice. We put $C=C_1+C_2$ and then at resonance $\omega_0^2=1/LC$ so if we substitute for $C_2$ we get (and with $R_2+j\omega L\equiv Z$):

$$\frac{V_o}{V_i}=\frac{C_1 Z}{C_1R_2+C_1R_1+CR_2-C_1R_2-\dfrac{CC_1LR_1}{LC}+\dfrac{C_1^2LR_1}{LC}} \tag{3.5.33}$$

$$=\frac{CC_1Z}{C^2R_2+C_1^2R_1}$$

Now we can differentiate this with respect to $C_1$ to find the value for maximum $V_o$ ($V_i$ is fixed) since this will also give the maximum value of current in $L$. Thus:

$$\frac{\partial(V_o/V_i)}{\partial C_1}=\frac{CC_1Z(0+2C_1R_1)-(C^2R_2+C_1^2R_1)(CZ)}{(C^2R_2+C_1^2R_1^2)}=0, \qquad \text{for a maximum}$$

$$\text{so}\quad 2CC_1^2ZR_1-C^3ZR_2-CC_1^2ZR_1=0 \tag{3.5.34}$$

$$C_1=C\left(\frac{R_2}{R_1}\right)^{\frac{1}{2}}=137.7\times0.316=43.5\text{ p} \qquad \text{and hence} \qquad C_2=94.2\text{ p}$$

and for $V_i=3$ V in the simulation $V_o=20.14$ V with a current in $L$, $i_L=236$ mA and a drive current $i_{R1}=74$ mA. Better efficiency, i.e. the ratio of $i_L$ to $i_{R1}$, may be obtained by *increasing* $C_2$ (and hence decreasing $C_1$) at the expense of some decrease in $i_L$.

A famous resonance catastrophe, the Tacoma Narrows bridge, is discussed in Billah and Scanlan (1991) and a SPICE simulation in Irwin (1996).

## SPICE simulation circuits

| | |
|---|---|
| Fig. 3.5.4 | Resnant1.SCH |

## References and additional sources 3.5

Billah K. Y., Scanlan R. H. (1991): Resonance, Tacoma Narrows bridge failure, and undergraduate physics text books. *Amer. J. Phys.* **59**(2), 118–124.

Blanchard J. (1941): The history of electrical resonance. *Bell Syst. Tech. J.* **20**, 415–433.

Everitt W. L., Anner G. E. (1956): *Communication Engineering*, 3rd Edn, New York: McGraw-Hill.

Grimsehl E. (1933): *A textbook of Physics Vol. III: Electricity and Magnetism*, London: Blackie & Sons, Translated from 7th German Edn.

Irwin J. D. (1996): *Basic Engineering Circuit Analysis*, Englewood Cliffs: Prentice Hall. ISBN 0-13-397274-7. See pp. 652 and 818.

Terman F. E. (1950): *Radio Engineers' Handbook*, New York: McGraw-Hill.

# 3.6 Bandwidth and risetime

Some patients, conscious that their condition is perilous, recover their health simply through their contentment with the goodness of the physician.

Hippocrates

The bandwidth $\mathcal{B}$ of a system designates the range of frequencies that it will accommodate. This is not a sharply defined parameter in that the cut-off in frequency cannot be instantaneous; it must fall off at some finite rate rather than instantaneously, i.e. unlike the proverbial 'brick-wall that stops you dead'. If you examine the response of the simple *RC* low-pass filter in Fig. 3.3.3 you can see what is meant. Since there is a gradual fall off it is necessary to agree on some accepted definition as to what defines $\mathcal{B}$. For convenience and for the geometric reasons described there the −3 dB point is usually used. In discussing the Fourier transform the influence of bandwidth on waveforms was discussed. Here we consider the matter in more detail: the impossibility of a 'brick-wall' frequency cut-off and the relationship of bandwidth to risetime.

Suppose we have a signal $f(t)$ of fundamental frequency $\omega_1$ which we can describe in terms of its Fourier components:

$$f(t) = A_0 + \sum_n C_n \cos(n\omega_1 t - \phi_n) \tag{3.6.1}$$

If the wave is delayed by a time $\tau$ then:

$$\begin{aligned} f(t-\tau) &= A_0 + \sum_n C_n \cos[n\omega_1(t-\tau) - \phi_n] \\ &= A_0 + \sum_n C_n \cos(n\omega_1 t - \phi_n - n\omega_1 \tau) \end{aligned} \tag{3.6.2}$$

i.e. introducing a phase lag $n\omega_1\tau$, proportional to the frequency component, results in delaying the signal by time $\tau$. To shift a waveform in time we must have a phase lag proportional to frequency. An ideal low-pass filter will pass all frequencies up to the cut-off without distortion. There may be some time delay and an overall change of amplitude, but these do not produce distortion, i.e. a change of shape. That is, if the effect is:

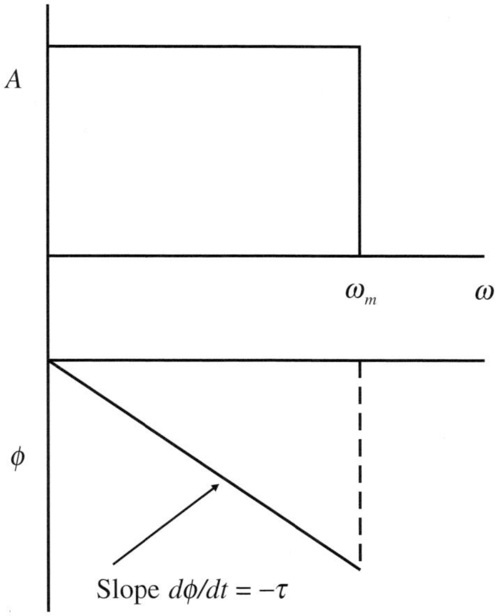


**Fig. 3.6.1** Brick-wall filter; amplitude and phase.

$$f(t) \to \text{filter} \to Af(t-\tau) \tag{3.6.3}$$

then the filter is 'ideal'. If the Fourier transform of $f(t)$ is $F(\omega)$ then the transform of $Af(t-\tau)$ is:

$$Af(t-\tau) \Leftrightarrow Ae^{-j\omega\tau}F(\omega) \tag{3.6.4}$$

A delay of $\tau$ in the time domain is equivalent to a multiplication by $\exp(-j\omega\tau)$ in the frequency domain. The necessary amplitude and phase response are shown in Fig. 3.6.1.

The filter will pass all frequencies up to the maximum $\omega_m$ and the phase response will have a slope $d\phi/d\omega = -\tau$ at least up to $\omega_m$. In terms of the Laplace transform we can say that the transfer function is given by:

$$H(\omega) = Ae^{-j\omega\tau} \tag{3.6.5}$$

and recall that the transfer function is the impulse response of the system (Section 1.12). We can now evaluate the inverse Fourier transform to find the impulse function in the time domain. Using the definition of Eq. (3.6.5), and noting that the system is now band-limited to $\omega_m$:

$$h(t) = \frac{1}{2\pi}\int_{-\infty}^{+\infty} H(\omega)e^{-j\omega t}d\omega$$

$$\begin{aligned} &= \frac{1}{2\pi}\int_{-\omega m}^{+\omega m} (Ae^{-j\omega\tau})e^{j\omega t}d\omega \\ &= \frac{A\sin[\omega_m(t-\tau)]}{\pi(t-\tau)} \\ &= \left(\frac{A\omega_m}{\pi}\right)\text{sinc}\,[\omega_m(t-\tau)] \end{aligned} \tag{3.6.6}$$

where the standard mathematical function sinc is as shown in Fig. 3.6.2(*a*) (Abramowitz and Stegun 1970, p. 231).

We are confronted by the result that in this case the response is present *before* the input is applied at $t=0$! This brings us to the idea of causality – the response to a cause cannot precede the cause. That is, the output of a causal system at any particular time depends only on the input prior to that time and not on any future input. In general, causal systems are physically realizable while non-causal systems are not.

In order for a linear system to be causal it is necessary and sufficient that the impulse response $h(t)$ be zero for $t<0$. It follows that the response due to an input $f(t)$ depends only on past values of $f(t)$. Above, the violation of causality arises from the independent definitions of $f(\omega)$ and $\phi(\omega)$ so we see that these cannot be independent. Let us consider further the response of the filter to a step input $Au(t)$. Generally, if an input $f(\tau)$ gives an output $g(\tau)$, then an input $\int f(\tau)$ gives an output $\int g(\tau)$, and similarly for a differential. Thus since the unit step $u(t)$ is the integral of the impulse function $\delta(t)$ (or $\delta(t)$ the differential of $u(t)$) we find using the impulse response found above for a step of amplitude $A$:

$$\begin{aligned} a(t) &= \int_{-\infty}^{t} \frac{A\omega_m}{\pi}\,\text{sinc}\,[\omega_m(t-\tau)]dt \\ &= \int_{0}^{\omega_m(t-\tau)} \text{sinc}\,(x)\frac{dx}{\omega_m}, \quad \text{where } x=\omega_m(t-\tau) \\ &= \int_{-\infty}^{0} \frac{A}{\pi}\,\text{sinc}\,(x)dx + \int_{0}^{\omega_m(t-\tau)} \frac{A}{\pi}\text{sinc}\,(x)dx \\ &= \frac{A}{2} + \frac{A}{\pi}\text{si}\,[\omega_m(t-\tau)] \end{aligned} \tag{3.6.7}$$

The integrals sinc and si are detailed in Abramowitz and Stegun (1970, p. 231) and the graph of si($x$) is shown in Fig. 3.6.2($b$). The response $a(t)$ is shown in Fig. 3.6.2($c$). The slope of $a(t)$ is by definition $h(t)$ for the impulse and it is a maximum when $t=\tau$ where it has the value $(A\omega_m/\pi)$. We may mention in passing that causality can cause difficulties in SPICE, for example, in handling some Laplace transforms. PSpice performs an inverse Fourier transform to obtain the impulse response and then carries out a convolution with the input function to determine the output. The impulse function can sometimes lead to a violation of causality and then a delay has to be introduced to reduce the violation to some negligible magnitude (MicroSim 1996).

A measure of the risetime given by $t_r$, defined by the tangent to the response at $t=\tau$ is then:

$$t_r = \frac{\pi}{\omega_m} = \frac{1}{2f_m} \tag{3.6.8}$$

This gives us a rough rule for the bandwidth required to transmit a reasonable facsimile of an input step of risetime $t_r$. Once again we see the response before the stimulus, which is not possible in a realizable system. The phase and amplitude response functions are not independent but are related, as they are in many other physical systems, through the Kramers–Kronig relations (Faulkner 1969, and Section 3.9). These apply in any case where one encounters ‘absorption and dispersion’ – just other names for $f(\omega)$ and $\phi(\omega)$. Absorption implies energy loss, while dispersion produces phase changes. You may be more familiar with the optical occurrence of dispersion, as for example in a prism, which deviates the light by amounts differing according to wavelength. In circuits, resistance causes energy loss and reactive elements produce phase changes. Here one has the real part of the impedance being responsible for the losses and the imaginary components for the phase change. The expression given in Eq. (2.10.1) for the permittivity of a dielectric indicates similar effects. However, because of the way $\varepsilon$ is defined it is the imaginary part $\varepsilon''$ that results in absorption and the real part $\varepsilon'$ which results in dispersion. It is arbitrary as to which way such relations are defined.

The relationship of gain and phase for electronic circuits was formalized by Bode (1940, 1947). The interaction is important for us in particular with regard to feedback systems. As we will see in Section 3.10, it is essential for stability against oscillation in negative feedback systems, that the fall off in gain, and hence phase change, is controlled to ensure that the phase shift does not reach 360° before the gain becomes $<1$.

Consider a step function applied to a simple low-pass $RC$ filter as shown in Fig. 3.6.3. The response will be the exponential rise of the form:

$$(1-e^{-t/RC}) \tag{3.6.9}$$

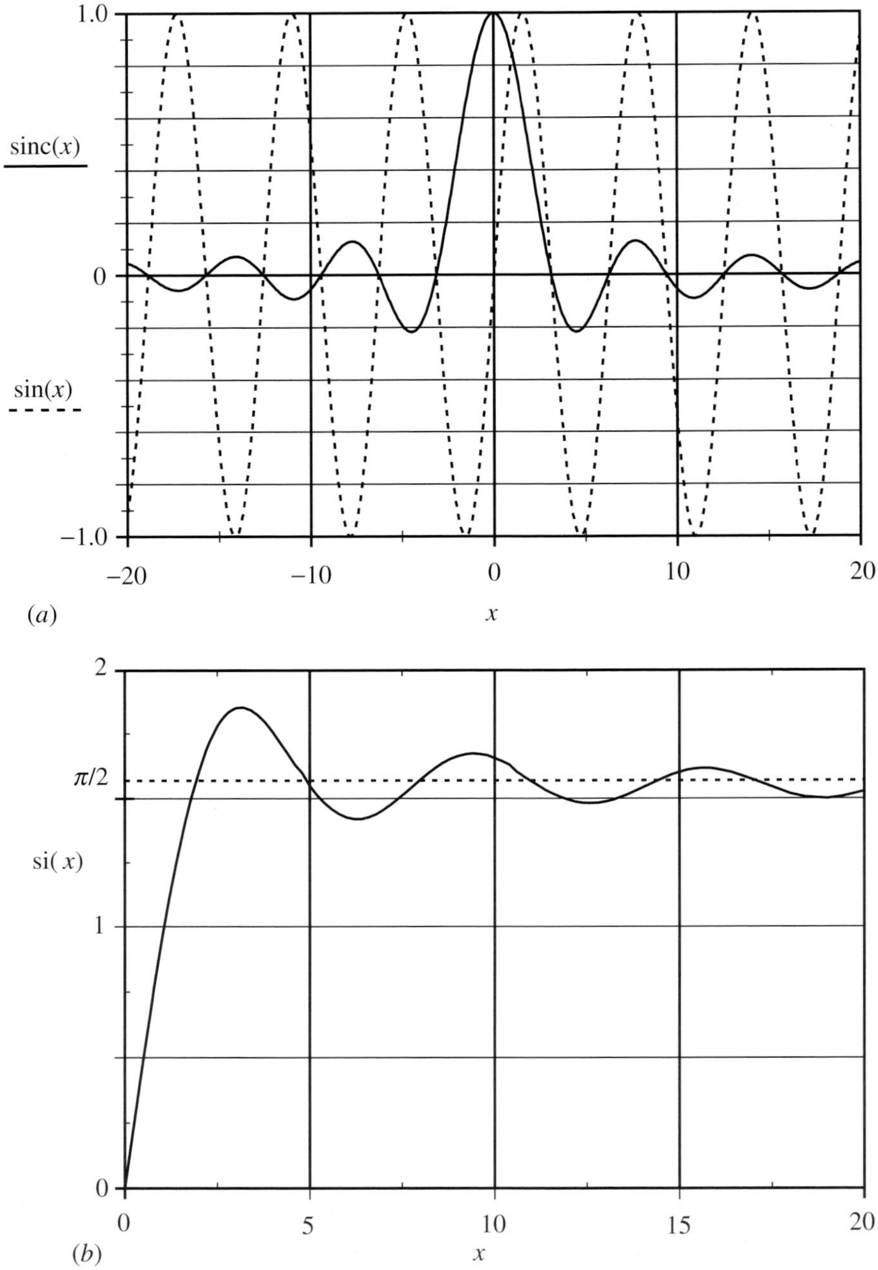


**Fig. 3.6.2** (*a*) Sinc function with corresponding sin function. (*b*) The si function.

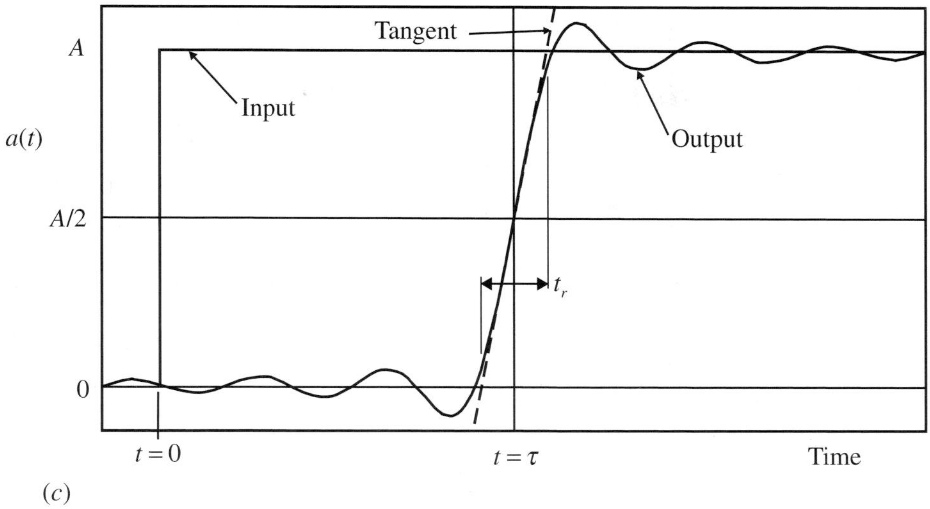


**Fig. 3.6.2** (*c*) The response $a(t)$ given by Eq. (3.6.7).

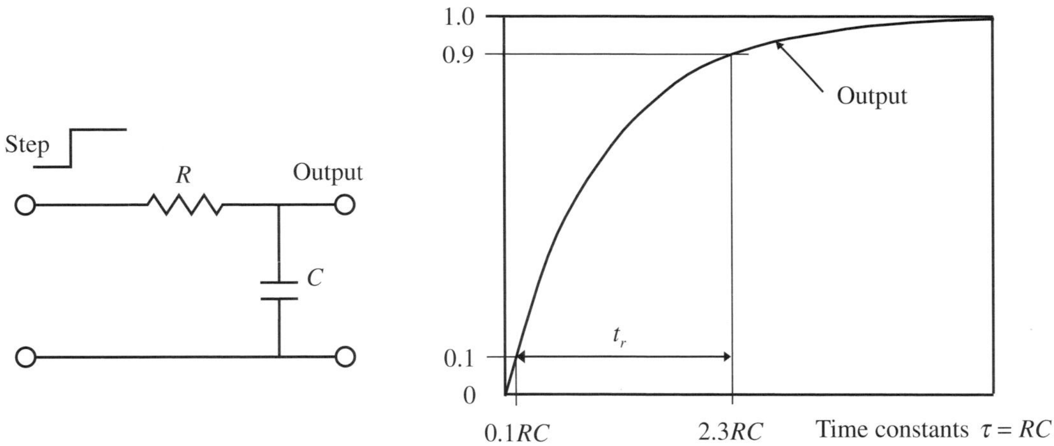


**Fig. 3.6.3** *RC* circuit risetime.

Since there is no clear end to the rise, the risetime $t_r$ is, by general agreement, commonly measured from 10 to 90%. The values of $t$ to reach these levels are:

$$t_{10} = 0.1RC \quad \text{and} \quad t_{90} = 2.3RC \quad \text{with bandwidth} \quad f_c = \frac{1}{2\pi RC}$$
$$\text{so} \quad t_r = 2.2RC = \frac{2.2}{2\pi f_c} = \frac{0.35}{\text{Bandwidth}} \tag{3.6.10}$$

e.g. a '100 MHz' oscilloscope will have a risetime of 3.5 ns. An oscilloscope will not necessarily have such a simple frequency response so the relationship may be a little different, but this relation gives a good guide. It is much the same as given in Eq. (3.6.8). These results hold for a smooth exponential response. If there are overshoots of say 10% then instead of 0.35 you should use about 0.45.

The relationship between bandwidth and response time is not precise and depends rather on how you define the various quantities. Early proposals were made by Elmore (Elmore and Sands 1949, p. 136; Chen 1995, p. 2138) and wider consideration is, for example, given in Siebert (1986, Chapter 16). Though we cannot expect close agreement the theory and practice do bear a useful relationship. If $h(t)$ is the impulse response of the low-pass amplifier circuit with little or no overshoot, then Elmore proposed that the delay time $\tau_D$ between input and output is given by the first moment in time as:

$$\tau_D = \int_0^\infty t\, h(t)\, dt \tag{3.6.11}$$

and that the risetime $\tau_R$ is given by the second moment (offset by $\tau_D$):

$$\tau_R = \left[ 2\pi \int_0^\infty (t-\tau_D)^2\, h(t)\, dt \right]^{\frac{1}{2}} \tag{3.6.12}$$

One attraction is that these two quantities can be fairly simply related to the coefficients of the transfer function $H(s)$ of $h(t)$:

$$H(s) = \frac{1 + a_1 s + a_2 s^2 + \cdots + a_m s^m}{1 + b_1 s + b_2 s^2 + \cdots + b_n s^n}, \text{ with } n > m \tag{3.6.13}$$

where $a_0$ and $b_0$ are put equal to 1 so that the step response tends to unity as $t \to \infty$. Expanding (3.6.12):

$$\begin{aligned}
\tau_R &= \left[ 2\pi \int_0^\infty (t-\tau_D)^2\, h(t)\, dt \right]^{\frac{1}{2}} \\
&= \left[ 2\pi \int_0^\infty (t^2 - 2t\tau_D + \tau_D^2) h(t)\, dt \right]^{\frac{1}{2}} \\
&= \left\{ 2\pi \left[ \int_0^\infty t^2\, h(t)\, dt - \int_0^\infty 2t\tau_D h(t)\, dt + \int_0^\infty \tau_D^2\, h(t)\, dt \right] \right\}^{\frac{1}{2}} && (3.6.14) \\
&= \left\{ 2\pi \left[ \int_0^\infty t^2\, h(t)\, dt - 2\tau_D^2 + \tau_D^2 \int_0^\infty h(t)\, dt \right] \right\}^{\frac{1}{2}} && \text{using (3.6.11)}
\end{aligned}$$

$$= \left\{ 2\pi \left[ \left[ \int_0^\infty t^2\, h(t)\, dt - \tau_D^2 \right] \right] \right\}^{\frac{1}{2}}, \quad \text{since} \int_0^\infty h(t)\, dt = 1$$

From the definition of the Laplace transform:

$$\begin{aligned} H(s) &= \int_0^\infty h(t) e^{-st}\, dt \\ &= \int_0^\infty h(t) \left( 1 - st + \frac{s^2 t^2}{2!} - \cdots \right) dt \\ &= \int_0^\infty h(t) dt - s \int_0^\infty t\, h(t) dt + \frac{s^2}{2!} \int_0^\infty t^2\, h(t) dt - \cdots \\ &= \int_0^\infty h(t) dt - s\tau_D + \frac{s^2}{2!} \left( \frac{\tau_R^2}{2\pi} + \tau_D^2 \right) - \cdots \quad \text{using (3.6.11) and (3.6.14)} \end{aligned} \tag{3.6.15}$$

Dividing the numerator into the denominator of Eq. (3.6.13) gives:

$$H(s) = 1 - (b_1 - a_1)s + (a_2 - b_2 - b_1 a_1 - b_1^2)s^2 + \cdots \tag{3.6.16}$$

and equating coefficients of the same power in $s$ in (3.6.15) and (3.6.16) gives:

$$\int_0^\infty h(t)\, dt = 1, \quad s\tau_D = (b_1 - a_1)s, \quad \frac{s^2}{2!} \left( \frac{\tau_R^2}{2\pi} + \tau_D^2 \right) = (a_2 - b_2 - b_1 a_1 - b_1^2)s^2$$

$$\text{or} \quad \tau_D = (b_1 - a_1) \quad \text{and} \quad \frac{\tau_R^2}{2\pi} + (b_1 - a_1)^2 = 2(a_2 - b_2 - b_1 a_1 - b_1^2) \tag{3.6.17}$$

$$\text{so} \quad \tau_R = \{2\pi[b_1^2 - a_1^2 + 2(a_2 - b_2)]\}^{\frac{1}{2}}$$

and the scaling factor is the cut-off frequency $\omega_0$. As an example we may examine the response of the 3-pole low-pass filter discussed in Section 5.13, and we choose the Paynter configuration as this has good pulse response and negligible overshoot. The coefficients (with a corner frequency of 10 kHz) are:

$$a_0 = 1, \quad a_1 = 0, \quad a_2 = 0, \quad b_0 = 1, \quad b_1 = 3.2, \quad b_2 = 4, \quad \omega_0 = 2\pi \times 10^4$$

$$\text{so} \quad \tau_D = \frac{3.2 - 0}{2\pi \times 10^4} = 51\ \mu\text{s} \quad \text{and} \quad \tau_R = \frac{\{2\pi[3.2^2 - 0 + 2(0 - 4)]\}^{\frac{1}{2}}}{2\pi \times 10^4} = 60\ \mu\text{s} \tag{3.6.18}$$

Simulating the circuit with PSpice with a LM6142 amplifier, and using a 1 μs risetime pulse and using the usual 10 to 90% measure for risetime, and to 50% for delay (Chen 1995) gave:

$$\tau_D = 50\ \mu\text{s} \quad \text{and} \quad \tau_R = 57\ \mu\text{s} \tag{3.6.19}$$

but one should not read too much into the reasonably close agreement. It does, however, under the appropriate conditions, give a reasonable indication of what to expect. For the same circuit it gave good agreement for the Bessel configuration but failed drastically for the Butterworth at least for risetime, where there was over 9% overshoot.

The delay defined as proportional to $\omega$ is known as the phase delay and in terms of the slope of the phase curve it is known as the group delay (Van Valkenburg 1982).

$$\text{Phase delay} \quad \theta = -\omega\tau, \quad \text{group delay} \quad \tau = \frac{-d\theta}{d\omega} \tag{3.6.20}$$

and for a transfer function $H(s)$ in terms of the real and imaginary parts:

$$H(j\omega) = R(j\omega) + jX(j\omega), \quad \text{so phase} \quad \theta = \tan^{-1}\left[\frac{X(\omega)}{R(\omega)}\right]$$

and differentiating

$$\begin{aligned}\frac{d\theta}{d\omega} &= \frac{1}{\left[1 + \dfrac{X(\omega)^2}{R(\omega)^2}\right]} \frac{d[X(\omega)/R(\omega)]}{d\omega} \\ &= \frac{1}{\left[1 + \dfrac{X(\omega)^2}{R(\omega)^2}\right]} \frac{R(\omega)\dfrac{dX(\omega)}{d\omega} - X(\omega)\dfrac{dR(\omega)}{d\omega}}{R(\omega)^2} \\ &= \frac{R(\omega)\dfrac{dX(\omega)}{d\omega} - X(\omega)\dfrac{dR(\omega)}{d\omega}}{R(\omega)^2 + X(\omega)^2} = -\tau\end{aligned} \tag{3.6.21}$$

where we have used Eqs. (1.8.1) and (1.8.2) for the differentiation.

The statement of the bandwidth of a system tells only part of the story. The rise-time–bandwidth relationship of Eq. (3.6.10) assumes that the system is able to deliver any rate of change that is necessary. For example, in an amplifier there are node capacities that must be charged through source resistances that limit the rate of change to some maximum value which leads to the definition of what is known as the slewing rate $SR$. For a sine wave the maximum rate of change is:

$$v = v_0 \sin(\omega t), \quad \text{so} \quad \frac{dv}{dt} = v_0\omega\cos(\omega t) \tag{3.6.22}$$

which has a maximum value for $\cos(\omega t) = 1$ or $\sin(\omega t) = 0$

Thus the maximum rate of change of the sine wave depends on both frequency and amplitude. For the output of the amplifier to be able to reproduce the amplified wave faithfully it must be able to slew at the required rate. If it cannot achieve the required rate then the output will be distorted. Since the internal charging mechanisms tend to act like current sources charging the node capacities, the output will be limited to a 'linear' ramp according to the slewing rate rather than the input wave. Specifications for the bandwidth of an op amp are usually in terms of a small enough amplitude of signal so that the available slewing rate is not exceeded. The large-signal bandwidth will, as a consequence of the maximum slewing rate, be substantially less. SPICE can be misleading in this respect since for an *AC SWEEP* to determine the frequency response it finds the bias point and then assumes the system is linear, i.e. without any amplitude limitations. You will have to run *TRANSIENT* simulations to see the limiting effects. Data sheets generally give a specification of the maximum slewing rate, usually in terms of V $\mu s^{-1}$. If the slewing rate is $S$ then:

$$(v_0 f)_{max} = \frac{S}{2\pi} \qquad (3.6.23) \qquad \text{(see Allen 1977).}$$

## SPICE simulation circuits

None

## References and additional sources 3.6


Abramowitz M., Stegun I. A. (1970): *Handbook of Mathematical Functions with Formulas, Graphs and Mathematical Tables*, Washington: National Bureau of Standards, Applied Mathematics Series.

Allen P. E. (1977): Slew induced distortion in operational amplifiers. *IEEE J.* **SC-12**, 39–44.

Bode H. W. (1940): Relation between attenuation and phase in feedback amplifier design. *Bell Syst. Tech. J.* **19**, 412–454.

Bode H. W. (1947): *Network Analysis and Feedback Amplifier Design*, New York: Van Nostrand.

Chen, Wai-Kai (Ed.) (1995): *The Circuits and Filters Handbook*, Boca Raton: CRC Press and IEEE Press. ISBN 0-8493-8341-2.

Elmore W. C., Sands M. (1949): *Electronics: Experimental Techniques*, New York: McGraw-Hill.

Faulkner E. A. (1969): *Introduction to the Theory of Linear Systems*, London: Chapman and Hall. ISBN 412-09400-2.

MicroSim (1996): *PSPICE User's Guide Ver. 7.1*, MicroSim Corp. Oct., pp. 6–45.

Siebert W. McC. (1986): *Circuits, Signals, and Systems*, New York: MIT Press/McGraw-Hill. ISBN 0-07-057290-9.

Skilling J. K. (1968): Pulse and frequency response. *General Radio Experimenter* November–December, 3–10.

Van Valkenburg M. E. (1982): *Analog Filter Design*, New York: Holt, Rinehart and Winston. ISBN 0-03-059246-1, or 4-8338-0091-3 International Edn.

# 3.7 Pulse and transient response

The object of this communication is to determine the motion of electricity at any instant after an electrified conductor, of given capacity, is put in connexion with the earth by means of a wire or other linear conductor of given form and resisting power.

William Thomson to the Glasgow Philosophical Society, 19 January 1853

In Section 3.2 we considered the analysis of circuits for the case of sinusoidal waveforms. In Section 1.5 the simple exponential responses were found for the cases of charging of a capacitor through a resistor or establishing the current in an inductor when the input is a step function. For waveforms other than sinusoidal we generally resort to the power of the Laplace transform (Section 1.12) to determine responses, but it is informative and beneficial to examine a few simple cases from the basic circuit equations as in Section 1.5. A good discussion of *RLC* waveshaping is given by Millman and Taub (1965).

Consider a voltage pulse rising from zero to $v_{in}$ applied to the *RC* circuit of Fig. 3.7.1(*a*).

Considering the pulse in two parts, it is evident from Eq. (1.5.2) that the initial rise will give an exponentially rising output as illustrated in (*b*). How close the output gets to the input $v_{in}$, depends on the relative values of the pulse length $t_p$ and the time constant $\tau = RC$. If we assume for the moment that $t_p \gg \tau$ then the falling edge of the input will cause a similar falling exponential from $v_{in}$ down to zero, the quiescent value of the input. If $t_p$ is decreased then the output will be as shown for the given ratios of $t_p$ to $\tau$. The peak values may be read from Table 1.5.1.

Inverting *R* and *C* gives Fig. 3.7.2(*a*) and we apply the same range of inputs to get the outputs shown in (*b*).

In this case, since the gain at z.f. is zero, the output must go negative so that the average value is zero, i.e. the 'positive' area between the waveform and common is equal to the 'negative' area. This can be shown by plotting the integral of $v_{out} = v_R$, which will tend to zero after the pulse (plot $S(V(V_{out})))$. We can consider this in another way. After the pulse, $v_{in} = 0$ and the quiescent value of $v_R$ must be zero; thus $v_C$ must also go to zero. If $i$ is the current at any instant and $q$ the charge on $C$, then:

$$v_R = iR = R\frac{dq}{dt}, \quad \text{where } q \text{ is the charge on } C$$

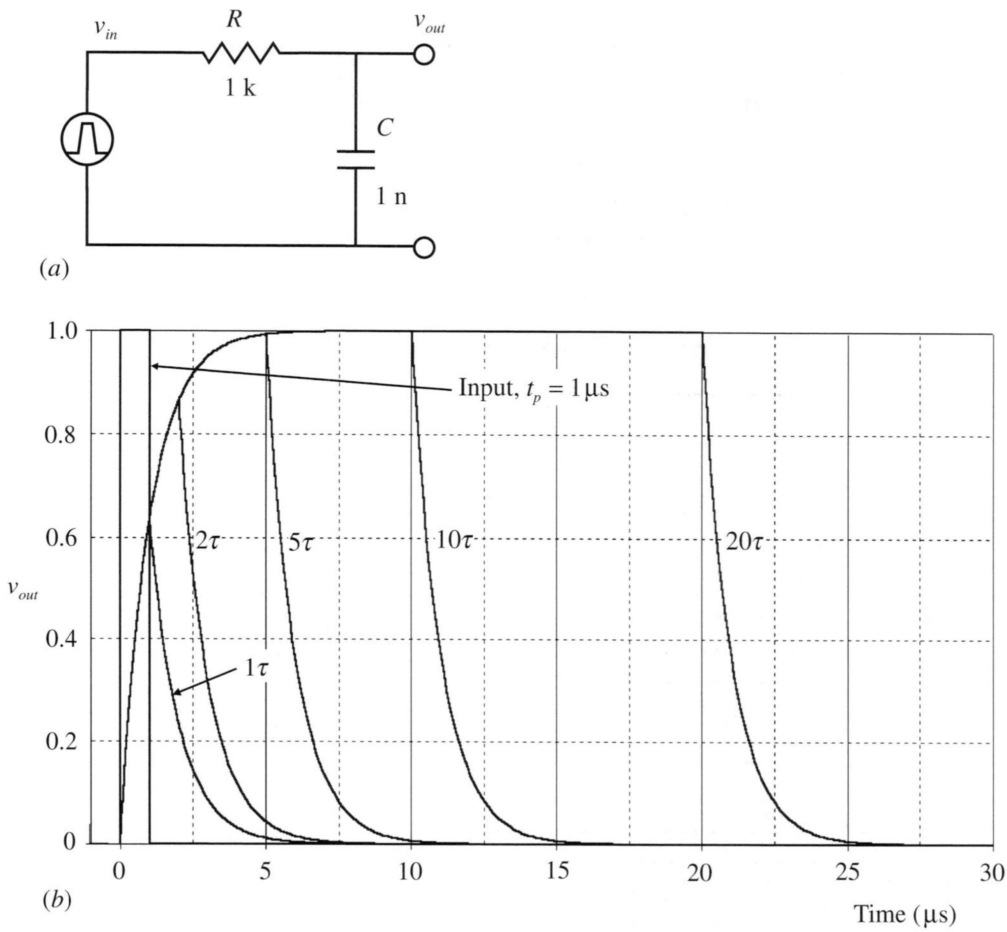


**Fig. 3.7.1** (*a*) Pulse applied to *RC* circuit with $\tau = 1$ μs. (*b*) Output responses for input pulse widths $t_p = 1$, 2, 5, 10 and 20 times $\tau$. Only the $t_p = 1$ μs input pulse is shown (amplitude 1 V, rise and fall times 10 ns).

$$\text{so} \quad \int_0^t v_R dt = R\int_0^t \frac{dq}{dt}dt = R\int_0^Q dq = RQ = RCv_C, \quad \text{since} \quad Q = Cv_C \qquad (3.7.1)$$

so if you plot $RCv_C$ you will find it equal to the integral of $v_R$ and that it eventually goes to zero. For repetitive pulses the average output level must be zero so the pulses will rise from some negative level dependent on the mark-to-space ratio, which presents a problem if the ratio changes since the base level will also change (Millman and Taub 1965, p. 30). The problem is even more difficult if the mark-to-space ratio or the pulse repetition rate is random since then the absolute levels cannot be predicted. In cases where this matters d.c. coupling must be used to maintain the absolute levels or more complex correction techniques implemented to provide base-line restoration (Chase 1961).

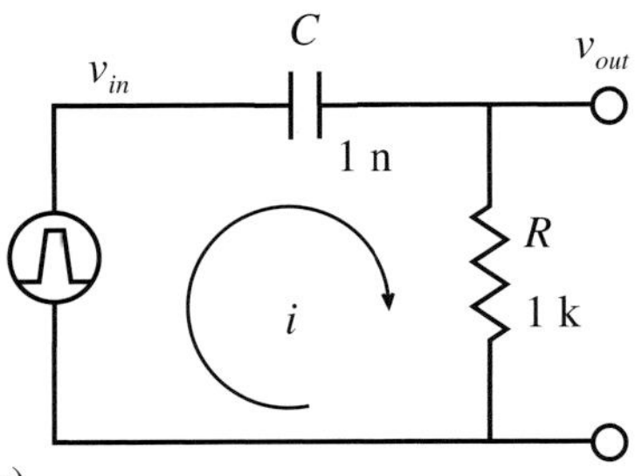


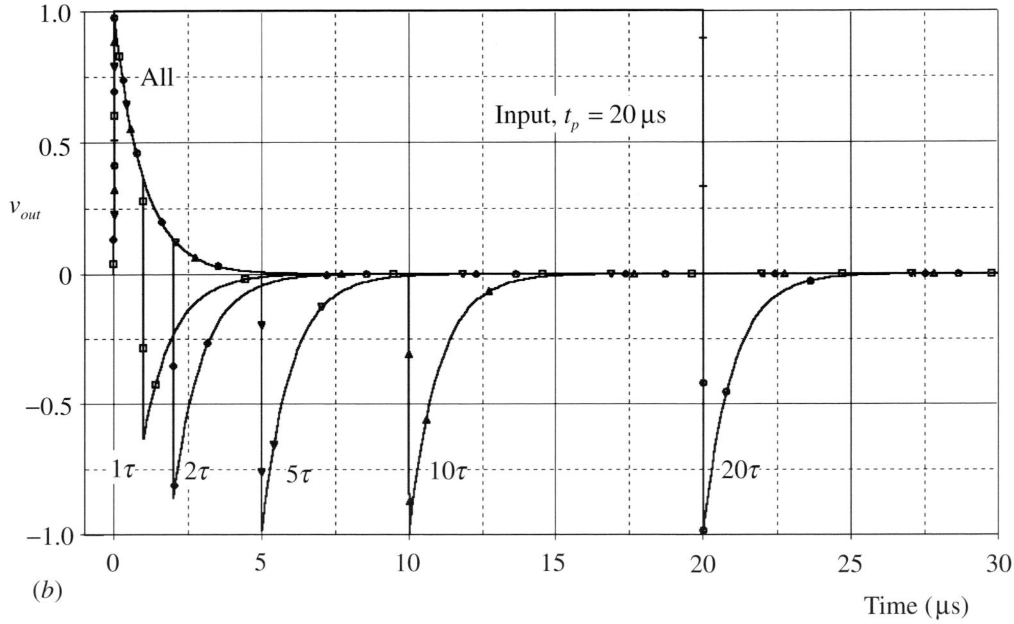


**Fig. 3.7.2** (*a*) Pulse applied to *CR* circuit with $\tau = 1$ μs. (*b*) Output responses for input pulse widths $t_p = 1$, 2, 5, 10 and 20 times $\tau$. Only the $t_p = 20$ μs input pulse is shown (amplitude 1 V, rise and fall times 10 ns).

If the input waveform is not a fast edged pulse then it is less easy to see what the output will be on the basis of simple physical arguments. Consider a ramp input as shown in Fig. 3.7.3(*a*).

The describing equation relating $v_{in}$ and $v_{out}$ is (where $q$ is the charge on $C$):

$$v_{in} = \frac{q}{C} + v_{out} \quad \text{with} \quad i = \frac{v_{out}}{R}$$
$$\text{so} \quad \frac{dv_{in}}{dt} = \frac{1}{C}\frac{dq}{dt} + \frac{dv_{out}}{dt} = \frac{i}{C} + \frac{dv_{out}}{dt} = \frac{v_{out}}{RC} + \frac{dv_{out}}{dt} \tag{3.7.2}$$

Since at time $t = 0$ the current $i$ is zero then initially:

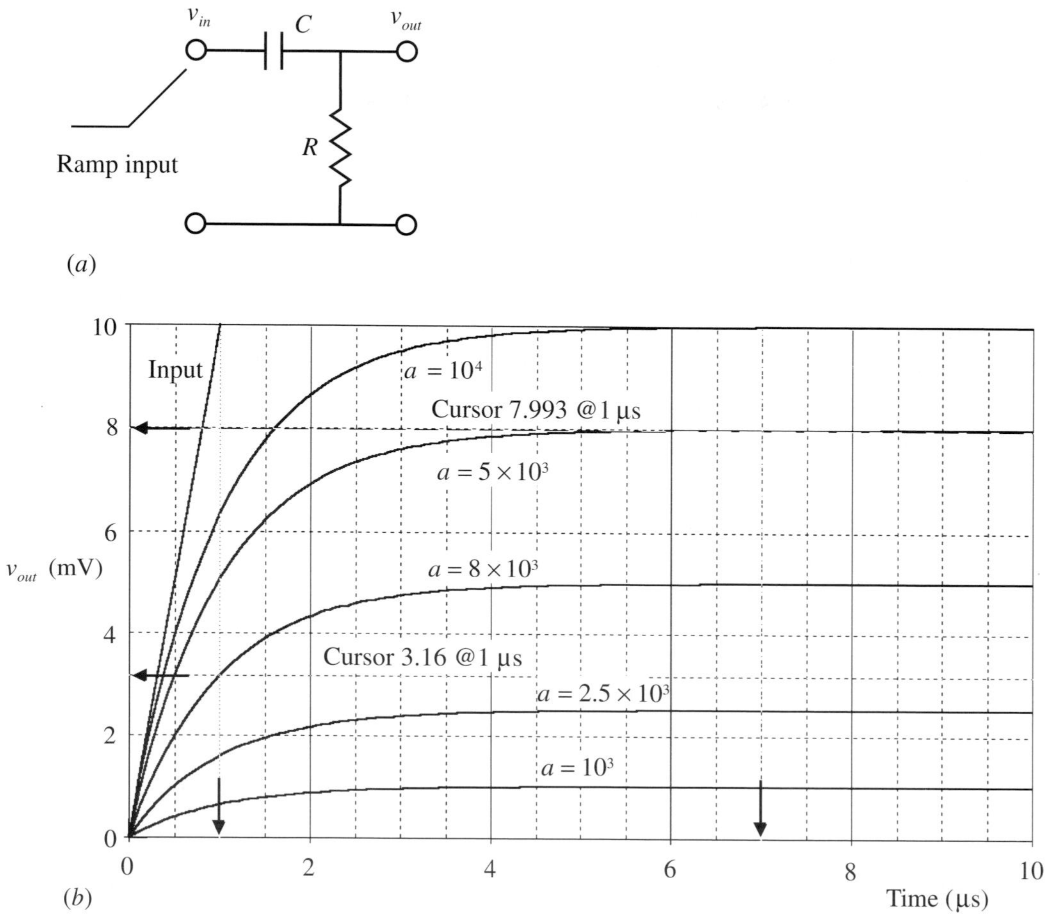


**Fig. 3.7.3** (*a*) Circuit with ramp input. (*b*) $v_{out}$ for a series of values of $a$ in V s$^{-1}$. The time constant $RC = 10^{-6}$ s.

$$\left.\frac{dv_{in}}{dt}\right|_{t=0} = \left.\frac{dv_{out}}{dt}\right|_{t=0} \tag{3.7.3}$$

so the initial rate of rise is the same. If the ramp input is zero before $t = 0$ and $v_{in} = at$ after then we have:

$$a = \frac{v_{out}}{RC} + \frac{dv_{out}}{dt} \quad \text{or} \quad RC\frac{dv_{out}}{dt} = aRC - v_{out} \tag{3.7.4}$$

which we must solve to find $v_{out}$. We make use of the approach of (1.9.5) and (1.9.7) by substituting:

$$x = aRC - v_{out} \quad \text{so that} \quad \frac{dx}{dt} = \frac{-dv_{out}}{dt} \quad \text{and with limits}$$
$$t = 0,\ v_{out} = 0,\ x = aRC \quad \text{and} \quad t = t,\ v_{out} = v_{out},\ x = aRC - v_{out} \tag{3.7.5}$$

so that (3.7.4) becomes:

$$\frac{dx}{dt}=\frac{-x}{RC} \quad \text{and integrating} \quad \int_{aRC}^{(aRC-v_{out})}\frac{dx}{dt}=\frac{-1}{RC}\int_0^t dt$$

$$\text{so } \ln(aRC-v_{out})-\ln(aRC)=\frac{-t}{RC} \quad \text{or} \quad \ln\left(\frac{aRC-v_{out}}{aRC}\right)=\frac{-t}{RC} \tag{3.7.6}$$

$$\text{then} \quad \frac{aRC-v_{out}}{aRC}=e^{\frac{-t}{RC}} \quad \text{or} \quad v_{out}=aRC\left(1-e^{\frac{-t}{RC}}\right)$$

so that after a long time the output should level out at $aRC$. Responses for a range of input rates $a$ with a fixed time constant $RC$ are shown in Fig. 3.7.3(*b*). The cursors, at 1 μs (= $RC$) and 7 μs (= $7RC$ for within 0.1% of final value, Table 1.5.1), show agreement with expected values.

For an exponential input waveform of time constant $\tau$, the input $v_{in}$ with a final amplitude $V_0$, and the describing equation and output $v_{out}$, are given by:

$$v_{in}=V_0\left(1-e^{\frac{-t}{\tau}}\right) \quad \text{and} \quad \frac{dv_{in}}{dt}=\frac{i}{C}+\frac{dv_{out}}{dt} \quad \text{so} \quad \frac{V_0}{\tau}e^{\frac{-t}{\tau}}=\frac{v_{out}}{RC}+\frac{dv_{out}}{dt}$$

$$\text{and letting} \quad n\equiv\frac{RC}{\tau} \quad \text{we get} \quad v_{out}=\frac{nV_0}{(n-1)}\left(e^{\frac{-t}{n\tau}}-e^{\frac{-t}{\tau}}\right) \quad \text{provided } n\neq 1 \tag{3.7.7}$$

$$\text{and for } n=1 \text{ we have the special case} \quad v_{out}=\frac{tV_0}{\tau}e^{\frac{-t}{\tau}}$$

The procedure used is much like that for ramp input, but here we find again the problem of a function of the form 0/0 if $n=1$ so we need a separate solution. Both may be proved by differentiating. Simulated waveforms are shown in Fig. 3.7.4.

The time at maximum for a given value of $n$ (other than $n=1$ which occurs at 1) is given by:

$$\left.\frac{t}{\tau}\right|_{\max}=\frac{n}{n-1}\ln(n) \tag{3.7.8}$$

The form of the output from the high-pass $CR$ circuit is approximately the differential of the input and hence the arrangement is often referred to as a differentiating circuit. Conversely, the $RC$ low-pass arrangement is often called an integrating circuit. Two high-pass $CR$ circuits in series, but separated by a buffer so they do not interact, provide double differentiation, which has some applications. The circuit shown in Fig. 3.7.5(*a*) produces the outputs shown in (*b*). The case illustrated has the two $CR$ time constants equal and the same as that for the input exponential, to match the results shown in Millman and Taub (1965, p. 42).

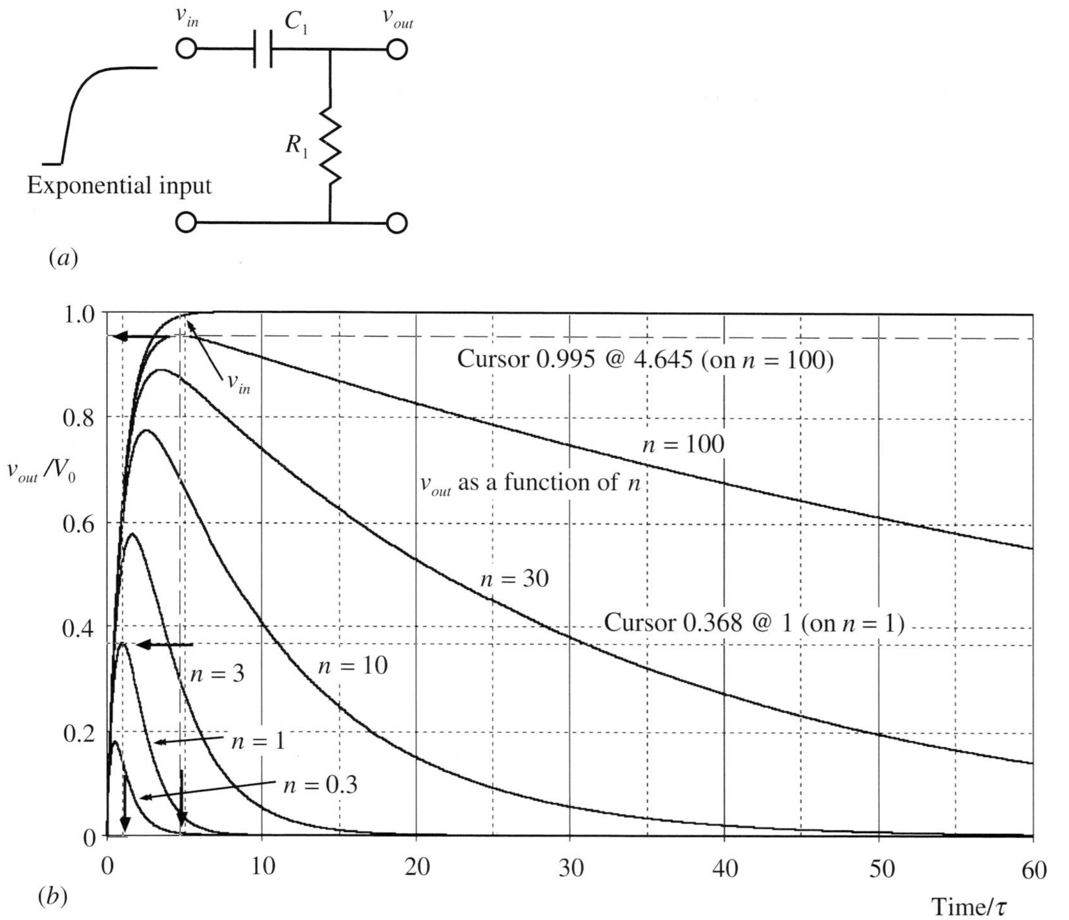


**Fig. 3.7.4** (*a*) High-pass *CR* circuit with exponential input of time constant $\tau$. (*b*) Output waveforms for various values of $n = RC/\tau$. The $x$-axis is scaled as $t/\tau$ and the $y$-axis as $v/V_0$.

The advantage now is that variations of any of the parameters can be quickly checked, whereas in the reference the simple case was used as otherwise the sums get very messy. The attraction of the circuit is that we get a sharpened pulse from a slower exponential waveform and a sharp zero crossing if timing is of interest.

One of the earliest papers on transients was written by William Thomson (1853) (subsequently the eminent Lord Kelvin). He considered the current flow when a 'discharger' was connected across a capacitor, and in particular some experiments carried out by Weber. The 'dischargers' used were various lengths of 'wet cord', which we may imagine were not exactly reproducible. As well as having resistance $k$, Thomson introduces a quantity $A$ which he calls 'the electrodynamic capacity of the discharger'. He relates that $A$ is a quantity which, when the length of discharger is bent back on itself, the 'energy of the current' goes to nil. He shows that

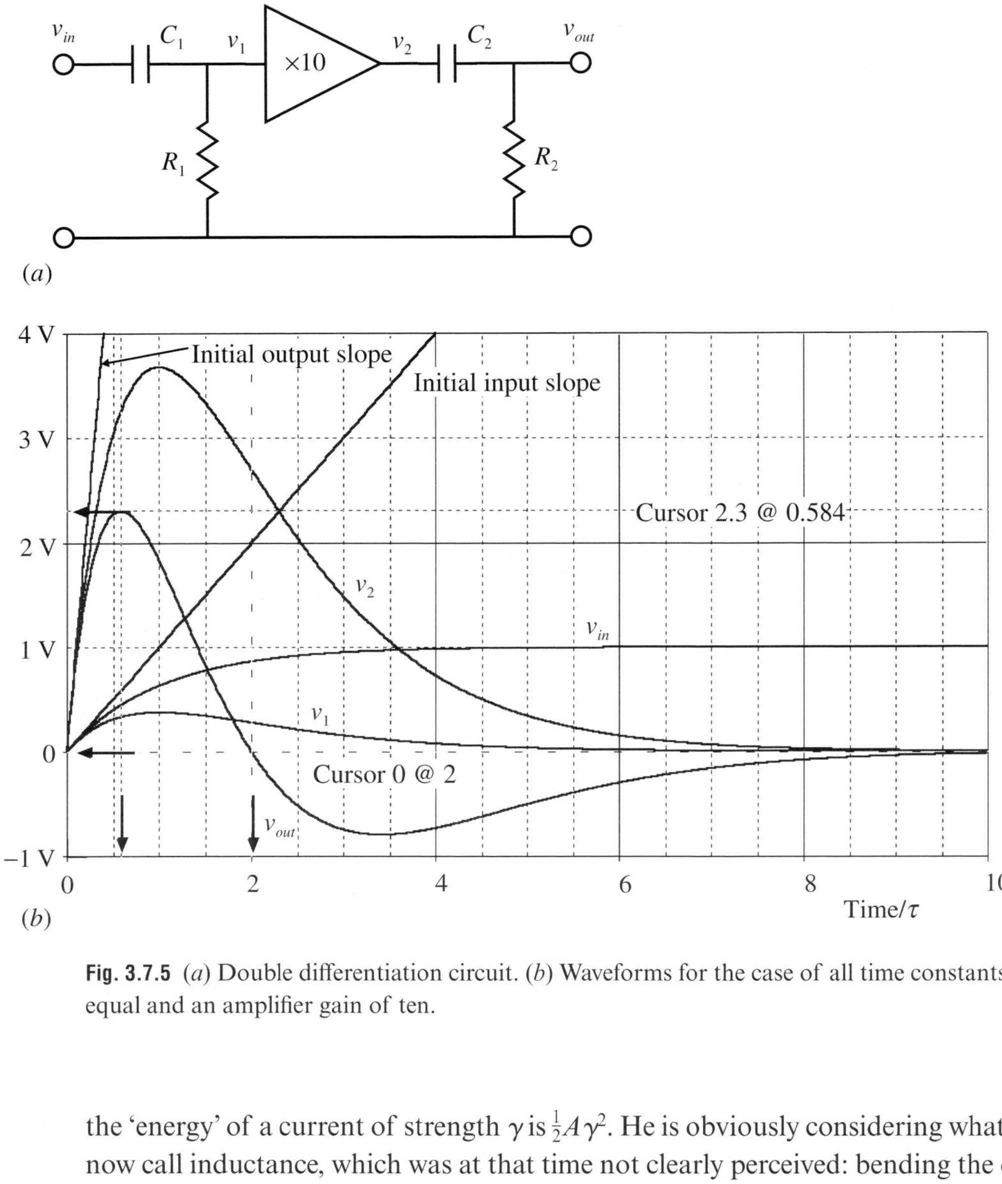


**Fig. 3.7.5** (*a*) Double differentiation circuit. (*b*) Waveforms for the case of all time constants equal and an amplifier gain of ten.

the 'energy' of a current of strength $\gamma$ is $\frac{1}{2}A\gamma^2$. He is obviously considering what we now call inductance, which was at that time not clearly perceived: bending the discharger back on itself is what we would do to have a non-inductive resistance, and the energy stored in an inductor carrying current $i$ is $\frac{1}{2}Li^2$ (Section 2.2). He then derives the equation for the charge $q$ on the capacitor as:

$$\frac{d^2q}{dt^2}+\frac{k}{A}\frac{dq}{dt}+\frac{q}{CA}=0, \quad \text{with} \quad \gamma=\frac{-dq}{dt}$$

$$\text{or} \quad \frac{d^2q}{dt^2}+\frac{R}{L}\frac{dq}{dt}+\frac{q}{CL}=0, \quad \text{with} \quad i=\frac{-dq}{dt} \tag{3.7.9}$$

which is exactly the equation we would now get for the circuit shown in Fig. 3.7.6.

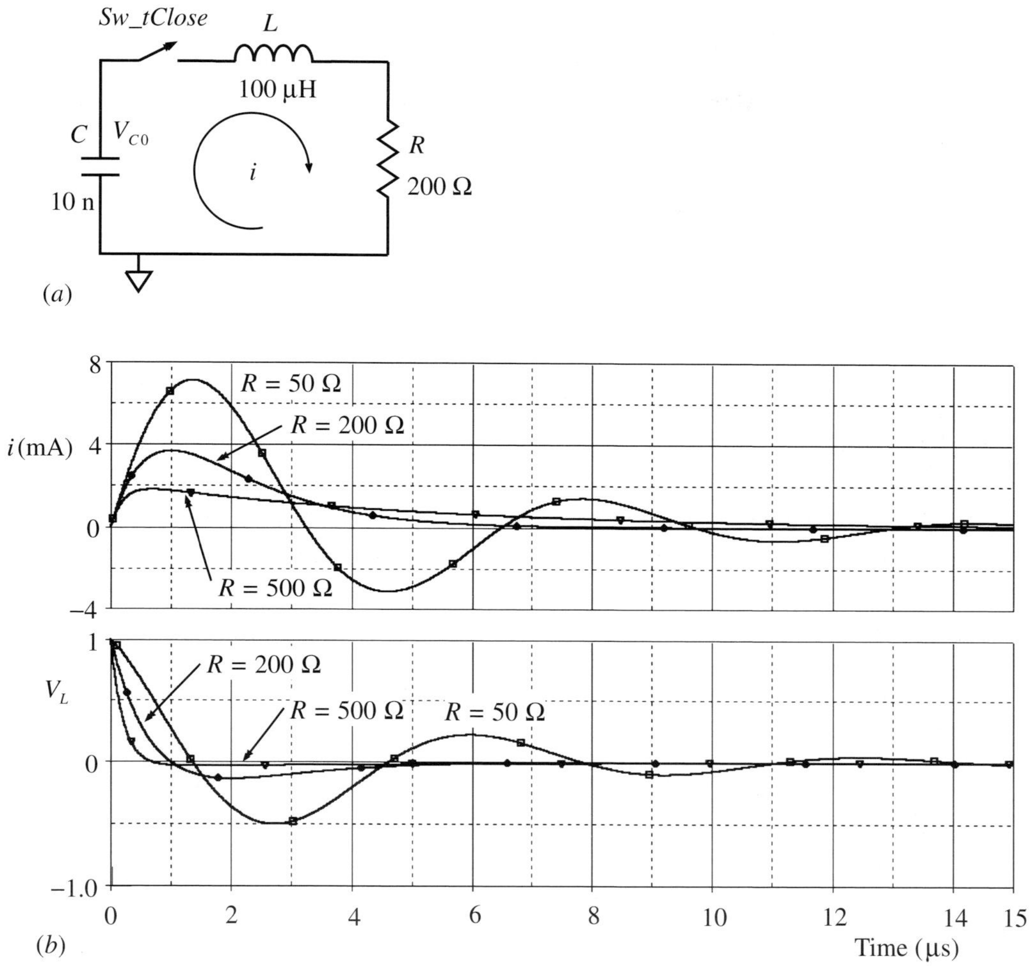


**Fig. 3.7.6** (*a*) Discharge of a capacitor through a resistor and an inductor. (*b*) Waveforms for under ($R=50\ \Omega$), critical ($R=200\ \Omega$) and overdamped ($R=500\ \Omega$) conditions showing voltage across $L$ and the current.

We would more usually derive the describing differential equation in terms of current or voltage but it can readily be transformed in terms of $q$:

$$v_L = L\frac{di}{dt},\quad v_R = iR,\quad v_C = \frac{q}{C}\quad \text{and}\quad v_L + v_R + v_C = 0$$

$$\text{so}\quad L\frac{di}{dt} + iR + \frac{q}{C} = 0\quad \text{and}\quad i = \frac{dq}{dt},\quad \frac{di}{dt} = \frac{d^2q}{dt^2}$$

$$\text{giving}\quad L\frac{d^2q}{dt^2} + R\frac{dq}{dt} + \frac{q}{C} = 0\quad \text{or}\quad \frac{d^2q}{dt^2} + \frac{R}{L}\frac{dq}{dt} + \frac{q}{CL} = 0 \qquad (3.7.10)$$

$$\text{and in terms of } i\text{:}\quad \frac{d^2i}{dt^2} + \frac{R}{L}\frac{di}{dt} + \frac{i}{CL} = 0$$

To find a solution for the differential equation we make a guess and see if it fits. Assume an exponential form for $i$, differentiate and substitute in Eq. (3.7.10) to get:

$$i = Pe^{\alpha t}, \quad \text{with} \quad \frac{di}{dt} = \alpha Pe^{\alpha t} \quad \text{and} \quad \frac{d^2i}{dt^2} = \alpha^2 Pe^{\alpha t}$$

$$\text{so} \quad L\alpha^2 + R\alpha + \frac{1}{C} = 0 \quad \text{and } \alpha \text{ may have two values} \tag{3.7.11}$$

$$\alpha_{1,2} = \frac{-R \pm \left(R^2 - \dfrac{4L}{C}\right)^{\frac{1}{2}}}{2L} = \frac{-R}{2L} \pm \frac{1}{2}\left(\frac{R^2}{L^2} - \frac{4}{LC}\right)^{\frac{1}{2}}$$

and depending on whether $\alpha$ is real or complex we find simple exponential or oscillatory decay. In fact we are just back to the results of (1.12.12) and (1.12.13), as we should be since the circuits are in effect identical. The difference of the sign for the first term $R/2L$ arises from our use of positive $k$ here in the form assumed for $i$ rather than negative in the other. The $\alpha$'s must be negative so that we get a decaying response as there is no continuing source of energy present. The equation for the response can now be written:

$$i = Me^{\alpha_1 t} + Ne^{\alpha_2 t} \tag{3.7.12}$$

and, since the current through the inductor cannot be changed instantaneously, then at time $t = 0$ we must have $i = 0$; thus we find $M = -N$. The capacitor is initially charged to say a voltage $V_{C0}$. At $t = 0$, $i = 0$ and all the voltage across $C$ must be the same as that across $L$. Thus we have for $V_{C0}$:

$$V_{C0} = L\frac{di}{dt} = k_1 Me^{\alpha_1 t} - k_2 Me^{\alpha_2 t} = \alpha_1 M = \alpha_2 M$$

$$\text{or} \quad M = \frac{V_{C0}}{\alpha_1 - \alpha_2} \tag{3.7.13}$$

$$\text{and} \quad i = \frac{V_{C0}\, e^{\alpha_1 t}}{L(\alpha_1 - \alpha_2)} - \frac{V_{C0}\, e^{\alpha_2 t}}{L(\alpha_1 - \alpha_2)} = \frac{V_{C0}(e^{\alpha_1 t} - e^{\alpha_2 t})}{L(\alpha_1 - \alpha_2)}$$

Weber and Thomson were faced with great problems in trying to detect any oscillation as well as in measuring the 'electrodynamic capacity' or inductance. They had to infer oscillation from such phenomena as the varying magnetization of fine steel needles next to the conductor or the decomposition of water in an electrolysis apparatus in that 'both descriptions of gases are exhibited at both electrodes'.

There is, however, no difficulty in our simulating the circuit to check the oscillations and the transition from damped oscillation to exponential decay. In the circuit of Fig. 3.7.6 the switch *Sw_tClose* is set up for an off resistance of 10 MΩ and an on resistance of 0.01 Ω. The switch is closed at $t = 0$ and the transition time $ttran = 1$ ns. This finite transition time for the switch to act, i.e. to make the linear variation of resistance from the open to closed values, is necessary to avoid convergence problems with SPICE. The open resistance also means that there is not a floating node at the switch terminals. If you make the values of the components such that the response times are comparable with *ttran* you will see small discrepancies with predicted times. The initial condition for $C$ is set to 1 V.

For the critically damped case, where $\alpha_1 = \alpha_2 = -R/2L$ we find a difficulty with Eq. (3.7.13) in that we land up with $i = 0/0$, which is undefined. This may be resolved by the use of l'Hôpital's rule (Section 1.3) or by series expansion of the terms to see how the functions approach the limit rather than what they do exactly at the limit. If $\alpha$ is the common value of $\alpha_1$ and $\alpha_2$ at the limit, then we can write for the values near the limit:

$$\alpha_1 = \alpha + \delta, \quad \alpha_2 = \alpha - \delta \quad \text{so} \quad \alpha_1 - \alpha_2 = 2\delta \quad \text{and let} \quad \delta \to 0 \tag{3.7.14}$$

and writing (3.7.13) in terms of these expressions gives:

$$\begin{aligned} i &= \left(\frac{V_{C0}}{L}\right)\frac{(e^{\alpha t}e^{\delta t} - e^{\alpha t}e^{-\delta t})}{2\delta} = \left(\frac{V_{C0}}{L}\right)\frac{e^{\alpha t}(e^{\delta t} - e^{-\delta t})}{2\delta} \\ &= \left(\frac{V_{C0}}{L}\right)\frac{e^{\alpha t}[1 + \delta t - (1 - \delta t)]}{2\delta} = \left(\frac{V_{C0}}{L}\right)\frac{e^{\alpha t}\, 2\delta t}{2\delta} = \left(\frac{V_{C0}}{L}\right) t e^{\alpha t} \end{aligned} \tag{3.7.15}$$

where the exponentials have been approximated by the series expansion (Section 1.3), keeping only the first two terms since $\delta$ is very small and tending to zero. The response will rise according to the factor $t$ but then fall again, since $\alpha = -R/2L$, as the decaying exponential comes to dominate. The time at the maximum is found by differentiating and equating to zero:

$$\frac{di}{dt} = \exp\left(-\frac{Rt}{2L}\right)\left(\frac{V_{C0}}{L}\right) + \left(\frac{V_{C0}t}{L}\right)\left(\frac{-R}{2L}\right)\exp\left(-\frac{Rt}{2L}\right) = 0$$

$$\text{or} \quad t = \frac{2L}{R} = \frac{2 \times 10^{-4}}{200} = 10^{-6}\ \text{s} \quad \text{for the values shown} \tag{3.7.16}$$

$$\text{since at critical damping} \quad R = \left(\frac{4L}{C}\right)^{\frac{1}{2}} = \left(\frac{4 \times 10^{-4}}{10^{-8}}\right)^{\frac{1}{2}} = 200\ \Omega$$

and the peak is found to occur at a time $t = 2L/R = \tau$, the time constant of the circuit. The value of the peak current is then $i_p = 10^{-2}\exp(-1) = 3.679$ mA, which agrees with the simulation. Note also that the overdamped response ($R = 500\ \Omega$) takes longer to settle than the critical response, as it should do.

## SPICE simulation circuits

| | |
|---|---|
| Fig. 3.7.1(*b*) | Plsresp 1.SCH |
| Fig. 3.7.2(*b*) | Plsresp 2.SCH |
| Fig. 3.7.3(*b*) | Plsresp 5.SCH |
| Fig. 3.7.4(*b*) | Plsresp 6.SCH |
| Fig. 3.7.5(*b*) | Plsresp 7.SCH |
| Fig. 3.7.6 | Plsresp 3.SCH |

## References and additional sources 3.7

Chase R. L. (1961): *Nuclear Pulse Spectrometry*, New York: McGraw-Hill.

Millman J., Taub H. (1965): *Pulse, Digital and Switching Waveforms*, New York: McGraw-Hill. Library of Congress Cat. No. 64-66293.

Oliver B. M. (1966): *Square Wave Testing of Linear Systems*, Hewlett-Packard Application Note 17, April.

Skilling J. K. (1968): Pulse and frequency response. *General Radio Experimenter* November–December, 3–10.

Thomson W. (1853): On transient electric currents. *Philosophical Magazine* **5**, 393–405.

# 3.8 Equivalent circuits

All generalisations are dangerous, even this one.

Ralph Waldo Emerson

It is sometimes convenient to view a circuit in a different but equivalent topology. Thévenin's equivalent circuit (Section 3.1) is an example that leads to much simplification in circuit analysis. We will examine a few of the more commonly used simple circuit transformations to see where the advantages may lie, and the limitations. Figure 3.8.1 shows a series *RC* circuit that we wish to transform to a parallel arrangement so that the terminal impedances are the same.

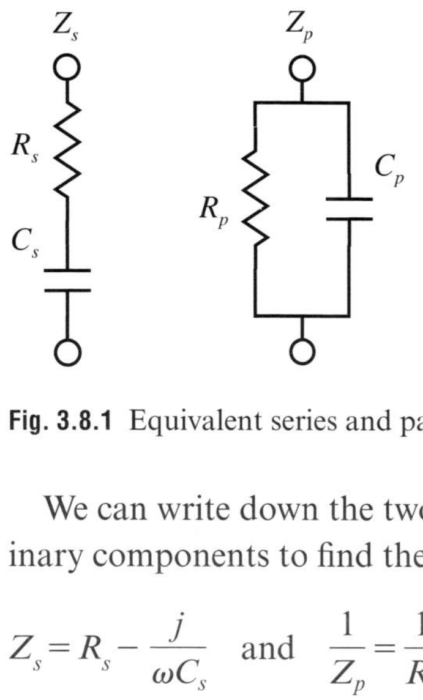


**Fig. 3.8.1** Equivalent series and parallel *RC* circuits.

We can write down the two impedances and then equate the real and the imaginary components to find the necessary relationships.

$$Z_s = R_s - \frac{j}{\omega C_s} \quad \text{and} \quad \frac{1}{Z_p} = \frac{1}{R_p} + j\omega C_p \quad \text{or} \quad Z_p = \frac{R_p(1 - j\omega C_p R_p)}{1 + (\omega C_p R_p)^2}$$

and equating the real and the imaginary components

$$R_s = \frac{R_p}{1 + (\omega C_p R_p)^2} \quad \text{and} \quad \frac{1}{\omega C_s} = \frac{\omega C_p R_p^2}{1 + (\omega C_p R_p)^2}$$

$$\frac{R_p}{R_s} = 1 + (\omega C_p R_p)^2, \quad \frac{C_p}{C_s} = \frac{(\omega C_p R_p)^2}{1 + (\omega C_p R_p)^2} \quad \text{and using } \alpha \equiv \omega C_p R_p \tag{3.8.1}$$

$$\omega C_s R_s = \omega \frac{C_p(1+\alpha^2)}{\alpha^2} \frac{R_p}{(1+\alpha^2)} = \frac{1}{\omega C_p R_p}$$

Since the transformation relations include $\omega$ then the transformation is only valid at that frequency.

A second useful transformation is the Tee-Pi or Star-Delta (or Y-delta), which are illustrated in Fig. 3.8.2. The former name is usually used in electronic circuits while the latter is more commonly used in discussing three-phase transformers. It is evident that the two are identical.

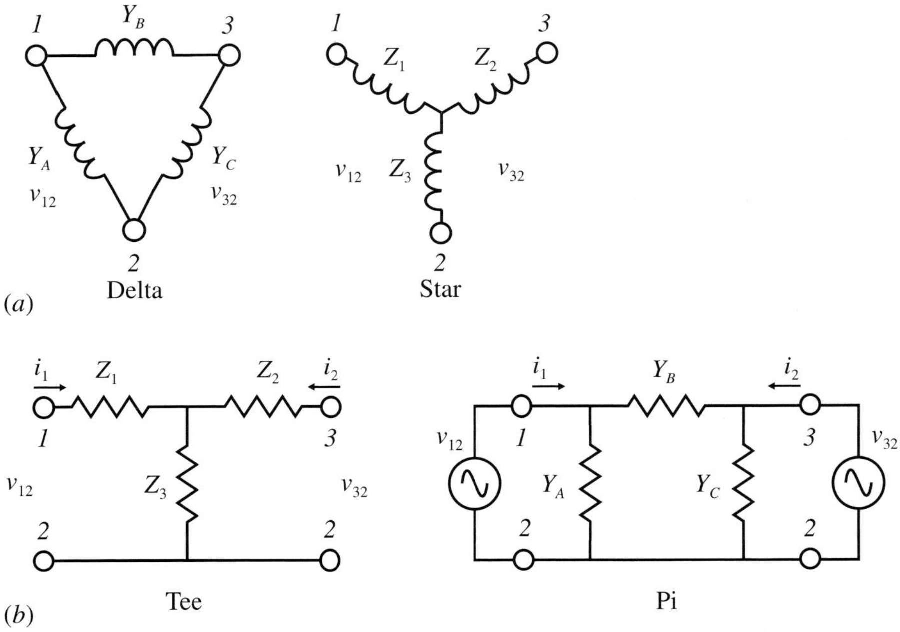


**Fig. 3.8.2** (*a*) Star-Delta transformation. (*b*) Tee-Pi circuit transformation.

The analysis is readily carried out by means of the superposition theorem (Section 3.1). For the Pi circuit, considering the two voltage sources separately (and putting the other equal to a short circuit), we can write (admittances make the sums neater):

$$i_1 = v_{12}(Y_A + Y_B) - v_{32} Y_B \quad \text{and} \quad i_2 = -v_{12} Y_B + v_{32}(Y_B + Y_C) \tag{3.8.2}$$

and solving for $v_{12}$ from the second and substituting in the first gives:

$$i_1 Y_B = -i_2(Y_A + Y_B) + v_{32}(Y_B + Y_C)(Y_A + Y_B) - v_{32} Y_B^2$$

$$\text{or} \quad v_{32} = \frac{i_1 Y_B + i_2(Y_A + Y_B)}{(Y_B + Y_C)(Y_A + Y_B) - Y_B^2}$$

$$=\frac{i_1 Y_B + i_2(Y_A + Y_B)}{Y^2}, \quad \text{where } Y^2 \equiv Y_A Y_B + Y_B Y_C + Y_C Y_A$$

and similarly

$$v_{12}=\frac{i_1(Y_B + Y_C) + i_2 Y_B}{Y^2} \tag{3.8.3}$$

A similar process for the Tee circuit, but this time using impedances, gives:

$$v_{32}=i_1 Z_3 + i_2(Z_2 + Z_3) \quad \text{and} \quad v_{12}=i_1(Z_1 + Z_3) + i_2 Z_3 \tag{3.8.4}$$

As the circuits must be equivalent and independent of current, the coefficients of $i_1$ and of $i_2$ must be equal, giving:

$$Z_3=\frac{Y_B}{Y^2}, \qquad Z_2 + Z_3=\frac{Y_A + Y_B}{Y^2}, \qquad Z_1 + Z_3=\frac{Y_B + Y_C}{Y^2}$$

or substituting for $Z_3$

$$Z_1=\frac{Y_C}{Y^2}, \qquad Z_2=\frac{Y_A}{Y^2}, \qquad Z_3=\frac{Y_B}{Y^2} \tag{3.8.5}$$

which has a pleasing symmetry to aid memory. For the reverse transformations we may note from (3.8.3) and (3.8.5) that:

$$Y^2=Z_2 Y^2 Z_3 Y^2 + Z_3 Y^2 Z_1 Y^2 + Z_1 Y^2 Z_2 Y^2 = Y^4(Z_2 Z_3 + Z_3 Z_1 + Z_1 Z_2) = Y^4 Z^2$$

$$\text{or} \quad Z^2=(Z_2 Z_3 + Z_3 Z_1 + Z_1 Z_2)=\frac{1}{Y^2}, \quad \text{so that} \tag{3.8.6}$$

$$Y_A=\frac{Z_2}{Z^2}, \quad Y_B=\frac{Z_3}{Z^2}, \quad Y_C=\frac{Z_1}{Z^2}$$

An application of this type of transformation arises in providing a variable attenuator in the classic long-tailed pair amplifier illustrated in Fig. 3.8.3(*a*).

Substituting the resistor values appropriately, and noting that commonly $R_1 \gg R_2$, the Pi values will be as shown. Varying $R_3$ alone now provides a gain control.

In an operational feedback network it is sometimes required to include a small capacitor to allow adjustment of the transient response and to ensure stability, as for example in the transimpedance circuit (see Sections 5.3 and 5.12). Small adjustable capacitors typically have a minimum capacity of a pF or two, so to obtain a very small adjustable capacitor of less than 1 pF it is necessary to resort to an 'attenuator' style, typically in the form of a Tee as shown in Fig. 3.8.4 (see e.g. Horowitz and Hill 1989; Burr-Brown 1995, p. 4; Graeme 1996).

The requirement now is to determine the relationship between the Tee capacitors and the equivalent single capacitor, say $C_E$. In the Tee a current $i_1$ will flow in

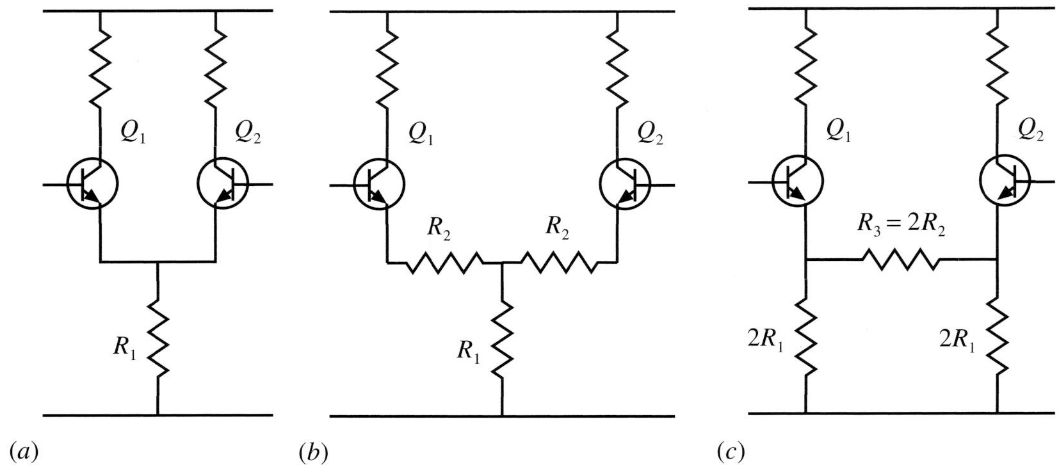


**Fig. 3.8.3** (*a*) Long-tailed pair circuit. (*b*) Introduction of a symmetrical attenuation via $R_2$. (c) A Tee-Pi transformation so that there is a single variable resistor $R_3 = 2R_2$.

$C_1$ and we wish the same current to flow in the equivalent single capacitor $C_E$. Due to the virtual common, $C_1$ and $C_3$ are effectively in parallel so the impedance of the Tee is:

$$Z = \frac{1}{sC_2} + \frac{1}{s(C_1 + C_3)} = \frac{C_1 + C_2 + C_3}{sC_2(C_1 + C_3)} \quad \text{and since} \quad i_2 = i_1 + i_3 \quad \text{we have}$$

$$\frac{i_1}{i_3} = \frac{C_1}{C_3} \quad \text{or} \quad \frac{i_1}{i_1 + i_3} = \frac{C_1}{C_1 + C_3} = \frac{i_1}{i_2} \quad \text{so} \quad i_2 = \frac{i_1(C_1 + C_3)}{C_1}$$

and considering a voltage $v_{AT}$ as shown (3.8.7)

$$i_2 = \frac{v_{AT}}{Z} = \frac{v_{AT}sC_2(C_1 + C_3)}{(C_1 + C_2 + C_3)} = \frac{i_1(C_1 + C_3)}{C_1}$$

and for the equivalent capacitor $C_E$: $i_1 = v_{AT}sC_E$

$$\frac{v_{AT}sC_2(C_1 + C_3)}{(C_1 + C_2 + C_3)} = \frac{v_{AT}sC_E(C_1 + C_3)}{C_1} \quad \text{so} \quad C_E = \frac{C_1 C_2}{(C_1 + C_2 + C_3)}$$

For design we would probably make $C_1 = C_2 = C$ and consider $C_3$ as a variable trimmer (connect the adjuster side to common). We will choose a maximum value for $C_E = 2$ pF and a typical small trimmer of 7–50 pF and calculate the range of variation of $C_E$. With the minimum value of $C_3$ (7 p) we get the maximum value of $C_E$ (2 p):

$$2 = \frac{C^2}{2C + 7} \quad \text{or} \quad C^2 - 4C - 14 = 0 \quad \text{with solution} \quad C = 6.2\text{ p}$$

(the other root is negative)

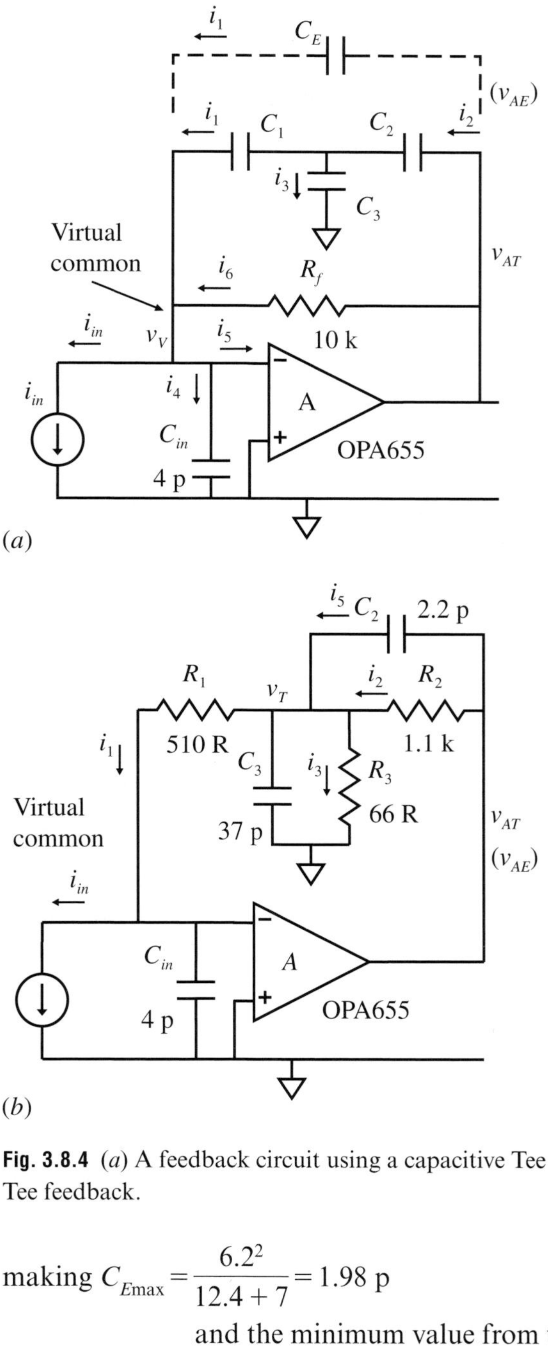


**Fig. 3.8.4** (*a*) A feedback circuit using a capacitive Tee compensator. (*b*) Compensated resistor Tee feedback.

$$\text{making } C_{E\max} = \frac{6.2^2}{12.4+7} = 1.98\ \text{p} \tag{3.8.8}$$

and the minimum value from the maximum $C_3 = 50$ p

$$C_{E\min} = \frac{6.2^2}{12.4+50} = 0.62\ \text{p}$$

so the controlling factor is the max/min ratio for $C_3$. You may of course need to iterate to get the range you need and in this case it would be necessary to make $C$

a standard value. We may now test (3.8.7) on the circuit shown in Fig. 3.8.4(*a*). With a single feedback capacitor $C_E$ it was determined that the optimum value was 0.8 p. Using (3.8.7) and the value of $C=6.2$ p from Eq. (3.8.8) the value of $C_3$ was found to be 36 p. With $C_3=36$ p there was significant overshoot, so for comparison it has been reduced to 24 p to eliminate this. The resulting waveforms are shown in Fig. 3.8.5(*a*) (you will need a high value resistor from the junction of the Tee to common to avoid a floating node).

The risetime has increased from 13.4 to 16.9 ns, about 25%. It is evident that the Tee is not equivalent to the single capacitor $C_E$, and the traces for the voltage $v_V$ (shown $\times 10$ for visibility) at the 'virtual common' is an indication of why they are not. We had assumed that $C_1$ and $C_3$ were in parallel on the basis that the 'virtual was common', but the delay in the transient feedback means that it is not. Thus SPICE has alerted us to the error and allows us to adjust the capacitors to see if we can find an effective equivalent. We may guess that the closer we come to the single effective feedback capacitor the more nearly there would be a correspondence, i.e. to make the correction as small as possible. On this basis $C_1$ was decreased and $C_3$ adjusted to get optimum response and a good fit was obtained for $C_1=2.2$ p and $C_3=8$ p, giving an accessible value for $C_1$. To make the correspondence more clear only $v_{AT}$ is shown together with the difference $v_{AT}-v_{AE}$. It may be noted in passing that the actual range of variable capacitors can be significantly different from their nominal values.

Similar equivalent circuits may be achieved by using a resistive Tee, primarily for achieving high effective resistance with more modest resistors (see e.g. Burr-Brown 1995, p. 4). The arrangement in the reference is illustrated in Fig. 3.8.4(*b*) (but without $C_3$). The reference supplies an equation for the effective resistance $R_E$ as:

$$R_E = R_1 + R_2 + (R_1 R_2 / R_3) \tag{3.8.9}$$

but this is also derived on the basis of the virtual common. For our purposes it can be adapted for our $R_E$ of 10 k but the compensation is much improved by addition of $C_3$ in effect to make the 'attenuator' ratio independent of frequency (see Fig. 5.2.1, p. 399). The sums are left as an exercise, but for a particular set of values that provide an effective 10 k we can get a response $v_{AT}$ that matches very closely the direct response $v_{AE}$ as found in Fig. 3.8.5. The responses are shown in Fig. 3.8.6.

The component values were found in part from Eq. (3.8.9) followed by simple cut and try. It will be noted that the ratio $C_3/C_2$ is effectively equal to $R_2/R_3$. The references following should be consulted for discussion of the effects of the Tee configurations on noise performance.

Note that the model listing for the OPA655 recommends that the SPICE parameter *ITL4* be set to 40 to assist convergence. This can be found under *ANALYSIS/OPTIONS*. (The model does not include input capacity; you may wish to add 1 p between inputs and 1 p from each to common.)

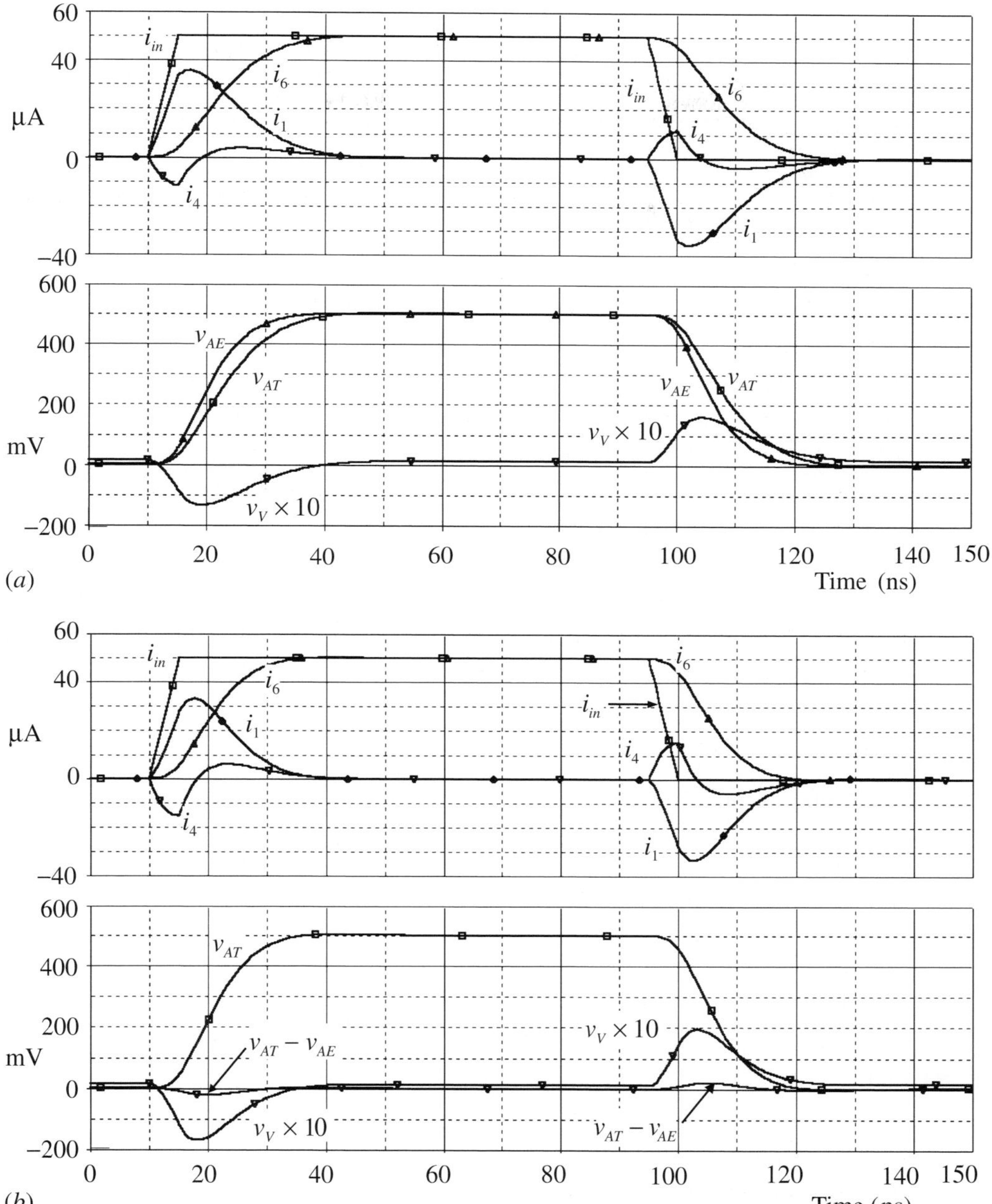


**Fig. 3.8.5** (*a*) Simulation waveforms with single feedback capacitor $C_E = 0.8$ p and with equivalent Tee configuration. $C_3$ has been reduced from 36 to 24 p to remove any overshoot. (*b*) Comparison, with adjusted capacitor values: $C_1 = 2.2$ p, $C_2 = 6.2$ p, $C_3 = 8$ p.

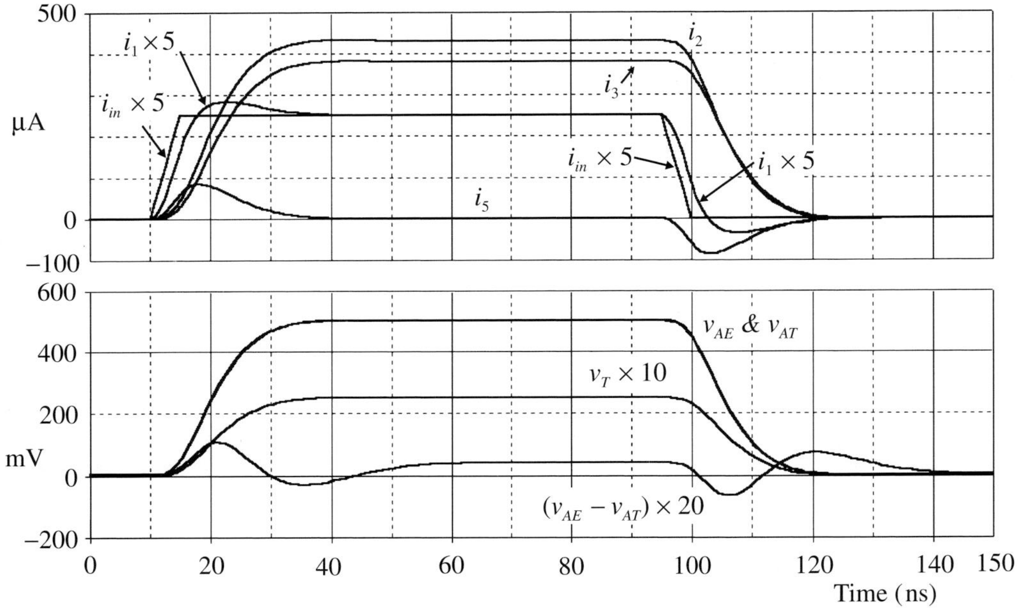


**Fig. 3.8.6** Responses with compensated resistor Tee feedback network with values given in Fig. 3.8.4(*b*). Some small offsets have been removed and some responses multiplied to improve clarity.

## SPICE simulation circuits

| | |
|---|---|
| Fig. 3.8.5(*a*) | Tcapfbk1.SCH |
| Fig. 3.8.5(*b*) | Tcapfbk1.SCH |
| Fig. 3.8.6 | Tcapfbk2.SCH |

# References and additional sources 3.8

Burr-Brown (1995): *Photodiode Monitoring with Op Amps*, Burr-Brown Application Note AB-075, January.

Graeme J. (1995): *Photodiode Amplifiers: Op Amp Solutions*, New York: McGraw-Hill. ISBN 0-07-024247-X. See pp. 59 and 130.

Horowitz P., Hill W. (1989): *The Art of Electronics*, 2nd Edn Cambridge: Cambridge University Press. ISBN 0-521-37095-7. See p. 1040.

# 3.9 Cauchy's dog bodes well

*Essentia non sunt multiplicanda praeter necessitatum* – hypotheses are not to be multiplied without necessity. The tenet now known as Occam's Razor.

William Occam 1324

Dispersion is probably a familiar term as applied to the variation of refractive index $n$ with visible wavelength in optical systems. A glass prism disperses white light into the familiar rainbow of colours because the velocity of light in the glass varies with wavelength or, more properly, with frequency since the wavelength depends on the material. The dispersion curve seen in optics books is only a small segment of a wider picture Fig. 3.9.1.

The full dispersion curve, in the simplest case, would look somewhat like (*b*). The reason why the full curve is not seen optically is because of the accompanying absorption, which increases strongly so that no light propagates through the glass. The dispersion and absorption arise from a resonance in the material (though for a glass there are usually many 'resonances', which means that the glass does not transmit again until the X-ray region) and in the simple case the responses are just like those in a resonant electronic circuit (Section 3.5).

This form of response in materials is found in many circumstances that concern us – in magnetic materials and in dielectrics of which refractive index is in effect the extension into the region of the spectrum where we usually refer to light rather than radio or microwaves. Circuits too also show such responses even though we are not in close proximity to a resonance. The effects are usually represented by considering the relevant quantity, e.g. dielectric constant or permittivity $\varepsilon$, the magnetic permeability $\mu$, the impedance $Z$ or the conductivity $\sigma$, as complex quantities:

$$\varepsilon=\varepsilon'-j\varepsilon'', \quad \mu=\mu'-j\mu'', \quad Z=R+jX, \quad \sigma=\sigma'-j\sigma'' \tag{3.9.1}$$

We need to be aware of the differing meaning of the symbols. For $Z$ we consider that the real part $R$ corresponds to the in-phase components while the imaginary part $X$ refers to the out-of-phase components. For permittivity $\varepsilon$, the real part $\varepsilon'$ is that which we would in simple terms associate with a capacitor and hence $\varepsilon''$ with the in-phase components, as is also the case with $\mu$.

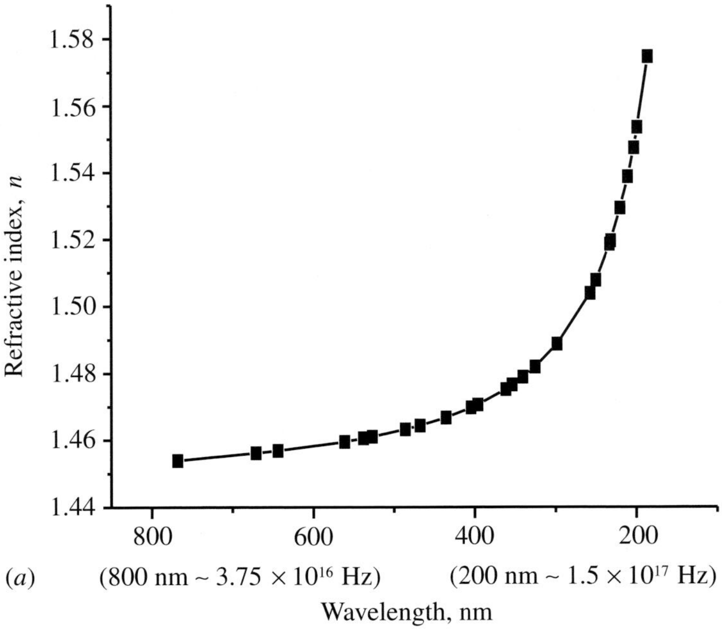


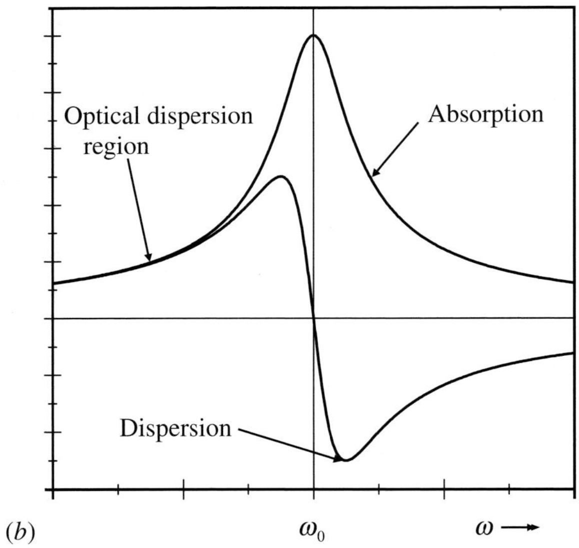


**Fig. 3.9.1** (*a*) Variation of the refractive index of quartz in the visible region. (*b*) Full dispersion curves including the region of strong absorption.

The primary point which we wish to make here is that there is an inescapable relation between the absorption and the dispersion – if one varies with frequency then the other must also and the variations are not independent. These connections were formalized by Kramers (1924) and Kronig (1926) and are now known as the Kramers–Kronig relations. The derivation of the relations requires some complex contour integration using Cauchy's theorem (e.g. Thomason 1955; Panofsky and Phillips 1962; Kittel 1967; Lipson and Lipson 1969; Faulkner 1966), which would be out of place here. However, in carrying out the integrations around a contour it is necessary to make small detours around any poles of the function (where the function goes off to infinity) and then determine the contributions or *residues* arising from these detours, separately. As an antidote to Schrödinger's (quantum) cat (see for example Gerry and Knight 1997) we have Cauchy's dog, which leaves a small residue at every pole (see Van Valkenburg 1982). In terms of a quantity $\chi = \chi' + \chi''$ the relations are:

$$\chi'(\omega) - \chi'(\infty) = \frac{2}{\pi}\int_0^\infty \frac{u\chi''(u)du}{u^2 - \omega^2}$$

$$\chi''(\omega) = \frac{-2}{\pi}\int_0^\infty \frac{\chi'(u) - \chi'(\infty)du}{u^2 - \omega^2} \tag{3.9.2}$$

which give the values of $\chi'$ or $\chi''$ at frequency $\omega$ in terms of one another ($u$ is a frequency dummy variable).

For the moment we wish to examine the relevance of the Kramers–Kronig relations to circuit theory. The quantities equivalent to absorption and dispersion are gain and phase, so that we must accept that if gain changes with frequency then so must phase. Bode (1940, 1947) deduced the corresponding relationships and gave them in a form that is more informative for circuit analysis. The phase shift $\theta(\omega_c)$ (in radian) at some given frequency $\omega_c$ is related to the amplitude response $A(\omega)$ as a function of frequency by (Terman 1943, Thomason 1955, Hakim 1966 – coth is the hyperbolic cotangent):

$$\theta(\omega_c) = \frac{\pi}{12}\left(\frac{dA}{d\mu}\right)\Bigg|_{\omega_c} + \frac{1}{6\pi}\int_{-\infty}^{\infty}\left(\frac{dA}{d\mu} - \frac{dA}{d\mu}\Bigg|_{\omega_c}\right)\ln\left(\coth\frac{|\mu|}{2}\right)d\mu$$

where $\dfrac{dA}{d\mu}$ = slope of the response in dB/oc (3.9.3)

$$\mu = \ln\left(\frac{\omega}{\omega_c}\right)$$

and $\coth\dfrac{|\mu|}{2}$ is the real part since it is complex if $\mu$ is negative

This does not look like a function you would like to deal with on an everyday basis but it does provide some very important information. The phase shift at any frequency, $\omega_c$ here, depends on the gain slopes over the whole frequency range but weighted by the function:

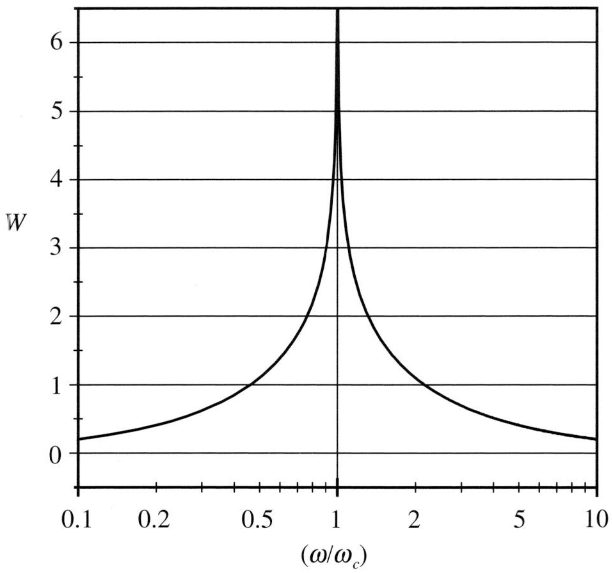


**Fig. 3.9.2** Bode's weighting function.

$$W = \ln\left(\coth\frac{|\mu|}{2}\right), \quad \text{with} \quad \mu = \ln\left(\frac{\omega}{\omega_c}\right) \tag{3.9.4}$$

which is plotted in Fig. 3.9.2. This shows a very peaked response indicating that it is the slope close to $\omega_c$ that makes the major contribution to the phase shift there.

A related expression gives the integrated phase shift in terms of the extreme gains (Hakim 1966, p. 224):

$$\int_{-\infty}^{\infty} \theta(\omega)d\mu = \frac{\pi}{2}[A(\infty) - A(0)] \tag{3.9.5}$$

i.e. the area under the logarithmic phase plot is $\pi/2$ times the difference in gain between infinite and zero frequency. These phase-amplitude relations are not of particular use in the design process as they are clearly difficult to handle and in any case we may so much more readily use SPICE. They do, however, provide a more general constraint, rather like the conservation of energy. The variation in gain will be associated with a related change in phase but it does not prescribe exactly where, as a function of frequency, the phase changes must occur. That is, there is some latitude which allows one, within limits, to tailor the distribution of the phase change as required, rather like squeezing the toothpaste in a tube.* In terms of feedback systems this may allow one to put off the onset of positive feedback until the gain

* There is a longstanding battle between those that squeeze the tube at the top and those that, more properly, squeeze it from the bottom. It would be wise not to become involved in such domestic conflicts.

falls below unity and hence achieve a stable system. Simple circuits like the phase advance and phase retard (Section 3.3) are examples that are commonly used (see e.g. Section 5.10).

## SPICE simulation circuits

None

## References and additional sources 3.9


Bode H. W. (1940): Relation between attenuation and phase in feedback amplifier design. *Bell Syst. Tech.* J. **19**, 412–454.

Bode H. W. (1947): *Network Analysis and Feedback Amplifier Design*, New York: Van Nostrand.

Faulkner E. A. (1966): *Principles of Linear Circuits*, London: Chapman and Hall.

Gerry C. C., Knight P. L. (1997): Quantum superposition and Schrödinger's cat states in quantum optics. *Amer. J. Phys*. **65**, 964–974.

Hakim S. S. (1966): *Feedback Circuit Analysis*, London: Iliffe Books.

Kittel C. (1967): *Elementary Statistical Physics*, New York: John Wiley. Library of Congress Cat. No. 58-12495. See p. 206.

Kramers H. A. (1924): The law of dispersion and Bohr's theory of spectra. *Nature* **113**, 673–674; and *Atti. Congr. Intern. Fis. Como* **2**, 545–557.

Kronig R. de L. (1926): Theory of dispersion of X-rays. *J. Opt. Soc. Amer*. **12**, 547–557.

Lipson S. G., Lipson H. (1969): *Optical Physics*, Cambridge: Cambridge University Press. ISBN 521-06926-2.

Panofsky W. K. H., Phillips M. (1962): *Classical Electricity and Magnetism*, 2nd Edn, Addison-Wesley. Library of Congress Cat. No. 61-10973.

Terman F. E. (1943): *Radio Engineers' Handbook*, New York: McGraw-Hill.

Thomason J. G. (1955): *Linear Feedback Analysis*, London: Pergamon Press.

Van Valkenburg M. E. (1982): *Analog Filter Design*, New York: Holt, Rinehart and Winston. ISBN 0-03-059246-1, or 4-8338-0091-3 International Edn. (See footnote p. 69: The author provides guidance on the pronunciation of Bode, to sound as *Boh-dah*, with which my Dutch friends concur, but I am afraid that the battle has been lost and *Bohd* has become the norm.)

# 3.10 Feedback

> But the history of science is not restricted to the enumeration of successful investigations. It has to tell of unsuccessful enquiries, and to explain why some of the ablest men have failed to find the key to knowledge. . . .
>
> James Clerk Maxwell in his introductory lecture on appointment as the first Cavendish Professor, Cambridge, in 1871

Feedback is one of the most important concepts in electronics so it is most desirable that one understands the reasons for and benefits of using it, the techniques of applying it and the consequences, drawbacks and pitfalls. We can consider the two possible regimes of positive and negative feedback separately since the intent is usually quite different but it must be remembered that one can merge into the other in many circumstances. In terms of the general principle negative feedback was 'invented' first, in particular as applied to governors for early steam engines (Maxwell 1868). However, in the field of electronics positive feedback was the first to be used (see Tucker 1972) and it was rather later that the great benefit of negative feedback was recognized (Black 1934, 1977; Bennett 1993; Klein 1993). We will consider negative feedback first.

The general arrangement of negative feedback is that part of the output signal of an amplifier (or a more complex system) is added to the input signal in such a way as to oppose it. If a significant amount of feedback is applied then to a substantial degree the performance of the system is found to depend on the feedback components rather than on the active amplifying elements. The immediate benefit is that since the feedback components are (generally) passive elements, which can have much greater temporal and temperature stability than the active amplifying elements, the gain may be much more accurately determined and maintained. The general arrangement is illustrated in Fig. 3.10.1.

If the proportion of the output signal that is fed back is $\beta$ then the signal input to the amplifier is given by:

$v_A = v_{in} + \beta v_{out}$, so we have $v_{out} = -A(v_{in} + \beta v_{out})$

so the net gain is $$G = \frac{v_{out}}{v_{in}} = \frac{-A}{1 + A\beta} \tag{3.10.1}$$

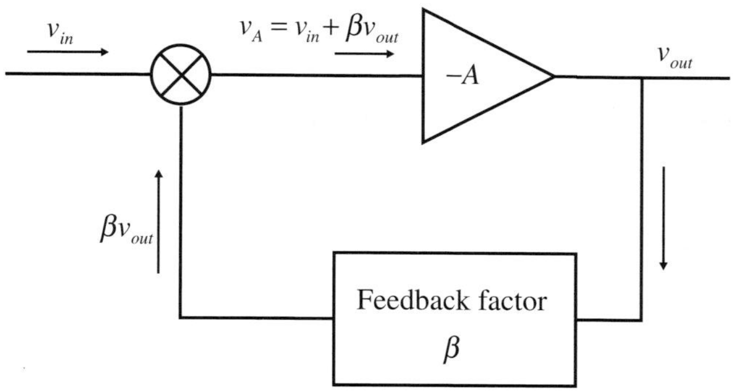


**Fig. 3.10.1** General negative feedback arrangement.

the deceptively simple relation that has had such a wide and beneficial effect. The quantity $L = A\beta$ is called the loop gain and is the arbiter of the performance of a feedback system.

The impetus for Black's work was the urgent need to reduce distortion greatly so that the many amplifiers required for telephone links across the USA did not make the output unintelligible. There are other significant benefits as we shall see, as well as some drawbacks, but overall the improvement in performance can be astonishing. When Black first described his idea, gain was relatively expensive and vacuum tubes had considerable limitations. Now gain is cheap and in comparison has vastly improved performance in terms of bandwidth and stability, and this means that we can obtain even greater improvement from the use of negative feedback.

The statement about taking part of the output signal and adding it to the input hides important details as to how these two operations are configured. There are four possible ways of applying feedback. The output voltage or the output current may be sampled and the feedback signal can be connected in series or in parallel with the input signal. The combinations are illustrated in Fig. 3.10.2. The classifications are named in a number of ways by different authors but the schematics make clear the four cases. The consequences are detailed in Table 3.10.1.

The configuration that we will most commonly encounter in the present work is the node–node (or shunt–shunt) of Fig. 3.10.2(*a*). Since this is the standard form of the operational amplifier the analysis is included in Section 5.3, together with the non-inverting amplifier which is an example of the loop–node configuration. If the output current, rather than voltage, is to be controlled then we must sample the output current and hence use output loop-sampling, i.e. configurations (*b*) or (*c*), which as is indicated increase the output impedance to emulate a current source. An example of this is given in Section 5.14. In some circumstances it is not too evident what the topology is – it may be something of a mixture. A number of example cases are examined in, for example, Grey and Searle (1969), Ghausi (1971) or Roberge (1975).

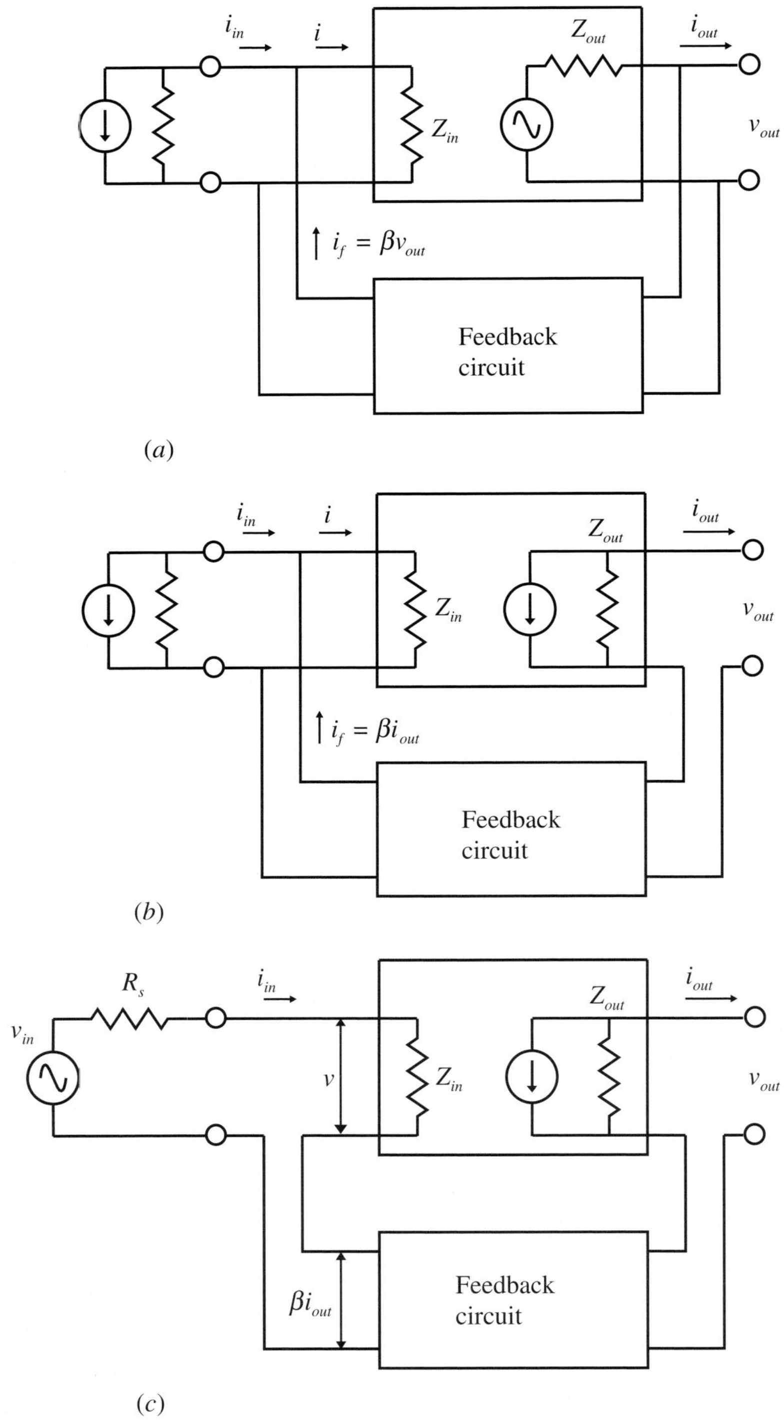


**Fig. 3.10.2** Negative feedback configurations.

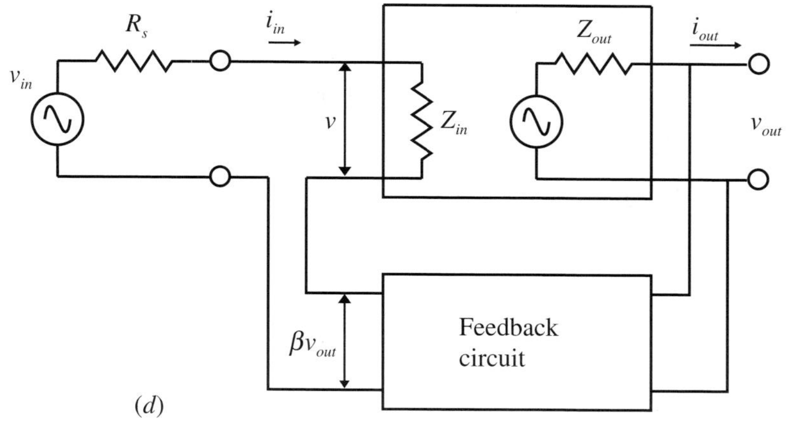


**Fig. 3.10.2** (cont.)

Table 3.10.1 *Effects of negative feedback for the configurations of Fig. 3.10.2*

| Topology: input–output[a] | Input impedance | Output impedance | Voltage gain | Current gain |
|---|---|---|---|---|
| (a) Node–node | Low/decrease | Low/decrease | No change | Decrease |
| (b) Node–loop | Low/decrease | High/increase | No change | Decrease |
| (c) Loop–loop | High/increase | High/increase | Decrease | No change |
| (d) Loop–node | High/increase | Low/decrease | Decrease | No change |

*Note:*
[a] Under topology the first term refers to the form of the input comparison and the second to the output sampling. Loop is otherwise referred to as series and node as shunt.

The benefits of negative feedback are to be seen in many sections of this book. The driving requirement which motivated Black was the need to reduce distortion in the then available amplifiers by a very large factor of some 50 dB ($\times 10^5$ in power terms) and this was simply not possible by trying to linearize the amplifiers. His idea of negative feedback achieved all this and had additional advantages, but the idea was disbelieved by many and it took some ten years to obtain the patent on it. It may not be immediately evident why the feedback reduces distortion so we will examine a simple simulation in which we can see what occurs. In the circuit of Fig. 3.10.3 the resistor–diode network at the output is added to produce some severe distortion which we may imagine as an internal part of the amplifier.

In (*a*) the distortion is outside the feedback loop so the feedback has no effect on it. Thus $v_{A1}$ is an inverted but good reproduction of $v_{I1}$, and $v_{O1}$ is severely distorted on the positive going half cycles. The negative going half cycles are not distorted, since the diode is not conducting, only reduced in amplitude according to the ratio of $R_3$ to $R_4$ (but see below). In (*b*) the distortion is placed inside the feedback loop so now it will be affected. $v_{O2}$ is an inverted replica of $v_{I2}$ while $v_{A2}$ is

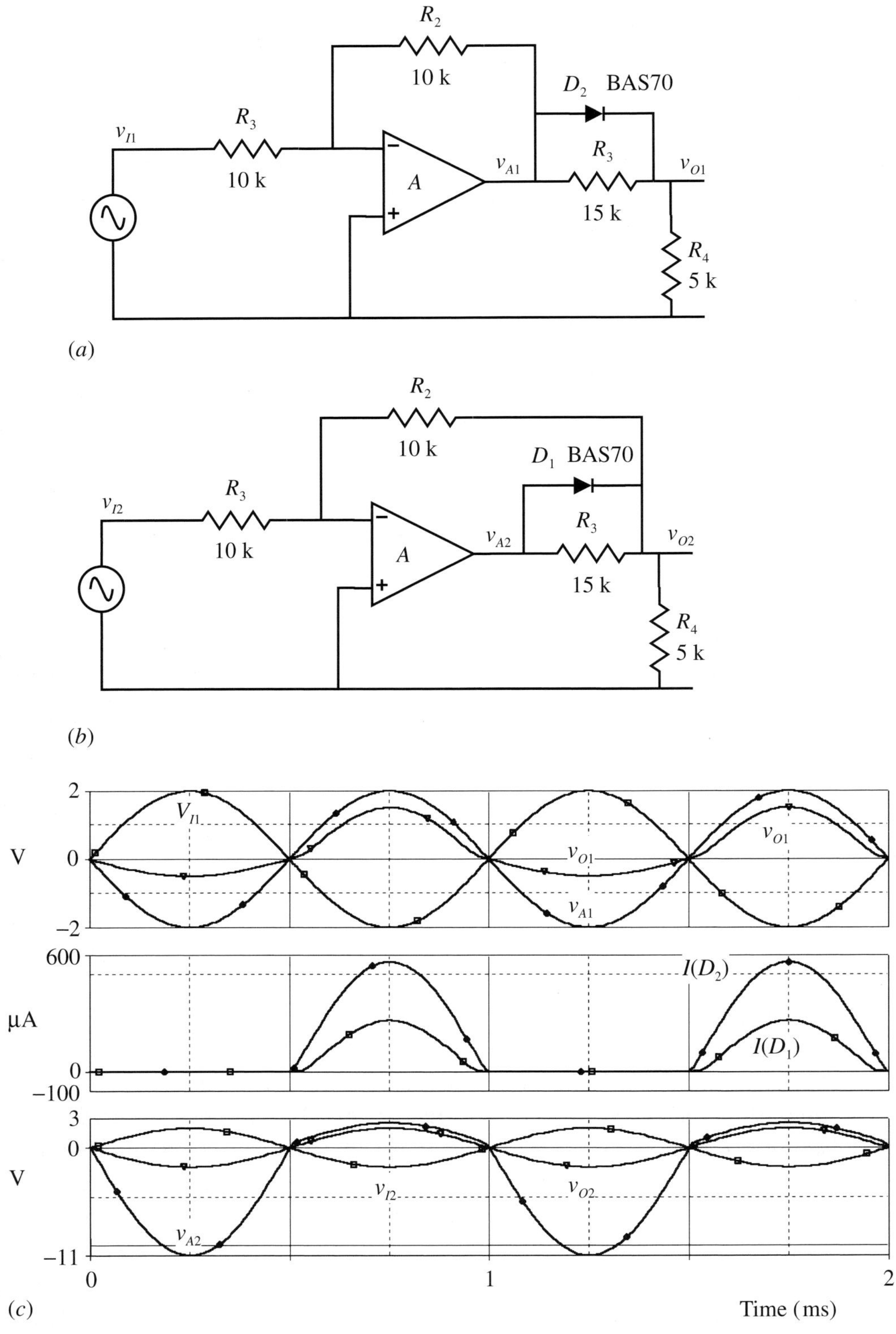


**Fig. 3.10.3** (*a*) Circuit with distortion outside the feedback loop. (*b*) Circuit with distortion inside the feedback loop. (*c*) Simulated waveforms for the two cases.

hugely distorted. But $v_{A2}$ is such that when passed through the distorting network it is 'redistorted' back to the original waveform! Part of $v_{O2}$ is passed back to the input summing node and amplified and inverted by the high gain of $A$ to just the shape and amplitude to compensate. This coincidence is precise, to the accuracy determined by $A$, because the distorting network is itself setting the overall gain. The apparent discrepancy in the 11 V peak of $v_{A2}$ arises because $R_2$ is in effect in parallel with $R_4$, giving a feedback attenuation of 5.5 rather than 4 considering $R_3$ and $R_4$ only. This self-correcting property was just what Black needed. He recounts how for previous amplifiers 'every hour on the hour – 24hours a day – somebody had to adjust the filament current to its correct value. In doing this, they were permitting plus or minus $\frac{1}{2}$- to 1-dB in amplifier gain, whereas, for my purpose, the gain had to be absolutely perfect. In addition, every six hours it became necessary to adjust the $B$ battery voltage, because the amplifier gain would be out of hand.' (Black 1977).

The difficulty with the concept of negative feedback at that time is described by Belevitch (1962).

> The elementary, but erroneous, physical reasoning based on the round and round circulation of the signal in the feedback loop yielded the geometric series $\mu(1+\mu\beta+\mu\beta^2+\cdots)$ for the effective amplification. This gave the impression that instability must occur for all $|\mu\beta|>1$ for the series then diverges. Stable behavior with negative feedback exceeding 6 dB was thus apparently forbidden, and such low values have little practical interest. It was soon recognized that the theory based on the geometric series was contradicted by experience, and the expression $\mu\beta/(1-\mu\beta)$ was used even for $|\mu\beta|>1$ without theoretical justification until Nyquist (1932) proved his famous stability criterion and showed the error of the older theory – the physical reasoning based on the loop circulation is only correct when transients are taken into account, the terms in the geometric series then becoming convolution products; the steady-state formula is obtained as the limit after infinite time; the transient series is convergent, but not always uniformly, so that the limit operation and the summation cannot be interchanged.

Reproduced by permission of the IEEE, © IEEE 1962

It must have been of considerable help to Nyquist to know what answer he had to land up with. See also the review by Bennett (1993).

The work of Nyquist, and soon after by Bode (1940, 1947) collaborating with Black set the sound mathematical foundations for understanding and successful application of negative feedback. Before this, Black (1977)* records 'In the course of this (previous, on oscillators) work, I had computed and measured transfer factors around feedback loops and discovered that for self oscillations to occur the loop transfer factor must be real, positive and greater than unity at some frequency. Consequently I knew that in order to avoid self-oscillation in a feedback amplifier it would be sufficient that at no frequency from zero to infinity should $\mu\beta$ be real, positive and greater than unity.' It is of interest that the math-

* Reproduced by permission of the IEEE, © IEEE 1977.

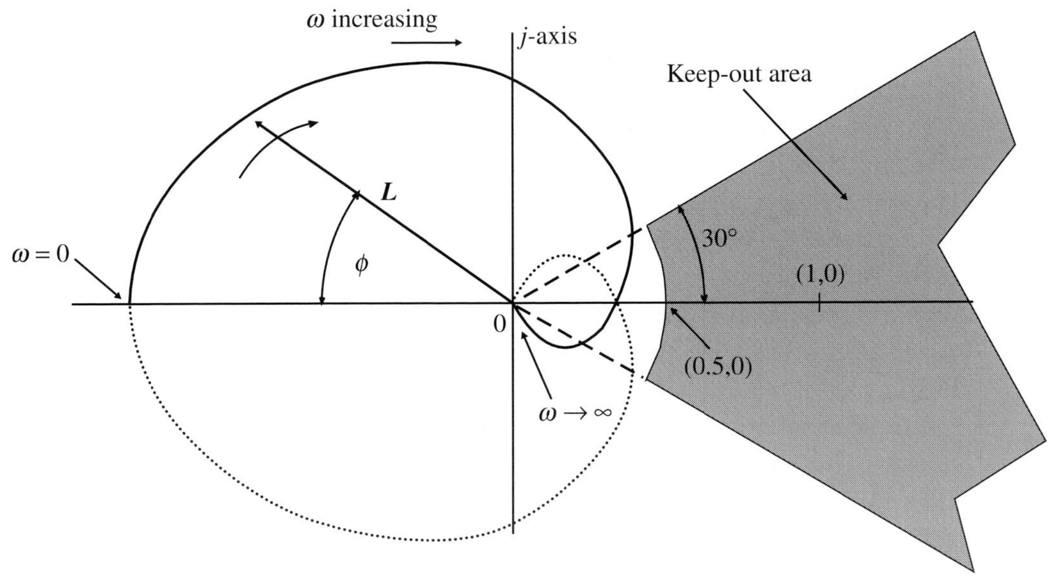


**Fig. 3.10.4** Nyquist plot of loop gain.

ematical techniques that Nyquist used are the same as those used to prove a fundamental theorem of algebra that a polynomial of order $n$ must have $n$ roots (Thomason 1955; Hakim1966). In dealing with transfer functions and poles and zeros this is of vital importance. Though Nyquist showed that what Black concluded, and what we would expect on a straightforward argument was not the complete story, the exceptions are generally not of consequence. But it is as well to know that the theory is properly understood so we need not stumble on some inexplicable behaviour. Nyquist's criterion is most readily demonstrated by his diagram shown in Fig. 3.10.4. This is a vector plot of the loop gain $L$ as a function of frequency.

The locus of $L$ as a function of frequency will follow a path somewhat as shown. $L$ is a complex quantity so it is plotted on a complex plane. The angle $\phi$ is of course the additional phase shift on top of the basic 180° inherent in the feedback being negative. In this example the loop gain extends to zero frequency and like all systems it must eventually become zero at infinite frequency. Nyquist showed that for stability the locus must not enclose the point (1,0); 'enclosure' for a system as shown requires that we imagine a conjugate locus as shown dotted, though in circumstances where the low frequency response goes to zero as with a.c. coupling, this will arise naturally. It also illustrates that the enclosure can occur at the low frequency end and result in oscillation there. The somewhat odd significance of the point (1,0) arises since the quantity that matters is, from Eq. (3.10.1), actually $(1-L)$ and the enclosure is of the origin (0,0). By plotting just $L$ the effective origin is shifted to (1,0). The enclosure is just an indication of whether there are any poles in the right-half plane, i.e. whether $(1-L)$ has any roots there. It is the poles and zeros of $G$ that are now controlling the response.

The control of phase and amplitude around a feedback loop – and remember that the two are not independent entities (see Section 3.4) – is a complex matter and requires much discussion. Many books discuss the techniques, e.g. Hakim (1966), Roberge (1975), Gray and Meyer (1977), Hamilton (1977), Siebert (1986), Chen (1995), and many others. With SPICE, systems can be readily examined and schemes for stabilizing a loop tested. The technique discussed in Section 5.14 makes it relatively easy to determine both open- and closed-loop gain and phase responses as a function of frequency and hence to see clearly the effects of any stabilisation measures. Section 5.10 illustrates some techniques for controlling the stability of a regulator, and Section 5.6 for the improvement of the transient behaviour of the operational differentiator. Consideration of parasitic capacity in op amp circuits is given by Karki (1999).

The prescription for stability that the loop gain must be less than unity before the phase shift reaches 360° is correct but it does not mean that the system will have an acceptable response. If the system is just stable it implies at least a pole very near to the $j\omega$-axis and from Fig. 1.12.4, p. 67, it will be seen that the output will ring in response to any disturbance. Experience has shown that margins for gain and phase are necessary and it is generally accepted that the loop gain $L$ should avoid the shaded region shown in Fig. 3.10.4. Again, with SPICE, it is easy to test the system with a pulse or step input to view the transient response.

The Nyquist diagram is useful to illustrate the stability conditions but it is not commonly used to assess a system. It is much more common to infer the performance from Bode plots where the close relationship between slopes of the response and the phase shifts using the fairly good approximations as illustrated in Section 3.3. It is essential, however, to be careful to differentiate between signal phase shift and loop phase shift. It is the latter that counts. The loop gain response is often referred to as the 'noise gain' (Graeme 1991) since the circuit noise is amplified by the loop gain.

In the early days of electronics it was discovered that the input impedance of valves or tubes was more restricted than expected or hoped for. This turned out to be one of the earliest recognized cases of (inadvertent) feedback and was first described by Miller (1919). Though he was of course dealing with triodes we may equally well consider a transistor as shown in Fig. 3.10.5.

If the transistor gain $v_C/v_B=-K$ and the impedance of the collector–base capacity $C_{CB}$ is $Z_{CB}$ then the current flowing in $C_{CB}$ is given by:

$$i=\frac{v_B-v_C}{Z_{CB}}=\frac{v_B(1+K)}{Z_{CB}} \qquad (3.10.2)$$

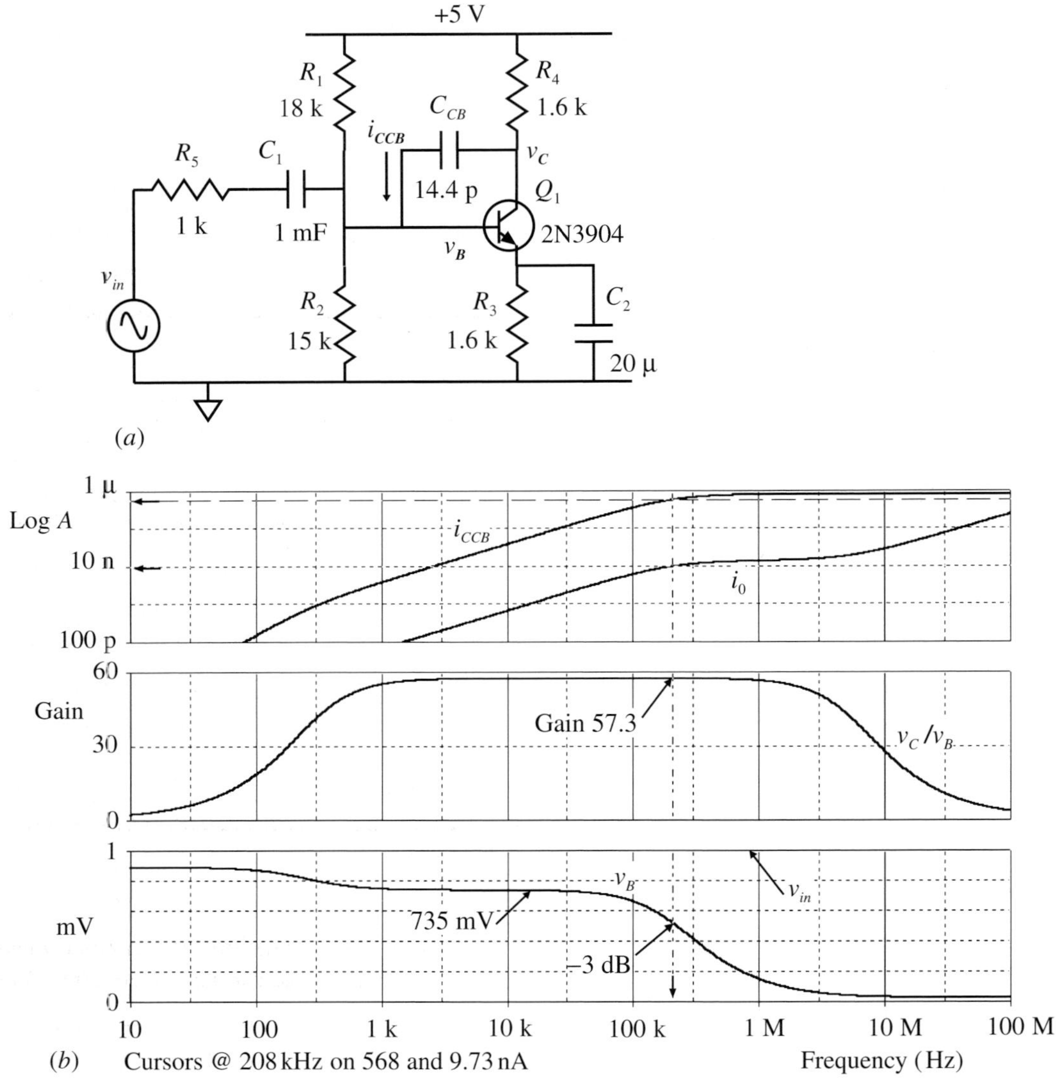


**Fig. 3.10.5** (*a*) Illustrating the Miller effect. (*b*) Simulation waveforms.

so that the effective value of the feedback capacity has been increased by $(1+K)$ times. This was the reason for the introduction of the screen grid between the control grid and the plate in valves so as to reduce the feedback capacity. We can illustrate the effect by simulating the circuit shown in (*a*) with the results shown in (*b*). This is effectively the same circuit as shown in Fig. 4.6.3, see p. 354, and the origin of the changes at low frequencies is described there. Since we do not have access to the collector–base capacitor of the transistor model ($C_{INT}=3.6$ p from the SPICE model) we have added an external capacitor of value four times the internal. The curve $i_0$ is the current that would flow in $C_{CB}$ if there was no Miller effect and is the function given in (3.10.3):

$$i_0 = v_B \times 2\pi \times \text{Frequency} \times 14.4\ \text{E} - 12 \qquad (3.10.3)$$

The cursors are at the −3 dB point for the mid-band value of $v_B$ (735 mV). The effective gain here is $K = v_C / v_B = 57.3$. The currents in $C_{CB}$ are given in the figure and have a ratio of 58.5, i.e. $K+1$ as expected (as can be seen from the parallel '*i*' curves the ratio is not dependent on this particular frequency but holds across the range of constant gain). The effective signal source resistance $R_s = 735\ \Omega$ and capacity $C_{IN}$ at the base is (the base–emitter capacity $C_{BE}$ is 4.5 p, again from the model):

$$C_{IN} = (C_{CB} + C_{INT}) \times \text{Gain} + C_{BE} = (14.4 + 3.6) \times 57.3 + 4.5 = 1035\ \text{p}$$

so the −3 dB frequency for the base voltage $v_B$ is given by

$$f_{-3\,\text{dB}} = \frac{1}{2\pi R_S C_{IN}} = \frac{1}{2\pi \times 735 \times 1035 \times 10^{-12}} = 209\ \text{kHz} \qquad (3.10.4)$$

again in agreement with the simulation.

In the earliest days of electronics it was positive, rather than negative, feedback that was of primary interest (Tucker 1972). The interest was twofold: one to make oscillators and the other to make near oscillators, i.e. to use what was called regeneration to increase the gain, and the selectivity, of amplifiers, since the available gain from the early amplifying tubes was low (de Forest's audion is said to have had a gain of about 3). Tucker covers the development of the ideas and the extensive litigation surrounding the application, a salutary story about keeping audited accounts of your work. The application of positive feedback to obtain hysteresis is covered in Section 3.12, and examples of oscillators in Sections 5.7 and 5.8. Applications producing negative resistance are treated in Sections 5.26 and 5.16.

## SPICE simulation circuits

| | |
|---|---|
| Fig. 3.10.3(*c*) | Distorn1.SCH |
| Fig. 3.10.5(*b*) | Miller 1.SCH |

## References and additional sources 3.10


Belevitch V. (1962): Summary of the history of circuit theory. *Proc. IRE* **50**, 848–855.

Bennett S. (1993): *A History of Control Engineering 1930–1955*, London: Peter Peregrinus and IEE. ISBN 0-86341-299-8.

Black H. S. (1934): Stabilised feedback amplifiers. *Bell Syst. Tech. J.* **13**, 1–18. See also reprint in *Proc. IEEE* **72**, 715–722, 1984.

Black H. S. (1977): Inventing the negative feedback amplifier. *IEEE Spectrum* **14**, 54–60.

Bode H. W. (1940): Relation between attenuation and phase in feedback amplifier design. *Bell Syst. Tech. J.* **19**, 412–454.

Bode H. W. (1947): *Network Analysis and Feedback Amplifier Design*, New York: Van Nostrand.

Bode H. W. (1960): Feedback – the history of an idea. In *Proceedings of the Symposium on Active Networks and Feedback Systems*, X, Polytechnic Press of the Polytechnic Institute of Brooklyn.

Chen, Wai-Kai (Ed.) (1995): *The Circuits and Filters Handbook*, Boca Raton: CRC Press and IEEE Press. ISBN 0-8493-8341-2. See Section V, p. 799 onwards.

de Sa A. (1990): *Principles of Electronic Instrumentation*, 2nd Edn: London; Edward Arnold. ISBN 0-7131-3635-9.

Ghausi M. S. (1971): *Electronic Circuits*, New York: Van Nostrand Reinhold. Library of Congress Cat. No. 70-154319. See Chapter 6.

Graeme J. G. (1991): *Feedback Plots Define Op Amp AC Performance*, Burr-Brown Application Note AB-028A. See also *EDN* 19 Jan. and 2 Feb. 1989, and Burr-Brown Applications Handbook LI459, 1994, p. 194.

Gray P. R., Meyer R. G. (1993): *Analysis and Design of Analog Integrated Circuits*, New York: John Wiley. ISBN 0-471-01367-6. (3rd Edn 1993, ISBN 0-471-57495-3).

Grey P. E., Searle C. L. (1969): *Electronic Principles. Physics, Models, and Circuits*, New York: John Wiley. ISBN 471-32398-5. See Chapter 18.

Hakim S. S. (1966): *Feedback Circuit Analysis*, London: Iliffe Books.

Hamilton T. D. S. (1977): *Handbook of Linear Integrated Electronics for Research*, London: McGraw-Hill. ISBN 0-07-084483-6.

Karki J. (1999): *Effect of Parasitic Capacitance in Op Amp Circuits*, Texas Instruments Literature Number SLOA013, February. (Stability analysis.)

Klein R. (1993): Harold Black and the negative-feedback amplifier. *IEEE Control Systems* **13**(4), 82–85.

Mancini R. (1999): *Feedback Amplifier Analysis Application Report*, Texas Instruments Literature Number SLOA017, April.

Mancini R. (1999): *Stability Analysis of Voltage-Feedback Op Amps Application Report*, Texas Instruments Literature Number SLOA020, July.

Maxwell J. C. (1868): On governors. *Proceedings Royal Society* **16**, 270–283.

Miller J. M. (1919): *Dependence of the Input Impedance of a Three-Electrode Vacuum Tube upon the Load in the Plate Circuit, National Bureau of Standards Scientific Papers* **15** (351), 367–385.

Millman J., Taub H. (1965): *Pulse, Digital and Switching Waveforms*, New York: McGraw-Hill. Library of Congress Cat. No. 64-66293.

Nyquist H. (1932): Regeneration theory. *Bell System Tech. J.* **13**, 126–147.

Roberge J. K. (1975): *Operational Amplifiers. Theory and Practice*, New York: John Wiley. ISBN 0-471-72585-4.

Siebert W. McC. (1986): *Circuits, Signals, and Systems*, Cambridge: MIT Press/McGraw-Hill. ISBN 0-07-057290-9.

Siegel Barry (1997): *High Frequency Amplifier Instability*, Élantec Application Note 22.

Smith J. I. (1971): *Modern Operational Circuit Design*, New York: Wiley-Interscience. ISBN 0-471-80194-1.

Thomason J. G. (1955): *Linear Feedback Analysis*, Oxford: Pergamon Press.

Tucker D. G. (1972): The history of positive feedback. *Radio and Electronic Engineer* **42**, 69–80.

Tustin A. (1952): Feedback. *Scientific American* **187**, 48–55.

# 3.11 Noise in circuits

Somehow the wondrous promise of the Earth is that there are things beautiful in it, things wondrous and alluring, and by virtue of your trade you want to understand them.

Mitchell Feigenbaum

In Section 2.13 we considered the fundamental sources of noise and derived expressions for their contributions. Here we will consider some examples of the effects of noise in circuits and how these effects may be minimized when necessary. Some readable accounts of how to compute the noise in circuits are given in the references, e.g. Smith and Sheingold (1969), Bryant and Counts (1990); and SPICE provides facilities for determining and plotting noise spectra (Vladimirescu 1994; Tuinenga 1995). In general it is not the magnitude of the noise signals that is of prime importance but rather the signal-to-noise ratio (*SNR*), which is often enshrined in a measure called the noise figure *F*. This expresses the amount by which the circuit diminishes the *SNR* and is usually specified in dB (remember though that here we are dealing with power ratios so to convert a ratio to dB you must take the log and multiply by 10 and not 20). Thus if the circuit was perfect and had no effect so that the *SNR* of the output was the same as *the SNR* of the input, then the noise figure would be 0 dB. It is important to understand that applying negative feedback in op amps, for example, does not of itself change the *SNR*: the gain and the noise are affected equally. It may, however, change impedances so that there is a better match which could improve matters.

In specifying the noise performance of amplifiers it is usual to refer all the noise contributions to equivalent values at the input. It is also necessary to use both voltage noise sources and current noise sources to represent these equivalent inputs effectively and then the amplifier itself can be considered as noiseless. These generators may in most cases be considered to be uncorrelated, so that all the different contributions can be added in the usual root-sum-of-squares manner.

To carry out calculation of the output noise rising from the various 'input' noise generators it is necessary to know the gain for these signals, which may well be different from the usual signal gain since they are input at different points. Consider the simple amplifier with frequency rolloff as shown in Fig. 3.11.1.

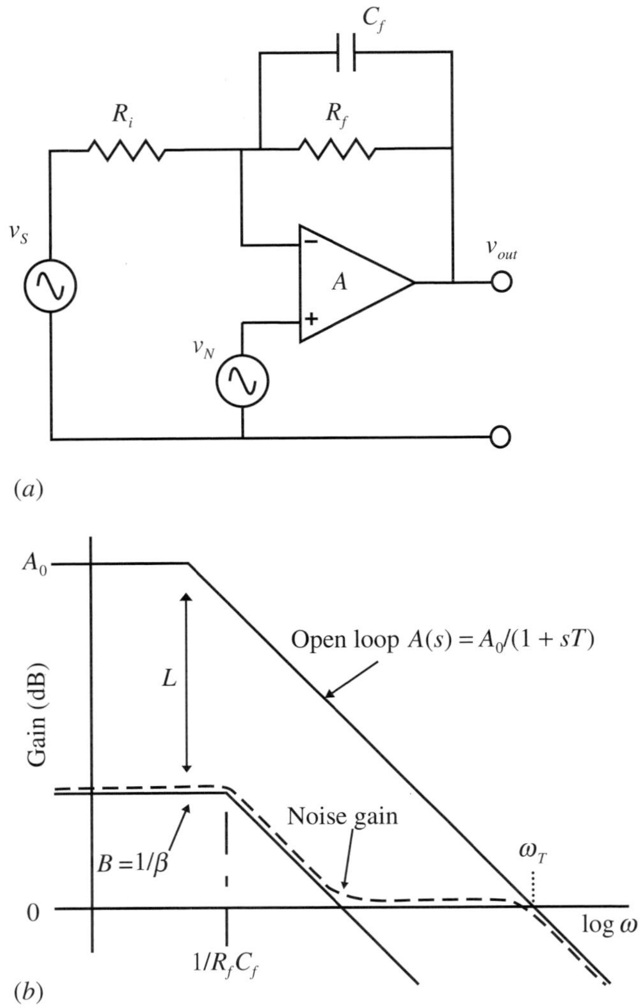


**Fig. 3.11.1** (*a*) Amplifier with rolloff and showing input noise generator. (*b*) Frequency responses.

For a non-inverting system the signal will be input as at $v_S$ while the noise $v_N$ appears at the amplifier non-inverting input. The amplifier output due to $v_S$ will follow the $B = 1/\beta$ curve while the output from $v_N$ will follow $B+1$ until the unity gain point, then follow unity gain until meeting the $A$ curve. For a non-inverting signal configuration, $v_N$ can equally represent the signal, and the signal and noise gain will be the same. Thus the 'noise gain' is not necessarily the same as the signal gain and, as can be seen from the responses, a lot more noise bandwidth is available resulting in a larger noise output than you might have expected from the lower frequency cut-off introduced by $C_f$ (see e.g. Smith and Sheingold 1969; Fredericksen 1984; Bryant 1990; Graeme 1991; Burr-Brown 1994). These references also spell out the steps in calculating the overall output noise.

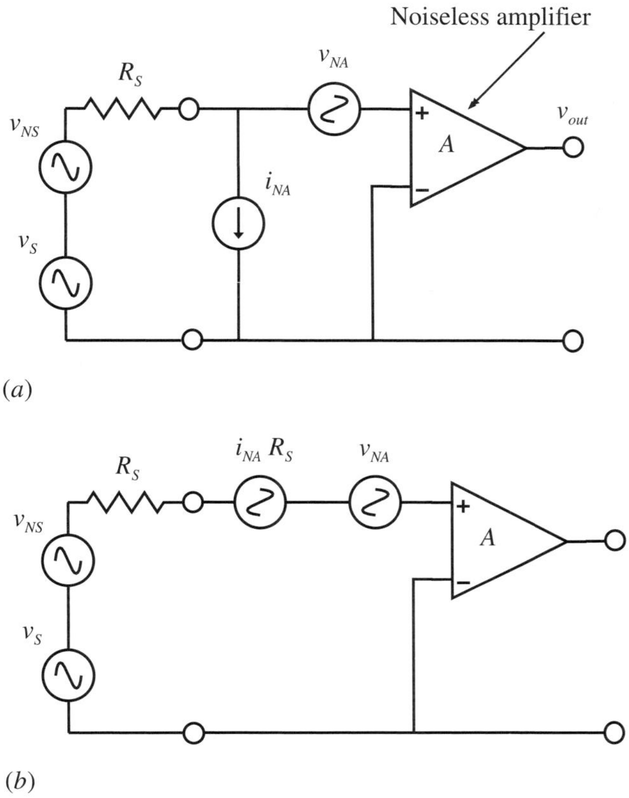


**Fig. 3.11.2** (*a*) Equivalent circuit for a noisy amplifier. (*b*) Transformed circuit.

We can now determine the noise factor $F$. If $v_S$ is the signal, $v_{NS}$ the noise in the signal and $v_{NA}$ is the noise at the output contributed by the amplifier itself, then $F$ is given by:

$$
\begin{aligned}
F &= \frac{\text{signal-to-noise ratio of the input}}{\text{signal-to-noise ratio of the output}} \\
&= \frac{\langle v_S^2\rangle / \langle v_{NS}^2\rangle}{A^2\langle v_S^2\rangle / (A^2\langle v_{NS}^2\rangle + \langle v_{NA}^2\rangle)}, \quad \text{where } A \text{ is the gain of the amplifier} \\
&= \frac{A^2\langle v_{NS}^2\rangle + \langle v_{NA}^2\rangle}{A^2\langle v_{NS}^2\rangle} \\
&= 1 + \frac{\langle v_{NA}^2\rangle}{A^2\langle v_{NS}^2\rangle}, \quad \text{so that it is always } \geq 1 \text{ (or 0 dB).}
\end{aligned}
\tag{3.11.1}
$$

To illustrate the effect of the signal source–circuit configuration on the noise performance we will follow the treatment of Faulkner (1968, 1975). Figure 3.11.2 shows an amplifier with input equivalent voltage and current noise generators, together with a signal source $v_S$ with an effective source resistance $R_S$.

The noise generators arising from the amplifier itself are $v_{NA}$ and $i_{NA}$ referred to the input, and the amplifier is now assumed to be noiseless. $v_S$ represents the input signal and $v_{NS}$ the noise from $R_S$, the signal source resistance. The current noise $i_{NA}$ flowing in $R_S$ is equivalent to a voltage generator $i_{NA}R_S$, which can then be placed in series with $v_{NA}$ as shown in (*b*). Two noise resistances related to the noise generators may be defined. $R_{NV}$, in series with the input, will generate a noise voltage equal to $V_{NA}$, and $R_{NI}$, in parallel with the input, will generate an effect equivalent to $i_{NA}$:

$$R_{NV}=\frac{\langle v_{NA}^2\rangle}{4k_BT\,\Delta f}\equiv\frac{\langle v_{NA}^2\rangle}{N} \quad \text{and} \quad R_{NI}=\frac{4k_BT\,\Delta f}{\langle i_{NA}^2\rangle}\equiv\frac{N}{\langle i_{NA}^2\rangle}, \quad \text{where} \quad N\equiv 4k_BT\Delta f \tag{3.11.2}$$

We may now determine the noise figure from (3.11.1) assuming there is no correlation between the noise sources:

$$\begin{aligned} F&=\frac{\langle v_S^2\rangle}{NR_S}\times\frac{NR_S+\langle v_{NA}^2\rangle+\langle i_{NA}^2\rangle R_S^2}{\langle v_S^2\rangle} \\ &=\frac{NR_S+NR_{NV}+(NR_S^2/R_{NI})}{NR_S} \quad \text{using Eq. (3.11.2)} \\ &=1+\frac{R_{NV}}{R_S}+\frac{R_S}{R_{NI}} \end{aligned} \tag{3.11.3}$$

which shows how the noise figure depends on the three resistances. For given $R_{NV}$ and $R_{NI}$ we can find the optimum source resistance $R_S$ by differentiating with respect to this:

$$\frac{\partial F}{\partial R_S}=0-\frac{R_{NV}}{R_S^2}+\frac{1}{R_{NI}}=0 \quad \text{for a minimum } F$$

$$\text{so} \quad R_{Sopt}=(R_{NV}\,R_{NI})^{\frac{1}{2}} \tag{3.11.4}$$

which from (3.11.3) will give a minimum noise figure $F_{min}$:

$$\begin{aligned} F_{min}&=1+\frac{R_{NV}}{(R_{NV}\,R_{NI})^{\frac{1}{2}}}+\frac{(R_{NV}\,R_{NI})^{\frac{1}{2}}}{R_{NV}} \\ &=1+2\left(\frac{R_{NV}}{R_{NI}}\right)^{\frac{1}{2}} \end{aligned} \tag{3.11.5}$$

As an example we can examine the noise performance of a JFET amplifier as these are commonly used to achieve a low noise factor. The noise generators (Robinson 1970) are:

(*i*) thermal noise in the channel, equivalent to a drain current $i_{ND}$ (with $g_{fs}$ the JFET transconductance) and translated to the gate as $v_{ND}$:

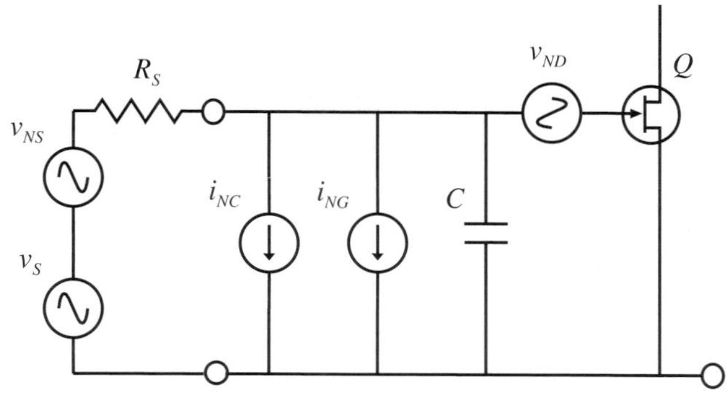


**Fig. 3.11.3** Equivalent noise circuit for a JFET amplifier.

$$\langle i_{ND}^2\rangle = \tfrac{2}{3}4k_BT\,\Delta f g_{fs} \quad \text{or} \quad \langle v_{ND}^2\rangle = \frac{8k_BT\,\Delta f}{3g_{fs}} \quad \text{in the gate circuit} \tag{3.11.6}$$

(*ii*) shot noise due to reverse gate leakage current $I_{GSS}$, and which is particularly important at low frequencies:

$$\langle i_{NG}^2\rangle = 2q_eI_{GSS}\,\Delta f \tag{3.11.7}$$

(*iii*) fluctuations in the channel under the gate (with $C$ the input capacity), important at higher frequencies:

$$\langle i_{NC}^2\rangle = \frac{\omega^2C^2}{4g_{fs}} \times 4k_BT\,\Delta f \tag{3.11.8}$$

which gives us the equivalent circuit shown in Fig. 3.11.3.

At low frequencies we may neglect $i_{NC}$ while at high frequencies we can neglect $i_{NG}$. Considering the low frequency regime for simplicity, the equivalent noise resistances are found to be:

$$\langle v_{ND}^2\rangle = \frac{8k_BT\,\Delta f}{3g_{fs}} = 4k_BTR_{NV}\,\Delta f \quad \text{or} \quad R_{NV} = \frac{2}{3g_{fs}}$$

$$\langle i_{NG}^2\rangle = 2q_eI_{GSS}\,\Delta f = \frac{4k_BTR_{NI}\,\Delta f}{R_{NI}^2} \quad \text{or} \quad R_{NI} = \frac{2k_BT}{q_eI_{GSS}} \tag{3.11.9}$$

so from (3.11.3)–(3.11.5) we have:

$$F = 1 + \frac{2}{3g_{fs}R_S} + \frac{q_eR_SI_{GSS}}{2k_BT}$$

$$F_{min} = 1 + \left(\frac{4q_eI_{GSS}}{3k_BT\,g_{fs}}\right)^{\frac{1}{2}} = 1 + 7.2\left(\frac{I_{GSS}}{g_{fs}}\right)^{\frac{1}{2}} \tag{3.11.10}$$

$$R_{Sopt}=\left(\frac{4k_BT}{3k_BT\,g_{fs}}\right)^{\frac{1}{2}}=\frac{0.185}{(g_{fs}\,I_{GSS})^{\frac{1}{2}}}$$

where we have put in the values of the constants $q_e$ and $k_B$ and let $T=300$ K. For a U310 JFET (Siliconix 1989) with conservative values $I_{GSS}=10^{-10}$ A and $g_{fs}=10$ mA $V^{-1}$ we find:

$$R_{NV}=67\ \Omega,\quad R_{NI}=517\ \text{M}\Omega,\quad F_{min}=1.00072=0.003\ \text{dB},\quad R_{Sopt}=185\ \text{k}\Omega \tag{3.11.11}$$

so it is a very good low noise amplifier. For low source resistances bipolar transistors are more appropriate, but if the source resistance for the U310 is only 10 kΩ then we find $F$ is still only 0.03 dB so the minimum is a shallow one. These figures indicate the sort of performance that can be achieved, but it should be remembered that there will be an increase in $1/f$ noise at low frequency and from channel fluctuations ((*iii*) above) at high, and increases at higher temperatures. Horowitz and Hill (1989) discuss a range of bipolar and FET low noise transistors. See Jefferts and Walls (1989) for an example of a very low-noise JFET amplifier.

PSpice provides facilities for noise measurements when an *AC SWEEP* is performed. The total output noise *ONOISE*, at a specified output, is calculated by determining the individual contributions from all the sources, e.g. resistors, transistors, etc., which are then multiplied by the appropriate gain and then summed as root-sum-of-squares to give the output in terms of a noise density with units of volt per root Hz. This value may be divided by the overall gain to give an *INOISE* equivalent at a specified input. As an example of this simulation we will consider the simple amplifier circuit shown in Fig. 3.11.4 (Griffiths 1970).

Under *ANALYSIS/SETUP/AC SWEEP* you need to check the box for *NOISE ANALYSIS* to enable it and set *OUTPUT VOLTAGE* to *V*(*VOUT*) and *I/V* to *V*3 for this circuit, leaving *INTERVAL* (which concerns printouts of noise values) blank. This setting defines $V_{out}$ as the place to determine *ONOISE* and *V*3 the place to refer *INOISE* to. The gain is set by $R_f$ with values 15 k, 3k3, 1 k, 300*R* giving nominal gains of 3, 10, 30, 100. Setting $R_f=300\ \Omega$ for a gain of $\approx 100$ and varying $R_S$ from 1 to 100 k a set of simulations were run with results shown in Table 3.11.1.

$V(ONOISE)=A\langle v_{NS}^2\rangle+\langle v_{NA}^2\rangle$ in Eq. (3.11.1), is plotted by SPICE and $V_{NS}^2=4k_BTR_S$ is the source resistance noise. An example of a run is shown in Fig. 3.11.5 together with a plot of the variation of $F$ as a function of $R_S$ and the measured results as shown in the reference. The curves differ somewhat but that is not surprising since the characteristics of transistors can vary widely (Faulkner and Harding 1968) and we have not made any allowance for additional low and high frequency noise.

If you wish to measure the noise from an isolated resistor then Fig. 3.11.4(*b*) is

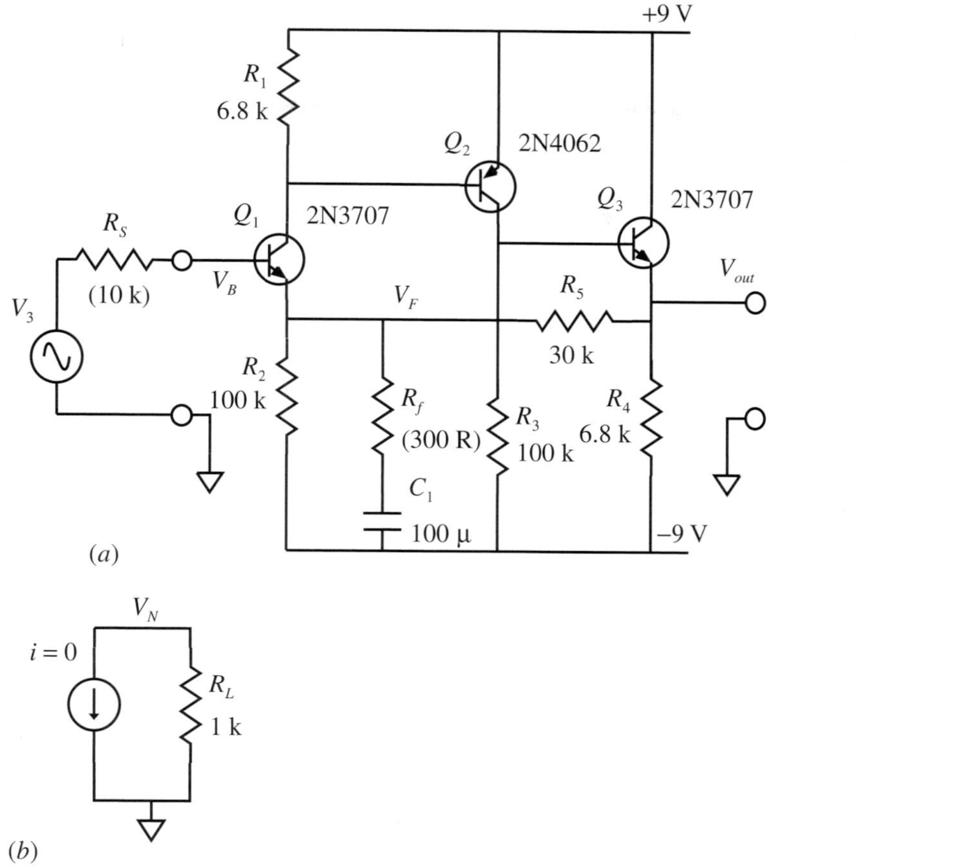


**Fig. 3.11.4** (*a*) Amplifier circuit for noise measurements. (*b*) Noise measurement for an isolated resistor.

Table 3.11.1 *Measured values from noise simulations. All values taken at* 1 kHz *and refer to unit bandwidth* ($4k_BT = 1.565\text{E}-20$)

| $R_S$ | Gain $A$ | $V(ONOISE)$ | $V(ONOISE)^2 = X$ | $A^2 V_{NS}^2 = Y$ | $F = X/Y$ ratio: (dB) |
|---|---|---|---|---|---|
| 1 k | 98.05 | 4.94E−7 | 24.4E−14 | 15.92E−14 | 1.53 : (1.85) |
| 3k3 | 98.0 | 7.92E−7 | 62.73E−14 | 52.48E−14 | 1.20 : (0.77) |
| 10 k | 97.8 | 1.35E−6 | 1.82E−12 | 1.58E−12 | 1.15 : (0.62) |
| 33 k | 97.3 | 2.61E−6 | 6.81E−12 | 5.17E−12 | 1.32 : (1.2) |
| 100 k | 95.8 | 5.40E−6 | 29.16E−12 | 15.2E−12 | 1.92 : (2.83) |

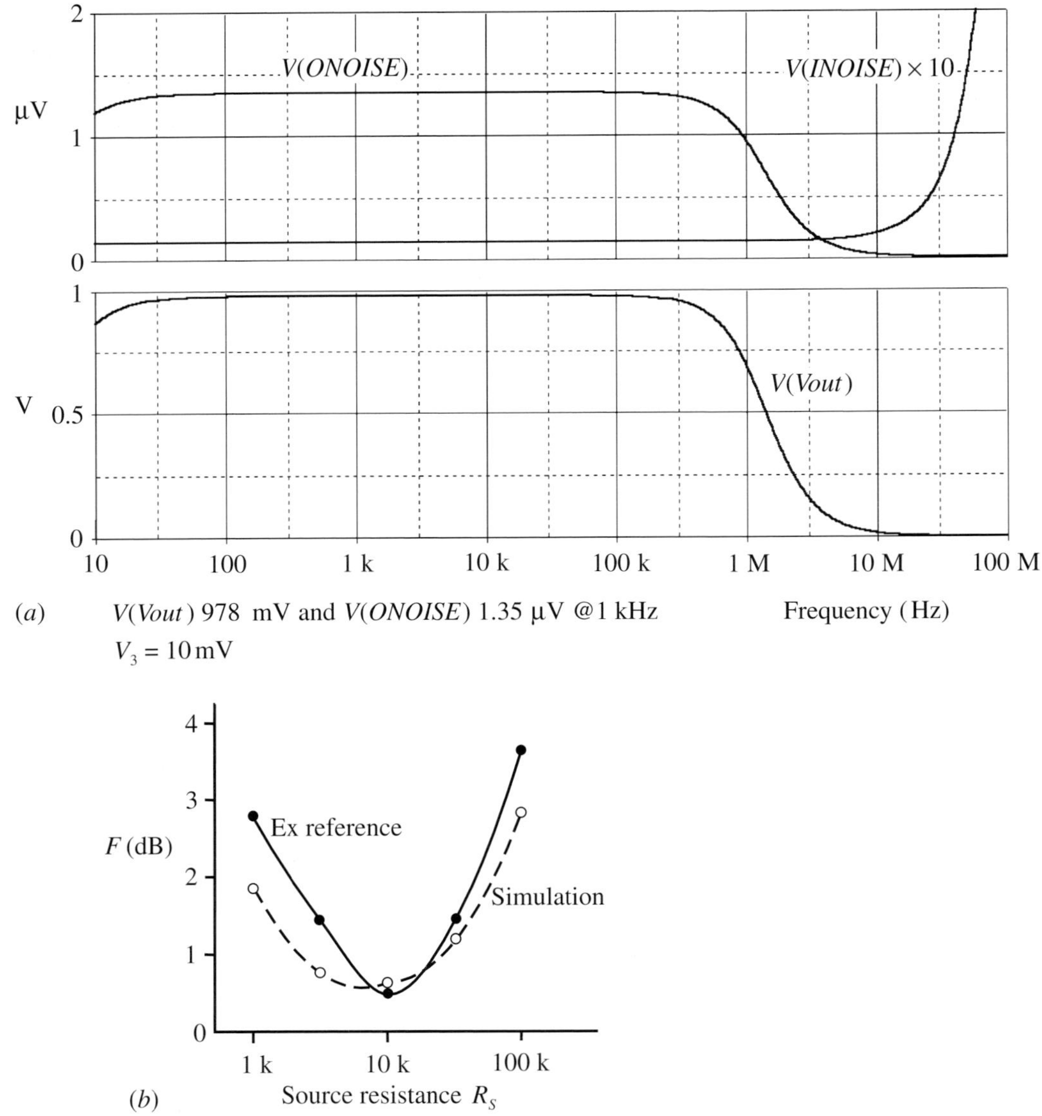


**Fig. 3.11.5** (*a*) Noise and gain as a function of frequency for $R_S = 10$ k and $R_f = 300\ \Omega$. (*b*) Noise figure $F$ as a function of source resistance $R_S$.

one way of doing it – set the current source for zero current, adjust the names under *ANALYSIS* and plot *V*(*ONOISE*). The current source copes with the dangling node and has infinite resistance so no current will flow. When plotting noise spectra in *PROBE* you will find listed under *ADD TRACE* a long list of noise variables, such as *NRE*(*Q1*), etc., which is the noise from the emitter resistance of transistor *Q*1. Under *PROBE HELP/INDEX* you will find *Noise, syntax and variables* which gives a listing of such variables. There is no entry for noise in *SCHEMATICS HELP*.

When using models of more complex devices, e.g. an op amp, you should check whether the model implements internal noise sources or not. For example, the Analog Devices library header in the PSpice Version 8 model library contains the statement 'Unique to ADI, the following models contain noise sources, allowing SPICE analysis of the total noise of a circuit:' (release K, 1/95).

A noise generator is available for transient simulation of circuits where, for example, you may wish to examine the effect of filtering a noisy signal. This is described by Hageman (1996) and the program is available from MicroSim/OrCAD/Cadence. An example of its use is given in Section 3.16.

### SPICE simulation circuits

| | |
|---|---|
| Fig. 3.11.5(*a*) | Nseamp 1.SCH |

## References and additional sources 3.11

Anderson P. T., Pipes P. B. (1974): A low noise amplifier with application to noise thermometry between 300 and 4 K. *Rev. Sci. Instrum.* **45**, 42–44.

Anson M., Martin S. R., Bayley P. M. (1977): Transient CD measurements at submillisecond time resolution – Application to studies of temperature-jump relaxation of equilibria of chiral biomolecules. *Rev. Sci. Instrum.* **48**, 953–962.

Baxandall P. J. (1968): Noise in transistor circuits. *Wireless World* **74**, Nov., pp. 388–392, and Dec. 1968, pp. 454–459.

Blake K. (1996): *Noise Design of Current-Feedback Op Amp Circuits*, National Semiconductor Application Note OA-12, Rev. B, April.

Bryant J. (1990): Op-amp issues – Noise. *Analog Dialogue* **24**(3), 24–25.

Bryant J., Counts L. (1990): Op-amp issues – Noise. *Analog Dialogue* **24**(2), 20–21.

Buckingham M. J., Faulkner E. A. (1974): The theory of inherent noise in p-n junction diodes and bipolar transistors. *Radio and Electronic Engineer* **44**(3), March, 125–140.

Burr-Brown (1994): *Noise Analysis of FET Transimpedance Amplifiers*, Burr-Brown Application Bulletin B-076: also *Photodiode Monitoring with Op Amps*, Bulletin AB-075 (both in Applications Handbook LI459).

Buxton J. (1992): Improve noise analysis with op-amp macromodel. *Electronic Design*, 2 April,73, 74, 76–81.

Chang Z. Y., Sansen W. M. C. (1991): *Low Noise Wide-Band Amplifiers in Bipolar and CMOS Technologies*, Amsterdam: Kluwer.

Faulkner E. A. (1968): The design of low-noise audio-frequency amplifiers. *Radio and Electronic Engineer* **36**, 17–30.

Faulkner E. A. (1975): The principles of impedance optimization and noise matching. *J. Phys. E Sci. Instrum.* **8**, 533–540.

Faulkner E. A., Harding D. W. (1968): Some measurements on low-noise transistors for audio-frequency applications. *Radio and Electronic Engineer* **36**, July, 31–33.

Fredericksen T. M. (1984): Intuitive op amps. *National Semiconductor*, p. 123.

Graeme J. G. (1991): *Feedback Plots Define Op Amp AC Performance*, Burr-Brown Application Note AB-028A. See also *EDN* 19 Jan. 1989 and 2 Feb. 1989.

Gray P. R., Meyer R. G. (1977): *Analysis and Design of Analog Integrated Circuits*, New York: John Wiley. p. 607. ISBN 0-471-01367-6.

Griffiths D. (1970): Low noise, constant loop gain, ac amplifier details. *J. Phys. E Sci. Instrum.* **3**, 243–245.

Gupta M. S. (1977): *Electrical Noise: Fundamentals and Sources*, New York: IEEE Press: Reprints of important papers on noise.

Hageman S. C. (1996): *Create Analog Random Noise Generators for PSpice Simulation*, MicroSim Application Notes Ver.6.3, p. 61, MicroSim Corp., April.

Halford D. (1968): A general model for spectral density random noise with special reference to flicker noise 1/*f*. *Proc. IEEE* **56**, March, 251.

Hamilton T. D. S. (1977): *Handbook of Linear Integrated Electronics for Research*, London: McGraw-Hill. See Sections 1.8 and 4.1, with many references.

Hamstra R. H., Wendland P. (1972): Noise and frequency response of silicon photodiode operational amplifier combination. *Applied Optics* **11**, 1539–1547.

Horowitz P., Hill W. (1989): *The Art of Electronics*, 2nd Edn, Cambridge: Cambridge University Press. See Chapter 7.

Jefferts S. R., Walls F. L. (1989): A very low-noise FET input preamplifier. *Rev. Sci. Instrum.* **60**, 1194–1196. (Uses the 2SK117.)

Keshner M. S. (1982): 1/*f* Noise. *Proc. IEEE* **70**, March, 212–218.

Morrison R. (1992): *Noise and Other Interfering Signals*, New York: John Wiley. ISBN 0-471-54288-1.

Motchenbacher C. D., Fitchen F. C. (1973): *Low-Noise Electronic Design*, New York: John Wiley.

Motchenbacher C. D., Connelly J. A. (1993): *Low Noise Electronic System Design*, New York: John Wiley. ISBN 0-471-57742-1.

National Semiconductor (1993): *A Tutorial on Applying Op Amps to RF Applications*, National Semiconductor Application Note OA-11, September.

O'Dell T. H. (1991): *Circuits for Electronic Instrumentation*, Cambridge: Cambridge University Press. ISBN 0-521-40428-2.

Ott H. (1988): *Noise Reduction Techniques in Electronic Systems*, 2nd Edn, New York: John Wiley. ISBN 0-471-85068-3.

Radeka V. (1988): Low-noise techniques in detectors. *Ann. Rev. Nuc. Part. Sci.* **38**, 217–277.

Robinson F. N. H. (1962): *Noise in Electrical Circuits*, Oxford: Oxford University Press.

Robinson F. N. H. (1970): Noise in transistors. *Wireless World* **76**, July, 339–340.

Ryan A., Scranton T. (1969): D-c amplifier noise revisited. *Analog Dialogue* **3**(1), 3–10.

Schouten R. N. (1998): A new amplifier design for fast low-noise far-infrared detectors using a pyroelectric element. *Meas. Sci. Tech.* **9**, 686–691.

Sherwin J. (1974): *Noise Specs Confusing?* National Semiconductor Application Note AN-104.

Siliconix (1989): *Low Power Discretes Data Book*. See also Watson B. *Audio-Frequency Noise Characteristics of Junction FETs*, LPD-5 or AN74-4 Application Note, 1974.

Smith L., Sheingold D. H. (1969): Noise and operational amplifier circuits. *Analog Dialogue* **3**(1), March, 5–16.

Spectrum Software (1997): Noise source macro. *Spectrum Newsletter* Fall Issue, 6–8. www.spectrum-soft.com

Steffes M. (1993): *Noise Analysis for Comlinear's Op Amps*, Comlinear Application Note OA-12, January.

Steffes M. (1993): *Improving Amplifier Noise Figure for High 3rd Order Intercept Amplifiers*, Comlinear Application Note OA-14, January.

Texas Instruments (1999): *Noise Analysis in Operational Amplifier Circuits*, Texas Instruments Literature Number SLOA019, March.

Tuinenga P. W. (1995): *SPICE: A Guide to Circuit Simulation and Analysis Using PSPICE*, Englewood Cliffs, NJ: Prentice Hall. 1988 (1st Edn). ISBN 0-13-834607-0. 3rd Edition ISBN 0-13-158775-7.

Usher M. J. (1974): Noise and bandwidth. *J. Phys. E Sci. Instrum.* **7**, 957–961.

Van der Ziel A. (1954): *Noise*. Englewood Cliffs: Prentice-Hall.

Van der Ziel A. (1958): Noise in junction transistors. *Proc. IRE* **46**, 1019–1038.

Van der Ziel A. (1970): *Noise Sources, Characterization, Measurement*, Englewood Cliffs: Prentice-Hall. Library of Congress Cat.No. 71-112911.

Van Driessche W., Lindemann B. (1978): Low-noise amplification of voltage and current fluctuations arising in epithelia. *Rev. Sci. Instrum.* **49**, 52–57.

Vladimirescu A. (1994): *The Spice Book*, New York: John Wiley, p. 149. ISBN 0-471-60926-9.

Williams J. (1993): *A Broadband Random Noise Generator*, Linear Technology Design Note 70.

Wilmshurst T. H. (1985): *Signal Recovery from Noise in Electronic Instrumentation*, Bristol: Adam Hilger. ISBN 0-85274-783-7.

# 3.12 Hysteresis

Nothing holds up the progress of science so much as the right idea at the wrong time.

De Vignaud, biochemist

A system is said to exhibit hysteresis if the response to a stimulus depends on the sense or direction of the stimulus. A well known case is that of magnetic materials where the magnitude of the induction $\boldsymbol{B}$ for a given field $\boldsymbol{H}$ depends also on the sense of $\boldsymbol{H}$, i.e. on how that value of $\boldsymbol{H}$ was reached. In electronic systems another form of hysteresis is often used to clean up noisy signals and is introduced by applying positive feedback. How this acts to produce hysteresis raises an interesting philosophical problem as we shall see.

Consider a simple comparator circuit. A comparator is a device somewhat similar to an operational amplifier with high gain, but designed rather for fast response without consideration of the need to be able to apply negative feedback. An operational amplifier can also be used but the response time may be slower. One input will be connected to a reference voltage and the other to the signal. The output will be at one limit or the other depending on whether the signal is positive or negative relative to the reference. Thus over a small range of input, e.g. 10 mV, the output changes very rapidly. However, if the signal is noisy then if it remains in the vicinity of the reference the output will switch with each significant noise variation. To defeat this effect hysteresis is introduced by applying a degree of positive feedback as shown in Fig. 3.12.1. This form of circuit is commonly referred to as a Schmitt trigger (Schmitt 1938).

By taking particular values for the resistors it is easier to see the consequences. Assume the limiting values of the output are ±10 V, then if the signal is negative relative to the reference input (point *B*) the output will be at +10 V (point *U*) and the reference input will be at +0.909 V (point *C*) determined by the potential divider. As the input moves from *B*, the output remains constant from *U* to *V*, and nothing happens until the input reaches *C*; as it passes, the output will change from point *V* to point *W* and any further increase of input to *D* will cause the output to stay at −10 V (point *X*). Now the reference voltage is −0.909 V (point *E*) so if the input moves back from *D* nothing will happen until point *E* is reached when the

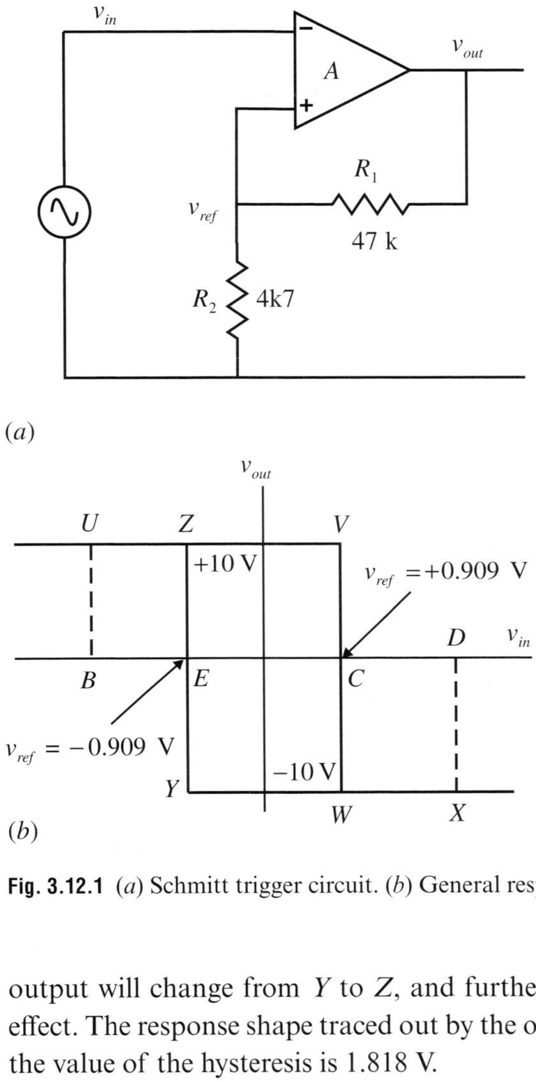


**Fig. 3.12.1** (*a*) Schmitt trigger circuit. (*b*) General response with positive feedback.

output will change from $Y$ to $Z$, and further change of input to $B$ will have no effect. The response shape traced out by the output is a classic hysteresis curve and the value of the hysteresis is 1.818 V.

We now examine this in more detail. The gain of a system with negative feedback is given by Eq. (3.10.1), so that for positive feedback we need only change the sign of '$A$' to get:

$$G = \frac{A}{1 - A\beta} \tag{3.12.1}$$

The open-loop response of the amplifier is shown as curve *I* in Fig. 3.12.2.

If $A$ is say $10^4$ then for ±10 V output, as above, the linear input region will extend from about −1 to +1 mV. The slope of this region is given by the gain. For inputs greater than 1 mV the output saturates and the gain (slope) tends to zero. If a small amount of positive feedback, given by $\beta$, is applied then the gain must increase and we get say curve *II* (this technique was used in the early days of radio to increase

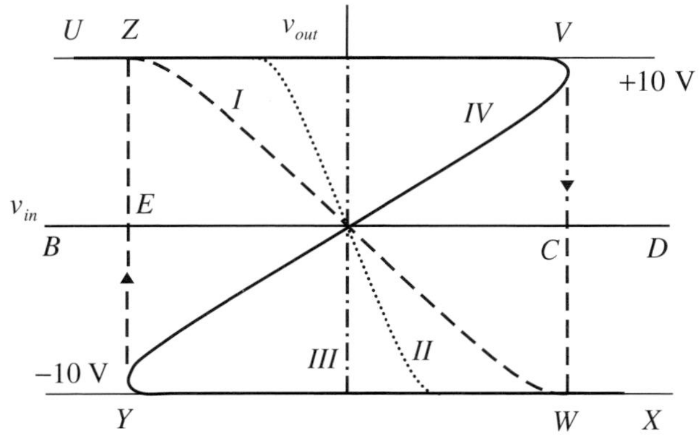


**Fig. 3.12.2** Responses as a function of magnitude of positive feedback.

the gain of amplifiers and was known as regeneration (Tucker 1972). If $\beta$ increases further so that the loop gain $A\beta = 1$ then the gain is infinite and we have curve *III*. It may be a little difficult to think of an infinite gain but you can consider the signal going round and round the loop being multiplied each time. For curves *I* and *II* the rate of change of the output is dependent on the rate of change of the input. However, for curve *III* we now have the situation where an infinitesimal change of input causes a change between limits in a time independent of the rate of change of the input – the circuit is said to have been triggered. This idea was first described by Schmitt (though with a valve circuit) and such systems are now commonly referred to by his name (Schmitt 1938; Williams 1946). The hysteresis of curve *III* is evidently zero. If now the feedback is increased still further the gain must be considered to be even greater than infinite, and changes sign (slope) as shown at curve *IV*. We can now consider again the sequence of inputs as above, starting with an input at *B* and output at *U*. As $v_{in}$ moves towards *C* then $v_{out}$ moves towards *V*. Now the amplifier starts to come out of saturation and the gain (slope) becomes finite. Small further change brings the gain to infinity (slope vertical) after which the positive feedback takes over. The system is now in an unstable condition and cannot remain on this part of the response. For the given input at *C* the stable state is at *W* – in effect the system makes the vertical transition from *V* to *W*. Further change of $v_{in}$ to *D* moves $v_{out}$ to *X* so there is no further change. Reversing $v_{in}$, moving from *D* to *E* causes a similar change at *E* where the output changes from *Y* to *Z* and the hysteresis is the difference between $v_{in}$ at *E* and *C*.

When the critical points at *E* or *C* are reached the positive feedback takes control and drives the system as fast as it will go, i.e. limited by the slewing rate, between the two limiting output states. When it reaches either of these the gain decreases to zero as the output is saturated and so we have a stable state again. This is why the

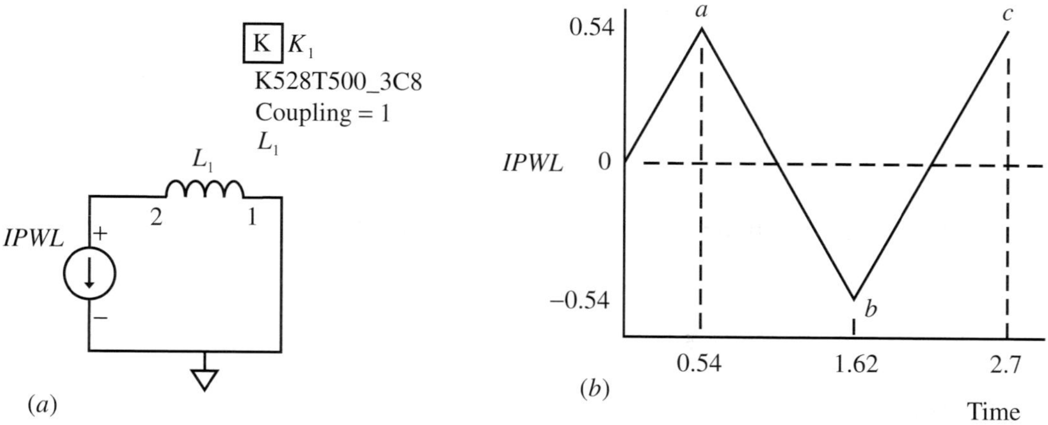


**Fig. 3.12.3** (*a*) Simulation circuit for hysteresis curves. (*b*) Current waveform for the *IPWL* generator.

system is said to be triggered in that once the critical point is reached the input has no further influence, just like pulling the trigger on a gun. Although the original Schmitt trigger referred to a particular circuit the name is now commonly given to any system of this kind. Circuit symbols for devices which incorporate hysteresis are often shown with an enclosed hysteresis response curve to remind you (e.g. the logic device 74XXX14 Hex Schmitt Inverter).

Hysteresis has traditionally been associated with the behaviour of magnetic materials subjected to alternating fields, as exemplified by the original work of J. A. Ewing, who introduced the idea in 1881 (Ewing 1885), and of Steinmetz (1892). The facilities for magnetic material modelling in PSpice are not extensive but there is an example for drawing a hysteresis curve hidden in the header of the *Magnetic.lib*. Guidance and instruction as to how to determine the parameters of a model for a material on the basis of the Jiles–Atherton (1986) model used by PSpice is given by Prigozy (1993) which we will outline, with modification for the newer version of PSpice. The requirement is to be able to draw, as a function of one of the parameters, several *B–H* curves on the same display. Since PSpice will only do this if the variable is time, a means of persuading it to do it as a function of a different variable is required. Figure 3.12.3(*a*) shows the simple circuit for simulation and you should orient the components as shown to get the senses correct. Here we can again make the suggestion that you edit your standard symbols, in this case the inductor, to make the pin numbers visible. Figure 3.12.3(*b*) shows the form of the current to be defined for the *IPWL* current source. Since we will use the model for the ferrite 3C8 from the *Magnetic.lib* rather than that for the grain-oriented steel used by Prigozy, the currents have been changed to suit.

The definitions of the circuit components are:

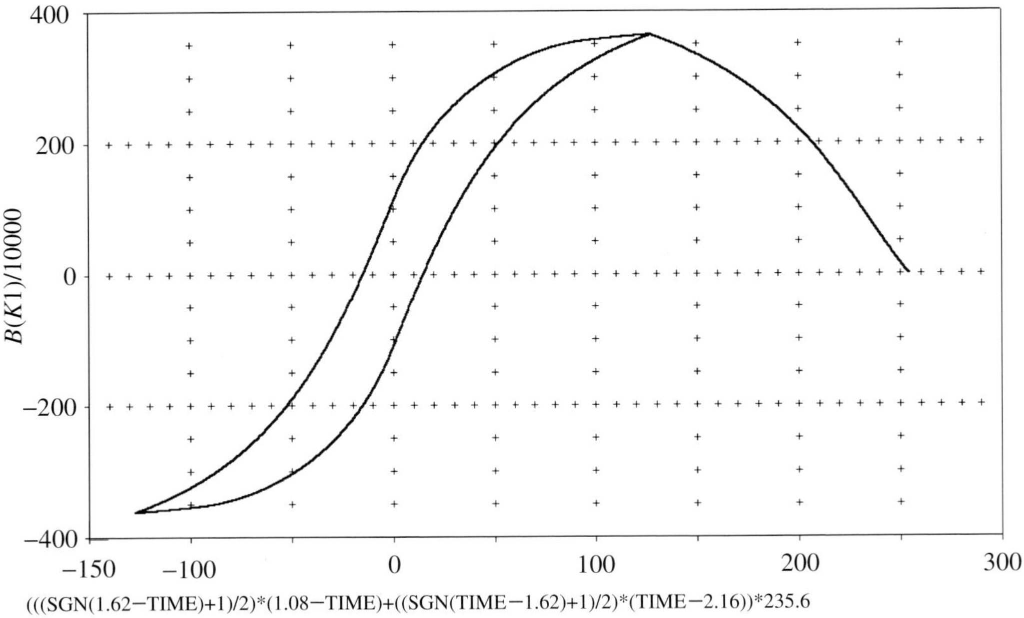


**Fig. 3.12.4** Hysteresis curve from Fig. 3.12.3 circuit.

For *IPWL*, in the form $(T_n, I_n)$: (0, 0; 0.54, 0.54; 1.62, −0.54; 2.7, 0.54)
For the inductor *L*1: *Value* = 20 (note that this will be interpreted as number of turns by PSpice)
For *Kbreak*: *L*1 = *L*1; *Coupling* = 1; *Model* = K528T500_3C8

and you must use the *Kbreak* parameter, rather than the simple *K*, to cause the simulator to interpret the *Value* 20 as turns and to access the material (3C8) and form (pot core) model. The apparently random values for the *IPWL* generator are Prigozy's values divided by 5 to suit the different material. The important consideration is the 1:1 relation between time and current. The conversion to enable multiple plots is to make the *x*-axis the function:

$$x = f(time) = (((SGN(3*I_{max} - TIME) + 1)/2)*(2*I_{max} - TIME)) + (((SGN(TIME - 3*I_{max}) + 1)/2)*(TIME - 4*I_{max})) \qquad (3.12.2)$$

where in our case $I_{max} = 0.54$ and *SGN* is a function available in *PROBE* that returns +1, 0 or −1 depending on the value of the argument. *TIME* is also a standard variable available in *PROBE*. The mapping function Eq. (3.12.2) does not readily spring to mind but the hard work has been done for us. If you run the simulation and plot trace *I*(*L*1) you will get Fig. 3.12.3(*b*). Now alter the *x*-axis to match Eq. (3.12.2) and you will get a diagonal line. When PSpice sees a magnetic model it automatically determines *H*(*K*1) (in oersted) and *B*(*K*1) (in gauss, the *K*1

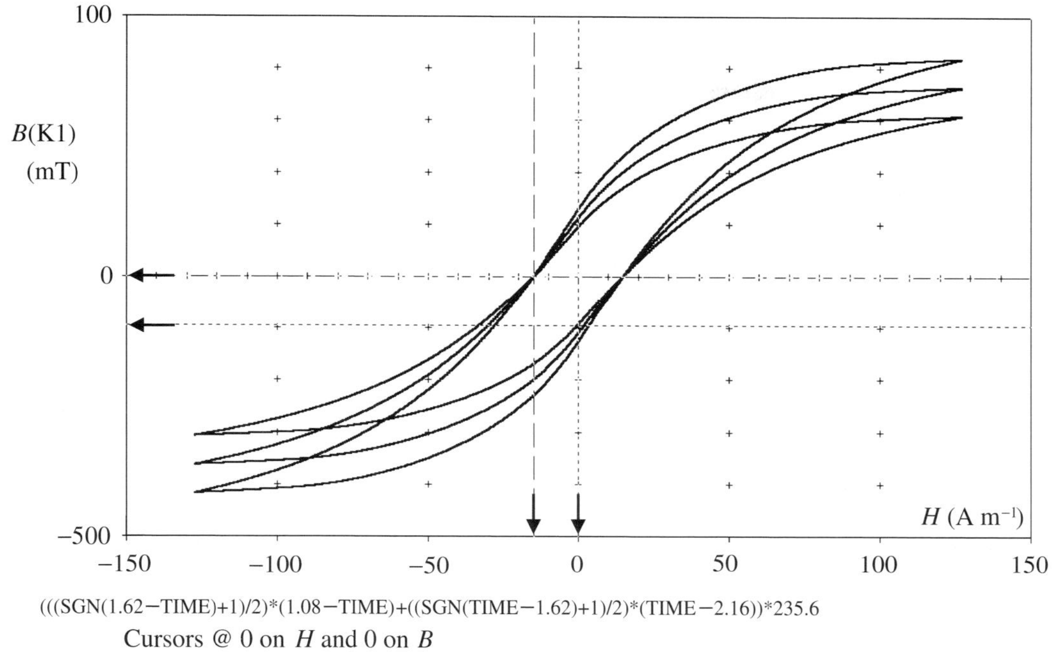


**Fig. 3.12.5** Example of hysteresis curves for variation of parameter *MS*.

being the *Kbreak* above), so if you replace trace *I*(*L*1) with *B*(*K*1) you will get the hysteresis curve as in Fig. 3.12.4.

The tail on the right-hand side arises from the portion of the current in Fig. 3.12.3(*b*) from zero to point *a* and is in effect the initial magnetization response but displaced and reflected by the mapping. If you wish to suppress this then in the *TRANSIENT* setup set the *NO PRINT DELAY* to 0.54 s. If you now wish to plot multiple curves you need to make one of the model values a parameter using the *PARAM* device. It would be wise to make a copy of the model to a user library and give it a variant name before making alterations in case you forget to restore the original. Under *ANALYSIS/SETUP/PARAMETRIC* select *Model Parameter* and say *Value List*. The *Model Type* is *Core*, the *Model Name* is K528T500_3C8, and the *Parameter Name* is *MS* (or *A* or *C* or *K* as required). Fill in the values in the value list and run the simulation. An example, with *MS*=352 k, 414 k, 478 k, is shown in Fig. 3.12.5. Dividing *B*(*K*1) by $10^4$ gives a scale of mT and multiplying the *x*-axis function by number of turns/path length (20/0.00849=253.6 in this case) gives a scale of A $m^{-1}$.

## SPICE simulation circuits

Consult the SimCmnt.doc file on the CD before running.

| | |
|---|---|
| Fig. 3.12.4 | Hystmag 2.SCH |
| Fig. 3.12.5 | Hystmag 2.SCH |

## References and additional sources 3.12


Ewing J. A. (1885): Experimental researches in magnetism. *Phil. Trans. Roy. Soc.* **176**(2), 523–640.

Hirasuna B. (1999): *Using Coupled Inductors and Inductor Cores*, Cadence/MicroSim Application Note PSPA021, April.

Jiles D. C., Atherton D. L. (1986): Theory of ferromagnetic hysteresis. *J. Magnetism and Magnetic Materials* **61**, 48–60.

O'Dell T. H. (1991): *Circuits for Electronic Instrumentation*, Cambridge: Cambridge University Press. ISBN 0-521-40428-2.

Prigozy S. (1993): PSPICE computer modeling of hysteresis effects. *IEEE Trans*. **EDUC-36**, 2–5. I am indebted to the author for discussions.

Prigozy S. (1994): PSPICE computer modeling of power transformers with hysteresis effects. *Computers in Education J.* **IV**(2), April–June, 68–76.

Schmitt O. H. (1938): A thermionic trigger. *J. Sci. Instrum.* **15**, 24–26.

Steinmetz C. P. (1892): On the law of hysteresis. *American Institute of Electrical Engineers Trans.* **9**, 3–64. See also a reprint in *Proc. IEEE* **72**(2), 196–221, 1982.

Tucker D. G. (1972): The history of positive feedback. *Radio and Electronic Engineer* **42**, 69–80.

Williams F. C. (1946): Introduction to circuit techniques for radiolocation. *J. IEE* **93**, Part IIIA, No.1, 289–308. See his Fig. 33.

# 3.13 Bridges

There is no reason to assume that the truth, when found, will be interesting.

Bertrand Russel

The classic Wheatstone bridge (Christie 1833; Wheatstone 1843) was one of the earliest electrical instruments and was used for measuring resistance. The form of the circuit, Fig. 3.13.1, proved to be adaptable to other applications and many other forms evolved to enable measurement of a range of other parameters. As we will come across bridge circuits in several places it is well worth examining the qualities of the bridge that make it useful.

The basic idea behind the bridge is to have a null system that allows much increased sensitivity, and hence resolution, in the measurements. The bridge is adjusted so that two large signals are balanced out so that the small residual signal can be greatly amplified to provide the resolution. If we were trying to measure a small change in a large signal then the large signal would also be amplified so our dynamic range would have to be very large. We start with the original arrangement and determine the conditions for balance. To make the sums rather simpler we assume that the input impedance of the detector measuring the bridge output voltage $v_o$ is so high that it does not draw current. Since the same current $i_1$ must flow in $R_1$ and $R_2$, and the same current $i_2$ in $R_3$ and $R_4$ then we can write:

$$i_1 = \frac{v_{in}}{R_1 + R_2} \quad \text{and} \quad i_2 = \frac{v_{in}}{R_3 + R_4} \tag{3.13.1}$$

and since $v_o$ is the difference between the potentials across $R_2$ and $R_4$ we have:

$$\begin{aligned} v_o &= i_1 R_2 - i_2 R_4 \\ &= \frac{v_{in} R_2}{R_1 + R_2} - \frac{v_{in} R_4}{R_3 + R_4} \\ &= \frac{v_{in}(R_2 R_3 - R_4 R_1)}{(R_1 + R_2)(R_3 + R_4)} \end{aligned} \tag{3.13.2}$$

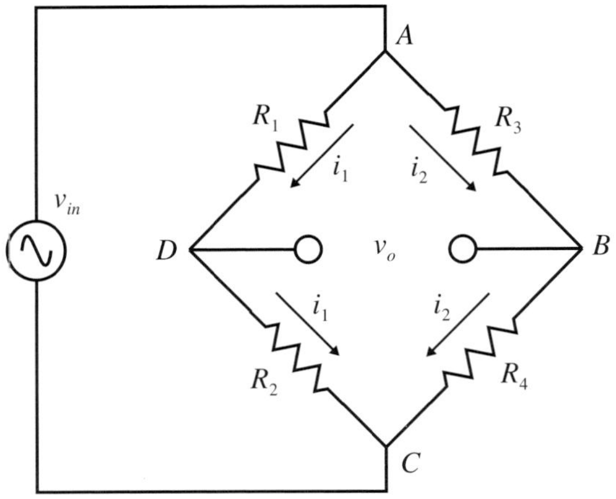


**Fig. 3.13.1** Basic Wheatstone bridge.

and for this to be zero we must have:

$$R_2R_3 - R_4R_1 = 0 \quad \text{or} \quad \frac{R_1}{R_2} = \frac{R_3}{R_4} \tag{3.13.3}$$

For a.c. voltages and arms with reactive elements then, there are two equations to be satisfied rather than one. The impedances are complex and the real and the imaginary components must be separately balanced, but the principle is just the same.

There is another question that must be answered. For maximum sensitivity, what relationship should there be between the resistances? This can be determined by considering a small change $\delta v_o$ in $v_o$ arising from a small change $\delta R_2$ in $R_2$ say. We have:

$$\delta v_o = \frac{\partial v_o}{\partial R_2}\delta R_2$$

$$= \frac{[v_{in}R_3(R_1+R_2)(R_3+R_4) - v_{in}(R_2R_3 - R_4R_1)(R_3+R_4)]\delta R_2}{[(R_1+R_2)(R_3+R_4)]^2}$$

$$= \frac{v_{in}R_1\,\delta R_2}{(R_1+R_2)^2} \tag{3.13.4}$$

so the relative change in $R_2$ is given by:

$$\frac{\delta R_2}{R_2} = \frac{\delta v_o}{v_{in}}\frac{(R_1+R_2)^2}{R_1R_2} \tag{3.13.5}$$

and differentiating the right-hand side and putting the result equal to zero to find the minimum gives $R_1 = R_2$. The same result will hold for $R_3$ and $R_4$, i.e. the

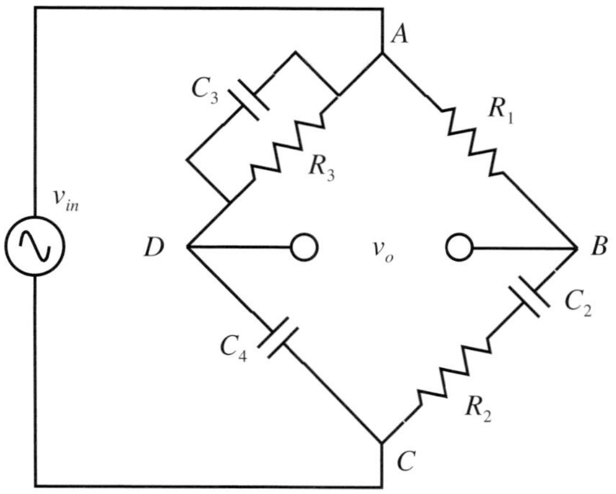


**Fig. 3.13.2** Schering bridge.

maximum sensitivity occurs for equal values of all the arms. A logarithmic null detector is most beneficial in that the large signal when far away from balance is of less interest than the small signals close to balance. An example of a simple resistive bridge is the commonly used strain gauge. Another is found in the Meacham-bridge oscillator (Meacham 1938; Strauss 1970) which provides greatly improved stabilization for a crystal oscillator.

A.C. bridges have the same basic topology but the requirement for balance means that there are two conditions, one for amplitude and one for phase. There are very many forms of a.c. bridge (Hague 1971) but here we will only consider one, the Schering bridge, Fig. 3.13.2 (Hague 1971, p. 342).

The conditions for balance may be readily derived from the form of Eq. (3.13.3):

$$\frac{R_2+\dfrac{1}{sC_2}}{R_1}=\frac{\dfrac{1}{sC_4}}{Z_3}\quad\text{with}\quad Z_3=\frac{R_3}{1+sC_3R_3}\quad\text{so}$$

$$\frac{1+sC_2R_2}{sC_2R_1}=\frac{1+sC_3R_3}{sC_4R_3}\quad\text{and equating real and imaginary parts}\tag{3.13.6}$$

$$\frac{sC_2R_2}{sC_2R_1}=\frac{sC_3R_3}{sC_4R_3}\quad\text{and}\quad\frac{1}{sC_2R_1}=\frac{1}{sC_4R_3}\quad\text{or}$$

$$\frac{R_2}{R_1}=\frac{C_3}{C_4}\quad\text{and}\quad\frac{R_3}{R_1}=\frac{C_2}{C_4}$$

so the two balance conditions can be adjusted in a non-interacting way by making say $R_2$ and $R_3$ the variables. Some examples of bridge applications are given in the references.

The Wien bridge is discussed in Section 1.12 and the application to an oscillator in Section 5.8, where it is shown how the characteristics of the bridge substantially improve the frequency stability.

### SPICE simulation circuits

None

## References and additional sources 3.13


Borchardt I. G., Holland L. R. (1975): Pseudo-bridge: a different way to compare resistances. *Rev. Sci. Instrum.* **46**, 67–70.

Christie S. H. (1833): Experimental determination of the laws of magneto-electric induction. *Phil. Trans* **123**, 95–142.

Giffard R. P. (1973): A simple low power self-balancing resistance bridge. *J. Phys. E Sci. Instrum.* **6**, 719–723.

Griffin J. A. (1975): An a.c. capacitance bridge temperature controller for use in strong magnetic fields at low temperatures; *Rev. Sci. Instrum.* **46**, 5–8.

Hague B. (Revised by Foord T. R.) (1971): *Alternating Bridge Methods*, 6th Edn, London: Pitman Publishing. ISBN 0-273-40291-9.

Jackman A. P., Noble R. D., Bonouvrie H. J. (1977): Inexpensive recording thermometer. *Rev. Sci. Instrum.* **48**, 865–869.

Kibble B. P., Rayner G. H. (1984): *Coaxial AC Bridges*, Bristol: Adam Hilger. ISBN 0-85274-389-0.

Meacham L. A. (1938): The bridge stabilized oscillator. *Bell System Tech. J.* **17**, 574–590. Also *Proc. IRE* **26**, 1278–1294, 1938.

Moody M. V., Paterson J. L., Ciali R. L. (1979): High resolution dc-voltage-biased ac conductance bridge for tunnel junction measurements. *Rev. Sci. Instrum.* **50**, 903–908.

Paesler M. A., Fritzsche H. (1974): Measurement of internal stress in thin films. *Rev. Sci. Instrum.* **45**, 114–115.

Strauss L. (1970): *Wave Generation and Shaping*, New York: McGraw-Hill. Library of Congress Cat. No. 74-90024.Reggio M. (1979): A linear ohmmeter for resistors having negative temperature coefficient. *J. Phys. E Sci. Instrum.* **12**, 173–174.

Tick P. A., Johnson D. (1973): A dynamic bridge measurement for lossy systems. *Rev. Sci. Instrum.* **44**, 798–799.

Wheatstone C. (1843): An account of several new instruments and processes for determining the constants of a voltaic circuit. *Philosophical Transactions* **133**, 303–327.

Wien M. (1891): Das Telephon als optischer Apparat zur Strommessung. *Ann. Phys.* **42**, 593–621, and **44**, 681–688, 1891. Also **44**, 689–712, 1891. (The first ac bridge.)

Williams J. (1990): *Bridge Circuits*, Linear Technology Application Note AN43, June.

# 3.14 Approximation

An approximate answer to the right question is worth a great deal more than a precise answer to the wrong question.

John Wilder Tukey, mathematician, 16 June 1915–26 July 2000

The design of filters involves approximation. We require a system that has the desired frequency or time response and seek a circuit that will provide this. There are, however, limits on what can be constructed with physically realizable elements in a causal system, i.e. time flows only one way. In a digital filter where you may go backward and forward in time there are additional possibilities but this is not an area visited in this book. Much of filter theory has revolved around finding approximations to the ideal response and is associated with names like Butterworth, Bessel, Tschebychev, Cauer, Thomson, elliptic and so on. The field of filters is covered by an enormous number of texts, a very few of which are listed in the references, but they will lead you on to others.

As an illustration of the approximation process we will choose a somewhat unusual approach which is both uncommonly treated and which has a familiarity in form to the usual transfer function. (In Stephen Potter terms, it may also serve to impress your friends!) A recent article (Smith 1997) describes a very simple circuit for obtaining a predictable analog time delay (Fig. 3.14.1). See also Roberge (1975) and National Semiconductor (1996).

The transfer function for a pure delay of time $\tau$ is given by (Table 1.12.1, No. 2):

$$H(s) = \exp(-s\tau) \tag{3.14.1}$$

which requires an infinite number of poles and zeros (like a transmission line) to implement exactly. The problem is then how to approximate this function to whatever degree is required. A classic technique is to use a polynomial series (Section 1.3) but some functions are not well approximated in this way. Rational functions, i.e. quotients of polynomials, are sometimes superior since they can model poles, as we are familiar with from the treatment of Laplace transforms (Section 1.12). A rational function has the form (Press et al. 1992; Sections 3.2 and 5.12):

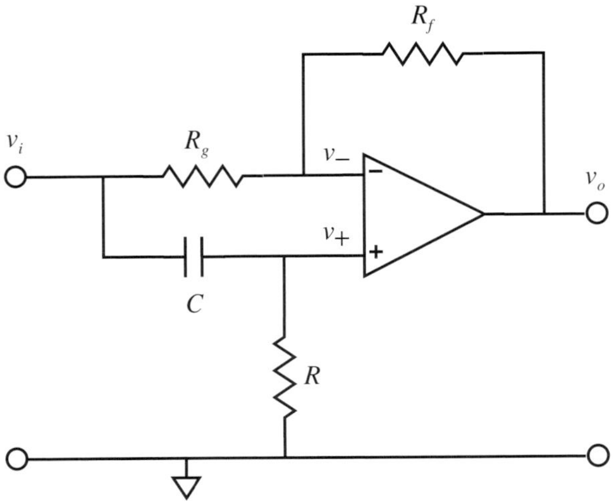


**Fig. 3.14.1** Delay circuit.

$$R(s)=\frac{a_0+a_1s+a_2s^2+\cdots+a_Ms^M}{b_0+b_1s+b_2s^2+\cdots+b_Ns^N}=\frac{\sum_{k=0}^{M}a_ks^k}{1+\sum_{k=1}^{N}b_ks^k} \tag{3.14.2}$$

just like the general form of transfer function Eq. (1.12.9) and where usually $M\leq N$ (in the sum form $b_0$ is arbitrary and put equal to 1). $R(s)$ is said to be a Padé approximant to the series:

$$f(s)=\sum_{k=0}^{\infty}c_ks^k \tag{3.14.3}$$

if the following conditions are true:

$$R(0)=f(0) \quad \text{and} \quad \frac{d^k}{ds^k}R(s)\bigg|_{s=0}=\frac{d^k}{ds^k}f(s)\bigg|_{s=0}, \quad \text{for} \quad k=1,2,\ldots,M+N \tag{3.14.3}$$

where the subscripts $s=0$ indicate that the differentials are evaluated there. The expansion for (3.14.1) is from Eq. (1.3.1):

$$f(s)=1-s\tau+\frac{s^2\tau^2}{2}-\frac{s^3\tau^3}{6}+\frac{s^4\tau^4}{24}-\frac{s^5\tau^5}{120}+\frac{s^6\tau^6}{720}-\cdots \tag{3.14.4}$$

so a simple power series expansion to first order would only use the first two terms and leave much to be desired as to the closeness of the approximation. Smith uses only the first order Padé approximant, but to see how the approximation is determined we will take it to second order in numerator and denominator, i.e. with $M=N=2$. The conditions of (3.14.3) lead to $M+N+1$ equations for the unknowns $a_0, \ldots, a_M$ and $b_1, \ldots, b_N$, which leads to $a_0 = c_0$ and the following relations:

$$\sum_{p=1}^{N} b_p c_{N-p+k} = -c_{N+k}, \quad \text{for} \quad k = 1, 2, \ldots, N$$
$$\sum_{p=0}^{k} b_p c_{k-p} = a_k, \quad \text{for} \quad k = 1, 2, \ldots, N \tag{3.14.5}$$

and taking the first of Eq. (3.14.5) we get (since $N = 2$):

$$b_1 c_{1+k} + b_2 c_k = -c_{2+k}\,, \text{ with } \quad c_0 = 1, \quad c_1 = -\tau, \quad c_2 = \frac{\tau^2}{2}, \quad c_3 = \frac{-\tau^3}{6}, \quad c_4 = \frac{\tau^4}{24}$$

so for $k = 1$: $b_1 c_2 + b_2 c_1 = -c_3$ and for $k = 2$: $b_1 c_3 + b_2 c_2 = -c_4$

$$b_1 \frac{\tau^2}{2} + b_2(-\tau) = \frac{\tau^3}{6} \qquad b_1\left(\frac{-\tau^3}{6}\right) + b_2 \frac{\tau^2}{2} = \frac{-\tau^4}{24}$$
$$b_1 \tau^2 - 2b_2 \tau = \frac{\tau^3}{3} \quad \text{(i)} \qquad -b_1 \tau^3 + 3b_2 \tau^2 = \frac{-\tau^4}{4} \quad \text{(ii)} \tag{3.14.6}$$

and multiplying (i) by $\tau$ and adding to (ii) gives

$$0 + b_2 \tau^2 = \frac{\tau^4}{12} \quad \text{or} \quad b_2 = \frac{\tau^2}{12}, \text{ and so from (i)} \quad b_1 = \frac{\tau^2}{2}$$

Using the second of Eq. (3.14.5) gives:

$$\text{for } k = 1: \quad b_0 c_{1-0} + b_1 c_0 = a_1 \quad \text{or} \quad 1(-\tau) + \frac{\tau}{2} 1 = a_1 \quad \text{so} \quad a_1 = \frac{-\tau}{2}$$
$$\text{for } k = 2: \quad b_0 c_2 + b_1 c_1 + b_2 c_0 = a_2 \quad \text{or} \quad 1\frac{\tau^2}{2} + \frac{\tau}{2}(-\tau) + \frac{\tau^2}{12} = a_2 \quad \text{so} \quad a_2 = \frac{\tau^2}{12} \tag{3.14.7}$$

and the final approximation to the transfer function Eq. (3.14.1) is:

$$H(s) = \frac{a_0 + a_1 s + a_2 s^2}{b_0 + b_1 s + b_2 s^2} = \frac{1 - \dfrac{s\tau}{2} + \dfrac{s^2 \tau^2}{12}}{1 + \dfrac{s\tau}{2} + \dfrac{s^2 \tau^2}{12}} \tag{3.14.8}$$

which, to first order, agrees with Smith. As the order of approximation is increased the sums get more extensive and one can resort to computational assistance as provided by Press et al. (1992). We now have to synthesize a circuit that will have a transfer function of the form required. The configuration of Fig. 3.14.1 does for first order as we now show for an ideal amplifier.

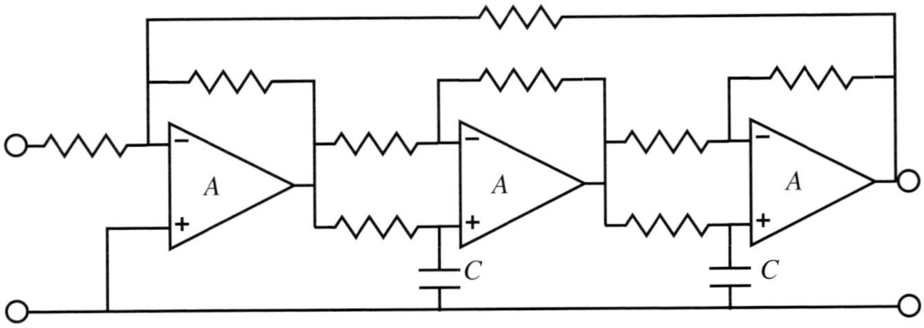


**Fig. 3.14.2** All-pass biquad circuit.

$$v_+ = \frac{v_i R}{R + 1/sC} = \frac{v_i sCR}{1 + sCR} \quad \text{and} \quad \frac{v_i - v_-}{R_g} = \frac{v_- - v_o}{R_f} \quad \text{or} \quad v_- = \frac{v_o R_g + v_i R_f}{R_f + R_g}$$

$$\text{so} \quad \frac{v_i sCR}{1 + sCR} = \frac{v_o R_g + v_i R_f}{R_f + R_g} \quad \text{since} \quad v_- = v_+ \quad \text{and putting} \quad R_f = R_g$$

$$\frac{v_o}{v_i} = \frac{sCR(R_f + R_g) - R_f(1 + sCR)}{R_g(1 + sCR)} = \frac{-(1 - sCR)}{(1 + sCR)} = \frac{-\left(1 - \dfrac{s\tau}{2}\right)}{\left(1 + \dfrac{s\tau}{2}\right)} \quad \text{so} \quad \tau = 2CR \tag{3.14.9}$$

Though Smith (inconsequentially) omits the negative sign in front of the transfer function it is of significance in one respect. Some other references (e.g. Chen 1995, p. 2407; Moschytz 1972) exchange the positions of $R$ and $C$ as shown in Fig. 3.14.2, and if you do so in (3.14.9) ($sC \rightarrow 1/R$ and $R \rightarrow 1/sC$) the result is the same function but without the negative sign, i.e. a non-inverting output, so you have a choice. You should note that the response of the circuit depends on the rate of change of the input signal. If the rise is too rapid for the bandwidth of the circuit then you will get overshoots and a distorted output so check carefully when using or simulating these types of circuit (Smith 1997, 1999).

We may compare the fit of the second order Padé approximation to expansion for the ideal delay equation (3.14.4) by dividing out (3.14.8) (Roberge 1975, p. 530, and see p. 555). If you do not relish doing this by hand you can resort to Mathcad and the Maple symbolic processor. Enter the expression using a single variable symbol for convenience, load the Symbolic Processor, Select the variable $s$ and choose Expand to Series. You will be requested to enter the order required and enter 9, to give:

$$f(s) = 1 - s\tau + \frac{s^2\tau^2}{2} - \frac{s^3\tau^3}{6} + \frac{s^4\tau^4}{24} - \frac{s^5\tau^5}{144} - \frac{s^7\tau^7}{1728} + \frac{s^8\tau^8}{5040} - O(s^9) \cdots \tag{3.14.10}$$

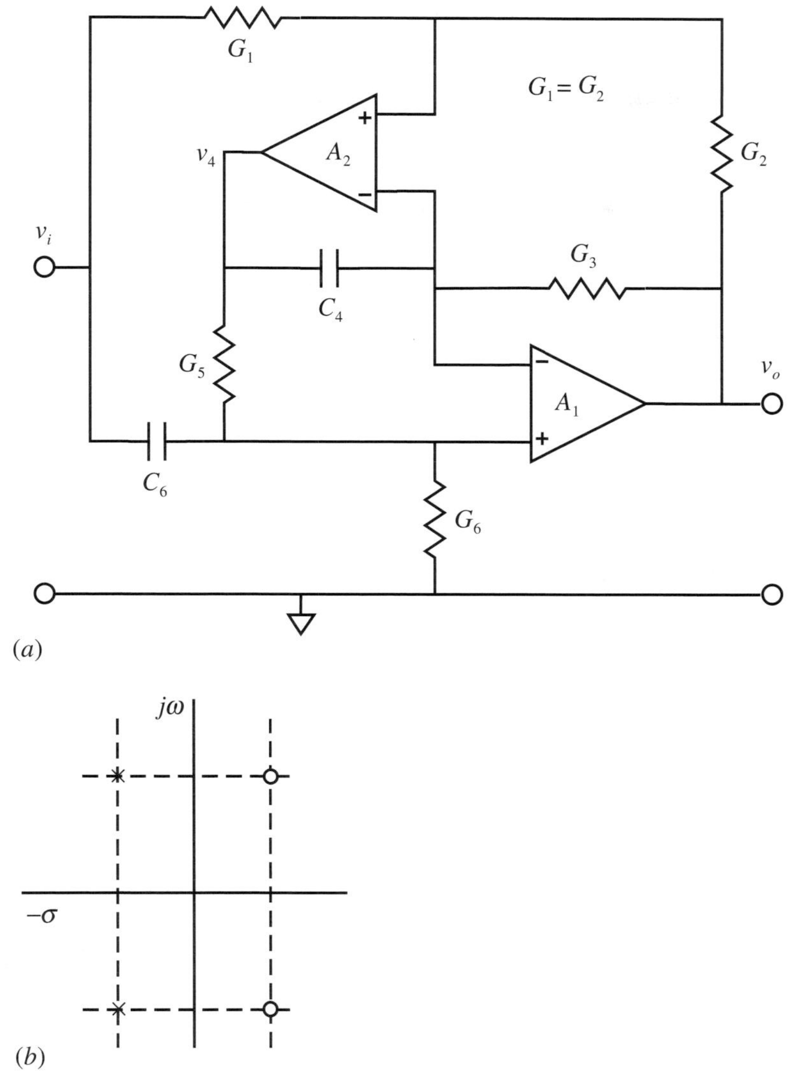


**Fig. 3.14.3** (*a*) Second order all-pass delay circuit. (*b*) Pole-zero diagram.

noting that there is no term in $s^6$, and that the final term just indicates the order of the 'error' arising from the limit on the number of terms selected for the expansion. The fit is exact up to fourth order and close up to fifth, so it is a pretty good approximation. See also Truxal (1955), Hakim (1966) and Van Valkenburg (1982).

A second order all-pass circuit using a generalized immittance converter (GIC, Section 5.16) is shown in Fig. 3.14.3 (Chen 1995, p. 2397).

This has a transfer function:

$$H_{AP}(s) = \frac{s^2 C_6 C_4 G_2 - s C_4 G_1 G_6 + G_1 G_3 G_5}{s^2 C_6 C_4 G_2 + s C_4 G_2 G_6 + G_1 G_3 G_5} \quad \text{and with} \quad G_1 = G_2 \tag{3.14.11}$$

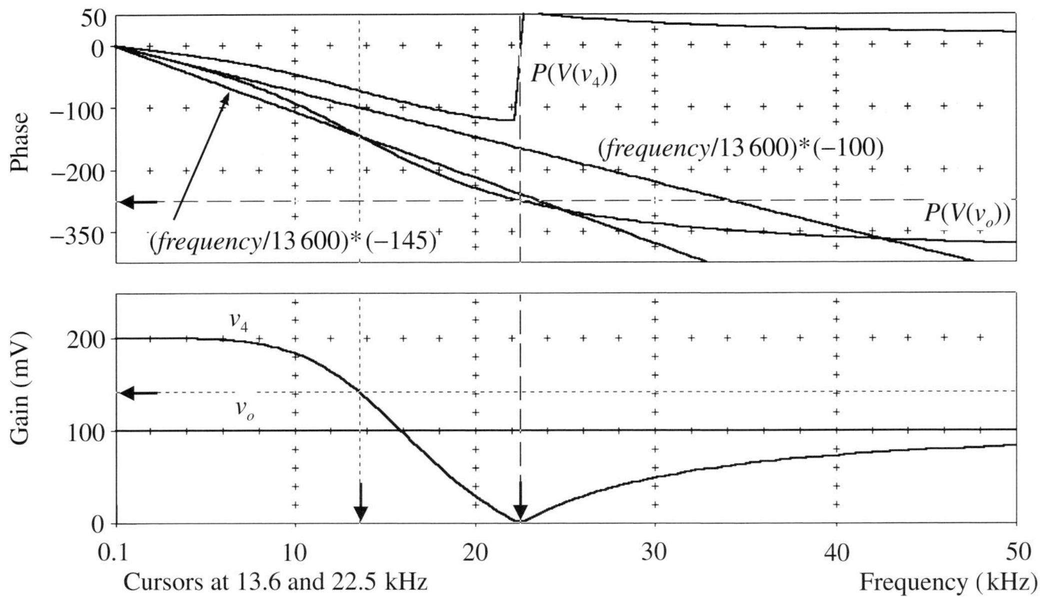


**Fig. 3.14.4** Frequency response of Fig. 3.14.3 circuit.

$$= \frac{s^2 - s/T_{66} + 1/T_{43}T_{65}}{s^2 + s/T_{66} + 1/T_{43}T_{65}}, \quad \text{where} \quad T_{uv} \equiv \frac{C_u}{G_v} \equiv C_u R_v$$

$$= \frac{1 - \dfrac{sT_{43}T_{65}}{T_{66}} + s^2 T_{43}T_{65}}{1 + \dfrac{sT_{43}T_{65}}{T_{66}} + s^2 T_{43}T_{65}} \quad \text{in the form of Eq. (3.14.8)} \qquad (3.14.11\ \text{cont.})$$

so the delay is given by $\tau = \dfrac{2T_{43}T_{65}}{T_{66}}$

In a PSpice simulation of this circuit with all conductances equal and of value 1/(10 k), and all capacitances equal and of value 1 n, we would predict a delay of $\tau = 2 \times 10^4 \times 10^{-9} = 20$ μs. With LM6142 amplifiers this is just what you get. It is worth noting that the output $v_4$ from $A_2$ is of twice the amplitude of $v_o$ and has half the delay. A suitable pulse has a risetime of ~100 μs in this case. Plots of the frequency responses are shown in Fig. 3.14.4 (note the linear frequency scale).

The dip in $v_4$ indicates the range over which the phase is an approximation to the linear function of frequency that is ideal for a true delay. To show the deviation from linearity we have chosen the 3 dB point on $v_4$, which occurs at 13.6 kHz, and drawn a straight line from zero phase at z.f. to the phase (−145°) at the 3 dB frequency (plot (*frequency*/13600)*(−145)) which gives the approximation shown with about equal positive and negative deviations. The consequential delay is found from Section 3.6 as:

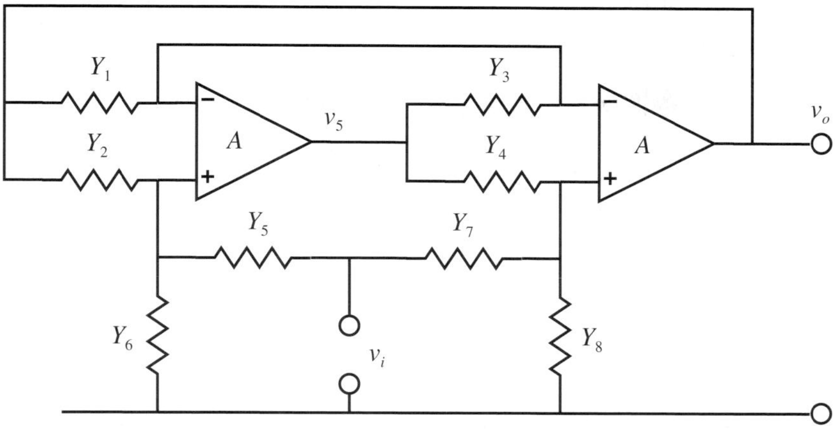


Fig. 3.14.5 Alternative second order all-pass delay circuit.

$$\tau = \frac{1}{360}\frac{d\phi}{df} = \frac{1}{360}\frac{145}{13\,600} = 29.6\ \mu\text{s} \quad (3.14.12)$$

which is rather longer than the measured value. To get closer agreement it is necessary to use an angle of $-100°$ which will lead to a line that is about tangential to the low frequency phase curve (0 to ~4 kHz). Testing the response with a pulse input can show the effects of input signal risetime. From (3.6.8) we can estimate the allowed risetime as a consequence of the limited bandwidth:

$$t_r = \frac{1}{2f_m} = \frac{1}{2 \times 13\,600} \approx 37\ \mu\text{s} \quad (3.14.13)$$

so carry out runs using a pulse with risetimes of say 10, 40 and 100 μs and confirm that the prediction is about correct.

An alternative GIC based all-pass circuit is shown in Fig. 3.14.5 (Chen 1995, p. 2412). The responses for this circuit are similar to those for Fig. 3.14.4 but the linearity of the phase response appears to be better and the calculated, measured and slope-derived delays agree.

### SPICE simulation circuits

Consult the SimCmnt.doc file on the CD before running.

| | |
|---|---|
| Fig. 3.14.4 | Apasgic2.SCH |

## References and additional sources 3.14


Chen, Wai-Kai (Ed.) (1995): *The Circuits and Filters Handbook*, Boca Raton: CRC Press and IEEE Press. ISBN 0-8493-8341-2.

Hakim S. S. (1966): *Feedback Circuit Analysis*, London: Iliffe Books.

Moschytz G. S. (1972): High-Q factor insensitive active RC networks, similar to the Tarmy–Ghausi circuit but using single-ended operational amplifiers. *Electronics Letters* **8**, 458–459.

National Semiconductor (1996): *Data Sheet CLC428.*

Press W. H., Teukolsky B. P., Vetterling W. T., Flannery B. P. (1992): *Numerical Recipes in Fortran*, 2nd Edn, Cambridge: Cambridge University Press. Sections 5.12 and 3.2. Versions are available that use other programming languages, e.g. C.

Roberge J. K. (1975): *Operational Amplifiers. Theory and Practice*, New York: John Wiley. ISBN 0-471-72585-4. See p. 536.

Smith S. O. (1997): An accurate analog delay circuit. *Electronic Design* 1 December, 148, 150. (Also private communication 1999.)

Truxal J. G. (1955): *Automatic Feedback Control System Synthesis*, New York: McGraw-Hill.

Van Valkenburg M. E. (1982): *Analog Filter Design*, New York: Holt, Rinehart and Winston. ISBN 0-03-059246-1, or 4-8338-0091-3 International Edn.

# 3.15 Control systems

Chance favours only the prepared mind.

Louis Pasteur

Control or servo systems are used in many forms to control to ensure that the output of a system is maintained at some desired value in spite of external influences that would cause the output to vary. We may for instance wish to keep the temperature of an enclosure constant, independent of variations in ambient temperature, or to keep the speed of a motor constant independent of the loading. Voltage regulators (Section 5.10) are a type of servo system that keeps the output constant independent of the current demand or the ambient temperature. In such systems there will be a demand input and some reference signal against which the output may be compared. Any deviation is amplified and the system acts to return the output to the desired value. These are therefore negative feedback systems with both the advantages and the problem of maintaining stability, with regard to both the deviation of the output relative to the demand and to the prevention of oscillation. A survey of the development of control systems is given by Bennett (1993).

The general arrangement of a control system may be illustrated by Fig. 3.15.1.

The required, or demand, input is shown as $r$. The controlled output $c$ is sampled by the feedback network and the signal $b$ is compared with $r$. If there is a difference $e$ then this will be amplified in the feedforward block to produce the necessary correction in $c$. In terms of Laplace variables we can write the equations:

$$B(s)=H(s)C(s),\quad C(s)=G_F(s)E(s),\quad E(s)=R(s)-B(s)=R(s)-H(s)C(s) \tag{3.15.1}$$

and we can define the overall $G_O$ and error $G_E$ transfer functions as:

$$\begin{aligned} G_O&=\frac{C}{R}=\frac{G_FE}{E+HC}=\frac{G_FE}{E+HG_FE}=\frac{G_F}{1+HG_F} \\ G_E&=\frac{E}{R}=\frac{E}{E+HC}=\frac{E}{E+HG_FE}=\frac{1}{1+HG_F} \end{aligned} \tag{3.15.2}$$

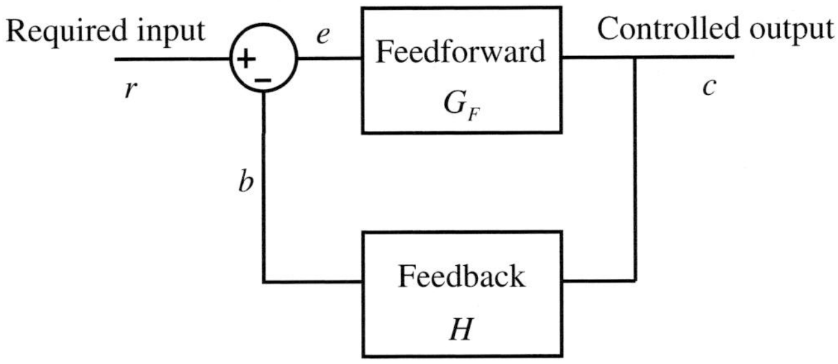


**Fig. 3.15.1** Schematic control system.

In general terms, a transfer function can be written in the form of (1.12.9), so for $G_F$:

$$G_F = \frac{K_N(q_m s^m + q_{m-1}s^{m-1} + \cdots + 1)}{s^N(p_n s^n + p_{n-1}s^{n-1} + \cdots + 1)} \tag{3.15.3}$$

and the order of the exponent $N$ of $s$ in the denominator designates the type of servo. While $N$ may in principle be greater than 2, in practice this is not made use of. The type determines the steady-state error $e_{ss}(t)$, the error after any transients have settled or as $t \to \infty$. The magnitude of the error depends on the form of the input $r$. The two standard forms are the step $r = r_0\, u(t)$, for which $R = r_0/s$, and the ramp $r = r_1(t)$, for which $R = r_1/s^2$. It is straightforward to determine the final steady-state error by way of the final-value theorem as given in Table 1.12.2 (No. 6) (e.g. Hamilton 1977).

$$e_{ss}(t) = \lim_{t\to\infty} e(t) = \lim_{s\to 0} sE(s) = \lim_{s\to 0} sG_E(s)R(s) \tag{3.15.4}$$

which gives for the two input functions:

$$\begin{array}{lll}
 & \text{Input step} \quad R = \dfrac{r_0}{s} & \text{Input ramp} \quad R = \dfrac{r_1}{s^2} \\[2ex]
 & e_{ss}(t) = \lim\limits_{s\to 0} \dfrac{r_0}{\left(1 + \dfrac{K_N}{s^N}\right)} & e_{ss}(t) = \lim\limits_{s\to 0} \dfrac{r_1}{\left(1 + \dfrac{K_N}{s^N}\right)} \cdot \dfrac{1}{s} \\[3ex]
\text{Type 0 } (N=0) & e_{ss}(t) = \dfrac{r_0}{1 + K_0} & e_{ss}(t) = \infty \\[2ex]
\text{Type 1 } (N=1) & e_{ss}(t) = 0 & e_{ss}(t) = \dfrac{r_1}{K_1} \\[2ex]
\text{Type 2 } (N=2) & e_{ss}(t) = 0 & e_{ss}(t) = 0
\end{array} \tag{3.15.5}$$

So for a type 0 servo there is always an offset between the demand input $r$ and the actual output $c$, finite for a step input and tending to $\infty$ for a ramp. This type of servo is usually called a regulator, as for example a voltage regulator, and is said to have proportional control. A type 1 servo will have zero error for the step, but a finite error for a ramp. For a type 2 servo both errors are zero. To achieve type 1 performance requires the addition of an integrator for the error signal so that the error is summed over time until it eventually becomes zero. Type 2 performance is obtained by adding additionally a differential term that can reduce the rate errors to zero. This leads to the concept of the three-term controller to include proportional, integral and differential terms, commonly referred to as a PID servo. Such a system can evidently provide much better response, but the difficulty is in knowing what to make the relative magnitudes of the three terms and at the same time maintaining stability. In the early days of servo systems some rule-of-thumb conditions were developed, deduced by the Ziegler–Nichols method which involves measurements on open- and closed-loop settings (Ziegler and Nichols 1942, 1943; Franklin et al. 1994). Now with simulation it is much easier to investigate the response to various inputs and to see the stability margins.

There are very many substantial books on the theory and practice of servomechanisms, e.g. Chestnut and Mayer (1951); James et al. (1947); Franklin et al. (1994), D'Azzo and Houpis (1995), Ogata (1997). It is not our intention to cover the field here, but we will choose a reasonably simple example which will also allow us to demonstrate the assistance that SPICE can provide. SPICE provides a number of what are called 'control system parts', which are in effect function blocks defined by mathematical expressions just like ABMs. We will make use of these to examine the response of a servo system as they somewhat simplify the simulation and in the frequency region in which we will operate these ideal elements involve minimal compromise. The elements we will use are:

SUM: the output is the algebraic sum of the two inputs.
DIFF: the output is the algebraic difference of the two inputs.
DIFFER: the output is the differential of the input with respect to time; the gain must be defined.
INTEG: the output is the integral of the input with respect to time; the gain must be defined.
GAIN: the output is the input multiplied by the gain, which may be positive or negative.

For the differentiator and the integrator the significance of the gain attribute is to define the unity-gain frequency. For example, the two gains specified in Fig. 3.15.4 ($RC=900$ for the differentiator and $1/RC=10$ for the integrator) give unity-gain frequencies:

$$f_I=\frac{10}{2\pi}=1.59\text{ Hz}\quad\text{and}\quad f_D=\frac{1}{2\pi\times 900}=0.177\text{ mHz} \qquad (3.15.6)$$

and you can run a simulation to check these.

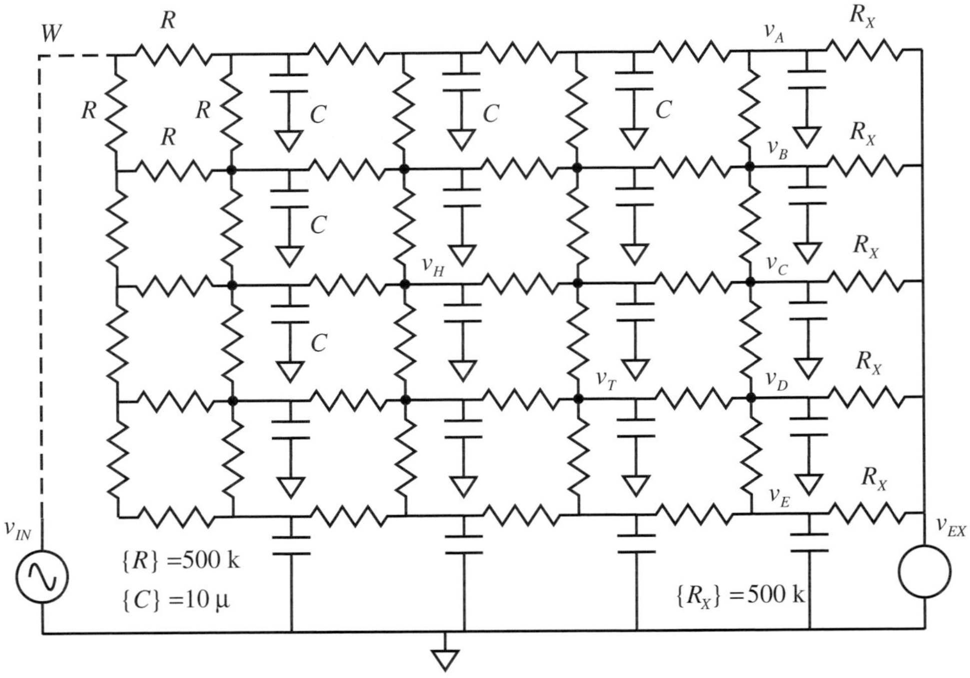


**Fig. 3.15.2** Resistor–capacitor network as an analog of the volume within an oven. All capacitors are equal and all resistors ($R$) are equal. The resistors $R_X$ represent the contact with the external ambient surroundings. The generator $v_{IN}$ represents power input and $v_{EX}$ represents input from ambient. The sensing thermistor may be at position $v_T$ and the point to be controlled at $v_H$.

As the example we choose a temperature control system as discussed by Stanbury (1965). He outlines a resistor–capacitor network as an analog of the oven, as illustrated in Fig. 3.15.2. It should be three-dimensional but we will make do with two dimensions as it is a bit difficult to draw such a network and the simulation time will increase considerably. This is of course a naive model of the oven but it will serve our purpose.

Such a network is too complex for reasonable analysis so Stanbury derives a simplified circuit as shown in Fig. 3.15.3.

The time constant $R_1C_1$ represents the response of the heater, $R_2C_2$ that of the enclosure, $R_3C_2$ the cooling of the enclosure and $R_4C_4$ the thermistor sensor. Taking some appropriate values for the time-constants Stanbury examines the possible values of the gain $A$ to avoid oscillation. Note that $A$ is positive since the feedback signal $v_F$ is inverted in *DIFF*. With the aid of SPICE we can include the array of Fig. 3.15.2 in place of $R_2C_2R_3$ and also introduce an integral and a differential term as shown in Fig. 3.15.4.

The values of the array components are as shown in the figure and it is convenient to express them as parameters, e.g. $\{C\}$, so that they may be more readily

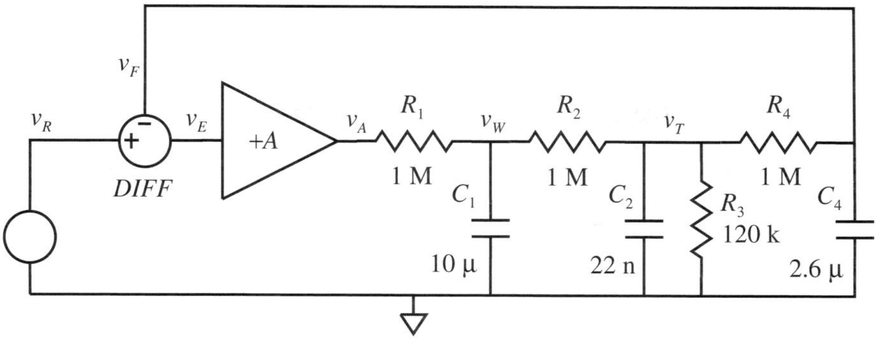


**Fig. 3.15.3** Simplified circuit for oven controller.

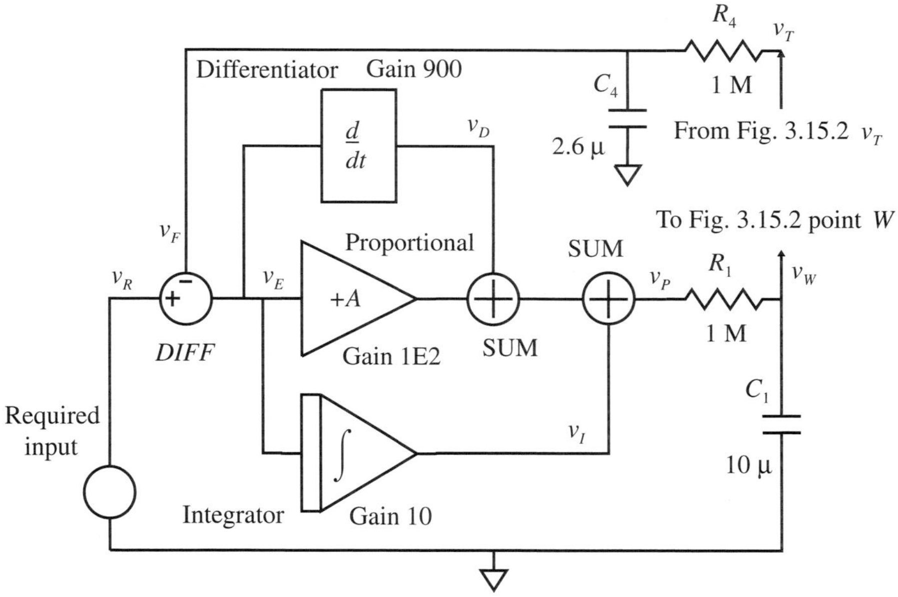


**Fig. 3.15.4** Introduction of the array to represent the volume of the oven, and integral and differential feedback terms.

changed. The values for the various time constants bear some sort of relation to those in the reference but we are primarily interested in the technique rather than any particular system. An example of a detailed investigation of the design and analysis of a high precision regulator is given, for example, by Dratler (1974), and in a different application by Yin and Usher (1988), though the latter is realized digitally.

The system is sensitive to (proportional) gain – if it is too high the system will be unstable and oscillate. The several series connected time constants produce a large phase shift which gives a small stability margin. We use a step input signal of 1 V (0.1 s risetime) to test the response. For small gain, and with no derivative and

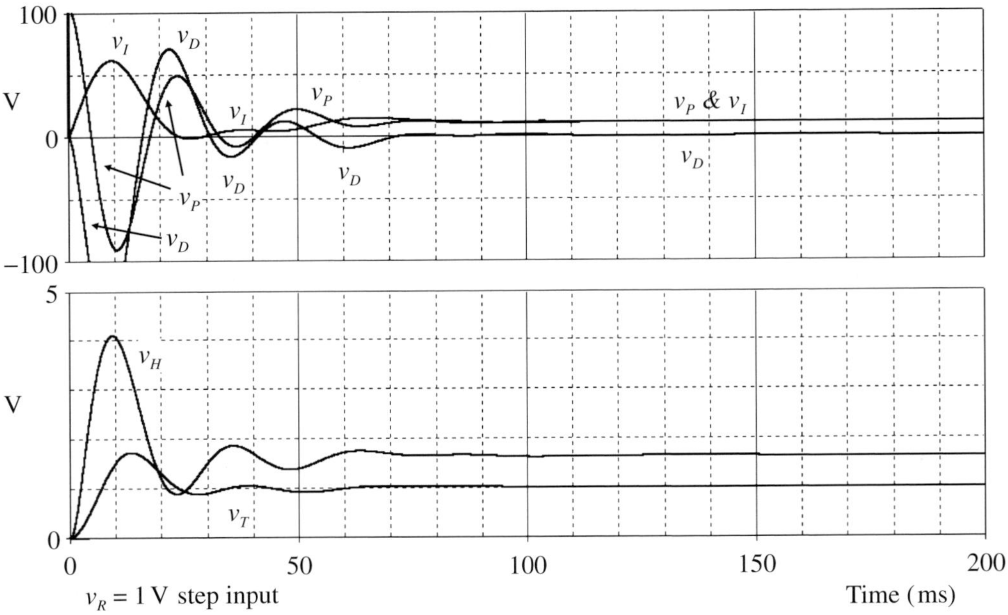


**Fig. 3.15.5** Responses of circuit of Fig. 3.15.4.

integral gain, the system is stable but the difference between the output $v_T$ and the demand input $v_{IN}$ is significant. As the proportional gain is increased ringing grows and at a gain of 100 there is nearly continuous oscillation. Now, increasing derivative gain progressively damps the ringing with an optimum value in the vicinity of 900. Introducing integral gain, up to a value of 10, has some small effect on the approach to equilibrium but more particularly drives $v_T$ to be closely equal to $v_R$. The final result at $v_T$ is shown in Fig. 3.15.5, which also shows the response at $v_H$, the integral $v_I$, the differential $v_D$ and the sum $v_P$. The difference between $v_H$ and $v_T$ illustrates the importance of placing the sensor as close as possible to the object whose temperature you are trying to control. Input of a pulse at $v_{EX}$ to represent an ambient variation can be used to test the reaction at $v_T$ and $v_H$.

## SPICE simulation circuits

Consult the SimCmnt.doc file on the CD before running.

| | |
|---|---|
| Fig. 3.15.5 | Ovenctl4.SCH |

## References and additional sources 3.15

Bennett S. (1993): *A History of Control Engineering 1930–1955*, London: Peter Peregrinus and IEE. ISBN 0-86341-299-8.

Chestnut H., Mayer R. W. (1951): *Servomechanisms and Regulating Systems Design*, New York: John Wiley. (2 Vols) Vol. 2 (1955) Library of Congress Cat. No. 51-10512.

D'Azzo J. J., Houpis C. H. (1995): *Linear Control System Analysis and Design: Conventional and Modern*, $4^{th}$ Edn, New York: McGraw-Hill. ISBN 0-07-016321-9.

Dratler J. (1974): A proportional thermostat with 10 microdegree stability. *Rev. Sci. Instrum.* **45**, 1435–1444.

Franklin G. F., Powell J. D., Emami-Naeini A. (1994): *Feedback Control of Dynamic Systems*, $3^{rd}$ Edn, Reading, Mass: Addison-Wesley. ISBN 0-201-53487-8.

Hamilton T. D. S. (1977): *Handbook of Linear Integrated Electronics for Research*, London: McGraw-Hill. ISBN 0-07-084483-6. See Section 4.5.

James H. M., Nichols N. B., Phillips R. S. (Eds) (1947): *Theory of Servomechanisms*, New York: McGraw-Hill.

Levine W. S. (Ed.) (1995): *The Control Handbook*, Boca Raton: CRC Press and IEEE Press. ISBN 0-8493-8570-9.

Miller C. A. (1967): Temperature control using commercially available dc amplifiers. *J. Phys. E Sci. Instrum.* **44**, 573–574.

Ogata K. (1997): *Modern Control Theory*, Upper Saddle River; Prentice-Hall International. ISBN 0-13-261389-1.

Stanbury A. C. (1965): Some design criteria for proportional temperature control systems. *J. Sci. Instrum.* **42**, 787–790.

Yin Z., Usher M. J. (1988): A high resolution wideband digital feedback system for seismometers. *J. Phys. E Sci. Instrum.* **21**, 748–752.

Ziegler J. G., Nichols N. B. (1942): Optimal settings for automatic controllers. *Trans. ASME* **64**, 759–768.

Ziegler J. G., Nichols N. B. (1943): Process lags in automatic control circuits. *Trans. ASME* **65**, 433–444.

# 3.16 Filters

Could Holy Writ be just the first literate attempt to explain the universe and make ourselves significant within it? Perhaps science is a continuation on new and better-tested ground to attain the same end. If so, then in that sense science is religion liberated and at large.

Edward O. Wilson, Consilience (Little Brown) 1998

The subject of filters is vast and multitudinous books on their design exist. Here we will examine only some of the basic ideas and in particular how SPICE may be employed to test and modify designs mostly derived from standardized configurations. A comprehensive book in tune with our approach is Van Valkenburg (1982). In early days, filters were primarily passive and considerable effort was put into realization of various forms that could be constructed using only $L$, $C$ and $R$. Nowadays we have the advantage of active forms that can generally produce the required response more conveniently and efficiently, but the data accumulated for passive filters provide a most useful resource of starting points for active designs. An example of this is given in Section 5.16. Other examples of active filters are given in Sections 3.14 and 5.13.

Here we can only consider a few of the more general characteristics and how SPICE may be of considerable use in both initial design and testing as well as providing a means of tweaking a circuit. All the standard types such as low-pass, high-pass, all-pass, etc. have been exhaustively analysed so design is generally a case of looking up the appropriate reference book and carrying out the calculations or referring to graphical charts (e.g. Hilburn and Johnson 1973). A more detailed understanding requires some appreciation of the relationship between the responses and the pole-zero configurations of the circuits. We will restrict ourselves to the commonly used second order (or biquadratic) responses which form the backbone of commonly used circuits. In these the poles and zeros can be complex, and hence can generate cut-off slopes approaching 40 dB/decade, in contrast to first order systems where they are all on the real axis and the slopes can only approach 20 dB/decade. The general transfer function for this group of filters is given by Eq. (3.16.1):

$$H(s) = K\frac{s^2 + (\omega_z/Q_z)s + \omega_z^2}{s^2 + (\omega_p/Q_p)s + \omega_p^2} \tag{3.16.1}$$

Table 3.16.1 *Transfer functions for second order filters*

| Filter form | Transfer function | Pole-zero configuration |
|---|---|---|
| Low-pass | $K\dfrac{\omega_0^2}{s^2+(\omega_0/Q)s+\omega_0^2}$ | 2 poles |
| High-pass | $K\dfrac{s^2}{s^2+(\omega_0/Q)s+\omega_0^2}$ | 2 poles, 2 zeros at origin |
| Band-pass | $K\dfrac{(\omega_0/Q)s}{s^2+(\omega_0/Q)s+\omega_0^2}$ | 2 poles, 1 zero at origin |
| Band-stop, (band-reject) | $K\dfrac{s^2+\omega_0^2}{s^2+(\omega_0/Q)s+\omega_0^2}$ | 2 poles, 2 zeros on $j\omega$-axis |
| All-pass (delay) | $K\dfrac{s^2-(\omega_0/Q)s+\omega_0^2}{s^2+(\omega_0/Q)s+\omega_0^2}$ | 2 poles, 2 zeros image of poles in $j\omega$-axis |

and the transfer function forms and pole-zero configurations for the various forms are shown in Table 3.16.1 and Fig. 3.16.1.

To illustrate the form of the gain and the phase we can simulate a 'universal' second order filter, generally referred to as a biquad (Tow 1968; Thomas 1971; Van Valkenburg 1982, p. 123). This arrangement, Fig. 3.16.2(*a*), allows the realization of all the five above filter forms.

The output $v_4$ can be configured to give one of the three indicated functions as indicated in Table 3.16.2. The references should be consulted for the full design equations. $R_1$ controls the $Q$ of the filter (increasing the resistance gives higher $Q$) and the values given in the table will allow a simple simulation as illustrated in Fig. 3.16.2(*b*), where it is evident that the $Q$ is a bit high for the low- and high-pass responses leading to some peaking.

The LM4162 amplifiers used have a good bandwidth but eventually their open-loop response limits the performance of the filter as can be seen at high frequency on the simulated responses. After initial design it is now easy to investigate the effect of the various design parameters or the effect of component tolerance.

All the treatments of filters give great attention to the attenuation and phase responses but do not say too much about prediction of the transient response. The sums involved are complex and the results difficult to parameterize (van Vollenhoven et al. 1965; Hansen 1963; Henderson and Kautz 1958). This is where SPICE can be so useful and informative. To demonstrate some results for the biquad we can input a 50 mV pulse with a fast risetime (10 μs) and a slow fall time (500 μs) with the results as shown in Fig. 3.16.3.

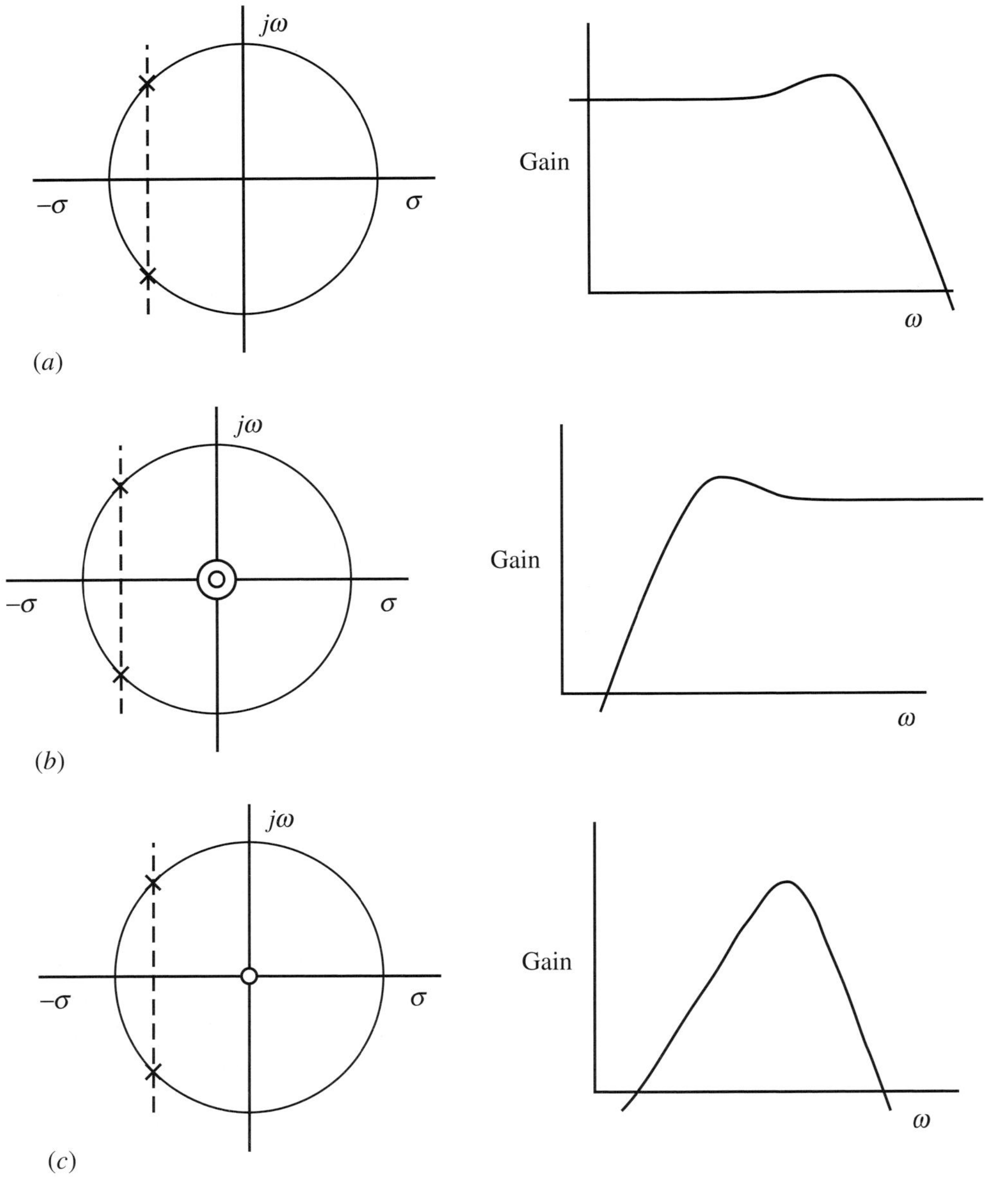


**Fig. 3.16.1** Pole-zero locations and response shapes. (*a*) Two poles. (*b*) Two poles, and two zeros at the origin. (*c*) Two poles, and one zero at the origin. (*d*) Two poles, and two zeros on the $j\omega$-axis. (*e*) Two poles, and two zeros images in the $j\omega$-axis.

These are not the kind of responses one would like to determine from first principles. IC devices for constructing such filters are commonly available, e.g. Linear Technology (2000).

Earlier designs of active filter were rather restricted by the limited bandwidth of available operational amplifiers. With the more recent substantial extension of available bandwidths to much higher frequencies the restrictions have been consid-

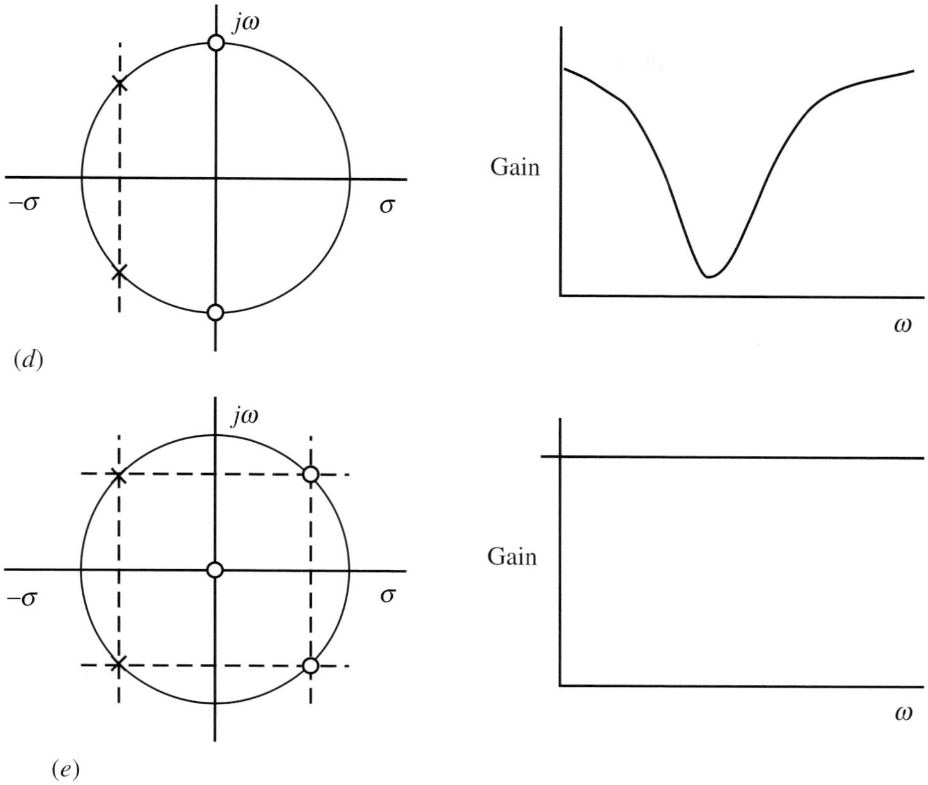


**Fig. 3.16.1** (cont.)

erably eased and operation to say 100 MHz is possible. See, for example, Steffes (1993), Henn and Lehman (1994), Gehrke and Hahn (1999), Karki (1999).

There are times, particularly in research applications, when the signals sought are small and overwhelmed by noise. The problem of filtering to extract the signal is acute but can be met by trading on the random nature of noise and being willing to trade time for improvement in the signal-to-noise ratio. This can be done digitally by signal averaging or by means of the so called lock-in, coherent or phase-sensitive detection (PSD) technique. They are equivalent techniques and in effect rely on time averaging the random noise, which tends to zero as time tends to infinity. A wide-ranging treatment of these techniques is given by Wilmshurst (1985) and many references to the literature are given by Hamilton (1977). The lock-in technique is of long standing (Ayrton and Perry 1888) and we will examine how SPICE can illuminate the operation. Application of the technique usually requires that the experiment is under full control of the experimenter so that the quantity to be determined can be modulated at a chosen carrier frequency, say $f_C$. This frequency is usually chosen to be above that at which the pink noise becomes negligible relative to the white noise (Sections 2.13 and 3.11). This is illustrated in Fig. 3.16.4.

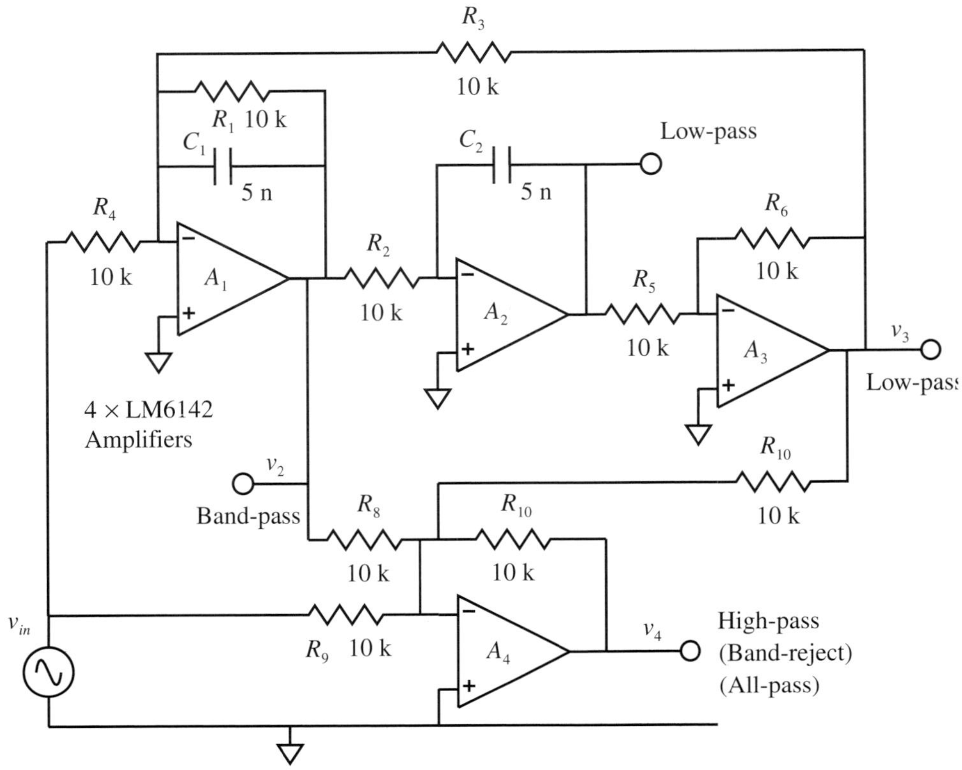


(*a*)

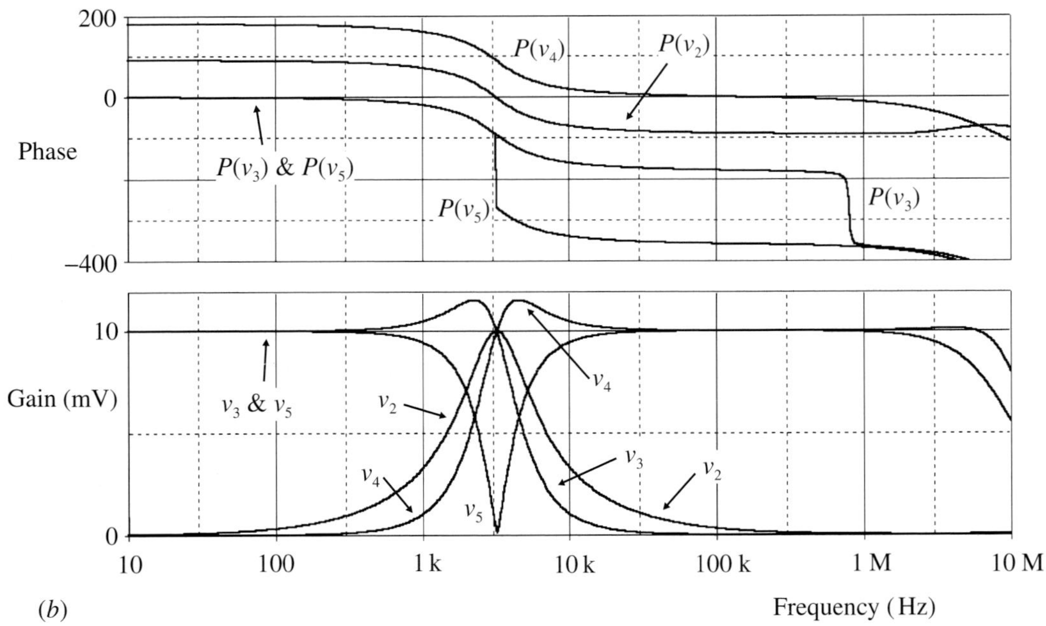


(*b*)

**Fig. 3.16.2** (*a*) The biquad filter circuit. (*b*) Simulation responses, omitting the all-pass configuration.

Table 3.16.2 *Conditions for filter forms at $v_4$ output*

| Filter form at $v_4$ | Conditions |
|---|---|
| High-pass | $R_7 = R_8 = R_9 = R_{10} = 10$ k, and $R_1 = R_3 = R_4$ |
| Band-stop | $R_7 = \infty$, and $R_1 = R_3 = R_4$ |
| All-pass | $R_7 = \infty$, and $R_4 = R_1/2$ |

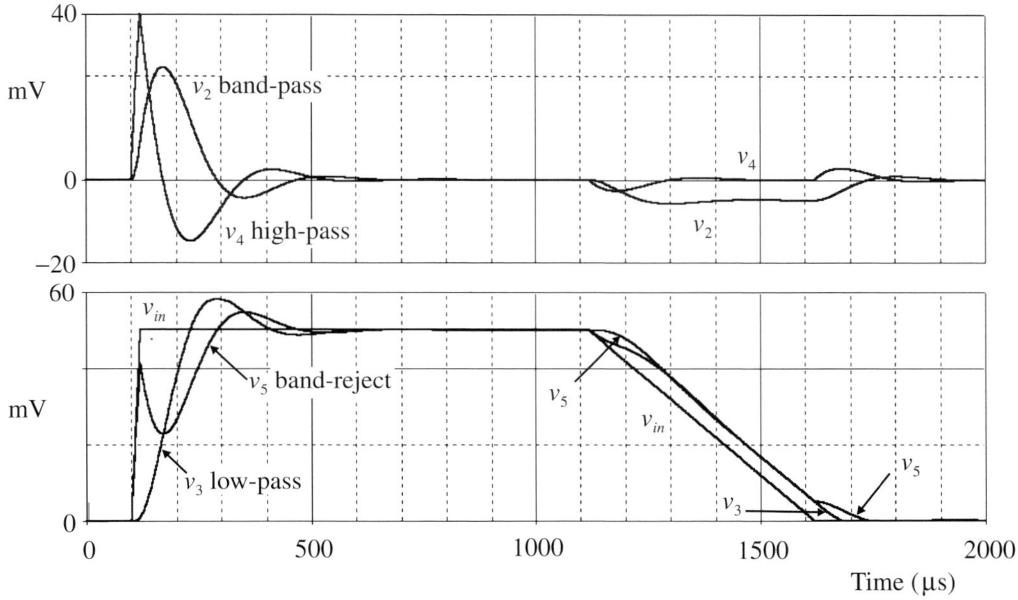


**Fig. 3.16.3** Response of biquad filter of Fig. 3.16.2 to a pulse input.

The straightforward approach would be to use a band-pass filter centred at $f_C$ to extract the signal and discriminate against the noise. However, as we make the pass-band narrower to improve the signal-to-noise ratio we encounter two difficulties: that of creating a narrow band filter, which may for example require to be say 0.01 Hz wide at 10 kHz, and the relative drift between the signal frequency and the filter centre frequency. The answer is to frequency shift the effective band-pass to zero frequency by multiplying the output signal coherently by the same driving frequency. Considering for a moment a clean experiment output signal $v_S$ then the output of the multiplier, allowing for any incidental phase difference $\phi$, is:

$$\begin{aligned} v_X &= v_S\cos(\omega_C t)\times v_R\cos(\omega_C t+\phi) \\ &= \tfrac{1}{2}v_S v_R[\cos(2\omega_C t+\phi)+\cos\phi] \quad \text{using Eq. (1.1.2(i))} \end{aligned} \tag{3.16.2}$$

and since the low-pass $RC$ filter will suppress the signal at $2\omega_C$ but pass the signal at zero frequency:

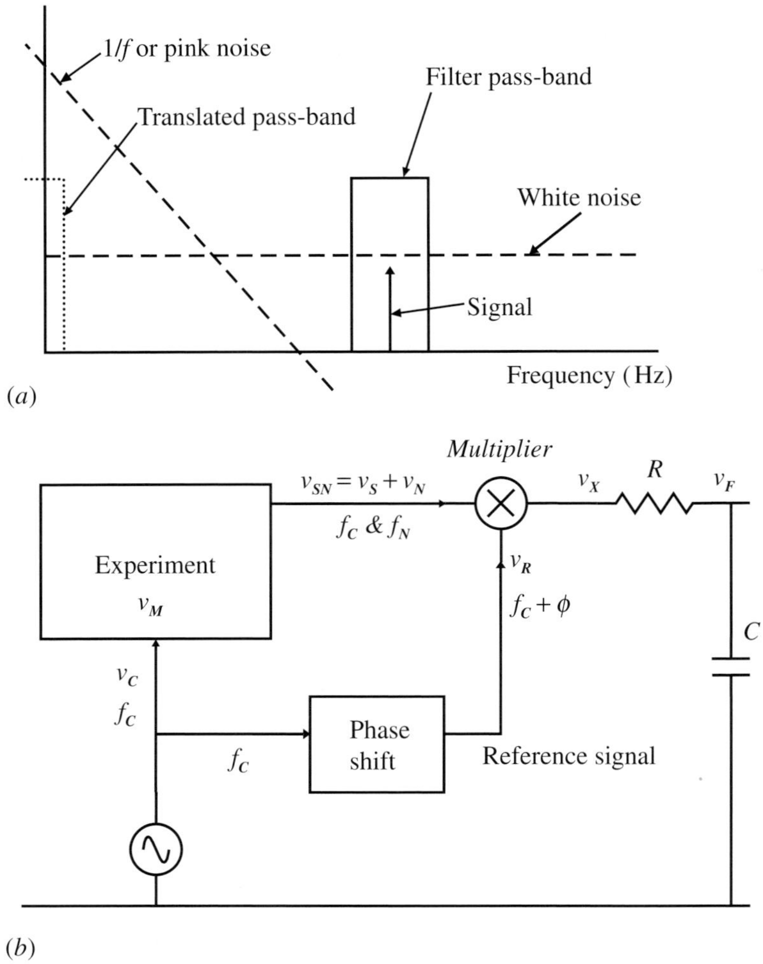


**Fig. 3.16.4** (*a*) Signal and noise spectra for lock-in detection. (*b*) Schematic of lock-in detector.

$$v_F = \tfrac{1}{2} v_S v_R \cos\phi \qquad (3.16.3)$$

It is important that the phase shift is near zero so that $\cos\phi \approx 1$ to give the maximum output, so we may need a phase shifter as shown. The dependence on $\phi$ is the origin of the name phase-sensitive detector. There is now no difficulty with frequency changes since the signal is locked to the drive $f_C$ and the band-pass, determined by *RC*, is locked to zero frequency. As far as the noise is concerned any noise component at a frequency that differs from $f_C$ by more than the width of the low-pass filter will be eliminated. For a noise signal represented by $v_N\cos(\omega_N t)$ (and not caring about any phase effects which are immaterial) we have:

$$\begin{aligned} v_X &= v_S\cos(\omega_C t) \times v_N\cos(\omega_N t) \\ &= \tfrac{1}{2} v_S v_N[\cos(\omega_C + \omega_N)t + \cos(\omega_C - \omega_N)t] \end{aligned} \qquad (3.16.4)$$

so, as before, the sum term will be deleted and the difference term will need $\omega_N \approx \omega_C$ to have any effect, and even then the phase will be random. The price that you pay for this improvement is in the increased time it takes to record the signal. If the signal is a single value you must wait for many time constants of the filter to achieve the final value (Table 1.5.1, p. 18). If, as is more common, you are looking for a varying signal produced as a consequence of some other variable, then you must sweep this variable very slowly for the same reason.

To demonstrate the operation of this form of filtering we need a source of noise that we can add into the system and see if we can extract a signal. SPICE does not offer such a *Transient* as against the computed *AC Sweep* noise, but there is a convenient model provided by Hageman (1996). A copy of the program is available from MicroSim/Cadence and allows the generation of a file of random noise voltages with a chosen time point interval, overall duration and r.m.s. magnitude. When the program is run it will generate the set of voltages and output the number of points in the table, the 3 dB bandwidth of the source set by an internal *RC* filter adjusted to prevent aliasing, and a slew rate in V $s^{-1}$ proportional to the bandwidth. There is also a direct unfiltered output and an implied ground connection.

The simulation circuit is shown in Fig. 3.16.5 and makes use of behavioural models. The carrier signal $v_C$ is modulated by the data signal $v_M$ to give $v_S$ which is then summed with the noise $v_N$ to give $v_{SN}$. This is then multiplied by the reference signal $v_R$ to give $v_X$ which is then filtered to produce the outputs $v_{F1}$ and $v_{F2}$. Since a 100 ms run time of 26 μs interval noise points was available for the *PWL* noise generator (there is a maximum allowed number of points) $v_C$ and $v_M$ were adjusted to suit and Fig. 3.16.5(*b*) illustrates the results. The unfiltered noise output was used. The two gain blocks are for convenient adjustment. An expanded portion of $v_{SN}$ is shown in the top graph together with $v_C$ for comparison to show how the sought signal is buried in the noise. The two traces are offset for clarity. The middle graph shows the modulation $v_M$ and $v_{F2}$ for comparison, and the lower $v_{F1}$ the first filtered recovered modulation. The initial condition of both $C_1$ and $C_2$ was set to zero to save simulation time getting past the initial transient. The phase and frequency of the two *VSIN* generators should be the same, and having separate generators makes it convenient to change the relative phase to examine the effect.

Switched-capacitor filters are available in IC form and cover a wide range of filter types. The frequency range of most of these can be programmed by means of an external clock which allows them to be dynamically tuned if required. The basis of this type of filter is the use of MOSFET analog switches to make a capacitor appear to be a resistor with the effective value of the resistor depending in particular on the switching frequency. The capacitors may be integrated within the IC to allow the construction of complex filters in which the tuning of a filter may be changed by varying the switching frequency. Though we will not examine these filters it is instructive to understand the mechanism involved (Van Valkenburg

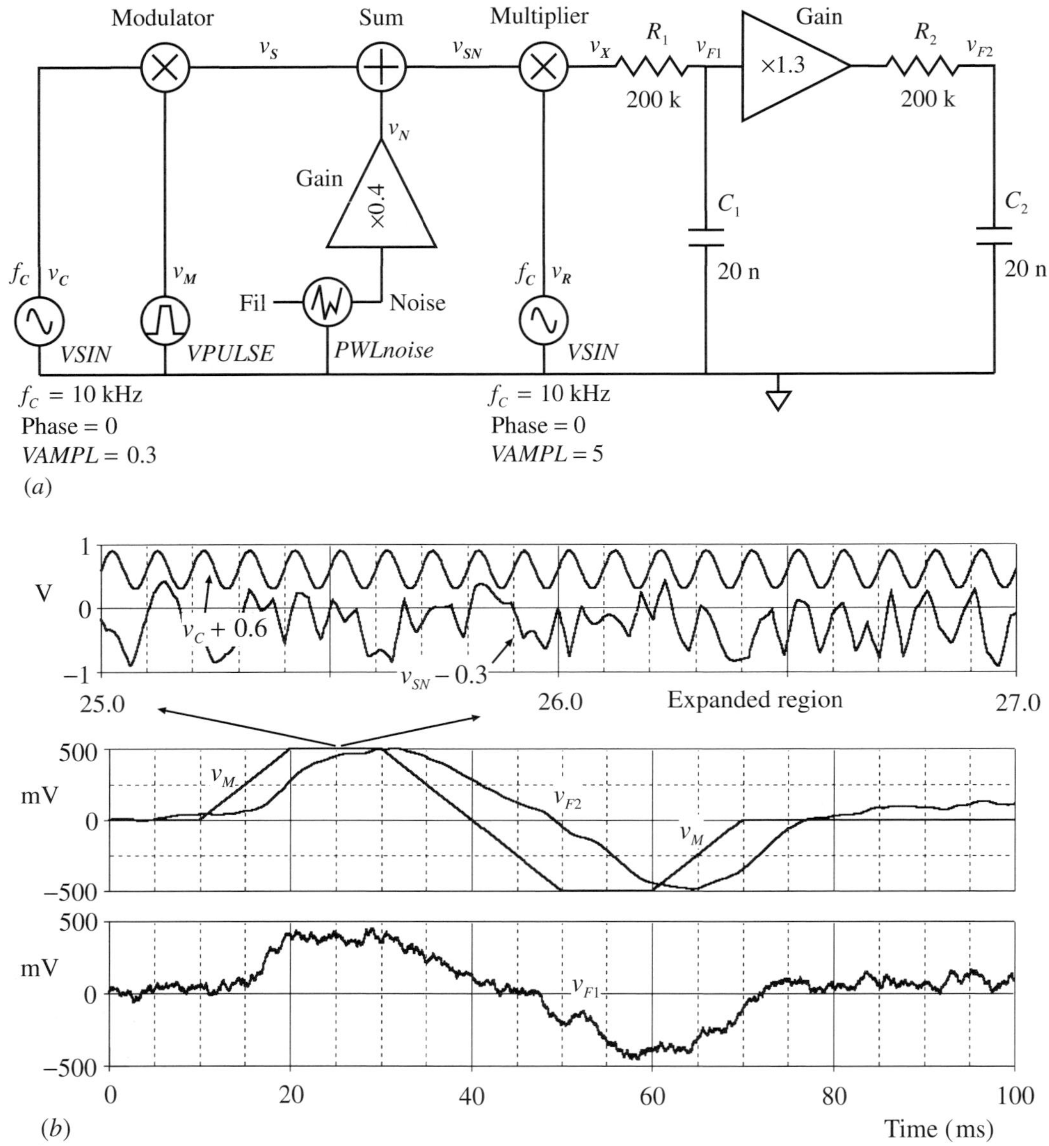


**Fig. 3.16.5** (*a*) Circuit for simulation of a lock-in detector. (*b*) Simulation results.

1982). Though this properly takes us into the realm of sampled-data with the limitations imposed by another of Nyquist's criteria, we can illustrate the operation by a more conventional approach. Consider a capacitor and a switch as shown in Fig. 3.16.6(*a*), which shows appropriate component values, or the equivalent in (*b*), where the switches operate in antiphase.

We must assume that the switching period is much shorter than significant variations in the input signal. In practice the switches will have some resistance and we

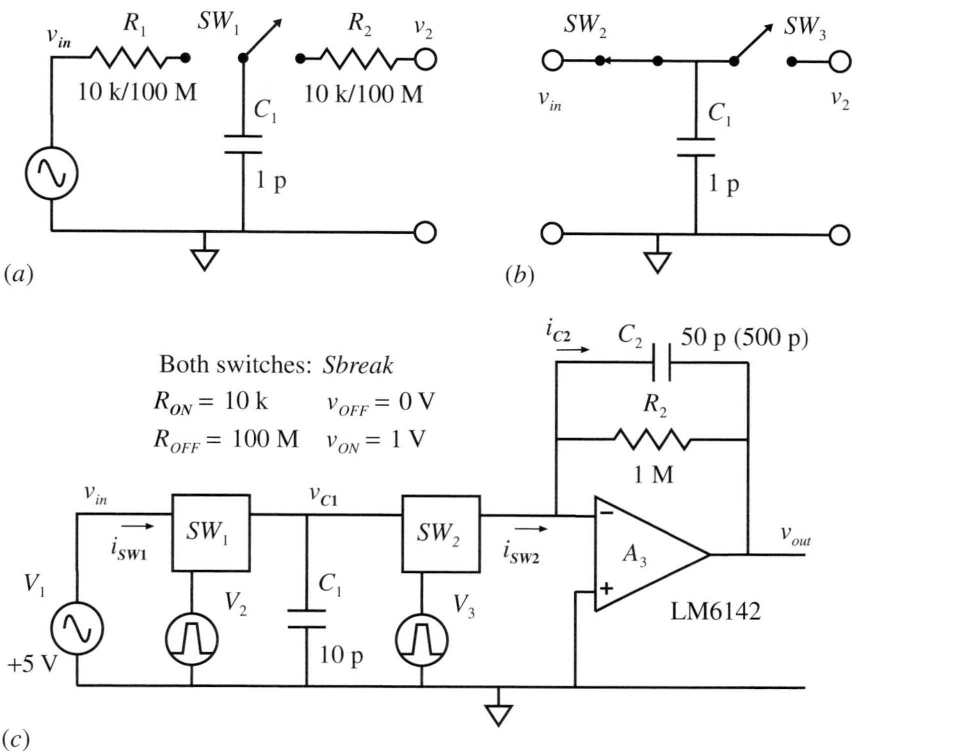


**Fig. 3.16.6** (*a*) Switched capacitor. (*b*) Equivalent circuit if switches operate in antiphase. (*c*) Circuit for simulation. $V_2$ and $V_3$ parameters: $T_R = T_F = 1$ μs, $PW = 10$ μs, $PER = 30$ μs, $v_1 = 0$ V, $v_2 = 1$ V. $T_D = 0$ for $V_2$ and $T_D = 15$ μs for $V_3$ to give non-overlapping switching. The switches are *Sbreak* devices.

assume that the ratio of $R_{OFF}$ to $R_{ON}$ is large. With $SW_1$ 'on' the capacitor $C_1$ will charge rapidly towards $v_{in}$, say to voltage $v_a$. $SW_1$ then opens and $SW_2$ closes and the capacitor discharges through some load to a voltage $v_b$. The charge transferred from the input to the output is thus:

$$Q = C_1(v_a - v_b) \tag{3.16.5}$$

and this is repeated for each cycle of the switching frequency $f$, say with a period $T$. The average current flow is then:

$$i_{av} = \frac{\partial Q}{\partial t} = \frac{C_1(v_a - v_b)}{T}, \quad \text{giving an equivalent resistance}$$

$$R = \frac{(v_a - v_b)}{i_{av}} = \frac{T}{C_1} = \frac{1}{fC_1} \tag{3.16.6}$$

For the simulation circuit as shown in Fig. 3.16.6(*c*) we have a period $T = 30$ μs and $C_1 = 10$p, so that we expect the effective resistance to be:

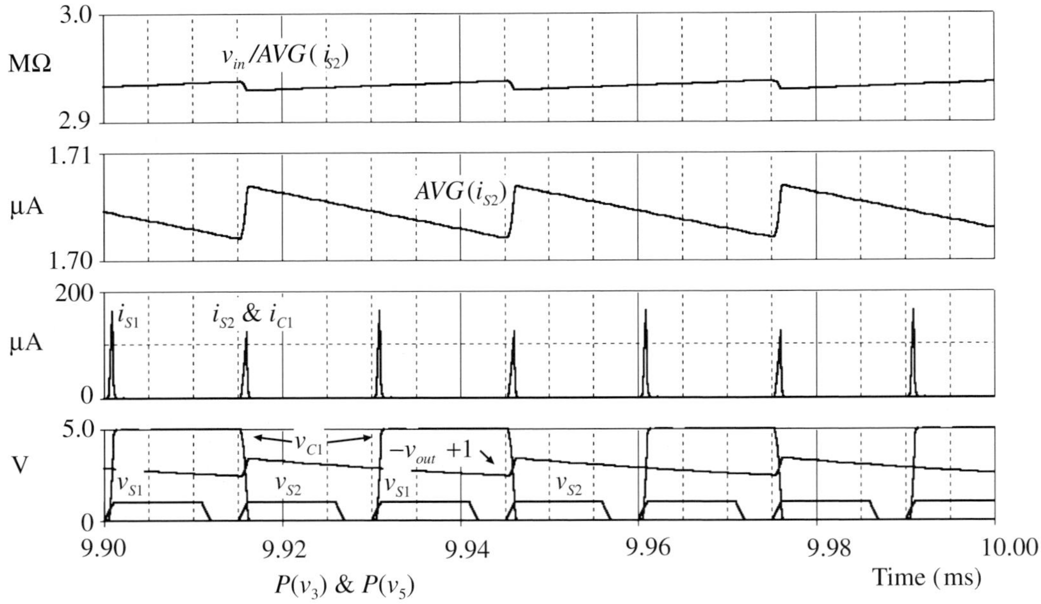


**Fig. 3.16.7** Simulation results for circuit of Fig. 3.16.6($c$).

$$R_{eff} = \frac{30 \times 10^{-6}}{10 \times 10^{-12}} = 3\ \text{M}\Omega \tag{3.16.7}$$

The simulation is run for 10 ms to allow any initial transients to decay and Fig. 3.16.7 shows the results for a short interval at the end of the run.

The lower graphs show the switching waveforms and the currents in the switches. The second graph shows the $AVG(i_{S2})$ and the top shows the effective resistance $R_{eff} = v_{in} / AVG(i_{S2})$ which is seen to be very nearly 3 MΩ as predicted.

To check that the system acts as a low-pass filter presents a problem. We cannot use the PSpice *AC SWEEP* facility since this does not allow for the dynamic switching upon which the operation depends. The $-3$ dB frequency is determined by $R_2$ and $C_2$, but to make the frequency rather lower so that the chopping ripple is less significant, we make $C_2 = 500$ p which gives a frequency of 318 Hz. Making $V_1$ an *AC* source of say 1 V at 100 Hz allows us to determine the in-band gain which is found to be $R_2/R_{eff} = 1/3$ as expected. To examine the frequency response we may use a linearly varying frequency input to see where the amplitude of the output falls to the $-3$ dB point which we can determine from the in-band gain we have just found, i.e. for a 1 V input we find that the amplitude at the corner should be $1 \times 0.333 \times 0.707 = 236$ mV. To generate a varying frequency input we can use an ABM device to give a sin output but with $\omega = kt$ instead of just a constant as for a normal sin function. Thus we can write the expression *EXP*1, for a 1 V amplitude

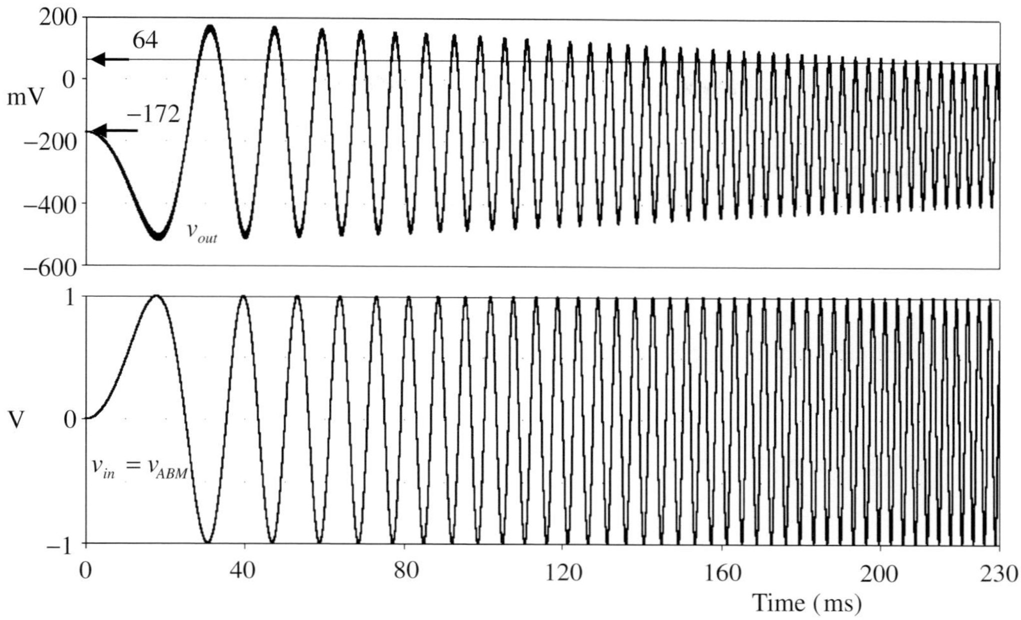


**Fig. 3.16.8** Frequency response of the circuit of Fig. 3.16.6(*c*) with $C_2 = 500$ p and input $v_{in}$ from an ABM defined by Eq. (3.16.8).

and using a suitable multiplier to cover the frequency range required (we will see below how this is arrived at):

$$\text{EXP1} = \sin(5\text{E}3{*}TIME{*}TIME) \tag{3.16.8}$$

so that the effective $\omega$ is a function of the SPICE variable *TIME*. This form has a somewhat unexpected consequence (which means that I did not think of it until I saw the simulation output). The ABM output is shown in the lower part of Fig. 3.16.8, which looks as we intended but closer examination shows that the frequency at any given time is twice that found from 5E3×*TIME*; this would predict $\omega = 5\text{E}3 \times 0.230 = 1150\ \text{s}^{-1}$ or $f = 183$ Hz at the end on the run at 230 ms whereas the simulation shows twice this. The reason for this is that the rate of change of the argument in (3.16.8) is proportional to $d(TIME^2)/dt = 2 \times TIME$, which determines when the sin function crosses zero. The time when the frequency equals our expected −3 dB frequency of 318 Hz is then found from:

$$TIME = \frac{318 \times 2\pi}{5\text{E}3 \times 2} = 200\ \text{ms} \tag{3.16.9}$$

and this form of calculation prompted the choice of the multiplier 5E3 and time of the run so as to exceed the required frequency. As the quiescent output of the filter was at –172 mV and the −3 dB amplitude was 236 mV, a trace at

(236 −172) = +64 mV is shown on the upper graph, which does indeed fit with the 200 ms *TIME* ≡ 318 Hz (the switching fuzz makes accurate determination difficult). Note that since we also have the much higher switching frequency to contend with, the maximum step size will be short so that simulation times will be long. We have demonstrated that the switched capacitor circuit does operate as described. The basic switch can be used to construct very complex filter devices and a wide range is available from several manufacturers.

PSpice has an associated filter design software package (MicroSim 1995) and a number of IC semiconductor manufacturers provide design packages particularly for their own products (see software references) and these are generally available.

### SPICE simulation circuits

Consult the SimCmnt.doc file on the CD before running.

| | |
|---|---|
| Fig. 3.16.2(*b*) | Biquadf1.SCH |
| Fig. 3.16.3 | Biquadf2.SCH |
| Fig. 3.16.5(*b*) | Lockin2.SCH |
| Fig. 3.16.7 | Swcapf1.SCH |
| Fig. 3.16.8 | Swcapf2.SCH |

## References and additional sources 3.16

Allen P. E., Blalock B. J., Milam S. W. (1995): Low-gain active filters. in Chen, Wai-Kai (Ed.): *The Circuits and Filters Handbook*, Part 75, pp. 2339–2371, Boca Raton: CRC Press and IEEE Press. ISBN 0-8493-8341-2. Section XV, Huelsman L. P. (Ed.), covers many active filters.

Ayrton W. E., Perry J. (1888): Modes of measuring the coefficients of self and mutual inductance. *J. Soc. Telegr. Engrs.* **16**, 292–243.

Gehrke D., Hahn A. (1999): *10 MHz Butterworth Filter Using the Operational Amplifier THS4001*, Texas Instruments Application Report SLOA032, October.

Hageman S. C. (1996): *Create Analog Random Noise Generators for PSpice Simulation*, MicroSim Application Notes Ver. 6.3, p. 61, MicroSim Corp. April. (The program is called PWL Noise.)

Hamilton T. D. S. (1977): *Handbook of Linear Integrated Electronics for Research*, London: McGraw-Hill. Section 4.4 Active filters, and references. Section 10.2 Lock-in or coherent detection, and references.

Hansen P. D. (1963): New approaches to the design of active filters. *The Lightning Empiricist* **13** (1 + 2), January–July. Philbrick Researches Inc. Part II in **13** (3 + 4), July–October 1965.

Henderson K. W., Kautz W. H. (1958): Transient response of conventional filters. *IEEE Trans.* **CT-5**, 333–347.

Henn C., Lehman K. (1994): *Designing Active Filters with the Diamond Transistor OPA660. Part 1*, Burr-Brown Application Bulletin AB-190.

Hilburn J. L., Johnson D. E. (1973): *Manual of Active Filter Design*, New York: McGraw-Hill, ISBN 0-07-028759-7.

Karki J. (1999): *Analysis of the Sallen-Key Architecture*, Texas Instruments Application Report SLOA024A, July.

Kuo F. F. (1966): *Network Analysis and Synthesis*, New York: John Wiley. ISBN 0-471-51118-8. See Section 13.5.

Lacanette K. (1994): *A Basic Introduction to Filters – Active, Passive, and Switched Capacitor*, National Semiconductor Application Note AN-779. See also *Linear Applications Handbook*.

Linear Technology (2000): *Active RC, 4th Order Lowpass Filter Family LTC1563-2/LTC1563-3*, Linear Technology Data Sheet, January.

Mancini R. (1998): Crystal filter for pure signals. *Electronic Design* 13 May, 136.

Markell R. (1994): *"Better than Bessel" Linear Phase Filters for Data Communications*, Linear Technology Application Note 56, January.

MicroSim (1995): *MicroSim Filter Designer, Filter Synthesis User's Guide*, MicroSim Corp., April.

MicroSim (1996): *Filter Models Implemented with ABM*, MicroSim Application Notes Ver. 6.3, p. 105, MicroSim Corp., April.

Sauerwald M. (1997): *Designing High Speed Active Filters*, National Semiconductor Application Note OA-26, September.

Scott H. H. (1938): A new type of selective circuit and some applications. *Proc. IRE* **26**, 226–236.

Steffes M. (1993): *Simplified Component Value Pre-Distortion for High Speed Active Filters*, Comlinear Application Note OA-21, March.

Stout D. F., Kaufman M. (1976): *Handbook of Operational Amplifier Circuit Design*, New York: McGraw-Hill. ISBN 0-07-061797-X.

Tavares S. E. (1966): A comparison of integration and low-pass filtering. *IEEE Trans.* **IM-15** March-June, 33–38.

Thomas L. C. (1971): The biquad: Part I – some practical design considerations. *IEEE Trans.* **CAS-18**, 350–357, and The biquad: Part II – a multipurpose active filtering system. *IEEE Trans.* **CAS-18**, 358–361.

Tow J. (1968): Active RC filters – a state-space realization. *Proc. IEEE* **56**, 1137–1139.

Van Valkenburg M. E. (1982): *Analog Filter Design*, New York: Holt, Rinehart and Winston. ISBN 0-03-059246-1, or 4-8338-0091-3 International Edn. Appendix 8 is a substantial bibliography of filter books.

Van Vollenhoven E., Reuver H. A., Somer J. C. (1965): Transient response of Butterworth filters. *IEEE Trans.* **CT-12**, 624–626.

Williams A. B., Taylor F. J. (1995): *Electronic Filter Design Handbook*, New York: McGraw-Hill.

Williams J. M. (Ed.) (1997): *Linear Applications Handbook, Vol. III: Linear Technology*. See subject index under filters/active RC/switched capacitor.

Wilmshurst T. H. (1985): *Signal Recovery from Noise in Electronic Instrumentation*, Bristol: Adam Hilger. ISBN 0-85274-783-7.

Yager C., Laber C. (1988): *High Frequency Complex Filter Design Using the ML2111*, Micro Linear Application Note 4, November.

Zverev A. I. (1967): *Handbook of Filter Synthesis*, New York: John Wiley.

**Series articles**

**A**

Mittleman J. (1968): Active filters: Part 1, The road to high Q's. *Electronics* 27 May, 109–114.

de Pian L. (1968): Active filters: Part 2, Using the gyrator. *Electronics* 10 June, 114–120.

de Pian L., Meltzer A. (1968): Active filters: Part 3, Negative impedance converters. *Electronics* 2 September, 82–93.

de Pian L., Meltzer A. (1968): Active filters: Part 4, Approaching the ideal NIC; *Electronics* 16 September, 105–108.

Moschytz G. S., Wyndrum R. W. (1968): Active filters: Part 5, Applying the operational amplifier. *Electronics* 9 December, 98–106.

Welling B. (1969): Active filters: Part 6, The op amp saves time and money. *Electronics* 3 February, 82–90.

Salerno J. (1969): Active filters: Part 7, Analog blocks ensure stable design. *Electronics* 17 February, 100–105.

Hurtig G. (1969): Active filters: Part 8, Positive results from negative feedback. *Electronics* 31 March, 96–102.

Marsocchi V. A. (1969): Active filters: Part 9, Applying nonlinear elements. *Electronics* 14 April, 116–121.

Aaronson G. (1969): Active filters: Part 10, Synthetic inductors from gyrators. *Electronics* 7 July, 118–125.

Mullaney J. W. (1969): Active filters: Part 11, Varying the approach. *Electronics* 21 July, 86–93.

Shepard R. R. (1969): Active filters: Part 12, Short cuts to network design. *Electronics* 18 August, 82–91.

**B**

Girling E. J., Good E. F.: Active filters. A practical approach on feedback amplifier theory: *Wireless World*,

(1) Survey of circuits. August 1969, 348–352.
(2) Basic theory: $1^{st}$ and $2^{nd}$ order responses. September 1969, 403–408.
(3) Properties of passive and non-feedback CR networks. October 1969, 461–465.
(4) Basic theory: active circuits. November 1969, 521–525.
(5) An integrator and a lag in a loop. December 1969, 568–572.
(6) Lead-lag network and positive gain. January 1970, 27–31.
(7) The two integrator loop. February 1970, 76–80.
(8) The two integrator loop, continued. March 1970, 134–139.
(9) Synthesis by factors. April 1970, 183–188.
(10) Uses of the parallel-T network. May 1970, 231–234.
(11) More on the parallel-T network. June 1970, 285–287.
(12) The leapfrog or active ladder synthesis. July 1970, 341–345.

**Filter design software**

Linear Technology: *FilterCAD Ver. 3.01*, July 1999. www.linear-tech.com

Maxim Integrated Products: *Filter Software for Max 274, Max 275, Max 260–268.*

Molina J., Stitt R. M. (1994): *Filter Design Program for the UAF42 Universal Active Filter*, Burr-Brown Application Bulletin AB-035C. www.burr-brown.com

Trump B., Stitt R. M. (1994): *MFB low-pass filter design program*, Burr-Brown Application Bulletin AB-034B. www.burr-brown.com

# 3.17 Transmission lines

I am aware of the deficiencies of the above diagrams. Perhaps some electrical student who possesses the patient laboriousness sometimes found associated with early manhood may find it worth his while to calculate the waves thoroughly and give tables of results, and several curves in each case. It should be a labour of love, of course; for although if done thoroughly there would be enough to make a book, it would not pay, and most eminent publishers will not keep a book in stock if it does not pay, even though it be a book that is well recognised to be a valuable work, and perhaps to a great extent the maker of other works of a more sellable nature. Storage room is too valuable.

Oliver Heaviside (1899): 'On transmission lines.' *Electromagnetic Theory*, April 10, Vol. II, p. 433

At high frequencies or for very fast signals involving short time intervals our normal view of components as lumped elements in which the propagation time is short has to be changed. Signals now take a significant time to travel from $A$ to $B$ so that we can no longer assume that the current is the same at every point in a conductor. The small inductance and capacitance of the interconnections and the incidental parasitics of our components are now no longer negligible and significantly affect the operation of the system. As in many circumstances the region of changeover from one regime to another is the most difficult to deal with and we will consider later some ideas on the conditions which suggest that we have to make the transition. We will not be concerned with conditions where it is necessary to solve Maxwell's field equations but there is one area where propagation times are commonly significant and that is where we have long interconnections (at least with respect to the time scale of interest), as for example with coaxial cables. We will examine the response of a transmission line where the electrical parameters are uniformly distributed along the length. The commonest form of transmission line is the coaxial cable, though other geometries are sometimes more convenient. The justification for using the concepts of inductance and capacitance when dealing with what is more properly a field application is outlined in Section 4.9. In any case there are two conductors which we can represent as shown in Fig. 3.17.1. Though we show the resistive and reactive components of the small section $\delta x$ of the line as lumped and unsymmetrical, it should be understood that they are effectively

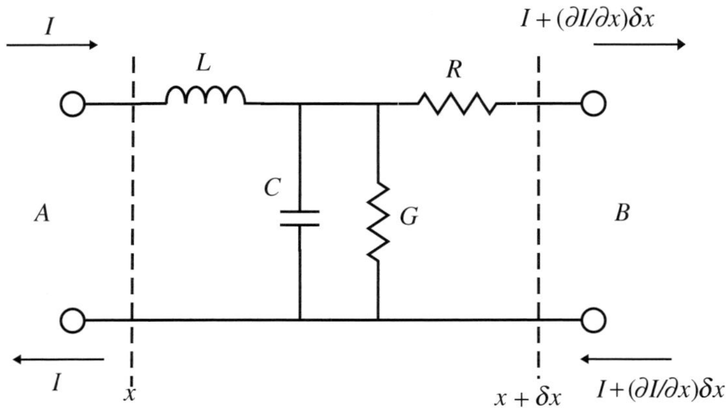


**Fig. 3.17.1** Transmission line segment.

distributed in both conductors. The components shown are to be taken as the value per unit length.

For the sort of lengths involved in interconnections between local systems the effect of the loss components $R$ (series) and $G$ (shunt) are negligible so we will ignore these. When a signal is applied to the end of the cable the voltages and currents along the length will not be uniform since there is a finite time required to propagate along the line. If the line is in air then the velocity of propagation will, as we shall show, be $c$, the velocity of light, i.e. the delay will be $3.3\times10^{-9}\,\text{s m}^{-1}$. If the space surrounding the wires is a dielectric with $\varepsilon\times1$ then the velocity will be less. For this small length $\delta x$ of line we can write the equations:

$$V_B - V_A = \frac{\partial V}{\partial x}\delta x = -L\delta x\frac{\partial I}{\partial t} \quad \text{or} \quad \frac{\partial V}{\partial x} = -L\frac{\partial I}{\partial t}$$

$$\text{so} \quad \frac{\partial^2 V}{\partial x^2} = -L\frac{\partial^2 I}{\partial x \partial t} \quad \text{by differentiating with respect to } x \tag{3.17.1}$$

and similarly for the shunt current flowing in $C$:

$$I_B - I_A = \frac{\partial I}{\partial x}\delta x = -C\delta x\frac{\partial V}{\partial t} \quad \text{or} \quad \frac{\partial I}{\partial x} = -C\frac{\partial V}{\partial t}$$

$$\text{so} \quad \frac{\partial^2 I}{\partial t \partial x} = -C\frac{\partial^2 V}{\partial t^2} \quad \text{by differentiating with respect to } t \tag{3.17.2}$$

For the partial differentials the order of differentiation is not significant, i.e. $\partial x \partial t = \partial t \partial x$, so we have from (3.17.1) and (3.17.2):

$$\frac{1}{L}\frac{\partial^2 V}{\partial x^2} = \frac{-\partial^2 I}{\partial x \partial t} = \frac{-\partial^2 I}{\partial t \partial x} = C\frac{\partial^2 V}{\partial t^2} \quad \text{or} \quad \frac{\partial^2 V}{\partial x^2} = LC\frac{\partial^2 V}{\partial t^2} \tag{3.17.3}$$

which we can recognize as a wave equation (Section 1.13). The solution to this equation is:

$$V(x, t) = f_1(ut - x) + f_2(ut + x) \tag{3.17.4}$$

where $f_1$ represents a wave proceeding in the direction $A \rightarrow B$ and $f_2$ a wave from $B \rightarrow A$, both with velocity $u = (LC)^{-\frac{1}{2}}$, which will be evaluated below for particular transmission lines. The direction follows from the sense of increasing $x$ as shown in Fig. 3.17.1. You can show that this is a solution of the wave equation by differentiating it and seeing whether it agrees. Each of $f_1$ or $f_2$ will separately be a solution, so taking $f_1$ for example (remember that you must differentiate as a function of a function; Section 1.8):

$$V = f_1(ut - x), \qquad \frac{dV}{dx} = f_1'(0 + 1), \qquad \frac{d^2V}{dx^2} = -f_1''(-1) = f_1''$$

$$\text{and} \quad \frac{dV}{dt} = f_1'(u - 0), \quad \frac{d^2V}{dt^2} = f_1''(u^2) \tag{3.17.5}$$

In a similar manner to the derivation of (3.17.3) we could also obtain a wave equation for $I$:

$$\frac{\partial^2 I}{\partial x^2} = LC\frac{\partial^2 I}{\partial t^2} \tag{3.17.6}$$

which would have corresponding solutions. Consider a sinusoidal input voltage given by:

$$V_x = V_0\cos[k(ut - x)] = V_0\cos[(\omega t - kx)], \quad \text{where} \quad k = \frac{\omega}{u} = \omega(LC)^{\frac{1}{2}} \tag{3.17.7}$$

i.e. a wave equivalent to $f_1$, and $k$ is the phase constant or wavenumber part of the propagation constant (see end of section). We can determine the corresponding equation for $I$ from (3.17.1) by integration (we may ignore the constant of integration as this will represent some steady-state initial condition):

$$-L\frac{\partial I}{\partial t} = \frac{\partial V}{\partial x} = -V_0\,k\sin[k(ut - x)]$$

so

$$I = \frac{-V_0\,k}{L}\int \sin[k(ut - x)]dt = \frac{-V_0\,k}{L}\left\{\frac{\cos[k(ut - x)]}{ku}\right\} \tag{3.17.8}$$

$$= V_0\left(\frac{L}{C}\right)^{\frac{1}{2}}\cos[k(ut - x)], \qquad \text{since} \quad Lu = \frac{L}{(LC)^{\frac{1}{2}}} = \left(\frac{L}{C}\right)^{\frac{1}{2}}$$

so that at any point on, say, an infinite line (so there will be no return wave) we have:

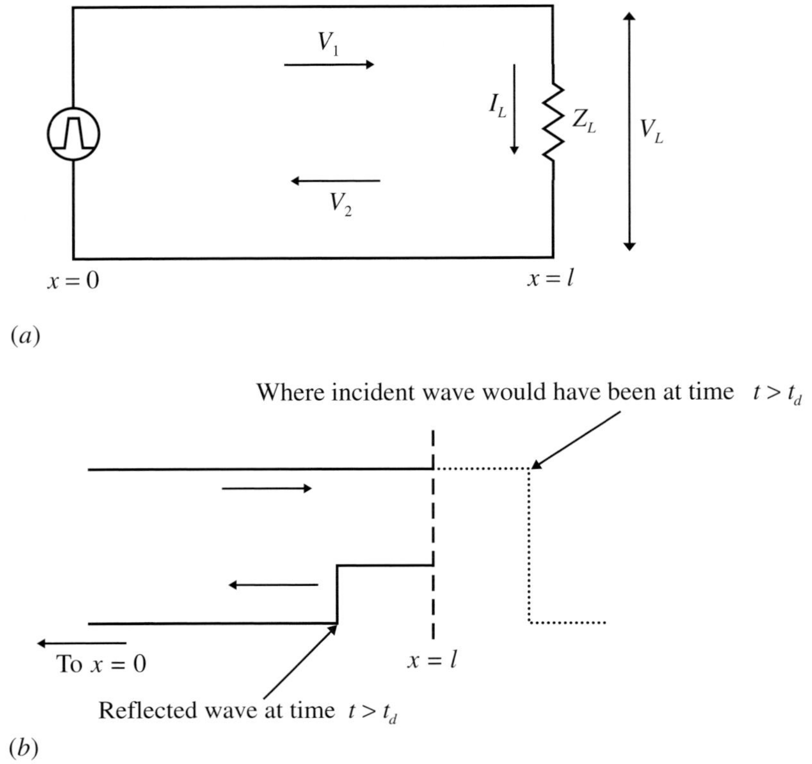


**Fig. 3.17.2** (*a*) Transmission line waves. (*b*) Transmission line reflections.

$$\frac{V}{I} = \left(\frac{L}{C}\right)^{\frac{1}{2}} = Z_0 \qquad (3.17.9)$$

which is resistive and is called the characteristic impedance of the line. $Z_0$ depends only on $L$ and $C$, and so is just a function of the geometry of the line as shown by (3.17.20) and (3.17.21) below. If we had used the wave corresponding to $f_2$ instead, then the procedure of Eq. (3.17.8) would lead to the same result as (3.17.9) for $Z_0$ except that it would have a minus sign which indicates the phase difference.

Now consider a line of length $l$ terminated by an impedance $Z_L$ as shown in Fig. 3.17.2.

At $x = l$ the voltage and current due to the two waves add to give the voltage $V_L$ and current $I_L$:

$$V_L = V_1 + V_2, \qquad I_L = I_1 + I_2, \qquad \frac{V_1}{I_1} = Z_0, \qquad \frac{V_2}{I_2} = -Z_0$$

$$\text{so} \quad Z_L = \frac{V_L}{I_L} = \frac{V_1 + V_2}{I_1 + I_2} = \frac{(I_1 - I_2)Z_0}{(I_1 + I_2)} \quad \text{or} \quad \frac{Z_L}{Z_0} = \frac{(I_1 - I_2)}{(I_1 + I_2)} \qquad (3.17.10)$$

$$\text{thus} \quad \frac{Z_0 - Z_L}{Z_0 + Z_L} = \frac{I_2}{I_1} \quad \text{or} \quad \frac{I_L}{I_1} = \frac{2Z_0}{Z_0 + Z_L} \quad \text{and} \quad \frac{I_2}{I_L} = \frac{Z_0 - Z_L}{2Z_0} \qquad \text{(3.17.10 cont.)}$$

$$\text{and similarly} \quad \frac{V_L}{V_1} = \frac{2Z_L}{Z_L + Z_0} \quad \text{and} \quad \frac{V_2}{V_L} = \frac{Z_L - Z_0}{2Z_L}$$

so that if $Z_L = Z_0$ then $V_L = V_1$, $V_2 = 0$ and $I_L = I_1$, $I_2 = 0$. Thus when the line is terminated with its characteristic impedance there is no reflected wave and all the power is absorbed by the termination. The line is said to be matched and a terminated line of any length, viewed from the sending end, looks identical.

If the line is not terminated by $Z_0$ then there will be reflections of magnitude depending on the load. From the equations above we can write:

$$\frac{V_2}{V_1} = \frac{Z_L - Z_0}{2Z_L} \frac{2Z_L}{Z_L + Z_0} = \frac{Z_L - Z_0}{Z_L + Z_0} = \rho, \text{ the voltage reflection coefficient}$$

$$\text{and} \quad \frac{I_2}{I_1} = \frac{Z_0 - Z_L}{Z_L + Z_0} = -\rho \qquad \text{(3.17.11)}$$

If $Z_L > Z_0$ then $\rho$ is positive so $V_2$ will have the same sense as $V_1$ and the voltage at the load is:

$$V_L = V_1 + V_2 = V_1 + \rho V_1 = (1+\rho)V_1$$

$$\text{and} \quad I_L = I_1 + I_2 = (1-\rho)I_1 \qquad \text{(3.17.12)}$$

and the two waves on the line at a time greater than $t = l/u$ will be as shown in Fig. 3.17.2(*b*) where we have shown an incident signal of unity magnitude and imagined the two waves in isolation.

The initial current will be $V/Z_0 = 1/Z_0$ and the effect of the combined waves will then be as shown in Fig. 3.17.3.

When the reflected waves reach the sending end then of course they will be reflected again depending on the impedance they meet, so that if there is a reflection at the far end then it is necessary to make the sending end impedance $Z_0$ to ensure there are no further reflections. The to-and-fro reflections (usually with attenuation) that can ensue are often seen in high speed digital systems. If the line is open circuited then the reflected wave will arrive back at the sending end at twice the initial amplitude.

If the load $Z_L < Z_0$ then $\rho$ is negative and the senses of the reflections will be inverted and the waves appear as in Fig. 3.17.4.

If the line is short-circuited then the reflected wave will be equal and opposite to the incident wave and the voltage will go to zero.

For reactive loads the consequences may also be determined. Consider the arrangement shown in Fig. 3.17.5(*a*) where we now include a source impedance $R_0$ equal to $Z_0$ so that there will be no further reflections at that end. The model used is the lossless T-line.

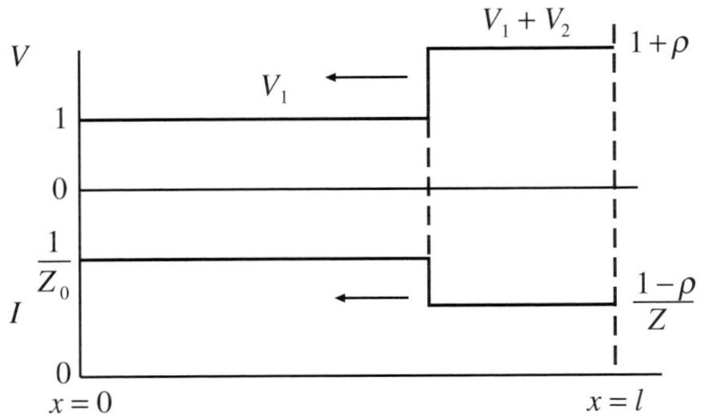


**Fig. 3.17.3** Reflections for resistive load $Z_L > Z_0$.

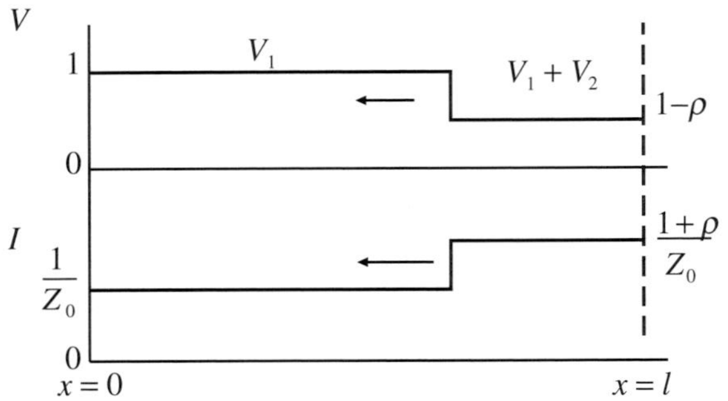


**Fig. 3.17.4** Reflections for resistive load $Z_L < Z_0$.

We assume a unit step input of amplitude $v_g(t)$ so that the input to the line itself will be half this since the line looks like $R_0$. The reflection coefficient $\rho$ is given by:

$$\rho(s)=\frac{\dfrac{1}{sC}-R_0}{\dfrac{1}{sC}+R_0}=\frac{\dfrac{1}{CR_0}-s}{\dfrac{1}{CR_0}+s} \tag{3.17.13}$$

then using the Laplace transform we have:

$$V_2(s)=V_1(s)\rho(s)=\frac{1}{2s}\frac{\left(\dfrac{1}{CR_0}\right)-s}{\left(\dfrac{1}{CR_0}\right)+s} \tag{3.17.14}$$

and the inverse transform is from Table 1.12.1, No. 18

$$v_2(t)=\frac{1}{2}-\exp(-t'/R_0C) \tag{3.17.15}$$

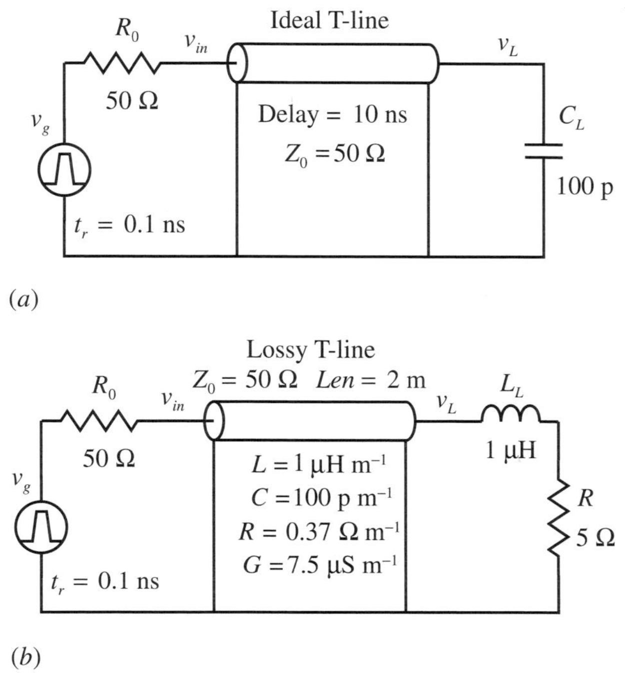


**Fig. 3.17.5** (*a*) Capacitive load with a generator matched to the line. (*b*) Inductive load, including self-resistance *R*, with a generator matched to the line. *R* has been exaggerated to show its effect.

where $t'$ is the time after the line delay $t_d$, i.e. $t' = t - t_d$. Thus at the load the response is given by:

$$v_L = v_1 + v_2 = 1 - \exp(-t'/R_0 C) \tag{3.17.16}$$

and the response will be as shown in Fig. 3.17.6 (lower) with the final voltage tending to one as it should for an open circuit at z.f. The voltage at the input end will be half for time $2t_d$ when the return wave reaches it and then follows the output response form.

If the termination is an inductor $L$, and we will include a resistance $R$ for the inductor, then the reflection factor is:

$$\rho(s) = \frac{sL + R - R_0}{sL + R + R_0} = \frac{s + \left(\dfrac{R - R_0}{L}\right)}{s + \left(\dfrac{R + R_0}{L}\right)} \tag{3.17.17}$$

and proceeding as before we find:

$$v_2(s) = \frac{1}{2}\left\{\left(\frac{R - R_0}{R + R_0}\right) + \left(\frac{2R_0}{R + R_0}\right)\exp\left[-\frac{t'(R + R_0)}{L}\right]\right\} \tag{3.17.18}$$

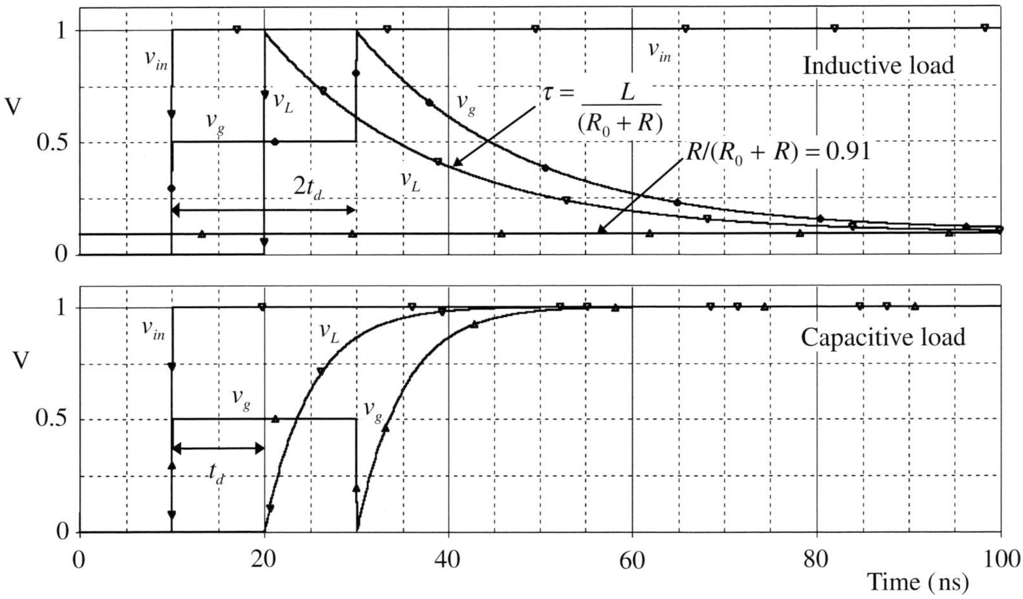


**Fig. 3.17.6** Reflections for capacitive (lower) and for inductive plus resistive load (upper).

so that at time $t_d$ (i.e. $t' = 0$) and for very long times we have for $V_L$:

$$v_L = \frac{1}{2} + \frac{1}{2}\left[\left(\frac{R - R_0}{R + R_0}\right) + \left(\frac{2R_0}{R + R_0}\right)\right] = \frac{1}{2} + \frac{1}{2}\left(\frac{R + R_0}{R + R_0}\right) = 1, \quad \text{for } t' \text{ short}$$

$$v_L = \frac{1}{2} + \frac{1}{2}\left(\frac{R - R_0}{R + R_0}\right) = \frac{R_0 + R + R - R_0}{2(R_0 + R)} = \frac{R}{R_0 + R}, \quad \text{for } t' \text{ long} \qquad (3.17.19)$$

which produces the responses shown in Fig. 3.17.6 (upper) with the exponential time constant $\tau = L/(R + R_0)$. The line model used is *TLOSSY*, but the effect for this length of line is insignificant.

Using these responses it is possible to deduce the form of the mismatch using a fast step function input (Hewlett-Packard 1964, 1966; Oliver 1964).

For general use there are two common types of transmission line, coaxial and balanced pair. Their performance (as far as we are concerned with short lengths) depends primarily on their inductance and capacity per unit length. These depend on geometric factors together with the properties of the dielectric in the vicinity of the conductors. For coaxial cables we have (dimensions as in Fig. 3.17.7):

$$C = \frac{2\pi\varepsilon\varepsilon_0}{\ln\left(\frac{D}{d}\right)} \ \text{Fm}^{-1}, \qquad L = \frac{\mu\mu_0}{2\pi}\ln\left(\frac{D}{d}\right) \ \text{Hm}^{-1}$$

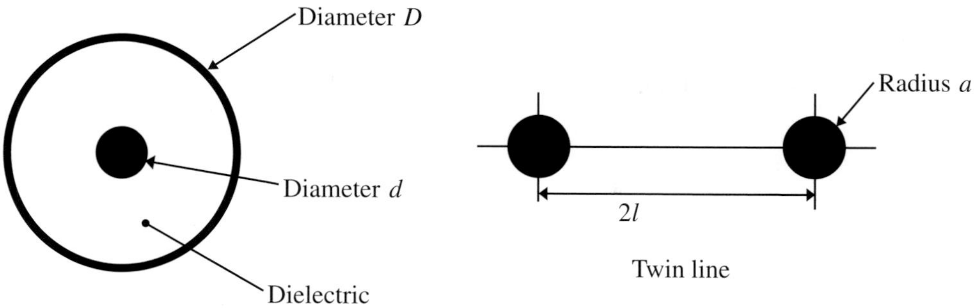


**Fig. 3.17.7** Transmission line cross-sections.

$$\text{so} \quad Z_0 = \left(\frac{L}{C}\right)^{\frac{1}{2}} = \left(\frac{\mu\mu_0}{4\pi^2\varepsilon\varepsilon_0}\right)^{\frac{1}{2}} \ln\left(\frac{D}{d}\right) \cong \frac{60}{\sqrt{\varepsilon}} \ln\left(\frac{D}{d}\right) \tag{3.17.20}$$

$$\text{and} \quad u = (LC)^{-\frac{1}{2}} = (\mu\mu_0\varepsilon\varepsilon_0)^{-\frac{1}{2}} = \frac{c}{\sqrt{\varepsilon}} \approx 0.66c, \quad \text{typical for } \varepsilon = 2.3$$

$$\text{giving} \quad t_d = 1/u \approx 5 \text{ ns m}^{-1}$$

The range of values for $D$ and $d$ is limited for mechanical reasons and a wide range of characteristic impedance is further constrained by the logarithmic dependence on the ratio. For a 50 Ω cable the ratio is about 3.6 and for a 75 Ω cable it is 6.7.

For two parallel conductors, say two wires of radius $a$ and separation $2l$:

$$C = \frac{2\pi\varepsilon_0}{\ln\left(\dfrac{l+p}{l-p}\right)}, \quad \text{where} \quad p^2 = l^2 - a^2 \quad \text{or for } l \gg a \quad C = \frac{2\pi\varepsilon_0}{\ln\left(\dfrac{2l}{a}\right)}$$

$$L = \frac{\mu\mu_0}{\pi} \ln\left(\frac{2l}{a}\right) \cong 120 \ln\left(\frac{2l}{a}\right) \tag{3.17.21}$$

$$Z_0 = \left(\frac{L}{C}\right)^{\frac{1}{2}} = \left(\frac{\mu\mu_0}{\pi^2\varepsilon\varepsilon_0}\right)^{\frac{1}{2}} \ln\left(\frac{2l}{a}\right) \cong 120 \ln\left(\frac{2l}{a}\right)$$

$$u = (LC)^{-\frac{1}{2}} = \frac{c}{\sqrt{\varepsilon}}$$

and it is evident that the propagation velocity $u$ is independent of the geometry of a transmission line and depends only on the dielectric constant. Alternative expressions for twin lines are given by Ramo and Whinnery (1953):

$$C = \frac{\pi\varepsilon\varepsilon_0}{\cosh^{-1}\left(\dfrac{2l}{d}\right)}, \quad L = \frac{\mu\mu_0}{\pi} \cosh^{-1}\left(\frac{2l}{d}\right), \quad Z_0 = \left(\frac{\mu\mu_0}{\pi^2\varepsilon\varepsilon_0}\right)^{\frac{1}{2}} \cosh^{-1}\left(\frac{2l}{d}\right) \tag{3.17.22}$$

which may be quicker to use than the full expressions in Eq. (3.17.21). Using any of these relations to obtain the parameters of a twisted pair, as for example if you consider two enamelled wires as being 'in contact' so that $2l = d$, then the $\cosh^{-1}$ term becomes zero. As a more general example take $2l = 3a$, i.e. a gap between the wires equal to the radius, then the 'exact' formula gives 28.9 pF $m^{-1}$ while the approximate formula intended for $l \gg a$ gives 25.3 pF $m^{-1}$, so the error is not great. The corresponding characteristic impedance is 115 Ω, typical of many twisted pairs.

Parameters for other forms of line are given in Chen (1995, Chapters 39 and 40). These include microstrip lines (a signal track on a PCB with a ground plane on the reverse side), a coplanar waveguide (a signal track on say a PCB with ground tracks adjacent on both sides and all on the same side as the signal track) and striplines (a signal track embedded in say a PCB with ground planes on both outer surfaces). The expressions are complex but readily computable.

If the transmission lines are lossy then there will be attenuation along the line and the characteristic impedance $Z_0$ and propagation constant $\gamma$ will be of the form (e.g. Ramo and Whinnery 1953; Everitt and Anner 1956):

$$Z_0 = \left(\frac{R + j\omega L}{G + j\omega C}\right)^{\frac{1}{2}} \quad \text{and} \quad \gamma = a + jk = [(R + j\omega L)(G + j\omega C)]^{\frac{1}{2}} \tag{3.17.23}$$

and a wave on the line will now be of the form (compare with Eq. (3.17.7)):

$$V_x = V_0 \exp(-ax)\cos[(\omega t - kx)] \tag{3.17.24}$$

with both $k$ and $a$ complicated functions of the line parameters. The exponential term shows the amplitude decaying with distance. Approximations may be made for low-loss lines, but for the sort of lengths commonly used losses may be ignored.

In driving transmission lines it is often necessary to back-terminate (at the driver) with the characteristic impedance. This means that the signal is attenuated by a factor of two, which may be a problem with the trend to amplifiers working off supplies of only a few volts. An approach to limiting the loss is presented by Steele (1995).

## SPICE simulation circuits

| | |
|---|---|
| Fig. 3.17.6 | Tlnterm1.SCH |

## References and additional sources 3.17

Boorom K. (1998): *Modeling of a High-Speed Digital Bus with OrCAD Pspice*, OrCAD Technical Note No. 80.

Chen, Wai-Kai (Ed.) (1995): *The Circuits and Filters Handbook*, Boca Raton: CRC Press and IEEE Press. ISBN 0-8493-8341-2.

Everitt W. L., Anner G. E. (1956): *Communication Engineering*, 3rd Edn, New York: McGraw-Hill. Library of Congress Cat. No. 55-12099.

Grivet P. (1970): *The Physics of Transmission Lines at High and Very High Frequencies*, Vols 1 and 2, New York: Academic Press. ISBN 12-303601-1.

Hewlett-Packard (1964): *Time Domain Reflectometry*, Hewlett-Packard Application Note 62.

Hewlett-Packard (1966): *Selected Articles on Time Domain Reflectometry Applications*, Hewlett-Packard Application Note 75, March.

Matick R. E. (1969): *Transmission Lines for Digital and Communication Networks*, New York: McGraw-Hill. Library of Congress Cat. No. 68-30561.

Oliver B. M. (1964): Time domain reflectometry. *Hewlett-Packard Journal* 15, 1–7.

Ramo S., Whinnery J. R. (1953): *Fields and Waves in Modern Radio*, 2nd Edn, New York: John Wiley.

Ramo S., Whinnery J. R., van Duzer T. (1965): *Fields and Waves in Communication Electronics*, New York: John Wiley.

Steele J. (1995): Positive feedback terminates cables. *Electronic Design* 6 March, 91–92. See update 10 July 2000, pp. 141, 142, 144 for correction and comment. (In my view there are some matters for concern in this circuit proposal.)

# Part 4

# Circuit elements

One ought, every day at least, to hear a little song, read a good poem, see a fine picture, and, if it were possible, to speak a few reasonable words.

Johannes Wolfgang Goethe (1749–1832)

In this part, a number of the more common basic circuit devices are discussed. There are of course many types of component and many variants of each, but we will be concerned with general matters rather than the detailed specifics of a particular device. In particular it is worth being aware of the limitations of real devices as these can significantly affect the performance of your circuit. As our present interests are very much involved with SPICE, some attention will be given to the parameters of the models available for simulation. Even for such apparently simple devices as resistors, capacitors and inductors there are many parasitic effects which in one way or another may affect their operation. Parasitic capacity; temperature coefficients; self-resonance; skin effect; permittivity and permeability changes with voltage, current or temperature; and others make the task of reliable design more difficult.

Again it should be emphasized that we are not doing a survey of anything like all the different types of component or even examining those that are discussed in great detail. We cover basic ideas and operation, record some of the less evident matters that should be taken into account and introduce the models that have been developed to allow simulation with reasonable accuracy. The variability of (particularly active) components means that you will in general only get approximations to the performance of your actual circuits. For bipolar and for FET transistors, SPICE is used to demonstrate its convenience in displaying characteristics and carrying out the iterative process of finding operating points.

The section on thermistors may appear to be giving prominence to a minor or less significant component but it does provide an opportunity to discuss a slightly different type of model which involves both internal and external thermal effects, and is made use of in Section 5.8.

# 4.1 Resistors

Uncritical egalitarianism poses a threat to excellence, seen by democratic man as an easily removable cause of envy and exclusion.

Alexis de Tocqueville

Resistors are available in a vast range of values, types and forms so it is improbable that you will not be able to find what you need. It is, however, necessary to consider carefully the properties of any resistor to make sure that it is appropriate for the application. Such aspects as stability (with respect to both temperature and time) and tolerance, power dissipation, parasitic effects of capacity and inductance, and physical size may be relevant. To minimize many undesirable effects, especially at higher frequencies, and for use around ICs where scales are small, surface mount types are much to be preferred. From my experience, dating from the days of solid carbon rods with wires wound round and soldered to the ends, SMD devices can be strongly recommended. One word of warning, however; too many soldering and desolderings of the device can result in removal of the end plating and a hidden dry joint.

PSpice presents a simple resistor model without any incidental parasitics, but it does provide a means of including allowance for a temperature coefficient of first or second order, or as an exponent. This is not specifically indicated in the *ATTRIBUTES* display but you may type in for the *VALUE* attribute a statement including the keyword *TC* such as:

$$Value\ TC = TC1,\ TC2; \quad \text{for example} \quad 10\text{ k}\ TC = 0.001,\ 0.0003 \tag{4.1.1}$$

where $TC1$ is the linear coefficient, per °C and $TC2$ is the quadratic coefficient, per $(°\text{C})^2$. The temperature is reckoned as the difference between the actual temperature and *TNOM*, which is set by SPICE to be 27 °C by default, though this can be set as you require (under *ANALYSIS/OPTIONS*). The resulting resistance, with *TEMP* being the PSpice temperature variable, will therefore be given by:

$$R = Value[1 + TC1(TEMP - TNOM) + TC2(TEMP - TNOM)^2] \tag{4.1.2}$$

Temperature coefficients of resistors are usually small and second order coefficients smaller still, but if using this facility for some other purpose you need

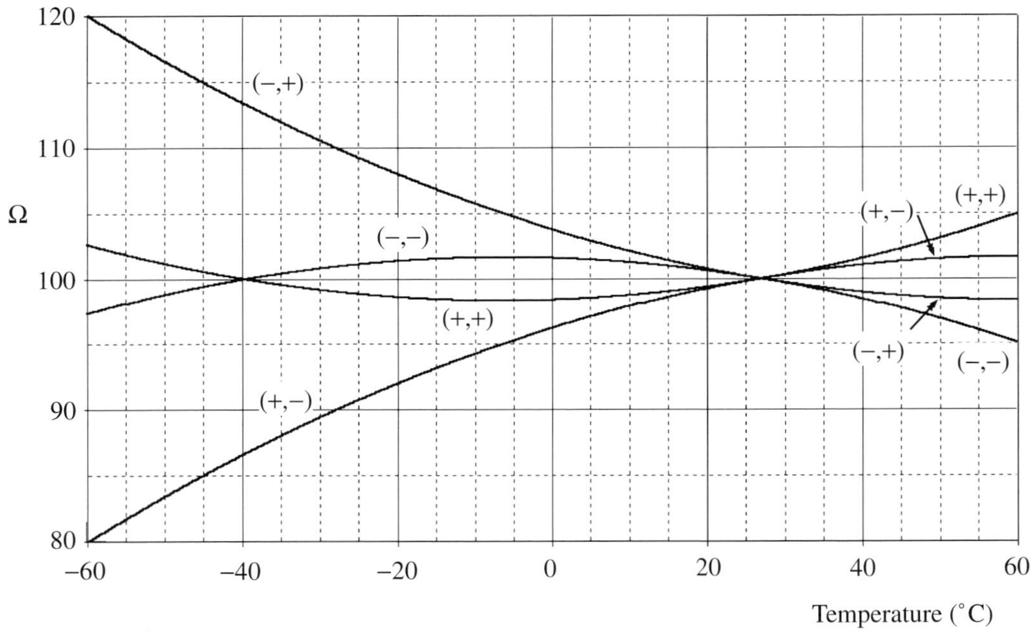


**Fig. 4.1.1** Variation of resistance with linear plus quadratic temperature coefficients. The nominal value is 100 Ω and $|TC1| = 1E{-}3$, $|TC2| = 15E{-}6$. The curves are labelled with the signs of ($TC1$, $TC2$).

to be careful in choosing the coefficients so as to avoid the resistance passing through zero. PSpice will flag an error as $R = 0$ is not allowed. An example of the variation in shape of the function is shown in Fig. 4.1.1 for the possible set of signs of $TC1$ and $TC2$.

The modelling of other forms of resistor can be performed in a number of ways. A technique for voltage-controlled resistors is described by Hirasuna (1999). The resistor is represented by a *VCCS* (or a *GVALUE* device in this case) as shown in Fig. 4.1.2.

If a table of values is appropriate then the *ABM* expression would look like that shown in the figure. Pairs of values (voltage, resistance) define the fixed points and linear interpolation is used between these points. Outside the defined range of voltage the terminal values of resistance are taken as constant. If you do a *DC Sweep* of the applied voltage $V1$ then plotting $V1/I(V1)$ will display the resistance. (Note, however, that the segments adjoining 0 V may be misleading as this depends on the *INCREMENT* specified for the sweep: this is an error in Versions 8 and 9.) If you specify a sweep including a negative voltage you do not get a negative resistance since the current simply changes direction. Note that if $R$ is a linear function of $V$ as given in the reference, $I$ will be constant (within the defined limits).

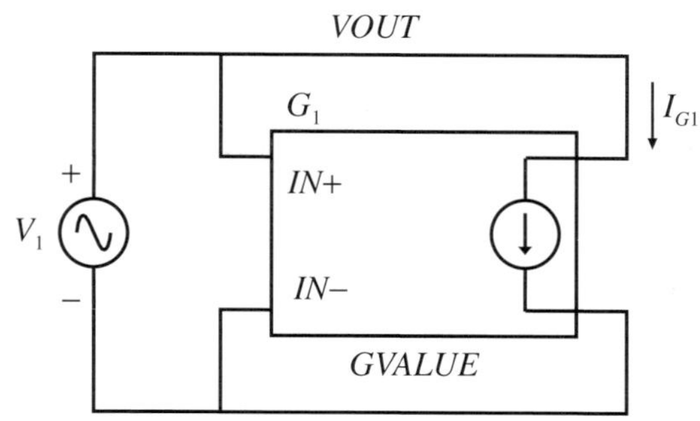


V(%IN+,%IN-)/TABLE(V(%IN+,%IN-),0.5, 25, 1.0, 10, 2.0, 90)

**Fig. 4.1.2** Model for a voltage-controlled resistor.

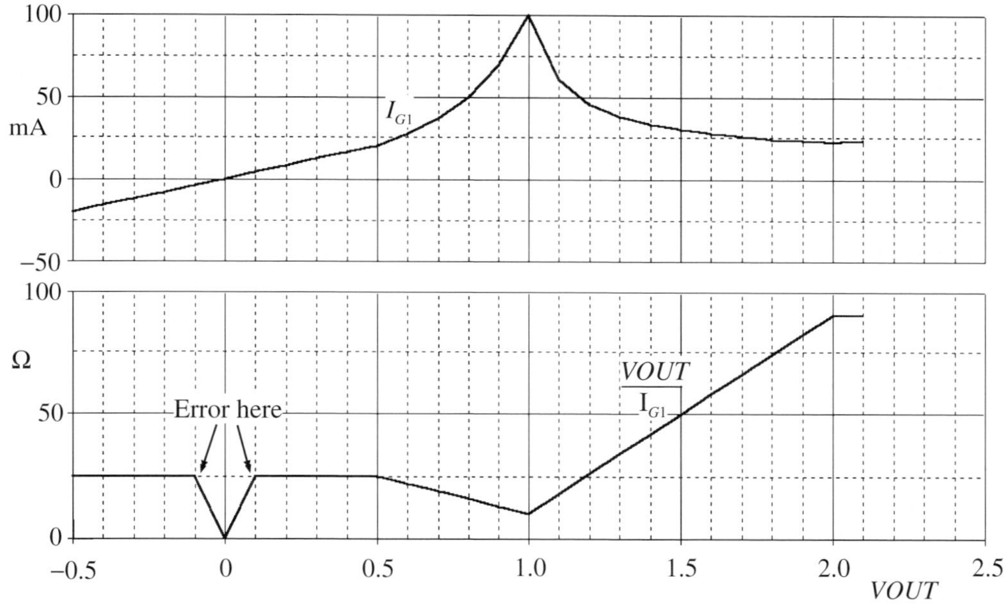


**Fig. 4.1.3** Simulation results for the voltage-controlled resistor model.

$$\text{If} \quad R = bV, \quad \text{where } b \text{ is a constant then} \quad I = \frac{V}{R} = \frac{V}{bV} = \frac{1}{b} \tag{4.1.3}$$

If $R$ is a non-linear function of $V$ then $I$ will be the inverse of this function. Figure 4.1.3 shows an example for a non-linear set of points and also illustrates the glitch in the resistance plot adjacent to zero volts.

The application note also illustrates a similar configuration for use as a temperature-dependent resistor as shown in Fig. 4.1.4. Thermistors will be examined in Section 4.8 but the model there is not sensitive to self-heating so we include it here.

The relation between resistance $R$ and absolute temperature $T$ for thermistors (n.t.c.) can be represented by (Hyde 1971, p. 13):

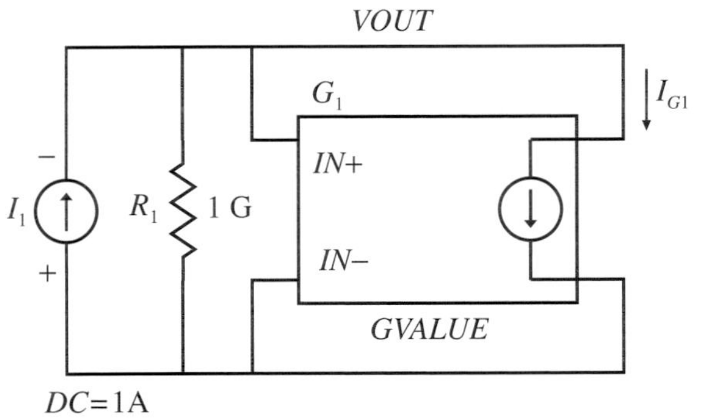


**Fig. 4.1.4** Model for a temperature-dependent resistor.

$$R = R_a \exp\left[B\left(\frac{1}{T} - \frac{1}{T_a}\right)\right] = R_a \exp\left[-B\left(\frac{T - T_a}{TT_a}\right)\right] \tag{4.1.4}$$

where $R_a$ is the resistance at absolute temperature $T_a$ and $B$ is a constant. For SPICE the temperature is represented by *TEMP* and is in celsius, so we must convert to the absolute kelvin scale by adding 273.15. The correspondence with Hirasuna's parameters and the expression for the *ABM* are:

*rho* ≡ $R_a$ = 10 k   *beta* ≡ $B$ = 3435   $T_a$ = *TEMP*0 = 298

*V*(%*IN*+, %*IN*-)/(*rho**exp(-*beta**(*TEMP* + 273.15 - *TEMP*0))/ ((*TEMP* + 273.15)**TEMP*0))) (4.1.5)

The parameters are set with a *PARAM* statement, the current source is set to say 1A and a *DC SWEEP* with temperature as the variable is run. Plotting *V*(*VOUT*)/*I*(*I*1) gives the resistance *R* as shown in Fig. 4.1.5. A too small step size gives a rather noisy *B* trace.

The coefficient of resistance $\alpha$ is given by (see Eq. (1.4.3) and use Eq. (4.1.4)):

$$\alpha = \frac{1}{R}\frac{dR}{dT} = \frac{d}{dT}\ln\left(\frac{R}{R_a}\right) = \frac{-B}{T^2} \quad \text{so plot the function}$$

*D*(log(*V*(*VOUT*)/*I*(*I*1)*1*E*4)   for $\alpha$, and for $B$ plot (4.1.6)

(*TEMP* + 273.15)*(*TEMP* + 273.15)**D*(log(*V*(*VOUT*)/*I*(*I*1)*1*E*4)

and the plots for $\alpha$ and $B$ are also shown in Fig. 4.1.5. As indicated by the cursors $\alpha = 0.0387$ and $B = 3435$ at a temperature of $T = 298 - 273.15 = 24.85$ °C, as they should.

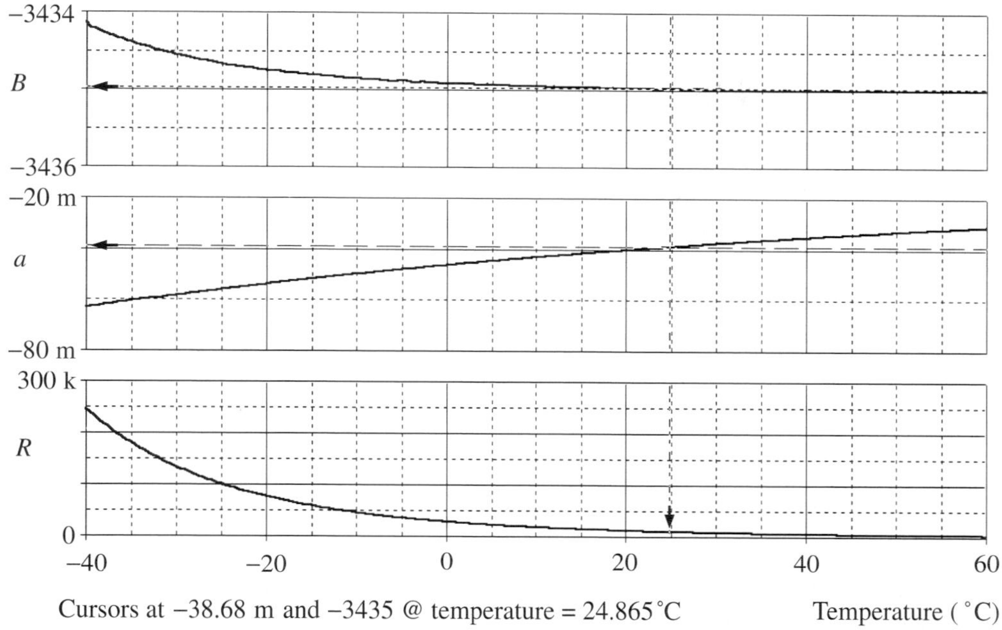


**Fig. 4.1.5** Variation with temperature of resistance of model of Fig. 4.1.4.

## SPICE simulation circuits

| | |
|---|---|
| Fig. 4.1.1 | Restcf.SCH |
| Fig. 4.1.3 | Vctl_res.SCH |
| Fig. 4.1.5 | Tctl_res.SCH |

# References and additional sources 4.1

Bryant J. (1991): Ask the applications engineer – 10 (on resistors). *Analog Dialogue* **25-2**, 24–25. See also: Ask the applications engineer. *30th Anniversary Bonus* 1997, 16–19.

'Cathode Ray' (1953): Ohm's law. Do we know what it really means? *Wireless World* August, 383–386.

Daire A. (1998): *Improving the Repeatability of Ultra-High Resistance and Resistivity Measurements*, Keithley Instruments White Paper.

Grant D., Wurcer S. (1983): Avoiding passive-component pitfalls. *Analog Dialogue* **17-2**, 6–11. See also: The best of *Analog Dialogue* 1967–1991. *Analog Devices* 1991, 143–148.

Hirasuna B. (1999): *Modeling Voltage-Controlled and Temperature-Dependent Resistors*, OrCAD Technical Note, No. 90.

Hyde F. J. (1971): *Thermistors*, London: Iliffe Books. ISBN 0-592-02807-0.

Mancini R. (1999): *Understanding Basic Analog – Passive Devices Application Report*, Texas Instruments Literature Number SLOA027, July.

# 4.2 Capacitors

There is only one good, knowledge, and one evil, ignorance.

Socrates (469–399 BC)

In electronic circuits charges are stored in a capacitor formed by two conductors. If the capacitor is isolated, i.e. there are no other external fields other than that arising from the charges on the two conductors, the charges on the two conductors must be equal in magnitude and opposite in sign since by Gauss' theorem the total charge must be zero. The charges must reside on the surface of the conductors and there can be no electric field parallel to the surface. We may find the potential difference between the conductors by determining the work that must be done to take a test charge from one conductor to the other. To make the derivation simple we choose a parallel plate capacitor of large area $S$ and separation $d$. Since capacity is a geometric constant it does not matter what form we use. The field $E$ between the plates is given in terms of the surface charge density $\sigma$, which in this case is related to the charge on a plate $Q$ and its area $S$, as $\sigma S = Q$:

$$E = \sigma/\varepsilon_0 \tag{4.2.1}$$

so that the work done in moving a charge $q$ from plate to plate between which there is a potential difference $V$, is $U = qV$, which will be given by integrating the force acting ($qE$) over the distance $d$:

$$qV = \int_0^d Eq\,dl = \frac{q\sigma}{\varepsilon_0}\int_0^d dl = \frac{q\sigma d}{\varepsilon_0} = \frac{qQd}{S\varepsilon_0}$$
$$\text{or} \quad V = \frac{Q}{C}, \quad \text{where} \quad C = \frac{S\varepsilon_0}{d} \tag{4.2.2}$$

and $C$, a geometrical factor, is called the capacity. The SI unit of capacity is the farad (F). As an example, two plates, each 10 mm square and separated by 1 mm will have a capacity according to (4.2.2) of:

$$C = \frac{10^{-2} \times 10^{-2} \times 8.85 \times 10^{-12}}{10^{-3}} = 0.9\ \text{pF} \tag{4.2.3}$$

The capacity of two concentric spheres of inner radius $R_1$ and outer radius $R_2$ is found to be (Corson and Lorrain 1962):

$$C = 4\pi\varepsilon_0 \left( \frac{R_2 R_1}{R_2 - R_1} \right) \quad \text{or for an isolated sphere } (R_2 = \infty) \quad C = 4\pi\varepsilon_0\, R_1 \tag{4.2.4}$$

For an isolated 10 mm radius sphere we find $C = 1.1$ pF and for the Earth of radius 6400 km, $C = 716$ μF. If $R_1$ is very small, as for example for an atomic-sized point on the end of a needle (say $10^{-8}$ m), then the electric field can readily exceed $10^8$ V m$^{-1}$ above which field emission occurs and electrons are emitted. This is the basis of the field-emission microscope, for example.

The energy $U$ stored in the capacitor is given by:

$$U = \int_0^Q V\, dQ = \int_0^Q \frac{Q}{C} dQ = \frac{Q^2}{2C} = \tfrac{1}{2} C V^2 \tag{4.2.5}$$

and the unit of energy is the joule (J).

The capacity may be substantially increased by using a suitable dielectric rather than free space. The property of the material is called the dielectric constant or relative permittivity ε, which can vary from unity (free space) to many thousand. The capacity, assuming that the dielectric fills the space between the plates, multiplies the value by ε, a complex quantity and which is written:

$$\varepsilon = \varepsilon' - j\varepsilon'' \tag{4.2.6}$$

At low frequency, $\varepsilon''$ is generally negligible and the capacitor behaves as we would expect. At high frequency, $\varepsilon''$ becomes significant and results in a component of the current through the capacitor in-phase with the applied voltage and so to power loss.

There are many types of capacitor and it is necessary to appreciate their characteristics so that they can be used effectively. The type of dielectric used and the physical structure determine their performance particularly as a function of frequency. We can divide them into three categories: ceramic dielectric, polymer film dielectric and electrolytic.

There are a wide range of ceramic materials with a wide range of dielectric constant, temperature coefficient and loss characteristic. Ceramics with a low temperature coefficient have a relatively low dielectric constant (NPO) and typically range from zero to say 1 nF. Other ceramics have a very high dielectric constant of the order of thousands so allowing substantial capacity in a small volume, but since their temperature coefficient is large they cannot be used where close tolerance is required. However, they are very good for decoupling applications.

Polymer dielectric capacitors have greater stability particularly for higher capacitance values but since their dielectric constants are low, around three, they have

Table 4.2.1 *Properties of some dielectrics*[a]

| Dielectric | Permittivity, $\varepsilon$ | Dissipation factor, *DF* | Absorption, *DA* % | Refractive index, *n* |
|---|---|---|---|---|
| Air or free space | 1 | 0 | 0 | 1 |
| Polyethylene | 2.3 | $<0.0002$ | – | 1.5 |
| Polystyrene (PS) | 2.5 | 0.0002 | 0.01 | 1.6 |
| Polysulphone | 3.1 | 0.0006 | – | 1.76 |
| Polyimide | 3.4 | 0.005 | – | 1.84 |
| Polyester/mylar (PET) | 3.3 | 0.006 | 0.2 | – |
| Polycarbonate (PC) | 2.8 | 0.002 | 0.06 | – |
| Polypropylene (PP) | 2.2 | 0.0002 | 0.01 | – |
| Polyethylene naphthalate (PEN) | 3.0 | 0.004 | 1.2 | – |
| Teflon | 2.1 | $<0.0002$ | – | – |
| Quartz | 3.8 | 0.0006 | – | – |
| Silicon | 11.8 | – | – | 3.44 |
| Silicon dioxide | 3.9 | – | – | 1.97 |
| Germanium | 16.0 | – | – | 4 |
| FR4 Epoxy fibreglass | 5 | 0.02 | – | 2.24 |
| Mica | 7 | 0.0014 | – | – |
| COG/NPO ceramic | 6 to 500 | 0.001 | – | – |
| Y5V ceramic | 3000 to 10000 | 0.01 | – | – |
| X7R ceramic | 500 to 4000 | $<0.025$ | – | – |

*Note:*
[a] The values must be taken as approximate as they depend on source of the material, temperature and frequency. They are indicative values culled from many manufacturers' data. $n$ is the refractive index in the optical region of the spectrum and also varies as a function of frequency.

larger volume. These types have low leakage, significant variation in soakage and of loss tangent (see below) with frequency. Some typical dielectric constants are given in Table 4.2.1.

Appropriate frequency ranges for capacitors using the various types of dielectric are indicated in Figs. 4.2.1 and 4.2.2. The latter has been adapted from Ott (1976) with some slight modification. They may all be used at lower frequency than suggested but the efficiency in terms of size and cost will probably be less.

A concise discussion of the characteristics of capacitors and their application is given by Guinta (1966). A résumé of characteristics is given in Table 4.2.2.

A 'real' capacitor can have a fairly complex equivalent circuit if we are to model all its variability, including changes in the dielectric as functions of temperature and frequency. An approximation to the actual circuit is shown in Fig. 4.2.3.

$C$ is the nominal capacity, $R_L$ represents the leakage resistance, $R_{ESR}$ is the

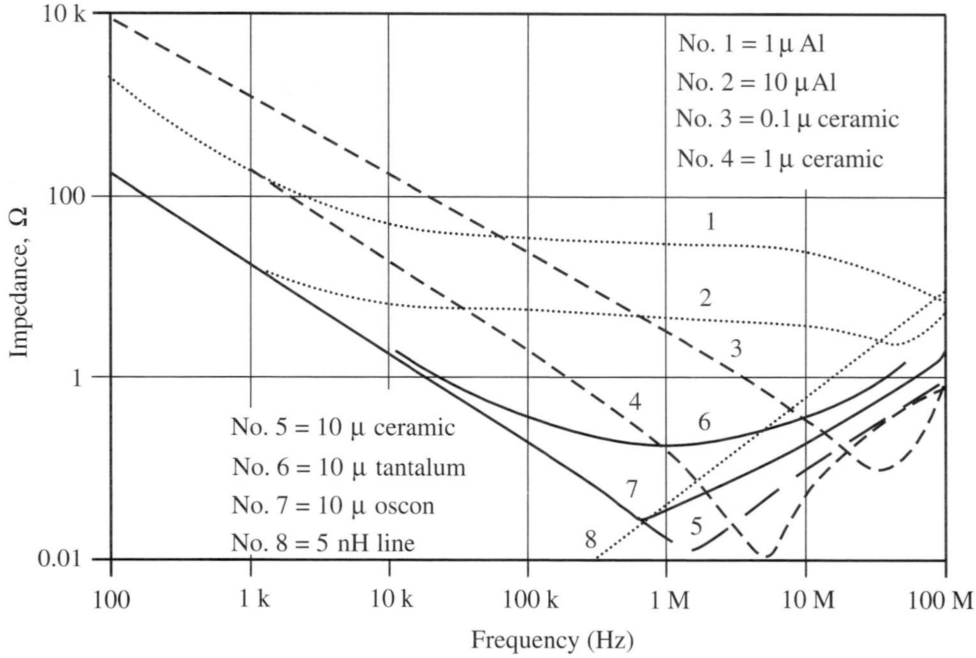


**Fig. 4.2.1** Impedance as a function of frequency.

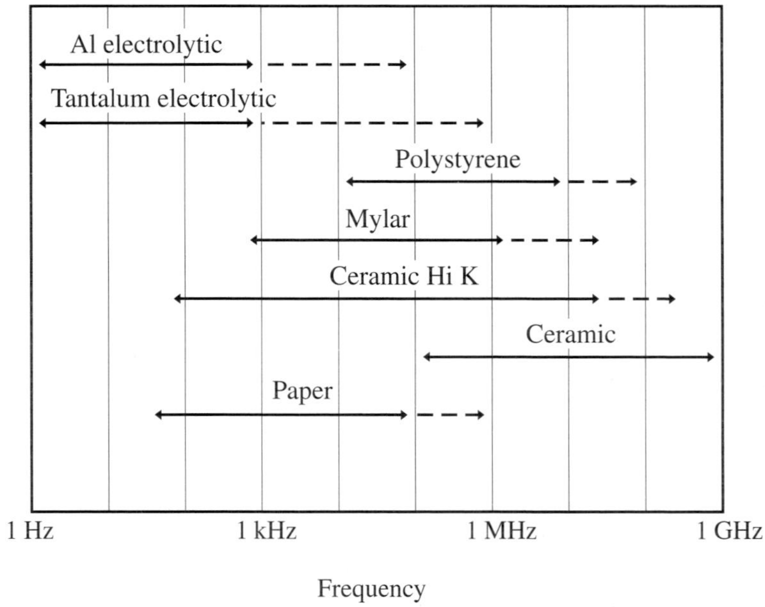


**Fig. 4.2.2** Appropriate frequency ranges for various types of capacitor.

Table 4.2.2 *Characteristics of capacitors (after Guinta 1996; adapted with permission of Analog Devices)*

| Capacitor type | *DA*[a] | Advantages | Disadvantages |
|---|---|---|---|
| NPO ceramic | <0.1% | Small case size<br>Inexpensive, good stability<br>Wide range of values<br>Many vendors, low inductance | *DA* low but not specified<br>Up to about 10 n only |
| Hi K ceramic | >0.2% | Low inductance<br>Wide range of values | Poor stability and *DA*<br>High voltage coefficient |
| Polystyrene | 0.001% | Inexpensive, low *DA*<br>Wide range of values<br>Good stability | Temperature max. 85 °C<br>High inductance |
| Polypropylene | 0.001% | Inexpensive, low *DA*<br>Wide range of values | Temperature max. 105 °C<br>High inductance |
| Teflon | 0.003% | Good stability, low *DA*<br>Wide range of values<br>Temperature to above 125 °C | Relatively expensive<br>Large size<br>High inductance |
| Polycarbonate | 0.1% | Good stability, low cost<br>Wide temperature range | Large size, poor *DA*<br>High inductance |
| Polyester | 0.3% | Moderate stability, low cost<br>Wide temperature range<br>Low inductance for stacked film | Large size, poor *DA*<br>High inductance |
| Mica | 0.003% | Low loss at HF, very stable<br>Low inductance | Quite large, expensive<br>Low values (<10 n) |
| Aluminium electrolytic | High | Large values, high currents<br>High voltages, small size | High leakage, poor stability<br>Poor accuracy, inductive<br>Usually polarized |
| Tantalum electrolytic | High | Small size, large values<br>Medium inductance | Quite high leakage, expensive<br>Poor stability and accuracy<br>Usually polarized |
| MOS | 0.01% | Small size, good *DA*, low inductance, use to above 125 °C | Limited availability<br>Low capacity values |

*Note:*
[a] *DA*, dielectric absorption.

equivalent series resistance, $L_{ESR}$ is the equivalent series inductance, and $R_{DA}$ and $C_{DA}$ represent the dielectric absorption. The last arises from a polarization effect in the dielectric with a rather long time constant. If, after charging, the capacitor is shorted for a short time so that there should be no remaining charge, then when it is open circuited a voltage develops due to the long relaxation time-constant of the dielectric (Dow 1958; Stata 1967; Demrow 1971; Pease 1998). $R_L$ and $R_{ESR}$ are

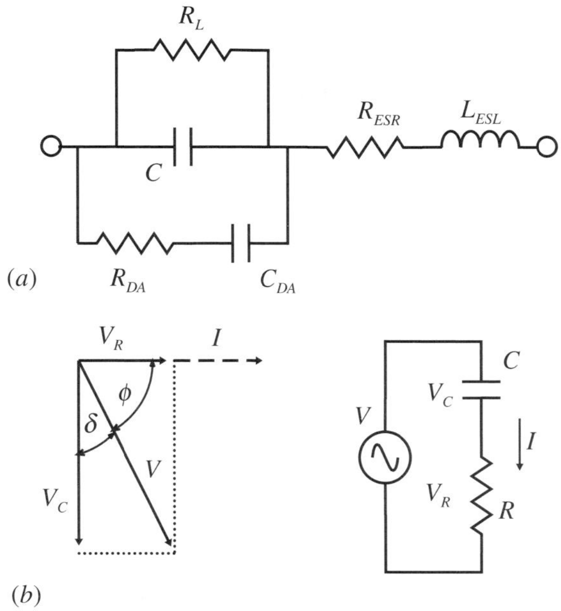


**Fig. 4.2.3** (*a*) Approximate equivalent circuit for a capacitor. (*b*) Illustrating capacitor losses and dissipation factor *DF*.

important for the electrolytic capacitors (see Section 5.10, for example), and for capacitors used at high frequencies $C$ and $L_{ESR}$ will eventually form a series resonant circuit, which is responsible for the minimum in the impedance curves shown in Fig. 4.2.1. At higher frequency the impedance becomes inductive so the capacitor becomes less effective in fulfilling its purpose.

An ideal capacitor has a 90° phase difference between the applied voltage and the current. Any deviation from ideal results in a change in the phase angle and the magnitude of the difference is a measure of the deviation. The capacitor now has a component of the current in phase with the voltage and hence power will be dissipated. Taking a simple effective circuit as shown in Fig. 4.2.3(*b*) and applying a voltage $V$, then if the current flowing is $I$ we may draw the phasor diagram as shown. For a good capacitor, $V_R$ will be very much smaller than $V_C$. The tangent of angle $\delta$ is called the loss tangent and the cosine of angle $\phi$, or sin of angle $\delta$, is called the power factor. The $Q$ of the capacitor is defined as the ratio of reactive to resistive impedance.

$$Z = R + j\omega C, \quad V_C = \frac{I}{j\omega C}, \quad V_R = IR$$

$$\text{so} \quad \tan\delta = \frac{V_R}{V_C} = \omega CR = DF \quad \text{and} \quad \cos\phi = \sin\delta = \frac{V_R}{V} = \frac{R}{Z} = \text{Power factor} \tag{4.2.7}$$

$$\text{Power dissipation} = VI\cos\phi, \quad \text{which tends to zero as } R \to 0$$

$$Q = \frac{X}{R} = \frac{1}{\omega CR} = \frac{1}{\tan\delta} = \frac{1}{DF}, \quad \text{which tends to infinity as } R \to 0$$

The small surface mount ceramic capacitors are the most effective since their size gives them the lowest inductance and allows them to be placed very close to the point to be decoupled so minimizing the inductance of any connecting lead. Larger capacity values have lower resonant frequencies so it is often desirable or necessary to use an additional smaller value in parallel to maintain the low impedance to high enough frequency. It should be remembered, however, that there are two terminals and the lead to common is equally important. Thus the strong encouragement to use ground (common) planes to reduce inductance. Normally of course the voltage supply line will be strongly decoupled to common, in which case, from the signal point of view, the supply and common are the same. For high frequency applications very careful PCB layout, power planes and miniature surface mount components are almost mandatory.

A convenient design guide that allows the SPICElike modelling of both ceramic and tantalum capacitors is available and may be downloaded from the web (Prymak 2000). It allows selection of specific devices, displays their performance with regard to temperature, frequency and voltage, and draws an equivalent circuit. The last entails at least 13 elements in the equivalent circuit (Fig. 4.2.4) for a tantalum capacitor! For higher frequencies (1 MHz+) there is a design aid from Johanson (1999).

Voltage-dependent capacitors, or varactors (e.g. Chadderton 1996), are commonly used in tuning at higher frequencies and so a model for such devices is desirable. Some specific types are available in the SPICE libraries and one is used in Section 5.26 to provide both non-linearity and time delay for a chaotic circuit. One realization has been outlined in Section 5.27(f) and another is given by Hirasuna (1999). In such a capacitor Eqs. (4.2.2) and (4.2.5) no longer hold so we need to re-examine the relationship. If we represent the capacity as a function of voltage by $C(V)$, then at any particular voltage $V$ the current $I$ will still be given by:

$$I = C(V)\frac{dV}{dt} \tag{4.2.8}$$

and the charge $Q$ and energy $U$ stored in the capacitor by (assuming $V=0$ at $t=0$, and $V=V$ at $t=t$):

$$\begin{aligned} Q &= \int_0^t I\,dt = \int_0^t C(V)\frac{dV}{dt}dt = \int_0^V C(V)dv \\ U &= \int_0^t VI\,dt = \int_0^t VC(V)\frac{dV}{dt}dt = \int_0^V VC(V)dv \end{aligned} \tag{4.2.9}$$

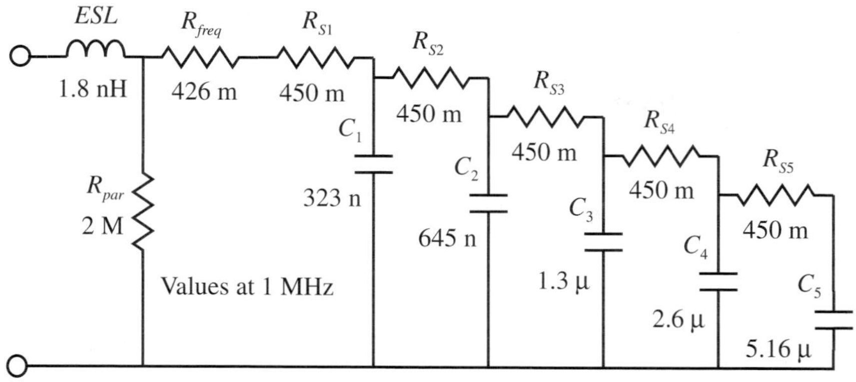


**Fig. 4.2.4** Equivalent circuit for a 10 μ tantalum capacitor at a frequency of 1 MHz (after Prymak 2000).

The form of the relation between capacity and voltage is now required to evaluate these quantities. For SPICE we may use either a table of values or a polynomial relation. To follow the example of Hirasuna we can take a second order polynomial:

$$C(V) = C_0 + C_1 V + C_2 V^2$$

then $$Q = \int_0^V C(V)dV = C_0 V + \tfrac{1}{2} C_1 V^2 + \tfrac{1}{3} C_2 V^3$$

and $$U = \int_0^V VC(V)dV = \tfrac{1}{2} C_0 V^2 + \tfrac{1}{3} C_1 V^3 + \tfrac{1}{4} C_2 V^4 \qquad (4.2.10)$$

and the first term in each expression matches those in (4.2.2) and (4.2.5), as they should. To compare with the example of Hirasuna we may calculate from his table example to find the equivalent values of the polynomial coefficients to be:

$$C_0 = 1\ \mu \qquad C_1 = 10\ \text{n} \qquad C_2 = 100\ \text{p} \qquad (4.2.11)$$

The SPICE model, using an *ABM VCCS* device, is then as shown in Fig. 4.2.5. The output current is based on Eq. (4.2.8) and is specified by the *GVALUE* statement as shown, where *DDT* is the differential with respect to time. A word of warning, however. It is necessary to ensure that the applied voltage is always positive as otherwise the term in $C_1$ will change sign and give a quite different variation, though the $C_2$ term will of course be unaffected. This would be consistent with the practical implementation of a varactor, which is in effect a reverse biased

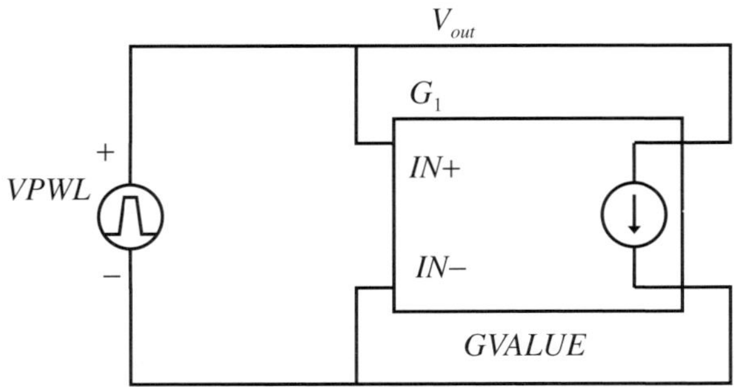


(C0+C1*V(%IN+,%IN-)+C2*PWRS(V(%IN+,%IN-), 2)*DDT(V(%IN+,%IN-)))

(C0/(PWR(1+0.462*V(%IN+,%IN-), M))*DDT(V(%IN+,%IN-)))

**Fig. 4.2.5** PSpice model for a varactor. The specifications for the two examples are shown.

diode, and which would normally not be used in the forward direction. If the model were to be used in this sense then an additional diode component will need to be added.

There are several different styles of varactor with greater or lesser rates of variation of capacity and an alternative form of representation is often used:

$$C(V) = \frac{C_{J0}}{\left(1 + \frac{V}{VJ}\right)^{M}} \tag{4.2.12}$$

where $C_{J0}$ is the capacity at zero applied volts $V$, $VJ$ is the barrier potential and $M$ a grading coefficient that indicates the type of diffusion profile. As an example we will examine the ZC830 hyper-abrupt varactor for which there is a model. Examining its model in the library (place a ZC830/ZTX on your schematic, select it, select *Edit Model* and then *Edit Instance Model* (*Text*) to get the listing), we find the parameters:

$$CJ0 = 19.15\text{E}-12, \quad VJ = 2.164, \quad M = 0.9001 \tag{4.2.13}$$

Using (4.2.8) we may now write the expression that defines the current output of the *G* device (a *VCCS*) as:

(CJ0/(PWR(1-(1/VJ)*V(%IN+, %IN-), M)))*DDT(V(%IN+, %IN-)) (4.2.14)

where *DDT* is the differential with respect to time, and declare the *Param*(eters) of Eq. (4.2.13). To get a suitable *dV/dt*, and to get the correct sign of current, we set the *VPWL* generator to produce a ramp of 0 to −30 V in 3 μs, i.e. a rate of $10^7$ V s$^{-1}$. This also accounts for the change of sign before the (1/*VJ*) term com-

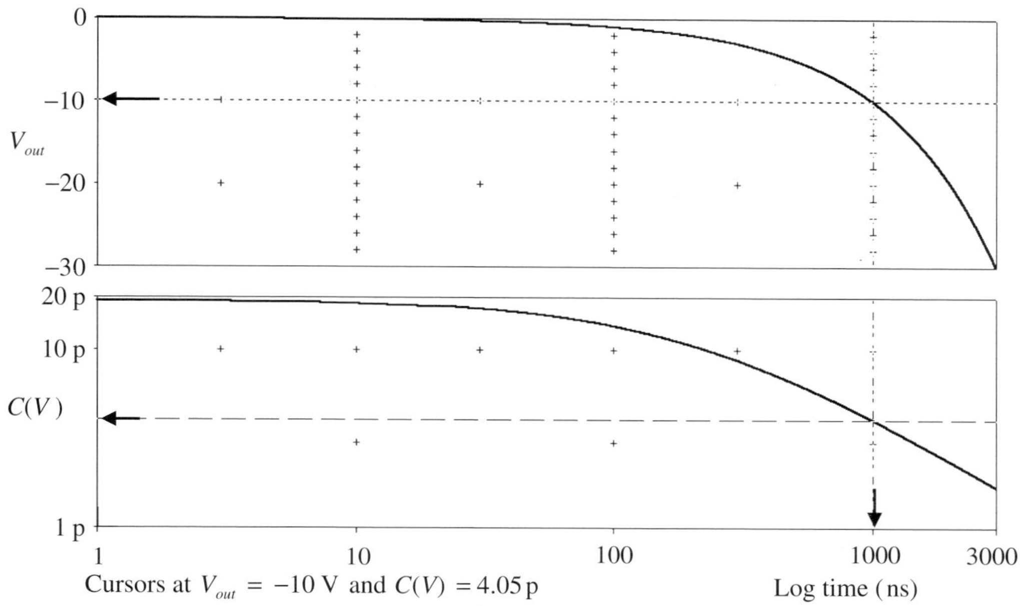


**Fig. 4.2.6** Simulation of a ZC830 using the circuit of Fig. 4.2.5.

pared with Eq. (4.2.12). The results of the PSpice simulation are shown in Fig. 4.2.6. The logarithmic time scale makes the linear voltage ramp appear curved and if we plot $I(G1)/D(V(Vout)) = C(V)$ to get the effective capacity the scale in pA will be equivalent to pF.

The cursor value agrees with the value directly calculated from Eq. (4.2.12).

## SPICE simulation circuits

| | |
|---|---|
| Fig. 4.2.6 | Vctlcap.SCH |

# References and additional sources 4.2

AVX Corp.: *Capacitor Design Software Allspiap.exe*; www.avxcorp.com

Chadderton N. (1996): *Zetex Variable Capacitance Diodes*, Zetex Application Note 9, January, www.zetex.com

Corson D. R., Lorrain P. (1962): *Introduction to Electromagnetic Fields and Waves*, San Francisco: W. H. Freeman. Library of Congress Cat. No. 62-14193. See p. 61.

Demrow R. (1971): Op amps as electrometers or – the world of fA; *Analog Dialogue* **5**, February/March, 6–7.

Dow P. C. (1958): An analysis of certain errors in electronic differential analyzers. II Capacitor dielectric absorption. *IRE Trans.* **EC-7**, 17–22.

Guinta S. (1966): Ask the applications engineer – 21, Capacitance and capacitors. *Analog Dialogue* **30**(2), 18–21. See also: Ask the applications engineer. *30th Anniversary Bonus* 1997, 46–49.

Hirasuna B. (1999): *A Nonlinear Capacitor Model for use in PSpice*, OrCAD Technical Note No. 88, www.orcad.com

Johanson Technology (1999): *MLCsoft*, www.johanson-caps.com

Ott H. W. (1976): *Noise Reduction Techniques in Electronic Systems*, New York: John Wiley.

Pease R. A. (1998): What's all this soakage stuff, anyhow? *Electronic Design* May 13th, pp. 125, 126, 128.

Prymak J. (2000): *Kemet Spice Simulation Software*, Kemet Electronics Corp. www.kemet.com

Schaffner G. (1965): *High-Power Varactor Diodes. Theory and Application*, Motorola Application Note AN-147.

Stata R. (1967): Operational integrators. *Analog Dialogue* **1** (1), April, 6–11.

# 4.3 Inductance

Information is endlessly available to us; where shall wisdom be found.

Harold Bloom in *How to Read and Why* (Fourth Estate)

The requirement to create a magnetic field around any conductor carrying a current means that there will be some impedance to the flow of current. This impedance is called inductance which acts to oppose any change in current by generating an opposing e.m.f. and is a manifestation of Faraday's law (Section 2.4). The fields surrounding a conductor have been examined in Section 2.5. Energy is stored in the magnetic field and for an ideal inductor with no resistance all the energy supplied to the inductor will be stored in the field. If there is a voltage $V$ across the inductor when a current $i$ is flowing producing a magnetic field flux $N$, then the rate of supply of energy $U$ is (the signs are not relevant here):

$$\frac{dU}{dt} = iV = i\frac{dN}{dt} = Li\frac{di}{dt} \tag{4.3.1}$$

Then the total energy $U$ (in joule) supplied to the inductor to establish the final current $I$ will be:

$$U = \int_0^I Li\frac{di}{dt}dt = L\left(\frac{i^2}{2}\right)_0^I = \tfrac{1}{2}LI^2 \tag{4.3.2}$$

Inductances may be computed from knowledge of the current distribution and the geometry of the inductor but the sums are complex other than in very simple configurations. As will be seen from the following examples there are usually correction factors to be applied depending on geometry. Most computations of the inductance of circuits assume that the currents are filamentary, that is that the cross-section dimension of the conductors is in effect zero. This does lead to conceptual difficulties since the field $B$ would tend to infinity (see Section 2.5; Corson and Lorrain 1962, p. 233). For finite sized circuits it is often assumed that the conductor cross-section is negligible but with close wound conductors, for example, the size is important. It is also worth noting that there is inductance (internal inductance $L_i$) related to the magnetic flux within the conductor which is effectively independent of the shape of the circuit (Ramo and Whinnery 1944, p. 255; see also Ramo et al. 1965). However, at higher frequencies, where the current is concentrated at the surface, the internal contribution will diminish. At these frequencies it is found that $\omega L_i \approx R_{hf}$, which is given in (2.8.12).

Many books provide formulae and graphs for the inductance of a wide range of winding shapes and wire forms (Terman 1950; Grover 1946 and 1973; Langford-Smith 1954; Snelling 1969 and 1988; Postupolski 1995). The inductance of an isolated straight wire of length $l$ and diameter $d$ (both in m) is given by (Terman 1950, p. 48):

$$L = 0.2l\left[\ln\left(\frac{4l}{d}\right) - 0.75\right]\ \mu\text{H} \tag{4.3.3}$$

e.g. a 25 cm length of 0.5 mm diameter wire has an inductance of 0.35 μH. If the wire is turned back on itself so that the field from one part can couple with another more effectively, i.e. a coil, then the inductance is increased. As more turns are added the inductance increases approximately by the square of the number $n$ of turns. It should be proportional to $n^2$, but since the wire is of finite size there are geometric factors that reduce the coupling. The inductance for a single turn is given by:

$$L = 0.628D\left[\ln\left(\frac{8D}{d}\right) - 2\right]\ \mu\text{H} \tag{4.3.4}$$

The same length of wire as above gives an inductance of 0.26 μH, suggesting that a largish diameter ($D \approx 8$ cm in this case) does not increase inductance. These measurements were checked using an *RLC* bridge at 400 kHz. The straight wire gave 0.34 μH while the single turn loop gave 0.2 μH, so the formulae are about right; but for such low inductances stray effects make accurate measurement difficult. With one, or only a few turns, end effects become significant and one wonders about the question of when does a 'turn' become a turn. For a multiturn single-layer close-wound solenoid of $n$ turns, the inductance is given by the Nagaoka formula (1909):

$$L = \frac{C(d/2)^2\, n^2}{l} K \tag{4.3.5}$$

where $C$ is a numerical constant and $K$ a constant depending on the ratio of diameter $d$ to winding length $l$. The formula is commonly used in the form:

$$L = Fn^2 d \tag{4.3.6}$$

where $F$ (which includes the factor $d/l$) is given in tabular or graphical form (Terman 1950, pp. 54–55). The values there must be multiplied by 39.37 (conversion of inch to metre) to give the value when $d$ is in m. Taking the same diameter and wire size as used for the single turn coil we find for two turns $L = 1.06\ \mu$H, i.e. $n^2 = 4$ times the single turn value. Formulae for multilayer solenoids, either short or long, become more difficult to derive but in general the inductance will be somewhat less than you would expect from a simple application of Eq. (4.3.6) with the requisite number of turns. A simpler relation for cases where $l \geq 0.8\ r$, where $r$ is the radius, is given by ($r$ and $l$ in m):

$$L = \frac{39.37 r^2 n^2}{9r + 10l}\ \mu\text{H} \tag{4.3.7}$$

For a *long multilayer* solenoid of length $b$, mean winding radius $a$ and radial winding thickness $c$, the inductance is given by (dimensions in m):

$$L = Fn^2 d - \frac{1.286 n^2\, ac(0.693 + B_s)}{b}\ \mu\text{H} \tag{4.3.8}$$

where the correction quantity $B_s$ depends on the ratio $b/c$ (Terman 1950, p. 60). For $b/c = 5$, 10, 20, $B_s = 0.23$, 0.28, 0.31. The inductance is evidently less than given by Eq. (4.3.6), demonstrating the decrease in coupling. An alternative form is given by Grover (1946), (or Snelling 1969, p. 348), but this also requires reference to graphical relationships. An approximate equivalent to Eq. (4.3.8) that does not require graphical reference is (dimensions in m):

$$L = \frac{31(an)^2}{6a + 9b + 10c}\ \mu\text{H} \tag{4.3.9}$$

In practice you should not expect very close correspondence between calculated and actual inductance, and at high frequencies even less so. In the more distant past of radio, the calculation of inductance for all sorts of shape and type of wire was an important matter, so there is a huge literature covering this and you should be able to find something appropriate (see Terman 1950; Langford-Smith 1954 and many references therein). If possible, measure the inductance at the frequency at which it will be used. Good modern *RLC* bridges have the facility to set the measurement frequency in quite close steps.

A further consideration, particularly at higher frequencies, is the inherent self-capacity of the winding so that the effective equivalent circuit is now of an inductor and resistor in series with a capacitor overall in parallel, i.e. just a parallel resonant circuit (Section 3.5). Thus the 'inductor' will resonate at some frequency and hence substantially change the effective impedance and phase change. Calculation of this stray capacity is very difficult but some discussion is given by Snelling (1969, pp. 350–354). It is as well to measure the self-resonant frequency to ensure that it is well above the region of interest. This can be readily done by connecting the inductor in series with a resistor (a few k$\Omega$) across a signal generator and measuring the voltage across the inductor as a function of frequency. If the stray capacity is small some correction should be made for the oscilloscope probe capacity.

A general drawback with inductors is that they do produce a surrounding magnetic field which can couple with other parts of the circuit. Conversely they will also pick up fields from some other source. Inductors must therefore be used with caution. To obtain higher inductance and/or to screen the fields, inductors are often wound on magnetic materials which have substantial permeability. This enables the required inductance to be obtained with fewer turns and, if the magnetic circuit is closed, to diminish the stray coupling effects greatly. Shapes such as pot cores or toroids keep the fields almost completely contained within them so that coupling in either direction is largely eliminated. Pot cores can also be provided with adjusters which allow the inductance to be set closely to the required value in-circuit. The properties of some magnetic materials are discussed in Section 2.11. An extensive discussion of the use of ferrites is given by Snelling (1969, 1988).

The design of inductors using closed ferrite structures, such as pot cores or the various forms of transformer core, is fairly straightforward as the manufacturers provide the appropriate parameters and design information for the particular core. Many types and values of inductor both fixed and variable and through hole or surface mount are nowadays available as standard components (see references).

When winding an inductor it is desirable to try to estimate the number of turns required. A test winding will give a first approximation of the 'constant' for the configuration and then the estimate can be made on the basis that the inductance is proportional to the square of the number of turns. This will always be an underestimate owing to geometrical factors affecting the coupling between turns. As can be seen from Fig. 2.5.1 the field surrounding a wire falls off as $1/r$, i.e. a steeper fall off at small distances ($dB/dr \propto 1/r^2$), so that even small gaps can have a significant effect. At higher frequencies the skin effect changes the distribution of current in the wire and hence the coupling to adjacent wires which changes the inductance. For a bunched winding, personal experience suggests that you should increase the calculated number of turns by about 7 to 10% so that you can then remove turns to get the desired inductance. If too low, having to join on extra wire is a bit of a pain.

In PSpice the standard inductor $L$ is ideal, i.e. it has no resistance (d.c. or skin) and no self-capacity. In using such a device in simulation you will get results that may deviate significantly from reality and it is necessary to add components to represent the deviations from the ideal. The primary addition should be a series resistance that prevents the inductor having an infinite $Q$. At higher frequencies the resistance becomes frequency dependent owing to the skin effect (Section 2.8), and if the inductor includes magnetic material, such as ferrite, then allowance should also be made for losses in this. The self-capacity of the inductor means that it will be self-resonant at some frequency, above which it will become capacitive rather than inductive. On similar lines to the SPICE models described for voltage-dependent resistors or capacitors, it is possible to construct a model for either r.f. inductors (coil core losses negligible) and for power inductors (Coilcraft 1999a,b). The model circuit is shown in Fig. 4.3.1(*a*) for r.f. (1MHz and up) and in (*b*) for power inductors (up to 100 kHz). *EXPR* and *XFORM* define the *GLAPLACE* device and the factor 0.159 is just $1/2\pi$.

The *GLAPLACE* device generates an effective resistance that is proportional to $\sqrt{f}$ to represent the increase of resistance due to the skin effect. Taking as an example the type 1812CS-333 surface mount device, which has the parameters $R_1=200\ \Omega$, $R_2=0.001\ \Omega$, $L_1=33\ \mu\text{H}$, $C_1=1.1$ p, $Q>20$ at 10 MHz, $SRF>20$ MHz, we obtain Fig. 4.3.2 from simulation.

This shows the total current $I(R2)$ and that in each of the two arms. The upper response is the effective impedance showing the self-resonant frequency at about 26.2 MHz with a $Q$ of 19. If you plot the impedance of $G_1$ you will find a value of 111 Ω at the resonant frequency and the phase of this as 45°. Adding this vectorially to $R_1=200$ gives an effective resistance of 278 Ω and thence a calculated $Q\approx 19$. From the resonance curve the $Q$ derived from the width at $1/\sqrt{2}$ of the peak will be found to be the same.

When two (or more) inductances are close enough we have mutual coupling, with the field of each linking with the other leading to an additional inductive impedance. The relationship between the individual inductances and the new mutual inductance can be derived for conductors of any shape, but we will consider a particular geometry that will make the sums simpler. Consider two solenoidal windings as shown in Fig. 4.3.3.

If the coil $P$ has $N_1$ turns per unit length and carries a current $i_1$ then the flux $\Phi_1$ produced is from (2.5.12) and (2.5.13):

$$\Phi_1=\mu_0 N_1 i_1 A, \quad \text{where the cross-section area } A=\pi r^2 \tag{4.3.10}$$

so the total flux linked to the secondary is then:

$$\Phi_{12}=\Phi_1 N_2 h \tag{4.3.11}$$

The e.m.f. induced in coil $S$ is then from Eq. (2.4.1):

$$\begin{aligned} v_2 &= -\frac{d\Phi_{12}}{dt} = -\mu_0 N_1 N_2 h A\frac{di_1}{dt} \\ &= -M_{12}\frac{di_1}{dt}, \quad \text{where} \quad M_{12}=\mu_0 N_1 N_2 h A \end{aligned} \tag{4.3.12}$$

If we reverse the procedure and pass a current $i_2$ through coil $S$, then the e.m.f. induced in coil $P$ will be:

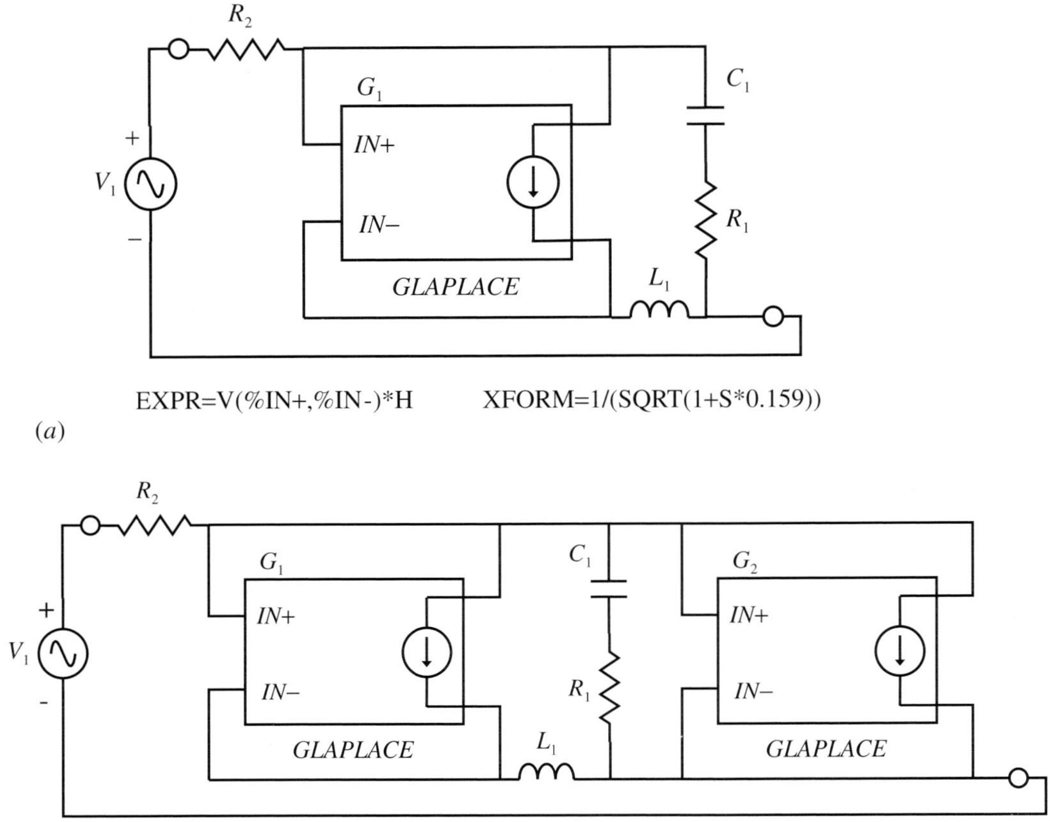


EXPR=V(%IN+,%IN-)*H XFORM=1/(SQRT(1+S*0.159))

(*a*)

(*b*) *EXPR=V(%IN+,%IN-)*H1* *XFORM=1/(SQRT(1+S*0.159))* *EXPR=V(%IN+,%IN-)*H2* *XFORM=1/(SQRT(S*0.159))*

**Fig. 4.3.1** (*a*) Model circuit for r.f. inductors. (*b*) Model circuit for power inductors.

$$v_1 = -M_{21}\frac{di_2}{dt}, \quad \text{where} \quad M_{21} = \mu_0 N_1 N_2 hA$$
$$\text{so} \quad M_{12} = M_{21} = M, \quad \text{say} \tag{4.3.13}$$

$M$ is the mutual inductance and depends on geometry. It is equal to the flux through one coil arising from unit current in the other. From (2.5.14) the two self-inductances $L_1$ and $L_2$ are given by:

$$L_1 = \mu_0 N_1^2 hA, \quad \text{and} \quad L_2 = \mu_0 N_2^2 hA$$
$$\text{so} \quad L_1 L_2 = \mu_0^2 N_1^2 N_2^2 h^2 A^2 = M^2 \quad \text{or} \quad M = (L_1 L_2)^{\frac{1}{2}} \tag{4.3.14}$$

We have assumed all the flux generated by one coil passes through the other. In practice this will not necessarily be true so we can write, using a coupling factor $k$ $(0 \le k \le 1)$:

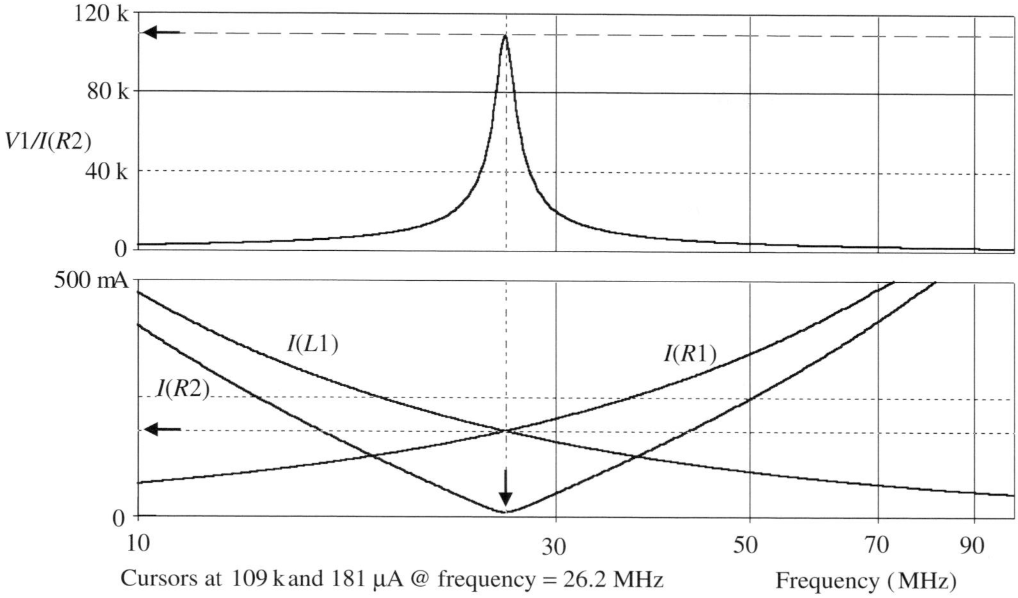


**Fig. 4.3.2** Resistance, reactance and impedance of inductor as a function of frequency.

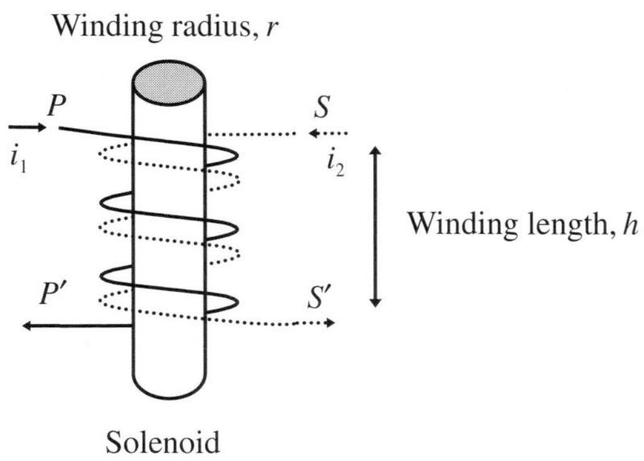


**Fig. 4.3.3** Mutually coupled solenoidal coils.

$$M = k(L_1 L_2)^{\frac{1}{2}} \tag{4.3.15}$$

and this is the form used in PSpice coupling (see Section 4.4).

## SPICE simulation circuits

| | |
|---|---|
| Fig. 4.3.2 | Indctrrf.SCH |

## References and additional sources 4.3


Breen B. N., Goldberger C., Talalaevsky L. (2000): *The Accu-L Multi-Layer Inductor for High Frequency Applications*, AVX Technical Information, www.avxcorp.com

Coilcraft (1999a): *Modeling Coilcraft RF Inductors*, Coilcraft Document 158-1, www.coilcraft.com

Coilcraft (1999b): *Modeling Coilcraft Power Inductors*, Coilcraft Document 209–1, www.coilcraft.com

Corson D. R., Lorrain P. (1962): *Introduction to Electromagnetic Fields and Waves*, San Francisco: W. H. Freeman. Library of Congress Cat. No. 62-14193.

Fair-Rite (1998): *Soft Ferrites Databook*, www.fair-rite.com

Fish G. E. (1989): Soft magnetic materials. *Proc. IEEE* **78**, 947–972.

Giacoletto L. J. (1977): *Electronic Designers Handbook*, New York: McGraw-Hill.

Grover F. W. (1946): *Inductance Calculations*, New York; Van Nostrand.

Grover F. W. (1973): *Inductance Calculations*, New York; Dover.

Langford-Smith F. (1954): *Radio Designer's Handbook*, London: Illife and Sons.

MicroSim (1996): *Application Notes Handbook Ver. 6.3*: Frequency-Domain Modeling of Real Inductors. Model Ferrite Beads in SPICE. Use Ferrite Bead Models to Analyze EMI Suppression.

MMG Neosid (1995): *Soft Ferrites Databook 1A*, MMG Neosid, Letchworth, www.mmg-neosid.com

Nagoaka H. (1909): The inductance coefficients of solenoids. *J. Coll. Sci. Imp. Univ. Tokyo*, **27**(6).

Neosid Pemetzrieder (2000): *Soft Ferrite Components*, www.neosid.de

O'Hara M. (1996): Improved resonant inductor SPICE model. *Electronic Product Design* August, 16, 19.

O'Meara K. (1997): Simple Spice file models all powdered inductors. *Electronic Design* 15 September, 186.

Philips (1998): *Magnetic Products, Soft Ferrites Databook*; Philips Components, www.acm.components.philips.com

*Soft Ferrites and Accessories, 2000 Data Handbook*, www.ferroxcube.com

Postupolski T. W. (1995): Inductor. In Chen, Wai-Kai (Ed.) *The Circuits and Filters Handbook*, Boca Raton: CRC Press and IEEE Press. ISBN 0-8493-8341-2.

Ramo S., Whinnery J. R. (1944): *Fields and Waves in Modern Radio*, New York: John Wiley. (2nd Edn 1953.)

Ramo S., Whinnery J. R., van Duzer T. (1965): *Fields and Waves in Communication Electronics*, New York: John Wiley.

Randall C. (1996): Comment – Resonant inductor modelling in Spice. *Electronic Product Design* December, 19.

Reeves R. (1988): Calculating the field functions of circular coils by direct numerical integration. *J. Phys. E Sci. Instrum.* **21**, 31–39.

Rostek P .M. (1974): Avoid wiring-inductance problems. *Electronic Design* 6 December, 62–65.

Siemens Matsushita (1999): *Ferrites and Accessories Databook*, www.siemens.de/pr/index.htm

Snelling E. C. (1969): *Soft Ferrites. Properties and Applications*, London: Iliffe Books. Second edition Butterworths 1988.

Terman F. E. (1950): *Radio Engineers' Handbook*, New York: McGraw-Hill.

Toko (1998): *Coils and Filters Catalogue C-027001*, Toko Inc, www.tokoeurope.com

Vacuumschmelze GMBH (1998): *Vitrovac 6025Z Tape-Wound Cores for Magnetic Amplifier Chokes*, Vacuumschmelze Application Note PK-002. VAC Magamp Calculator 99 Software, www.vacuumschmelze.de

Wyatt M. A. (1992): Model ferrite beads in Spice. *Electronic Design* 15 October, 76.

# 4.4 Transformers

There seems to be a natural law that regulates the advance of Science. Where only observation can be made, the growth of knowledge creeps; but where laboratory experiments can be carried on, knowledge leaps forward.

Michael Faraday

Transformers can play several roles:

1. Galvanic isolation of one circuit from another while preserving coupling, e.g. to isolate the input high voltage part of a mains driven switching power supply for feedback from the output stage.
2. A low loss means of changing voltage and power levels, e.g. voltage reduction from the mains to obtain a suitable voltage for a system power supply.
3. Impedance matching between circuits for maximum power transfer.

We will analyse a basic circuit for a transformer to determine the relationships between input and output and consider the departures from the ideal in practical realizations. Though transformers can have more than two effective windings we need only analyse the simple case (Fig. 4.4.1). The reference to effective windings is to cover the case of a tapped single winding or auto transformer (Fig. 4.4.2), e.g. the most useful mains Variac™. The coupling between the windings depends on Faraday's law (Section 2.4) resulting in mutual inductance $M$ and can only couple of course for a.c. signals and not for d.c. For signals at least, rather than power, a d.c. optocoupler is available using photons.

We consider a simple transformer circuit as shown in Fig. 4.4.1. For sinusoidal inputs the voltage $v_2$ induced in the secondary by the current $i_1$ in the primary is given by:

$$\begin{aligned} v_2 &= M\frac{di_1}{dt} \\ &= j\omega M i_1, \quad \text{since for} \quad i_1 = i_0 e^{j\omega t} \quad \text{then} \quad \frac{di_1}{dt} = j\omega i_0 e^{j\omega t} = j\omega i_1 \end{aligned} \tag{4.4.1}$$

and the current $i_2$ in the secondary will in an equivalent way produce a voltage $j\omega M i_2$ in the primary. The mesh equations for the two circuits are therefore (we now use $s \equiv j\omega$ for convenience, Section 1.12):

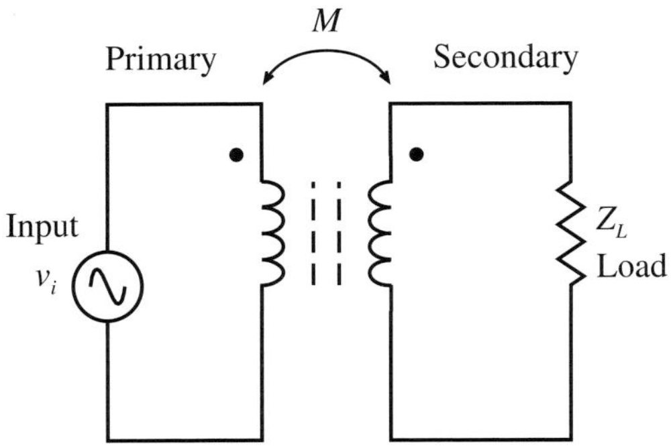


**Fig. 4.4.1** Basic transformer.

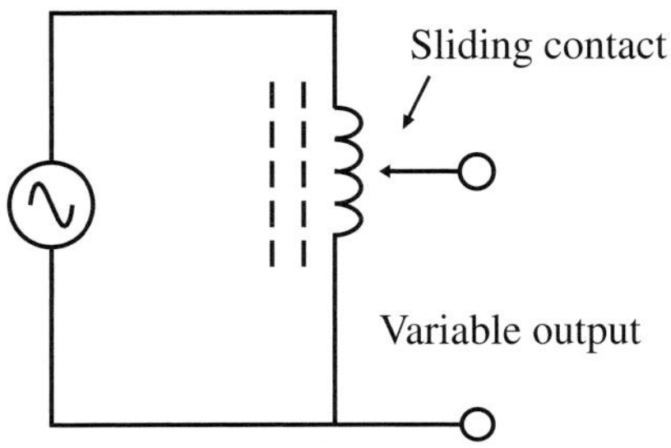


**Fig. 4.4.2** Basic auto transformer.

$v_1 = sL_1 i_1 + sMi_2$ (for the primary)

$0 = sMi_1 + sL_2 i_2 + Z_2 i_2$ (for the secondary)

so $\quad i_2 = \dfrac{-sMi_1}{(sL_2 + Z_2)}$

$$\text{and} \quad v_1 = sL_1 i_1 - sM\left(\frac{sMi_1}{sL_2 + Z_2}\right) \tag{4.4.2}$$

$$= sL_1 i_1 + \left(\frac{\omega^2 M^2 i_1}{sL_2 + Z_2}\right)$$

$$= i_1\left[\frac{\omega^2(M^2 - L_1 L_2) + sL_1 Z_2}{sL_2 + Z_2}\right]$$

which we can represent in terms of the equivalent circuit as shown in Fig. 4.4.3.

The corresponding value of $i_2$ can now be determined:

$$i_2 = \frac{-sM}{(sL_2 + Z_2)} \frac{v_1(sL_2 + Z_2)}{\omega^2(M^2 - L_1 L_2) + sL_1 Z_2} \tag{4.4.3}$$

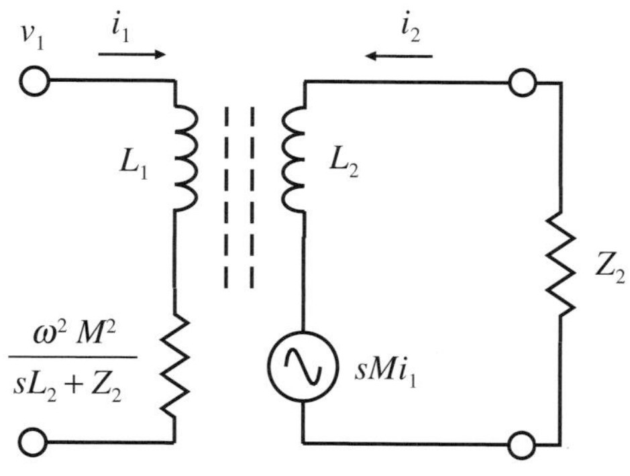


**Fig. 4.4.3** Equivalent circuit for Eq. (4.4.2).

$$=\frac{-sv_1M}{\omega^2(M^2-L_1L_2)+sL_1Z_2} \qquad (4.4.3\text{ cont.})$$

$$=\frac{-v_1M/L_1}{sL_2+Z_2-\dfrac{sM^2}{L_1}} \quad \text{(by dividing through by } sL_1\text{)}$$

which gives an alternative equivalent circuit showing the secondary voltage in terms of the primary (Fig. 4.4.4).

If the source has internal impedance $Z_1$ then we can add this to $sL_1$ to give the equivalent circuit shown in Fig. 4.4.5. Letting $Z_1 \to 0$ gives the configuration of Fig. 4.4.4.

If the transformer is considered as ideal then from Eq. (4.3.15) we have $M^2 = L_1L_2$ and we can define $n$:

$$\frac{M}{L_1}=\frac{L_2}{M}=n \quad \text{so} \quad \frac{L_2^2}{M^2}=n^2 \quad \text{or} \quad \frac{L_2^2}{L_1L_2}=\frac{L_2}{L_1}=n^2 \qquad (4.4.4)$$

so that (4.4.2) becomes:

$$v_1=i_1\left(\frac{sL_1Z_2}{sL_2+Z_2}\right) \quad \text{or} \quad i_1=v_1\left(\frac{sL_2}{sL_1Z_2}+\frac{Z_2}{sL_1Z_2}\right)=v_1\left(\frac{L_2}{L_1Z_2}+\frac{1}{sL_1}\right)=v_1\left(\frac{n^2}{Z_2}+\frac{1}{sL_1}\right) \qquad (4.4.5)$$

and we can find $i_2$ from Eq. (4.4.3):

$$i_2=\frac{-sM}{(sL_2+Z_2)}\frac{v_1(sL_2+Z_2)}{sL_1Z_2}=\frac{-L_2v_1}{nL_1Z_2}=\frac{-v_1n}{Z_2} \qquad (4.4.6)$$

which indicates an equivalent circuit as shown in Fig. 4.4.6.

From Section 4.3 it is seen that the inductance of a coil is proportional to the square of the number of turns so in this case we can put:

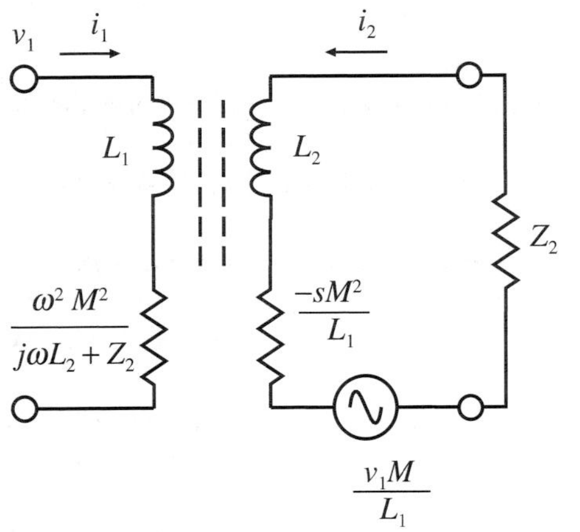


**Fig. 4.4.4** Equivalent circuit for Eq. (4.4.3).

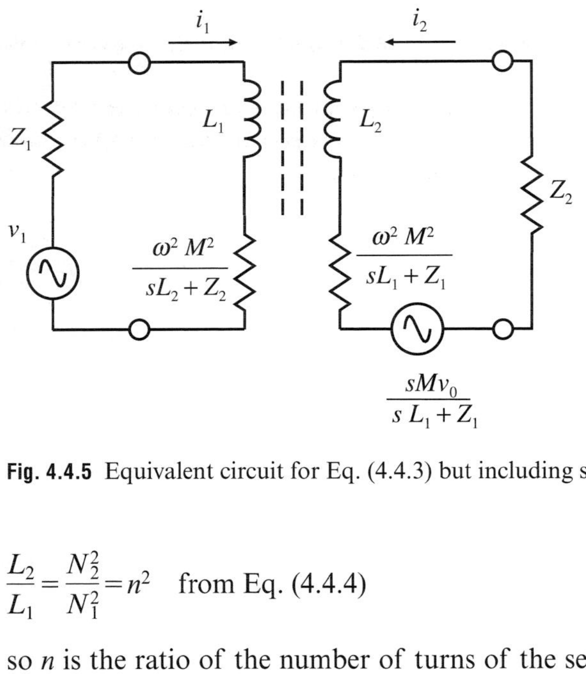


**Fig. 4.4.5** Equivalent circuit for Eq. (4.4.3) but including source impedance $Z_1$.

$$\frac{L_2}{L_1} = \frac{N_2^2}{N_1^2} = n^2 \quad \text{from Eq. (4.4.4)} \tag{4.4.7}$$

so $n$ is the ratio of the number of turns of the secondary to the primary. If, as is usual, $sL_1 \gg Z_2/n^2$ then:

$$i_1 = ni_2 \quad \text{and} \quad v_2 = nv_1 \quad \text{or} \quad v_1 i_1 = v_2 i_2 \tag{4.4.8}$$

as must be the case for an ideal transformer since we must have conservation of energy. We can have either a step up or a step down transformer depending on whether $n$ is greater or less than one. We can also see that for the conditions stated the source generator sees an impedance $Z_2/n^2$ so that the transformation ratio can be selected to match an impedance $Z_2$ to the generator (Section 3.1).

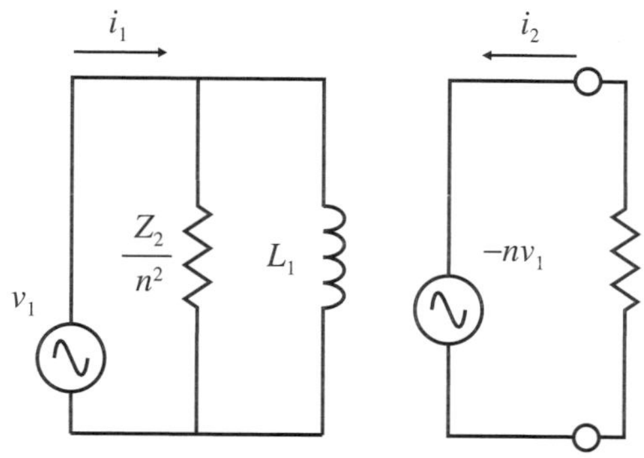


**Fig. 4.4.6** Equivalent circuit for Eq. (4.4.6).

A real transformer can depart from the ideal in a number of ways. The main sources of deviation at low frequencies are:

1. The magnetic coupling is not complete, i.e. all the flux from the primary does not couple with the secondary.
2. The windings have resistance causing energy loss and voltage drops.
3. Eddy currents in the magnetic core cause energy loss.
4. Hysteresis in the magnetic core causes energy loss.
5. At higher magnetic fluxes the core magnetization is non-linear.

At higher frequencies there are additional effects. Resistance of the windings changes due to the skin effect and there are significant capacities between primary and secondary and self-capacity across each of the windings. These deviations from the ideal can be represented by adding appropriate elements to the previous equivalent circuit to give Fig. 4.4.7.

Here Eq. (4.4.4) has been modified to allow for incomplete coupling between primary and secondary so that there is now a leakage inductance on each side. We write $M^2 = k^2 L_1 L_2$, where $k$ is the coupling factor and must lie between 0 and 1 (see Eq. (4.3.15)). Then the leakage inductances have the values shown; $L_1(1-k^2)$ in the primary and $k^2 L_2$ on the secondary. The factor $k$ is just that you will find in the SPICE Linear transformer or the *K_Linear* coupling parameter (see the PSpice reference manual on 'Inductor or transmission line coupling', MicroSim 1996; Hirasuna 1999). $R_1$ and $R_2$ represent the resistances of the two windings and allowance must be made for the increase as a result of the skin effect. Hysteresis losses may be represented by an additional resistor $H_1$ in parallel across $L_1$ and $H_2$ across the secondary. At higher frequencies the capacities $C_1$ and $C_2$ must be included, which now presents a very complex circuit to analyse. We can examine the equivalence using SPICE for the two circuits shown in Fig. 4.4.8, where we are ignoring the effects of capacity and losses.

Taking $k = 0.9$ and $n = 0.5$ together with source and load resistances and primary

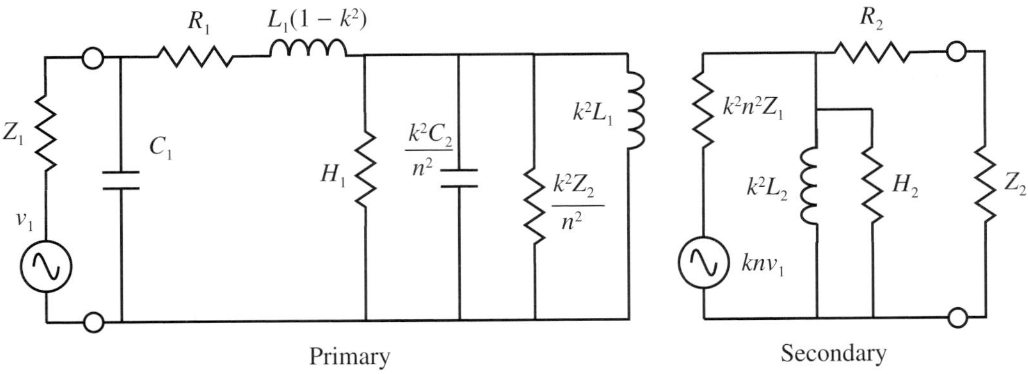


**Fig. 4.4.7** Equivalent circuit for Eq. (4.4.6) but allowing for leakage, hysteresis and stray capacity.

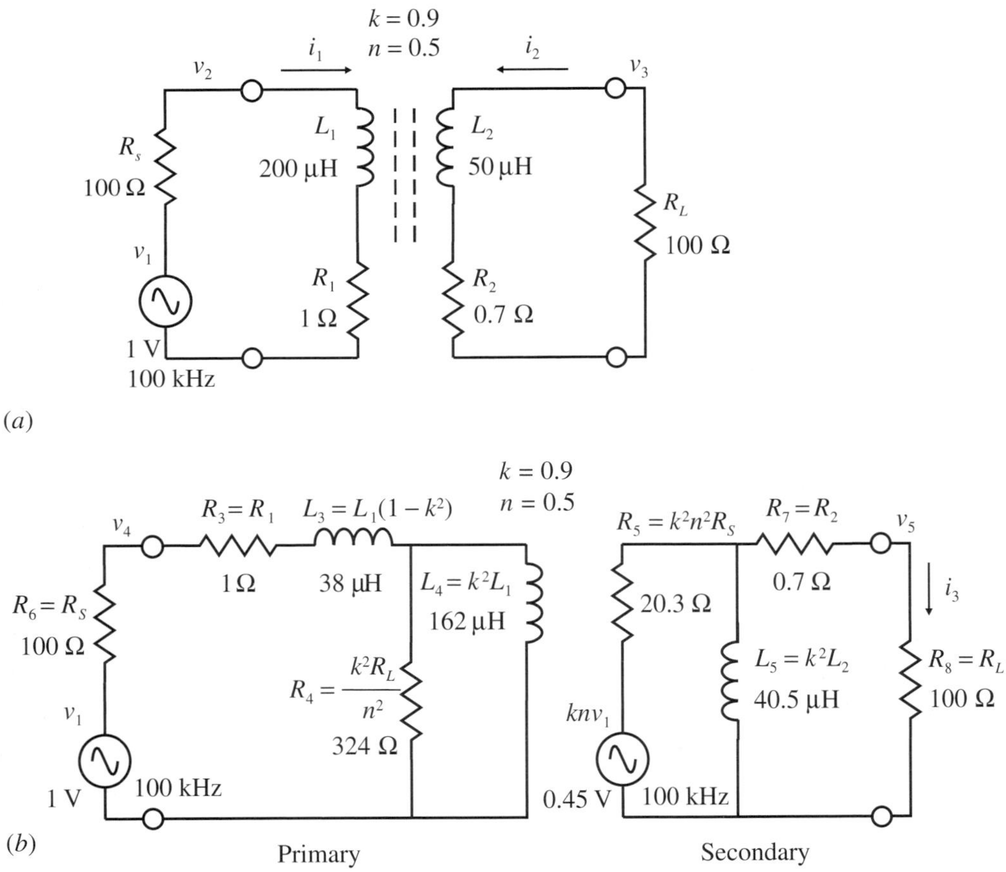


**Fig. 4.4.8** (*a*) Simple transformer schematic. (*b*) Equivalent circuit based on Fig. 4.4.7.

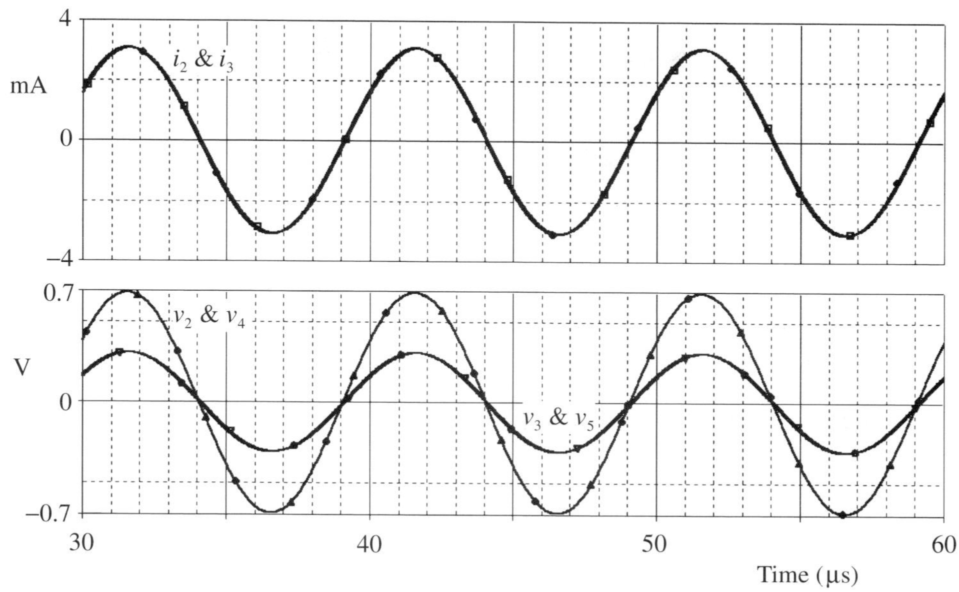


**Fig. 4.4.9** Comparison of voltages and currents for the circuits of Fig. 4.4.8.

inductance the equivalent circuit values are as shown in the figure. If we run a simulation we can compare the primary and secondary voltages and the secondary currents to see if the circuits are equivalent. The results are shown in Fig. 4.4.9 and it is evident that there is a good correspondence.

To achieve close coupling it is necessary to use magnetic materials to guide the flux from one winding to the other. At low frequency, e.g. for mains transformers and up to a few tens of kilohertz, various types of iron or other magnetic metals are used. To avoid eddy currents these must be laminated and insulated but eventually it becomes impossible to make the laminations thin enough. Above this ferrite materials are used which consist of compounds of iron, zinc and oxygen together with manganese, nickel or magnesium. These are polycrystalline sintered materials and are very hard, brittle and ceramic like. They have a wide range of permeabilities and frequency range and have high to very high intrinsic resistivities that largely avoid the problem of eddy currents. Since the permeability depends on orientation effects in the microcrystalline domains it also depends on the frequency of the applied magnetic field and hence must be represented as a complex quantity (see Section 2.11).

Pulse transformers present considerable difficulties since for preservation of the pulse shape they must have a very wide bandwidth and we have mentioned above the additional effects at high frequency. A description of the techniques of design is given Section 5.20, together with examples of some other applications.

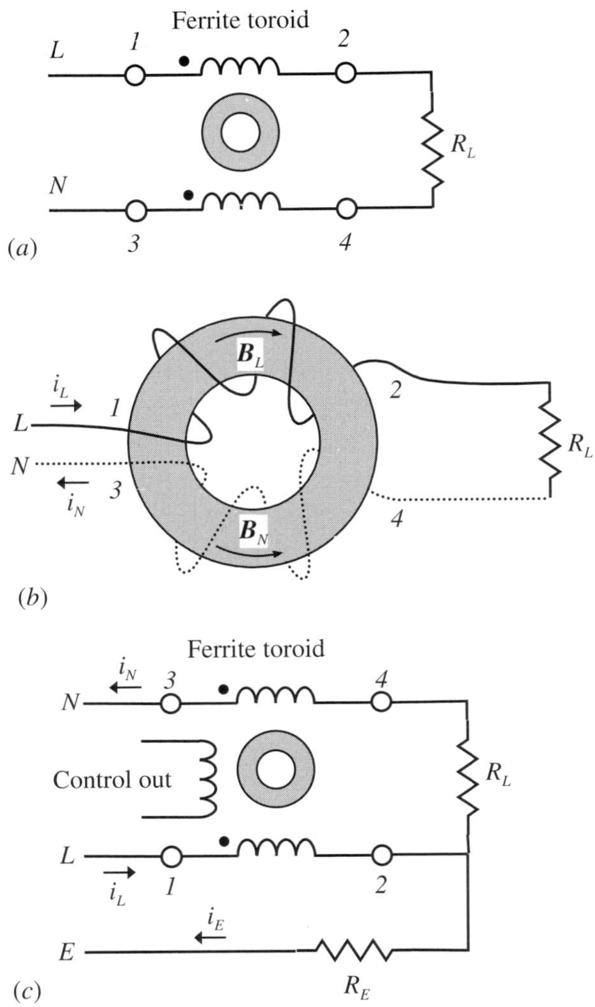


**Fig. 4.4.10** (*a*) Common-mode choke. (*b*) Winding on a toroid. (*c*) Earth-leakage circuit breaker.

So far we have not been concerned with the phasing of the primary and the secondary but in some cases this is important. As illustrated in Fig. 4.4.1 corresponding phase is indicated by a dot at the appropriate terminal. The PSpice transformer symbols do not show this so it is worth editing the symbol to make the terminal numbers visible on the schematic. A simple test will reveal the corresponding sense of the windings; pin 1 and pin 3 have the same phase. The use of the *K_Linear* coupling symbol, used for coupling separate inductors, should also be considered. In this case corresponding pin numbers are in phase and conventional current flows from pin 1 to pin 2.

Another special application of a form of transformer is the common-mode choke. As the name implies the intention is to inhibit common-mode signals while allowing differential mode signals to pass unaffected. Such chokes are often used on mains leads, for example, to decrease EMC radiations and the general configuration is shown in Fig. 4.4.10.

Differential currents $i_L$ and $i_N$ flowing as indicated will produce magnetic inductions $\boldsymbol{B}$ in opposite senses so there will be no added inductance ((*a*) and (*b*)). For common-mode currents, which both flow in the same direction, the $\boldsymbol{B}$ fields will add so that there will be considerable additional inductance causing a decrease in their magnitude. This arrangement is also the basis of a type of earth-leakage circuit breaker. If there is some earth leakage as represented by $R_E$ in (*c*) drawing a current $i_E$ then the currents $i_L$ and $i_N$ will no longer be equal so there will be some net $\boldsymbol{B}$ flux and the consequent signal from a separate winding can be used to activate the protective trip mechanism. For higher frequencies, as often seen for example on interconnections in computer applications, clip-on ferrite forms may be used for common-mode suppression.

## SPICE simulation circuits

| | |
|---|---|
| Fig. 4.4.9 | Trnsfmr 1.SCH |

## References and additional sources 4.4


Dean D. (2000): Gain confidence in network designs by simulating the LAN magnetics. *Electronic Design* 20 March, 121, 122, 124, 128.

Faulkner E. A. (1966): *Principles of Linear Circuits*, London: Chapman and Hall.

Giacoletto L. J. (1977): *Electronic Designers Handbook*, New York: McGraw-Hill.

Grant I., Philips W. R. (1975): *Electromagnetism*, London: John Wiley. ISBN 0-471-32246-6. (2nd Edn 1990.)

Grossner N. R. (1967): *Transformers for Electronic Circuits*, New York: McGraw-Hill. Library of Congress Cat. No. 66-23276. (2nd Edn 1983.)

Hirasuna B. (1999): *Using Coupled Inductors and Inductor Cores*, Cadence/MicroSim Application Note PSPA021, April.

MicroSim (1996): *PSpice A/D Reference Manual Ver. 7.1*, MicroSim, October.

Snelling E. C. (1969): *Soft Ferrites. Properties and Applications*, London: Iliffe Books. (2nd Edn Butterworths 1988.)

Tse C. K. (1998): *Linear Circuit Analysis*, Harlow: Addison Wesley. ISBN 0-201-34296-0.

Woodward W. S. (1995): The many analog uses for optical isolators. *Electronic Design* 17 April, 101–102, 104–106,108.

# 4.5 Diodes

The desire to know is natural to good men.

Leonardo da Vinci

The relationship between the voltage $V_D$ applied to a p-n junction diode and the current flow $I_D$ is given by the Shockley equation:

$$I_D = I_S \left[ \exp\left( \frac{V_D}{nV_J} \right) - 1 \right] \tag{4.5.1}$$

$I_S$ is called the saturation current and is ideally of the order of $10^{-14}$ A at room temperature and increases with increase in temperature. In practice there are other current generation effects which considerably increase the current and for the common silicon signal diodes it is of the order of nanoamps or more. $I_S$ is not the measured reverse leakage current, which is generally much larger and around room temperature doubles for about a 10 °C increase. $n$ is a factor that depends on the details of the construction of the p-n junction and varies between 1 and 2. The thermal voltage $V_J$ is given in terms of the absolute temperature $T$, the Boltzmann constant $k_B$ and the electronic charge $q_e$ by:

$$\begin{aligned} V_J &= \frac{k_B T}{q_e}, \quad \text{with } k_B = 1.38 \times 10^{-23}\ \text{J K}^{-1} \quad \text{and} \quad q_e = 1.602 \times 10^{-19}\ \text{C} \\ &= 25.9, \quad \text{say} \quad 26\ \text{mV} \quad \text{at 300 K} \quad (\text{approximately } \tfrac{1}{40} V) \end{aligned} \tag{4.5.2}$$

The inverse form of Eq. (4.5.1) is:

$$V_D = nV_J \ln\left( \frac{I_D}{I_S} + 1 \right) \tag{4.5.3}$$

This logarithmic dependence is made use of to construct mathematical processing circuits and for logarithmic amplifiers. It may be noted that diode connected transistors can provide closer conformance to the ideal response, and the so-called transdiode can do even better (Gibbons and Horn 1964; Sheingold and Pouliot 1974; Hamilton 1977).

The characteristics of diodes as described by the SPICE models are determined by the parameters listed in Table 4.5.1 (see for example Vladimirescu 1994).

Table 4.5.1 *Main SPICE diode parameters*[a]

| SPICE parameter | Equation symbol | Significance | Example values for 1N4148[b] |
|---|---|---|---|
| *N* | $n$ | Emission coefficient | 1.84 |
| *RS* | $R_S$ | Ohmic series resistance | 0.56 Ω |
| *IS* | $I_S$ | Saturation current | 2.68 nA |
| *CJO* | $C_{j0}$ | Depletion capacity at zero bias | 4 pF |
| *TT* | $\tau_T$ | Transit time | 11.5 ns |
| *BV* | | Breakdown voltage | 100 V |
| *EG* | | Energy gap (e.g. for Si or Ge) | 1.11 |
| *VJ* | $\phi_0$ | Barrier potential | 0.5 V |
| *M* | $m$ | Grading coefficient (for capacity) | 0.3333 |

[a] SPICE component libraries vary in their use of case for symbols.
[b] The values for the 1N4148 are taken from a model library.

The variation of junction capacity with reverse voltage is modelled by:

$$C_J = \frac{C_{j0}}{\left(1 - \dfrac{V_D}{\phi_0}\right)^m} \tag{4.5.4}$$

where it should be noted that $V_D$ is negative for reverse bias so that $C_J$ decreases with increasing reverse bias.

If a forward biased diode is suddenly switched to reverse bias it will conduct in the reverse direction for a short interval called the reverse recovery time $t_{rr}$. This is made up from two parts, the storage time $t_s$ and the transition time $t_t$. The storage time arises from the delay in extracting all the minority carriers from the junction region and is related to the carrier transit time $\tau_T$ (see Table 4.5.1) by:

$$t_s = \tau_T \ln\left(\frac{I_F - I_R}{-I_R}\right) \qquad (I_R \text{ is a negative quantity. Hambley 1994, p. 660}) \tag{4.5.5}$$

where $I_F$ is the forward current before reversal and $I_R$ the reverse current during the transition. When the carriers have been removed the junction capacity $C_J$ has to be charged (approximately exponentially), the time $t_t$ taken being dependent on the external circuit conditions. The sum $t_s + t_t = t_{rr}$. This allows an estimate of the reverse recovery time from the model value of the transit time $\tau_T$.

For forward currents such that we may neglect unity relative to the exponential term in Eq. (4.5.1), and this is applicable for quite small currents as we go round the cut-in region, then we can derive an expression for the incremental resistance of the diode at a given current. Taking $n = 1$:

$$I = I_S \exp\left(\frac{q_e V_D}{k_B T}\right) \quad \text{and hence} \quad \frac{dI}{dV_D} = \frac{q_e}{k_B T} I_S \exp\left(\frac{q_e V_D}{k_B T}\right) = \frac{q_e I}{k_B T}$$

$$\text{so} \quad r = \frac{dV_D}{dI} = \frac{k_B T}{q_e I} \cong \frac{\frac{1}{40}}{I} \Omega \approx 25\ \Omega \text{ at } 1\text{ mA} \tag{4.5.6}$$

When considering the operation of diodes in a circuit it is usual for one to allow the commonly used value of 0.6 V as the forward voltage drop before conduction becomes significant. This is often a good starting point but it should be remembered that there is a significant variation in the cut-in point for the wide range of diodes available. It is very simple to plot the characteristics using SPICE and Fig. 4.5.1 shows the characteristics of a number of different types. They are shown in two groups, signal diodes and rectifiers, the division being primarily on current range, together with reverse recovery time and junction capacity.

Signal diodes:

| | |
|---|---|
| 1N4148 | 'Standard' fast silicon diode, $V_{BR}=75$ V, 200 mA average |
| BAS70 | Schottky diode, SOT-23, $V_{BR}=70$ V, 70 mA average |
| BAT68 | Schottky diode, SOT-23, $V_{BR}=8$ V, $V_F=395$ mV at 10 mA, 130 mA max, low capacity |
| 1N5711 | Schottky barrier for UHF applications, $V_{BR}=70$ V |
| 1N5712 | Schottky barrier for UHF applications, $V_{BR}=20$ V |

Rectifiers:

| | |
|---|---|
| 1N4007 | Standard 1000 V, 1 A |
| 1N5817 | Schottky 20 V, 1 A |
| 1N5822 | Schottky 40 V, 3 A |

Schottky diodes are formed from the junction of a metal and a semiconductor. The benefits are that the reverse recovery time is generally much shorter than for a p-n junction owing to conduction by majority carriers, and the absence of minority carriers. The forward cut-in voltage is lower as can be seen in Fig. 4.5.1, but the reverse breakdown voltage is also generally lower than for silicon diodes.

When reverse biased sufficiently a diode will break down and usually be destroyed. The usual mechanism causing this is the acceleration of charge carriers which, if they can gain enough energy, are able to cause ionization of atoms by collision creating additional charge carriers. If these can also be accelerated sufficiently to cause further ionization then an avalanche breakdown occurs with the generation of considerable heat which can destroy the junction. However, by suitable design of the junctions it is possible to construct rectifiers which will not be damaged by reverse breakdown (within limits) so providing some protection for transient voltages. By appropriate doping it is also possible to produce diodes that have a breakdown voltage that is to first order independent of current, the so called

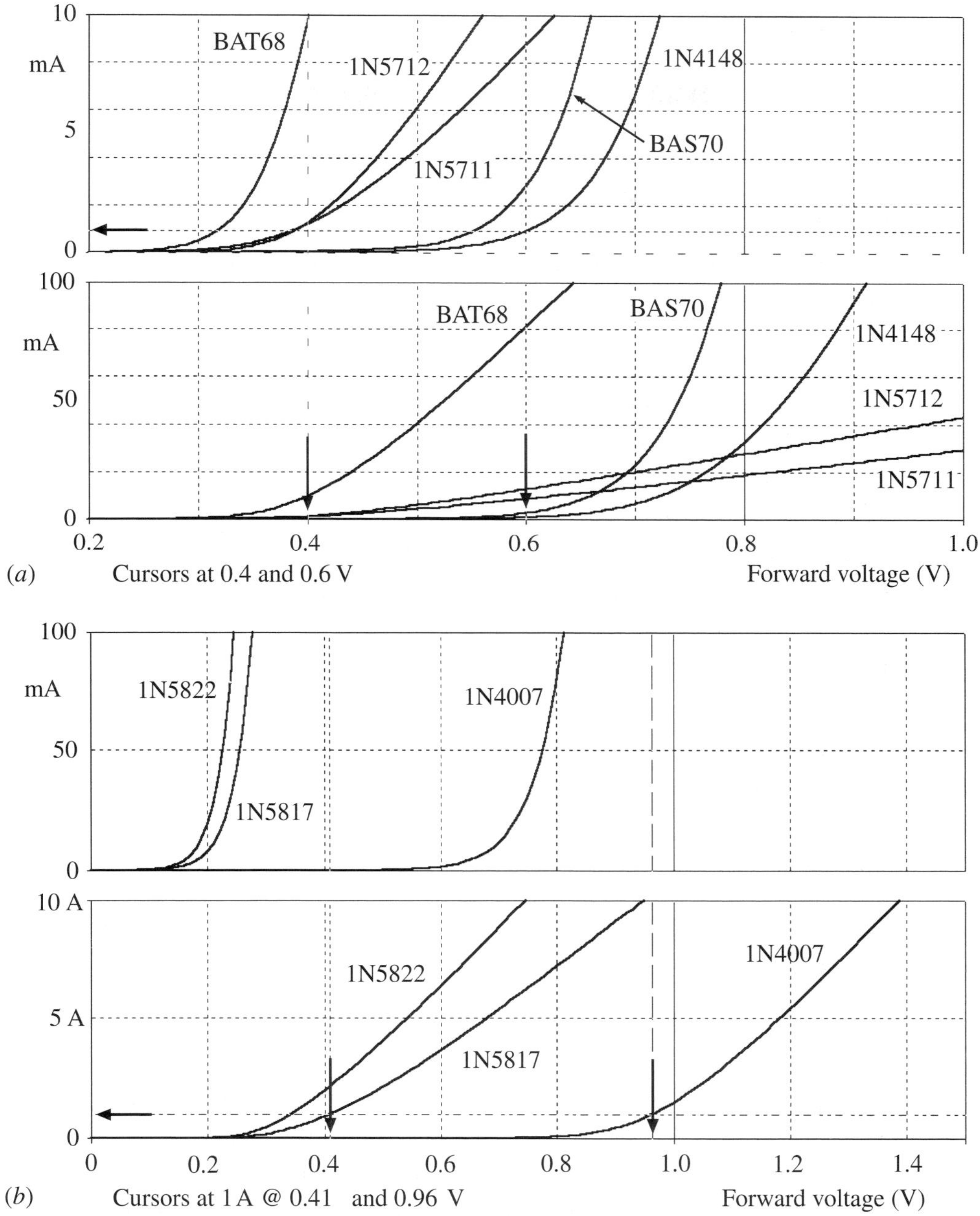


**Fig. 4.5.1** (*a*) Forward characteristics of some signal diodes (note that the voltage scale shown does not start at zero). (*b*) Forward characteristics of some rectifiers.

Zener diode. At rather low reverse voltages, say less than about 6 V, a different mechanism, first elucidated by Zener, can also produce breakdown. Nowadays both types of regulator diode are referred to as Zener diodes even though the higher voltage types do not rely on the Zener mechanism. The temperature coefficient of breakdown voltage for the Zener mechanism is negative, while that for avalanche breakdown is positive. Thus in the vicinity of 5–6 V the regulator diodes have minimum temperature coefficient. Though this is used to create stable voltage regulators, high precision types are more predictably made from a Zener diode in series with one or more ordinary diodes to obtain low temperature coefficients. In more recent times other types of voltage reference, e.g. the so called bandgap references based on Eq. (5.9.11) have often replaced Zeners.

An important factor with regard to Zeners, which is seldom referred to in data-sheets or textbooks, is their large junction capacity. It is not uncommon to see Zeners used inappropriately, where the large capacity could have significant effect. Since Zeners are so often used as voltage stabilizers, where the capacity is of advantage, manufacturers seem to find it unnecessary to provide the capacity specification. If you examine a SPICE library you will find, for example, that for a 1N756 (8.2 V) 400 mW Zener the junction capacity is 90 p, while for a 1N746 (3.3 V) it is 220 p. Higher power Zeners, having larger junction areas, will have correspondingly larger capacity. The breakdown mechanisms in Zener diodes makes them electrically rather noisy, so if you wish to use one for a reference voltage they should be well bypassed. Low voltage Zeners are in fact used as predictable noise sources as they are effectively shot noise limited (Motchenbacher and Connelly 1993). In using Zener diodes it is essential to include a series resistor to limit the current otherwise the device will probably die a sudden death. The current through the device should be greater than the knee values to obtain a stable voltage independent of current. Figure 4.5.2(*a*) illustrates the basic circuit arrangement and (*b*) shows a simulated characteristic for a 4.7 V and a 3 V Zener.

It is evident that the knee region can differ significantly between types so it is necessary to check your operating currents carefully. The current $i_R$ through the series resistor $R_S$ will be the sum of the current $i_Z$ required in the Zener and $i_L$ for the load $R_L$. The value of $R_S$ is found from:

$$R_S = \frac{V_{in} - v_Z}{i_Z + i_L} \tag{4.5.7}$$

making allowance for any variation in $V_{in}$ or $i_L$.

The output stage of many op amps and power amplifiers make use of a circuit arrangement that can be tailored to yield what is effectively an adjustable Zener. This arrangement is also known as a $V_{BE}$ multiplier. The circuit is shown in Fig. 4.5.3(*a*).

We assume that $i_B \ll i_A$ so that we can write:

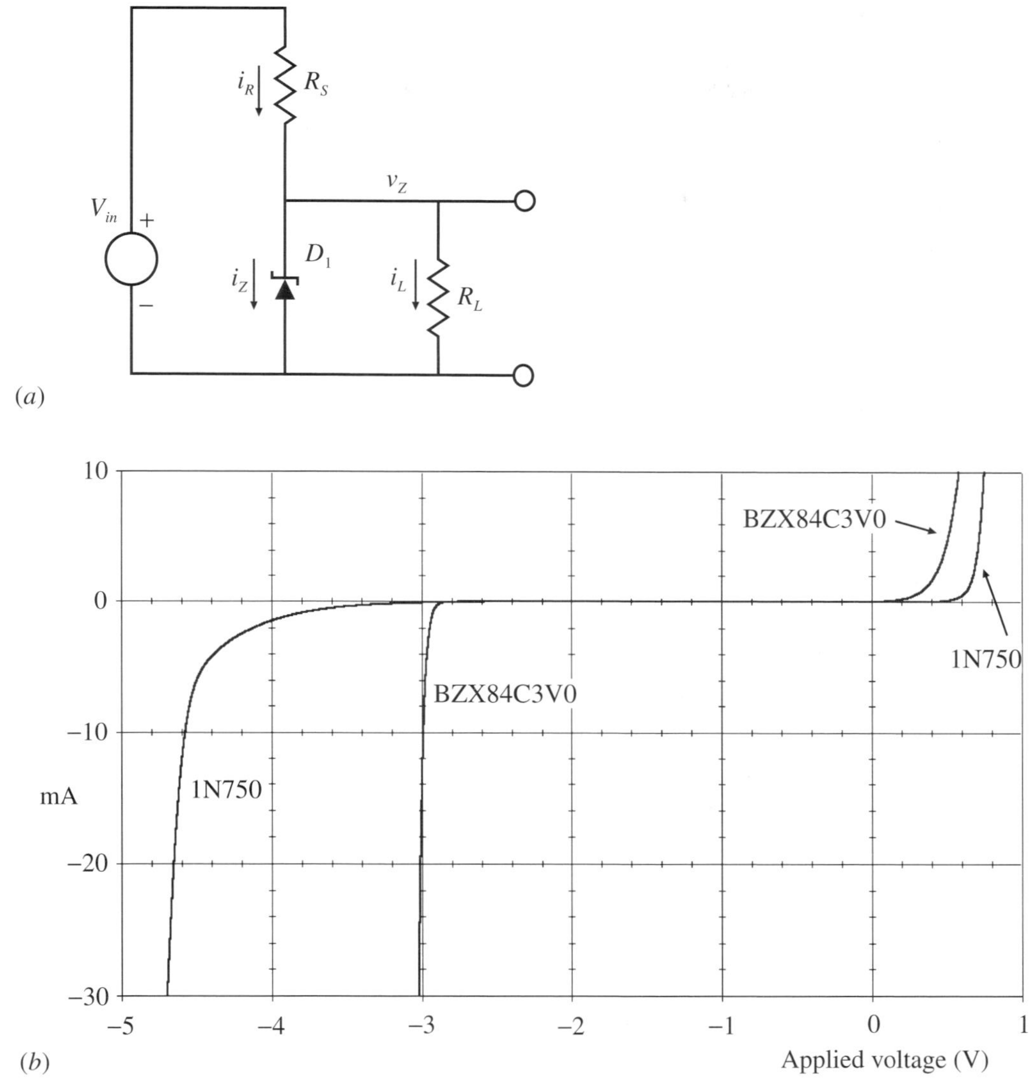


**Fig. 4.5.2** (*a*) Typical Zener circuit arrangement. (*b*) Example Zener characteristics.

$$V_{CE} = V_{BE}\frac{(R_1 + R_2)}{R_2} = \frac{(R_1 + R_2)}{R_2} V_J \ln\left(\frac{i_C}{I_S}\right) \quad \text{using Eq. (4.5.2)}$$

$$= \frac{(R_1 + R_2)}{R_2} 60 \text{ mV} \quad \text{per decade ratio of } i_C/I_S \text{ using } V_J = 26 \text{ mV and } \ln(10) = 2.3026 \qquad (4.5.8)$$

and so $V_{CE}$ is only logarithmically dependent on $i_C$ and hence is essentially constant, only depending on the resistive divider. As an example we choose $V_{CE} = 1$ V so the ratio $R_1/R_2 \approx (V_{CE} - V_{EB})/V_{EB} \approx 0.4/0.6$ and choose a total resistance of a few kΩ to

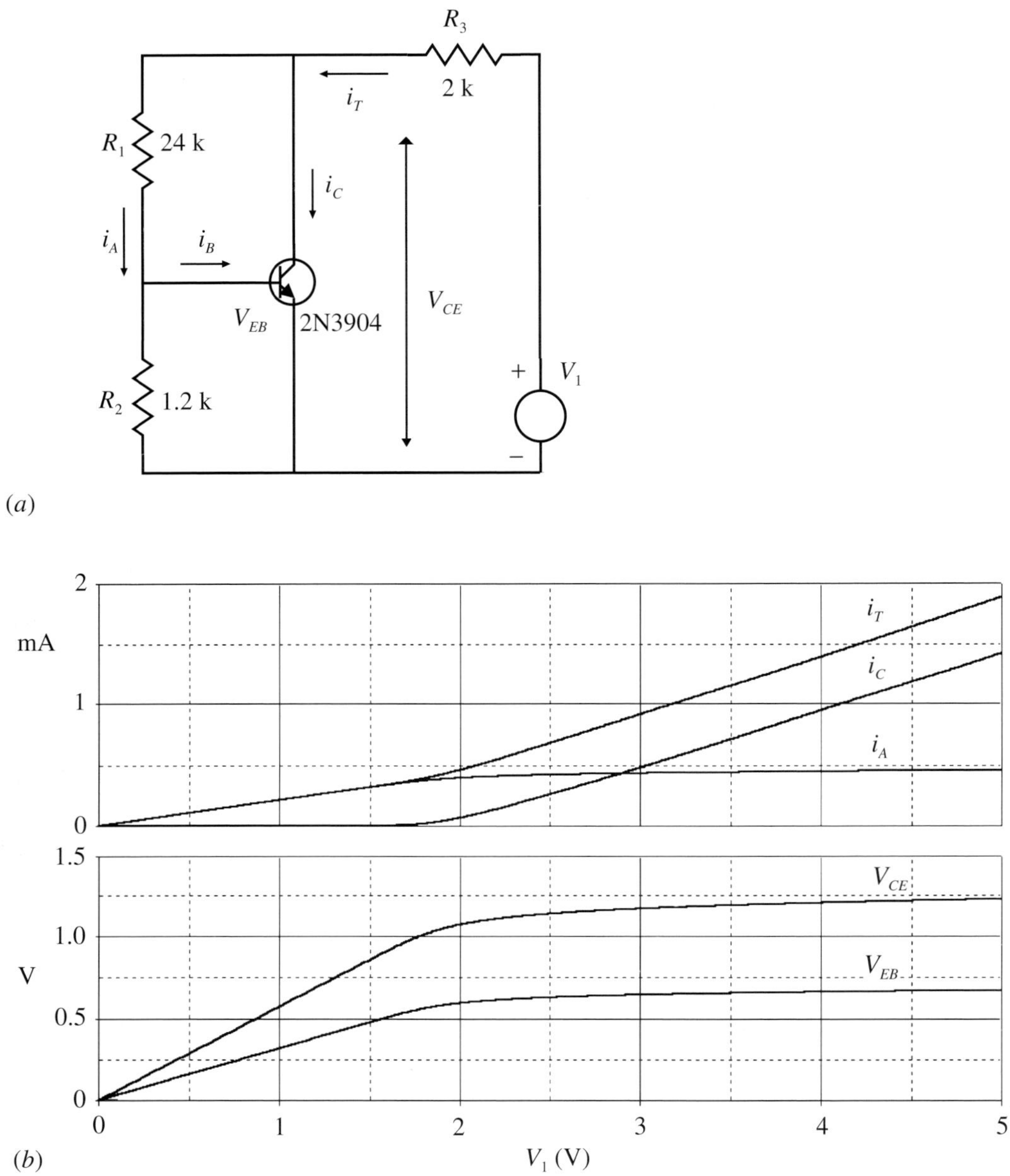


**Fig. 4.5.3** (*a*) Adjustable Zener circuit. (*b*) Characteristic curves.

keep $i_B \ll i_A$, and a 2N3904 transistor ($\beta \approx 400$). Sweeping the supply voltage $V_1$ over the range 0–5 V produced the simulation results shown in Fig. 4.5.3(*b*). It is seen that $V_{CE}$ is substantially constant, independent of the current $i_T$. The base current $i_B$ is about 10 μA and $i_A \approx 450$ μA.

Current regulator diodes are also available and are constructed from FETs with the gate connected to the source (Botos 1969; Watson 1970; Hamilton 1977).

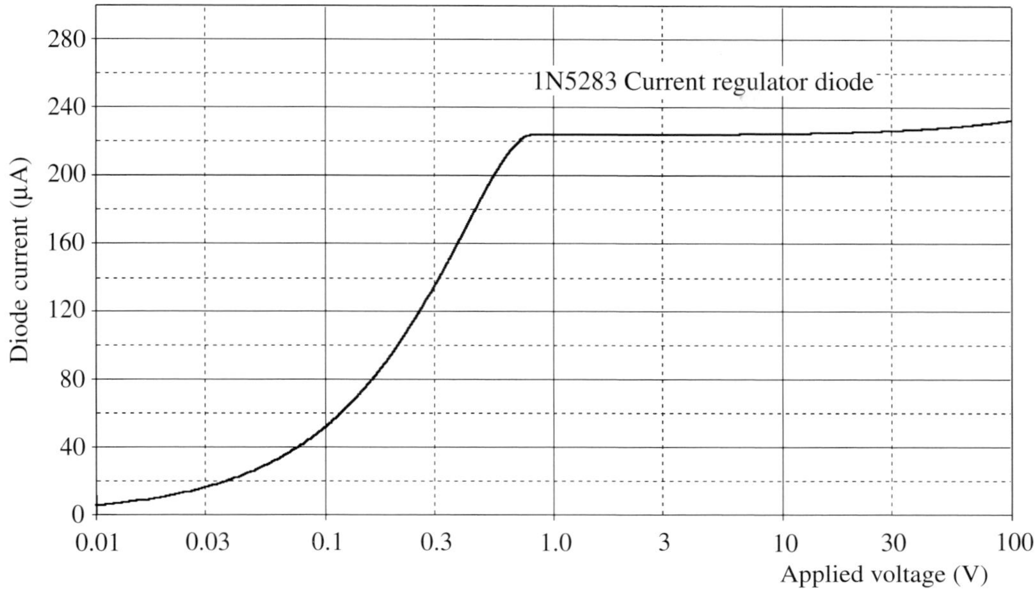


**Fig. 4.5.4** Characteristic of a 1N5283 current regulator diode. The voltage scale is logarithmic to show more clearly the low voltage region.

Within their breakdown and knee voltages these give a fixed current independent of applied voltage. Figure 4.5.4 shows the simulated characteristic for a 1N5283, specified at 220 μA with minimum voltage 1 V and maximum 100 V.

Diodes are also available for use as voltage controlled capacitors, particularly useful for r.f. tuning applications. Models for these are discussed in Section 4.2.

## SPICE simulation circuits

| | |
|---|---|
| Fig. 4.5.1(*a*) | Diode1.SCH |
| Fig. 4.5.1(*b*) | Diode1.SCH |
| Fig. 4.5.2(*b*) | Zenereg.SCH |
| Fig. 4.5.3(*b*) | Adjzenr1.SCH |
| Fig. 4.5.4 | Curdiod.SCH |

## References and additional sources 4.5


Botos B. (1969): *FET Current Regulators – Circuits and Diodes*, Motorola Application Note AN-462.

Chen, Wai-Kai (Ed.) (1995): *The Circuits and Filters Handbook*, Boca Raton: CRC Press and IEEE Press. ISBN 0-8493-8341-2. See Section 9.5 'Semiconductor diode' by B. M.Wilamowski, pp. 333–344.

Getreu I. (1974): Modeling the bipolar transistor. *Electronics* 19 September, 114–120; 31 October, 71–75; 14 November, 137–143.

Gibbons J. F., Horn H. S. (1964): A circuit with logarithmic transfer response over 9 decades. *IEEE Trans.* **CT-11**, 378–384.

Gray P. E., Searle C. L. (1969): *Electronic Principles. Physics, Models, and Circuits*, New York: John Wiley. ISBN 471-32398-5.

Gray P. E., Searle C. L. (1969): *Electronic Principles. Physics, Models, and Circuits. Instructors Manual*, New York: John Wiley. ISBN 471-32399-3.

Hambley A. R. (1994): *Electronics. A Top-Down Approach to Computer-Aided Circuit Design*, New York: Macmillan. See Chapter 4: Introduction to diodes, and Chapter 11: Internal physics and circuit models of semiconductor devices.

Hamilton T. D. S. (1977): *Handbook of Linear Integrated Electronics for Research*, London: McGraw-Hill. ISBN 0-07-084483-6. See Section 4.3 for many logarithmic references. See Section 7.10 for Zener and current diodes.

Motchenbacher C. D., Connelly J. A. (1993): *Low Noise Electronic System Design*, New York: John Wiley. ISBN 0-471-57742-1.

Sheingold D., Pouliot F. (1974): The hows and whys of log amps. *Electronic Design* 1 February, 52–59.

Siliconix (1989): *The FET Constant-Current Source*, Siliconix Application Note LPD-15.

Todd C. D. (1970): *Zener and Avalanche Diodes*, New York: Wiley-Interscience.

Vladimirescu A. (1994): *The Spice Book*, New York: John Wiley. ISBN 0-471-60926-9. See Section 3.2.

Watson J. (1970): *An Introduction to Field-Effect Transistors*, Siliconix.

# 4.6 Bipolar transistors

It is as important to present new knowledge in a form in which it can be assimilated and its essential import realized as it is to discover it, and this presentation is an art akin to poetry and literature.

W. L. Bragg

With such a wide range of integrated circuits to choose from the use of discrete transistors is nowadays considerably less than of yore. There are very many books that discuss the details of the internal operation, characteristics and small signal equivalent circuits (Gray and Searle 1969; Gronner 1970; Hambley 1994; Schubert and Kim 1996). Here we will largely restrict discussion to some practical matters and the significance of the model parameters used by SPICE.

Transistors are normally considered as current controlled devices with a current gain $\beta = I_C/I_B$ typically of several hundred. $\beta$ is roughly constant over a fairly wide range of current but at both low and high collector current decreases substantially. The value is rather variable even among the same type of transistor so you cannot rely on the specification sheet value for accurate calculation. In the simplest application, such as a switch which is either on or off, it is only necessary to ensure that the base current is high enough for the transistor to become saturated, when the collector voltage falls to the base voltage or even a little lower. $V_{CE}$ saturation voltages can be as low as say 0.2 V which, although the collector–base junction appears to be forward rather than reverse biased, does not cause significant forward current flow since it is still less than the cut-in voltage. Ensuring that significant saturation occurs allows for changes with time and temperature. This advantage, together with the minimum power dissipation in this state, must be balanced against the drawback in that the recovery time when the switch is turned off is much increased when recovering from saturation. It is possible to reduce this effect by clamping the collector to the base using a Schottky diode which can limit the degree of saturation and this is done, for instance, in high-speed logic circuits.

The common use of a transistor for switching an inductive load is illustrated in Fig. 4.6.1($a$). We simulate this circuit to provide a reminder of the need to protect the transistor against the high induced voltage that arises when the load is switched

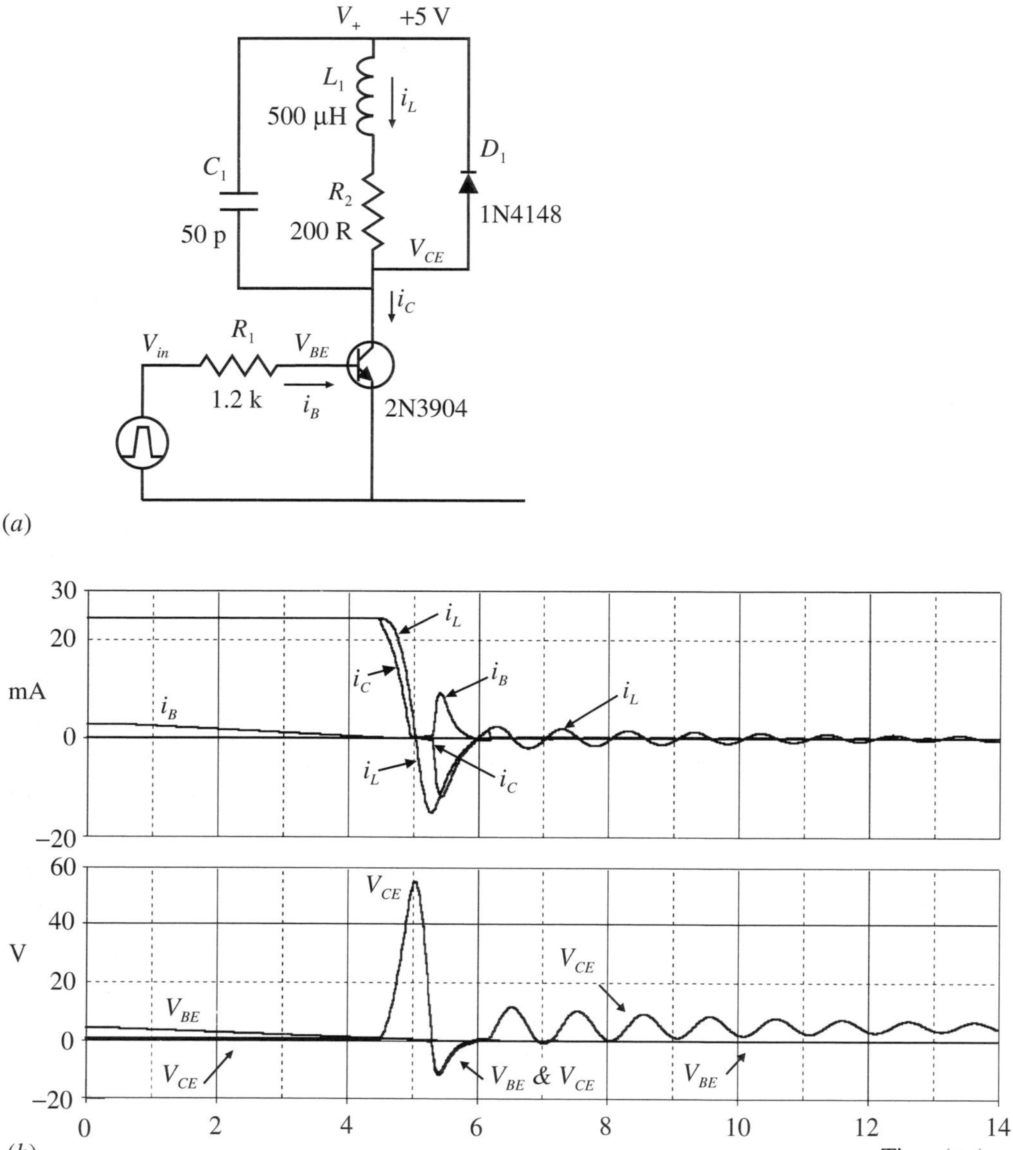


**Fig. 4.6.1** (*a*) Transistor switch for an inductive load. (*b*) Responses without a catcher diode. (*c*) Responses with catcher diode $D_1$. The current $i_D$ is shown as negative to make $i_C$ and $i_L$ more clear.

off. Figure 4.6.1(*b*) shows the waveforms for the circuit without the catcher diode $D_1$ and (*c*) when it is included.

The voltage transient in this case reaches about 54 V, which this particular transistor could withstand, but the value is very dependent on switch-off time, series resistance $R_2$ and capacity at the collector. The last is in part represented by $C_1$ but

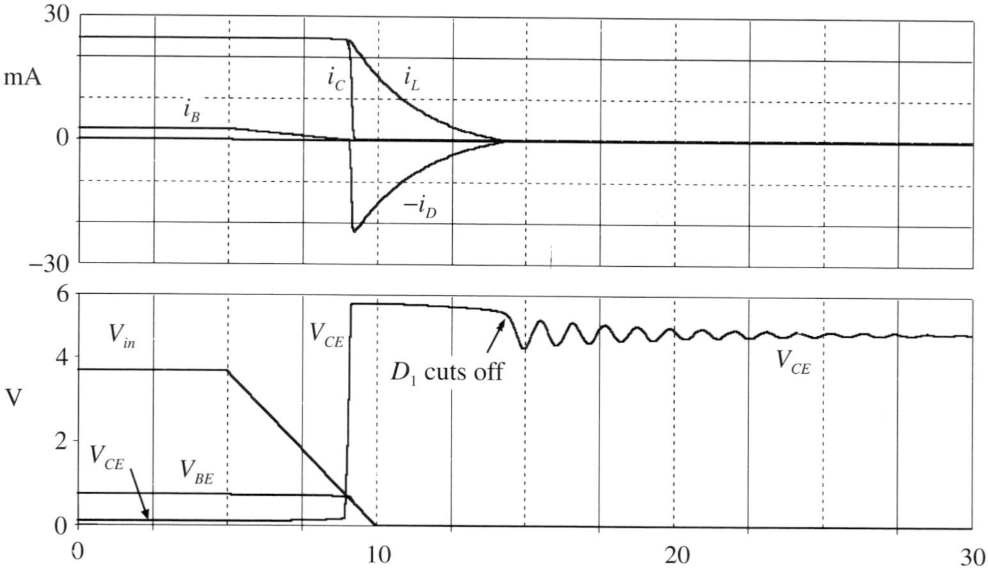


(c)

**Fig. 4.6.1** *(cont.)*

there is also the collector–base (and $D_1$) capacity which also contribute. The negative going swing of $V_{CE}$ following the positive peak also means that $V_{BE}$ goes to about −10 V which would well exceed the typical base–emitter breakdown voltage of −5 V and possibly damage the transistor. This negative excursion is also the reason for the sharp peak in $i_B$ at this time (base–collector forward biased). When the catcher diode is included the transient voltage is limited to one diode drop above the supply voltage and the energy stored in the inductor field is largely dissipated in $R_2$. When the diode is no longer conducting the remaining energy can cause oscillations at the resonant frequency which take some time to die out. Measurement of the ringing frequency gives an equivalent capacity of just over 50 p.

To obtain a linear amplifier it is necessary to bias the transistor so that the output quiescent voltage is such that it can swing positively and negatively relative to this by the desired amount. The variability of the current gain and the substantial effects of temperature variation make it very difficult to obtain a stable operating point simply by biasing the base directly. As in many other applications we can make use of negative feedback to make the operating point substantially independent of variations in the transistor itself, the penalty being of course that you pay for this in terms of a reduction in gain. However, gain is generally easy to come by so this is seldom a serious drawback. The feedback is introduced by means of an emitter resistor $R_E$ and the typical circuit is shown in Fig. 4.6.2(*a*).

The biasing resistors $R_1$ and $R_2$ establish a voltage $V_B$ so the emitter voltage $V_E$ will be one diode drop $V_{BE}$ below this. If the current $i_A$ in the bleeder is much greater

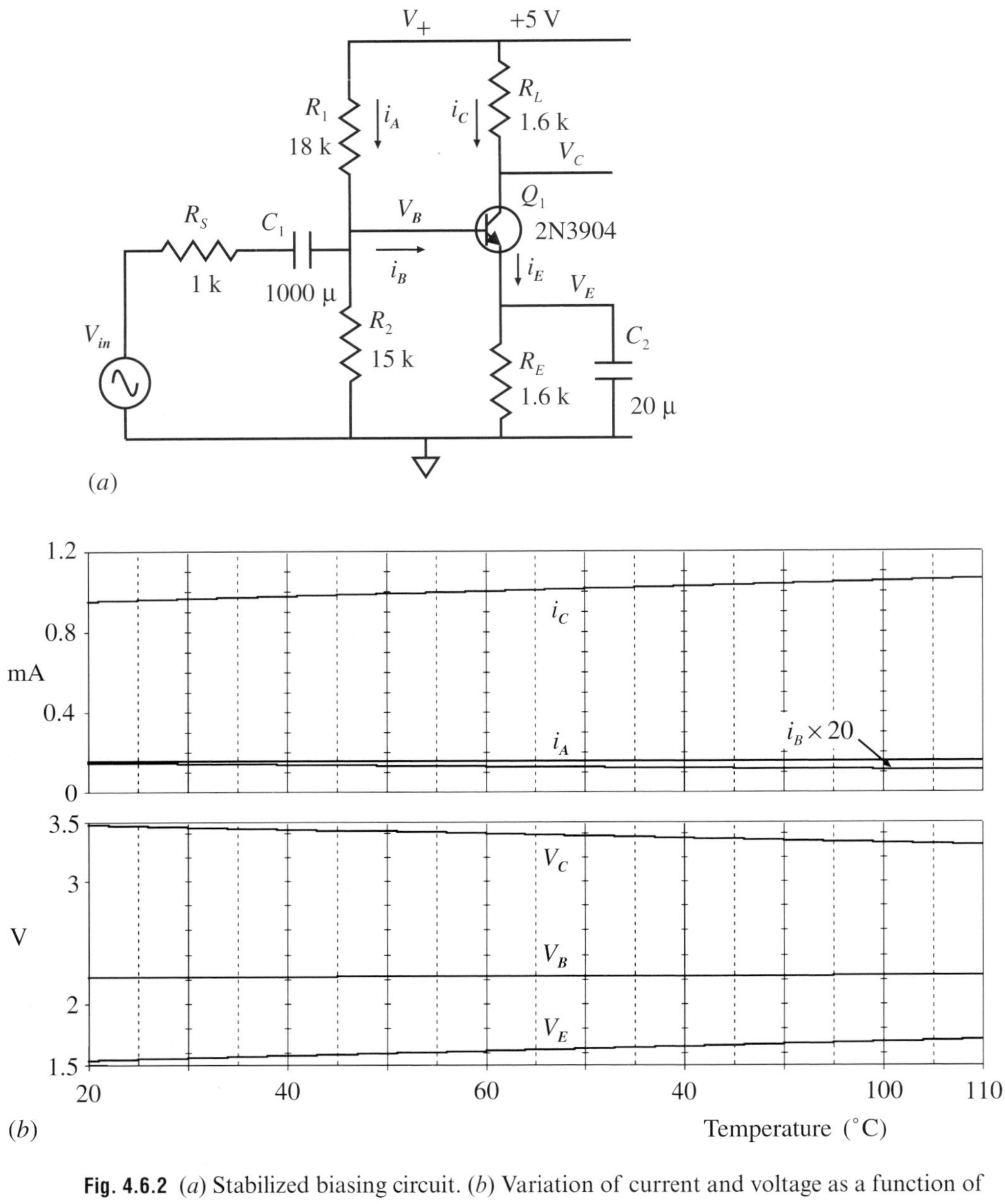


**Fig. 4.6.2** (*a*) Stabilized biasing circuit. (*b*) Variation of current and voltage as a function of temperature. Note that the base current is shown as $i_B \times 20$ for comparison with the bleeder current $i_A$ and to make it more visible.

than the base current $i_B$ then we may neglect $i_B$ in calculating $V_B$. The immediate aim is to set $V_C$ to a suitable value so that with an input signal from $V_{in}$ the output may swing symmetrically. $V_{in}$ is capacitively coupled so that it does not upset the biasing arrangement. The design involves an iterative process. At least for low voltage $V_+$ supplies, and since we want a reasonable amount of negative feedback

Table 4.6.1 *Currents and voltages from simulation of the circuit of Fig. 4.6.2(a). The precision shown is spurious in practice but it is needed here to show the small differences*

| Temp. (°C) | $i_A$ (μA) | $i_B$ (μA) | $i_C$ (μA) | $V_B$ | $V_E$ | $V_C$ | $R_1$ | $R_2$ |
|---|---|---|---|---|---|---|---|---|
| 27 | 51.6 | 6.737 | 898 | 2.109 | 1.447 | 3.546 | 56 k | 47 k |
| 27 | 155 | 7.170 | 961 | 2.214 | 1.550 | 3.462 | 18 k | 15 k |

from $R_E$, we may make a first pass choice to have about equal voltages across $R_E$, $Q_1$ and $R_L$. Since $\beta$ is large the current $i_C \cong i_E$ so that $R_E = R_L$. The choice of $i_C$ must now be made. The basis for this is not direct or exact. The current should not be too small so that $\beta$ is falling off and it should not be larger than necessary to minimize power dissipation and hence temperature effects. A significant factor is frequency response, since the current available, the load resistance and the effective capacity at the collector will set the upper frequency limit. Using the values shown, derived on the basis of a collector current of 1 mA and the SPICE model value of $\beta \equiv BF \approx 400$ for the 2N3904 giving a base current $i_B = 2.5$ μA, the simulated values shown in Table 4.6.1 were obtained.

The first pass may be acceptable but it is seen that the base current is about three times what was expected so that the bleeder current $i_A$ is now not 20 times $i_B$ as originally designed. The reason for this discrepancy is that we are operating in the region where the base current is of the same order as the leakage and recombination currents so that the effective value of $\beta$ is now about 133. This matter is described by Tuinenga (1995, Section 16.5) who illustrates the interaction of the various transistor model parameters. Decreasing the values of the bleeder resistors by a factor of about three gives the results shown in the second line of the table, which are closer to our original choice. The test is now to see how the quiescent values vary with temperature so we can run a simulation with temperature as the sweep variable. A linear sweep from 20 to 110 °C gives the curves of Fig. 4.6.2(*b*). The performance is quite satisfactory, though we have of course not considered any temperature effects for the resistors. However, these will be small and will in any case substantially cancel out if they have similar coefficients. For the basic circuit the large feedback will make the gain $\approx R_L/R_E$, i.e. effectively unity here, so that for greater gain at higher frequencies $R_E$ must be bypassed with $C_2$ so that the negative feedback is reduced or becomes negligible. A judicious selection of $C_1$ and $C_2$ allows us to display both the low frequency (effectively z.f.) and the high frequency responses as shown in Fig. 4.6.3. Measurements at 1 Hz and 100 kHz are given in Table 4.6.2.

The potential divider drop from $V_{in}$ to $V_B$ allows calculation of the effective input

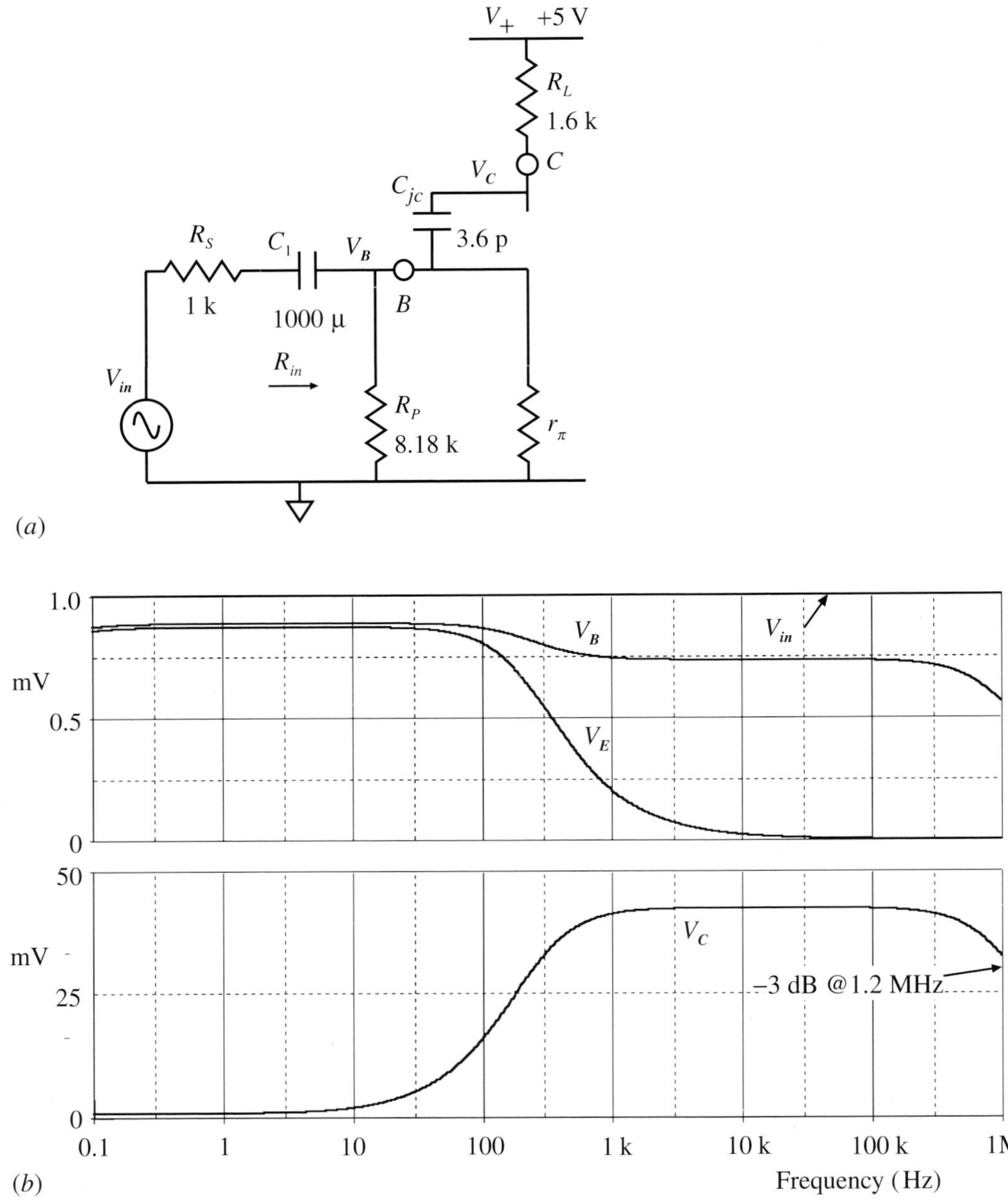


**Fig. 4.6.3** (*a*) Equivalent circuit for determination of input resistance. (*b*) Frequency response of circuit of Fig. 4.6.2(*a*).

resistance $R_{in}$. Using the value of $R_P$ from the resistors $R_1$ and $R_2$ in parallel allows the calculation of the input resistance $r_\pi$ of the transistor itself (Gray and Searle 1969, p. 414; Hambley 1994, p. 199). In other references $r_\pi$ is referred to as $r_{be}$ or $h_{ie}$, e.g. Gronner (1970). The large value at 1 Hz illustrates the effect of the emitter resistor $R_E$ in providing large negative feedback (as in the op-amp voltage follower, Eq. (5.3.14)). At 100 kHz, the feedback has largely disappeared as $C_2$ becomes

Table 4.6.2 *Measured values from Fig. 4.6.3 (voltages are in μV)*

| Frequency | $V_{in}$ | $V_B$ | $V_E$ | $V_C$ | Gain | $R_{in}$ | $r_\pi$ |
|---|---|---|---|---|---|---|---|
| 1 Hz | 1000 | 888 | 872 | 884 | 0.995 | 7.93 k | 259 k |
| 100 kHz | 1000 | 733 | 2 | 42000 | 57.3 | 2.74 k | 4.12 k |

Table 4.6.3 *Bipolar transistor SPICE model parameters*

| Spice parameter | Significance | Example for 2N3904 |
|---|---|---|
| BF | Forward current gain | 416 |
| IS | Saturation current | 6.7 fA |
| RB | Base resistance | 10 Ω |
| CJC | BC zero-bias capacity | 3.6 pF |
| VJC | BC built-in potential | 0.75 V |
| CJE | BE zero-bias capacity | 4.5 pF |
| VJE | BE built-in potential | 0.75 V |
| TF | Forward transit time | 301 ps |
| TR | Reverse transit time | 240 ns |

effective. $V_E$ decreases, $V_B$ dips and the voltage gain increases. The fall off in gain at high frequency arises from the increasing influence of the collector–base capacity which is multiplied by the Miller effect (Section 3.10). The model for the 2N3904 gives $C_{jc} = 3.6$ p, which when multiplied by the gain (57.3) gives 205 p. This has an impedance of 1 k, to match $R_S$, at about 800 kHz so we expect the –3 dB point to be near here. The simulation shows it just off the graph at 1.2 MHz. A condensed account of models and some guidance on determining parameters for operating points different from those given in the datasheets is given by Hamilton (1977, Section 1.9).

Though the bipolar transistor is almost invariably characterized as a current-controlled device this is not always appropriate. An alternative approach in terms of a voltage-controlled device is given by Faulkner (1969). Despite its long existence and the design of many thousands of different transistor types, and it is difficult to think of good reasons for there being so many, there is still room for innovation. The introduction of techniques from the ICs have been used, for example, to construct multiple or distributed bases to improve performance in terms of saturation voltage, current gain and power dissipation in tiny packages (Bradbury 1995).

Some of the more important SPICE model parameters are shown in Table 4.6.3 with some values for a commonly used transistor. The role of the parameters is discussed in Vladimirescu (1994) and Hambley (1994).

## SPICE simulation circuits

| | |
|---|---|
| Fig.4.6.1(*b*) | Transtr 1.SCH |
| Fig. 4.6.1(*c*) | Transtr 1.SCH |
| Fig. 4.6.2(*b*) | Transtr 2.SCH |
| Fig. 4.6.3(*b*) | Transtr 2.SCH |

## References and additional sources 4.6


Bradbury D. (1995): *Features and Applications of the FMMT618 and 619 (6A SOT23 Transistors)*, Zetex Application Note 11, October.

Dean K. J. (1964): *Transistors, Theory and Practice*, New York: McGraw-Hill.

Faulkner E. A. (1969): The bipolar transistor as a voltage-operated device. *Radio Electronic Engnr.* **37**, 303–304.

Faulkner E. A., Dawnay J. C. G. (1965): Characteristics of silicon transistors. *Electronics Lett.* **1**, 224–225.

Getreu I. (1974): Modeling the bipolar transistor. *Electronics* 19 September, 114–120; 31 October, 71–75; 14 November, 137–143.

Getreu I. E. (1976): *CAD for Electronic Circuits: Modeling the Bipolar Transistor*, Vol. 1, New York: Elsevier Scientific. ISBN 0-444-41722-2.

Gray P. E., Searle C. L. (1969): *Electronic Principles. Physics, Models, and Circuits*, New York: John Wiley. ISBN 471-32398-5.

Gray P. E., Searle C. L. (1969): *Electronic Principles. Physics, Models, and Circuits. Instructors Manual*, New York: John Wiley. ISBN 471-32399-3.

Gronner A. D. (1970): *Transistor Circuit Analysis*, New York: Simon and Schuster. Revised Edn.

Gummel H. K., Poon H. C. (1970): An integral charge-control model of bipolar transistors. *Bell Sys. Tech. J.* **49** May, 827–892.

Hambley A. R. (1994): *Electronics. A Top-Down Approach to Computer-Aided Circuit Design*, New York: Macmillan.

Hamilton T. D. S. (1977): *Handbook of Linear Integrated Electronics for Research*, London: McGraw-Hill. ISBN 0-07-084483-6.

Riordan M., Hoddeson L. (1997): *Crystal Fire. The Birth of the Information Age*, New York: W. W. Norton. ISBN 0-393-04124-7. (A 'popular' if somewhat breathless account of the discovery and development of the transistor.)

Schubert T., Kim E. (1996): *Active and Non-linear Electronics*, New York: John Wiley. ISBN 0-471-57942-4.

Tuinenga P. W. (1988): *SPICE: A Guide to Circuit Simulation and Analysis Using PSPICE*, Englewood Cliffs: Prentice Hall. ISBN 0-13-834607-0. (3rd Edn 1995, ISBN 0-13-158775-7, see Section 16.5.)

Vladimirescu A. (1994): *The Spice Book*, New York; John Wiley. ISBN 0-471-60926-9.

Whitfield G. R. (1968): Frequency response measurements on silicon planar transistors. *Radio Electronic Engnr.* **36**, 335–340.

Zetex (1995): *Temperature Effects on Silicon Semiconductor Devices*, Zetex Design Note DN-4, June.

# 4.7 Field effect transistors

Prediction is very hard . . . particularly of the future.

Neils Bohr

As with bipolar transistors the description of the inner action and the analysis of field effect transistors is covered by an extensive range of books. The primary difference between these and bipolar transistors is that field effect transistors are voltage controlled as against current controlled. The input resistance is very high, though this does not necessarily mean that they can be controlled with insignificant inputs since we usually have to contend with substantial input capacity, which in a sense means that they are in effect current controlled when working at any significant frequency. It is also possible to consider both bipolars and FETs on the same basis as charge controlled (Severns 1984, Section 2.2). There are two classes of FETs. Junction FETs use a reverse biased p-n junction to create the control gate and are of low power. Insulated gate MOSFETs use a very thin insulating layer to isolate the gate, and are also referred to variously according to their form of construction as VMOS, DMOS, etc., and are usually of high power. They can both be N-channel or P-channel, and for MOSFETs can be of enhancement or depletion mode, though the latter mode is uncommon. The symbols and polarities of these various types are illustrated in Fig. 4.7.1.

It should be noted that for JFETs the bias $V_{GS}$ should not cross into the forward region by any significant amount otherwise the junction diode will begin to conduct. For the MOSFETs the insulated gate allows bias of either polarity without gate conduction. There is a breakdown limit which is typically 15–20 V, so you must be careful not to exceed this. As the input resistance is very high, static charges can readily cause the limit to be exceeded with probable destruction. The arrow on the MOSFET symbol is in the opposite sense to that for the corresponding bipolar device since this represents the body-channel diode and is not directly related to the gate function. The body lead is brought out separately on some devices, but is more commonly internally connected to the source as the above symbols show. The alternative symbols illustrated show the body connection separately as well as indicating whether the device is an enhancement or depletion

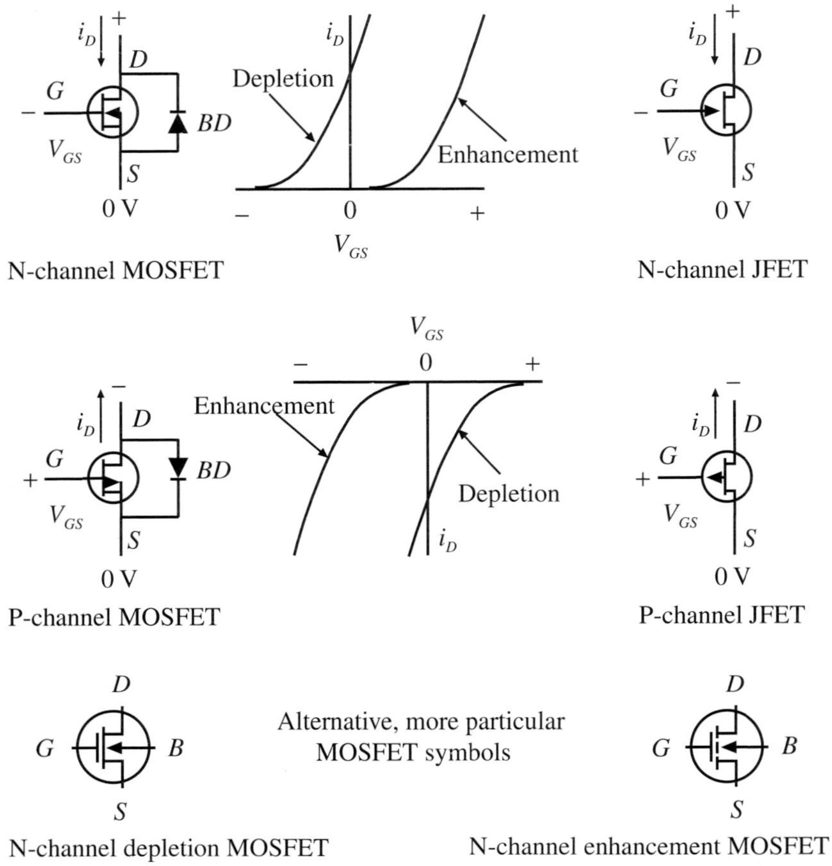


**Fig. 4.7.1** Symbols and operating polarities for the various forms of FET. For MOSFETs the body-diode *BD* is sometimes explicitly included, either within the transistor symbol or outside it as illustrated, but it is often omitted and taken as understood.

type. In MOSFETs there is in fact a reverse biased diode connected between source and drain arising from the usual form of construction. This is not shown on the basic symbols as shown in Fig. 4.7.1, but is sometimes included in the symbol or shown separately. If the FET source–drain voltage is reversed this diode will conduct, though the forward voltage is commonly somewhat higher than for a standard p-n diode. Many MOSFETs are 'avalanche rated' which means that they can withstand breakdown, within energy limits, when the limiting $V_{DS}$ is exceeded as for example from inductive kickback, so that a protection diode is not always necessary. FETs are majority carrier devices and so do not have the time delays associated with minority carriers in the base region of bipolar transistors. They are in principle capable of very fast switching (Oxner 1989), but there are difficulties as we will consider below.

To see the characteristics of a FET we can run a SPICE simulation using a con-

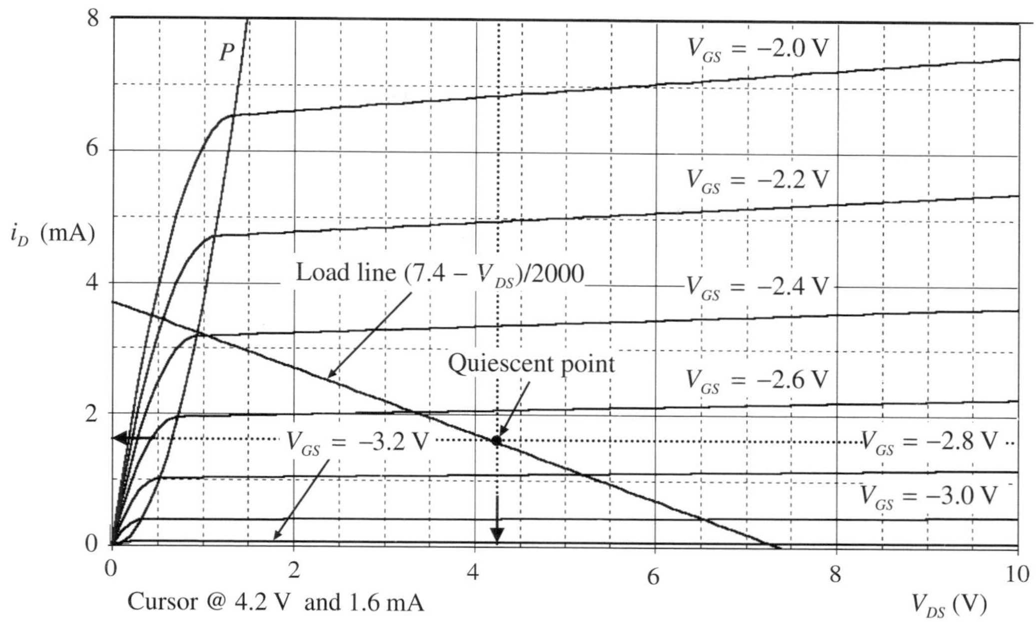


**Fig. 4.7.2** Characteristics of a type U310 JFET transistor.

venient model. We have chosen a N-channel JFET type U310, which is used as a low-noise example in Section 3.11. Some characteristics are shown in Fig. 4.7.2.

The bias voltages are negative with a cut-off value of about −3.4 V. The curve indicated by $P$ is the locus of the pinch-off voltage $V_P$, which indicates the boundary between the 'resistive' (or ohmic or triode) region to the left and the saturation or pinch-off region to the right. Normal amplifying operation is in the pinch-off region. In the resistive region the response is approximately linear near the origin and so can be used for devices such as voltage-controlled resistors or analog gates. The dividing line is given by the equation:

$$i_D = KV_{DS}^2 \quad \text{or, at the limit of zero bias } V_{GS}=0, \quad \text{by} \quad I_{DSS} \cong KV_P^2 \tag{4.7.1}$$

where $I_{DSS}$ and $V_P$ are values usually given in the data sheet for the device. Looking up the device model in the SPICE library we find:

$$\begin{aligned} &V_P \equiv VTO = 3.324 \quad \text{and} \quad K \equiv BETA = 3.688 \times 10^{-3} \quad \text{so} \\ &I_{DSS} = 40.7\ \text{mA} \quad \text{and for the pinch-off locus } i_D = 3.688 \times 10^{-3} \times V_{DS}^2 \end{aligned} \tag{4.7.2}$$

which is the pinch-off curve $P$ shown in the figure. $I_{DSS}$ is off scale but can be verified by simulating with $V_{GS}$ running from zero volts, which results in a value of 41.8 mA. In the saturation region the drain current is substantially independent of $V_{DS}$ but there is always some small increase as shown.

To construct an amplifier it is usually necessary to include a source resistor $R_S$ to generate the required negative gate bias as shown in Fig. 4.7.3($a$):

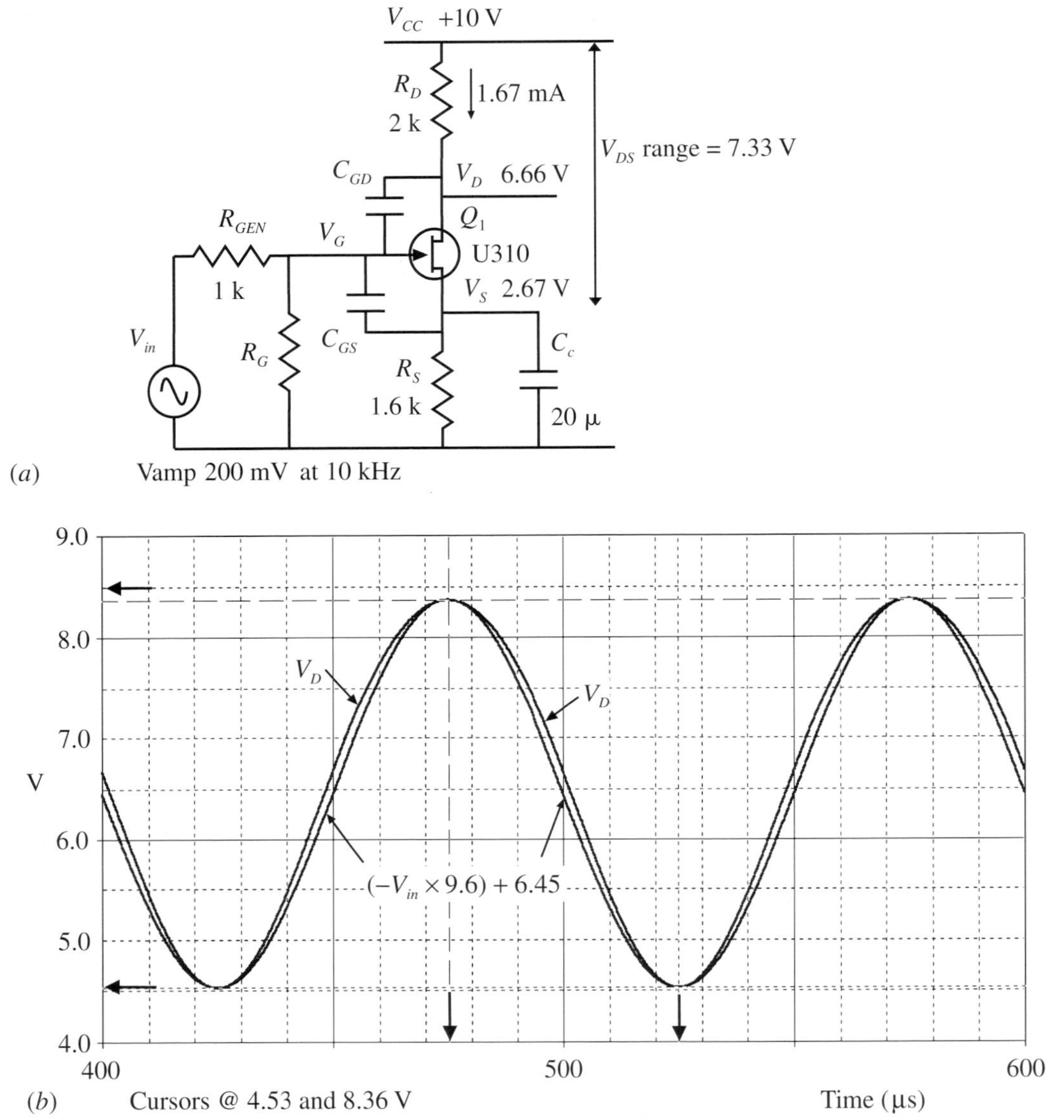


**Fig. 4.7.3** (*a*) Junction FET amplifier for a gain of ten. (*b*) Signals for maximum allowed input. The input signal is displayed as multiplied by the measured gain, and offset to allow comparison with the output, showing some distortion.

The capacitors $C_{GD}$ and $C_{GS}$ represent the internal capacities of the device and from the SPICE model the values are found to be the same at 7.4 p. $R_G$ is included as a reminder that the gate should not be left floating and for the simulations is set to a high value and so has no influence.

The choice of where to start in designing even as simple a circuit as this presents some difficulty. We will begin by specifying a gain of ten and a supply of +10 V.

From the characteristic curves of Fig. 4.7.2 you may determine the transconductance $g_m$ (some use $g_{fs}$) given by $\partial i_D/\partial V_{GS}$, from the vertical spacing of the $V_{GS}$ curves in the region of operation, as about 5 mS (or mA $V^{-1}$) so that we need a load resistor $R_D=2$ k for a gain of ten. This resistor will also determine the high frequency cut-off but we will see later what that turns out to be. As the spacing of the $V_{GS}$ curves is not equal for equal changes then $g_m$ will depend to some extent on the operating point. This also means that the output signal for any significant excursion must be distorted. To give maximum output voltage swing the drain current $i_D$ must be such as to place the quiescent drain voltage $V_D$ about half way between the available voltage range limits. To use the characteristic curves we must differentiate between the supply voltage (+10 V) and the accessible range of $V_{DS}$. Allowing 1 V ($\approx V_P$) minimum for the transistor and about 2 V for bias so that the accessible range for $V_{DS}=(10-3)$ V means that $i_D\approx(10-3)/2\times2$ k$=1.75$ mA. Looking at the characteristics we see that for this current we need a bias of about $-2.6$ V rather than 2 V, which reduces the current to about 1.6 mA. The applicable range of $V_{DS}$ is now 0 to 7.4 V so we can add a load line, given by $i_D=(7.4-V_{DS})/2$ k to the characteristics as shown in Fig. 4.7.2. This load line is the locus of the instantaneous operating point as a function of the input $V_{GS}$. To set the quiescent operating point at the centre of the accessible $V_{DS}$ range between 1 and 7.4 V, i.e. at 4.2 V, we need a bias of $\approx$2.6 V and with $i_D=1.6$ mA this makes $R_S=1.66$ k – say 1.6 k. We can now run the bias point simulation and find the values shown on the figure. The output signal can be allowed to swing about $\pm3$ V about the quiescent value of $V_{DS}\approx4.2$ (or $V_D\approx6.6$). As we expect the gain to be $\times10$ we can input a sinusoidal signal of amplitude 0.3 V. This results in substantial distortion, so the result for an input of 0.2 V is shown in Fig. 4.7.3(*b*). A few iterations will usually be necessary to obtain the desired results but SPICE makes this fairly simple. It is not worth working to high accuracy since even transistors of the same type can vary considerably. The large negative feedback at z.f. helps considerably in reducing the effects of such variations on the quiescent operating point.

Running a frequency sweep gives an upper $-3$ dB point of 4 MHz and a gain of 9.6. The device capacities given in the model are for zero bias voltage on the junctions and are $C_{GD}(0)=C_{GS}(0)=7.4$ p. The change in capacity as a function of reverse voltage is given by (Vladimirescu 1994, p. 97):

$$C_{GD}(V_{GD})=\frac{C_{GD}(0)}{(1-V_{GD}/PB)^{\frac{1}{2}}}=\frac{7.4}{(1+4/1)^{\frac{1}{2}}}=3.3\text{ p} \tag{4.7.3}$$

where from the model the potential barrier $PB=1$ V, and we have used the approximate quiescent value of $V_{GD}=-4$ V. Allowing for the Miller effect, the effective input capacity is then about ($C_{GD}\times9.6+C_{GS}$), say 36 p. With $R_{GEN}=1$ k, this gives a $-3$ dB point at 4.4 MHz.

MOSFETs have rather similar characteristics to JFETs as illustrated in

Fig. 4.7.2 except that the operating bias voltages, for an equivalent N-channel enhancement type, will be positive rather than negative. The gate–source voltage $V_{GS}$ must be greater than the threshold voltage $V_{TH}$ before any $i_D$ conduction occurs. The current capability will generally be much larger and the on-resistance $R_{DS(on)}$ very low, of the order of tens of milliohms for the higher power devices. Manufacturers vie with one another to have the lowest values of $R_{DS(on)}$ with a secondary aim to have low values of $V_{TH}$. Using MOSFETs as amplifiers at low frequencies is fairly straightforward, for example as discussed by Severns (1984, see Section 6.6). They are also popular for high power r.f. amplifiers (Motorola 1995) though the internal structure is usually somewhat different, but we will not consider this application. A larger proportion of applications use the device as a switch so that it is either on or off rather than operating in a linear mode. This form of operation is attractive in that the power dissipation in the device is low and makes particular use of the very low values of $R_{DS(on)}$ now available. To make effective use of this mode it is vital to switch the device very rapidly to minimize the time in the linear regime where the maximum power loss occurs, and this is where the problems arise. To obtain such very low resistance power, FETs consist in effect of a very large number of parallel connected devices, of the order of $10^5$ per cm$^2$, which, however, leads to very large device capacities which dominate the switching time. For large MOSFETs the input capacity is in the range 1–10 nF and this has to be charged very rapidly requiring a low impedance source. The position is complicated by the changing gain of the FET as it passes from non-conducting, through conducting and back to non-conducting. The gain has a major influence on the effective value of the capacity arising from the Miller effect so the charging regime is not linear.

To switch the FET on and off, the effective gate capacity must be charged and then discharged, so the driver circuit must be able to both source and sink large currents. Curves are commonly provided in the datasheet illustrating how the charge on the gate changes as the device is turned on, together with the change of $V_{DS}$. A circuit for deriving these curves is given by Pelly (1993b) and is shown in Fig. 4.7.4.

With the input $I_2$ at zero there will be no drain current and the current from the generator $I_1$ must flow through the freewheel diodes as $i_F$. Then at time 100 ns, $I_2$ changes to −20 mA (negative since $I_2$ flows internally from + to −, as shown similarly for the $I_1$ symbol) and the gate capacity starts charging up. When the threshold voltage is reached $i_D$ grows, and hence $i_F$ falls, until $i_D$ reaches the full set value of 1 A. $V_{DS}$ falls and the Miller effect via $C_{GD}$ causes the horizontal section of $V_{GS}$. When $V_{DS}$ bottoms, then $V_{GS}$ can resume its upward path. Integrating $i_G$ will give the total charge delivered, which is seen to be $Q=20\times10^{-3}\times500\times10^{-9}=10$ nC. This allows the inclusion of the additional horizontal scale in terms of gate charge as shown. Though the charge may seem small the coulomb is a large unit. One

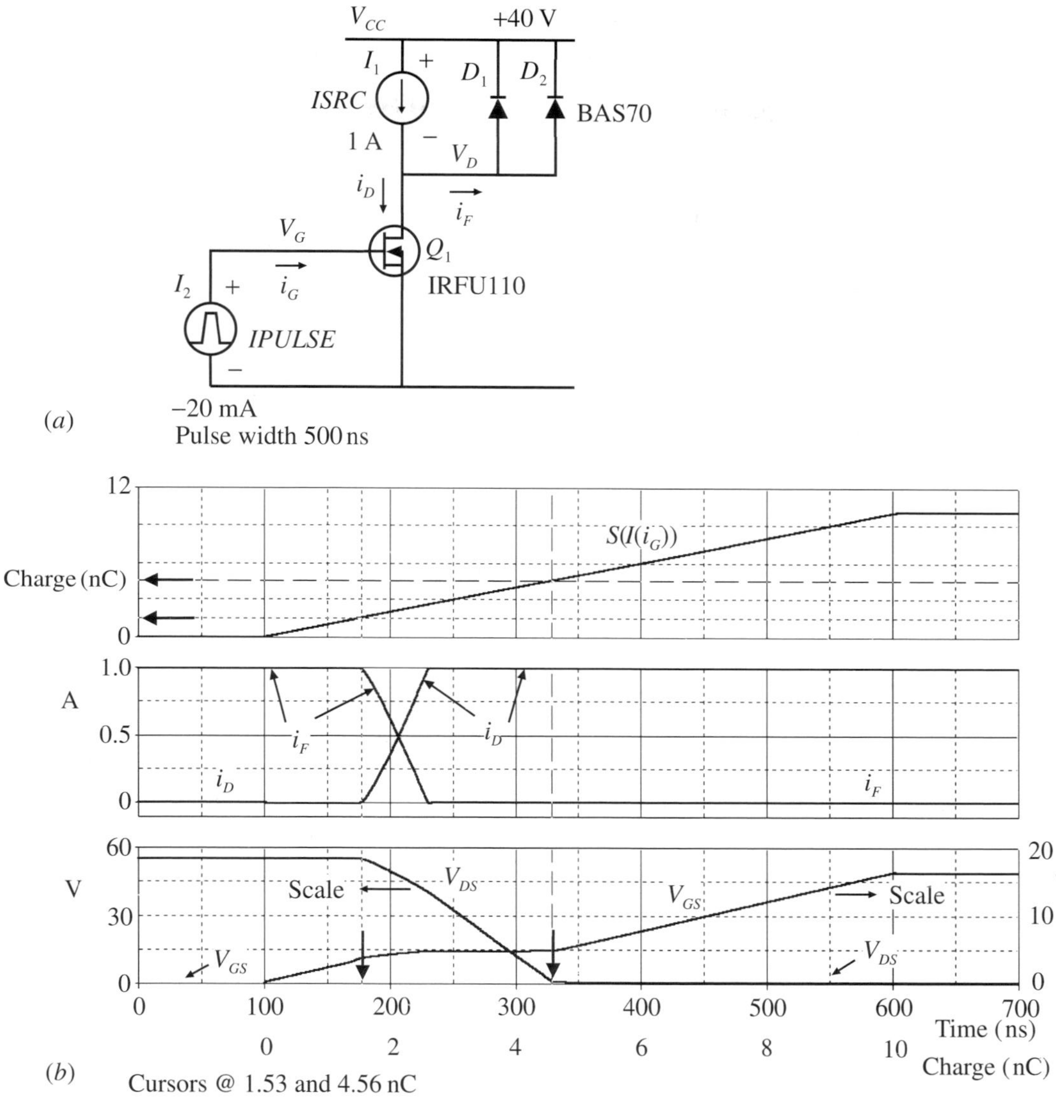


**Fig. 4.7.4** (*a*) Test circuit for gate capacity charging. (*b*) Simulation responses. The top curve is the integral as a function of time ($S(I(i_G))$ in PSpice terminology) and hence is a measure of the charge delivered to the gate as a function of $V_{GS}$.

coulomb placed on a sphere the size of the Earth (Eq. (4.2.4) and just below) would raise the potential by nearly 1400 V. To charge the gate in the very short times required will need large currents. The change in the effective value of $C_{GD}$ has two components: the dynamic change due to the Miller effect, and a rapid and substantial increase (see Eq. (4.7.3)) as the falling $V_{GD}$ passes through the rising $V_{GS}$ and in fact reverses the voltage between drain and gate (Severns 1984, pp. 3–10). Though this test allows the determination of the effective capacities it is not directly representative of normal operation since here the drain current is kept constant.

Capacities of FETs are recorded in two ways. They can be stated in a direct way as the capacity between gate and source or drain and are given the symbols $C_{GS}$ and $C_{GD}$. They are often measured with various combinations of terminal shorted and the terminology can be somewhat confusing since there are many subscripts. The subscripts $i$:input, $o$:output and $r$:reverse, together with $d$:$D$:drain and $g$:$G$:gate are the most obvious. The subscript $s$:$S$:source can be confused with $s$:$S$ for shorted, but the latter significance only applies as the third of three subscripts (e.g. Landon 1989). Measurements are usually made at some high frequency, e.g. 1 MHz, and the results are given in terms of three capacities (Gray and Searle 1969, p. 430; Severns 1984):

$C_{iss} = C_{gd} + C_{gs}$ is the common-source input capacity (drain-source shorted)

$C_{rss} = C_{gd}$ is the common-source reverse transfer capacity

$C_{oss} = C_{gd} + C_{ds}$ is the common-source output capacity

$$\text{so} \quad C_{gs} = C_{iss} - C_{gd} = C_{iss} - C_{rss} \tag{4.7.4}$$

and the small capacity $C_{ds}$ may usually be ignored. Allowance must be made for the variation of the capacities as illustrated by Eq. (4.7.3).

The common need for high current, high speed drivers has meant that many semiconductor manufacturers have produced suitable ICs that can source and sink say up to 2 A or more and can be driven from low power logic signals (see references). There are also very many more complex IC devices that include a range of control and protection features, or have level shifters to allow control of FETs floating at high voltages. There are also many combined FET devices with two or more transistors, of the same polarity or complementary pairs, suitable for half or full bridge circuits.

We will examine one of the simple push–pull configurations used for driving an inductive load such as a motor (Fig. 4.7.5($a$)). For convenience the load is connected to a supply fixed at half the main supply. The waveforms shown cover a small interval at one of the transitions so that we can see the cross-conduction that can occur when the two FETs are briefly on simultaneously. The lower graphs show the main voltages and currents and it is evident that at crossover there is a short period where current flows directly from $V_{CC}$ to common through both transistors. We have shown the source currents rather than the drain currents to eliminate any current through $C_{GD}$. The third graph shows the instantaneous power dissipation in $Q_2$ as the product of the source current and the output voltage $V_L$. The peak dissipation is 24 W! The top graph shows the integral (with respect to time) of the power dissipation, i.e. energy, with the scale in μWs. The total from the start of the input pulse (way off to the left) and up to point $X$ is about 0.293 while the increment between $X$ and $Y$, arising primarily from the cross-conduction, is about 0.160

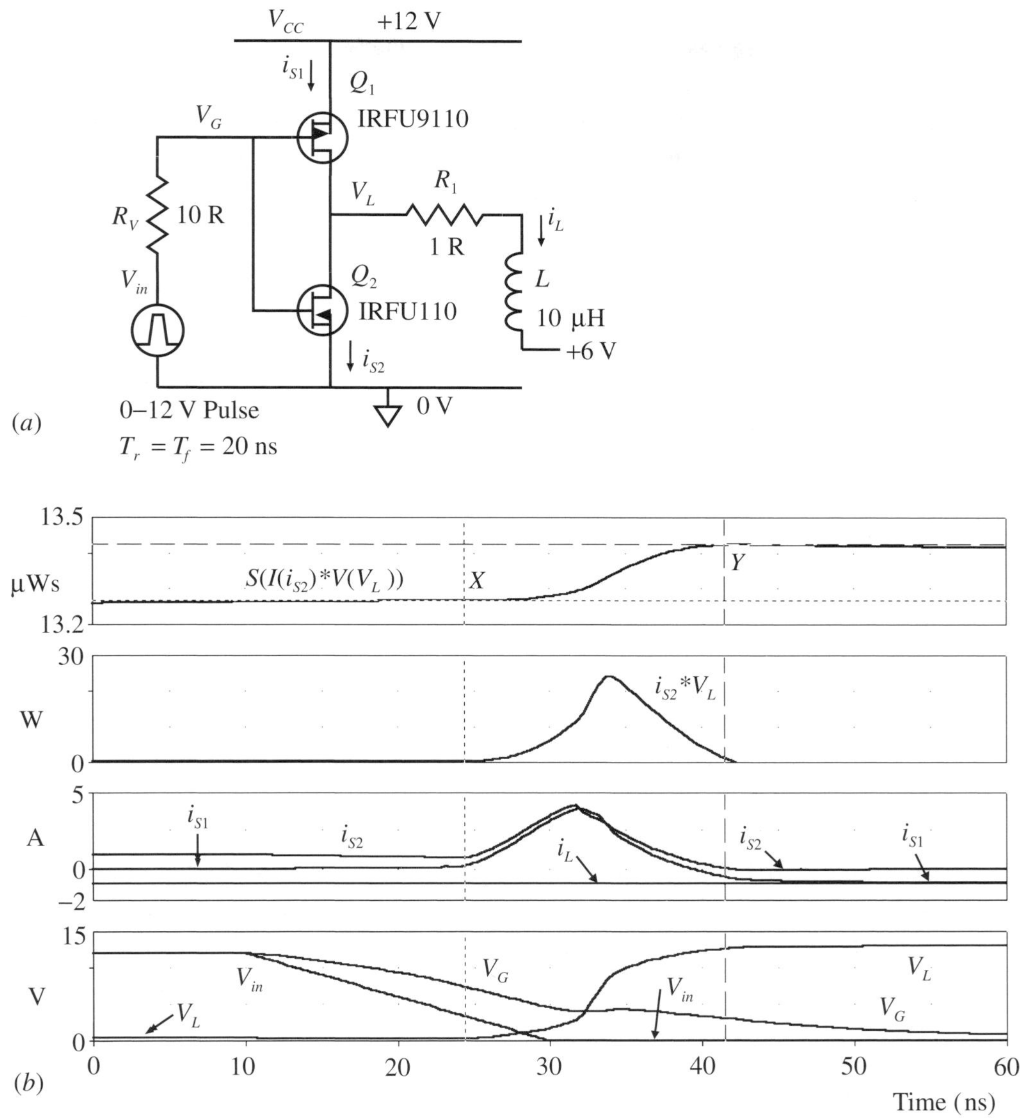


**Fig. 4.7.5** (*a*) Push–pull or half bridge power stage using complementary MOSFETs. (*b*) Simulation waveforms.

μWs, a ratio of nearly 55%. So, be warned and check your circuit carefully and take steps to decrease this conduction. The spikes can be very fast so you will need a good scope to see them. Not only do they substantially increase power dissipation but since they are fast they can cause considerable electromagnetic interference.

MOSFETs have one very useful advantage over bipolar transistors in that $R_{DS(on)}$ increases with increase in temperature, which makes it much less likely that thermal runaway can occur. The position is not quite as simple as sometimes made out, and is discussed in some detail by Severns (1984, see Sections 5.2 and 5.3). For

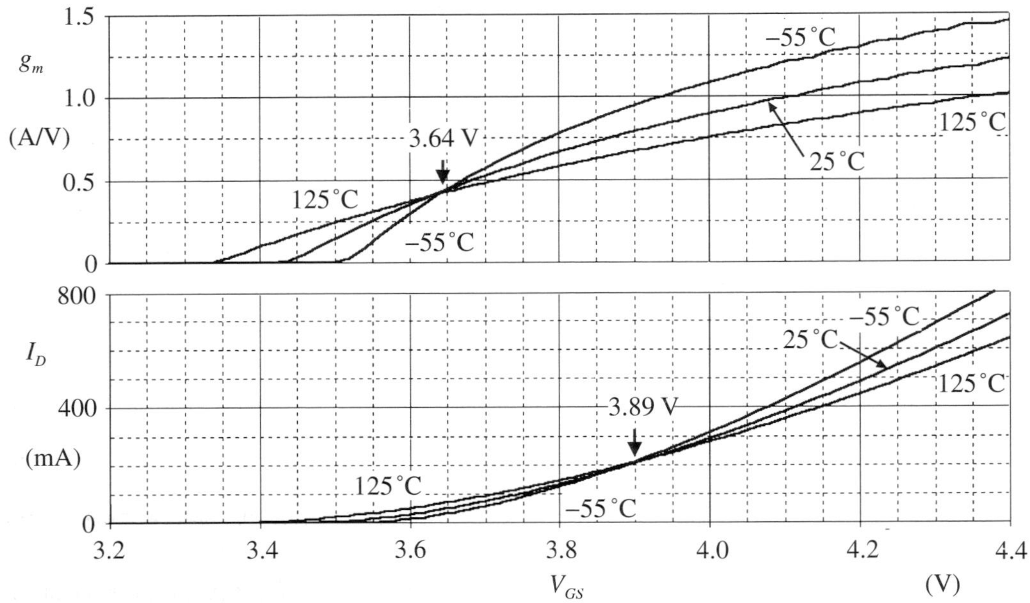


**Fig. 4.7.6** $I_D$ and $g_m$ *versus* $V_{GS}$ as a function of temperature for a FET.

operation as a switch where the FET is hard off or on then it is generally safe to expect that limiting will occur and runaway will be avoided. This does not mean that the device cannot be destroyed by excessive power dissipation, but that would signify a failure of basic design. The relationship between $V_{GS}$ and $I_D$ as a function of temperature is illustrated in Fig. 4.7.6, showing that both $V_{TH}$ and $g_m$ change with temperature.

The lower graph shows that the temperature coefficient (TC) changes sign and that there is a point of zero TC. The upper graph shows the differential of the lower set and hence represents $g_m$ which is given by $\partial I_D/\partial V_{GS}$. It also makes the change in $V_{TH}$ more evident. (Note: do not make the increment in $V_{GS}$ too small when running the simulation; 0.02 was found to be the minimum to avoid noise on the differentials.) For linear operation the quiescent operating point would tend to be to the left of the zero TC point, where the coefficient is positive, so that for a fixed $V_{GS}$ runaway could occur.

## SPICE simulation circuits

| | |
|---|---|
| Fig. 4.7.2 | Jfet1.SCH |
| Fig. 4.7.3(*b*) | Jfet2.SCH |
| Fig. 4.7.4(*b*) | Mfet3.SCH |
| Fig. 4.7.5(*b*) | Mfet4.SCH |
| Fig. 4.7.6 | Mfet5.SCH |

## References and additional sources 4.7


Andreycak, Bill (2000): *New Driver ICs Optimize High Speed Power MOSFET Switching Characteristics*, Texas Instruments/Unitrode Application Note U-118.

Clemente S. (1993): *Gate Drive Characteristics and Requirements for Power HEXFETs*, International Rectifier Application Note 937B.

Fairchild Semiconductor (1999): *MOSFET Basics*, Fairchild Korea Semiconductor Application Note AN9010, April. www.Fairchildsemiconductor.co and go to application notes.

Gray P. E., Searle C. L. (1969): *Electronic Principles. Physics, Models, and Circuits*, New York: John Wiley. ISBN 471-32398-5. Also *Instructors Manual*, ISBN 471-32399-3.

Hambley A. R. (1994): *Electronics. A Top-Down Approach to Computer-Aided Circuit Design*, New York: Macmillan. See Chapter 6.

Hamilton T. D. S. (1977): *Handbook of Linear Integrated Electronics for Research*, London: McGraw-Hill. ISBN 0-07-084483-6.

Harnden J. (1992): *Designing with Complementary Power MOSFETs in Surface-Mount (SO-8) Packages*, Siliconix Application Note AN-90-4 (AN801).

Landon B. (1989): *Understanding JFET Parameters*, Siliconix Application Note LPD-2. Note his eqn. 9 is in error.

Malouyans S. (1984): *SPICE Computer Models for HEXFET Power MOSFETs*, International Rectifier Application Note 975B.

Motorola Semiconductor (1995): *RF Application Reports*, Motorola Inc. Handbook HB215/D.

Oxner E. (1989): *Subnanosecond Switching with DMOS FETs*, Siliconix Application Note LPD-11.

Oxner E. (1989): *P-Channel MOSFETs, The Best Choice for High-Side Switching*, Siliconix Application Note LPD-17.

Oxner E., Bonkowski R. (1989): *Depletion-Mode MOSFETs Expand Circuit Opportunities*, Siliconix Application Note LPD-18.

Pelly B. R. (1993a): *The Do's and Don'ts of Using Power HEXFETS*, International Rectifier Application Note 936A.

Pelly B. R. (1993b): *A New Gate Charge Factor Leads to Easy Drive Design for Power MOSFET Circuits*, International Rectifier Application Note 944A.

Ronan H. R., Wheatley C. F. (1992): *Power MOSFET Switching Waveforms: A New Insight*, Harris Semiconductor Application Note 7260.1, May.

Schubert T., Kim E. (1996): *Active and Non-linear Electronics*, New York: John Wiley 1996. ISBN 0-471-57942-4.

Severns R. (Ed.) (1984): *MOSPOWER Applications Handbook*, Siliconix Inc. ISBN 0-930519-00-0.

Siliconix (1989): *FET Biasing*, Siliconix Application Note LPD-3.

Tuinenga P. W. (1988): *SPICE: A Guide to Circuit Simulation and Analysis Using PSPICE*, Englewood Cliffs: Prentice Hall. ISBN 0-13-834607-0. (3rd Edn, 1995, ISBN 0-13-158775-7. See Section 16.3.)

Vladimirescu A. (1994): *The Spice Book*, New York: John Wiley. ISBN 0-471-60926-9.

Watson J. (1970): *An Introduction to Field-Effect Transistors*, Siliconix.

FET driver examples: Texas Instruments TPS28XX; Maxim MAX62X; Elantec EL71XX, 72XX; Teledyne TSC442X; Motorola MC3415X. There are many variants in each series (XX).

# 4.8 Temperature dependent resistors

First get your facts; and then you can distort them at your leisure.

Mark Twain

In normal circumstances we would wish our resistors to have values independent of temperature. There are, however, times when we particularly want the resistance to vary with temperature to provide, for example, sensors for measuring temperature or for protection in case of a fault. In practice, suitable materials for temperature sensing have negative temperature coefficients (NTC) and those for protection have positive temperature coefficients (PTC). This differentiation is essentially one of availability of materials rather than any inherent properties. The name thermistor is often used to refer to both kinds, though nowadays it is taken to refer more commonly to the NTC types while the PTC types are referred to as posistors. PTC types have low resistance for the normal range of currents with a rather sharp increase at some critical value, hence providing protection against excessive current and possible damage. Hyde (1971) provides a general treatment of all these types of device. For our purposes we will only consider the NTC types so that we can make use of their characteristics for purposes of stabilization in circuits. The use for direct temperature measurement is relatively straightforward and will not detain us (see e.g. Williams 1997).

To enable us to use thermistors in simulations it is necessary to have a model. It is fairly straightforward to set up one for a temperature-dependent resistor (Hirasuna 1999) but we require a model that will reflect both the effect of ambient temperature as well as the internal power dissipated in the device due to the current through it. A suitable arrangement can be found on the Web (Epcos/Siemens 1999) and we will examine how this functions. The circuit of the model is shown in Fig. 4.8.1.

The model consists of two parts, an isolated controlled resistance between nodes No. *1* and No. *2*, and a separate thermal controller referred to common, node No. *0*. The model description is also in two subcircuit parts: *.subckt NTC*, which describes the common structure, and the other for the particular device which provides the values of the actual parameters. *.subckt C619_10000* illustrated is for a thermistor of value 10 k at 25 °C.

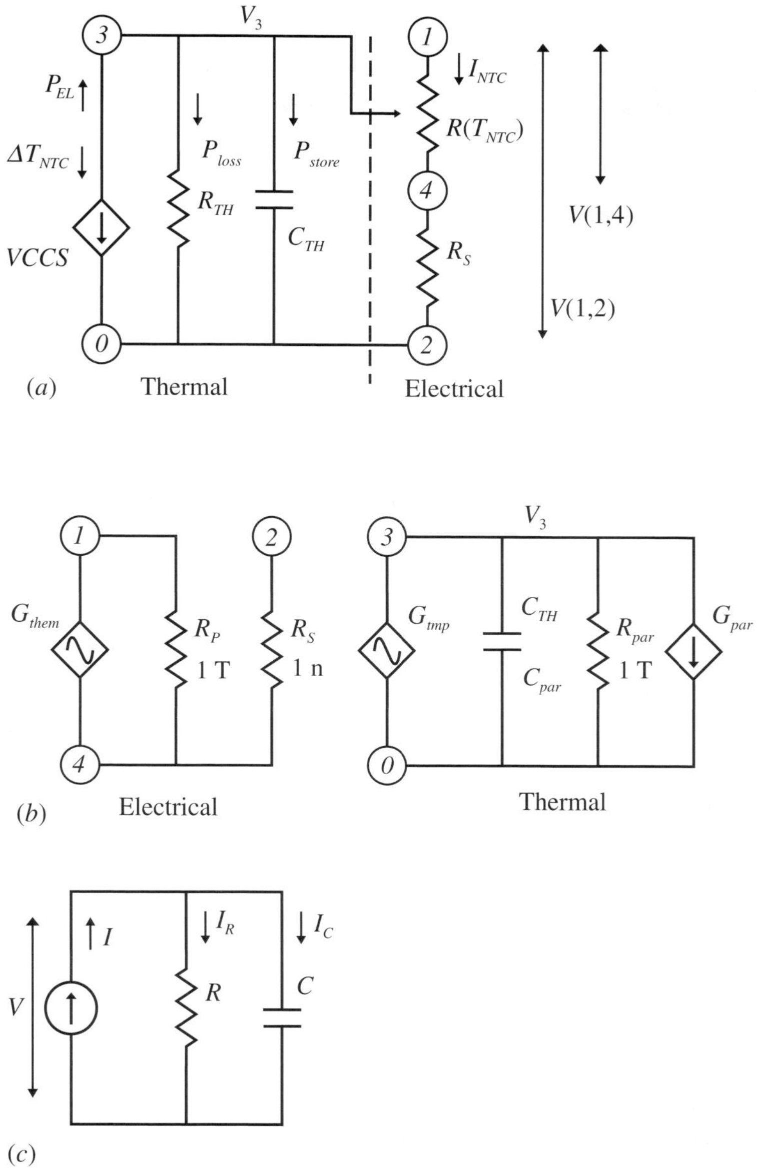


**Fig. 4.8.1** (*a*) Model configuration as given by Epcos. (*b*) Circuit derived from the model description. (*c*) An electrical equivalent to the thermal generator section.

```
.subckt NTC 1 2 Params:B0=1, B1=1, B2=1, B3=0, B4=0, R25=10,
+CTH=1, GTH0=1, GTH1=1, T=1,
+TK=273.15
*
Gthem 1 4 Value={V(1,4)/(R25*T*(exp((((B4/(V(3)+TK+TEMP)+B3)/(V(3)+TK+TEMP)+B2)
+/(V(3)+TK+TEMP)+B1)/(V(3)+TK+TEMP)+B0)))}
*
RP 1 4 1T
RS 4 2 1n
*
Gtmp 0 3
Value={V(1,4)*V(1,4)/(R25*T*(exp((((B4/(V(3)+TK+TEMP)+B3)/(V(3)+TK+TEMP)+B2)
+/(V(3)+TK+TEMP)+B1)/(V(3)+TK+TEMP)+B0)))}
C_par 3 0 {CTH}
R_par 3 0 1T
Gpar 3 0 Value={V(3)*(GTH0+(GTH1*(V(3)+TEMP)))}
.ends
*
*
.subckt C619_4700 1 2 PARAMS: TOL=0
X1 1 2 NTC Params:
+B0=-9.83085430743211
+B1=7716.66739328665
+B2=-1419289.76278818
+B3=216177153.122501
+B4=-14030842059.0076
+R25=4700
+T={1+TOL/100}
+CTH=0.0012   GTH0=0.0034586   GTH1=3.11E-6   TK=273.15
.ends
*
*
*
.subckt C619_10000 1 2 PARAMS: TOL=0
X1 1 2 NTC Params:
+B0=-11.9847664230072
+B1=3069.94526980822
+B2=418296.198088036
+B3=-102412371.979498
+B4=6690146322.502
+R25=10000
+T={1+TOL/100}
+CTH=0.0012   GTH0=0.0034586   GTH1=3.11E-6   TK=273.15
.ends
*
```

(Note: the minus signs have been written as hyphens (-) as that is what you would have to type).

Note that '*T*' used in the expressions for the resistance, (as in *R25*T**...) refers to the tolerance on *R25* as may be specified in the statement *T*={1 + *TOL*/100}, and has no connection with temperature. As *TOL* is specified as zero, *T* reduces to unity. The '*T*' used in the values of *RP* and *R_par* is the unit 'tera' and again has no connection with temperature. There are too many *T*'s around, so be careful. The *B* values look impractically precise, but since SPICE works to such precision it is no matter.

Following Epcos we can consider the operation of the model. The thermal circuit consists of a *VCCS* controlled by the voltage $V(1, 4)$ across the thermistor ($R_S$ is too small to be significant), $C_{TH}$ represents the heat capacity and $R_{TH}$ the heat loss to the surroundings. The dissipation factor $G_{TH} = 1/R_{TH}$ is used instead, which allows a linear dependence on temperature of conduction to the surroundings according to:

$$G_{TH} = G_{TH0} + G_{TH1} * T_{NTC} \tag{4.8.1}$$

with the *G* factors being specified in the model. These factors are of course significantly dependent on the physical mounting circumstances of the device. The time constant that $C_{TH}$ provides, represents the delay in heating up and the inability of the device to respond to rapid changes in the electrical input. The temperatures used are:

$T_{amb}$ is the ambient temperature in °C, and is represented by *TEMP* in SPICE

$T_{EL}$ is the temperature rise in °C owing to the electrical input and is given by $V(3)$ in the model

$T_{NTC}$ is the actual temperature of the device in °C and $= T_{EL} + T_{amb} = V(3) + TEMP$ (4.8.2)

$TK$ is the absolute temperature equivalent of 0 °C and = 273.15K

$T_{abs}$ is the absolute temperature in degrees K and $= V(3) + TEMP + TK$

We can see the relationship between the electrical input and the thermal consequence as follows. If the input electrical power is $P_{EL}$, the stored power is $P_{store}$ and the power loss to the surroundings is $P_{loss}$, then the energy, or more properly power, balance gives:

$$P_{EL} = P_{store} + P_{loss}$$

$$\text{or} \quad I_{NTC} \times V(1, 4) = C_{TH} \times \frac{dT_{NTC}}{dt} + G_{TH}(T_{NTC} - T_{amb}) \tag{4.8.3}$$

and we may compare this with the circuit shown in Fig. 4.8.1(*c*):

$$I = I_C + I_R = C\frac{dV}{dt} + \frac{V}{R} \quad \text{so we have the equivalences}$$

$$T_{NTC} - T_{amb} = T_{EL} \Rightarrow V, \quad C_{TH} \Rightarrow C, \quad G_{TH} \Rightarrow 1/R \tag{4.8.4}$$

$$P_{EL} = I_{NTC} \times V(1,4) \Rightarrow I, \quad \frac{dT_{NTC}}{dt} \Rightarrow \frac{dV}{dt}$$

and so from the circuit (*b*) the 'voltage' $V_3 = T_{EL}$ and the actual temperature $T_{NTC} = V_3 + T_{amb} = V_3 + TEMP$ which can be displayed in *PROBE*.

The variation of resistance with temperature must be given in terms of absolute temperature $T_{abs}$ otherwise we run into difficulties at 0 °C where we would be dividing by zero:

$$R(T) = R_{25} \exp\left( B_0 + \frac{B_1}{T_{abs}} + \frac{B_2}{T^2_{abs}} + \frac{B_3}{T^3_{abs}} + \frac{B_4}{T^4_{abs}} \right) \tag{4.8.5}$$

The voltage-controlled current sources $G_{therm}$ and $G_{tmp}$ can now be defined as shown in the model statements, but it may be noted that the resistance formula used is a sneaky but effective variation of Eq. (4.8.5). If we replace the absolute temperature $T_{abs} = (V(3) + TK + TEMP)$ by $(x)$ for convenience, then what is written in the exponential factor of the model is:

$$\left[\left\{\left[ [B_4/(x) + B_3]/(x) + B_2 \right]/(x) + B_1 \right\}/(x) + B_0 \right]$$

$$= \left[\left\{\left[ \left( \frac{B_4 + xB_3}{x} \right)/(x) + B_2 \right]/(x) + B_1 \right\}/(x) + B_0 \right]$$

$$= \left\{\left[ \left( \frac{B_4 + xB_3}{x^2} + B_2 \right)/(x) + B_1 \right]/(x) + B_0 \right\} = \left\{\left[ \left( \frac{B_4 + xB_3 + x^2B_2}{x^2} \right)/(x) + B_1 \right]/(x) + B_0 \right\}$$

and so on until $= \dfrac{B_4 + xB_3 + x^2B_2 + x^3B_1 + x^4B_0}{x^4} = \dfrac{B_4}{x^4} + \dfrac{B_3}{x^3} + \dfrac{B_2}{x^2} + \dfrac{B_1}{x} + B_0$

as required (4.8.6)

and the reason for this arrangement is that the divisions are all by the same factor $(x) = (V(3) + TK + TEMP)$. This makes for faster computation than the calculation of all the powers of $(x)$.

$G_{tmp}$ is of the form $V^2/R$ and so depends on power with both direct and alternating signals and represents the effects of electrical dissipation $P_{EL}$. $G_{par}$ represents the effect of the ambient conditions and has a sense opposite to $G_{tmp}$ since this represents power input while $G_{par}$ represents power loss. $G_{therm}$ produces a current $i = V(1,4)/R(T)$ or in effect a resistance $R(T) = V(1,4)/i$ as required. This model for a thermistor is used in Section 5.8 as the amplitude stabilizer for a Wien-bridge oscillator. In this application the values of the parameters given in the model above

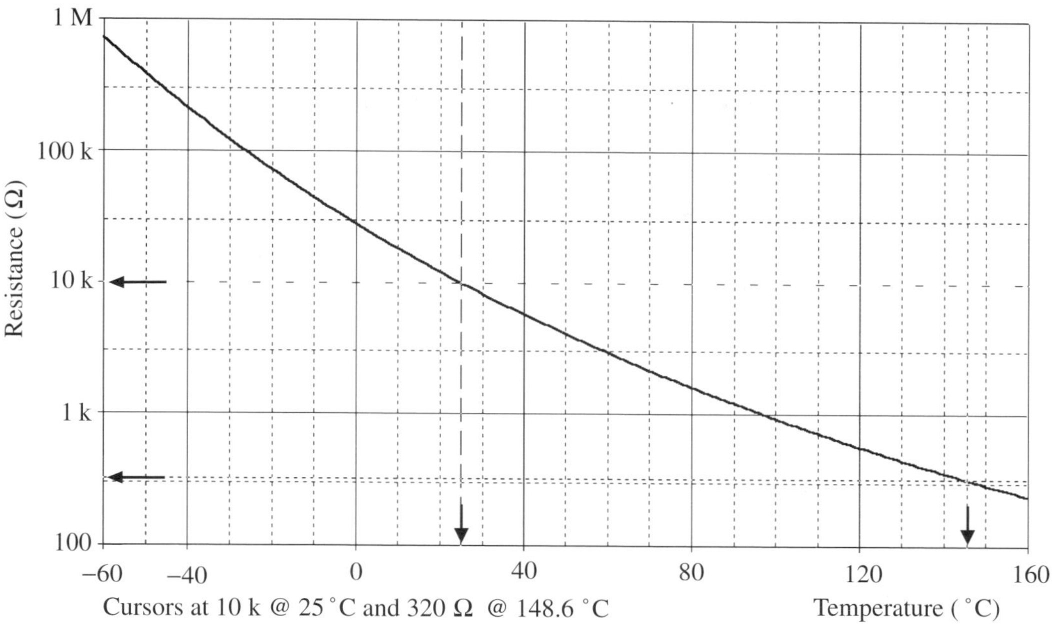


**Fig. 4.8.2** Variation of resistance with ambient temperature.

Table 4.8.1 *Parameter values for Fig. 4.8.3*

| Curve | *CTH* | *GTH0* | *GTH1* |
|---|---|---|---|
| A | 1.2E−3 | 3.459E−3 | 3.11E−6 |
| B | 5.0E−7 | 1.759E−3 | 1.00E−6 |
| C | 5.0E−7 | 1.759E−3 | 3.11E−7 |
| D | 5.0E−7 | 1.759E−3 | 3.11E−8 |

are not suitable, both from the point of view of the time response and the isolation from the ambient environment. After a little experimentation a suitable set of values was realized but it should be remembered that it does not represent any actual device. Variation of resistance with ambient temperature is shown in Fig. 4.8.2. In the simulation the current was so small that there was negligible internal power dissipation.

Some effects of variation of *GTH*0 and *GTH*1 are shown in Fig. 4.8.3. Curve A represents the original Siemens *C619_10000* response. For case D, which is used later, the resistance at the cursor position is 406 Ω and the power dissipation $P_{EL}$ = 2.34 mW.

The response as a function of the parameter *CTH*, found using a simple current

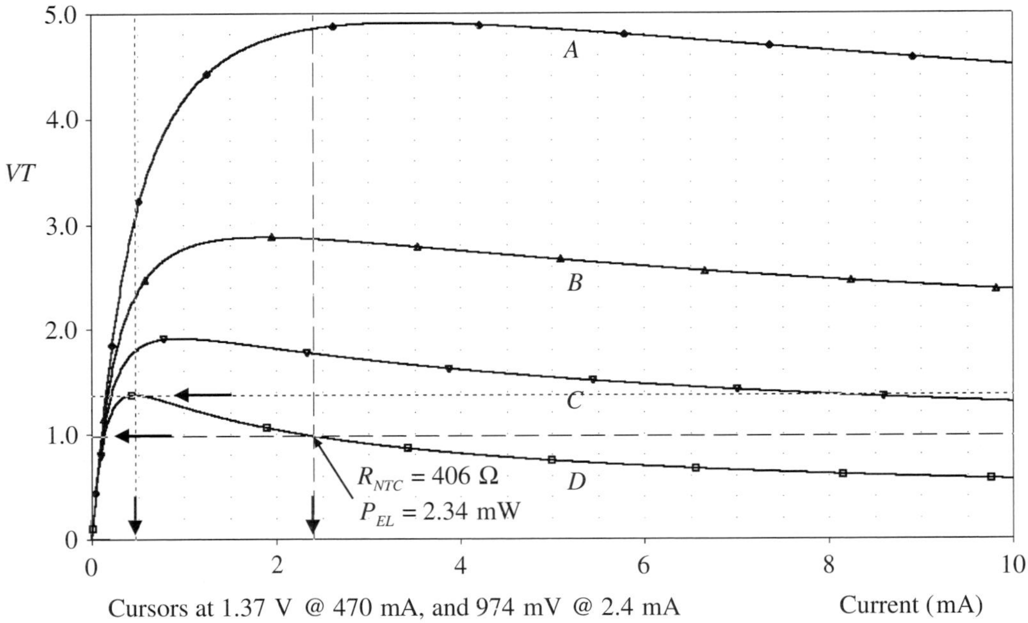


**Fig. 4.8.3** Effect of *GTH*0 and *GTH*1. The values used are given in Table 4.8.1.

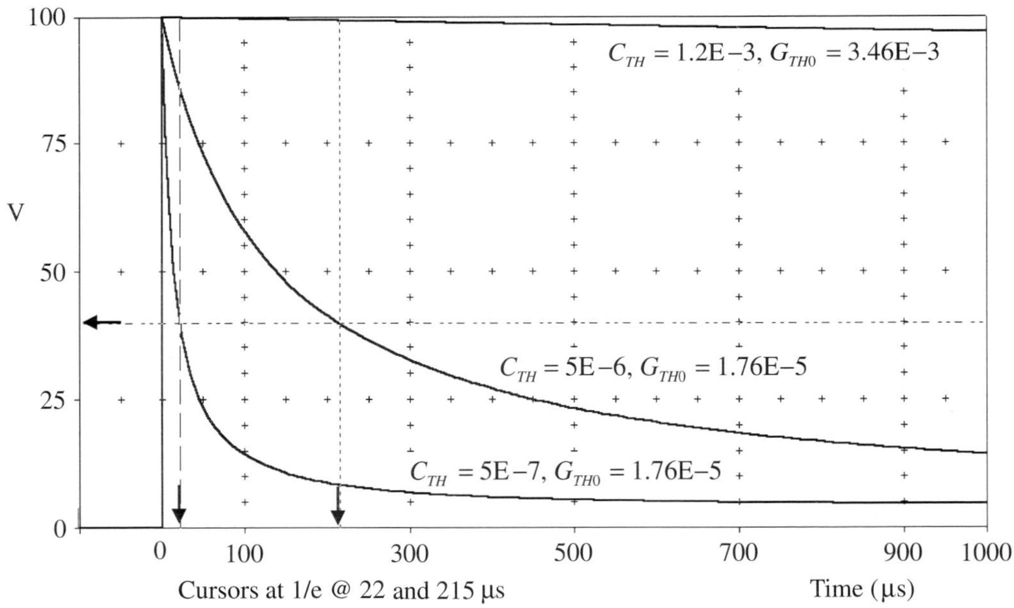


**Fig. 4.8.4** Effect of the parameter *CTH*. Resistance variation as a function of time for a 10 mA step current input. The parameters are shown with the responses.

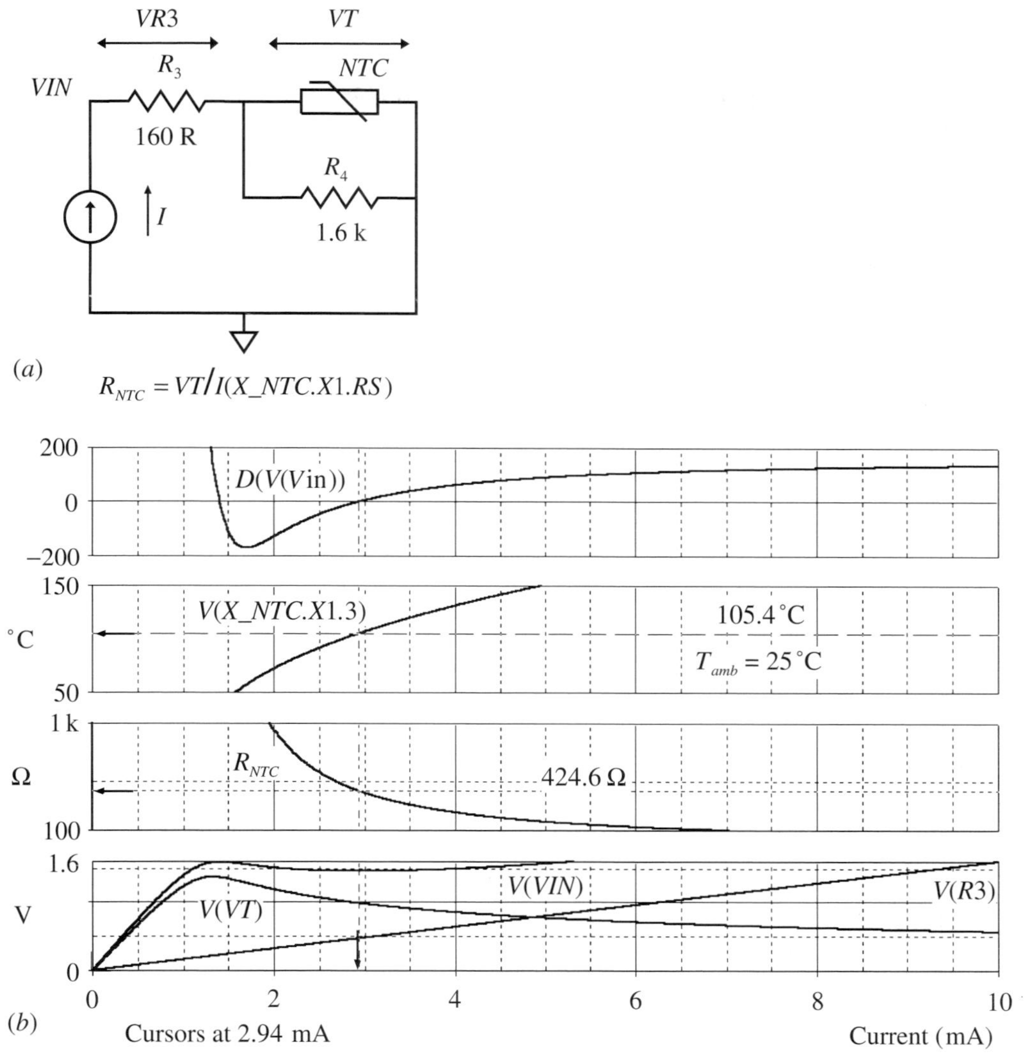


**Fig. 4.8.5** (*a*) Schematic of stabilization arms of a Wien-bridge oscillator. (*b*) Characteristics of the circuit showing region of constant output irrespective of the ratio of the two arms.

source step, is shown in Fig. 4.8.4 and it is seen that *CTH* governs the time and hence frequency response.

For use as an amplitude stabilizer in a Wien-bridge oscillator (Section 5.8) it has been proposed by Owen (1965) that the characteristics should be modified to enhance the constancy of the output. This involves adding a resistor $R_4$ in parallel with the thermistor so that at the operating point the slope of the thermistor characteristic is opposite to that of the resistor $R_3$ in series with the combination as shown in Fig. 4.8.5(*a*).

Using case D from Fig. 4.8.3, a little trial and error led to the resistor values shown in Fig. 4.8.5(*b*). The requirement is to have the operating current near the centre of the (near) constant region of *VIN*, found from the plot of the differential *D*(*V*(*VIN*)), and the ratio of *VIN*: *VT*: *V*(*R*3) = 3:2:1 to give the one-third attenuation required for balancing the full bridge (Sections 1.12 and 5.8). The resistance of the thermistor itself is seen to be 424.6 Ω, which together with $R_4$ gives an effective 335 Ω, i.e. ≈twice $R_3$ and the internal temperature of the thermistor is $T_{EL} = 105.4$ °C to give a net temperature of 105.4 + 25 = 130.4 °C, high enough for the effect of ambient variations to be small.

### SPICE simulation circuits

Consult the SimCmnt.doc file on the CD before running.

| | |
|---|---|
| Fig. 4.8.2 | Ntcchar.SCH |
| Fig. 4.8.3 | Ntcchar 5.SCH |
| Fig. 4.8.4 | Ntcchar 3.SCH |
| Fig. 4.8.5 | Ntcchar 4.SCH |

## References and additional sources 4.8


Epcos /Siemens: NTC thermistor data website at www.epcos.com . Information reproduced and used with permission.

Hirasuna B. (1999): *Modeling Voltage-Controlled and Temperature-Dependent Resistors*, OrCAD Technical Note No. 90. (www.orcad.com or www.cadence.com or www.pspice.com )

Hoge H. J. (1979): Comparison of circuits for linearizing the temperature indications of thermistors. *Rev. Sci. Instrum.* **50**, 316–320.

Hyde F. J. (1971): *Thermistors*, London: Iliffe Books. ISBN 0-592-02807-0.

Owen R. E. (1965): A modern, wide range, RC oscillator. *General Radio Experimenter* **39**(8), August, 3–12.

Sarid D., Cannell D. S. (1974): A ±15 Microdegree temperature controller. *Rev. Sci. Instrum.* **45**, 1082–1088. (Thermistor temperature sensor.)

Williams J. M. (Ed.) (1997): *Linear Applications Handbook, Vol. III, Linear Technology* (approx. 3 cms worth; also previous volumes in 1987(I) and 1990(II)). See p. 185 *et seq*. Signal conditioning – Temperature, for a very wide range of applications.

# 4.9 Coaxial cables

One cannot escape the feeling that these mathematical formulae have an independent existence and intelligence of their own, that they are wiser than we are, wiser than their discoverers, that we get more out of them than was originally put into them.

Heinrich Hertz on Maxwell's equations

Coaxial cables provide one of the most controlled, convenient and sanitary means of interconnecting electronic systems. The signals are well confined and the energy transfer is efficient. The theory of transmission, especially at high frequencies where the wavelengths involved are comparable with the length of the cable, is outlined in Section 3.17. We wish here to comment on the relationship of the electromagnetic field view and the circuit view (Ramo and Whinnery 1953, p. 299). To examine the propagation of waves guided by a uniform system of conductors properly it is necessary to find an acceptable solution to Maxwell's equations that satisfies the given boundary conditions at the conductors. In the case of the coaxial line the fields are contained within the outer shield, so that there is no external field arising from the currents in the cable, and fields from an external source do not affect the internal signals. The use of two conductors allows the propagation of what is termed a transverse electromagnetic wave (TEM), at least in the case of lossless conductors. In this mode the internal fields have components that are purely transverse relative to the direction of propagation along the cable. (This is also possible outside a single wire where the 'return' conductor is in effect at infinity, but such an open line has evident drawbacks.) Single and parallel pair lines were the basis of Hertz's experiments leading to the discovery of electromagnetic waves (O'Hara and Pricha 1987; Susskind 1995) but he was bedevilled by what turned out to be reflections from the walls of the laboratory, and it was only when he invented the 'screened cable' that he was able to make satisfactory measurements. His screening cage also allowed him to invent the 'slotted line' which permitted measurements on the waves travelling along the line without disturbance from the presence of the experimenter. How such a momentous discovery was made with such simple apparatus is a lesson to us all.

For a TEM wave in a coaxial cable the distribution of electric $\boldsymbol{E}$ and magnetic $\boldsymbol{B}$ fields are as shown in Fig. 4.9.1.

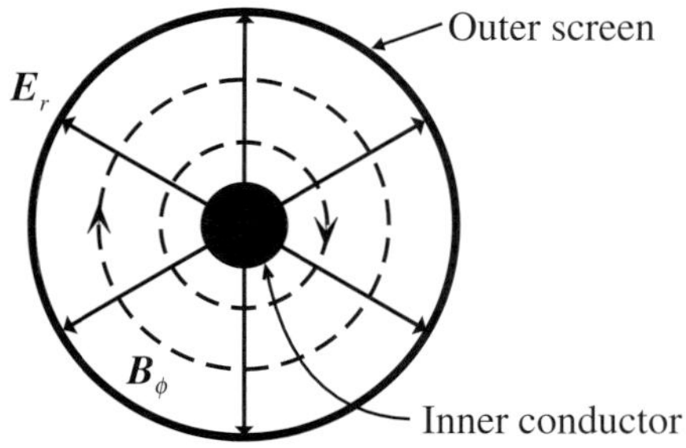


**Fig. 4.9.1** Electric and magnetic fields in a coaxial cable viewed along the direction of propagation.

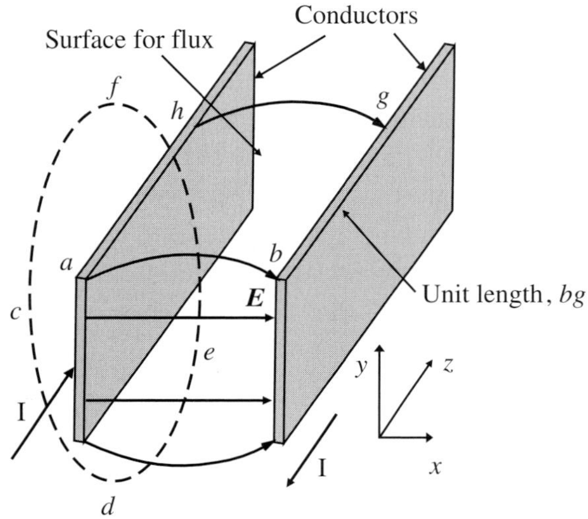


**Fig. 4.9.2** Fields in a planar guide.

The illustration is for the case of perfect conductors: the $\boldsymbol{E}$ field must be normal to the surface of the conductor and the $\boldsymbol{B}$ field must be tangential to the surfaces and normal to $\boldsymbol{E}$ (Sections 2.5, 2.7). It is possible in this case to show that the requirements of Maxwell's equations lead to the circuit equations used in solving for waves on a transmission line. Consider a twin line as shown in Fig. 4.9.2 where the cross-section shape is not of consequence except in that the line must be uniform in the $z$-direction.

A TEM mode requires that $H_z$ and $E_z$ must be zero. To determine the potential $V$ between the conductors we may integrate the electric field along any path between the conductors (say from $a$ to $b$) since in the plane of the diagram $E$ satisfies Laplace's equation and hence it is given by the gradient of the scalar potential, or inversely $V$ is the integral of $E$ over the path. Thus:

$$V = -\int_a^b E \cdot dl = -\int_a^b (E_x dx + E_y dy)$$

$$\text{so} \quad \frac{\partial V}{\partial z} = -\int_a^b \left( \frac{\partial E_x}{\partial z} dx + \frac{\partial E_y}{\partial z} dy \right) \tag{4.9.1}$$

From Eqs. (2.7.1(II)) and (1.6.15) we have with $E_z = 0$:

$$\nabla \times \boldsymbol{E} = \frac{-\partial \boldsymbol{B}}{\partial t} \quad \text{so} \quad \frac{\partial E_y}{\partial z} = \frac{\partial B_x}{\partial t} \quad \text{and} \quad \frac{\partial E_x}{\partial z} = \frac{-\partial B_y}{\partial t} \tag{4.9.2}$$

so that Eq. (4.9.1) becomes:

$$\frac{\partial V}{\partial z} = -\frac{\partial}{\partial t} \int_a^b (-B_y dx + B_x dy) \tag{4.9.3}$$

and the integral is the magnetic flux through the surface defined by the integration path (*ab*)and a unit length in the $z$-direction (i.e. area *abgh*). By definition (Section 4.3), this flux is equal to the inductance $L$ per unit length times the current $I$ so we have:

$$\frac{\partial V}{\partial z} = -\frac{\partial (LI)}{\partial t} = -L \frac{\partial I}{\partial t} \tag{4.9.4}$$

which agrees with Eq. (3.17.1) (except that there propagation was in the $x$-direction). If we now integrate around the path *cdef* then since the enclosed current is $I$ we have (Eq. (2.5.1)):

$$I = \oint \boldsymbol{H} \cdot d\boldsymbol{l} = \oint (H_x dx + H_y dy)$$

$$\text{so} \quad \frac{\partial I}{\partial z} = \oint \left( \frac{\partial H_x}{\partial z} dx + \frac{\partial H_y}{\partial z} dy \right) \tag{4.9.5}$$

From (1.6.15), as above for $E$, and with $H_z = 0$, we have:

$$\nabla \times \boldsymbol{H} = \frac{\partial \boldsymbol{D}}{\partial t} \quad \text{so} \quad \frac{\partial D_y}{\partial t} = \frac{\partial H_x}{\partial z} \quad \text{and} \quad \frac{\partial D_x}{\partial t} = \frac{-\partial H_y}{\partial z} \tag{4.9.6}$$

so that (4.9.5) becomes:

$$\frac{\partial I}{\partial z} = -\frac{\partial}{\partial t} \oint (D_x dy - D_y dx) \tag{4.9.7}$$

and in this case the integral is the electric displacement flux per unit length between the conductors. By definition it is the product of the capacity $C$ per unit length and the voltage between the conductors:

$$\frac{\partial I}{\partial z} = -\frac{\partial (CV)}{\partial t} = -C\frac{\partial V}{\partial t} \tag{4.9.8}$$

which is again in agreement, this time with (3.17.2). Thus the approach in Section 3.17 using the low frequency measures of $L$ and $C$ to examine the propagation of fields along conductors is seen to be justified at least in the case of perfect conductors, but this is not necessarily so for other configurations. In the case we have been considering the fields are just those we would deduce at zero frequency so the general conclusions are perhaps not unexpected.

If we now consider a more realistic line in which the conductors have finite resistance then it is evident that the fields must be somewhat changed since we now need a component of $\boldsymbol{E}$ along the conductors to cause the current to flow – perfect conductors need no field! This also means that the approach used in Section 3.17 is also in doubt and the problem becomes more complex to solve. A direct approach is difficult to apply but using perturbation theory, in effect successive approximation, shows that for low resistance conductors the deviation from the circuit approach is not significant (Schelkunoff 1934; Carson 1914).

For PSpice the definition of a lossy transmission line requires the specification of the values of $L$, $C$, $R$ and $G$ per unit length. The values of $L$ and $C$ are usually readily available or one can be calculated from the other together with $Z_0$. For most ordinary applications we may ignore $G$, but $R$ is sometimes required if we wish to calculate the attenuation of a signal. In some applications this is necessary for short lengths, as for example in fast logic circuits using ECL (Blood 1988). There is, however, a considerable difficulty in specifying an appropriate value for $R$ since it varies, in particular due to the skin effect, with the square root of the frequency (Section 2.8). For a sine wave you may calculate the proper value but for pulses this is not appliable and you should make $R$ an appropriate function of the frequency. For a cable using copper (or copper-plated) conductors with inner conductor radius $d$ and inner radius of the outer conductor $D$ (both in m):

$$R = 4.2 \times 10^{-8} f^{\frac{1}{2}} \left( \frac{1}{d} + \frac{1}{D} \right) \Omega \text{ m}^{-1} \tag{4.9.9}$$

Design handbooks for ECL applications give extensive coverage of high speed transmission line techniques and show the sort of performance one may achieve. Blood (1988, p. 70) illustrates a further consequence of the skin effect on the risetime of fast pulses. The waveform rises sharply for the initial 50% and then slows considerably with the 10–90% time being 30 times the 0–50% time! Skin effect is discussed in Section 2.8. The effect in braided screens differs from that of an equivalent solid tube and a step function input to the cable does not result in the usual exponential type rise at the output (Botos 1966), but rather more like $1 - \text{erf}$ (the

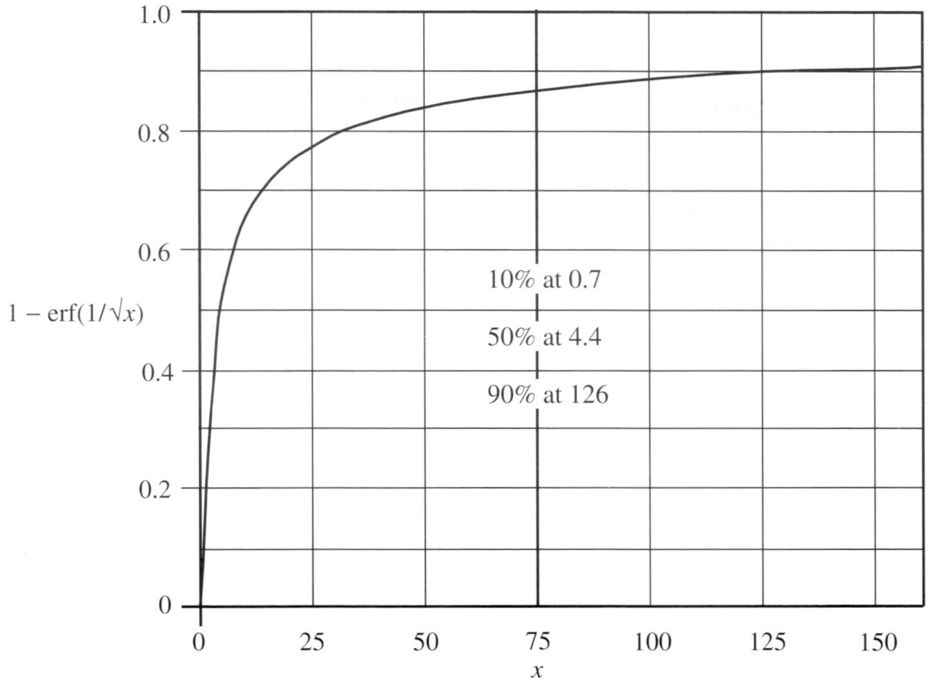


**Fig. 4.9.3** Risetime function of coaxial cable for fast edges.

error function erf is discussed in many texts, e.g. Boas 1966; Abramowitz and Stegun 1970):

$$1 - \mathrm{erf}\left[\frac{l\alpha_0}{2(\pi f_0 t)^{\frac{1}{2}}}\right] \tag{4.9.10}$$

where $l$ is the length of cable, $\alpha_0$ is the attenuation at frequency $f_0$, and $t$ is the time excluding the normal delay down the cable. A plot of this function (Fig. 4.9.3) shows the fast initial rise to about 50% followed by a very slow rise to the final level, as found in actual measurements. At high frequency, in the region of 1 GHz the dielectric loss is comparable with the skin effect loss further complicating any calculation (Nahman 1962). Wigington and Nahman (1957), Turin (1957), Magnusson (1968), Nahman and Holt (1972), and Cases and Quinn (1980) provide further examination of the responses. An extensive treatment of transmission lines at high frequencies is given by Grivet (1970).

## SPICE Simulation Circuits

None

## References and additional sources 4.9


Abramowitz M., Stegun I. A. (1970): *Handbook of Mathematical Functions with Formulas, Graphs and Mathematical Tables*, National Bureau of Standards, Applied Mathematics Series, p. 297.

Amsel G., Bosshard R., Rausch R., Zadje C. (1971): Time domain compensation of cable induced distortions using passive filters for the transmission of fast pulses. *Rev. Sci. Instrum.* **42**, 1237–1246.

Blood W. R. (1988): *MECL System Design Handbook*, 4th Edn, Motorola Semiconductor Products Inc.

Boas M. L. (1966): *Mathematical Methods in the Physical Sciences*, New York: John Wiley. Library of Congress Cat. No. 66-17646. See p. 404.

Botos Bob (1966): *Nanosecond Pulse Handling Techniques in I/C Interconnections*, Motorola Application Note AN-270.

Brookshier W. K. (1969): Noise signal pickup in coaxial cables. *Nuc. Instrum. Meth.* **70**, 1–10.

Carson J. R. (1914): The guided and radiated energy in wire transmission. *J. Amer. IEE* October, 906-913.

Cases M., Quinn D. M. (1980): Transient response of uniformly distributed RLC transmission lines. *Proc. IEEE* **CAS-27**, 200–207.

Grivet P. (1970): *The Physics of Transmission Lines at High and Very High Frequencies*, Vols 1 and 2, New York; Academic Press. ISBN 12-303601-1.

Ishii T. K. (1995): Transmission lines. in Chen, Wai-Kai (Ed.) *The Circuits and Filters Handbook*, Boca Raton: CRC Press and IEEE Press. ISBN 0-8493-8341-2. Chapter 39. See also De Zutter D., Martens L. Multiconductor transmission lines. Chapter 40, and references therein. (Equation (39.43) is missing $\times 10^{-2}$ for '*a*' and '*b*' in m.)

Kirsten F. (undated): *Pulse Response of Coaxial Cables*, Lawrence Radiation Laboratory Counting Note CC2-1A.

Magnusson P. C. (1968): Transient wavefronts on lossy transmission lines – Effect of source resistance. *Trans. IEEE* **CT-15**, 290–292.

MicroSim Corporation (1993): *Circuit Analysis Reference Manual*, July, p. 163.

Nahman N. S. (1962): A discussion on the transient analysis of coaxial cables considering high-frequency losses. *IRE Trans.* **CT-9**, 1962.

Nahman N. S., Holt D. R. (1972): Transient analysis of coaxial cables using the skin effect approximation $A + B\sqrt{s}$. *IEEE Trans.* **CT-19**, 443-451.

National Semiconductor (1989): *F100K ECL Logic. Databook and Design Guide*, National Semiconductor.

O'Hara J. G., Pricha W. (1987): *Hertz and the Maxwellians*, London: Peter Peregrinus. ISBN 0-86341-101-0.

Schelkunoff S. A. (1934): The electromagnetic theory of coaxial transmission lines and cylindrical shields. *Bell Sys. Tech. J.* **13** October, 532.

Susskind C. (1995): *Heinrich Hertz: A Short Life*, San Francisco: San Francisco Press. ISBN 0-911302-74-3.

Turin G. L. (1957): Steady-state and transient analysis of lossy coaxial cables. *Proc. IRE* **45**, 878–879.

Wigington R. L., Nahman N. S. (1957): Transient analysis of coaxial cables considering skin effect. *Proc. IRE* **45**, 166–174.

# 4.10 Crystals

The aim of science is not to open a door to infinite wisdom but to set a limit to infinite error.

Galileo, from the pen of Brecht

Though quartz crystals are not used in anything like the numbers of, say, resistors or capacitors we discuss them here as they present some difficulties when used in simulated circuits. This arises from their very high $Q$, which means that we must work with very small frequency increments and must expect long simulation times since it will take very many cycles to reach a steady state.

A simple equivalent circuit is shown in Fig. 4.10.1(*a*), and (*b*) shows the response as a function of frequency for the component values shown.

Such a circuit is only a rough approximation since a crystal can have many modes of oscillation, particularly at harmonic frequencies, in addition to that at the primary frequency. A little subterfuge is required to persuade PSpice to draw the reactance curves shown. The reactances are considered to be positive quantities and *PROBE* will only plot them as such even if you include a negative sign. The negative version of $X_C$ is obtained by taking the modulus *M*( ) (or absolute value *ABS*( )) of $X_C$ and then taking the negative. If you 'add' $X_C$ and $X_L$ you get the one-sided curve with the zero where the two are equal and which is the series resonant frequency $f_S$ (which we will discover below to be 10.2734 MHz). To get the curve as usually illustrated going from negative to positive reactance (e.g. Hambley 1994, p. 966) we cannot use the ruse applied above to get $-X_C$ as this would just reflect the $X_C + X_L$ curve in the $x$-axis. The ruse is to take the absolute value and multiply by *SGN*(*Frequency*−10.2734E6) which will produce −1 for frequencies below $f_S$ and +1 above, hence only reflecting the lower frequency portion. Note that since we are working with an expanded frequency axis and need to read small frequency differences, you will need to increase the cursor resolution. Set a large number of points in *ANALYSIS/SETUP* and under *TOOLS/OPTIONS* in *PROBE* set the number of cursor digits to 7.

At the fundamental there are two resonances possible, series involving $L$ and $C$, and parallel involving $L$ and $C_P$. We may determine the two frequencies by analysing the impedance to find where the phase is zero. To make the algebra a little more

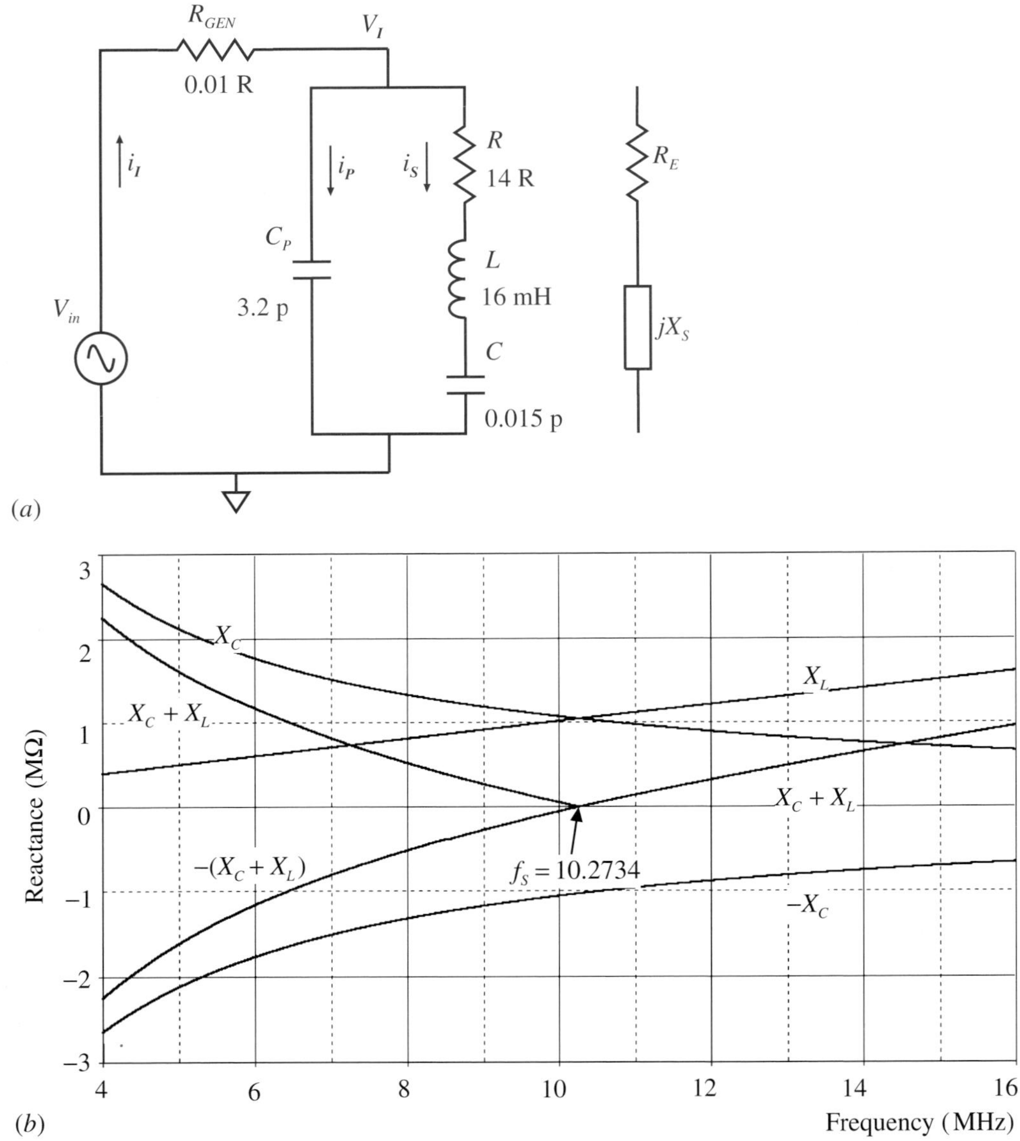


**Fig. 4.10.1** (*a*) Crystal equivalent circuits. (*b*) Reactances as a function of frequency.

convenient we define a reactance magnitude $X$ as indicated and then write down the expression for the impedance $Z$:

$$\frac{1}{Z}=\frac{1}{R+jX}+j\omega C_P, \quad \text{where} \quad X=\left(\omega L-\frac{1}{\omega C}\right)$$

$$=\frac{1+j\omega C_P(R+jX)}{(R+jX)} \tag{4.10.1}$$

$$\text{so}\quad Z=\frac{(R+jX)}{1-\omega C_P X+j\omega C_P R}\quad \text{and rationalizing} \tag{4.10.1 cont.}$$

$$=\frac{(R+jX)(1-\omega C_P X-j\omega C_P R)}{(1-\omega C_P X)^2+(\omega C_P R)^2}$$

$$=\frac{R+j[X(1-\omega C_P X)-\omega C_P R^2]}{(1-\omega C_P X)^2+(\omega C_P R)^2}$$

For the phase to be zero the $j$ term must be zero. We will need to make some slight approximations that depend on the relative magnitude of terms. The values used are those shown in Fig. 4.10.1 and are appropriate for a crystal of about 10 MHz. They were provided by the manufacturer (CMAC 2000). To make the $j$ term zero we have:

$$X^2\omega C_p - X+\omega C_p R=0,\quad \text{a quadratic which has two roots}$$

$$X=\frac{1\pm(1-4\omega^2 C_P^2 R^2)^{\frac{1}{2}}}{2\omega C_P}\quad \text{and since}\quad 4\omega^2 C_P^2 R^2\ll 1,\quad \text{use the binomial expansion}$$

$$=\frac{1\pm(1-2\omega C_P R)}{2\omega C_P} \tag{4.10.2}$$

$$\text{so}\quad X=R\quad \text{or}\quad X=\frac{1-\omega C_P R}{\omega C_P}\cong\frac{1}{\omega C_P}\quad \text{since}\quad \omega C_p R\ll 1$$

Now inserting the expression for $X$, and noting that $CR\ll 2\sqrt{LC}$, or $\omega CR\ll 1$, we have:

$$\left(\omega L-\frac{1}{\omega C}\right)=R\quad \text{or}\quad \left(\omega L-\frac{1}{\omega C}\right)=\frac{1}{\omega C_P}\quad \text{giving}$$

$$\omega_S=\frac{1}{\sqrt{LC}}\quad \text{or}\quad \omega_P=\frac{1}{\sqrt{LC}}\left(1+\frac{C}{C_P}\right)\quad \text{and hence} \tag{4.10.3}$$

$$f_S=10.2734\ \text{MHz},\quad f_P=10.2970\ \text{MHz}\quad \text{with our values}$$

a series resonance at $\omega_S$ and a parallel resonance at $\omega_P$. As $C\ll C_P$ the separation is small, about 0.2% in this case. To check the sums we can run a PSpice simulation which yields the results shown in Fig. 4.10.2. The series resonance results in a low impedance of $R$ so that we record the current $i_S$. The parallel resonance produces a high impedance so we record $V_I/i_S$ to show this.

The simulation agrees closely with the calculated values of Eq. (4.10.3). The impedance at the two resonances can now be found from (4.10.1):

$$Z_S=\frac{R}{1+\omega^2 C_P^2 R^2}=R,\quad \text{since}\quad X=0\quad \text{and}\quad \omega^2 C_P^2 R^2\ll 1 \tag{4.10.4}$$

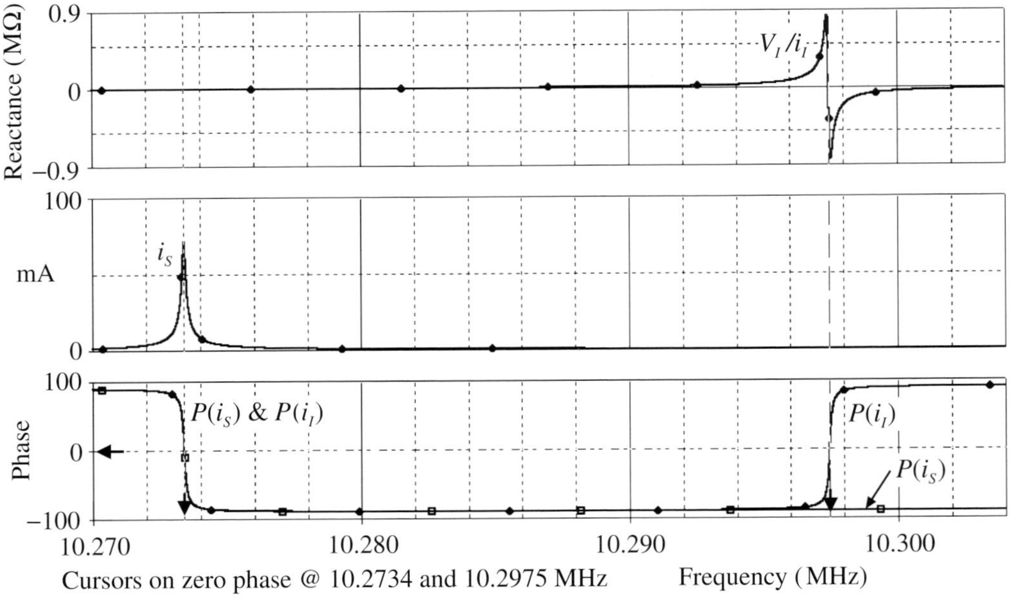


**Fig. 4.10.2** Simulation responses for the crystal circuit of Fig. 4.10.1(*a*).

$$Z_P = \frac{R}{\omega^2 C_P^2 R^2} = \frac{Q^2 C^2}{C_P^2}, \quad \text{since } X = \frac{1}{\omega C_P} \text{ so } (1 - \omega C_P X) = 0 \qquad (4.10.4\ \text{cont.})$$

and we have used $Q = \dfrac{\omega L}{R}$ so $Q^2 = \dfrac{L}{CR^2}$ with $\omega^2 = LC$

and it is seen that $Z_S$ is low and $Z_P$ is high as we would expect.

The determination of frequency response does not involve the $Q$ of the resonator in the sense that an impedance does not depend on a time delay to allow the response to build up. If we wish to carry out a transient simulation we will have to allow a run time long enough for the signal to grow to full amplitude. The conditions necessary are considered by MicroSim (1993) where it is recommended that the step ceiling (or time step) should be set to:

$$\textit{Step ceiling} = 4 \times \textit{resonance period} \times \left( \frac{RELTOL}{Q} \right)^{\frac{1}{2}} \qquad (4.10.5)$$

where *RELTOL* is found under *ANALYSIS/SETUP/OPTIONS*. For the crystal we have been considering this turns out to be 45 ps! The oscillations in a driven resonant circuit build up exponentially as given by Eq. (3.5.14). The equation for the growth, or time to reach within 0.1% of the steady state, given by MicroSim is in error and to demonstrate this we will simulate the resonator given as the example. The parameters given are $f_0 = 20$ kHz, $Q = 1000$, and we have deduced 'typical' circuit parameters to match, namely $L=8$ H, $C_p=20$ p, $C=7.915$ p, $R=1005\ \Omega$, to give a resonant frequency of 20.0008 kHz. The amplitude will grow according to the equation:

$$i=i_0\left[1-\exp\left(\frac{-\omega_0 t}{2Q}\right)\right],\quad \text{with time constant } \tau=\frac{2Q}{\omega_0}=\frac{2\times 1000}{2\pi\times 20\times 1000}=15.9\text{ ms} \tag{4.10.6}$$

so from Table 1.5.1 we may expect the time taken to reach within 0.1% to be about $15.9\times 7=111$ ms. From (4.10.6) we have:

$$\frac{i_0-i}{i_0}=\frac{1-\left[1-\exp\left(\frac{-\omega_0 t}{2Q}\right)\right]}{1},\quad \text{with}\quad \frac{i_0-i}{i_0}=0.001 \text{ at } t_{0.1} \text{ say}$$

$$\text{so}\quad 0.001=\exp\left(\frac{-\omega_0 t_{0.1}}{2Q}\right) \tag{4.10.7}$$

$$\text{or}\quad t_{0.1}=\frac{-2\times 1000\times \ln(0.001)}{2\pi\times 20\times 10^3}=110\text{ ms},\quad \text{where } \ln(0.001)=-6.908$$

in agreement with our estimate. Simulating the circuit presents some difficulty. If you drive the circuit close to its resonant frequency or make the step ceiling small you will tend to get beats which make our calculations inappropriate. A larger step size and a drive frequency a little way off will give the exponential form of growth we require. Using the same circuit as Fig. 4.10.1(*a*) but with our 20 kHz parameters produced the response shown in Fig. 4.10.3. The cycles are too compressed to show in the figure but the exponential rise is evident. It is difficult to make measurements of relative amplitudes to find $t_{0.1}$ but we can fit an exponential curve to the response and see whether the parameters fit. Starting with $\tau=15.9$ ms and a final amplitude of 0.942 determined from the value at 160 ms we can get *PROBE* to plot Eq. (4.10.6). A little adjustment showed that a time constant of 15 ms fitted best, and this is the curve shown in the figure.

The fit shows that the time to 0.1% of final level is as calculated. The calculated value in MicroSim is in error by a factor of $2\pi$ which would give $0.69/2\pi=0.11$ s.

For our original ~10 MHz oscillator a similar calculation gives a time of 15.8 ms, 162 k cycles and 350 M steps, which would take considerable time and storage memory. This is not an attractive prospect so you would have to make some compromise on step length and proximity to the final steady state. There is, however, an alternative approach which will get the resonator up to full oscillation almost immediately. By setting initial conditions we can make the resonator start at full amplitude, and has a parallel in the technique used in Section 5.23. You can either set the *SIN* generator phase to 90, i.e. a cosine wave starting at say 1 V

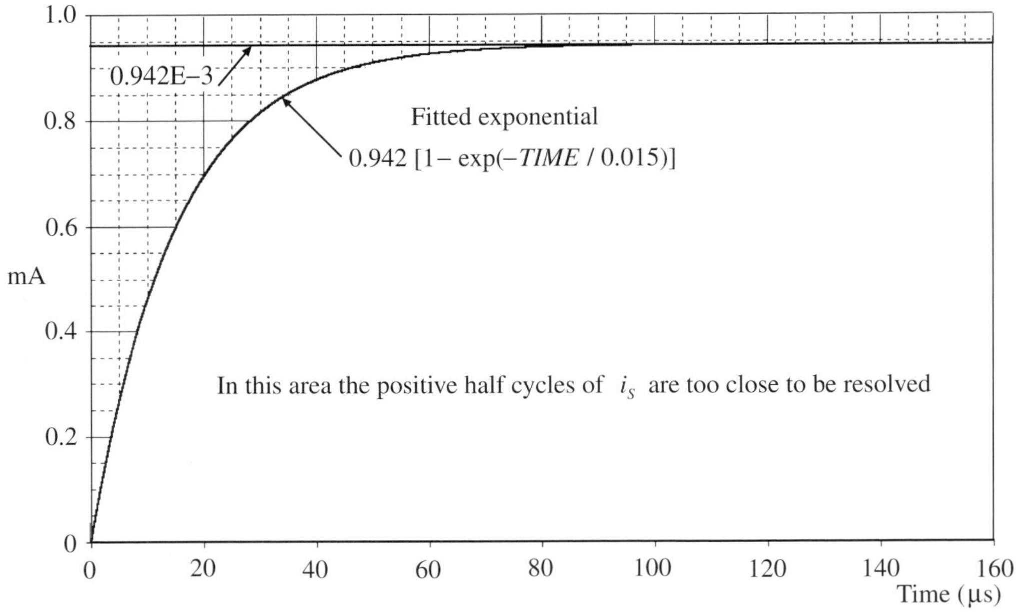


**Fig. 4.10.3** Growth of response of 20 kHz resonator to drive at 19.9 kHz with step size 2 μs.

together with the initial condition for $C$ to $IC=0$, or leave the phase at zero and set $IC=1$ for $C$. In either case you will find that the oscillation will start, and continue, at the full amplitude of 71.37 mA, and if you enable the bias point current display on the schematic you will find the same value.

A practical consideration is the power dissipated in the crystal. For stability the temperature should be kept stable and in such a high $Q$ resonator it is quite possible to overheat or fracture the crystal. As O'Dell (1988) points out a 1 MHz crystal dissipating 10 mW has stored energy of about 1 mJ, which he equates to the energy stored in a 1 μF capacitor charged to 50 V. Dissipations of microwatts should be aimed at for good stable oscillations. Connecting a crystal to a circuit inevitably adds parallel capacity so crystals are specified to work into a certain load capacity. The changes in frequency with this external capacity are not great but should be considered if frequency precision is required. Several of the references provide instruction on designing crystal oscillator circuits.

## SPICE simulation circuits

Consult the SimCmnt.doc file on the CD before running.

| | |
|---|---|
| Fig. 4.10.1(*b*) | Crystal2.SCH |
| Fig. 4.10.2 | Crystal1.SCH |
| Fig. 4.10.3 | Crystal4.SCH |

## References and additional sources 4.10


Baxandall P. J. (1965): Transistor crystal oscillators and the design of a 1-Mc/s oscillator of good frequency stability. *Radio Electr. Engnr.* **29**, 229–246.

CMAC Frequency Products (IQD Ltd.) (2000): *Crystal Product Databook*, Crewekerne, Somerset. http://cfpwww.com

Fairchild Semiconductor (1983): *HCMOS Crystal Oscillators*, Fairchild Semiconductor Application Note 340, May. www.Fairchildsemiconductor.co and go to application notes.

Ferking M. E. (1978): *Crystal Oscillator Design and Temperature Compensation*, New York: Van Nostrand Reinhold.

Hambley A. R. (1994): *Electronics. A Top-Down Approach to Computer-Aided Circuit Design*, New York: Macmillan. ISBN 0-02-349335-6.

Matthys R. J. (1983): *Crystal Oscillator Circuits*, New York: John Wiley.

MicroSim Corporation (1993): *Transient Analysis of High-Q Circuits*, MicroSim Application Notes, Ver. 5.4 July, p. 28.

MicroSim Corporation (1996): *Simulating High-Q Circuits Using Open Loop Response*, MicroSim Application Notes, Ver. 6.3 April, pp. 225–231. (But see Section 5.14.) Modeling quartz crystals, p. 179.

Nye J. F. (1957): *The Physical Properties of Crystals*, Oxford: Oxford University Press. See Chapters 4 and 10.

O'Dell T. H. (1988): *Electronic Circuit Design*, Cambridge: Cambridge University Press. ISBN 0-521-35858-2. (See pp. 107–111.)

Ramo S., Whinnery J. R. (1953): *Fields and Waves in Modern Radio*, 2nd Edn, New York: John Wiley.

Williams J. M. (Ed.) (1997): *Linear Applications Handbook, Vol. III, Linear Technology*. See 'Oscillators' in subject index.

# Part 5

# SPICE circuit applications

Read not to contradict and confute, nor to believe and take for granted, nor to find talk and discourse, but to weigh and consider.

Francis Bacon

In this part we cover a range of circuits that provide an instructive opportunity for making use of some of the many facilities provided by SPICE. Many of the circuits will be analysed mathematically where this may reasonably be done. In many cases it is necessary to make approximations to allow the analysis to be taken to a point where useful predictions may be made and then a SPICE simulation is performed to establish the validity of the analysis and particularly the approximations. Some are examined in detail while others are presented only briefly. The topics chosen cover a wide range; but as is evident to all, there are an infinite number of others that could have been selected. Those that have are ones that I have met in my own work or which I thought could be used to illustrate some point or technique of more general interest. The approach in Part 5 is therefore in no way comprehensive and is episodic rather than a coherent development. Within reason I have included component values and plots of representative results to demonstrate what you may expect. The intention is that you should be encouraged to experiment to develop confidence in some of the techniques of analysis and the use of SPICE. As mentioned much earlier a particular flavour of SPICE, i.e. PSpice, has been used, and though I do not have the means to check I expect other flavours will also be appropriate. For each topic there is appended a list of references to the literature which can supplement the text and provide additional instruction and insight. It is always helpful to have alternative approaches so that you are more likely to find one that suits your way of thinking. In some cases there is overlap between the references but this makes it more likely that you will be able to find one of them.

In Section 5.27 there are a series of comments on PSpice usage covering matters where I have encountered difficulties or ambiguities which have taken some time to resolve or required recourse to MicroSim/Cadence. Such, often small, matters can prove extremely annoying and time wasting so I hope some of the comments are illuminating. The comments refer in particular to Version 8 and have been vetted by MicroSim for technical accuracy, but the comments are entirely mine and are not necessarily endorsed by MicroSim. These software packages are huge structures and I have only gone as far as has been necessary, so there is no doubt much more available than referred to here. Any comments, observations or suggestions will be welcome. Version 9, which has appeared during the writing of this book, may well have attended to some of the problems encountered, but it seems a fact of life that new versions of software packages have a tendency to reintroduce problems from earlier versions.

Whatever conclusions you may reach as a result of SPICE simulations you will eventually have to construct the circuit and test it. Some authors are rather dismissive of the capabilities of SPICE but in my experience it is of very considerable benefit and will get you to the correct ball-park in which to start your practical investigations. There are not many books providing good practical advice on this phase but one such is Pease (1991), and Williams (1998) provides an inside look at the art of design.

## References

Pease R. A. (1991): *Troubleshooting Analog Circuits*, London: Butterworth-Heinemann. ISBN 0-7506-9499-8. His Appendix G, 'More on SPICE', refers to rather older versions and in my view the comments there are now largely inappropriate.

Williams J. (Ed.) (1998): *The Art and Science of Analog Circuit Design*, Boston: Butterworth-Heinemann. ISBN 0-7506-7062-2.

# 5.1 Absolute value circuit

If a man will begin with certainties, he shall end in doubts; but if he will be content to begin with doubts, he shall end in certainties.

Francis Bacon

The circuit is shown in Fig. 5.1.1 (Philbrick/Nexus 1968). The aim of this is to produce, from a bipolar input, an output that is the absolute value. What this means is that for a positive input the output will be positive, while for a negative input the output will also be positive. (We could also obtain the 'negative' absolute value if desired, though mathematically this is a contradiction. What is meant is that the signal is inverted.) In mathematical terms we can write:

$$v_{out} = |v_{in}| \tag{5.1.1}$$

This circuit has been chosen since it is useful in itself and also since it provides a very convenient case for a discussion of the effect of negative feedback on

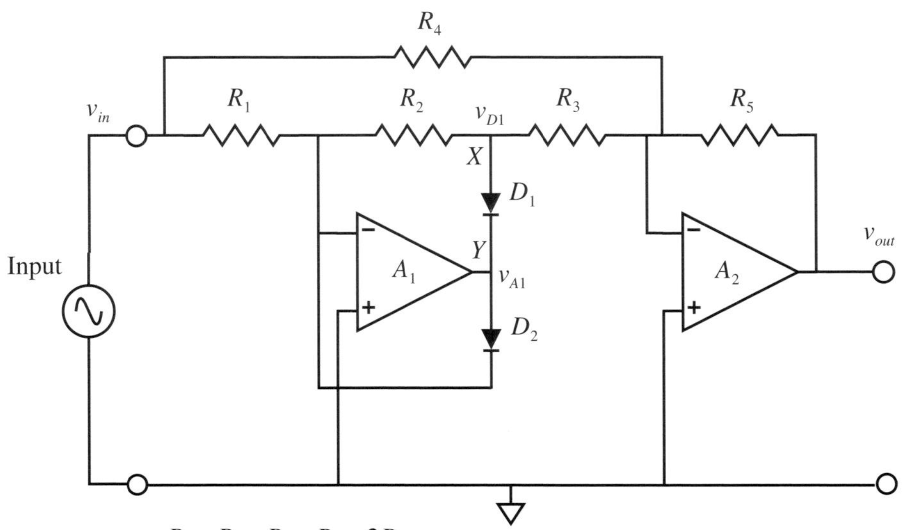


$R_1 = R_2 = R_4 = R_5 = 2R_3$

**Fig. 5.1.1** An absolute value circuit.

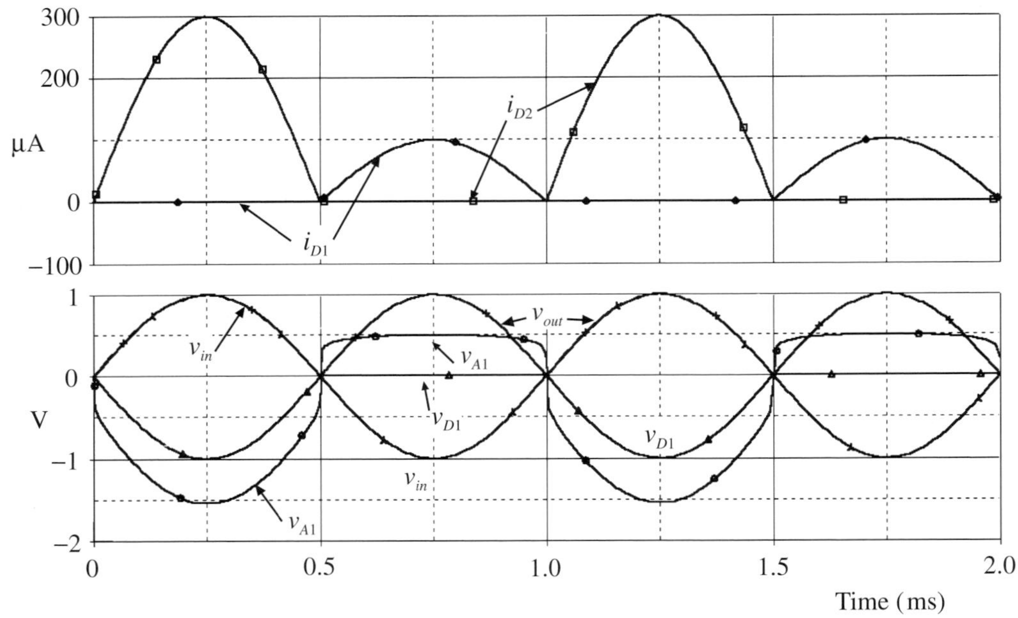


**Fig. 5.1.2** Waveforms for circuit of Fig. 5.1.1.

distortion. Here we have deliberately introduced distortion in the form of the diodes for which the form of distortion is well defined, so we know what effect it has.

The diodes serve as polarity selectors. As noted previously, diodes have a substantially non-linear response which we might expect to lead to considerable distortion in the output signal. Typical signals are shown in Fig. 5.1.2, and it can be seen that there is no evidence of distortion in the output.

The general principle of operation is as follows. For the positive portion of the input sine wave the output of amplifier $A_1$ will be negative causing $D_1$ to conduct to give a gain of 1. The input to amplifier $A_2$ via $R_4$ during this part of the cycle is amplified by –1 while the input to $A_2$ from $A_1$ is amplified by –2, with the output thus being equal to the original input. For the negative portion of the input, $D_2$ will conduct giving a gain of zero at point $X$ (Fig. 5.1.1) so the output from $A_2$ will just be the inversion of the input via $R_4$ and so we now have the absolute value output as shown. Since the circuit operates by algebraic addition it is important the gains be well defined, i.e. the resistor values must be well matched.

In the discussion above it was assumed that the diodes were ideal. Let us now see how the actual diode characteristics affect the operation. Start with zero volts input so we have zero volts at $Y$ (Fig. 5.1.1). The two diodes are therefore non-conducting so there is *no* feedback and the gain is the open-loop value which we will take to be say $10^4$ – it needs to be high to reduce distortion substantially. If now the input is 10 μV then the output would be 0.1 V so the appropriate diode would be conducting to a small degree, i.e. its resistance would be less than infinity (see Fig. 4.5.1 and Eq. (4.5.6)) and the gain would in fact now be less than $10^4$. To determine exactly what the gain is, we need to go round the loop in a series of successive approximations until we get a consistent answer. This is one of the great difficulties presented by non-linear elements, but it is what PSpice is good at and can rapidly find the new operating point. As the input signal departs from zero the feedback increases to control the gain to be exactly the value necessary to linearize the diode characteristic: it can do this, if the amplifier gain is substantial, since it is the diode itself that controls the gain. Thus at $Y$ we would expect to see a very rapid change of signal for the time when the input is small and the diode is starting to conduct but the closed-loop gain is still large, with a reversion to a good approximation of the input wave once the diode is strongly conducting and the closed-loop gain approaches 1. This region of rapid change can be clearly seen in $v_{A1}$, Fig. 5.1.2. The amplitude of this part is about 0.5 V, just what is expected for a silicon diode.

This is the basis by which distortion, within the feedback loop, is reduced. Of course, if the frequency is high and the loop gain is correspondingly low, the reduction in distortion will be less. It should be emphasized again that the distortion must be *inside* the loop: the feedback system cannot know anything about distortion that is without the loop.

A variation on this circuit has been used to construct a window comparator which has independent adjustment of the two levels (Mancini 1998).

An alternative absolute value circuit is shown in Fig. 5.1.3 (Tektronix). In this circuit, the output current is steered according to the polarity of the input so that it always flows in the same direction in the load $R_2$. $R_1$ sets the conversion factor of voltage to current, and the circuit has the attraction of a high input impedance. The requirement that the differential voltages at the amplifier inputs tend to zero means that $v_{E2}=0$ and $v_{E1}=v_{in}$ and so the voltage across $R_1$ is equal to $v_{in}$, and the current in the output will therefore be $v_{in}/R_1$. For positive inputs, current will flow from $v_{out}$ through $Q_1$, $R_1$ and $D_2$ to $A_2$. For negative inputs, the current will flow from $v_{out}$ through $Q_2$, $R_1$ and $D_1$ to $A_1$. The current in $R_2$ is thus unidirectional as required. As shown the circuit operates up to say 1 MHz with some distortion. The simulated waveforms shown in Fig. 5.1.4 are for a frequency of 10 kHz.

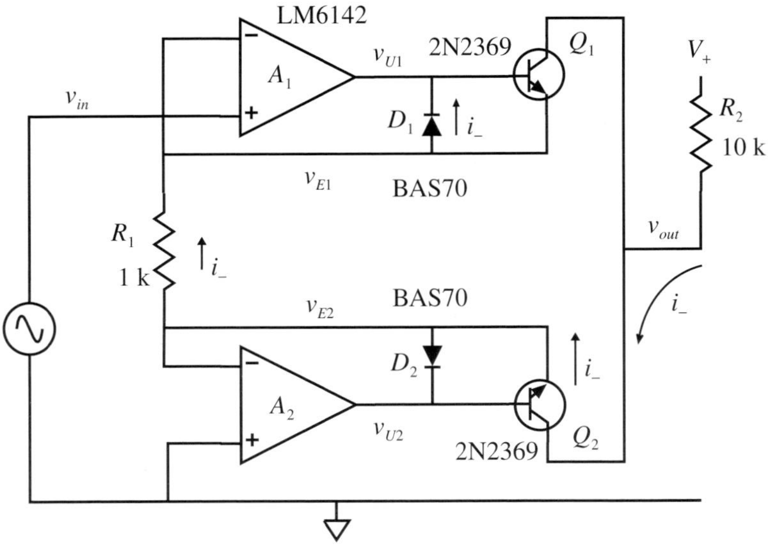


**Fig. 5.1.3** Another absolute value circuit.

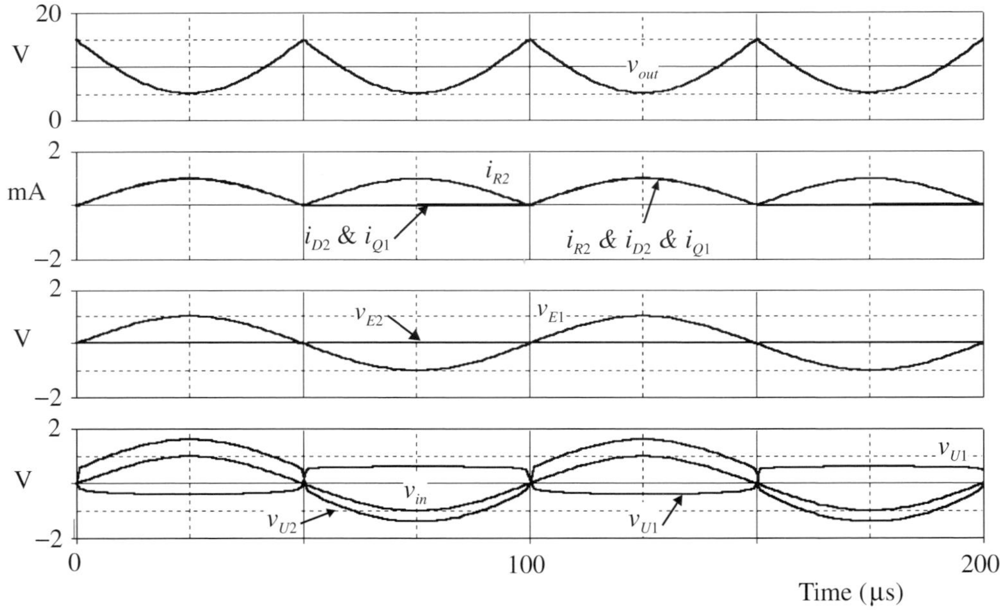


**Fig. 5.1.4** Waveforms for circuit of Fig. 5.1.3.

## SPICE simulation circuits

| | |
|---|---|
| Fig. 5.1.2 | Absval.SCH |
| Fig. 5.1.4 | Absvalb.SCH |

## References and additional sources 5.1

Jones H. T. (1993): Generate absolute value of difference. *Electronic Design* 24 June, 86.

Mancini R. (1998): Window comparator features independent adjustments. *Electronic Design* 9 February, 126.

Philbrick/Nexus (1968): *Applications Manual for Operational Amplifiers*, Philbrick/Nexus Research. See Section II.42. (Yes, I am one of the lucky ones to have a copy of this manual.)

Tektronix: This circuit appeared in a Tektronix magazine but they have not been able to identify it for me as I only have the single page.

# 5.2 Oscilloscope probes

How many a man has dated a new era in his life from the reading of a book.

Henry Thoreau

An oscilloscope probe guide (LeCroy 1994) asks whether oscilloscopes should carry a label stating 'connecting your oscilloscope probe to a circuit or device can distort the measured waveforms'. Any measurement requires interaction with the system and distorts the measurement to a greater or lesser extent. The aim is to make the disturbance as small as possible or to know how much it is disturbed and make a correction for it. Loading by the probe is usually decreased by means of a high impedance attenuator so that the improvement is bought at the expense of reduced gain. Figure 5.2.1(*a*) shows a 10:1 attenuator where $R_2$ and $C_2$ represent the input of the scope. A variable capacitor $C_1$ must be included to allow compensation so the attenuation is independent of frequency. Figure 5.2.2 shows some responses as a function of $C_1$.

In practice it is necessary to include a reasonable length of cable to allow flexibility in probing circuits and this introduces additional effects. A schematic of a probe is shown in Fig. 5.2.1(*b*), where we have represented the cable by a series of *RLC* sections rather than using one of the standard SPICE transmission line models. The main reason for this is that SPICE restricts its time steps to half the electrical time delay of the line ($\tau_D \approx 5$ ns) and so for a short line of, say, 1 m the steps will be small and the simulations take a very long time to run. The cables used are often somewhat different from normal coaxial cable in that the capacity is reduced and the resistance substantially increased. The reason for the latter is to help alleviate the major mismatch of the scope input impedance to the characteristic impedance of the cable.

The values used for the simulation as shown in the figure are approximate and do not represent any particular probe. It is first of all necessary to adjust the value of $C_1$ for optimum gain and transient response. This is normally done in practice using a low frequency squarewave output provided on the oscilloscope.

If you now include an inductor in the 'ground' lead to the source (at *G*) to represent the commonly used 'croc clip' connection you will see a significant amount

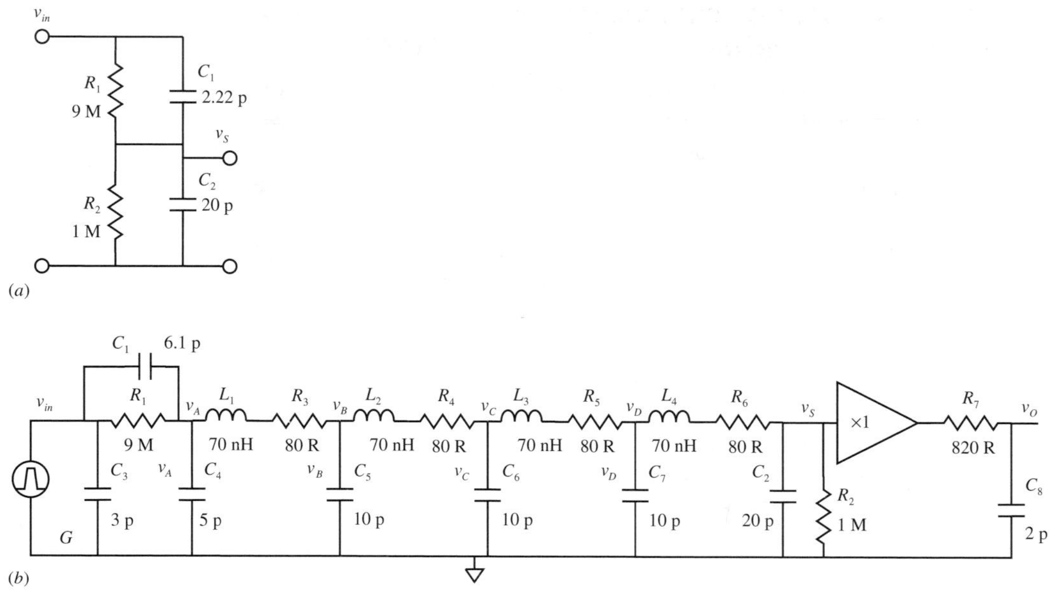


**Fig. 5.2.1** (*a*) Compensated attenuator. (*b*) Probe schematic.

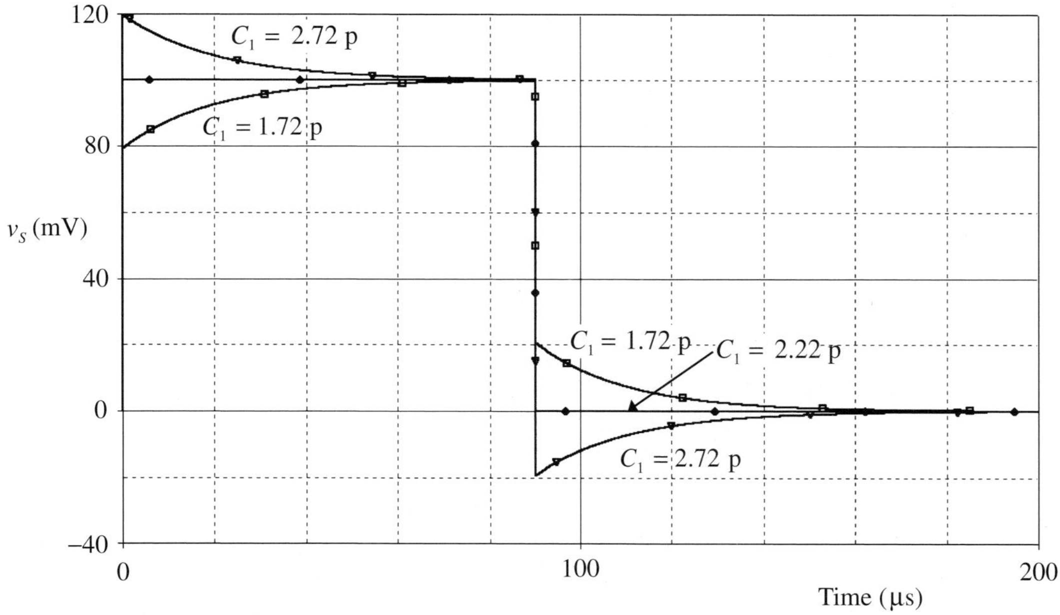


**Fig. 5.2.2** Response of the simple compensated attenuator of Fig. 5.2.1(*a*) with variation of $C_1$. The three curves show under- and over-compensation as well as critical compensation, which should occur at $C_1 = 20/9 = 2.22$ p as observed.

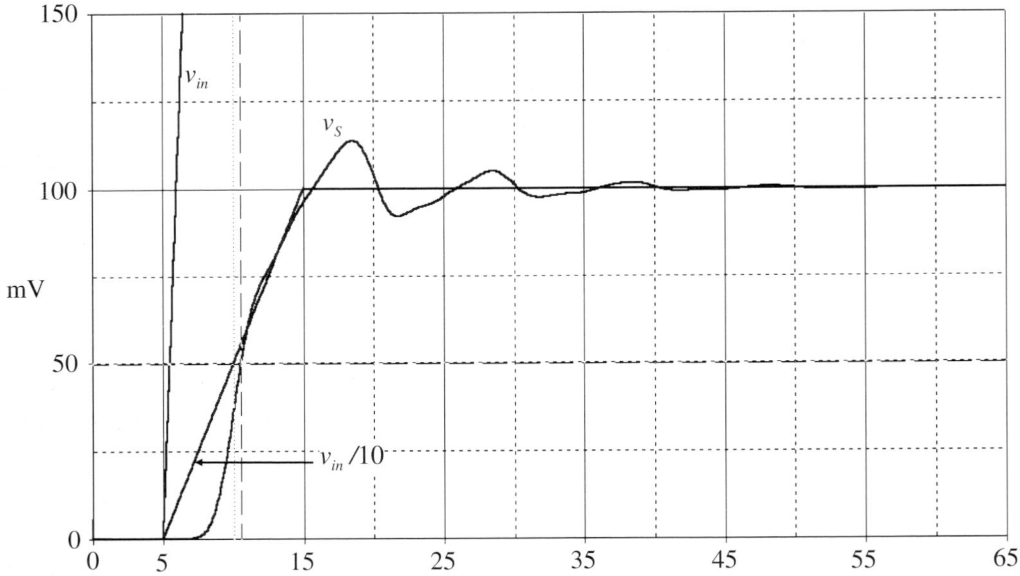


(*a*)

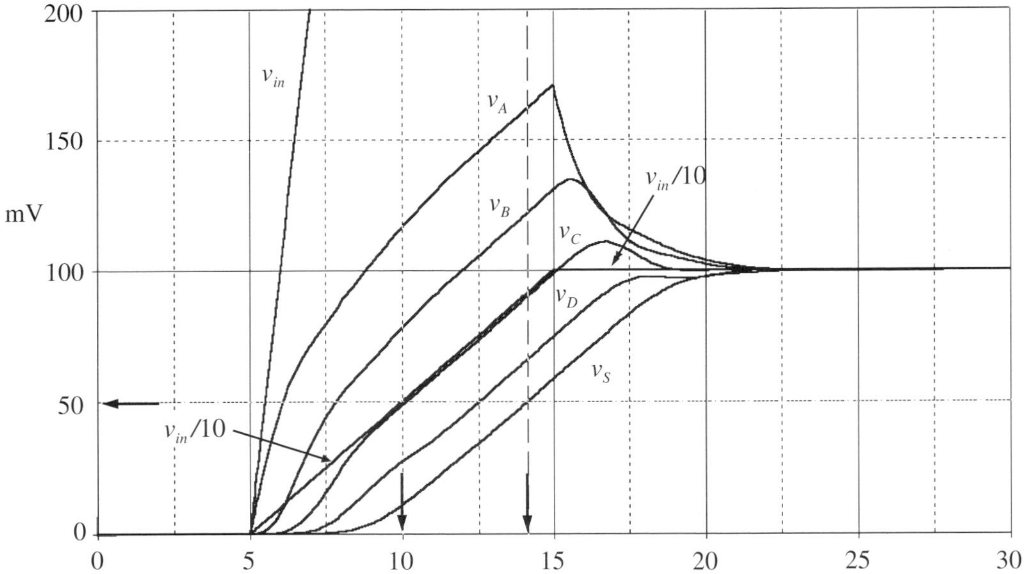


(*b*) Cursors at 10 and 14 ns

**Fig. 5.2.3** (*a*) Response of underdamped probe with cable resistor values 10*R* rather than 80*R*. The improvement may be followed by progressively increasing the resistor values and the optimum occurs at about 80*R*. (*b*) Responses for a compensated and damped probe. The input was 1 V with a risetime of 10 ns and the component values as shown in Fig. 5.2.1(*b*) for a 10 : 1 attenuation. The curve for $v_{in}/10$ is shown for comparison (and is nearly coincident with $v_C$ in the centre portion of the rise). The responses at intermediate points illustrate the progressive improvement in waveshape. The delay, $v_{in}$ to $v_S$ at 50%, is about 4 ns. The compensation capacitor $C_1$ was found to be 6.1 p for best transient response, significantly different from the value found for Fig. 5.2.1(*a*).

of ringing. A straight wire has an inductance of about 10 nH $cm^{-1}$ so for a 10 cm lead the inductance is 100 nH.

The input capacitor $C_3$ has no effect on the simulations above since we have used a voltage source. In practice of course the circuit being probed will have an output impedance and hence the responses found will differ to some degree from those shown. To find the input impedance the voltage generator can be replaced with a current source set, for example, to 1 A. Then running an a.c. sweep will give the input impedance as a function of frequency if you plot $v_{in}$ and read volts as ohms (you may plot $v_{in}/i_{in}$ to get ohms if you feel happier). The response has a corner at about 1 kHz, and at 10 MHz the impedance is only 1.88 k with a phase angle of $\approx 90°$, i.e. it is capacitive. Ignoring the line and thereafter, and only summing up the three input capacitors appropriately gives about 5.8 p, which has an impedance of 2.8 k at 10 MHz, so this is about correct.

The analysis above has been fairly straightforward and is appropriate for probes of say 10 or 20 ns risetimes. For faster response the conditions are somewhat more complex and are considered by McGovern (1977). The introduction of resistive cable, as used above, is shown to be the key to improved response at risetimes of 1 ns or less.

### SPICE simulation circuits

| | |
|---|---|
| Fig. 5.2.2 | Scopprb2.SCH |
| Fig. 5.2.3(*a*) | Scopprb1.SCH |
| Fig. 5.2.3(*b*) | Scopprb1.SCH |

## References and additional sources 5.2


Applebee P. D. (1999): Learn the ins and outs of probing those tricky differential signals. *Electronic Design* 5 April, 72, 74, 76, 79.

Bunze V. (1972): *Probing in Perspective*, Hewlett-Packard Application Note 152.

Feign E. (1998): High-frequency probes drive 50-ohm measurements. *RF Design* October, **21**(10), 66, 68–70.

Frost A., Whiteman D., Tsai J. (1999): Are you measuring your circuit or your scope probe? *EDN* 22 July, 53–58.

Heyberger C., Pryor M. (1981): *The XYZs of Using a Scope*, Tektronix Inc.

Johnson F. (1999): Simple 'homemade' sensors solve tough EMI problems. *Electronic Design* 8 November, 109, 110, 112, 114.

LeCroy (1994): *Probes and Probing*, LeCroy Corp.

McGovern P. A. (1977): Nanosecond passive voltage probes. *IEEE Trans.* **IM-26**, 46–52.

Murray J. K. (1965): Oscilloscope accessories. *Marconi Instrumentation*, **10**(1), April, 2.

O'Dell T. H. (1991): *Circuits for Electronic Instrumentation*, Cambridge: Cambridge University Press. ISBN 0-521-40428-2.

Parham J. (1998): *High-Speed Probing*, Tektronix Inc., June, Literature No. 55W-12107-0.

Roach S. (1998): Signal conditioning in oscilloscopes and the spirit of invention. Chapter 7 in Williams J. (Ed.) (1998): *The Art and Science of Analog Circuit Design*, Boston: Butterworth-Heinemann. ISBN 0-7506-7062-2. 107/1#.

Tektronix (1998): *ABC's of Probes*, Tektronix Inc., July, Literature No. 60W-6053-7.

Williams J. (1991): *High Speed Amplifier Techniques*, Linear Technology Application Note 47, August. Includes ABCs of probes, by Tektronix. *Linear Technology Applications Handbook*, Vol. II 1993. See also AN13, and AN35 for current probes.

# 5.3 Operational amplifier circuits

Give me a lever long enough and a fulcrum on which to place it, and I shall move the world.

Archimedes

The operational amplifier was originally developed as the active element of analog computers (Korn and Korn 1956). The name derives from their ability to carry out a range of mathematical operations which were essential to the solution of differential equations. These operations are such as addition, subtraction, integration and differentiation. They are now indispensable devices in electronic systems though they are not commonly used in analog computers as such, but the functions performed are still the same. The performance of the various functions depends not only on their inherent properties but particularly on the application of negative feedback. We will consider this in some detail both here and in Section 3.10 as this is an instructive application to consider the benefits and the inevitable penalties and difficulties faced in the use of feedback. We will first consider an ideal operational amplifier and then let some reality intrude to show what we have to make do with in practice.

The ideal operational amplifier requires the following properties:

(i) infinite open-loop gain;
(ii) infinite input impedance;
(iii) infinite bandwidth;
(iv) zero output impedance, offset, drift and noise.

The early vacuum tube operational amplifier had only a single input terminal owing to the nature of the drift-correcting chopper amplifier (Goldberg 1950). All standard operational amplifiers now have differential input terminals to give us the simple feedback configuration as shown in Fig. 5.3.1.

Here we show two resistors for simplicity, but as we shall see later there are many other configurations. The amplifier block amplifies the difference between the two inputs to give the output $v_{out}$. The two inputs are $v_+$, the non-inverting, and $v_-$ the inverting. This gives:

$$v_{out} = -A(v_- - v_+) \qquad (5.3.1)$$

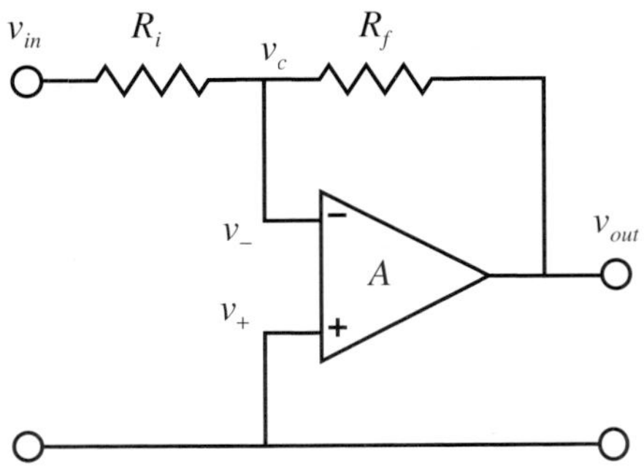


**Fig. 5.3.1** Basic operational amplifier.

the negative sign showing inversion (or what is equivalent to a 180° phase shift). It is shown this way as in many applications $v_+$ will be zero. If we now apply an input $v_{in}$ to $R_i$, then a current $i$ will flow. Since it has been assumed that the input resistance of the amplifier itself is infinite then the current must all flow through the feedback resistor $R_f$ to the output. Kirchhoff's law requires the current to be continuous, i.e. it *must* flow in a loop back to the source $v_{in}$. In drawing circuits it is usual to omit the power supply connections, but you should not forget them. In this case the current through $R_f$ will flow into or out of the amplifier output depending on the sense, then through the output stage of the amplifier, through the appropriate power supply and back to common and the input generator $v_{in}$. The current $i$ will be given by ($v_c \equiv v_-$):

$$i = \frac{v_{in} - v_c}{R_i} \quad \text{or} \quad i = \frac{v_c - v_{out}}{R_f} \tag{5.3.2}$$

and the actual input to the amplifier $v_c$ is related to the output $v_{out}$ by:

$$v_{out} = -Av_c \tag{5.3.3}$$

Now as we have assumed that $A$ tends to infinity, then $v_c$ must tend to zero. Thus from (5.3.2) we have:

$$\frac{v_{in}}{R_i} = \frac{-v_{out}}{R_f} \quad \text{or} \quad G_\infty = \frac{v_{out}}{v_{in}} = \frac{-R_f}{R_i} \tag{5.3.4}$$

The signal from the output is fed back to $v_c$ to *oppose* the input and the effect is to make $v_c \to 0$. This point is thus forced to the same potential as common or more correctly $v_+$. Most accounts refer to this as a *virtual ground* but in my view this is better referred to as a *virtual common* since ground has a quite separate meaning, i.e. common in a circuit is not necessarily at ground and in many applications $v_+$ is not necessarily at ground potential. The safest view to hold is that the differential input voltage $(v_- - v_+) \rightarrow 0$. The closed-loop gain is shown as $G_\infty$ to indicate the condition $A \rightarrow \infty$. We see that the gain depends on the external feedback components and is independent of the gain $A$ of the amplifier itself.

In practice $A$ is of course not infinite though it can be very large, say $10^6$, at least at low frequency, and falls off with frequency. It is necessary therefore to examine the effect of finite gain on the closed-loop gain. Using (5.3.2) and (5.3.3) and taking $A$ as finite gives:

$$\frac{v_{in} + \dfrac{v_{out}}{A}}{R_i} = \frac{\dfrac{-v_{out}}{A} - v_{out}}{R_f}$$

$$v_{out}\left(\frac{R_f}{A} + \frac{R_i}{A} + R_i\right) = -v_{in}R_f$$

$$\text{so} \quad G_A = \frac{v_{out}}{v_{in}} = \frac{-R_f}{R_i}\left(1 + \frac{R_f + R_i}{AR_i}\right)^{-1} \tag{5.3.5}$$

$$= \frac{-R_f}{R_i}\left(1 + \frac{1}{A\beta}\right)^{-1}, \quad \text{where} \quad \beta \equiv \frac{R_i}{R_i + R_f}$$

$$= \frac{-R_f}{R_i}\left(\frac{A\beta}{1 + A\beta}\right) \quad \text{or} \quad = \frac{-R_f}{R_i}\left(1 - \frac{1}{1 + A\beta}\right)$$

The quantity $\beta$ is called the feedback fraction, i.e. it is the fraction of $v_{out}$ that is fed back to $v_-$ (remember that $v_{in}$, as a voltage source, is in effect a short circuit). It is seen that the quantity that determines how close the performance is to the ideal is $A\beta$, which is called the loop gain $L$. This represents the effective gain in going around the feedback loop. We can demonstrate a convenient graphical relationship between $A$, $G_A$, $L$ and $\beta$ as shown in Fig. 5.3.2. If $G_A$ is not too small, i.e. if $R_i \ll R_f$ then $G_A \cong -1/\beta$ and:

$$\text{Loop gain} \quad L = A\beta = \frac{A}{G_A} = \frac{\textit{open-loop gain}}{\textit{closed-loop gain}} = A(\text{dB}) - G_A(\text{dB}) \tag{5.3.6}$$

The line $G_A$ represents the closed-loop gain and would indicate that the gain was independent of frequency. The curve $A(s)$ represents the gain of the amplifier itself showing the fall off in gain at higher frequencies (Section 1.2). The difference between these is, according to Eq. (5.3.6), the loop gain $L$. From (5.3.5), provided $L = A\beta$ is large then the gain follows $G_A$ but at the intersection with $A$ the value of $L$ is negligible so the feedback no longer controls the response and the actual gain must now follow the open-loop gain $A$. Thus with negative feedback the actual

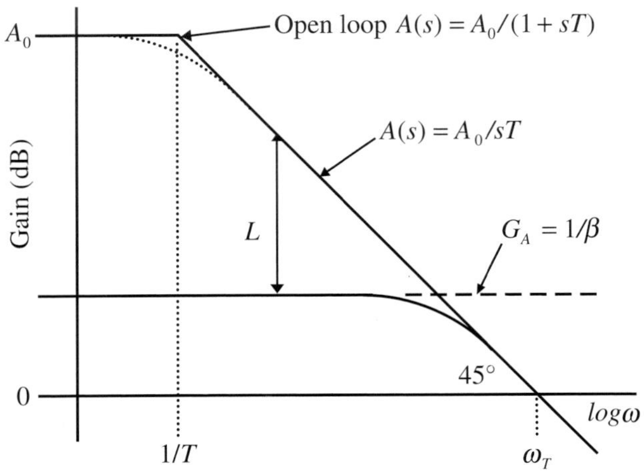


**Fig. 5.3.2** Amplifier frequency response and loop gain.

gain must be confined within the curve $A$. Equation (5.3.4) indicated that the actual gain was independent of $A$. From (5.3.5) we can find the sensitivity of $G_A$ to variations in $A$ by taking the logarithm as follows:

$$\ln G_A = \ln A - \ln(1 + A\beta) \quad \text{using} \quad \beta \cong R_i/R_f$$

and differentiating
$$\frac{dG_A}{G_A} = \frac{dA}{A} - \frac{\beta dA}{(1+A\beta)} \tag{5.3.7}$$
$$= \frac{1}{(1+A\beta)}\frac{dA}{A}$$

This shows that a fractional change $dA/A$ in the amplifier gain is reduced by effectively the loop gain $L = A\beta$ in the consequential fractional change in $G_A$. Since passive components like resistors or capacitors can be much more temperature and temporally stable as well as have higher precision relative to the amplifier characteristics, the benefit of negative feedback is evident. A drawback that is immediately evident from Fig. 5.3.2 is that the bandwidth available is inversely dependent on the gain. The point of intersection of $G_A$ with the $A$ curve is at (Section 1.2):

$$G_A = \frac{A_0}{sT} \quad \text{or} \quad G_A s = \frac{A_0}{T} \tag{5.3.8}$$

i.e. the gain–bandwidth product is a constant for this form of open-loop response.

We now examine the effect on the input and output impedances. Figure 5.3.3 shows a resistance $R_{out}$ to represent the output resistance of the amplifier itself. In practice this will be of the order of tens of ohm.

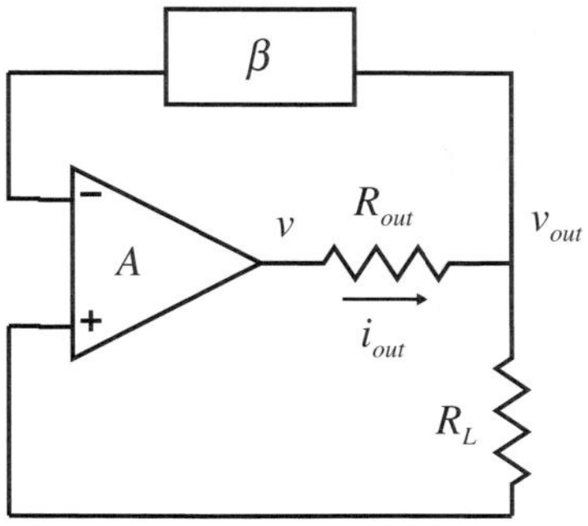


**Fig. 5.3.3** Determination of effective output impedance.

With no feedback connected we have:

$$v_{out} = v + i_{out} R_{out} \tag{5.3.9}$$

so the effective output impedance is:

$$\frac{\partial v_{out}}{\partial i_{out}} = R_{out} \tag{5.3.10}$$

as you would expect. Now connect the feedback and we have:

$$v_{out} = v + i_{out} R_{out} \quad \text{and} \quad v = -A\beta v_{out}$$

$$\text{giving} \quad v_{out} = \frac{i_{out} R_{out}}{1 + A\beta} \tag{5.3.11}$$

$$\text{so} \quad \frac{\partial v_{out}}{\partial i_{out}} = \frac{R_{out}}{1 + A\beta}$$

i.e. the output resistance has effectively been reduced by the loop gain. What of the input resistance of the amplifier? Ideally this is reduced to zero since we have a virtual common, but we can determine just how good an approximation this is. Consider Fig. 5.3.4, which shows an effective resistance at the virtual common of $R_c$. The generator $v_g$ will produce a current $i_c$:

$$i_c = \frac{v_c - v_{out}}{R_f} \quad \text{and} \quad v_{out} = -Av_c$$

$$\text{so} \quad R_c = \frac{v_c}{i_c} = \frac{R_f}{1 + A} \tag{5.3.12}$$

The input resistance of the system, i.e. from the point of view of $v_{in}$ in Fig. 5.3.1, is clearly just $R_i$. All these results correspond to the general expectation for a node–node configuration as given in Table 3.10.1.

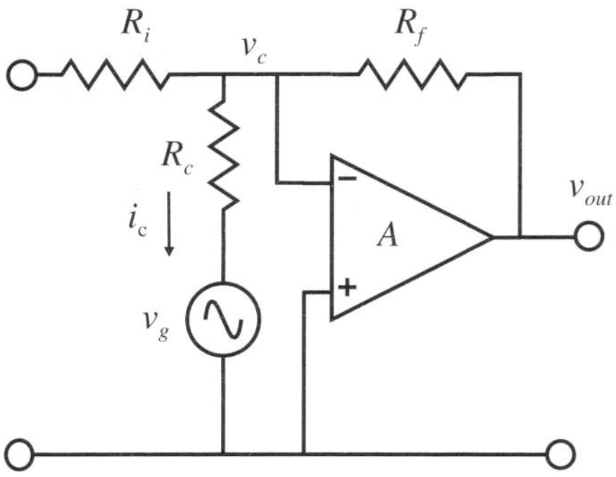


**Fig. 5.3.4** Determination of virtual common impedance.

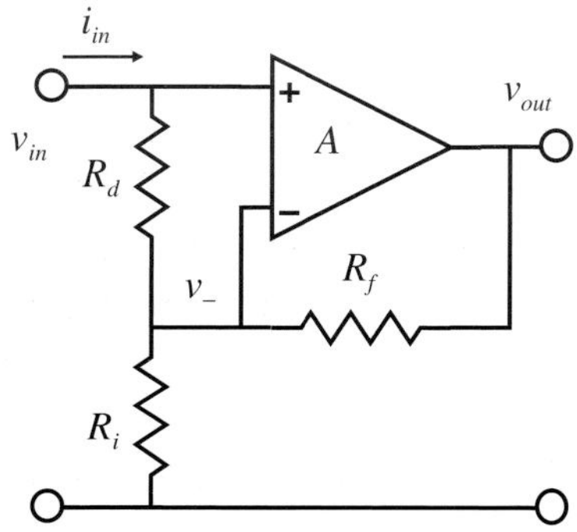


**Fig. 5.3.5** Non-inverting configuration. $R_d$ represents the input resistance of the amplifier itself.

## Non-inverting amplifier

A non-inverting configuration is shown in Fig. 5.3.5. If we proceed on the idealized basis the gain is readily determined. We expect the differential input voltage $v_d$ to be zero so that $v_- = v_{in}$. Thus:

$$v_- = \frac{v_{out} R_i}{R_i + R_f} = v_{in}$$

$$\text{so} \quad G = \frac{v_{out}}{v_{in}} = \frac{R_i + R_f}{R_i} = 1 + \frac{R_f}{R_i} = \frac{1}{\beta} \tag{5.3.13}$$

For high gains ($R_i \ll R_f$) the gain is essentially the same as for the inverting configuration. If $R_i = \infty$ the gain is unity, which may appear to be of little use. However, there is a very important difference between this configuration and the non-inverting with respect to input resistance. Ideally $v_- = v_{in}$ so there would be no current flowing in the amplifier input resistance $R_d$ and hence the effective system input resistance is infinite! Allowing for the actual value of $A$ we have:

$$i_{in} = \frac{v_{in} - v_-}{R_d}, \quad v_- = \frac{v_{out} R_i}{R_i + R_f} = v_{out}\beta, \quad v_{out} = A(v_{in} - v_-)$$

$$\text{so} \quad v_{out} = A(v_{in} - v_{out}\beta) = \frac{A v_{in}}{(1 + A\beta)} \quad \text{and} \quad v_- = \frac{A\beta v_{\text{in}}}{(1 + A\beta)}$$

$$\text{then} \quad i_{in} = \frac{v_{in} - \dfrac{A\beta v_{in}}{(1 + A\beta)}}{R_d} = \frac{v_{in}}{R_d}\left(1 - \frac{A\beta}{1 + A\beta}\right) = \frac{v_{in}}{R_d}\left(\frac{1}{1 + A\beta}\right) \tag{5.3.14}$$

$$\text{so} \quad R_{in} = \frac{v_{in}}{i_{in}} = R_d(1 + A\beta)$$

The effective input resistance is thus increased by the factor $(1 + A\beta)$. This is what is expected for the loop–node configuration of Table 3.10.1. This is not quite the whole story since there are additional common-mode resistances from each input to common which limit the increase. If the gain is unity the circuit is called a voltage follower similarly to the single transistor emitter follower. Notwithstanding the simple function of the voltage follower the application of maximum feedback makes it the most difficult case from the point of view of dynamic stability (Section 3.10).

## Summing and differencing amplifiers

The characteristics of the virtual common allow the construction of many useful circuits. Figure 5.3.6 shows a circuit for adding several voltages without any interaction between the inputs. Because of the virtual common the several input currents will be given by:

$$i_1 = \frac{v_1}{R_1}, \quad i_2 = \frac{v_2}{R_2}, \quad i_3 = \frac{v_3}{R_3} \tag{5.3.15}$$

$$\text{and} \quad i = i_1 + i_2 + i_3$$

Since each input will be amplified independently then:

$$v_{out1} = \frac{-R_f v_1}{R_1}, \quad \text{and} \quad v_{out2} = \frac{-R_f v_2}{R_2}, \quad \text{etc.}$$

$$\text{or} \quad v_{out} = -R_f\left[\frac{v_1}{R_1} + \frac{v_2}{R_2} + \frac{v_3}{R_3} + \cdots\right] \tag{5.3.16}$$

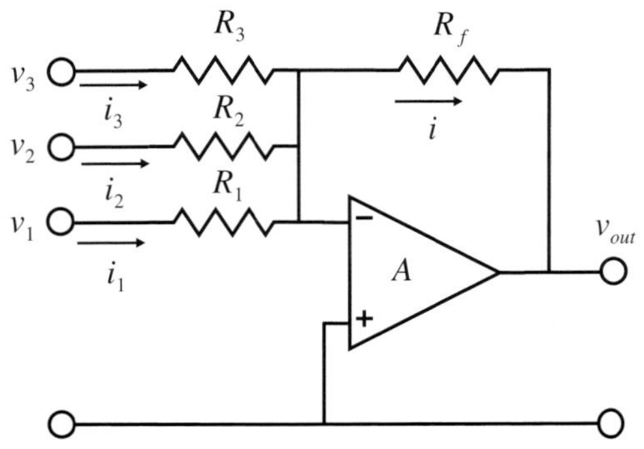


**Fig. 5.3.6** Summing amplifier.

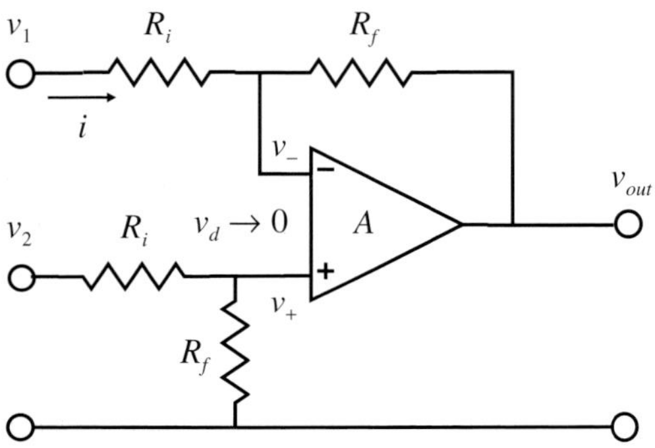


**Fig. 5.3.7** Difference amplifier.

so the inputs are added, scaled according to their particular input resistor. A connection to the non-inverting input as shown in Fig. 5.3.7 allows subtraction. This is a case where the name virtual ground causes difficulty since this is no longer true. However, taking the more appropriate condition that $v_d \to 0$ gives the relations:

$$i = \frac{v_1 - v_-}{R_i} = \frac{v_- - v_{out}}{R_f} \quad \text{and} \quad v_+ = \frac{R_f v_2}{R_i + R_f} = v_-$$

$$\text{so} \quad v_-\left(\frac{1}{R_i} + \frac{1}{R_f}\right) = \frac{v_{out}}{R_f} + \frac{v_1}{R_i}$$

$$\frac{v_2}{R_i}\left(\frac{R_f}{R_i + R_f}\right)\left(\frac{R_i + R_f}{R_f}\right) = \frac{v_{out}}{R_f} + \frac{v_1}{R_i} \tag{5.3.17}$$

$$\text{and} \quad v_{out} = \frac{R_f}{R_i}(v_2 - v_1)$$

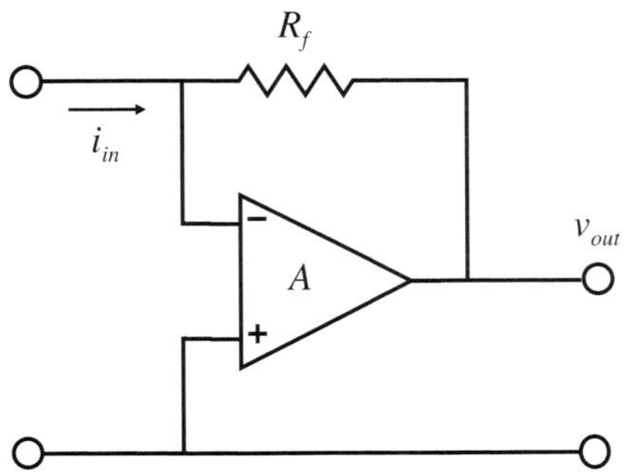


**Fig. 5.3.8** Current-to-voltage converter.

Combinations of addition and subtraction can be configured and the sums are easier to work out if the various resistors are considered as proportions of one of them.

The function of the input resistor $R_i$ in Fig. 5.3.1 is to convert the input voltage to a current. If the signal is a current source then $R_i$ may be dispensed with to give Fig. 5.3.8. The operation creating the virtual common is still effective so that the output is now given by:

$$v_{out} = -i_{in} R_f \tag{5.3.18}$$

i.e. a current-to-voltage converter without the voltage burden that would be present if a simple resistor was used. This configuration is commonly called a transimpedance amplifier since the gain has the unit volt/amp or resistance. This configuration is often used to measure very small currents in which case $R_f$ becomes very large, e.g. $10^9\ \Omega$. This raises a number of problems to which we shall return in Section 5.12. As an example of a simple application of this configuration see Nelson (1980), and for higher currents and wider bandwidth see Briano (2000) who uses a current feedback amplifier (CFA; Section 5.11).

## Operational integrator

So far we have only used resistances for the feedback elements but nothing that has been said precludes other impedances. If the feedback impedance is a capacitor as shown in Fig. 5.3.9 then, as we shall now show, the operation performed is that of integration as a function of time.

For ideal conditions we can directly write down the gain from Eq. (5.3.4) using the Laplace form:

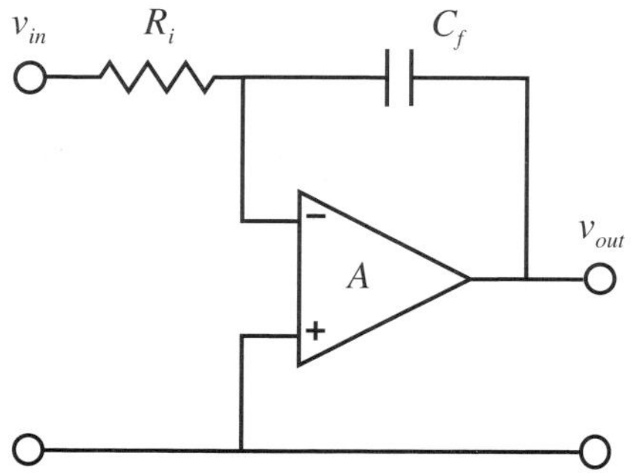


**Fig. 5.3.9** Operational integrator.

$$G = \frac{V_{out}(s)}{V_{in}(s)} = \frac{-Z_f(s)}{Z_i(s)} = \frac{-1}{R_i C_f s} \tag{5.3.19}$$

and using Table 1.12.1 to transform back to the time domain gives:

$$v_{out} = \frac{-1}{R_i C_f} \int_0^t v_{in} dt = \frac{-1}{C_f} \int_0^t i\, dt \tag{5.3.20}$$

where we have assumed no initial charge on $C_f$: if there were, there would be an additional constant term. The output is the integral of the input with a scaling factor $1/R_i C_f$. The circuit is essentially the same as that discussed in Section 3.10 and hence is often referred to as a Miller integrator. The circuit can be analysed in the same way as was done for the previous arrangement with two resistors. Since we have a virtual common, and if $Q$ is the charge on $C_f$, then we have:

$$i = \frac{v_{in}}{R_i} \quad \text{and} \quad v_{out} = \frac{Q}{C_f} \quad \text{and since} \quad Q = -\int_0^t i\, dt$$

$$\text{then} \quad v_{out} = \frac{-1}{R_i C_f} \int_0^t v_{in}\, dt \tag{5.3.21}$$

As discussed in Section 2.1 this emphasizes again the fact that currents flow through capacitors. A major concern with integrators is that they integrate everything. If there is an offset of any magnitude then this will be integrated and the output will sooner or later reach a limit, i.e. integrators whose response extends to z.f. will generally require some form of reset mechanism to keep them in the operating region.

We now consider the practical condition where the amplifier has finite gain and bandwidth. We take the simple single pole response given by Eq. (1.2.1):

$$A(s) = \frac{A_0}{1 + Ts} \tag{5.3.22}$$

Starting from Eq. (5.3.5) and inserting $1/C_f s$ for $R_f$ and Eq. (5.3.22) for $A$ we get:

$$\frac{V_{out}}{V_{in}} = \frac{-1}{R_i C_f s}\left(1 + \frac{\dfrac{1}{C_f s} + R_i}{\dfrac{R_i A_0}{1 + Ts}}\right)^{-1}$$

$$= \frac{-1}{R_i C_f s}\left(1 + \frac{\dfrac{1}{C_f s} + \dfrac{T}{C_f} + R_i + R_i Ts}{R_i A_0}\right)^{-1} \tag{5.3.23}$$

$$= \frac{-1}{R_i C_f s}\left(1 + \frac{1}{A_0 R_i C_f s} + \frac{T}{A_0 R_i C_f} + \frac{1}{A_0} + \frac{Ts}{A_0}\right)^{-1}$$

We must now make some approximations to get a more manageable expression. Since $A_0 \gg 1$ and $A_0 \gg T/R_i C_f$ then we have (and using $\omega_T = A_0/T$):

$$\frac{V_{out}}{V_{in}} = \frac{-1}{R_i C_f s}\left(1 + \frac{s}{\omega_T} + \frac{1}{A_0 R_i C_f s}\right)^{-1} \tag{5.3.24}$$

There are two frequency regions of interest. The high frequency response affects the response time, and the low frequency the linearity of the integration. At high frequency, defined by the condition $s \gg 1/A_0 R_i C_f$, Eq. (5.3.24) becomes:

$$\frac{V_{out}}{V_{in}} = \frac{-1}{R_i C_f s}\left(1 + \frac{s}{\omega_T}\right)^{-1} = \frac{-\omega_T}{R_i C_f s(s + \omega_T)} \tag{5.3.25}$$

The response of an ideal integrator to a unit step function input $u(t)$ would be a linear ramp, i.e. proportional to time, with a slope $1/R_i C_f$. For Eq. (5.3.25), using a negative step $v_{in} = -u(t)$ for convenience, we find a response (the transform of $u(t)$ is $1/s$):

$$V_{out} = \frac{\omega_T}{R_i C_f s^2 (s + \omega_T)}$$

$$\text{or} \quad v_{out} = \frac{1}{R_i C_f \omega_T}(e^{-\omega_T t} + \omega_T t - 1) \tag{5.3.26}$$

For $t$ small the exponential term is significant but this soon decays to zero and the response then becomes of the form $(t - 1/\omega_T)$ and there is thus a delay of $1/\omega_T$ in the response as shown in Fig. 5.3.10.

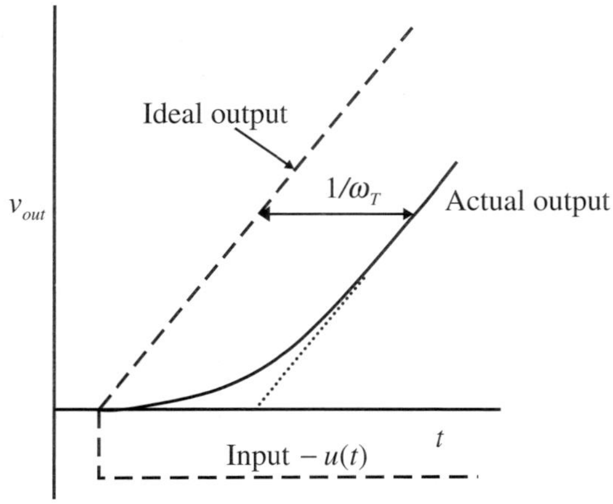


**Fig. 5.3.10** Integrator response to a step input.

If we examine the low frequency region where $s \ll \omega_T$ then Eq. (5.3.24) becomes:

$$\frac{V_{out}}{V_{in}} = \frac{-1}{R_i C_f s}\left(1 + \frac{1}{A_0 R_i C_f s}\right)^{-1} = \frac{-A_0}{1 + A_0 R_i C_f s} \tag{5.3.27}$$

A unit step function input in this case gives the time response (Table 1.12.1, No. 16 with $\beta = 0$):

$$\begin{aligned} v_{out} &= A_0[1 - \exp(-t/A_0 R_i C_f)] \\ &= A_0\left[1 - 1 + \frac{t}{A_0 R_i C_f} - \frac{t^2}{2A_0^2 R_i^2 C_f^2} + \cdots\right] \\ &\cong \frac{t}{R_i C_f} - \frac{t^2}{2A_0 R_i^2 C_f^2} \end{aligned} \tag{5.3.28}$$

The first term is just the ideal response, while the second shows that the error (at least for short times when we may ignore higher terms) increases as the square of the time. When $t \gg A_0 R_i C_f$ then the exponential part in Eq. (5.3.28) becomes negligible and $v_{out}$ tends to $A_0$ as it should. In effect, as far as z.f. is concerned, the system is operating open loop so with unit input you will get $A_0$ out. These predictions can be examined using SPICE by simulating the circuit of Fig. 5.3.9 using a μA741 amplifier with $R_i = 100\ \Omega$, $C_f = 100$ n and a resistor equal to $R_i$ in the amplifier (+) connection to common. The input step must be small to avoid limiting and a value of –10 μV was found satisfactory. The open-loop gain of the amplifier $A_0$ is given in Eq. (5.6.15) as about $2 \times 10^5$. If the simulation is run the

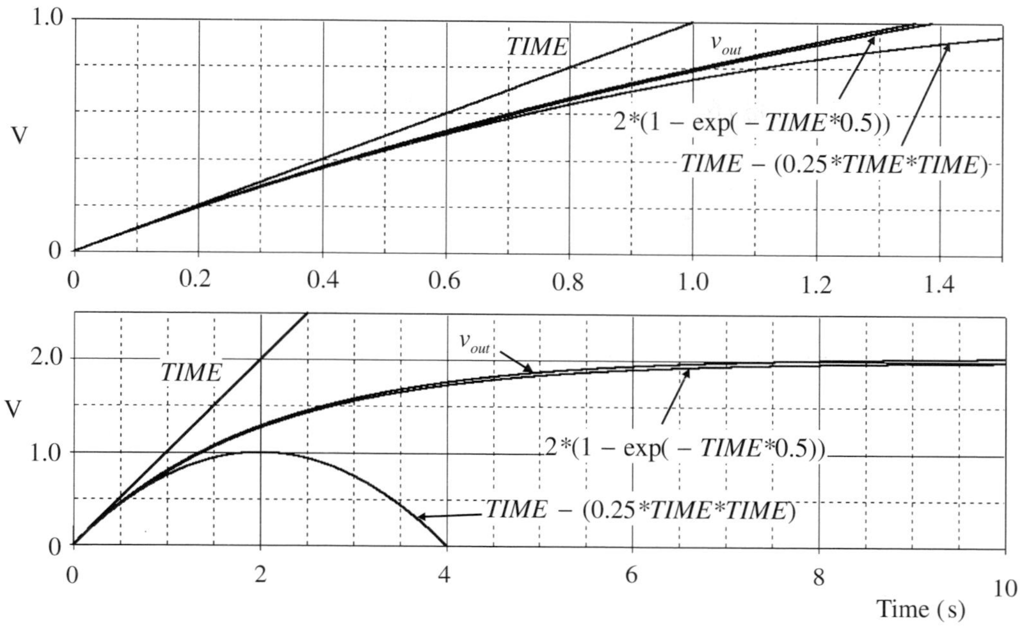


**Fig. 5.3.11** Integrator responses to a step input.

final output will not be 2 V as expected, but about 6 V. The reason for this is of course the amplifier offset voltage which will itself be integrated. A test with no applied input gave a final output of +3.8 V, which translates into an input offset of −19 μV so that an applied step of +9 μV will give us our required input of −10 μV. Setting the initial condition (*IC*) for $C_f$ to be zero we get the responses shown in Fig. 5.3.11 . Here we have also plotted the first (ideal) term of Eq. (5.3.28), the approximation including the $t^2$ term and the full exponential expression:

since $R_i C_f = 10^2 \times 10^{-7} = 10^{-5}$, $A_0 R_i C_f = 2$ and $v_{in}(2A_0 R_i^2 C_f^2)^{-1} = 0.25$

(1) Ideal: $\dfrac{TIME \times v_{in}}{R_i C_f} = \dfrac{TIME \times 10^{-5}}{10^{-5}} = TIME \times 1$

(2) Approximation: $TIME - (0.25 * TIME * TIME)$

(3) Exponential: $2*(1 - \exp(-TIME*0.5))$ (5.3.29)

It is seen that the final output is just $v_{in} \times A_0$ and that in the region where the circuit is at least approximately integrating, the $t^2$ correction is appropriate.

If precision integration is required the properties of the capacitor must be considered, with particular regard to leakage and memory effects. These are discussed in Section 4.2. Integrators are also examined in Section 5.5.

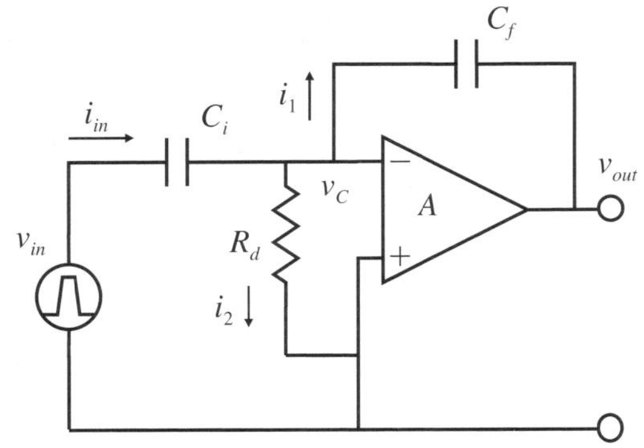


**Fig. 5.3.12** Charge amplifier.

## Differentiator

This circuit is analysed in some detail in Section 5.6.

## Charge amplifier

If both input and feedback impedances are capacitors then we have Fig. 5.3.12. For the ideal amplifier the gain will be given by:

$$G = \frac{-C_i}{C_f} \tag{5.3.30}$$

the same form as for two resistors, Eq. (5.3.4). The consequence of this is that the gain is constant within the limitations of the amplifier bandwidth down to z.f. This is clearly untrue as the impedance of the capacitors goes to infinity as frequency tends to zero. The improper proposition arises from the assumption of ideal conditions (as above) so we must examine the circuit in more practical terms. Assuming that the amplifier input resistance is not infinite but is given by $R_d$ as shown, and that $A$ has the form given by Eq. (5.3.22) then:

$$\frac{V_{in} - V_-}{1/sC_i} = \frac{V_- - V_{out}}{1/sC_f} + \frac{V_-}{R_d} \quad \text{and} \quad V_{out} = -AV_-$$

$$V_{out}\left(\frac{sC_i}{A} + \frac{sC_f}{A} + sC_f + \frac{1}{AR_d}\right) = -sC_iV_{in} \tag{5.3.31}$$

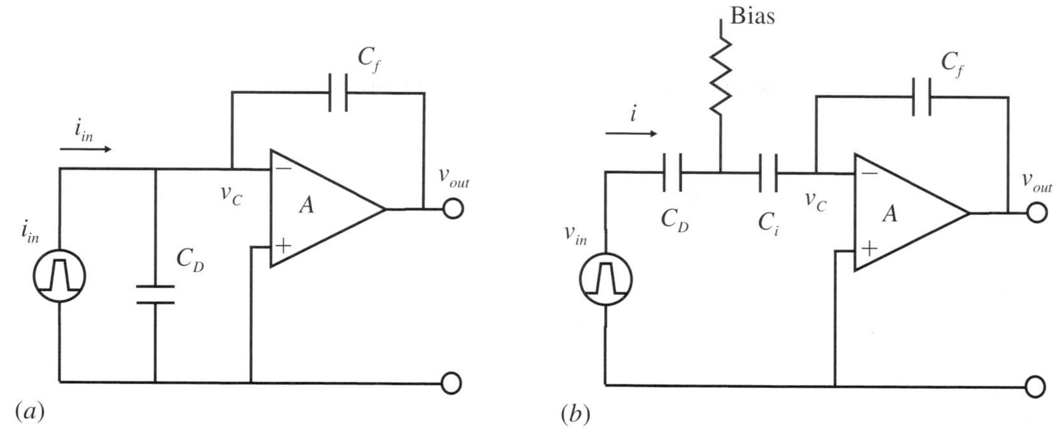


**Fig. 5.3.13** (*a*) Charge amplifier with detector. (*b*) Thévenin equivalent circuit.

$$\frac{V_{out}}{V_{in}} = \frac{-C_i}{C_f}\left[\frac{C_i}{AC_f} + \frac{1}{A} + 1 + \frac{1}{sAC_fR_d}\right]^{-1}$$

At low frequency, $s \to 0$, and since $A$ is large, then this reduces to:

$$\frac{V_{out}}{V_{in}} = \frac{-C_i}{C_f}\left[1 + \frac{1}{sAC_fR_d}\right]^{-1} \tag{5.3.32}$$

which tends to zero as we would expect.

The use of this configuration of amplifier is particularly relevant for signal sources that develop a charge rather than a voltage. Such devices include piezoelectric transducers, capacitor microphones and semiconductor nuclear particle detectors. We will consider the last as an example to see the benefits. This type of detector consists of a p-n junction reverse biased with a fairly high voltage. The ionizing particles produce hole–electron pairs in the bulk of the junction which are separated by the large electric field. For each particle a charge proportional to the energy of the particle appears on the capacity $C_D$ of the junction, so it is the charge that must be measured. $C_D$ is not fixed, depending on the applied voltage and external effects, so the voltage developed is not proportional to the charge. Figure 5.3.13(*a*) shows the circuit including the Norton equivalent of the detector, shown as a current source in parallel with the capacity $C_D$.

The virtual common ensures that there is zero voltage across $C_D$ so its value now no longer matters and the current $i$ flows to charge $C_f$. Thus:

$$v_{out} = \frac{-1}{C_f}\int i\,dt = \frac{-Q}{C_f} \tag{5.3.33}$$

so integrating over the interval of the current pulse produces an output voltage proportional to the charge produced and which is independent of the junction capacity. We may look at this in a different way by considering a Thévenin equivalent circuit as shown in Fig. 5.3.13(*b*). The capacitor $C_i$ in series with $C_D$ is now made large so that the input capacity is essentially $C_D$. $C_i$ now plays no part except that it is usually necessary to have a high voltage capacitor to isolate the high voltage bias on the junction from the amplifier. Now we have:

$$v_{out} = \frac{-C_D}{C_f} v_{in} \quad \text{and} \quad v_{in} = \frac{i}{sC_D}$$

$$= \frac{-i}{sC_f} \tag{5.3.34}$$

$$\text{so} \quad v_{out} = \frac{-1}{C_f} \int i\,dt = \frac{-Q}{C_f}$$

as before. In practice some resistance must be connected across $C_f$ to allow the charge to decay between pulses and to allow for any offset current. An example of the application of a charge amplifier is given in Cudney et al. 1972.

## Composite amplifiers

Composite amplifiers, consisting of two or more amplifiers in the feedback loop, are quite often proposed as a technique for improving performance (e.g. Wong 1987; Graeme 1993; Élantec 1997; Gerstenhaber et al. (and Brokaw) 2000). Though improvement may be obtained you have to proceed with considerable care. The 'update' by Brokaw discusses some of the problems. A much earlier approach is given by Buckerfield (1952). The gain is increased by a large factor, which is beneficial, but controlling the loop phase shift to maintain stability becomes very much more difficult. SPICE can be of considerable assistance in determining the performance and adjusting the compensation to achieve stability. Consider the circuit (Fig. 5.3.14) proposed by Graeme (1993, his Fig. 5).

There are two unusual features in this circuit. The first is the very large value of the 'load' capacitor of 30 nF. Reduction to just below 10 nF causes the circuit to oscillate so it is evident why it is there, though it seems a rather severe limitation on the use of the configuration unless you require an amplifier to drive a substantial capacity. The second is the magnitude of the signal at $v_3$ which is a differentiated form about ten times the amplitude of the output, so that the dynamic range is rather limited.

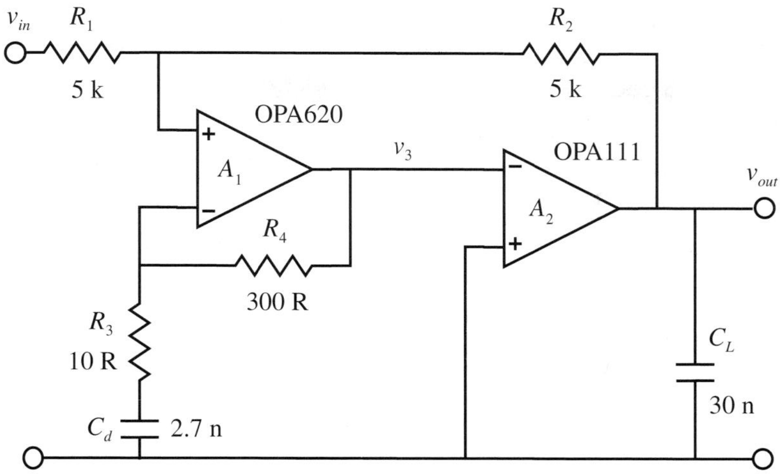


**Fig. 5.3.14** Composite amplifier with active zero compensation.

An application of a composite amplifier as a wideband integrator is described in Section 5.5.

## An Archimedian view

The basic operational circuit of Fig. 5.3.15 may be viewed in mechanical terms as a simple lever with the virtual common playing the role of the fulcrum. If the input $v_{in}$ is taken as the distance moved and the lengths of the two arms of the lever as $R_i$ and $R_f$, then the output $v_{out}$ will move by:

$$\text{(a)}\quad v_{out} = -v_{in}\frac{R_f}{R_i}, \qquad \text{(b)}\quad v_{out} = v_{in}\left(\frac{R_f + R_i}{R_i}\right) = v_{in}\left(1 + \frac{R_f}{R_i}\right) \tag{5.3.35}$$

with the signs indicating movement in opposite or the same sense. Archimedes would have appreciated this correspondence immediately.

### SPICE simulation circuits

| | |
|---|---|
| Fig. 5.3.11 | Intgrtr 1.SCH |

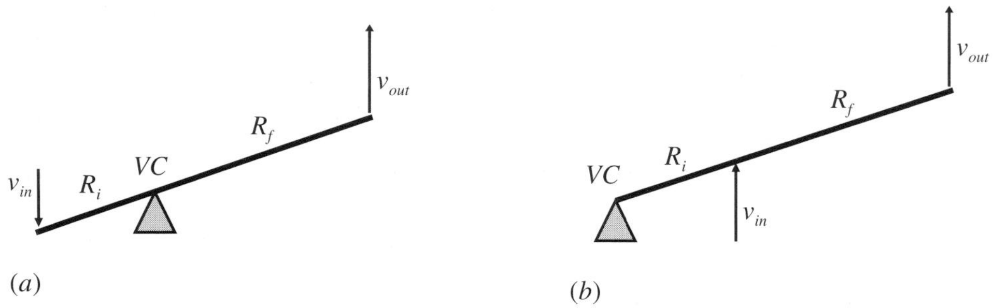


**Fig. 5.3.15** Operational amplifier as a lever: (*a*) Inverting. (*b*) Non-inverting.

## References and additional sources 5.3


Briano, Bob (2000): Improved photodiode pre-amp uses current-feedback amplifier. *Electronic Design* 10 July, 131.

Brokaw A. P. (1994): *Analog Signal Handling for High Speed and Accuracy*, Analog Devices Application Note AN-342.

Brokaw P., Barrow J. (1989): Grounding for low- and high-frequency circuits. *Analog Dialogue* **23-3**, 7–9.

Brokaw P. (1994): *An I.C. Amplifier User's Guide to Decoupling, Grounding, and Making Things Go Right for a Change*, Analog Devices Application Note AN-202.

Buckerfield P. S. T. (1952): The parallel-T d.c. amplifier: a low-drift amplifier with wide frequency response. *Proc. IEE 99*, Part II, October, 497–506.

Cudney R. A., Phelps C. T., Barreto E. (1972): An ungrounded electronic field meter. *Rev. Sci. Instrum.* **43**, 1372–1373.

Dostal J. (1993): *Operational Amplifiers*, 2nd Edn, London: Butterworth-Heinemann.

Élantec (1997): *The Care, Feeding, and Application of Unity Gain Buffers*, Élantec Application Note 19.

Franco S. (1988): *Design with Operational Amplifiers and Analog Integrated Circuits*, New York: McGraw-Hill.

Fredericksen T. M. (1984): *Intuitive Op Amps*, National Semiconductor.

Gerstenhaber M., Murphy M., Wurcer S. (2000): Composite amp has low noise, drift. *Electronic Design* 12 June, 142. Also Update by Brokaw P. 142, 144, 146, 148.

Goldberg E. A. (1950): Stabilization of wideband amplifiers for zero and gain. *RCA Review* **11**, 296–300.

Goodenough F. (Ed.) (1992): The magic of analog design. *Electronic Design* 25 November, 75, 76, 78–80, 81, 84, 86, 88, 90, 92, 93, 96.

Graeme J. (1993): Phase compensation perks up composite amplifiers. *Electronic Design* 19 August, 64–66, 68, 70, 72, 74, 76, 78.

Hamilton T. D. S. (1977): *Handbook of Linear Integrated Electronics for Research*, London: McGraw-Hill. ISBN 0-07-084483-6.

Intusoft (1999): Comparing op-amp models. *Intusoft Newsletter*, Issue 56, June, 8–10.

Karki J. (1999): *Understanding Operational Amplifier Specifications*, Texas Instruments White Paper SLOA011.

Karki J. (1999): *Effect of Parasitic Capacitance in Op Amp Circuits*, Texas Instruments Literature Number SLOA013, February. (There are a number of errors in the values of capacitors indicated in the text.)

Kester W. (Ed.) (1996): *High Speed Design Techniques*, Analog Devices Handbook. ISBN 0-916550-17-6.

Kitchen C., Counts L. (2000): *A Designer's Guide to Instrumentation Amplifiers*, Analog Devices Handbook.

Korn G. A., Korn T. M. (1956): *Electronic Analog Computers*, 2nd Edn, New York: McGraw-Hill.

Mancini R. (1996): *Feedback, Op Amps and Compensation*, Harris Semiconductor Application Note AN9415.3, November.

Mancini R. (1999): *Understanding Basic Analog – Ideal Op Amps Application Report*, Texas Instruments Literature Number SLOA068, July.

National Semiconductor (1993): *A Tutorial on Applying Op Amps to RF Applications*, National Semiconductor Application Note OA-11, September.

Nelson R. N. (1980): A capacitance and conductance adaptor for the DVM. *J. Phys. E Sci. Instrum.* **13**, 376–379. (Adaptor Sic.)

Peyton A., Walsh V. (1993): *Analog Electronics with Op-Amps*, Cambridge: Cambridge University Press. ISBN 0-521-33604-X.

Philbrick/Nexus (1968): *Applications Manual for Operational Amplifiers*, Philbrick/Nexus Research.

Smith J. I. (1971): *Modern Operational Circuit Design*, New York: John Wiley. ISBN 0-471-80194-1.

Sockolov S., Wong J. (1992): High-accuracy analog needs more than op amps. *Electronic Design* 1 October, 53, 54, 56–58, 60, 61.

Stout D. F., Kaufman M. (1976): *Handbook of Operational Amplifier Circuit Design*, New York: McGraw-Hill. ISBN 0-07-061797-X.

Williams J. (1991): *High Speed Amplifier Techniques*, LinearTechnology Application Note 47, August. Includes ABCs of Probes, by Tektronix. See also *The Contributions of Edsel Murphy to the Understanding of the Behaviour of Inanimate Objects*, by D. L. Klipstein, p. 130.

Wong J. (1987): Active feedback improves amplifier phase accuracy. *Electronic Design* 17 September, 179–186, 187, and Analog Devices Application Note AN-107.

# 5.4 Rectifier circuits

The highest pleasure to be got out of freedom, and having nothing to do, is labour.

Mark Twain

Rectifier circuits are arrangements for converting alternating power to d.c. power. A transformer is usually used to change the voltage levels and to provide safety isolation between the mains supply and your circuit. The three common rectifier configurations are shown in Fig. 5.4.1. Output voltages and currents, including smoothing capacitors and a load resistor are shown in Fig. 5.4.2.

The negative values of $I(C1)$ simply represent current flowing in the 'opposite' direction as $C_1$ supplies the output while the rectifiers are not conducting. The PSpice plot of $AVG(M(I(R2)))$ is the average of the modulus of the current through $R_2$ and this will be found to be equal to $AVG(I(R1)+I(C1))$ as it should.

The relationship between output current, ripple and smoothing capacity is usually derived from curves of the type produced by Schade (1943) and by Waidelich (1947), though it should be noted that these were originally derived for vacuum tube rectifiers which have substantial forward voltage drops. The defining parameter is the quantity $\omega R_L C$ as shown in Table 5.4.1. It is usual to have enough capacity so that there is at worst about 10% ripple on the output voltage. This means that with small error the exponential decay of the voltage between a.c. peaks can be considered linear with a rate of decay given by $1/R_L C$ (see Section 1.5). Thus if $T$ is the period of the rectified wave (i.e. 10 ms for a 50 Hz supply or 8.3 ms for 60 Hz) then the change $\Delta v_{out}$ in $v_{out}$ is given by:

$$\Delta v_{out} = \frac{v_{out} T}{R_L C} \tag{5.4.1}$$

where $R_L$ is the load resistance. For example, for the bridge rectifier the measured ripple in the figure is 1.03 V with $v_{out} = 7.2$ V. With the values shown in the figure, Eq. (5.4.1) gives $\Delta v_{out} = 1.44$ V, so it is somewhat pessimistic. The time for the decay is actually somewhat less than half the period, but since electrolytic capacitors have considerable tolerance this pessimism is desirable. This level of ripple is seldom

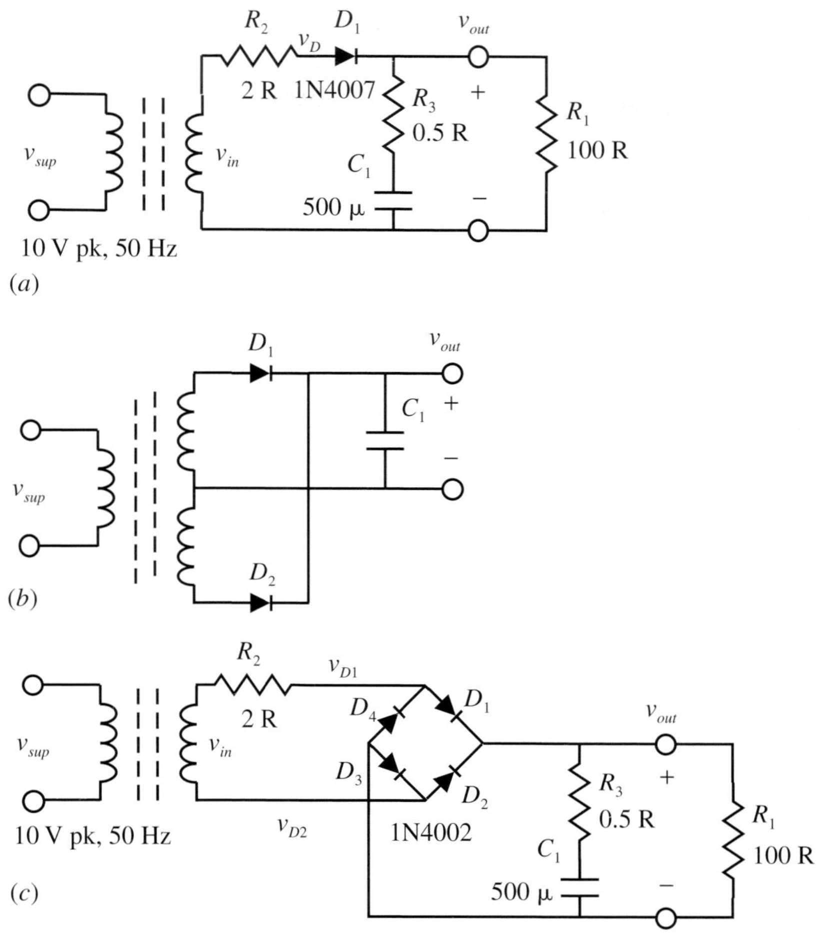


**Fig. 5.4.1** (*a*) Half-wave rectifier. (*b*) Full-wave rectifier. (*c*) Full-wave bridge rectifier.

acceptable for powering circuits so further stabilization is usually required. Remember that it is the minimum voltage at the maximum output current that is important. Nowadays regulation is mostly provided by IC regulators of which there are very many available. This general regulation function will be examined in Section 5.10.

In selecting the components for the rectifying circuit there are a number of considerations to be borne in mind:

(a) Rectifier diodes:

These must be selected to withstand the maximum peak-inverse voltage (PIV) with allowance for maximum mains voltage and any possible transients. It is usually possible to use devices with high PIV as the cost of these is low. If working close to the limit, avalanche rectifiers may be used, which can stand

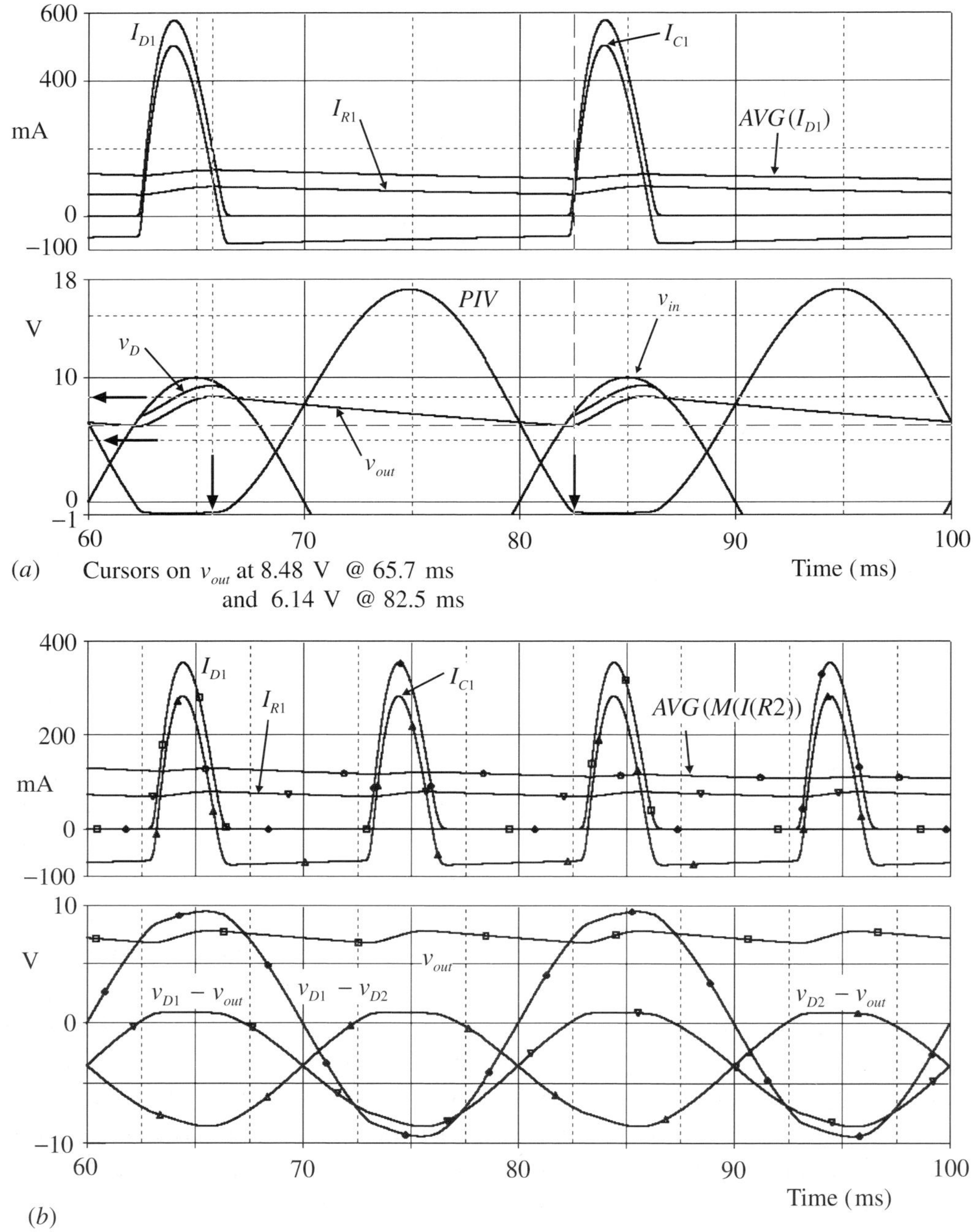


**Fig. 5.4.2** (*a*) Waveforms for half-wave rectifier. (*b*) Waveforms for full-wave bridge rectifier.

Table 5.4.1 *Parameters for full-wave rectifier*

| Voltage ripple (r.m.s., %) | $\omega R_L C$ |
|---|---|
| 10 | 7 |
| 5 | 15 |
| 1 | 70 |

$R_L$ = load resistance (Ω),
$C$ = smoothing capacity (F),
$\omega = 2\pi$ mains frequency

$$\frac{\text{average d.c. current per rectifier}}{\text{d.c. load current}} = 0.5$$

$$\frac{\text{repetitive peak current per rectifier}}{\text{d.c. load current}} = 4$$

$$\frac{\text{r.m.s. current per rectifier}}{\text{d.c. load current}} = 1.3$$

$$\frac{\text{r.m.s. ripple current through capacitor}}{\text{d.c. load current}} = 1.6$$

reverse voltage breakdown transiently without damage. It is also necessary to select a device which can carry the required current. The current in the diodes flows for only a short part of the cycle so the peak current will be many times the average d.c. output current (Fig. 5.4.2) and the rating of the diode must reflect this. Though not shown in Fig. 5.4.2 the initial peak current into an uncharged capacitor can be very large and there is also a limit on this. The common silicon rectifier diode is intended for low frequency operation. At frequencies above a few kilohertz their reverse recovery time is such that they will conduct effectively in *both* directions and so cease to rectify. Then fast recovery types must be used, or for lower voltages, Schottky rectifiers. The latter have somewhat lower forward voltages and much shorter reverse recovery times than the normal silicon rectifier, which reduces power dissipation in the device and also, in low voltage supplies, reduces the loss of voltage particularly in configurations like the bridge rectifier where two diodes conduct in series. However, an advantage in this configuration is that the PIV for each diode is halved.

(b) Smoothing capacitors:
Electrolytic capacitors are generally used when any significant capacity is required. These are polarized, have a maximum voltage and a maximum ripple current rating, with both of these temperature dependent. Such capacitors provide a large capacity for a given volume but have significant internal

resistance and inductance which limits their frequency response. A plot of impedance against frequency shows a minimum in the region from 10 to 100 kHz for aluminium types, and about 1 MHz for tantalum, with an inductive rise at high frequencies. Hence they should be used in parallel with more suitable capacitors if this presents a problem. The resistor $R_3$ shown in Fig. 5.4.1 represents the equivalent series resistance (ESR) though this does not represent the inductive impedance at higher frequency.

(c) Transformer:

The transformer must be selected for the required peak or r.m.s. voltage allowing for the voltage drops in the rectifiers and the average maximum output current which also determines the minimum voltage across the smoothing capacitor. The transformer will have its own winding resistance, which is often specified by a 'regulation' percentage figure, and which leads to lower voltages at higher currents. $R_2$ in Fig. 5.4.1 represents this. Transformers also generate significant leakage magnetic fields which may affect adjacent circuits. Toroidal types, though somewhat more expensive, can considerably reduce stray fields. But a word of warning as to how you mount a toroidal transformer: do not create a shorted turn around the toroid. When probing a rectifier circuit be careful where you clip your scope probe ground lead. In the bridge circuit, for example, the transformer secondary is floating and the negative terminal of the output may be grounded elsewhere.

For generating high voltages at low current, a voltage multiplying rectifier may be used. One form is shown in Fig. 5.4.3 and is often known as a Cockroft–Walton circuit after the originators who used it to generate high voltages to accelerate protons to produce the first artificial disintegration of the nucleus.

On the first half-cycle, with the secondary voltage at $W$ positive, $C_1$ will charge via $D_1$ so that $Y$ will be at the peak of $V_{in}$. On the second half-cycle, with $X$ now positive, the voltage applied to $D_2$ is the peak of $V_{in}$ plus the voltage across $C_1$, so $C_2$ will be charged to twice $V_{in}$ (at $Z$). The process is repeated up the stack until the output voltage becomes six times $V_{in}$. The assumption made so far is that there is only a very small output current so there is very little discharge of the capacitors. Waveforms for the circuit are shown in Fig. 5.4.3(*b*) for the parameters indicated. It is evident that an even number of stages must be used.

It is common for switching power supplies, for example, to be adjustable to operate from either 230 or 117 V a.c. supplies. In transformer input supplies this is usually achieved by tap changing on the transformer, but for switching supplies it is usual to operate direct off-line with the isolation transformer operating at the switching frequency rather than at mains frequency. A simple modification of the bridge rectifier allows a simple arrangement for dual mains operation and is shown in Fig. 5.4.4.

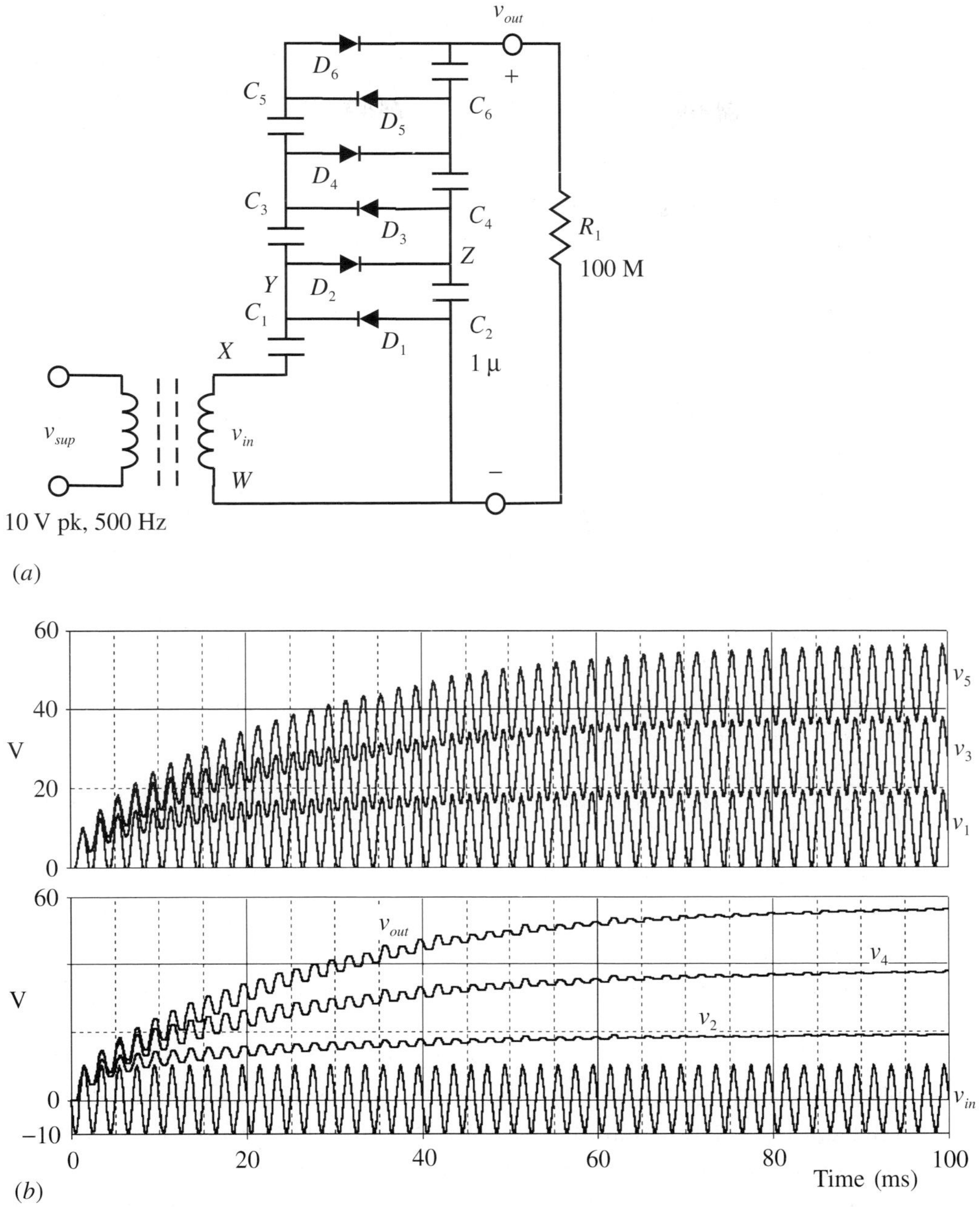


**Fig. 5.4.3** (*a*) Cockroft–Walton voltage multiplier. (*b*) Waveforms.

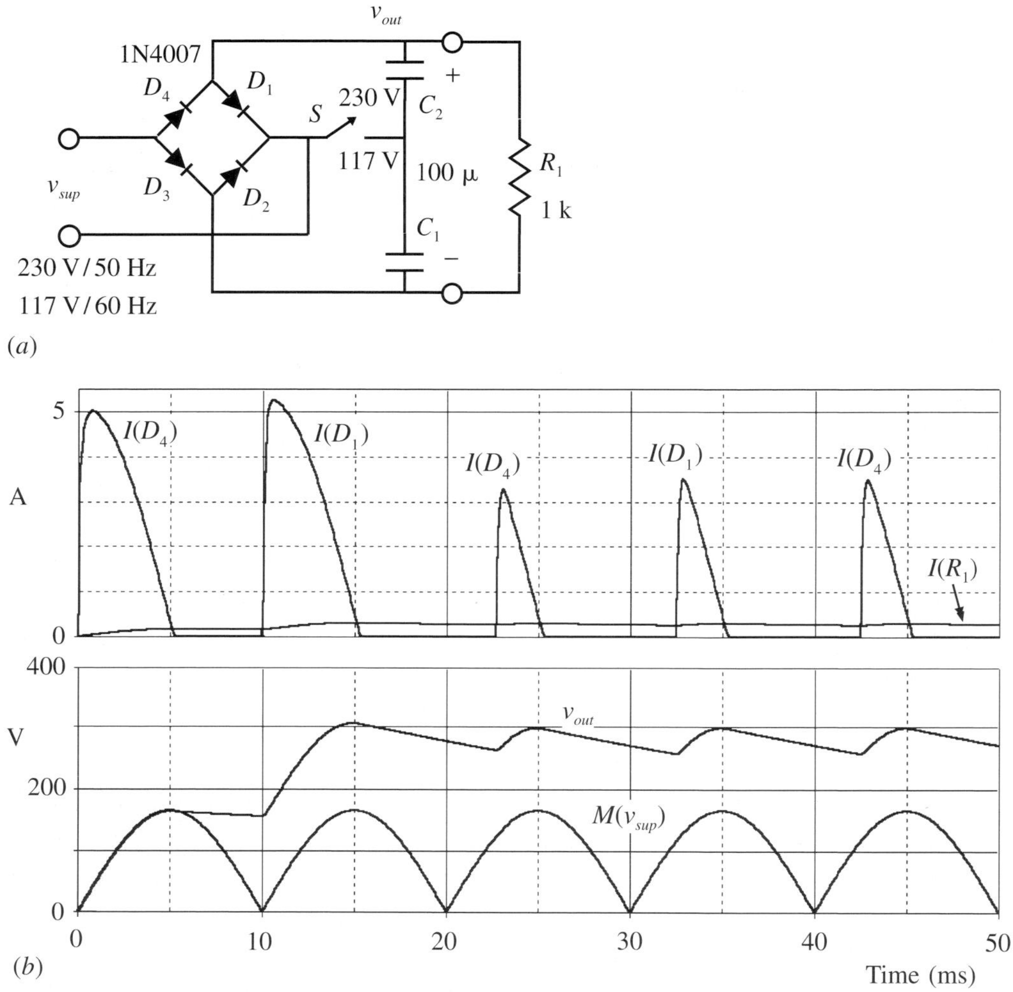


**Fig. 5.4.4** (*a*) Dual input voltage rectifier circuit. (*b*) Waveforms.

For 230 V input the switch $S$ is open and the circuit is a full-wave bridge rectifier. If the input is 117 V then $S$ is closed and the circuit acts as a voltage doubler to give the same rectified $V_{out}$ as the 230 V input case. The diodes $D_2$ and $D_3$ play no part with $S$ closed. The waveforms shown are for the 117 V configuration (165 V peak) and show the voltage doubling action and the substantial peak currents that flow even for a relatively low value of output current (about 280 mA here).

As mentioned above the normal type of power rectifier is not effective at higher frequencies. The half-wave circuit of Fig. 5.4.1(*a*) was simulated with the components indicated but at a frequency of 10 kHz and the results are illustrated in Fig. 5.4.5.

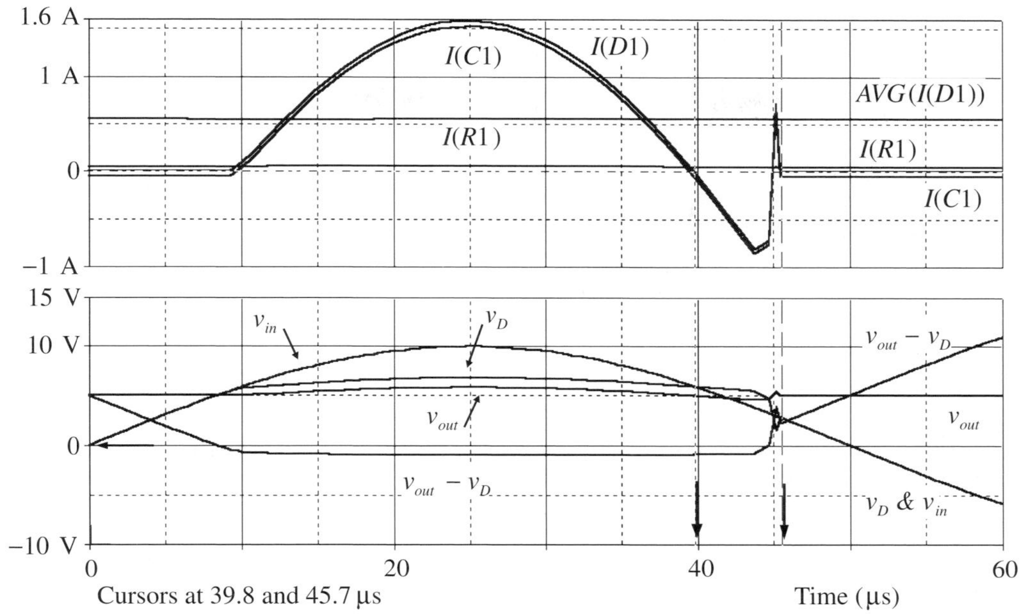


**Fig. 5.4.5** Waveforms for half-wave rectifier using 1N4007 rectifiers and a supply frequency of 10 kHz.

If you look up the model for the 1N4007 in the PSpice library (you will actually need to look at the model for the 1N4001 as the 1N4007 is given as 'a kind of' (ako) 1N4001 with only the breakdown voltage changed), you will see a parameter $Tt=$ 5.7 μs which gives the reverse recovery time. The cursors on the waveforms show that the recovery time has this value. The reverse current in the diode reaches a peak of about 850 mA and the sharp cut-off when the junction charges have been removed can lead to considerable wideband interference being generated. In conjunction with, say, leakage inductance in the transformer it can also lead to ringing waveforms on the output. If you add even a small inductor in series with the diode, say 1 μH, the simulation will produce considerable ringing.

## SPICE simulation circuits

| | |
|---|---|
| Fig. 5.4.2(*a*) | Halfwav1.SCH |
| Fig. 5.4.2(*b*) | Fwavbr 1.SCH |
| Fig. 5.4.3(*b*) | Voltmult.SCH |
| Fig. 5.4.4(*b*) | Dulvfwb1.SCH |
| Fig. 5.4.5 | Halfwav2.SCH |

## References and additional sources 5.4

'Cathode Ray' (1955): Voltage multiplying rectifier circuits. *Wireless World*, March, 115–119.

Dayal M. (1964): Power rectification with silicon diodes. *Mullard Technical Communication* **7**, 230–262, and **9**, 46–47, 1966.

Dayal M. (1964): Rectifier diode operation at kilocycle frequencies. *Mullard Technical Communications* **8**, 66–77.

Millman J., Halkias C. C. (1967): *Electronic Devices and Circuits*, New York: McGraw-Hill. See Chapter 20: Rectifiers and Power Supplies.

Noble P. G. (1978): Understanding rectifier diode data. *Electronic Components and Applications* **1**, 45–55.

Schade O. H. (1943): Analysis of rectifier operation. *Proc. IRE* **31**, 341–361.

Unvala B. A. (1968): *Power Supply Circuits Using Silicon Rectifiers*, Texas Instruments Application Report D4.

Waidelich D. L. (1947): Analysis of full-wave rectifier and capacitive input circuits. *Electronics* **20**(9), 121–123.

# 5.5 Integrators

Lesser artists borrow, great artists steal.

Igor Stravinsky

The integrator is one of the original operational circuits and formed the basis for the early analog computers or differential analysers. The basic integrator is discussed in Section 5.3. A general review of operational integrator circuits is given by Stata (1967). Here we will examine the augmenting and the non-inverting integrators. The augmenting integrator is shown in Fig. 5.5.1.

The transfer function, assuming an ideal amplifier, is:

$$H(s) = \frac{-R_f + \dfrac{1}{sC_f}}{R_i} = \frac{-R_f}{R_i}\left(1 + \frac{1}{sR_fC_f}\right) \tag{5.5.1}$$

so the response is a combination of simple gain and integration. Such a circuit is useful in the control of closed-loop servo systems which require both proportional and integral terms.

Using the non-inverting operational amplifier configuration does not lead to a non-inverting integrator, but addition of an input $RC$ as shown in Fig. 5.5.2 corrects the transfer function. The transfer function is now:

$$H(s) = \left(\frac{V_{out}}{V_+}\right)\left(\frac{V_+}{V_{in}}\right) = \left(1 + \frac{1}{sR_1C_1}\right)\left(\frac{1}{sR_2C_2 + 1}\right) = \left(\frac{sR_1C_1 + 1}{sR_1C_1}\right)\left(\frac{1}{sR_2C_2 + 1}\right) = \frac{1}{sRC}$$

$$\text{when} \quad R_1 = R_2 = R \quad \text{and} \quad C_1 = C_2 = C \tag{5.5.2}$$

which can be seen to be an integrator by comparison with Eq. (5.3.19).

A 'five decade integrator' is described in the datasheet for the CLC428 dual wideband low-noise voltage feedback amplifier (National Semiconductor 1997; Smith 1999). The circuit makes use of both amplifiers to give very high gain which results in an integrator response over a very wide bandwidth. There is of course a considerable problem in ensuring stability, both dynamic and of offset, with such

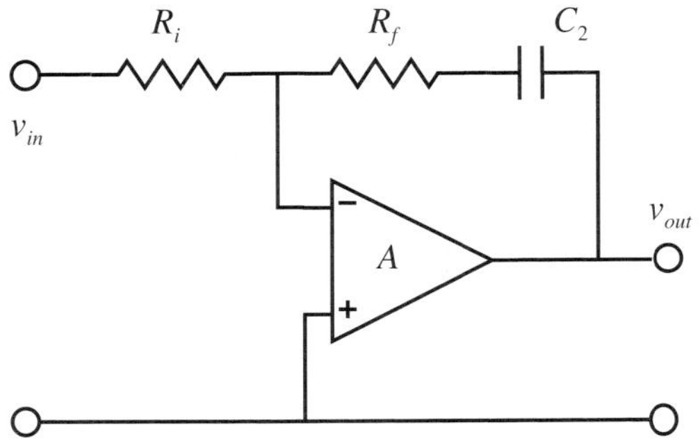


**Fig. 5.5.1** Augmenting integrator.

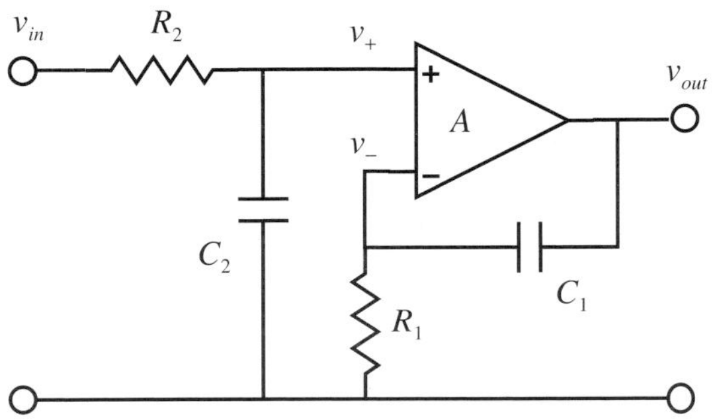


**Fig. 5.5.2** Non-inverting integrator.

high gain. The resistors $R_2$ and $R_4$ are, according to the reference, to 'reduce the loop gain and stabilize the network'. The circuit is shown in Fig. 5.5.3(*a*).

When simulated it is found that the very high z.f. gain makes the response comparator-like; small offsets drive the output to the supply rails. To avoid this, the modifications shown in (*b*) have been made: $C_2$ cuts off the z.f. gain and $R_3$ controls the gain of $A_1$, but these do not significantly affect the dynamic response where it matters. $R_4$ was increased to 390 Ω to adjust the output offset empirically to near zero, but it also helps the low frequency cut-off. The performance of the circuit as simulated is shown in Fig. 5.5.4.

This shows the frequency and phase response as well as the integrating performance for a fairly fast bipolar pulse. The frequency response should fall off at a constant rate of −20 dB/decade and the phase shift should be −90°, or more properly −270°, as shown. Taking an arbitrary level of −260° the integrating bandwidth is from about 40 Hz to 53 MHz. The integrated signal response is clean and returns to the same level after the symmetrical input pulse. The low frequency corner is the same as that for the original circuit of Fig. 5.5.3, and though that has response down to zero frequency it does not integrate in this region.

An earlier wide-band integrator, using positive feedback, is described in Jenkin

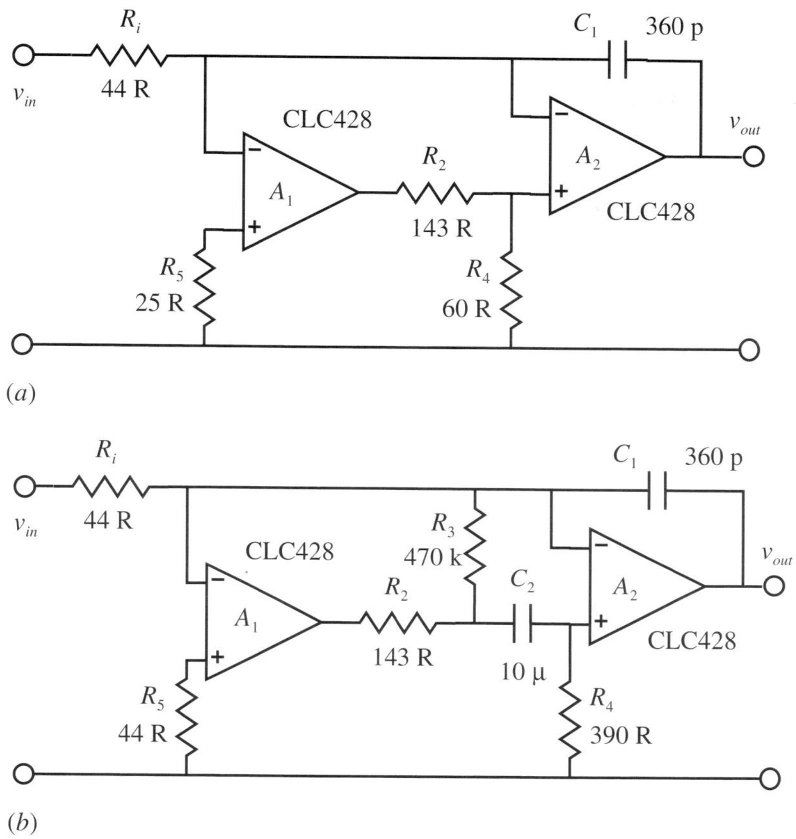


**Fig. 5.5.3** (*a*) Five decade integrator. (*b*) Modified circuit.

(1970). A low-noise integrator, also using positive feedback, is shown in the CLC425 ultra-low-noise wide-band op amp data sheet (National Semiconductor 1997) following deBoo (1967). The circuit is shown in Fig. 5.5.5.

The datasheet conditions are somewhat economical with the truth. It is necessary that $R_3/R_1 = R_5/R_4$ for a true integrator but this also puts the circuit on the edge of instability since then the positive feedback is just equal to the negative. Any source impedance or resistor tolerance would tip the balance. The resistor $R_2$ ($\gg R_1$) serves in this circumstance to decrease the positive feedback by a small amount to ensure stability. It should be noted that this circuit is just the Howland circuit (Smith 1971, p. 155).

## SPICE simulation circuits

Consult the SimCmnt.doc file on the CD before running

| | |
|---|---|
| Fig. 5.5.4(*a*) | Wbintgr 1.SCH |
| Fig. 5.5.4(*b*) | Wbintgr 2.SCH |

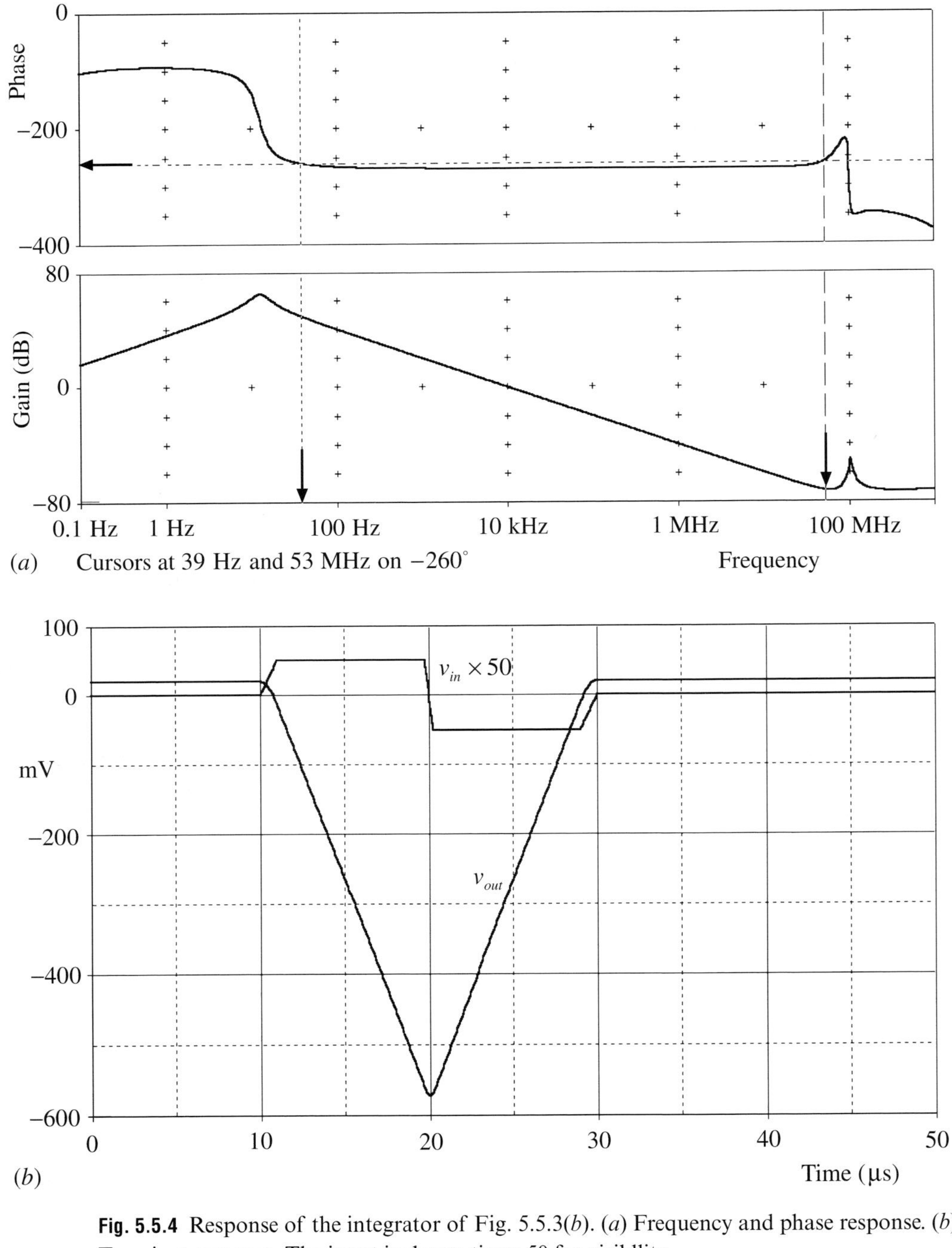


**Fig. 5.5.4** Response of the integrator of Fig. 5.5.3(*b*). (*a*) Frequency and phase response. (*b*) Transient response. The input is shown times 50 for visibllity.

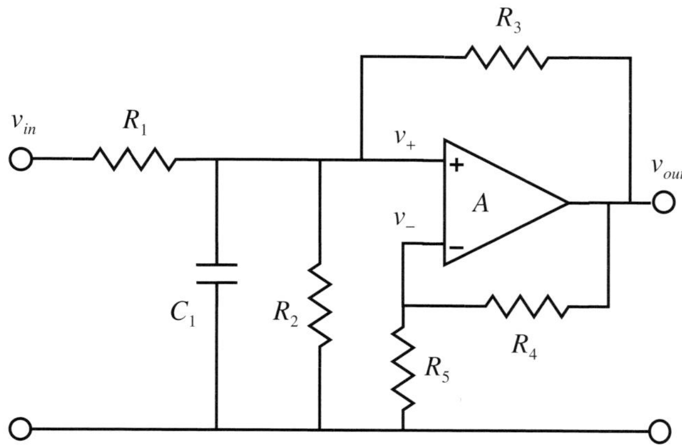


**Fig. 5.5.5** Low-noise integrator.

## References and additional sources 5.5

Alalaqui M. A. (1989): A novel approach to designing a noninverting integrator with built-in low frequency stability, high frequency compensation, and high Q, *IEEE Trans.* **IM-38**, 1116–1121.

deBoo G. J. (1967): A novel integrator results by grounding its capacitor. *Electronic Design* 7 June, 90.

Jenkin L. R. (1970): Signal integrators of wide bandwidth. *J. Phys. E Sci. Instrum.* **3**, 148–150.

National Semiconductor (1997): *Comlinear High-Speed Analog and Mixed Signal Databook.*

Smith J. I. (1971): *Modern Operational Circuit Design*, New York: John Wiley. ISBN 0-471-80194-1.

Smith S. O. (1999): Private communication. Originally at Comlinear and now at Kota Microcircuits.

Stata R. (1967): Operational integrators. *Analog Dialogue* **1**(1), April, 6–11.

# 5.6 Differentiator

Well done is better than well said.

Benjamin Franklin

Though the principle of the operational differentiator is well known its use is often avoided due to apparently undesirable properties (e.g. Tobey et al. 1971). With suitable analysis and understanding it is found that it is an amenable and designable circuit. The basic circuit is shown in Fig. 5.6.1(*a*). To analyse it effectively it is necessary to include the frequency response of the amplifier since this has a critical effect on the stability of the system.

We allow for a single pole frequency response of the amplifier given by:

$$A = \frac{A_0}{1 + Ts} \tag{5.6.1}$$

where $A_0$ is the z.f. gain and the corner frequency $\omega_1 = 1/T$ as shown in Fig. 5.6.1(*b*) (Hamilton 1974).

Using Eq. (5.3.5) for the closed-loop gain and inserting the appropriate values gives the transfer function:

$$\begin{aligned}\frac{V_{out}(s)}{V_{in}(s)} &= -R_f C_i \left[1 + \left(\frac{1 + Ts}{A_0}\right)(1 + R_f C_i s)\right]^{-1} \\ &= -s\omega_T \left[s^2 + s\left(\frac{1}{T} + \frac{1}{R_f C_i}\right) + \frac{\omega_T}{R_f C_i}\right]^{-1} \\ &= -s\omega_T \left\{\left[s + \left(\frac{1}{2T} + \frac{1}{2R_f C_i}\right)\right]^2 + \left[\frac{\omega_T}{R_f C_i} - \left(\frac{1}{2T} + \frac{1}{2R_f C_i}\right)^2\right]\right\}^{-1}\end{aligned} \tag{5.6.2}$$

where we have used $1/A_0 = 0$ and $\omega_T = A_0/T$. This is too complex to analyse readily, so we delete the last squared term since for typical values ($\omega_T \sim 10^7$, $A_0 \sim 10^6$, $R_f \sim 10^4$, $C_i \sim 10^{-8}$):

$$\frac{\omega_T}{R_f C_i} \gg \left(\frac{1}{2T} + \frac{1}{2R_f C_i}\right)^2 \tag{5.6.3}$$

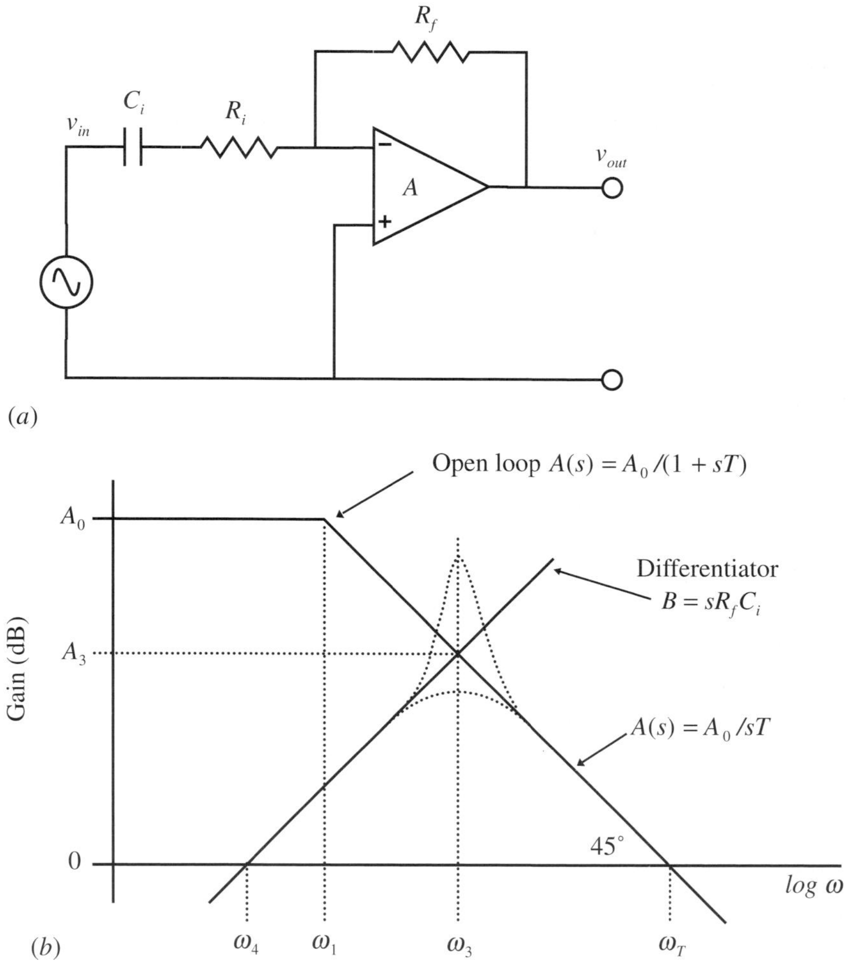


**Fig. 5.6.1** (*a*) Basic operational differentiator circuit. For the moment the resistor $R_i$ should be ignored. (*b*) Frequency responses.

To test the response we can input a ramp for which the output of an ideal differentiator would be a constant. Thus we have:

$$v_{in}(t) = kt \quad \text{and} \quad V_{in}(s) = \frac{k}{s^2} \tag{5.6.4}$$

which gives from Eq. (5.6.2) and condition (5.6.3):

$$V_{out}(s) = \frac{-k\omega_T}{s}\left\{\left[s + \left(\frac{1}{2T} + \frac{1}{2R_fC_i}\right)^2\right] + \frac{\omega_T}{R_fC_i}\right\}^{-1} \tag{5.6.5}$$

for which we find from Laplace transform tables:

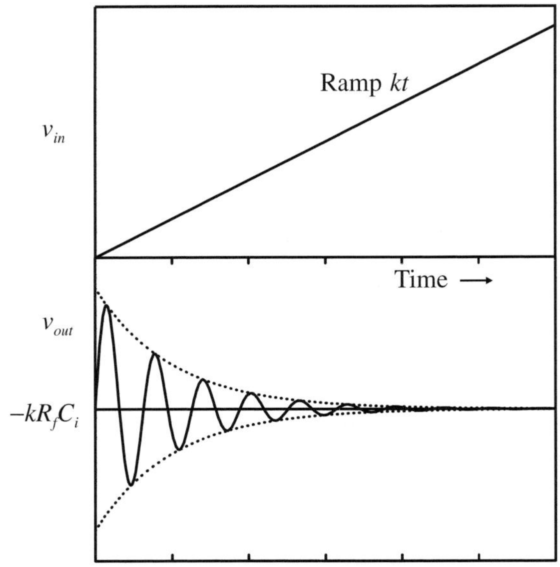


**Fig. 5.6.2** Waveforms of Fig. 5.6.1 for a ramp input.

$$v_{out}(t) = -kR_fC_i\left\{1 - \exp\left[-\left(\frac{1}{2T} + \frac{1}{2R_fC_i}\right)t\right]\cos\left(\frac{\omega_T}{R_fC_i}\right)^{\frac{1}{2}} t\right\} \tag{5.6.6}$$

Figure 5.6.2 shows the input ramp and the corresponding output, which is clearly far from ideal.

The decay time $\tau_0$ and frequency $\omega$ of the oscillation is given by:

$$\tau_0 = \left(\frac{1}{2T} + \frac{1}{2R_fC_i}\right)^{-1} \quad \text{and} \quad \omega = \left(\frac{\omega_T}{R_fC_i}\right)^{\frac{1}{2}} = (\omega_T\omega_4)^{\frac{1}{2}} = \omega_3 \tag{5.6.7}$$

where the last step is derived as follows. The intersection of the differentiator response $sR_fC_i$ with the amplifier response $A_0/sT$ occurs at the frequency $\omega_3$:

$$\omega_3R_fC_i = \frac{A_0}{\omega_3 T} \quad \text{or} \quad \omega_3^2 = \frac{A_0}{TR_fC_i}$$
$$\text{so} \quad \omega_3 = (\omega_4\omega_T)^{\frac{1}{2}} \quad \text{since} \quad \omega_T = \frac{A_0}{T} \quad \text{and} \quad \omega_4 = \frac{1}{R_fC_i} \tag{5.6.8}$$

The ringing indicates that the poles of the system are close to the $j\omega$-axis or, what is equivalent, that the loop phase shift is close to 360°, i.e. there is significant positive feedback. This can be seen if we consider the phase shifts around the loop (Fig. 5.6.3).

If we break the loop at $P$, say, then an input signal to the amplifier at $P$ will experience 180° shift due to inversion in the amplifier together with a lag tending to 90°

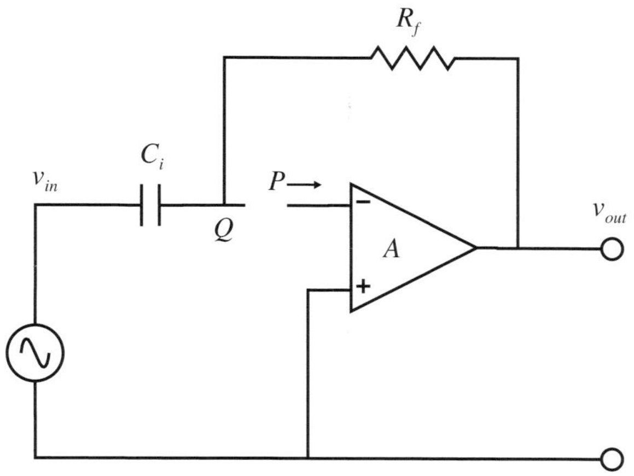


**Fig. 5.6.3** Origin of phase shift around the loop.

arising from the fall off of the amplifier gain. Then, since the signal source $v_{in}$ is a voltage source with zero internal impedance, we have a further lag approaching 90° from $R_f$ and $C_i$ at $Q$. A phasor diagram is shown in Fig. 3.3.6. The total shift around the loop thus approaches −360° and hence the ringing. If the gain is measured it will be found that the response has a peak as indicated in Fig. 5.6.5($a$), see p. 442, with gain greater than the open-loop gain, a sure sign of positive feedback. To remove the ringing, damping is introduced through a resistor $R_i$ in series with $C_i$(Fig. 5.6.1($a$)), and we now have to determine the appropriate value that will give the best response without overshoot. The transfer function including $R_i$ is:

$$\frac{V_{out}(s)}{V_{in}(s)}=\frac{-sR_fC_i}{(1+sR_iC_i)}\left[1+\left(\frac{1+sT}{A_0}\right)\left(1+\frac{sR_fR_i}{1+sR_iC_i}\right)\right]^{-1}$$

$$=\frac{-sR_fC_i}{(1+sR_iC_i)}\left[1+\frac{1}{A_0}+\frac{sR_fC_i}{A_0(sR_iC_i+1)}+\frac{sT}{A_0}+\frac{s^2TR_fC_i}{A_0(sR_iC_i+1)}\right]^{-1}$$

$$=\frac{-sR_fC_i}{(1+sR_iC_i)}\left\{\frac{TR_fC_i}{A_0(sR_iC_i+1)}\left[s^2+\frac{s}{T}+\frac{s(sR_iC_i+1)}{R_fC_i}+\frac{A_0(sR_iC_i+1)}{TR_fC_i}\right]\right\}^{-1},$$

$$\text{since}\quad\frac{1}{A_0}\approx 0$$

$$=-s\omega_T\left(s^2+\frac{s}{T}+\frac{s^2R_iC_i}{R_fC_i}+\frac{s}{R_fC_i}+\frac{sA_0R_iC_i}{TR_fC_i}+\frac{A_0}{TR_fC_i}\right)^{-1}$$

$$=-s\omega_T\left(s^2+\frac{s\omega_TR_i}{R_f}+\frac{\omega_T}{R_fC_i}\right)^{-1}\quad\text{using}\quad\omega_T=\frac{A_0}{T}\tag{5.6.9}$$

where for typical values we find that (in the second last line) the third term is much less than the first, and that the second and fourth are much less than the fifth. For critical damping the roots of the denominator must be equal (see Section 1.12), which requires:

$$\left(\frac{\omega_T R_i}{R_f}\right)^2 = \frac{4\omega_T}{R_f C_i} \quad \text{or} \quad R_{i(opt)} = \left(\frac{4R_f}{\omega_T C_i}\right)^{\frac{1}{2}} = \left(\frac{4TR_f}{A_0 C_i}\right)^{\frac{1}{2}} = \frac{2R_f}{A_3} \tag{5.6.10}$$

where $A_3$ is the gain at the intercept. Since the frequency is $\omega_{3,}$ we have:

$$A_3 = \omega_3 R_f C_i = \frac{\omega_3}{\omega_4} \tag{5.6.11}$$

In this case of critical damping the time response for a ramp input as before, will be given by:

$$v_{out}(t) = -kR_f C_i[1-(1+\omega_3 t)\exp(-\omega_3 t)] \tag{5.6.12}$$

For $t$ small the response will be dominated by the exponential to give a $(1-\exp(-t/\tau))$ form which eventually settles to the expected output $kR_f C_i$. If a small amount of overshoot is acceptable then the damping can be reduced to achieve a faster rise in the response. If we make $Q = 1/\sqrt{2}$ then the poles will lie on a locus at 45° to the axes so that the real and imaginary magnitudes (say $m$) are equal (Fig. 1.12.8). Roots of this form lead to a polynomial given by:

$$(s+m+jm)(s+m-jm) = s^2 + 2sm + 2m^2 \tag{5.6.13}$$

which when equated to the denominator of Eq. (5.6.9) gives:

$$\frac{\omega_T}{R_f C_i} = 2m^2 \quad \text{and} \quad \frac{\omega_T R_i}{R_f} = 2m \quad \text{or} \quad \frac{\omega_T^2 R_i^2}{2R_f^2} = 2m^2$$

$$\text{so} \quad \frac{\omega_T}{R_f C_i} = \frac{\omega_T^2 R_i^2}{2R_f^2} \quad \text{and} \quad R_i = \left(\frac{2R_f}{\omega_T C_i}\right)^{\frac{1}{2}} = R_{i(fr)} = \frac{R_{i(opt)}}{\sqrt{2}} \tag{5.6.14}$$

The circuit may be simulated in PSpice to check the predictions (Fig. 5.6.4). The second amplifier has been set up to produce a response for the open-loop gain and $R_i$ is set to be varied over a range of values. Using the common μA741 amplifier with $R_f = 470$ k, $C_i = 10$ n gave the following values:

$$\omega_1 = \frac{1}{T} = 32.7(5.2\text{ Hz}), \quad \omega_3 = 3.6\times10^4(5.75\text{ kHz}), \quad \omega_4 = 213.6(34\text{ Hz})$$

$$\omega_T = 5.57\times10^6(887\text{ kHz}), \quad A_0 = 1.95\times10^5, \quad A_3 = 156 \tag{5.6.15}$$

from which the optimum $R_{i(opt)}$ and fast response $R_{i(fr)}$ damping resistances are determined to be:

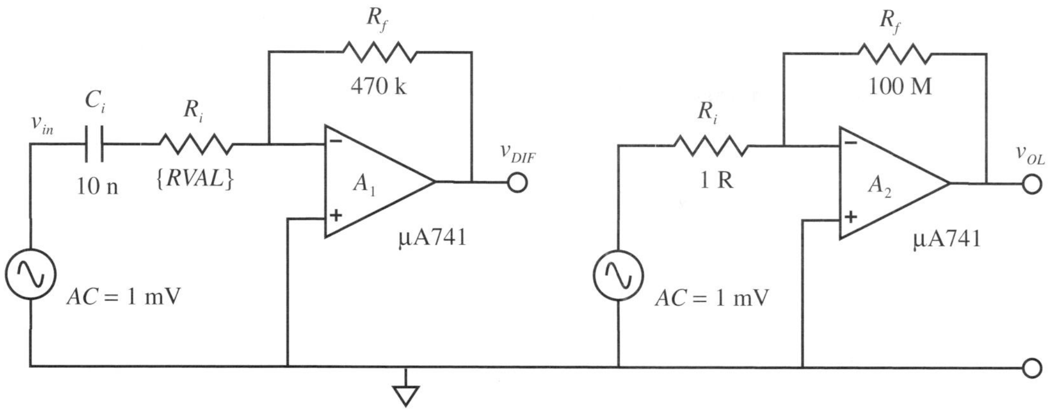


**Fig. 5.6.4** Circuits for PSpice simulation. The second circuit with $A_2$ presents the open-loop gain.

$$R_{i(opt)} = 5810\ \Omega \quad \text{and} \quad R_{i(fr)} = 4100\ \Omega \tag{5.6.16}$$

A 2 ms period triangular ramp of amplitude 0.1 V with appropriate values for the range of $R_i$ gives a suitable output response (Fig. 5.6.5). The ringing frequency is about 5820 Hz, which agrees with $\omega_3$. For critical damping, the poles are calculated to be at $34\,158 \pm j2271$ (using Mathcad), which is not quite the expected equal (real) roots. The difference is small and arises from the neglect of the small terms in Eq. (5.6.9), i.e. the damping resistors are slightly low (in this case a closer value is 5843 Ω, but the effect is negligible in practice). The poles for the fast-rise case are found to be at $24\,200 \pm j24\,286$, so the real and imaginary magnitudes are effectively the same. The fast-rise damping improved the time to reach 90% of final value by 27% with an overshoot of 5%. The frequency responses (Fig. 5.6.5(*a*)) show the resonant peak exceeding the open-loop gain, and the effect of the damping resistor. The signal phase responses show the same form as for a resonant circuit (Section 3.5). The $T$ technique (Section 5.14) can also be employed to examine the loop response and the phase shift.

## SPICE simulation circuits

| | |
|---|---|
| Fig. 5.6.5(*a*) | Opdiffr 1.SCH |
| Fig. 5.6.5(*b*) | Opdiffr 1.SCH |

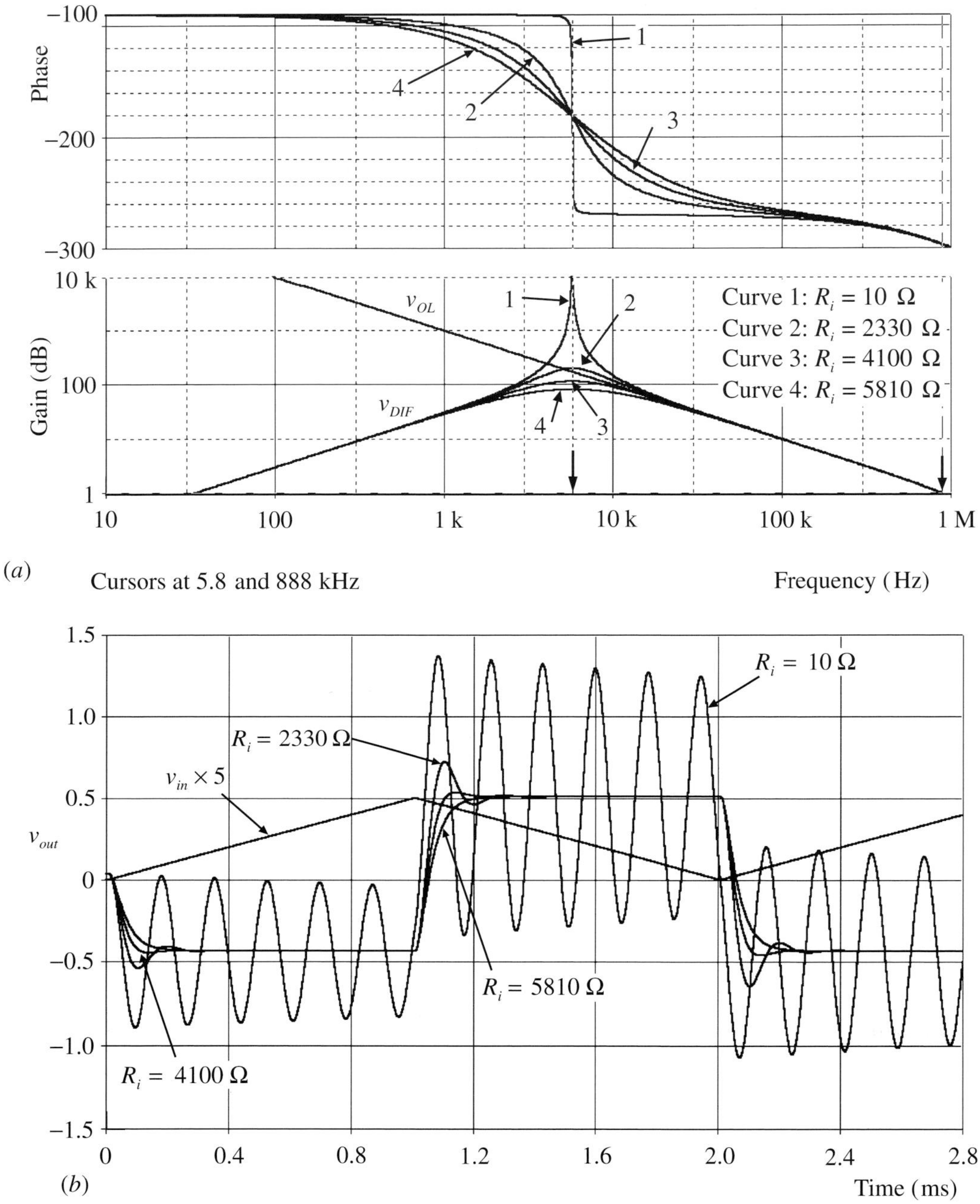


**Fig. 5.6.5** (*a*) Frequency and phase response of the differentiator. (*b*) Transient response. Note that the input signal has been multiplied by five to make it more evident.

## References and additional sources 5.6

Alalaqui M. A. (1989): A state variable approach to designing a resistive input, low noise, non-inverting differentiator. *IEEE Trans.* **IM-38**, 920–922.

Hamilton T. D. S. (1974): Operational differentiators. *Electronic Engng.* **46**, October, 53–55.

Hoft D. (1970): Applying the analog differentiator. *Analog Dialogue* June, 12–14.

Nordland D. R. (1972): High accuracy analog differentiator. *Rev. Sci. Instrum.* **43**, 1699–1700.

Schouten R. N. (1998): A new amplifier design for fast low-noise far-infrared detectors using a pyroelectric element. *Meas. Sci. Tech.* **9**, 686–691.

Tobey G. E., Graeme J. G., Huelsman L. P. (Eds) (1971): *Operational Amplifiers, Design and Applications*, New York: McGraw-Hill. Library of Congress Cat. No. 74-163297.

# 5.7 Two-phase oscillator

... getting to know, on all the matters that most concern us the best which has been thought and said in the world, and, through this knowledge, turning a stream of fresh and free thought upon our stock notions and habits.

Matthew Arnold

The aim of this circuit is to provide two outputs at the same frequency but in quadrature (Howe and Leite 1953; Good 1957; Earls 1962; Foord 1974). The circuit is shown in Fig. 5.7.1.

The standard operational integrator has been discussed in Section 5.5 on the basis of an ideal operational amplifier. The first section acts as a non-inverting integrator, as is shown below. As we need to know how the gain and phase vary with frequency it is necessary to examine the performance of each stage. Stage $A_1$ has the transfer function:

$$\frac{V_1}{V_3} = 1 + \frac{1/sC_1}{R_1} = \frac{sC_1R_1 + 1}{sC_1R_1} \tag{5.7.1}$$

which is clearly not that of an integrator. If the extra components $R_3$ and $C_3$ are included then:

$$\frac{V_3}{V_5} = \frac{\dfrac{1}{sC_3}}{R_3 + \dfrac{1}{sC_3}} = \frac{1}{sC_3R_3 + 1} \quad \text{and since} \quad \frac{V_1}{V_5} = \frac{V_1}{V_3}\frac{V_3}{V_5}$$

$$\text{then} \quad \frac{V_1}{V_5} = \left(\frac{sC_1R_1 + 1}{sC_1R_1}\right)\left(\frac{1}{sC_3R_3 + 1}\right) \tag{5.7.2}$$

In this application $R_1 = R_2 = R_3 = R$ and $C_1 = C_2 = C_3 = C$, so that (5.7.2) becomes:

$$\frac{V_1}{V_5} = \frac{1}{sCR} \tag{5.7.3}$$

which is the transfer function of a non-inverting integrator. At the oscillation frequency, $R_3C_3$ will produce a lag of 45° and $A_1$ will produce a further lag of 45° so

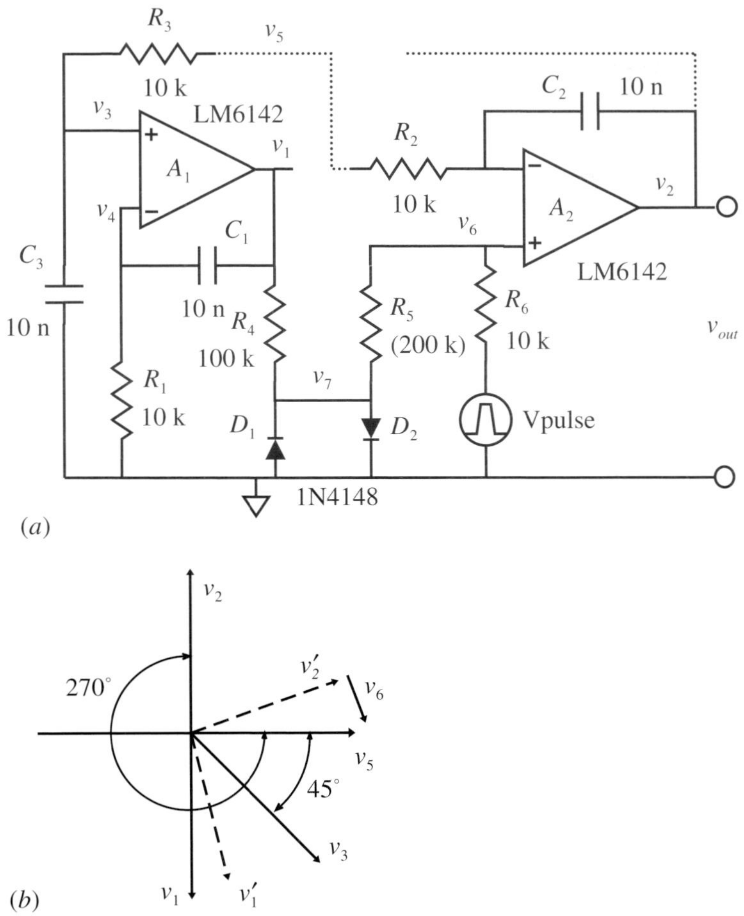


**Fig. 5.7.1** (*a*) Two amplifier two-phase oscillator. Normally $v_1$ would be connected to $R_2$ and $v_5$ to $v_2$. To show the separate phase shifts a signal may be input at $v_5$ with the connections as shown dotted. (*b*) Voltage phasors: $v_5$, $v_1$ and $v_2$ are for the separate stages; reconnecting $R_2$ to $v_1$ and allowing for non-ideal amplifiers results in $v_5$, $v_1'$ and $v_2'$. Adding $v_6$ corrects the phase shift and controls the amplitude of oscillation.

that the phase relation between $v_5$ and $v_1$ will be as shown in Fig. 5.7.1(*b*). The transfer function for the two stages $A_1$ and $A_2$ in series is then:

$$\frac{V_2}{V_5}=\frac{-1}{s^2C_2R^2} \quad \text{or} \quad \frac{V_2}{V_5}=\frac{1}{\omega^2C^2R^2} \tag{5.7.4}$$

where we have put $s=j\omega$ for sinusoids. For the particular value $\omega_0=1/RC$, $v_2=v_5$ so that the output can be connected back to the input to fulfil the Barkhausen criterion for oscillation. By breaking the feedback connection and that between stages and connecting a *VSIN* generator to both stages the responses can be simulated with the results as shown in Fig. 5.7.2.

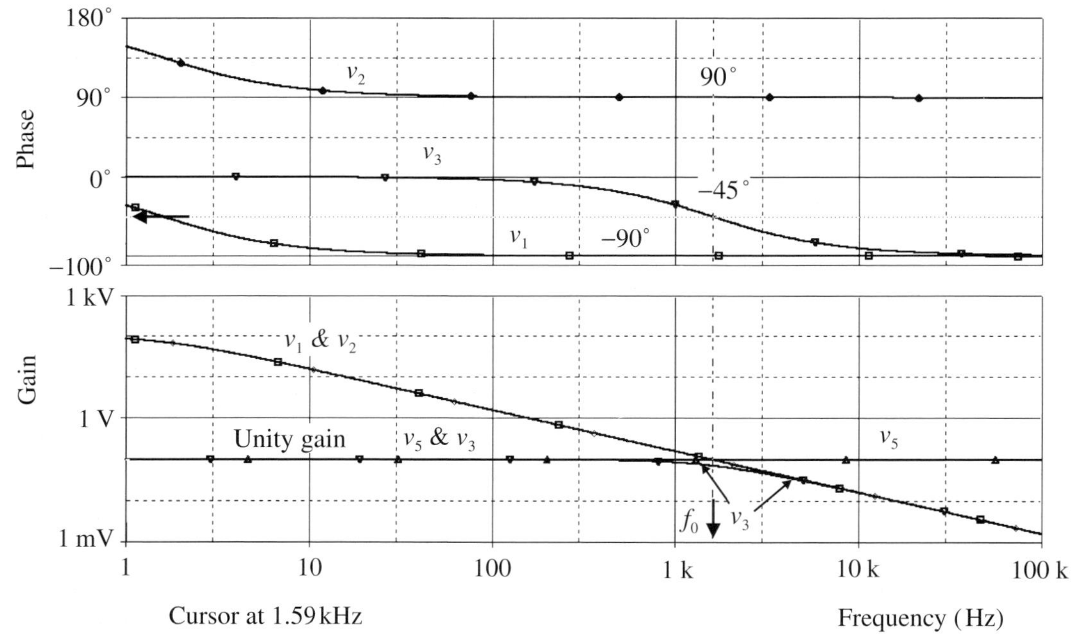


**Fig. 5.7.2** Responses for the separate stages of the oscillator.

The trace for $v_5$ represents unity gain and intercepts the integrator response at $f_0$, which for the values $R = 10$ k and $C = 10$ n used is 1.59 kHz. The phase shifts are also shown, including that for $R_3C_3$. For $A_1$ the overall phase shift is −90° and as shown for $A_2$ it is +90°. Though the latter is what most books give it is somewhat misleading in that it implies that (at the beginning, say) there is an output before any input (see Section 3.6). The actual shift is −270°, 90° for the integrator and 180° for the inversion, which does get us back to the same place but it is the correct way to look at it.

For the simulation of the oscillator it is usually necessary to give the circuit a prod to get it started (in practice switching transients would be sufficient).This is provided by *VPULSE* feeding the unused input of $A_2$. For the oscillation to build up it is necessary for the pole to be in the right half-plane (see Section 1.12) – how far into the plane, controls the rate at which the oscillations grow. This cannot continue, however, as otherwise the amplifiers would overload. It is therefore necessary that some form of limiter come into operation when the amplitude is large enough and this of course will also introduce some distortion. We need the gain to be slightly greater than 1 to allow the oscillations to start and build up until the limiter comes into operation. $R_3$ is therefore increased slightly above its nominal value since this increases the phase shift introduced by $R_3C_3$ so the system moves to a slightly lower frequency to keep the phase shift at 360°; but the gain, as can be seen from the frequency response, will now be slightly greater than 1. Changing $R_3$ thus controls the rate of growth. The feedback from the limiter provided by the diodes and $R_4$ comes into play at larger amplitudes and dynamically reduces the gain to bring the pole back to the $j\omega$-axis to give constant amplitude. If the simulation is run in transient mode to see the actual waveforms it will be found that the two outputs are indeed in quadrature but the frequency is significantly lower than $f_0$ given above. Setting $R_3 = 9.9$ k and $R_5 = 150$ k gives better amplitude control and closer oscillation frequency. Zener diodes have also been used for limiting (O'Dell 1988) but it should be remembered that they can have substantial self-capacity which makes them unsuitable for high frequencies.

At higher frequencies the gain $A$ of the amplifier will be less and for significant amplitudes there will be a limitation from the maximum slewing rate of the amplifier. Since the change of phase with frequency is slow it is evident that frequency prediction or stability is not good. This should be compared with the performance of the Wien bridge with its very rapid change of phase with frequency (Section 5.8).

An alternative configuration is shown in Fig. 5.7.3 which requires an extra amplifier but this is nowadays a small price to pay. There are two inverting integrators and a simple inverter to give the required 360° phase change. The inverter also allows for an extra feedforward to enable stabilization of the amplitude. The phasor diagram shows the relationship between the signals. Starting with $v_3$, then $v_1$ lags by 270° and $v_3'$ by a further 270°: $v_4$ is the output of the inverter, which should be in phase with $v_3$. Because of finite gains the net phase shift is slightly less than 360° so we add a small proportion of $v_1$, $v_5$, in quadrature with $v_3'$ to make up for the loss. The signal from $v_1$ is limited by the diodes so that $R_6$ controls the gain for $v_5$ and thus the output signals. The simulations have been run using LM6142 amplifiers which have a gain-bandwidth of 17 MHz to minimize any affects from the amplifier rolloff. If the ubiquitous 741 amplifier is used ($GB$ of 1 MHz) the stabilization conditions are more frequency dependent and the difference between calculated and actual frequency greater. For example, with 741's try the difference between $C = 10$ and 1 nF.

## SPICE simulation circuits

| | |
|---|---|
| Fig. 5.7.1(*b*) | 2phaosc3.SCH |
| Fig. 5.7.2 | 2phaosc3.SCH Frequency response |
| Fig. 5.7.2 | 2phaosc4.SCH Transient response |
| Fig. 5.7.3 | 2phseosc.SCH Transient response |

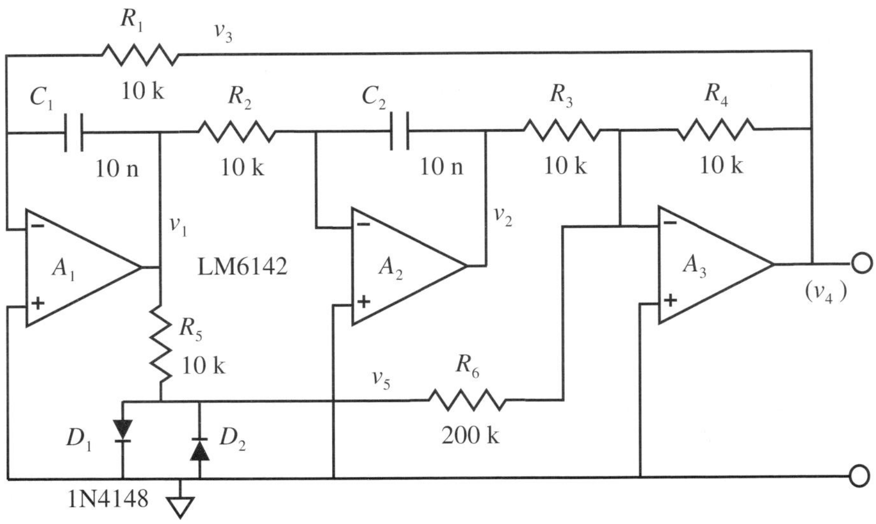


(a)

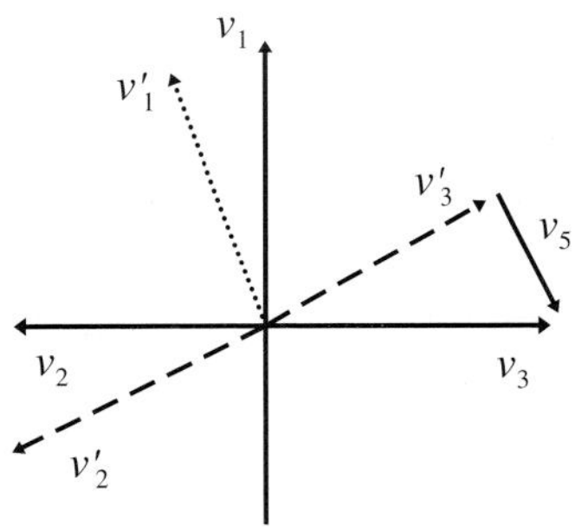


(b)

**Fig. 5.7.3** (*a*) Three amplifier two-phase oscillator. (*b*) Voltage phasors. $v_1'$, $v_2'$ and $v_3'$ show the effect of finite amplifier gain, and $v_5$ provides the correction.

## References and additional sources 5.7


Earls J. C. (1962): Generation of low frequency sinusoidal voltages of given phase, amplitude and frequency by means of a two-integrator loop. *J. Sci. Instrum.* **39**, 73–74.

Foord A. (1974): A two-phase low-frequency oscillator. *Electronic Engng.* **46**, December, 19, 21.

Good E. F. (1957): A two-phase low frequency oscillator. *Electronic Engng.* **29**, 164–169 and 210–213.

Howe R. M., Leite R. J. (1953): A low-frequency oscillator. *Rev. Sci. Instrum.* **24**, 901–903.

O'Dell T. H. (1988): *Electronic Circuit Design*, Cambridge: Cambridge University Press. ISBN 0-521-35858-2. See p. 105.

# 5.8 Wien-bridge oscillator

No man is an island, entire of itself; every man is a piece of the continent, a part of the main; if a clod be washed away by the sea, Europe is the less, as well as if a promontory were, as well as if a manor of thy friends or of thine own were; any man's death diminishes me, because I am involved in mankind; and therefore never send to know for whom the bell tolls; it tolls for thee.

John Donne (1572–1631)

The Wien-bridge oscillator was examined in Section 1.12 as an example of the use of Laplace transforms. Here we look at some additional aspects of the circuit and the use of SPICE simulations to provide some insight into frequency stability in oscillators. The frequency selectivity of the Wien network by itself is poor but its use in the bridge configuration, together with gain and feedback, can improve the characteristics considerably. The Wien network (Wien 1891) refers to the *RC* arms of the bridge and from Eq. (1.12.17) we have the transfer function for this:

$$Z_{par}=\frac{R}{1+sCR} \quad \text{and} \quad Z_{ser}=\frac{1+sCR}{sC}$$
$$\text{or} \quad H(s)=\frac{V_{out}(s)}{V_{in}(s)}=\frac{Z_{par}}{Z_{par}+Z_{ser}}=\frac{sCR}{s^2C^2R^2+3sCR+1} \tag{5.8.1}$$

and since we know the oscillation frequency will be $\omega_0=1/RC$, the transfer function may also be written (putting $j\omega$ for $s$):

$$\beta_+=\frac{V_{out}(s)}{V_{in}(s)}=\frac{1}{sCR+3+\dfrac{1}{sCR}}$$
$$=\frac{1}{3+j\left(\dfrac{\omega}{\omega_0}-\dfrac{\omega_0}{\omega}\right)} \tag{5.8.2}$$

so that at the oscillation frequency $\omega_0$ we will have $\beta_{+0}=1/3$ (Strauss 1960). Consider now the full bridge circuit as shown in Fig. 5.8.1. The additional arms $R_3$ and $R_4$ should also provide a transfer function of 1/3 to balance the bridge at resonance, but we will write it as deviating slightly from this so that we may examine

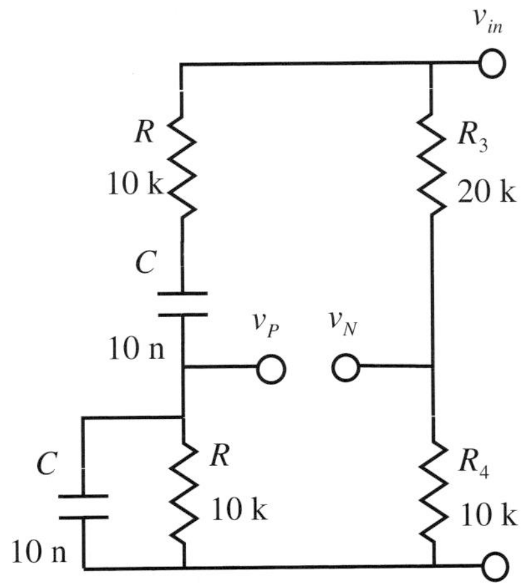


**Fig. 5.8.1** Wien-bridge circuit.

the effect of the negative feedback on selectivity and stability. By nearly balancing the bridge and using a large gain to amplify the small unbalance signal, with feedback to keep the balance, we will find that the frequency selectivity and stability may be considerably improved compared with the simple Wien network. Thus we write the transfer function for these arms as:

$$\beta_{-} = \frac{R_4}{R_3 + R_4} = \frac{1}{3} - \frac{1}{\delta} \tag{5.8.3}$$

so that the overall transfer function from the bridge input to the differential output is:

$$\beta = \beta_{+} - \beta_{-} = \beta_{+} - \left(\frac{1}{3} - \frac{1}{\delta}\right) \tag{5.8.4}$$

The phase shift $\phi$ of $\beta$ as a function of relative frequency $f/f_0$ is shown in Fig. 5.8.2 for several values of $\delta$. At the oscillation frequency we will have $\beta = 1/\delta$ and hence the gain of the amplifier required to sustain oscillation will be $A = \delta$ so that $A\beta = 1$. For an oscillator the total phase shift around the feedback loop will be zero (or 360°). If the frequency, and hence the phase, changes for any reason then the overall phase shift must readjust to bring it back to zero. If the phase can vary significantly with only a small frequency change then we will have a more frequency stable oscillator. A normalized stability factor $S_f$ may be defined:

$$S_f = \frac{\Delta\phi}{\Delta f/f_0} = f_0\frac{d\phi}{df} \equiv \omega_0\frac{d\phi}{d\omega} \tag{5.8.5}$$

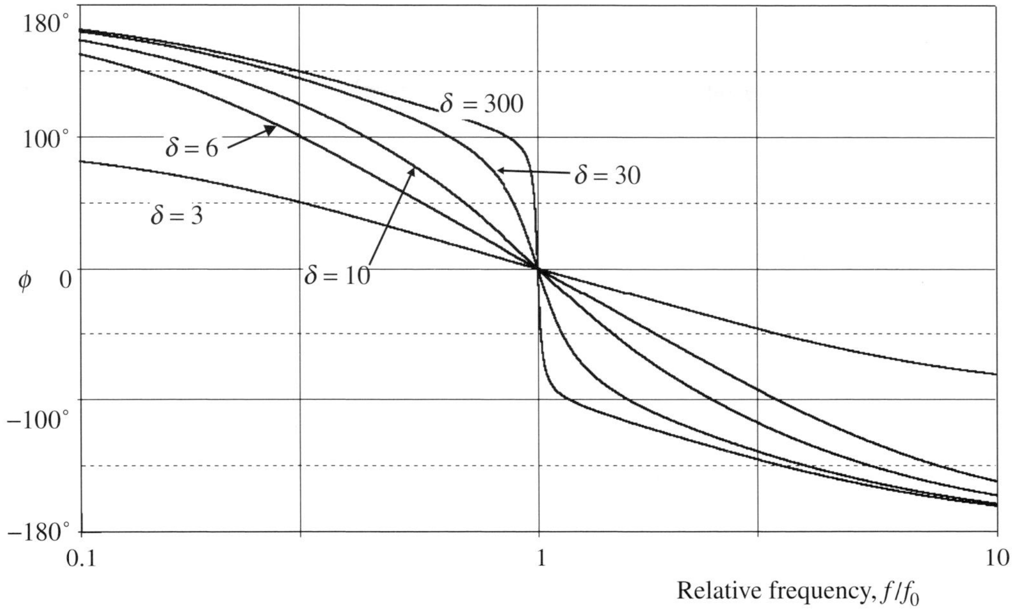


**Fig. 5.8.2** Phase shift $\phi$ as a function of relative frequency for values of $\delta$. The curve for $\delta=3$ is for the Wien network (i.e. $V_p$) alone.

so that a rapid change of phase with frequency is desirable. From Fig. 5.8.2 it is seen that a high value of $\delta$ is necessary. The phase relation for $\beta$ is found from Eqs. (5.8.2), (5.8.3) and (5.8.4) to be:

$$\phi=\tan^{-1}\left[\frac{(3-\delta)}{9}\left(\frac{\omega}{\omega_0}-\frac{\omega_0}{\omega}\right)\right]-\tan^{-1}\left[\frac{1}{3}\left(\frac{\omega}{\omega_0}-\frac{\omega_0}{\omega}\right)\right] \tag{5.8.6}$$

and differentiating this we get for $S_f$:

$$S_f=\frac{\dfrac{(3-\delta)}{9}\left[1+\left(\dfrac{\omega_0}{\omega}\right)^2\right]}{1+\left(\dfrac{3-\delta}{9}\right)^2\left(\dfrac{\omega}{\omega_0}-\dfrac{\omega_0}{\omega}\right)^2}-\frac{\dfrac{1}{3}\left[1+\left(\dfrac{\omega_0}{\omega}\right)^2\right]}{1+\dfrac{1}{9}\left(\dfrac{\omega}{\omega_0}-\dfrac{\omega_0}{\omega}\right)^2} \tag{5.8.7}$$

This rather complex expression (which as we shall see below can be readily plotted by PSpice) simplifies at $\omega=\omega_0$ to:

$$S_f=\frac{-2}{9}\delta=-66.67, \quad \text{for } \delta=300 \tag{5.8.8}$$

From (5.8.3) we have, with $\delta=300$ and $R_3=20$ k:

$$R_4=R_3\left(\frac{\delta-3}{2\delta+3}\right)=9.85\text{ k} \tag{5.8.9}$$

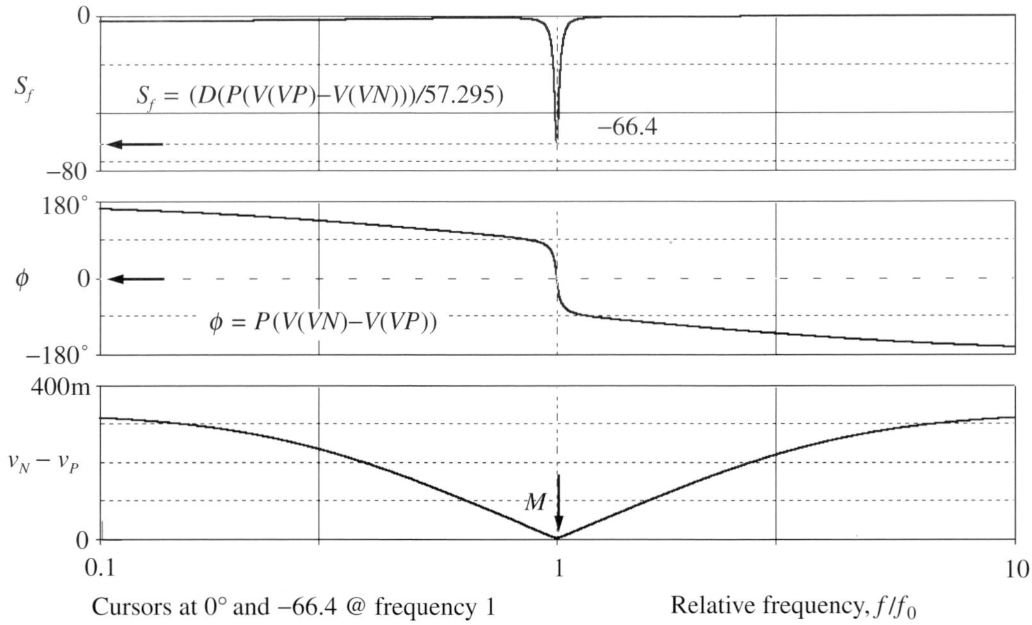


**Fig. 5.8.3** Simulation results for Wien bridge of Fig. 5.8.1. The value of $\delta = 300$.

If a simulation is run using the numerical values shown in Fig. 5.8.1 and $R_4$ given by Eq. (5.8.9) then the results shown in Fig. 5.8.3 are obtained. The 'resonance' frequency $f_0 = 1/2\pi RC = 1591.5$ Hz so the initial *PROBE* frequency scale is divided by this to normalize the scale and the extent of the scale is set to 0.1 to 10 (Hz, though it is now really just a number). For $S_f$ (using Eq. (5.8.5)) the value is divided by $360/2\pi = 57.295$ to convert degrees to radians for comparison with Eq. (5.8.8). As the function has such a sharp point it is necessary to make the maximum step size for the simulation small to get close to the minimum (a setting of 2001 points per decade was used). An input of 1 V peak was used so the value of $v_N - v_P$ at the minimum should be $1/\delta = 3.3$ mV, which is what is found at $M$. The value for $S_f$ also matches Eq. (5.8.8).

The shape of the amplitude response, $v_N - v_P$, is very broad and in itself would not provide good frequency stability. The phase change, however, is very rapid for high $\delta$ and this can provide good frequency stability. The effective value of $\delta$ will depend on the frequency response of the op amp, since the gain $A(f) = \delta$. Amplitude stability is usually achieved by making one of $R_3$ or $R_4$ power (or temperature) sensitive so that an increase of output amplitude results in increased negative feedback to give a stable amplitude. If $R_3$ is a negative temperature coefficient (NTC) thermistor, then increased current arising from higher amplitude causes a reduction in the value of the resistance and hence increased negative feedback. This does have a drawback in that the resistance is also affected by ambient tem-

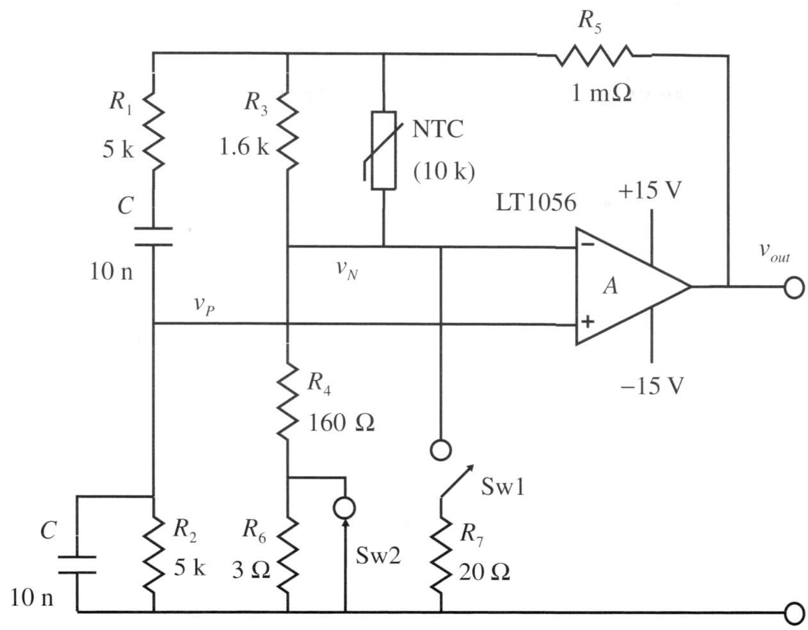


**Fig. 5.8.4** Wien-bridge oscillator for simulation. The NTC parameters are those of case D in Fig. 4.8.3.

perature changes and that at very low frequencies the thermal time constant is not long enough not to follow the oscillation. The characteristics of NTC thermistors are discussed in Section 4.8.

To see how the thermistor functions in an oscillator we will examine the performance of the circuit of Fig. 5.8.4. Some incidental additional components have been included to facilitate measurement and to ensure reliable startup.

Ensuring reliable startup of the oscillator proved to be somewhat difficult. After some experiment the arrangement using switch Sw1 was found to be effective. Sw1 is left closed for 100 μs after initiating the simulation to enable the NTC to get near to the operating region. When run the results found are shown in Fig. 5.8.5.

The output waveform $v_{out}$ was something of a surprise, but then a very distant bell was heard, a reminder of a publication of long ago. The Wien-bridge oscillator was the firstborn from a famous garage and no less a person than the vice-president for R&D analysed the circuit for just this form of output (Oliver 1960). In those days it was usual to use a PTC stabilizer in the form of a small tungsten lamp with a somewhat different temperature dependence function, but the effects are just the same. The squegging-type response (better known to oldtimers from the valve/tube days) does not matter too much at switch on, but it recurs whenever the oscillator is tuned to another frequency. Switch Sw2 is opened at 200 ms to include $R_6 = 3\ \Omega$ and cause a disturbance, and the results can readily be seen. The analysis

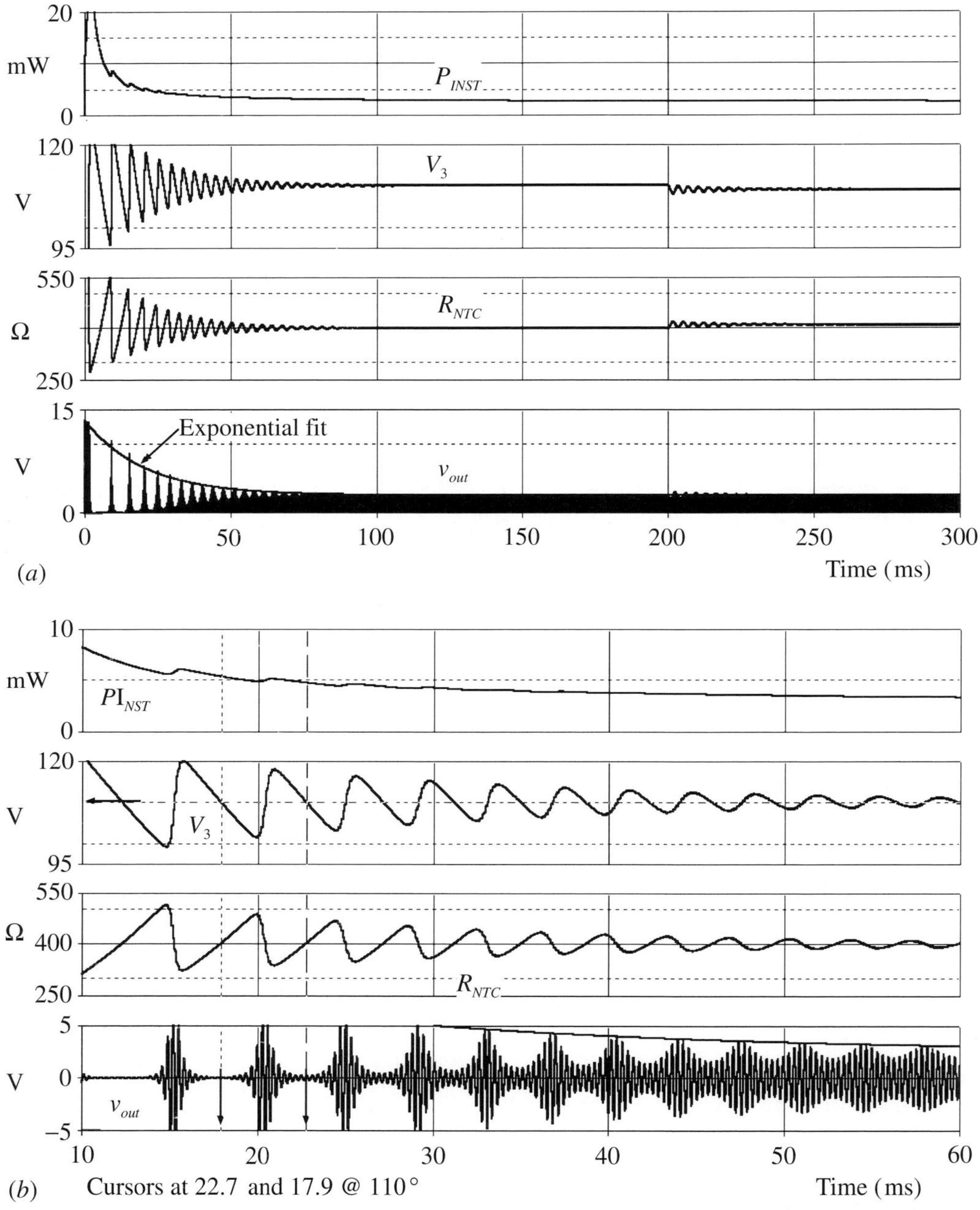


**Fig. 5.8.5** (*a*) Waveforms for circuit of Fig. 5.8.3. Sw2 opens at 200 ms to introduce a transient in the system. (*b*) Expanded segment of (*a*).

is too long to be examined here but a suggestion is made that introduction of some non-linearity into the amplifier-feedback loop will minimize the effect. This was tried using the PSpice *SOFTLIM* device which will introduce third order distortion as recommended (you can check the form of distortion using the Fourier facility of SPICE). There was a small effect, but at least in the present configuration it was not significant. The parameters for the NTC used here, in particular CTH, were made short to keep the simulation times more acceptable. *GTH*0 and *GTH*1 were reduced from the original device value to give small interaction with the external temperature, as if it was, say, a vacuum isolated device.

We have a considerable advantage in being able to see the fluctuations in the NTC temperature, and hence resistance, as well as the time delays in the response. Comparing the temperature or resistance variations with the oscillator output it is evident that there is a classic squegging-type of relaxation oscillation. Fitting an exponential to the peaks of the bounces (using the later rather than the earlier peaks since the latter do not have a consistent period) led to a decay time $\tau_b = 20.4$ ms and the resulting curve is plotted in the figure, offset by the steady state of 2.5 V. Using Oliver's analysis now allows us to derive a frequency response diagram corresponding to his Fig. 5. The thermistor transmission function $G(s)$ is given in terms of its 'gain' $S$ and corner frequency $\omega_L$, together with the bounce frequency $\omega_b$ and the oscillation frequency $\omega_0$:

$$G(s) = \frac{S\omega_L}{s + \omega_L}, \quad \text{with} \quad \omega_L = \frac{2}{\tau_b}, \quad \text{and} \quad \omega_b = (S\alpha\omega_0\omega_L)^{\frac{1}{2}}, \quad \text{with } \alpha = 2 \quad \text{here}$$

$$\text{so} \quad S = \frac{\omega_b^2}{\alpha\omega_0\omega_L} = \frac{f_b^2}{\alpha f_0 f_L}, \quad \text{with} \quad \omega_L = \frac{2}{20.4\times 10^{-3}} = 98.04\ \text{s}^{-1} \quad \text{or} \quad f_L = 15.60\ \text{Hz}$$

and using the measured frequencies

$f_b = 287.6$ Hz, $f_0 = 3170$ Hz

gives

$$S = \frac{(287.6)^2}{2\times 3170\times 15.60} = 0.836 \tag{5.8.10}$$

which may be compared with Oliver's lamp with $S = 0.25$.

The frequency response can now be drawn as shown in Fig. 5.8.6, where $F_0(s)$ is the system response and $H_0(s)$ is the total envelope gain. The 20 dB/dec slope of $H_0(s)$ above $f_L$ and the greater than ten times frequency interval before unity gain imply that the phase shift is close to 180° and hence that the response will be nearly unstable. Additional references below give much information on practical realizations.

As an aside it may be noted that for a value of $\delta = 6$ ($R_4 = 4$ k in Fig. 5.8.1) the

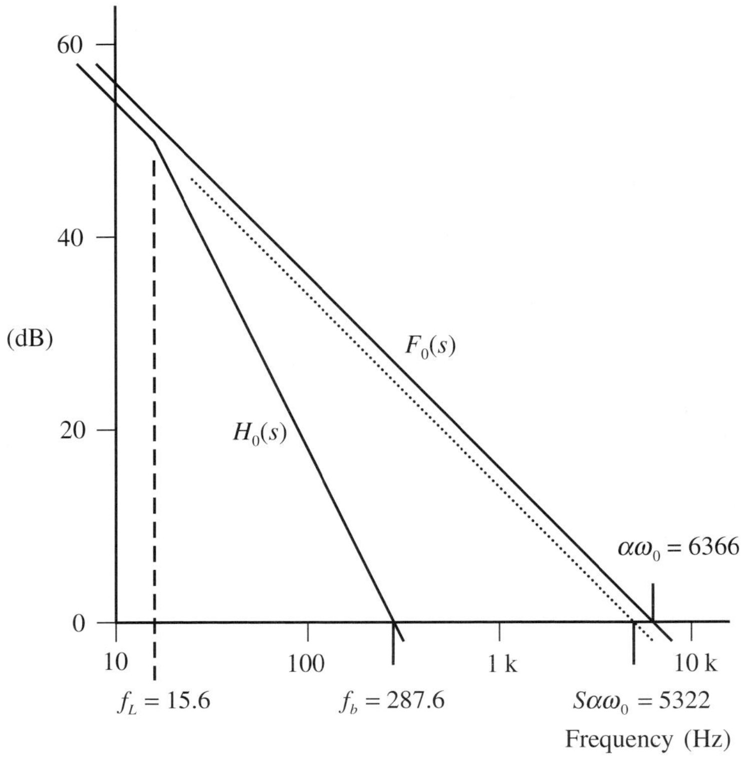


**Fig. 5.8.6** Version of Oliver's frequency response diagram.

differential output $v_N - v_P$ is independent of frequency. The outputs $|v_N - v_P| = |v_N| = |v_{in}/6|$, $v_N$ being in phase with $v_{in}$ and the phase of $v_N - v_P$ depending on frequency and the values of $R$ and $C$ in Fig. 5.8.1. The phase of $v_P$ is exactly half that of $v_N - v_P$.

## SPICE simulation circuits

Consult the SimCmnt.doc file on the CD before running

| | |
|---|---|
| Fig. 5.8.2 | Wienbrfr.SCH |
| Fig. 5.8.3 | Wienbrfr 2.sch |
| Fig. 5.8.5(*a*) | Wienbrg 7.SCH |
| Fig. 5.8.5(*b*) | Wienbrg 7.SCH (Expanded segment of (*a*)) |

## References and additional sources 5.8

Bradbury D. J. (1985): VMOS in a Wien bridge oscillator. *Electronic Product Design* February, 57–59. See also April 1985, 25. Note that Figs 1 and 2 in the paper are in error with the lower *RC* arm of the bridge.

Celma S., Martínez P. A., Carlosena A. (1994): Current feedback amplifier based sinusoidal oscillators. *IEEE Trans.* **CAS-41**, 906–908.

Kushnick E. (1994): Streamline your Wien bridge. *Electronic Design* 18 April, 94.

Oliver B. M. (1960): The effect of μ–circuit non-linearity on the amplitude stability of RC oscillators. *Hewlett-Packard* J. **11** (8–10), 1–8.

Owen R. E. (1965): A modern, wide range, RC oscillator. *General Radio Experimenter* **39** (8), August, 3–12.

Owen R. E. (1966): Solid-state RC oscillator design for audio use. *J. Audio Eng. Soc.* **14**, January, 53–57.

Sambrook J., Brady D., Rawlings K. C., van der Sluijs J. C. A. (1978): A low power Wien bridge oscillator for use in low-temperature thermometry. *J. Phys. E Sci. Instrum* **11**, 991–992.

Strauss L. (1960): *Wave Generation and Shaping*, New York: McGraw-Hill. Library of Congress Cat. No. 74-90024. See pp. 663–674, 688–691.

Wien M. (1891): Measung der Induction Constanten mit dern Optischen Telephon. *Annalen der Physik* **44**, 704–707.

Williams J. M. (Ed.) (1987): *Linear Applications Handbook, Linear Technology*, Vol. I, Vol. II 1990, Vol. III 1997. See in particular Vol. II, AN43, p. 29–33.

Williams P. (1971): Wien oscillators. *Wireless World* **77**, 541–547.

# 5.9 Current sources and mirrors

Scarcely anything in literature is worth a damn except what is written between the lines.

Raymond Chandler

The current mirror is one of the more common building blocks of analog integrated circuits. They are used to determine bias currents and voltages, provide level shifting, differential-to-single-ended converters for amplifier stages and variations allow the production of stable and predictable reference voltages (Brokaw 1975; Gray and Meyer 1977; Hambley 1994; Chen 1995, p. 1619).

The basic circuit is shown in Fig. 5.9.1. A benefit of integration is that the two transistors can be made to be nearly identical (i.e. matched), so that since $v_{BE}$ is the same then $i_{B1} = i_{B2}$. The matching also means that the current gain $\beta$ is also the same. Thus for $v_{CE2}$ greater than, say, 0.2 V (so $Q_2$ is in the active region) we have:

$$i_{C1} = \beta i_{B1} = \beta i_{B2} = i_{C2} \quad \text{and} \quad i_{REF} = i_{C1} + i_{B1} + i_{B2} = \beta i_{B1} + i_{B1} + i_{B1} = i_{B1}(2+\beta)$$

$$\text{so} \quad i_{C1} = i_{C2} = \frac{\beta i_{REF}}{(2+\beta)} = \frac{i_{REF}}{(1+2/\beta)} \tag{5.9.1}$$

and since $\beta$ is large (say much more than 100) we have nearly:

$$i_{REF} = i_{C2} \tag{5.9.2}$$

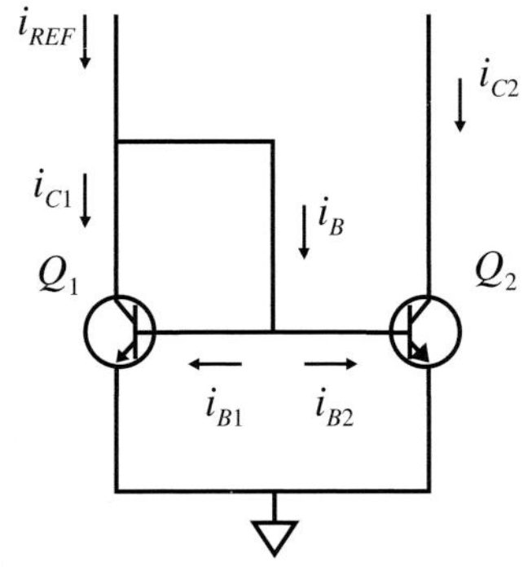


**Fig. 5.9.1** Basic current mirror.

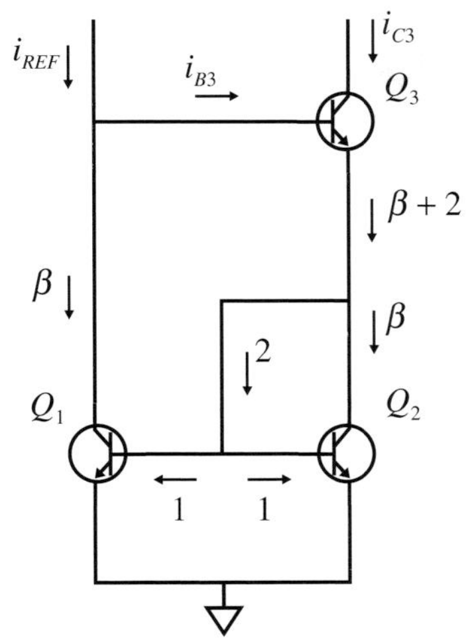


**Fig. 5.9.2** The Wilson current mirror.

i.e. the current $i_{C2}$ is a mirror of $i_{REF}$. A measure of the effectiveness of the current source is provided by the dynamic output resistance; an ideal current source has infinite output impedance. For this form of circuit at a current of, say, 1 mA the output resistance $R_0$ is about 130 k (Gray and Meyer 1977, p. 200). There are many variations on this type of circuit (Chen 1995, Section 57) but we will consider only two. The Wilson circuit is shown in Fig. 5.9.2. We again assume identical transistors as above. If we assign the base currents of $Q_1$ and $Q_2$ to be one unit, then the other currents will be as shown. We can then write:

$$i_{C3}=\beta+2-i_{B3}=\beta i_{B3} \quad \text{so} \quad i_{B3}=\frac{\beta+2}{\beta+1} \quad \text{and} \quad i_{C3}=\beta\left(\frac{\beta+2}{\beta+1}\right) \tag{5.9.3}$$

so we can now determine $i_{REF}$:

$$i_{REF}=i_{B3}+i_{C1}=\left(\frac{\beta+2}{\beta+1}\right)+\beta=\frac{\beta^2+2\beta+2}{\beta+1}\cong\beta\left(\frac{\beta+2}{\beta+1}\right) \tag{5.9.4}$$

and the ratio of currents:

$$\frac{i_{C3}}{i_{REF}}=\beta\left(\frac{\beta+2}{\beta+1}\right)\left(\frac{\beta+1}{\beta^2+2\beta+2}\right)=\beta\left(\frac{\beta+2}{\beta^2+2\beta+2}\right)\cong 1 \tag{5.9.5}$$

The matching is thus very close for any reasonable value for $\beta$. For this configuration we find similarly to the simple circuit, $R_0 \approx 6$ M. The Widlar circuit (Widlar 1965) is shown in Fig. 5.9.3 and is intended for use where very small currents are required.

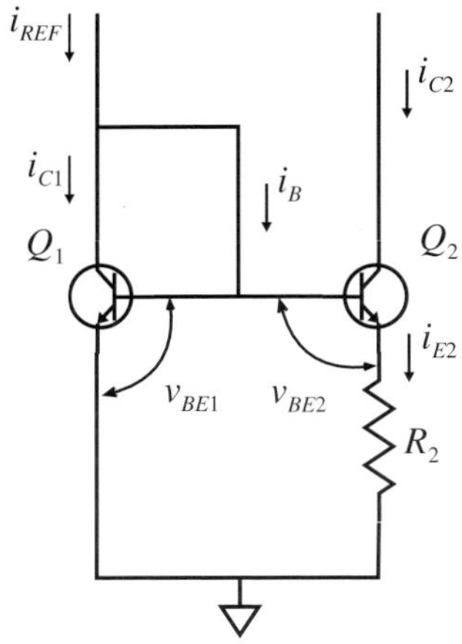


**Fig. 5.9.3** The Widlar current mirror.

The basic equation for a transistor relates the collector current to the base–emitter voltage:

$$i_C = i_S \exp\left(\frac{v_{BE}}{V_J}\right), \quad \text{where} \quad V_J = \frac{k_B T}{q_e} = 26 \text{ mV} \quad \text{at 300 K}$$

$$\text{or} \quad v_{BE} = V_J \ln\left(\frac{i_C}{i_S}\right)$$

$$\text{so that} \quad v_{BE1} - v_{BE2} = V_J\left[\ln\left(\frac{i_{C1}}{i_{S1}}\right) - \ln\left(\frac{i_{C2}}{i_{S2}}\right)\right] \tag{5.9.6}$$

$$= V_J \ln\left(\frac{i_{C1}}{i_{C2}}\right) \quad \text{since for identical transistors } i_{S1} = i_{S2}$$

and taking the emitter and base loop of the two transistors:

$$v_{BE1} - v_{BE2} - i_{E2}R_2 = 0 \quad \text{and for } \beta \text{ large} \quad i_{E2} \cong i_{C2}$$

$$\text{so} \quad i_{E2}R_2 = V_J \ln\left(\frac{i_{C1}}{i_{C2}}\right) \cong i_{C2}R_2 \tag{5.9.7}$$

$$\text{and} \quad \frac{i_{REF}}{i_{C2}} = \exp\left(\frac{R_2 i_{C2}}{V_J}\right) \quad \text{since} \quad i_{REF} \cong i_{C1}$$

This is a transcendental equation that must be solved by trial and error. The attraction of this circuit is that the resistor required for a given current is much lower than for the previous circuit, and hence the cost in chip area is correspondingly less. These circuits may readily be simulated, e.g. a d.c. sweep of the voltage

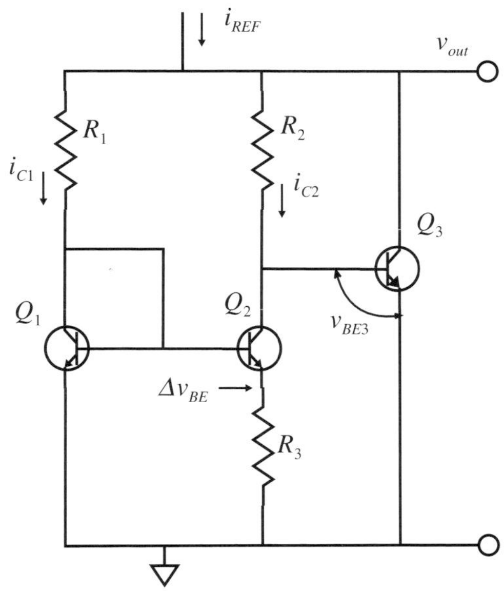


**Fig. 5.9.4** Voltage reference.

applied to the output transistor allows the slope of the response to be measured and hence the effective output resistance to be calculated.

A further common use is for the construction of voltage reference sources. In this application it is required to provide a predictable voltage with minimum temperature coefficient. A simple circuit is shown in Fig. 5.9.4 (Chen 1995, p. 1626).

In this circuit the two transistors are operated at different current densities $J$, say by making the areas of the junctions different. Using Eq. (5.9.7):

$$\Delta v_{BE} = V_J \ln\left(\frac{J_1}{J_2}\right) \quad \text{and} \quad i_{E2} = \frac{\Delta v_{BE}}{R_3} \cong i_{C2}$$
$$\text{so} \quad v_{out} = i_{C2}R_2 + v_{BE3} = V_J\left(\frac{R_2}{R_3}\right)\ln\left(\frac{J_1}{J_2}\right) + v_{BE3} \tag{5.9.8}$$

Now the temperature coefficient of $v_{BE}$ is $-2$ mV K$^{-1}$ and $k_B/q_e = +0.085$ mV K$^{-1}$, so we need to multiply $V_J$ by a factor $F = 2/0.085 = 23.5$ to give zero temperature coefficient; $F$ is thus given by:

$$F = \left(\frac{R_2}{R_3}\right)\ln\left(\frac{J_1}{J_2}\right) \quad \text{and} \quad v_{out} = 23.5 \times 26 \times 10^{-3} + 0.65 = 1.26\ \text{V} \tag{5.9.9}$$

assuming say a value of 0.65 V for $v_{BE}$. An improved circuit is shown in Fig. 5.9.5 (Brokaw 1975). The area of $Q_2$ is several times the area of $Q_1$ (in the reference the factor is 8), so that the current density $J_1 = 8J_2$ for equal collector currents. The amplifier controls the base voltages so that its input differential is zero, thus

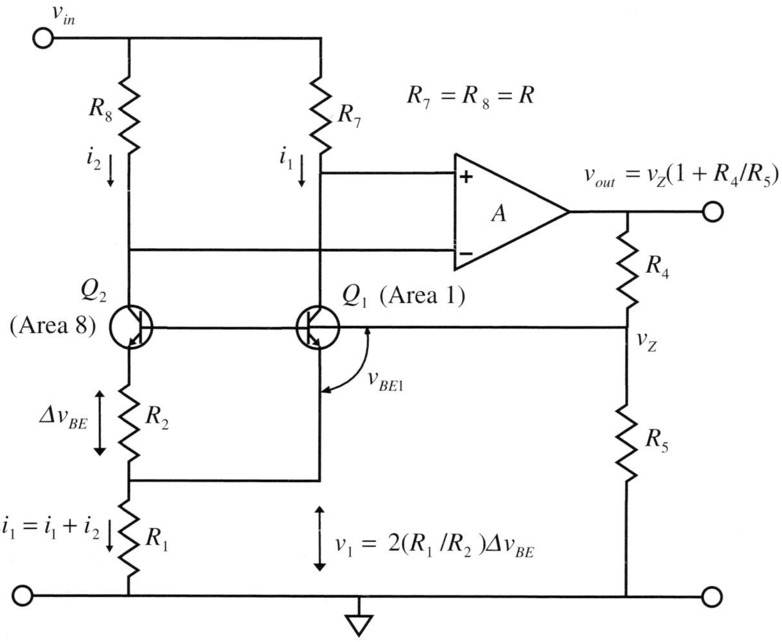


**Fig. 5.9.5** Improved voltage reference.

making $i_1 = i_2$ since the two resistors are equal. From above we can write for the difference $\Delta v_{BE}$ of the $v_{BE}$'s of the two transistors:

$$\Delta v_{BE} = V_J \ln\left(\frac{J_1}{J_2}\right) = V_J \ln(8) \quad \text{and} \quad i_2 = \frac{\Delta v_{BE}}{R_2} = \frac{V_J \ln(8)}{R_2} = i_1 \tag{5.9.10}$$

and since $i_1$ and $i_2$ both flow through $R_1$ the voltage $v_1$ is given by:

$$v_1 = \frac{2R_1 V_J \ln(8)}{R_2} \quad \text{and} \quad v_Z = v_1 + v_{BE1} = \frac{2R_1 V_J \ln(8)}{R_2} + v_{BE1} \tag{5.9.11}$$

which can be adjusted as before to give zero temperature coefficient. When this is done the value of $v_Z$ is found to be 1.205 V, effectively the bandgap voltage of silicon and hence the common name for these as bandgap references. If higher reference voltages are required, e.g. 2.5 or 5 V, then $R_4$ and $R_5$ are scaled to provide the required gain. The op amp output also provides low output impedance.

To simulate this circuit it is necessary to provide an appropriate area ratio. A simulation with four transistors in parallel for $Q_2$ (type 2N3904, $\beta \approx 100$) and an OP-07 amplifier worked well. $R_7$ and $R_8$ were set to 10 k and since $v_{BE}$ is about 0.6 V then $v_1$ should be the same to give the desired output voltage. Using the area ratio of 4, the ratio $R_1/R_2$ is calculated to be 8.32. For a supply of 5 V, the inputs to the amplifier were set to 4 V so the currents $i_1$ and $i_2$ were 100 μA and hence the current in $R_1$ is 200 μA. Thus $R_1$ must be ≈3 k to give $v_1$ = 0.6 V and hence $R_2$ is ≈360R. This was run to find the d.c. bias voltages and currents, which can be shown on the schematic, and with a few iterations resulted in $R_1$ = 2.75 k and $R_2$ = 330R (ratio = 8.33) for an output of 1.204 V. If the supply voltage is changed to 10 V, the output is found to be 1.203 V. The calculated value of $\Delta v_{BE}$ = 36 mV agrees closely with the simulated value.

## SPICE simulation circuits

| | |
|---|---|
| Fig. 5.9.5 | Voltref1.SCH |

## References and additional sources 5.9

Brokaw A. P. (1974): A simple three-terminal IC bandgap reference. *IEEE J. Solid-State Circuits* **SC-9**, December, 388–393.

Brokaw A. P. (1975): More about the AD580 monolithic IC voltage regulator. *Analog Dialogue* **9-1**, 6–7.

Burr-Brown (1988): *Dual Current Source REF200*, Burr-Brown Data sheet PDS-851.

Chen, Wai-Kai (Ed.) (1995): *The Circuits and Filters Handbook*, Boca Rator: CRC Press and IEEE Press. ISBN 0-8493-8341-2.

Damljanovic D. D. (1998): Current mirrors for higher currents. *Electronic Engineering* **70**, February, 16, 20.

Gray P. R., Meyer R. G. (1977): *Analysis and Design of Analog Integrated Circuits*, New York: John Wiley. ISBN 0-471-01367-6.

Hambley A. R. (1994): *Electronics. A Top–Down Approach to Computer-Aided Circuit Design*, New York: Macmillan. ISBN 0-02-349335-6.

Steele J., Green T. (1992): Tame those versatile current-source circuits. *Electronic Design* 25 October, 61, 62, 64, 66, 70, 72.

Widlar R. J. (1965): Some circuit design techniques for linear integrated circuits. *IEEE Trans. Circuit Theory* **CT-12**, December, 586–590.

# 5.10 Power supplies

Life can only be understood backwards; but it must be lived forwards.

Soren Kierkegaard

Power supplies are a somewhat overlooked area in many books on electronics. They are, however, essential in every application and can in themselves be quite complex systems. The performance of most circuits can be substantially affected or even negated by poor power supplies. The simplest source is probably a battery of which there are now many different types with a wide range of performance. Rechargeable-type batteries require careful consideration in their use and recharging, and many integrated circuit devices are now available to control the discharge and charge for best life and performance. Mains supplied regulators usually require rectifying circuits to give d.c. voltages from the input a.c. supplies. Nowadays switching supplies have become popular as they can be highly efficient, and again many integrated circuits are available to enable effective construction of these supplies. This type of supply can both reduce and increase the raw supply voltage as well as invert it to give the opposite polarity. Consideration should always be given to the switching frequency and its possible effects on the circuits it will power. Switched capacitor circuits can also transform voltages though they are usually only used for low power systems, but they are often useful for producing inverted voltages.

Rectifier circuits are discussed in Section 5.4. In selecting the components for the rectifying circuit there are a number of considerations to be borne in mind.

*Rectifier diodes*. These must be selected to withstand the maximum peak-inverse voltage (PIV) with allowance for maximum mains voltage and any possible transients. It is usually possible to use devices with high PIV as the cost of these is low. If working close to the limit, avalanche rectifiers may be used, which can stand reverse voltage breakdown transiently without damage. It is also necessary to select a device which can carry the required current. The current in the diodes flows for only a short part of the cycle so the peak current will be many times the average d.c. output current (Fig. 5.4.2) and the rating of the diode must reflect this. The common silicon rectifier diode is intended for low frequency operation. At frequencies above about 10 kHz their reverse recovery time is such that they will conduct effectively in *both* directions and so cease to rectify. Then fast recovery types must be used; or for lower voltages, Schottky rectifiers. The latter have somewhat lower forward voltages than the normal silicon rectifier which reduces power dissipation in the device and also, in low voltage supplies, reduces the loss of voltage particularly in configurations like the bridge rectifier where two diodes conduct in series.

*Smoothing capacitors.* Electrolytic capacitors are generally used when any significant capacity is required. These are polarized, have a maximum voltage and a maximum ripple current rating. Such capacitors provide a large capacity for a given volume but have significant internal resistance and inductance which limits their frequency response. A plot of impedance against frequency shows a minimum in the region from 10 to 100 kHz for aluminium types, and about 1 MHz for tantalum, with an inductive rise at high frequencies (see Fig. 4.2.1). Hence they should be used in parallel with more suitable capacitors if this presents a problem.

## Regulators

To reduce the ripple from the rectifier circuit and to provide a stable voltage independent of the mains supply, or of the load, it is necessary to use a regulator. As the name indicates these are members of a class of servo system referred to as type zero (Section 3.15). The characteristic of this type is that there is inherently an offset between the demand output and the actual output. The form of this system is in effect a voltage divider which can be controlled to maintain a constant output voltage.

There are three main parts in a regulator:

1. a series element whose conductance can be controlled;
2. a reference voltage against which to compare the output voltage;
3. an amplifier, that uses the error voltage to control the series element.

The general configuration is shown in Fig. 5.10.1. The output voltage, or a proportion of it, is compared with a reference voltage. If there is a difference then the voltage divider is adjusted to return to the required output. The series element of the divider is provided by a transistor and the error signal is amplified by a high-gain amplifier to control the transistor. There are many different circuit realizations of voltage regulators but we will consider one that has a number of advantages, and which, as we shall see, will be familiar (Fig. 5.10.2).

The box $v_u$ represents the unregulated input as obtained from the rectifier circuit.

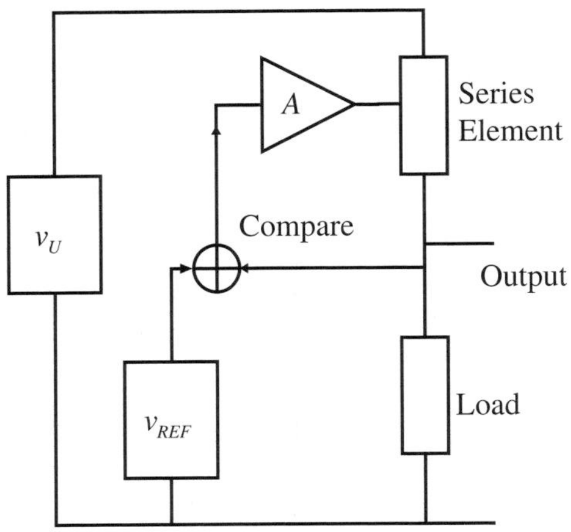


**Fig. 5.10.1** Basic regulator configuration.

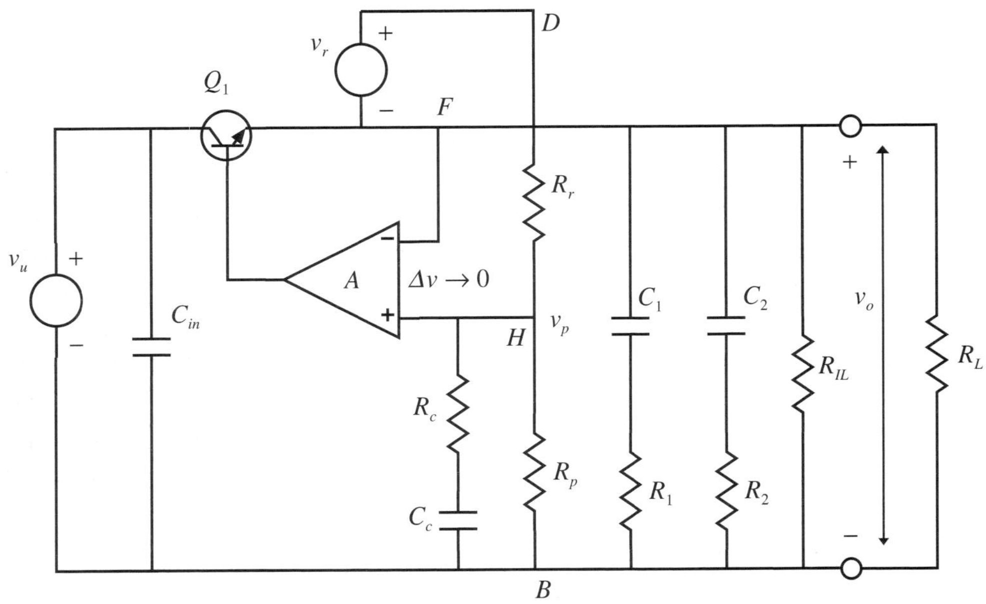


**Fig. 5.10.2** Bridge power supply regulator.

The series transistor $Q_1$ is the variable element of the voltage divider formed with the load $R_L$. $v_r$ represents a reference voltage that is independent of $v_u$ or $v_o$, and $A$ is a differential (operational) amplifier. The connection of the amplifier is such that we have negative feedback, which means that the differential input voltage to the amplifier $\Delta v \to 0$, i.e. the voltages at points $H$ and $F$ are the same. The consequence of this is that $v_r$ will cause a current $i_b = v_r / R_r$ to flow in $R_r$ and hence $R_p$. Thus $v_p = v_o$ and:

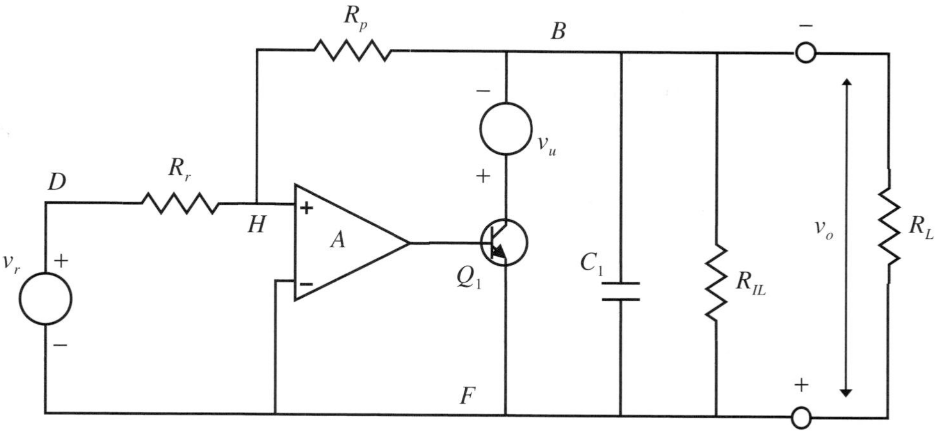


**Fig. 5.10.3** Bridge regulator showing operational amplifier form.

$$v_o = v_p = i_b R_p = \frac{v_r R_p}{R_r} \tag{5.10.1}$$

Thus the output voltage is proportional to $R_p$ and so is easily programmed from zero to the maximum allowable voltage. The latter is determined by the lowest value of $v_u$ and the minimum voltage across $Q_1$ for it to operate effectively. The minimum value of $v_u$ occurs at maximum load current, and minimum mains input and should be measured at the minimum of the rectifier ripple voltage, i.e. a d.c. meter will not give a proper reading as it shows the average input, so an oscilloscope must be used. If the voltage across $Q_1$ drops below saturation then the negative excursions of $v_u$ will appear on the output. It is important to appreciate the role of the load. The series element can only be turned on to source current as demanded by the load. If there were no load then the system cannot regulate as there is nowhere for any stray currents to flow – the series element cannot sink current. Thus there should always be some minimum load connected and this is normally provided internally. It is usually expected that the load will demand more or less current, which the regulator can cope with, but if for some reason the load should generate a voltage the regulator again cannot sink current and will tend to turn off the series element as it thinks the output is too high.

The circuit of Fig. 5.10.2 can be redrawn as shown in Fig. 5.10.3, where corresponding points are marked. It will now be recognized that the circuit is just a power op amp (compare Fig. 5.3.1). The apparent change in the use of the non-inverting input of the amplifier is explained by the extra inversion provided by the series transistor. The output is just an amplified version of the reference voltage, and since the gain is not infinite, there will be an offset between the expected and the actual voltage but this is usually of little consequence.

The regulator is a high-gain feedback loop and hence is prone to the problems of stability against oscillation. The roll-off of the gain with frequency must be controlled so that the feedback does not become positive until after unity gain is reached. This means that the response of the regulator to rapid changes in demand will be limited, becoming worse as frequency increases, so that fast transients will not be fully corrected. To reduce this effect it is necessary to shunt the output with substantial capacity to supply the currents demanded by the transient loads and hence maintain acceptable regulation to high frequencies. This usually requires the use of both electrolytic capacitors for low frequencies and of plastic and ceramic dielectric capacitors for high frequencies. It is also necessary to provide appropriate decoupling locally in your circuit since the long leads to the power supply will appear significantly inductive at high frequencies

For many circuit applications one of the very many types of IC regulator are satisfactory and are available in a range of fixed as well as variable voltage types. The standard models require about 3 V differential between input and output to operate satisfactorily, but there are also low-dropout types that can operate down to a few hundred millivolts differential. The latter usually have higher quiescent current consumption, i.e. at no load, but the introduction of FETs as series elements is also reducing this considerably. A number of types also have logic on–off control inputs that can be used to power-off systems giving flexibility in power sequencing and power saving (but see Section 5.22). Though IC regulators use somewhat different circuit configurations to that described above they are similar and have the same requirements. They must be decoupled at both input and output to ensure their own stability and the capacitors must be placed as close as possible to the IC. ICs usually have a number of protection facilities built in, such as current and temperature limiting, reverse voltage protection and transient input overvoltage protection. Those that are defined for automotive use are nearly indestructible and should be considered for hazardous applications.

The achievement of stability in the regulator is dependent on a number of factors, some of which may not be immediately evident. The datasheet of a recent MOSFET LDO driver/controller, the LP2975, provides a useful basis for discussion of stability since it provides the necessary information on its internal parameters (National Semiconductor 1997). We will investigate the circuit using the open-loop technique described in Section 5.14 to see the effects of the various parameters. The schematic circuit is shown in Fig. 5.10.4(*a*).

As there is no equivalent circuit for the amplifier, we have made a rough approximation using a LPC661 device for which a model is available. The gain is adjusted to suit by means of $R_1$, $R_2$ and $C_b$ as shown in (*c*). We can divide the circuit into two parts as shown in (*b*) and (*c*) for initial investigation. In (*b*), $R_a$ represents the output resistance of the feedback amplifier $A$ and the load components represent an output load of 1 A at 5 V together with the output capacitor $C_o$ and its equivalent series resistance (ESR) $R_e$. At first sight it may appear that it would be desirable to make $R_e$ as small as possible from the point of view of limiting the transient variation of $v_{out}$ due to changes in load current. In this respect it would be most desirable but we will see that from the point of view of loop stability a certain minimum value of $R_e$ is necessary. This may be just as well, as all electrolytic type capacitors inevitably have a significant ESR (Section 4.2). The offset of the generator $v_1$ was set to make $v_{out} \approx 5$ V. For the IRFU9110 used as the series element, the device capacities are about $C_{GS} = 185$ p and $C_{GD} = 50$ p. The model allows for these, so we have only shown them on the figure as a reminder as they play a role in the frequency response. Since the gain is not too high it is possible to measure the open-loop response directly by dividing the circuit into two portions, measuring the response of each and then multiplying the two outputs to get the overall result. If circuits (*b*) and (*c*) together are run in PSpice the results shown in Fig. 5.10.5 are obtained (remember, however, that the unity-gain value for the overall product of the individual gains is given by $v_1^2$). It is instructive to repeat the simulation with $R_e = 0.1\ \Omega$.

There are two poles and a zero associated with circuit Fig. 5.10.4(*b*): $f_p$ and $f_z$ from the output load and $f_{pg}$ from the FET input capacity and amplifier output resistance. The 0.16 is simply $1/2\pi$ and the numerical values are for the component values shown (or derived below for $C_{eff}$):

$$f_p = \frac{0.16}{(R_L + R_e)C_{out}} = \frac{0.16}{5.1 \times 180 \times 10^{-6}} = 174\ \text{Hz}$$

$$f_z = \frac{0.16}{R_e C_{out}} = \frac{0.16}{0.1 \times 180 \times 10^{-6}} = 8.9\ \text{kHz} \qquad (5.10.2)$$

$$f_{pg} = \frac{0.16}{R_a C_{eff}} = \frac{0.16}{1.2 \times 10^3 \times 240 \times 10^{-12}} = 556\ \text{kHz}$$

where $C_{eff}$ is the effective input capacity of the PFET gate. This capacity is determined from the datasheet values of capacities $C_{iss}$ and $C_{rss}$ and from the transconductance $G_m$ for the particular FET used. The effect of $C_{GD}$ is increased owing to the Miller effect and can be determined for our current $I_L$ from the datasheet value $G_{fs}$ at a current $I_D$. The relationships are:

$$C_{GS} = C_{iss} - C_{rss}, \quad C_{GD} = C_{rss}, \quad G_m = G_{fs}\left(\frac{I_L}{I_D}\right)^{\frac{1}{2}} \qquad (5.10.3)$$

$$C_{eff} = C_{GS} + C_{GD}[1 + G_m(R_L \| R_e)]$$

The datasheet values for the IRFU9110 are ($R_e \ll R_L$, so $R_L$ may be ignored):

$$C_{iss} = 230\ \text{p}, \quad C_{rss} = 45\ \text{p} = C_{GD}, \quad C_{GS} = 185\ \text{p}, \quad G_{fs} = 0.97\ \text{S@}1.9\ \text{A}$$
$$G_m = 0.7\ \text{S}, \quad C_{eff} = 185 + 45(1 + 0.7 \times 0.3) = 240\ \text{p} \qquad (5.10.4)$$

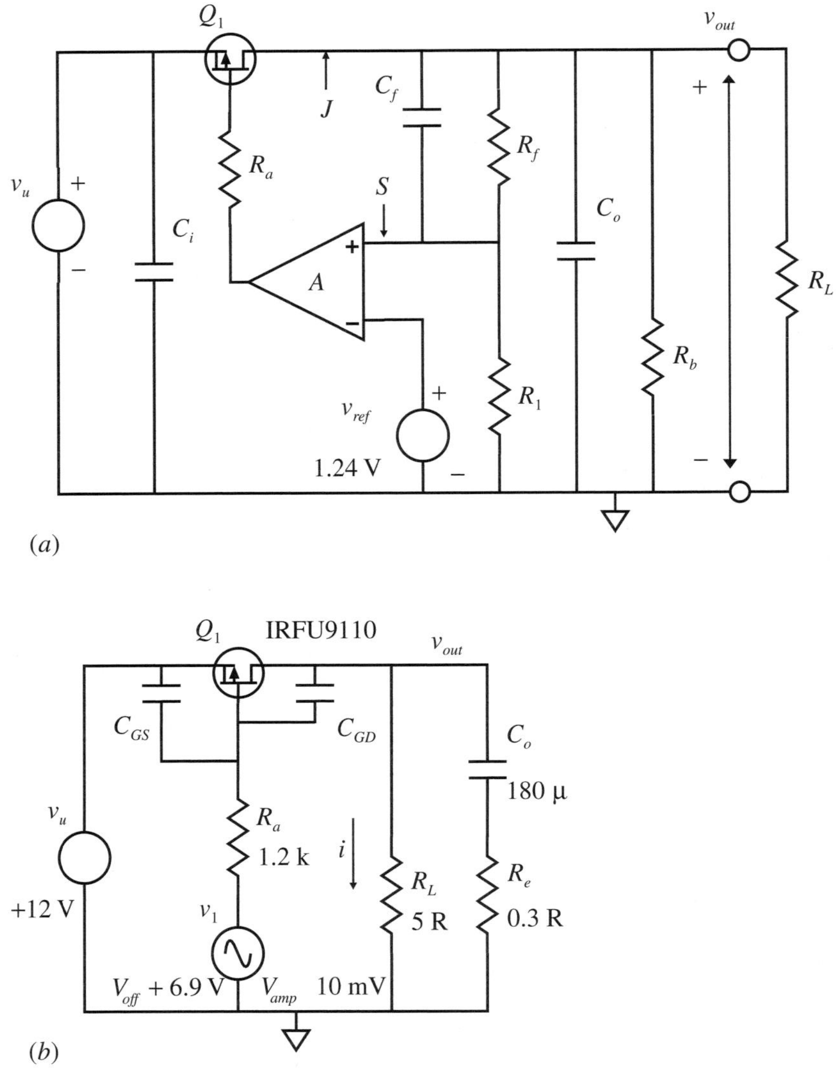


**Fig. 5.10.4** (*a*) Regulator circuit with external FET series regulator. (*b*) Circuit for investigation of FET response. (*c*) Circuit for investigation of feedback amplifier response.

It is difficult to pinpoint the various corner frequencies from the frequency response curves and we can only readily identify the unity-gain frequency:

$$f_U \approx 24\ \text{kHz} \quad \text{with phase margin} \approx 76° \tag{5.10.5}$$

$f_U$, for the full regulator circuit Fig. 5.10.4(*a*), is determined according to the data-sheet equation (the $1.24/v_{out}$ term is an equivalent way of referring to the resistors $R_i$ and $R_f$):

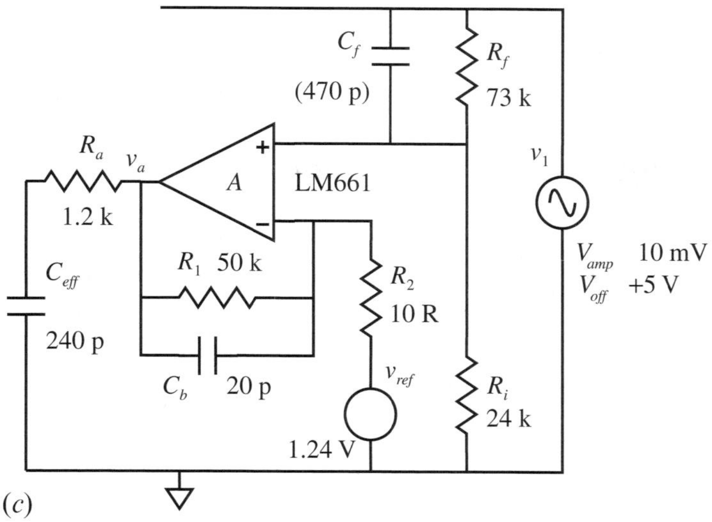


**Fig. 5.10.4** (*cont.*)

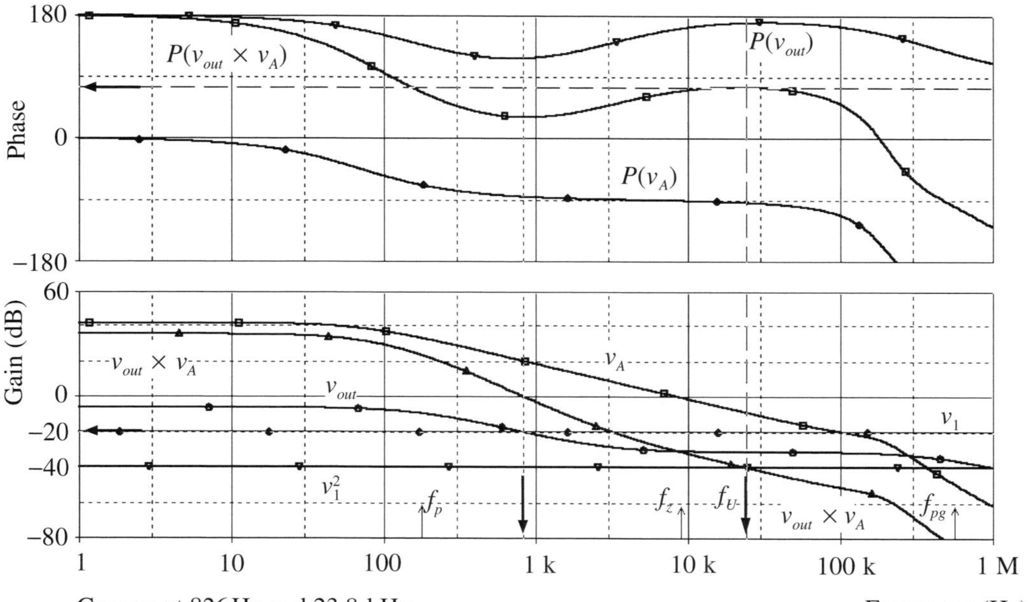


**Fig. 5.10.5** Gain and phase responses for Fig. 5.10.4(*b*) and (*c*) and for the product.

$$f_U = \left[\frac{3 \times 10^5 \times G_m(1.24 / v_{out})}{2\pi C_o}\right]^{\frac{1}{2}} = 6.8 \text{ kHz} \tag{5.10.6}$$

which differs considerably from the measured value, but it is very dependent on the load current via $G_m$ so you should not expect too close correspondence. In any case, our aim is to investigate the effect of the various components on the stability and the applicability of the *T* technique. The beneficial effect of the zero, which arises from $R_e$, is evident in that it reduces the phase lag in the region of $f_U$ giving improved phase margin. A lowish value of $f_U$ means that high frequency fluctuations will not be corrected so that $C_o$ will be important for this role.

To examine the performance of this type of circuit we need to be able to measure the open-loop performance, so we will make use of the *T* technique of Section 5.14. For the *T* technique, two example insertion points could be tried as shown in Fig. 5.10.6(*a*): insertion at *J* will result in *T* following $T_i$, and at *S*, *T* will follow $T_v$. If you run these two simulations, and the direct open-loop circuit of Fig. 5.10.4(*b*) and (*c*) but now with $C_f = 470$ p, you should find the unity-gain frequency as $\approx 112$ kHz and the phase margin about 36–46°, all in quite good agreement. You should experiment with variation of $R_e$ to see the importance of this and the effect of making this too small. Newer types of electrolytic capacitor (e.g. Oscon) have rather lower $R_e$, and though the improvement is not dramatic, say factors of two or three, this does provide additional flexibility.

In Fig. 5.10.6(*a*) a pulse controlled FET $Q_2$ can be added in series with $R_L$ to allow us to examine the transient response of the regulator. Typical waveforms, arising from a 1 ms pulse current of 1 A, are shown in Fig. 5.10.7. Figure 5.10.7(*a*) shows the waveforms as the load current $i_{RL}$ goes on. It is evident that the initial current supply to the load comes from $C_o$ and then the normal regulator action takes over after a short delay with the rise in current through $Q_1$, the transient decrease in voltage being about 100 mV. $i_{Q1}$ exceeds $i_{RL}$, the excess going to recharge $C_o$ as can be seen by the reversal in the direction of $i_{Co}$. $v_A$ decreases, making $v_{GS}$ greater, to switch $Q_1$ on.

When the current load is switched off the output voltage has instead a transient *rise* (of about 100 mV), which is equally undesirable. The origin of this rise can be seen from Fig. 5.10.7(*b*). The delay in the response of the feedback amplifier means that $Q_1$ is still on but the current now has nowhere to go so $C_o$ starts charging. When the amplifier output does eventually catch up and turns $Q_1$ off then $C_o$ can only discharge through the fixed load $R_b$ and so allow the output voltage to return towards normal. $v_A$ swings beyond $v_u$ (at +12 V) to more than cut-off $Q_1$, i.e. control has been lost until $v_{out}$ again comes within range. $v_A$ recovers control and there is a brief pulse of current $i_{Q1}$ into $C_o$ as control is re-established. There is thus a delay of about 1.2 ms after the end of the pulse before the system is settled. This brings home the desirability of having a wide frequency response, which has all the problems associated with trying to stabilize a system with many sources of phase

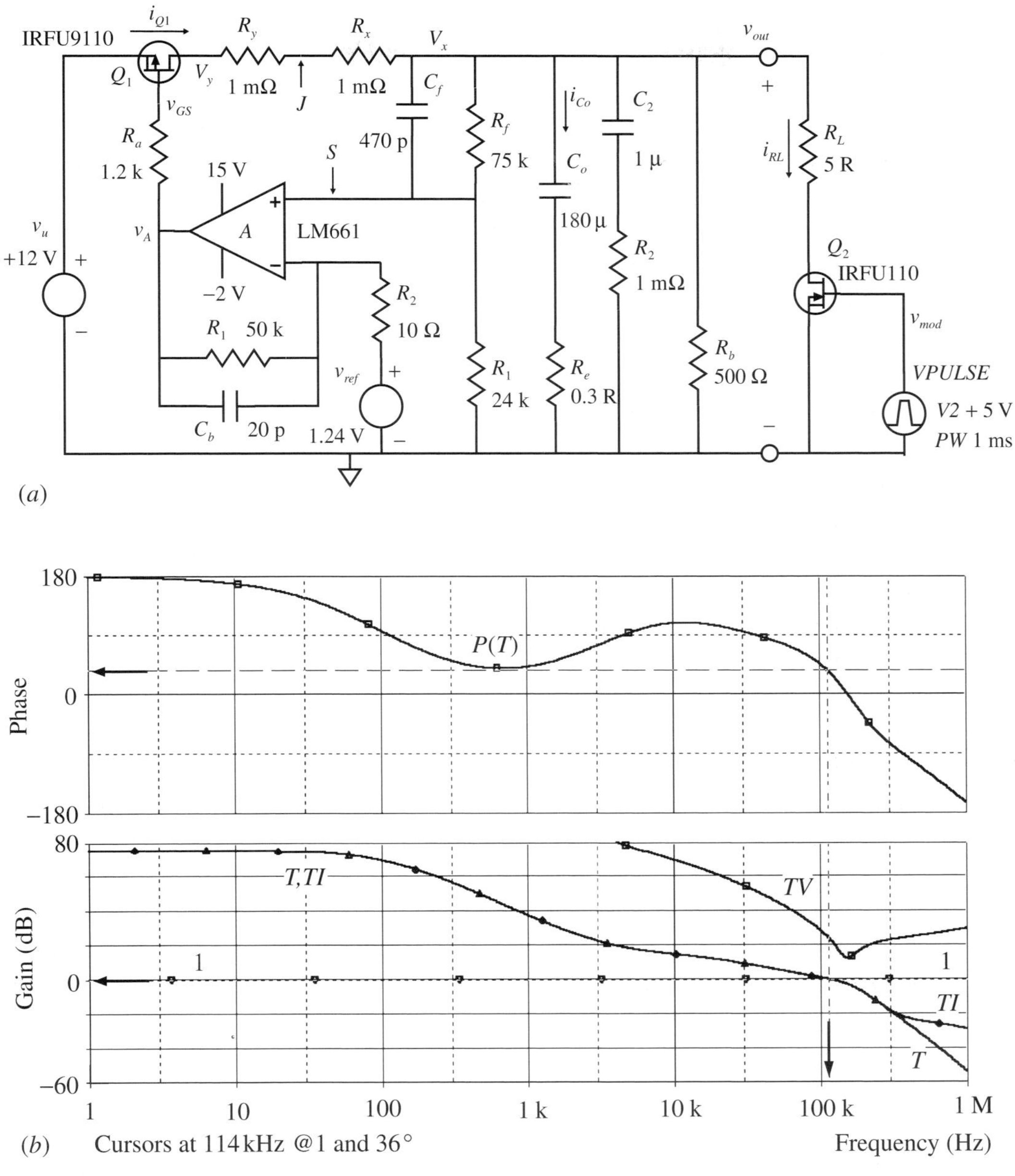


**Fig. 5.10.6** (*a*) Voltage regulator circuit for simulation. $Q_2$ is used to examine the transient behaviour. (*b*) *T* technique responses for injection at point *J* so that *T* will follow $T_i$. A notional resistor was inserted on either side of *J* to allow determination of the two currents.

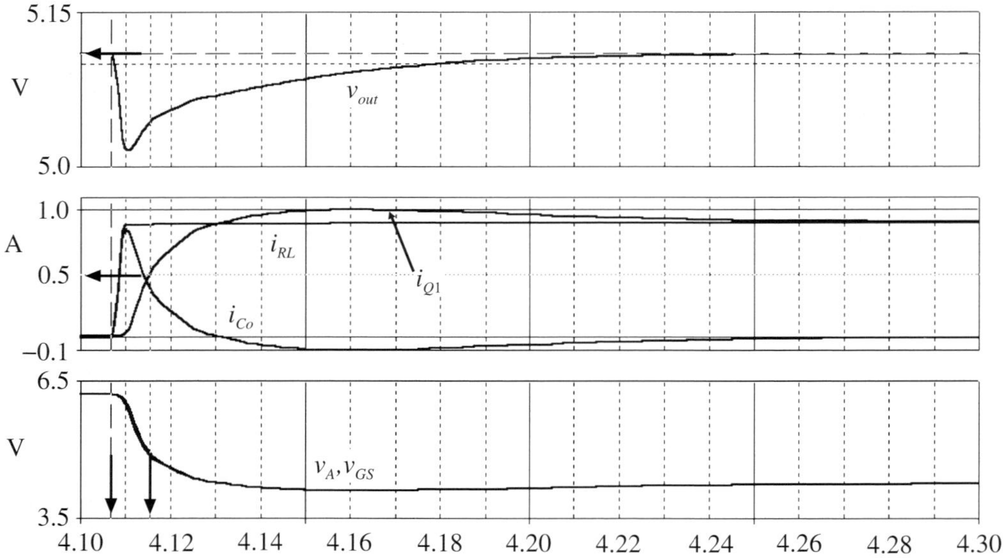


(*a*) Cursors at 4.107 and 4.115 ms Time (ms)

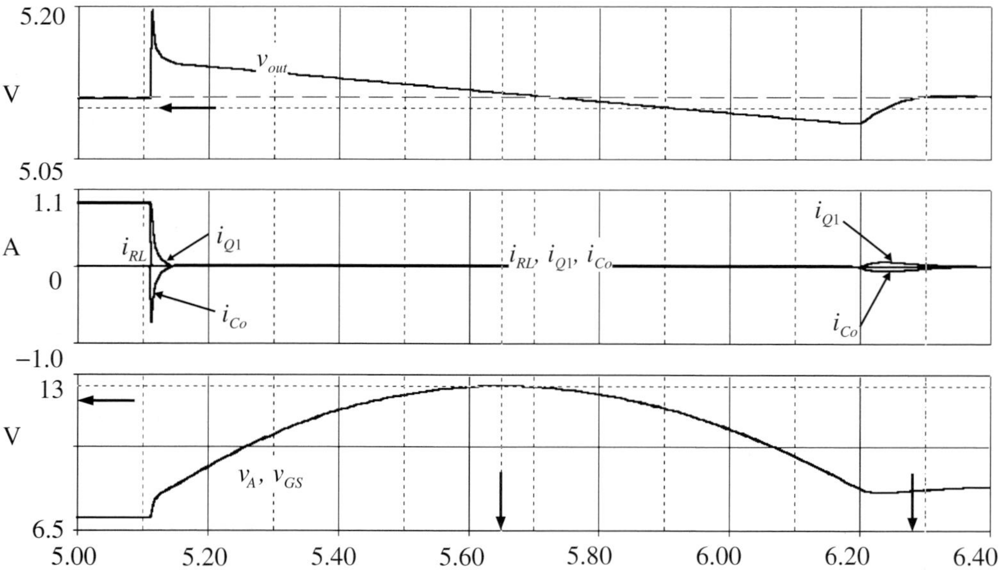


(*b*) Cursors at 5.65 and 6.3 ms Time (ms)

**Fig. 5.10.7** (*a*) Waveforms for $i_{RL}$ from zero to 1 A. (*b*) Waveforms for $i_{RL}$ from 1 A to zero. The quiescent current through $R_b$ is 10 mA.

shift, but it also suggests that the use of bridge-type power loads should be considered, in which the current is 'constant' and is switched between alternative paths.

It is of interest to examine the performance of a regulator using a very much wider bandwidth amplifier with good output capability to drive the FET gate capacity. We may try a 'what if' circuit as shown in Fig. 5.10.8(*a*), which is the same basic circuit but using an AD8012 current-feedback amplifier. Though this has a slew rate of 1200 V $\mu s^{-1}$ and a current capability of 100 mA the quiescent current is only 1.6 mA. The inverting input is actually a low impedance current output so we can apply the reference voltage here to also give maximum gain (Analog Devices 1997). As the bandwidth is so much greater it is necessary to adjust the location of the various poles and zeros to ensure stability, but there is much further room for experimenting on the various values. Components $C_3$, $R_4$ and $R_6$ are an approximation to a more realistic unregulated supply than the ideal voltage source. It is important to have a good high frequency characteristic for the unregulated supply as will be seen from the currents that must flow.

For a small load (about 10 mA through $R_b$) Fig. 5.10.8(*b*) shows the frequency response, with unity gain at 4.1 MHz and a phase margin of 38°. Increasing the load to 1 A gives a unity-gain frequency of 8.8 MHz and phase margin of 43° so stability should be ensured at the higher current. Figure 5.10.9 shows the transient response to a load change of 1 A.

The decrease of $V_{out}$ is now only 4 mV and there is no positive overshoot. It is evident that the primary control of the current is now via the active circuit and, if the current in the output capacitors is examined, that the output capacitors play a very much lesser role than previously. The figure shows the current supplied by $C_o$ and the currents through $R_6$ and $C_3$. Note that the current through $C_o$ has been multiplied by 100 so that it can be seen on the scale of the other currents. The sensitivity of the circuit to external load is now rather greater than for the slower regulator so the system should be designed with the 'external' load present. It should be said that this circuit has not been constructed to see the reality.

An integrated fast controller, with many additional facilities, is described by Craig Varga (1997).

## SPICE simulation circuits

Consult the SimCmnt.doc file on the CD before running

| | |
|---|---|
| Fig. 5.10.4(*c*) | Rpwsply6.SCH |
| Fig. 5.10.5 | Rpwsply6.SCH |
| Fig. 5.10.6(*b*) | Rpwsply7.SCH |
| Fig. 5.10.7(*a*) | Rpwsply1.SCH |
| Fig. 5.10.7(*b*) | Rpwsply1.SCH |
| Fig. 5.10.8(*a*) | Rpwsplyd.SCH |
| Fig. 5.10.8(*b*) | Rpwsplyd.SCH |
| Fig. 5.10.9 | Rpwsplye.SCH |

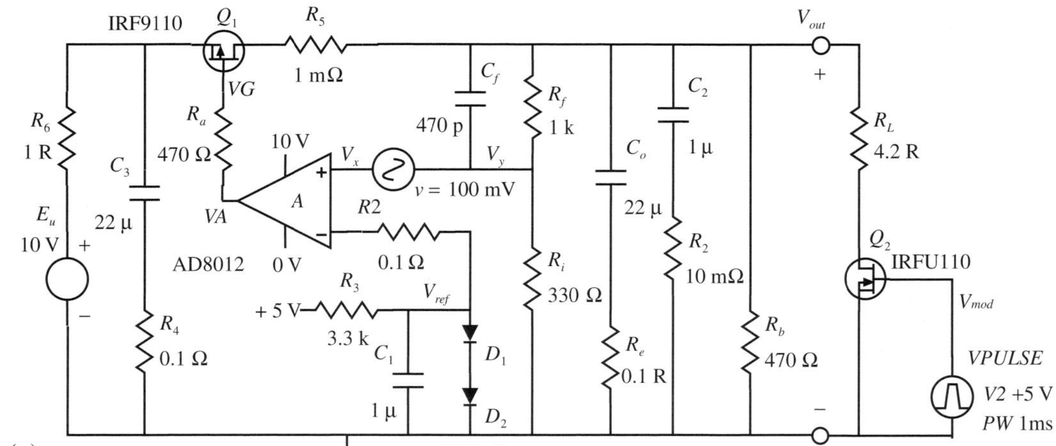

(*a*)

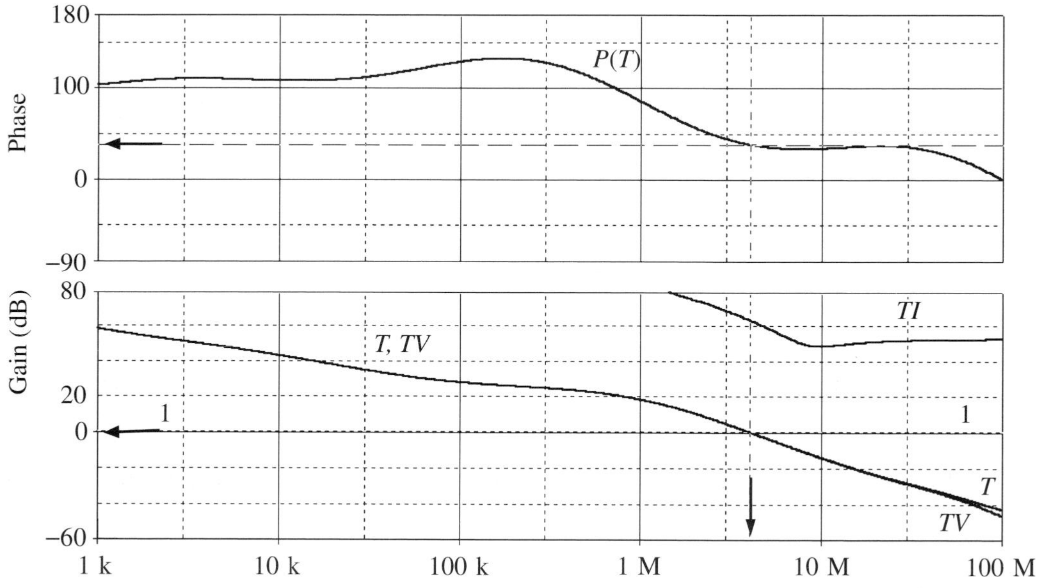

(*b*) Cursors at 4.08 MHz @ 0 dB and 38° Frequency (Hz)

**Fig. 5.10.8** (*a*) Regulator with fast feedback amplifier. (*b*) Frequency responses for low output current.

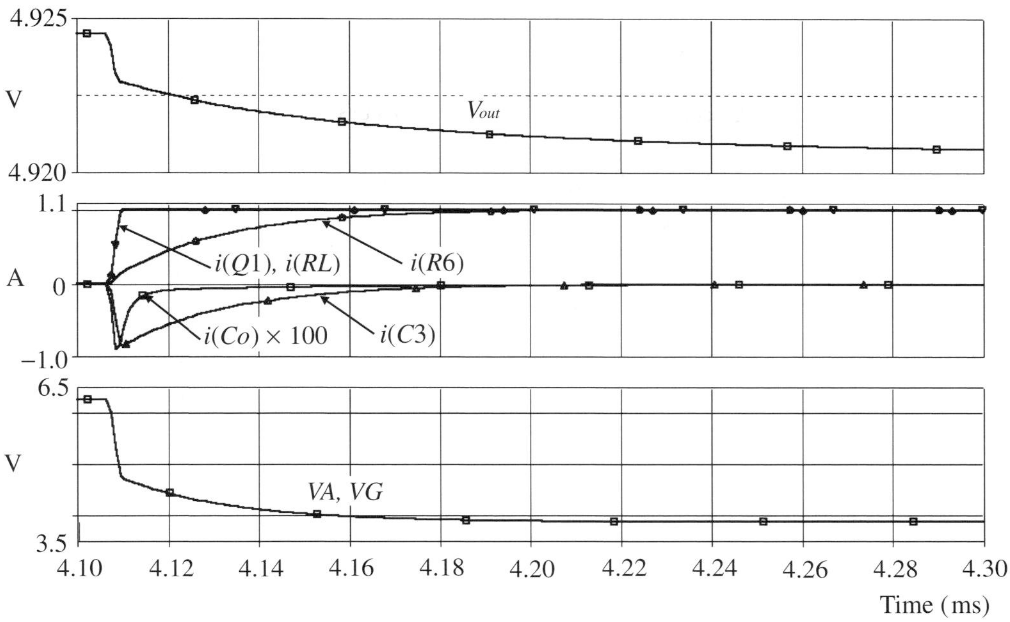


**Fig. 5.10.9** Transient response of the circuit of Fig. 5.10.8(*a*) to a change of 1 A output current.

## References and additional sources 5.10


Analog Devices (1997): *AD8012 Dual 350MHz Low Power Amplifier Datasheet.*

Craig Varga (1996): LT1575 ultrafast linear controller makes fast transient response power supplies. Application Note AN-69, September, in Williams J. M. (Ed.) (1997): *Linear Applications Handbook, Vol. III; Linear Technology.*

Hunt F. V., Hickman R. W. (1939): On electronic voltage stabilizers. *Rev. Sci. Instrum.* **10**, 6–21. (One of the earliest references on power supply regulators and possibly the origin of the cascode amplifier configuration.)

Kugelstadt T. (1999): *Fundamental Theory of PMOS Low-Dropout Voltage Regulators*, Texas Instruments Applications Report SLVA068, April.

Lee B. S. (1999): Understanding the stable range of equivalent series resistance of an LDO regulator. *Texas Instruments Analog Applications Journal*, November, 14–16. www.ti.com/sc/docs/products/msp/pwrmgmt/index.htm

Lee B. S. (1999): *Technical Review of Low Dropout Voltage Regulator Operation and Performance*, Texas Instruments Applications Report SLVA072.

National Semiconductor (1997): *LP2975 MOSFET LDO Driver/Controller*, Datasheet, September.

O'Malley K. (1994): Understanding linear-regulator compensation. *Electronic Design*, 22 August, 123, 124, 126, 128.

Rincon-Mora G. A., Allen P. E. (1998): Optimized frequency-shaping circuit topologies for LDOs. *IEEE Trans.* **CAS-45 Part II**, 703–708.

Rogers E. (1999): *Stability Analysis of Low-Dropout Linear Regulators with PMOS Pass Element*, Texas Instruments Analog Applications SLYT005, p. 10–12, August. www.ti.com/sc/docs/products/analog/tps 76433.html

Simpson C. (2000): *Linear Regulators: Theory of Operation and Compensation*, National Semiconductor Application Note 1148, May.

Wolbert, Bob (1996): *Designing with Low-Drop-out Voltage Regulators*, Micrel Semiconductor Application Guide.

Zendzian D. (2000): *A High Performance Linear Regulator for Low Dropout Applications*, Texas Instruments/Unitrode Application Note U-152.

# 5.11 Current-feedback amplifiers

Origin of things comes from union of yin and yang – passive and active principles.

Confucius

In more recent years a new form of operational amplifier has been developed, known as current-feedback amplifiers, and which differ significantly from the original voltage-feedback (VFA) types. Though, as discussed in Section 3.10, the traditional operational amplifier may use 'current feedback' by sampling the output current to obtain the feedback voltage, the new current-feedback amplifiers (CFA) have a quite different circuit topology that leads to rather different design techniques and performance. Newer semiconductor processes and circuit topologies have resulted in an order of magnitude improvement in performance (Kester 1996). Kester shows an illuminating plot of bandwidth against supply current for different processes and gives as an example the AD8011 with a bandwidth of 300 MHz at only 1 mA supply current.

A schematic form of the CFA amplifier is shown in Fig. 5.11.1.

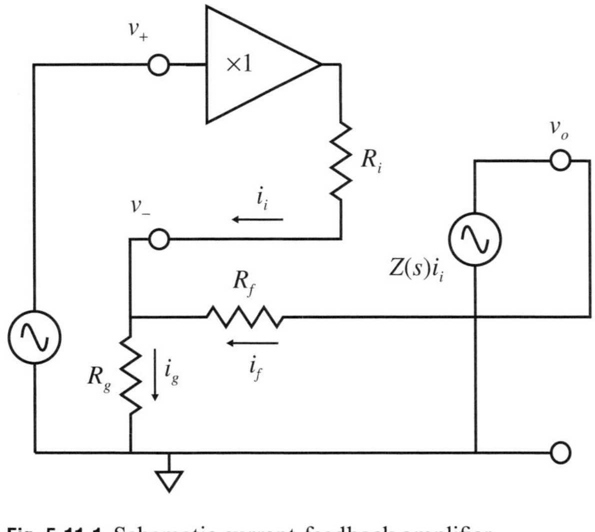


**Fig. 5.11.1** Schematic current-feedback amplifier.

We will generally follow the approach of Franco (1993) and Steffes (1993a). The usual non-inverting and inverting inputs are coupled by a very wideband unity-gain buffer with high input impedance and ideally zero output impedance. Here we include a resistor $R_i$ to represent a more realistic situation but the value will be say of the order of 50 Ω. Any current $i_i$ that flows between the buffer and the inverting input is converted by the transimpedance converter, with gain $Z(s)$, to an output voltage $V_o = Z(s)i_i$. $Z(s)$ has the units volt amp$^{-1}$, or ohm, and hence is called a transimpedance amplifier. It has high 'gain' and may have a value of many hundreds of kΩ at lower frequencies and a bandwidth of several hundred kHz (compare this with a few Hz for voltage-feedback amplifiers). We assume for now that the output impedance is low enough to be ignored. The feedback configuration with $R_f$ and $R_g$ looks familiar. We will analyse this non-inverting configuration. Applying an input signal $v_+$ would cause a current $i_i$ to flow to $v_-$. This is amplified by the transimpedance block to produce $v_o$ which causes a current $i_f$ to flow to cancel $i_i$ so that $v_-$ is forced to be equal to $v_+$ since the buffer gain is unity. Thus the differential input voltage tends to zero just as for the VFA. The particular difference here is that the CFA has a very low input resistance at the inverting input whereas the VFA has the same high input resistance at $v_-$ as at $v_+$.

If we define the response of the buffer by the function $\alpha(s)$ then we have the relationships:

$$i_f = \frac{v_o - v_-}{R_f}, \quad i_g = \frac{v_-}{R_g}, \quad i_i = \frac{v_o}{Z(s)}, \quad v_- = \alpha(s)v_+ - i_i R_i = \alpha(s)v_+ - \frac{v_o R_i}{Z(s)} \tag{5.11.1}$$

Summing the currents at the $v_-$ node:

$$i_i + i_f = i_g$$

$$i_i + \frac{(v_o - v_-)}{R_f} = \frac{v_-}{R_g}$$

$$R_f \frac{v_o}{Z(s)} + v_o = v_-\left(1 + \frac{R_f}{R_g}\right) = \left[\alpha(s)v_+ \frac{v_o R_i}{Z(s)}\right]\left(1 + \frac{R_f}{R_g}\right) \tag{5.11.2}$$

$$v_o\left[\frac{R_f}{Z(s)} + 1 + \frac{R_i}{Z(s)}\left(1 + \frac{R_f}{R_g}\right)\right] = \alpha(s)v_+\left(1 + \frac{R_f}{R_g}\right)$$

so for the overall gain $G$ we have:

$$G = \frac{v_o}{v_+} = \frac{\alpha(s)\left(1 + \dfrac{R_f}{R_g}\right)}{1 + \dfrac{R_f + R_i\left(1 + \dfrac{R_f}{R_g}\right)}{Z(s)}} \tag{5.11.3}$$

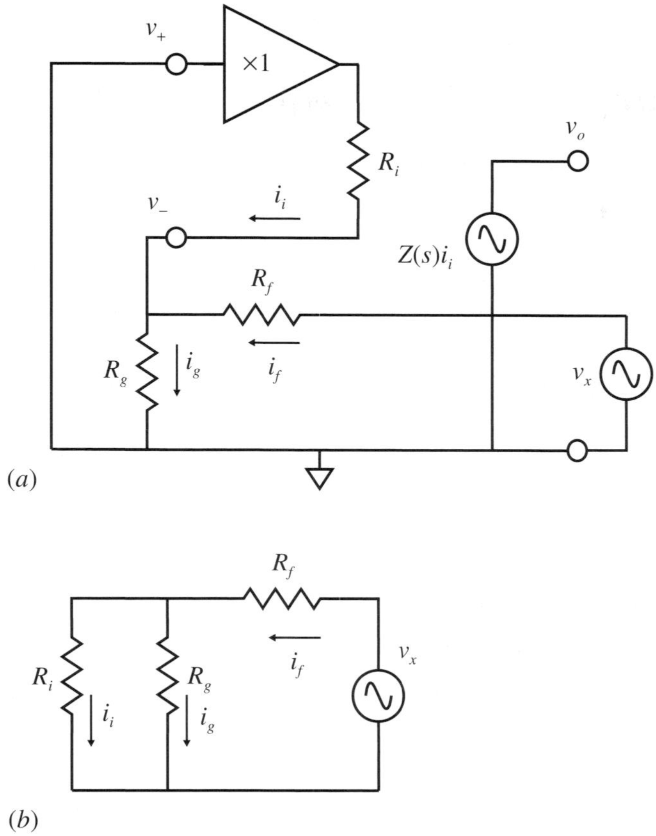


**Fig. 5.11.2** (*a*) Determination of gain. (*b*) Gain equivalent circuit.

As we approach the near-ideal condition where $R_i$ is small and at lower frequencies for the usual operating conditions where $Z(s) \gg R_f$ and $\alpha(s) = 1$, then the gain is:

$$G = 1 + \frac{R_f}{R_i} \tag{5.11.4}$$

just as for the VFA equivalent. We now determine the loop gain as this is the quantity that determines the stability of the amplifier. To do this we break the loop and add a generator $v_x$ as shown in Fig. 5.11.2.

For a voltage source, $v_+$ is effectively connected to common and since the buffer gain is unity and we assume $R_i$ is ideally zero, then $v_-$ is also at common. In this case the loop gain is given by:

$$i_i = i_f, \qquad i_f = \frac{v_x}{R_f}, \qquad v_o = i_i Z(s)$$

so the ideal loop gain $L_1 = \dfrac{v_o}{v_x} = \dfrac{i_i Z(s)}{i_i R_f} = \dfrac{Z(s)}{R_f}$ (5.11.5)

This gives us a guide to what to expect as we now consider a more practical derivation taking $R_i$ into account. In effect $R_i$ is now in parallel with $R_g$ so that the expression for $i_f$ is:

$$i_f = \frac{v_x}{R_f + \left(\dfrac{R_i R_g}{R_i + R_g}\right)} \tag{5.11.6}$$

Now for the two resistors in parallel we have:

$$\frac{i_i}{i_g} = \frac{R_g}{R_i} \quad \text{so} \quad \frac{i_i}{i_i + i_g} = \frac{R_g}{R_i + R_g} \quad \text{and hence} \quad \frac{i_i}{i_f} = \frac{R_g}{R_i + R_g} \tag{5.11.7}$$

and using (5.11.6) and the transimpedance generator relation gives:

$$\frac{v_o}{Z(s)} = i_i = \frac{i_f R_g}{R_i + R_g} = \frac{v_x R_g}{\left(R_f + \dfrac{R_i R_g}{R_i + R_g}\right)(R_i + R_g)}$$

$$= \frac{v_x R_g}{R_f(R_i + R_g) + R_i R_g} \tag{5.11.8}$$

$$= \frac{v_x}{R_f + \dfrac{R_i(R_f + R_g)}{R_g}}$$

so that we get for the loop gain:

$$L = \frac{v_o}{v_x} = \frac{Z(s)}{R_f + \dfrac{R_i(R_f + R_g)}{R_g}}$$

$$= \frac{Z(s)}{R_f + R_i\left(1 + \dfrac{R_f}{R_g}\right)} \tag{5.11.9}$$

which reduces to Eq. (5.11.5) for $R_i$ small as before. As frequency increases $Z(s)$ has a corner frequency at $f_C$ which may be say 300 kHz for example. The form of $Z(s)$ is thus:

$$Z(s) = \frac{Z(0)}{1 + sT}, \quad \text{where} \quad T = 1/\omega_C \tag{5.11.10}$$

to give a response as shown in Fig. 5.11.3. Here we plot the transimpedance gain in dB against log $\omega$ with the $x$-axis being the line $R_f$. The distance between the two

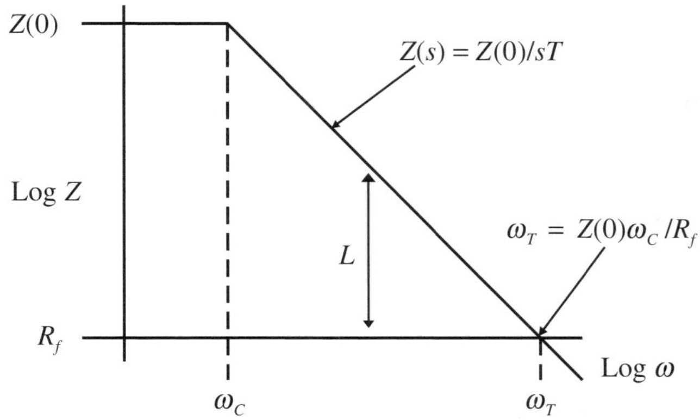


**Fig. 5.11.3** Current-feedback amplifier frequency response.

curves is the loop gain $L$. Where the two curves cross (i.e. $L = 1$) at $\omega_T = Z(0)/TR_f = Z(0)\omega_C/R_f$ defines the 3 dB bandwidth on the basis of the ideal loop-gain expression Eq. (5.11.5). The correction, according to Eq. (5.11.9), is small.

We can see the same result if we introduce (5.11.10) into (5.11.3), again taking the approximate condition of $R_i$ small. We then have:

$$G = \frac{\left(1 + \dfrac{R_f}{R_g}\right)}{1 + \left[\dfrac{R_f(1 + sT)}{Z(0)}\right]} = \frac{\left(1 + \dfrac{R_f}{R_g}\right)}{1 + \dfrac{R_f}{Z(0)} + \dfrac{sR_fT}{Z(0)}} = \frac{\left(1 + \dfrac{R_f}{R_g}\right)}{1 + \dfrac{s}{\omega_T}}, \quad \text{since} \quad R_f \ll Z(0) \quad \text{and} \quad \omega_T = \frac{Z(0)}{TR_f} \tag{5.11.11}$$

so the 3 dB frequency is $s = j\omega_T$ when the denominator becomes $(1 + j1)$. We can now draw some important conclusions regarding the CFA:

(i) The bandwidth is set by the choice of $R_f$ alone. Particular devices are designed with an optimum value of $R_f$ in mind.

(ii) The gain can be independently set, after $R_f$ is selected, by choice of $R_g$. Changing the gain does not change the bandwidth as happens for the VFA.

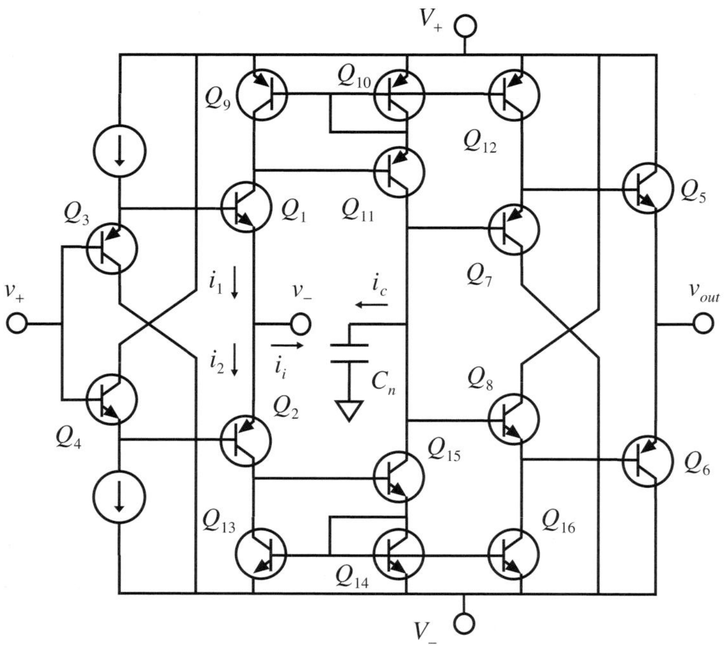


**Fig. 5.11.4** Current-feedback amplifier (CFA) internal structure.

However, for high gains our approximations become invalid and the CFA reverts to the VFA type fixed gain–bandwidth product.

(iii) The CFA is particularly appropriate for lower gains with wide bandwidth. This is where the VFA presents the greatest problems from the stability point of view because the feedback is large (see Section 3.10).

The large-signal transient response is determined by the bandwidth and the slewing rate (Section 3.6). The CFA has a considerable advantage over the VFA with regard to slewing rate which is particularly appropriate because of the large bandwidths possible. To see how this arises it is necessary to consider the circuit configuration of the CFA in a little more detail as shown in Fig. 5.11.4.

The left-hand portion forms the unity-gain buffer with the two inputs $v_+$ and $v_-$.The current $i_i$ is the difference between the currents $i_1$ and $i_2$. These two currents also flow in the collectors of $Q_1$ and $Q_2$ and are reflected by the two Wilson current mirrors (Section 5.9) to flow in the collectors of $Q_{11}$ and $Q_{15}$. The difference between these two currents gives $i_c$ to charge the node capacity $C_n$, the voltage across which is amplified to give the output $v_{out}$. Since the output amplifier has been designated as a transimpedance amplifier it is more correct to say that the difference between $i_{11}$ and $i_{15}$ is amplified to give $v_o$, but $C_n$ still has to be charged to reach the final output.

The transient magnitude of $i_i$ is determined by the magnitude of the input signal.

The action of the current mirrors makes $i_c = i_i$ so that the current available to charge $C_n$ depends also on the magnitude of the input signal. So within the limitations of the amplifier the slewing rate is proportional to the input amplitude and hence the CFA can cope with large fast signals. For an input $\delta v_{in}$ the current $i_i$ is given by (since $R_g$ and $R_f$ are in parallel):

$$i_i = \frac{\delta v_{in}}{[R_g R_f/(R_g + R_f)]} = i_c \tag{5.11.12}$$

The initial rate of charge is then (using (5.11.4)):

$$\frac{i_c}{C_n} = \frac{\delta v_{in}(R_g + R_f)}{R_g R_f C_n} = \frac{\delta v_{in}\left(1 + \dfrac{R_f}{R_g}\right)}{R_f C_n} = \frac{\delta v_o}{R_f C_n} \tag{5.11.13}$$

and from Section 3.7 this initial rate of rise is equivalent to an exponential time constant $\tau = R_f C_n$. Since $R_f$ is typically a few kΩ (or less) and $C_n$ a few pF we have risetimes in the region of say 10 ns so we can expect good fidelity large signal responses. However, we need to be aware that stray effects will be very much more important with such wideband systems. The layout and interconnection of 100 MHz systems is a much more difficult task than for 1 MHz bandwidth. Stray capacity can have a very significant effect; 1 pF at 1 MHz has an impedance of 160 k while at 100 MHz it is only 1.6 k. Load capacitance can cause response peaking even for ≈10 pF. If there is a problem the output should be decoupled with a small resistor of tens of Ω. Capacity at the inverting input can reduce phase margins and lead to ringing. Ground planes, which are recommended for high frequency systems, should be kept somewhat away from the output or inverting input pins. Since the capacity that can cause change is small, say 5 pF, probing with an oscilloscope probe, say 10 pF, can be very misleading. It is desirable to introduce some series resistance, say 100 Ω, to minimize any effects.

Operation of a CFA in the inverting mode is in general much the same as for the VFA so long as the feedback impedance is resistive. Reactive feedback components tend to cause reduced phase margins and are generally avoided.

A variation on the standard CFA configuration is the OPA622 (Henn and Sibrai 1993). The circuit is presented in a segmented form and includes an additional wideband buffer which allows the device to be configured as either a CFA or as a VFA. The buffer is used to feed the usual CFA inverting input to provide a high impedance input to match the non-inverting input. The feedback is now voltage rather than current but the system can retain the wideband 'constant' bandwidth of the CFA. If used in CFA configuration then the extra buffer can be used independently. Figure 5.11.5 shows the IC arrangement and the voltage feedback connection.

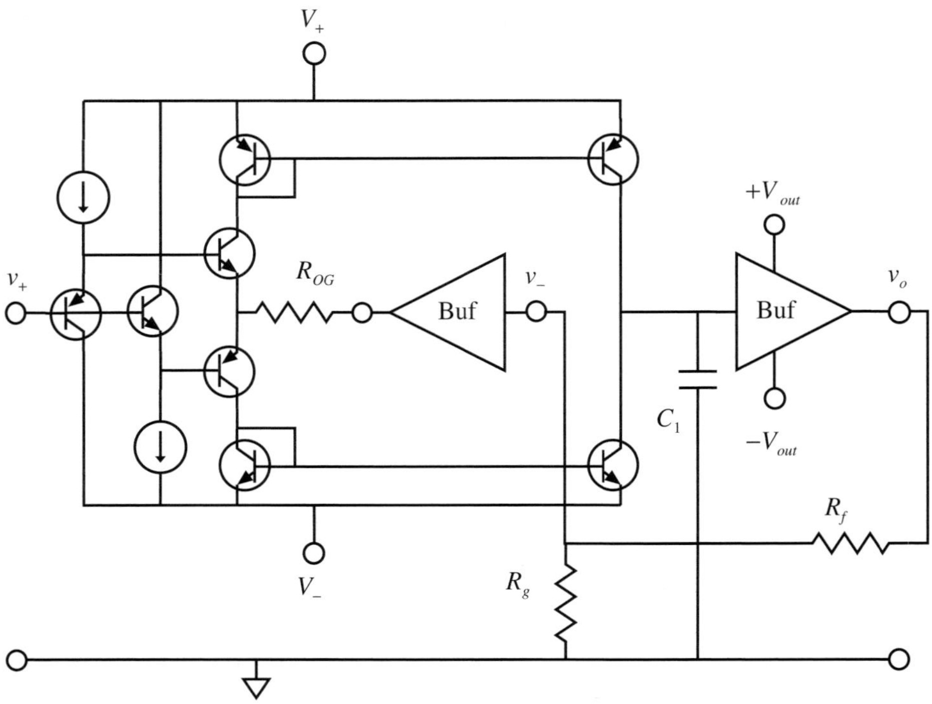


**Fig. 5.11.5** OPA622 configuration.

## SPICE simulation circuits

None

## References and additional sources 5.11


Blake K. (1996): *Noise Design of Current-Feedback Op Amp Circuits*, National Semiconductor Application Note OA-12, Rev. B, April.

Buck, Arne (1993): *Current-Feedback Myths Debunked*, Comlinear Application Note OA-20.

Franco, Sergio (1993): *Current-Feedback Amplifiers*, Comlinear Application Note.

Harold P. (1988): Current-feedback op amps ease high speed circuit design. *EDN* 7 July, 84–94

Harvey, Barry (1997): *Practical Current-Feedback Amplifier Design Considerations*, Élantec tutorial No. 3, or Application Note 23.

Henn C., Sibrai A. (1993): *Current or Voltage Feedback: The Choice is Yours with the New, Flexible Wide-Band Operational Amplifier OPA622*, Burr-Brown Application Note AN-186.

Karki J. (1998): *Voltage Feedback vs Current Feedback Op Amps Application Report*, Texas Instruments Literature Number SLVA051, November.

Kester W. (Ed.) (1996): *High Speed Design Techniques*, Analog Devices Handbook. ISBN 0-916550-17-6. Section 1: High speed operational amplifiers.

Lehmann K. (1991): *Current or Voltage Feedback? That's the Question Here*, Burr-Brown Application Note 110-E, May.
Mancini R., Lies J. (1995): *Current Feedback Amplifier Theory and Applications*, Harris Semiconductor Application Note AN9420.1, April.
Mancini, Ron (1996): *Comparison of Current Feedback Op Amp SPICE Models*, Harris Semiconductor Application Note AN9621.1, November.
Mancini, Ronald (1997): *Converting from Voltage-Feedback to Current-Feedback Amplifiers*, Harris Semiconductor Application Note AN9663, January.
Mancini R. (1999): *Current Feedback Analysis and Compensation*, Texas Instruments Application Report SLOA021, May.
National Semiconductor (1992): *Development of an Extensive SPICE Macromodel for 'Current-Feedback' Amplifiers*, Application Note 840, July.
National Semiconductor (1992): *Topics on Using the LM6181 a new Current Feedback Amplifier*, Application Note 813, March.
Palouda H (1989): *Current Feedback Amplifiers*, National Semiconductor Application Note 597, June.
Steffes M. (1993a): *Current-Feedback Loop Gain Analysis and Performance Enhancement*, Comlinear Application Note OA-13, January.
Steffes, Michael (1993b): *Frequent Faux-pas in Applying Wideband Current-Feedback Amplifiers*; Comlinear Application Note OA-15.
Tolley W. E. (1993): *AD9617/AD9618 Current-Feedback Amplifier Macromodels*, Analog Devices Application Note AN-259.
Wang A. D. (1994): *The Current-Feedback Op Amp. A High Speed Building Block*, Burr-Brown Application Bulletin AB193.
Wang, Tony (1994): *Voltage-Feedback Amplifiers vs. Current-Feedback Amplifiers: Bandwidth and Distortion Considerations*, Burr-Brown Application Bulletin AB091.

# 5.12 Fast operational picoammeter

'What is a "Designer"?' Is he/she a person who designs circuits? Wears flamboyant clothing and plans the décor for a home or an office – or a locomotive? Designs the transmission or the grille for a new car? Well, yes, a designer can be all or any of these things. But after you learn how to analyze things and prove the feasibility of a design, it's also of great value to be able to invent new circuits – new designs. To do that, you have to be familiar with lots of old designs. You have to know what each old design did well, and what it did badly. Basically, you just have to KNOW lots of old designs.

R. A. Pease (1995)

Perhaps the most convenient ammeter for very small currents is the operational current-to-voltage converter, which is often called a transimpedance amplifier (Fig. 5.12.1).

Using a suitable amplifier having negligible bias current relative to $i_{in}$ the current to be measured, then making ideal operational amplifier assumptions the transfer function is:

$$H(s) = \frac{v_{out}}{i_{in}} = -R_f \tag{5.12.1}$$

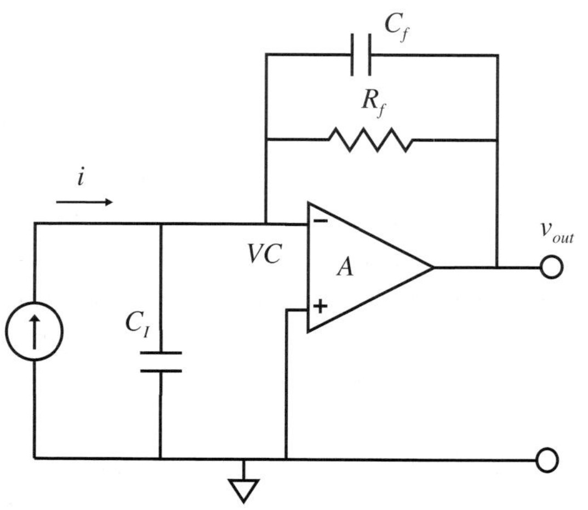


**Fig. 5.12.1** Basic transimpedance ammeter.

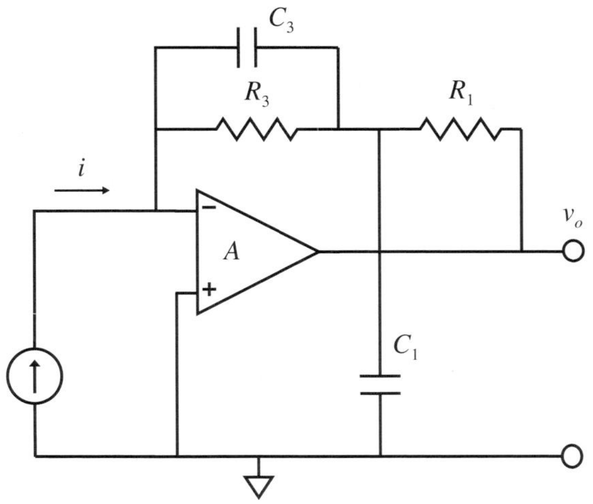


**Fig. 5.12.2** Modified picoammeter circuit.

The virtual common point $VC$ ensures there is no voltage burden applied to the current source and has the added attraction that source and interconnection capacity $C_I$ should have no effect (Keithley 1984, pp. 34, 38).

We are concerned here with the response time of the ammeter and how to decrease it. It is commonly stated that the primary factor which limits the bandwidth, and hence increases the response time, is the capacity $C_f$ shunting the feedback resistor $R_f$ and that $C_I$ has negligible effect; e.g. for $R_f = 1$ GΩ and $C_f = 1$ pF, the bandwidth would be 159 Hz. To improve bandwidth it is suggested that a modified feedback arrangement is used (Fig. 5.12.2) (Pelchowitch and Zaalberg van Zelst 1952; Praglin and Nichols 1960; Keithley 1984).

If the time constants $R_1C_1 = R_3C_3$ then it can be shown that the effect of $C_3$ will be compensated (Pelchowitch and Zaalberg van Zelst 1952; Keithley 1984). It is not now evident what determines the bandwidth. If such a circuit is constructed it will be found to have most undesirable characteristics. The frequency response and the output for a current pulse input is found to be as shown in Fig. 5.12.3.

It is evident that the phase shift is large and the system is on the verge of instability. There are two factors which have not been taken into account: the source capacity and the frequency response of the amplifier itself. If these are included it will be clear why the large resonance peak occurs – we now have the same circuit as for the differentiator (Section 5.6) and the system acts like a very lightly damped tuned circuit. What is required is a suitable damping device to reduce the $Q$ without decreasing the bandwidth.

A number of modifications to the basic circuit have been examined by several authors. For example, Kendall and Reiter (1974) examine an arrangement shown in Fig. 5.12.4 (see also Kendall and Zabielski 1970). In this case the dimensions of

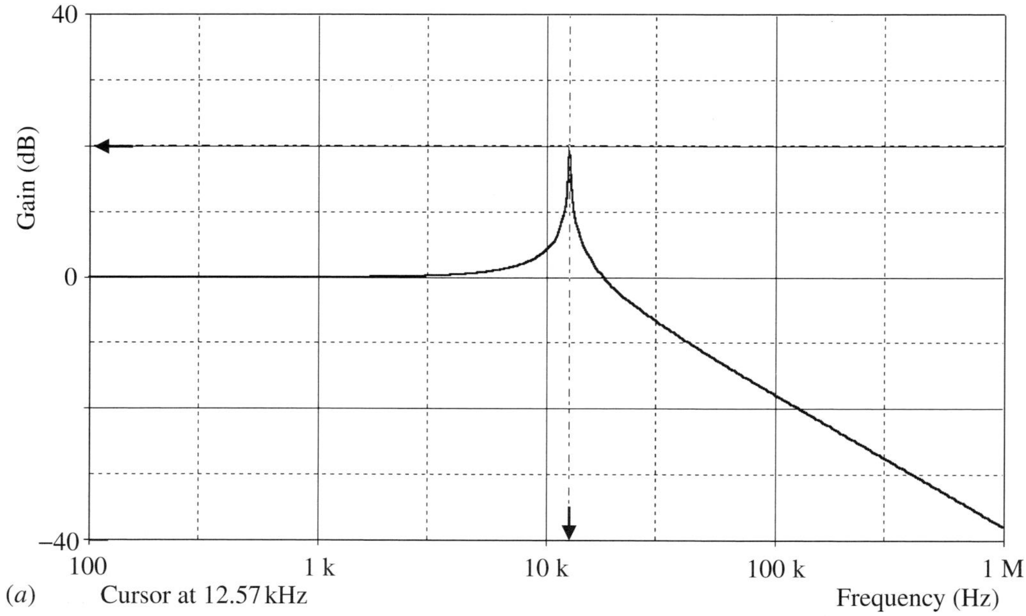


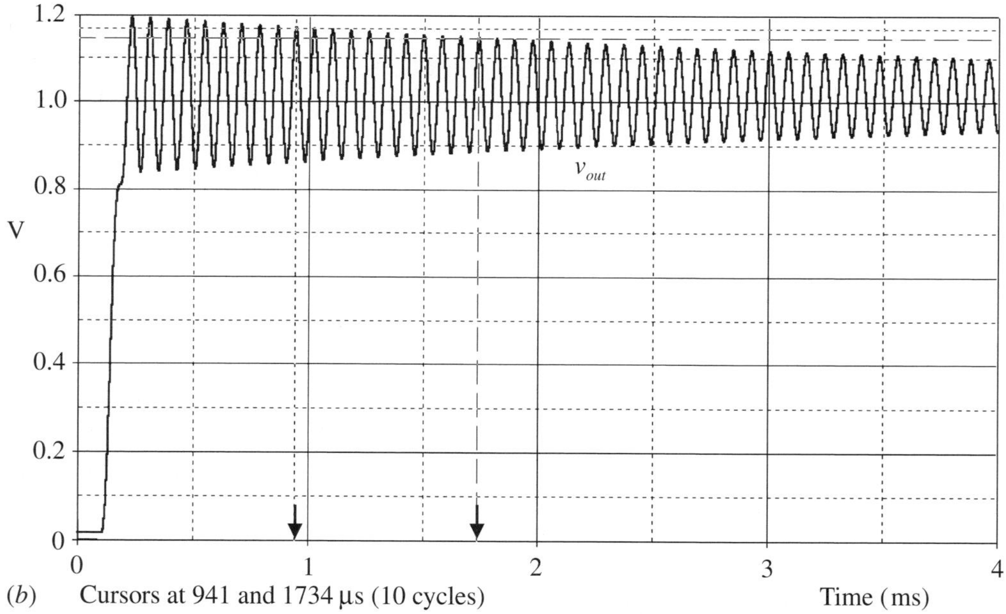


**Fig. 5.12.3** (*a*) Frequency response of Fig. 5.12.2. An input capacity of 1 pF was added and the peak is at 12.57 kHz. (*b*) Pulse response for 1 nA input pulse with rise- and falltimes 100 μs and pulse width 10 ms. The ringing frequency is ≈12.6 kHz.

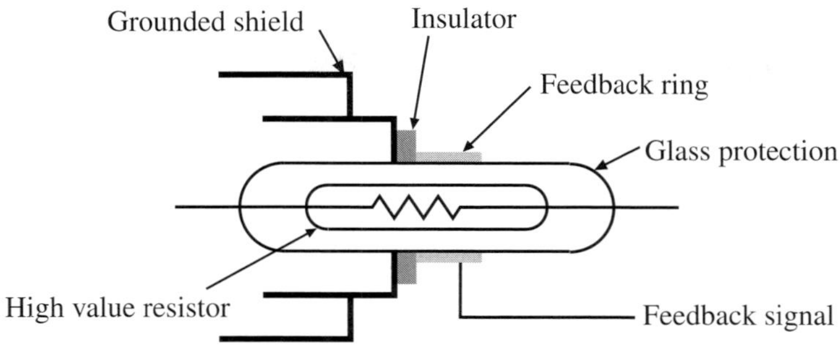


**Fig. 5.12.4** Feedback resistor compensation.

the feedback ring can only be determined by trial and error and is mechanically inconvenient.

We present here an analysis of the basic circuit of Fig. 5.12.2 but now taking account of the amplifier open-loop response and introducing a damping control ($R_2$, Fig. 5.12.5) which can be simply implemented and readily adjusted for optimum response. We also examine the effect of a non-ideal current source. It is difficult to know accurate values of all the components, e.g. $C_3$ and $C_2$, so we cannot expect very close agreement between simulation and experiment, but the simple damping control allows for quick and easy optimization. This does raise the problem of generating such low current pulses to carry out the adjustment. A convenient method is described by Praglin and Nichols and is illustrated in Fig. 5.12.6. A fuller analysis of this is given in 'Testing response time' below.

## Analysis of circuit with damping control

The circuit is shown in Figure 5.12.5. By choice $C_1(R_1+R_2)=C_3R_3$. As will be found later $R_2 \ll R_1$ and since the parallel impedance $Z_3$ of $R_3$ and $C_3$ is very much larger than the effective source resistance at point $B$ we assume that $v_2$ may be determined as a fraction of $v_o$ ignoring the loading effect of $C_3R_3$. Thus:

$$v_2 = \frac{v_o\left(R_2 + \dfrac{1}{sC_1}\right)}{R_1 + R_2 + \dfrac{1}{sC_1}} = \frac{v_o(sC_1R_2 + 1)}{1 + sC_1(R_1 + R_2)} \tag{5.12.2}$$

$C_2$ represents the capacity of the source connection and the amplifier input capacity. Since the current source is here assumed ideal its source resistance is infinite as is the input resistance of the amplifier. Thus:

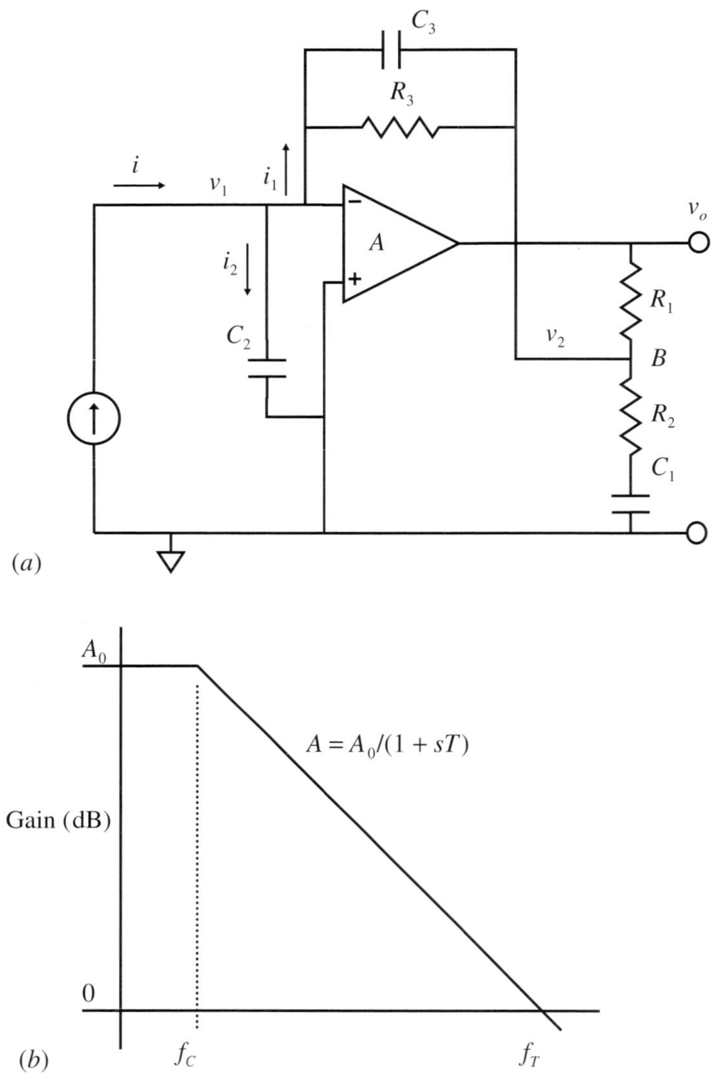


**Fig. 5.12.5** (*a*) Circuit for analysis, with damping control $R_2$. (*b*) Amplifier frequency response.

$$i = i_1 + i_2 = \frac{v_1 - v_2}{Z_3} + v_1 sC_2, \quad \text{with} \quad Z_3 = \left(\frac{R_3}{1 + sC_3R_3}\right) \tag{5.12.3}$$

It is also necessary to allow for the open-loop frequency response $A$ of the amplifier. Commonly used electrometer amplifiers have a single pole response with the corner frequency determined by a time constant $T$, i.e. $f_C = 1/(2\pi T)$ (Fig. 5.12.5(*b*)). Thus for a zero frequency gain $A_0$:

$$A = \frac{A_0}{1 + sT} \tag{5.12.4}$$

and the relation between $v_o$ and $v_1$ is:

$$v_o = -v_1 A \qquad \text{or} \qquad v_1 = \frac{-v_o}{A} \tag{5.12.5}$$

Then from (5.12.2):

$$\begin{aligned} I &= \frac{-V_o}{AZ_3} - \frac{V_o s C_2}{A} - \frac{V_o(1+sC_1R_2)(1+sC_3R_3)}{[1+sC_1(R_1+R_2)]R_3} \\ &= -V_o\left[\frac{1}{AZ_3} + \frac{sC_2}{A} + \frac{(1+sC_1R_2)}{R_3}\right], \quad \text{since} \quad C_3R_3 = C_1(R_1+R_2) \end{aligned} \tag{5.12.6}$$

The transimpedance transfer function is:

$$\begin{aligned} H(s) &= \frac{V_o}{I} = \frac{-1}{\dfrac{(1+sC_3R_3)}{AR_3} + \dfrac{sC_2}{A} + \dfrac{(1+sC_1R_2)}{R_3}} \\ &= \frac{-AR_3}{(1+sC_3R_3) + sC_2R_3 + A(1+sC_1R_2)} \end{aligned} \tag{5.12.7}$$

Since we are concerned with the response at high frequency well above $f_C$, we can write approximately:

$$A = \frac{A_0}{sT}, \quad \text{since} \quad sT \gg 1 \tag{5.12.8}$$

Then:

$$\begin{aligned} H(s) &= \frac{V_o}{I} = \frac{-A_0R_3/sT}{sC_3R_3 + 1 + sC_2R_3 + \dfrac{A_0(1+sC_1R_2)}{sT}} \\ &= \frac{-A_0R_3}{s^2TR_3(C_3+C_2) + sT + A_0(1+sC_1R_2)} \\ &= \frac{-A_0R_3}{s^2TR_3C_P + s(T+A_0C_1R_2) + A_0}, \quad \text{where} \quad C_P = C_3 + C_2 \end{aligned} \tag{5.12.9}$$

i.e. there are two poles given by:

$$s_{a,b} = \frac{-(T+A_0C_1R_2) \pm [(T+A_0C_1R_2)^2 - 4TC_PR_3A_0]^{\frac{1}{2}}}{2TC_PR_3} \tag{5.12.10}$$

Common electrometer amplifiers have a corner frequency $\approx 5$ Hz so $T = 0.03$, and with typical values of $A_0 = 10^6$, $C_1 = 100$ nF, $R_2 = 100\ \Omega$ then $A_0C_1R_2 \approx 10$ and we may neglect $T$ relative to this. For critical damping of the second order system the poles must be equal (and real), i.e. the condition required is:

$$(T + A_0C_1R_2)^2 = 4TC_pR_3A_0 \quad \text{or} \quad A_0^2C_1^2R_2^2 = 4TC_pR_3A_0 \quad (\text{neglecting } T) \tag{5.12.11}$$

$$\text{and} \quad R_2 = \left[\frac{4TR_3C_P}{A_0C_1^2}\right]^{\frac{1}{2}} \tag{5.12.11 cont.}$$

For example with the values above and say $C_P = 10$ pF and $R_3 = 1$ GΩ, then $R_2 = 346$ Ω. Since $A_0$, $T$ and $C_P$ will generally not be accurately known it will be necessary to make $R_2$ variable and tune the system. For an input current pulse, $R_2$ is adjusted so there is no ringing on $v_o$.

If critical damping conditions hold the two poles lie at:

$$s_1 = \frac{-(T + A_0C_1R_2)}{2TC_PR_3} \cong \frac{-A_0C_1R_2}{2TC_PR_3} \tag{5.12.12}$$

and using the above 'typical' values we find:

$$\omega_1 = 57\,667 \text{ s}^{-1} \quad \text{or} \quad f_1 = 9.2 \text{ kHZ} \tag{5.12.13}$$

The corresponding risetime $\tau_1$, assuming the standard 10 to 90% relationship, but allowing for two cascaded time constants (Millman and Taub 1965, p. 136), is given by:

$$\tau_1 = \frac{1.5 \times 0.35}{f_1} = \frac{1.5 \times 0.35}{9.2 \times 10^3} = 57 \text{ μs} \tag{5.12.14}$$

Note that since $f_1 \propto R_2/C_P$ and $R_2 \propto \sqrt{C_P}$, $f_1 \propto 1/\sqrt{C_P}$ and hence $\tau_1$ is proportional to $\sqrt{C_P}$.

## Testing response time

The production of fast low current pulses with large effective source resistance, i.e. a proper current source, presents some problems. The source must have a large source resistance as otherwise the analysis will be void since the feedback will be changed. A phototube in conjunction with an appropriate light source is most effective but may be inconvenient. An alternative approach has been proposed by Praglin and Nichols (1960) but their analysis is only partial and does not include the damping arrangement introduced above. The circuit arrangement is shown in Fig. 5.12.6.

A triangular voltage waveform $v_{in}$ is connected to the picoammeter via a small capacitor $C_4$ (say 1 pF). The current waveform of $i$ will ideally be the differential of the triangular voltage waveform, i.e. a square wave, the magnitude of $i$ being determined by the slope of the triangular ramp. A fuller analysis reveals that $i$ deviates somewhat from the expected square wave.

The starting equations are (using the same approximations as previously to determine $v_2$):

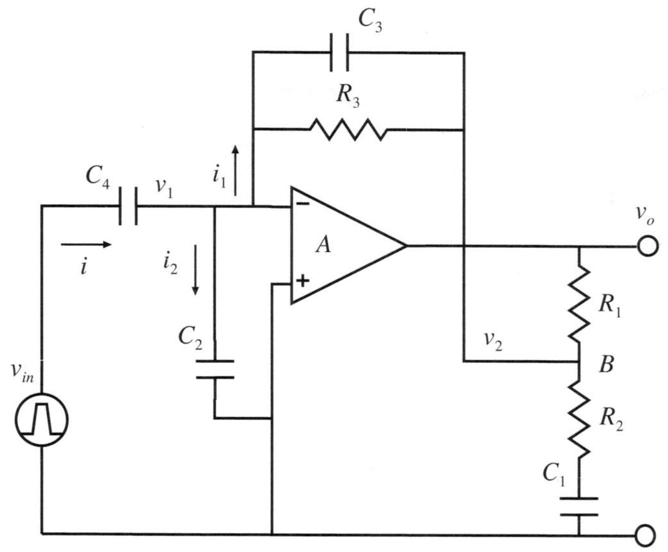


**Fig. 5.12.6** Circuit for generating fast very low current pulses.

$$I = sC_4(V_{in} - V_1), \quad I_1 = \frac{V_1 - V_2}{Z_3}, \quad I_2 = sC_2V_1, \quad I = I_1 + I_2$$

$$Z_3 = \frac{R_3}{1 + sC_3R_3}, \quad V_2 = \frac{V_o(1 + sC_1R_2)}{1 + sC_1(R_1 + R_2)}, \quad V_1 = \frac{-V_o}{A}, \quad A = \frac{A_0}{1 + sT} \tag{5.12.15}$$

This gives:

$$sC_4(V_{in} - V_1) = \frac{V_1 - V_2}{Z_3} + sC_2V_1$$

$$sC_4V_{in} = \frac{V_1}{Z_3} + V_1s(C_4 + C_2) - \frac{V_o(1 + sC_1R_2)(1 + sC_3R_3)}{R_3[1 + sC_1(R_1 + R_2)]} \tag{5.12.16}$$

and making $C_3R_3 = C_1\,(R_1 + R_2)$ as before and putting $C_4 + C_2 = C_S$:

$$sC_4V_{in} = \frac{-V_o}{A}\left[\frac{1}{Z_3} + sC_S + \frac{A(1 + sC_1R_2)}{R_3}\right] \tag{5.12.17}$$

The transfer function is then:

$$H(s) = \frac{V_o}{V_{in}} = \frac{-sC_4A}{\left[\dfrac{1}{Z} + sC_S + \dfrac{A(1 + sC_1R_2)}{R_3}\right]}$$

$$= \frac{-sC_4A}{\dfrac{(1 + sC_3R_3)}{R_3} + sC_S + \dfrac{A(1 + sC_1R_2)}{R_3}} \tag{5.12.18}$$

$$= \frac{-sAC_4R_3}{1 + sR_3(C_3 + C_S) + A(1 + sC_1R_2)} \qquad (5.12.18 \text{ cont.})$$

Again at high frequencies where $sT \gg 1$ we can use $A = A_0/sT$ and put $(C_3 + C_S) = (C_3 + C_4 + C_2) = C_U$ giving:

$$\frac{V_o}{V_{in}} = \frac{-sA_0C_4R_3}{s^2TC_UR_3 + s(A_0C_1R_2 + T) + A_0} \qquad (5.12.19)$$

so that for a ramp input $v_{in} = at$, $V_{in} = a/s^2$ and we get:

$$V_o = \frac{-aA_0C_4R_3}{s[s^2TC_UR_3 + s(A_0C_1R_2 + T) + A_0]} \qquad (5.12.20)$$

For critical damping the roots of the denominator must be equal and real, i.e.:

$$(A_0C_1R_2 + T)^2 = 4TC_UR_3A_0 \qquad (5.12.21)$$

and since in practice $T \ll A_0C_1R_2$ we get:

$$R_2 = \left(\frac{4TC_UR_3}{C_1^2 A_0}\right)^{\frac{1}{2}} \qquad (5.12.22)$$

The value of $R_2$ is as before except for the extra contribution from $C_4$ in $C_U$. Since $v_{in}$ is a voltage source it has zero impedance and hence $C_4$ is simply in parallel with $C_2$. To enable comparison between the calculation and the PSpice results we need a consistent set of component values. The simulation will be run using an OPA128 amplifier with the following additional component values ($A_0$, $f_C$, $T$ and $f_T$ were obtained from an 'open loop' measurement of the Burr-Brown SPICE macromodel).

OPA128: $A_0 = 2.29 \times 10^6$, $f_C = 0.88$ Hz or $T = 0.18$ s, $f_T = 1.94$ MHz,
$R_{in} = 10^{13}\ \Omega$, $C_{in} = 1$ p (5.12.23)
Other components: $C_1 = 100$ n, $C_2 = 10$ p, $C_3 = 1$ p, $C_4 = 1$ p, $R_3 = 1$ GΩ

which gives $R_2 = 614\ \Omega$ from Eq. (5.12.22) and $R_1 = 9386\ \Omega$ from just above Eq. (5.12.17).

The transform is now of the form:

$$V_o = \frac{-K}{s(s+\alpha)^2}, \quad \text{with} \quad K = \frac{aA_0C_4}{TC_U} \quad \text{and} \quad \alpha^2 = \frac{A_0}{TC_UR_3} \qquad (5.12.24)$$

The inverse transform gives a time function (Table 1.12.1, No. 21; Holbrook 1966):

$$v_o = K\left(\frac{1 - e^{-\alpha t} - \alpha t e^{-\alpha t}}{\alpha^2}\right) \qquad (5.12.25)$$

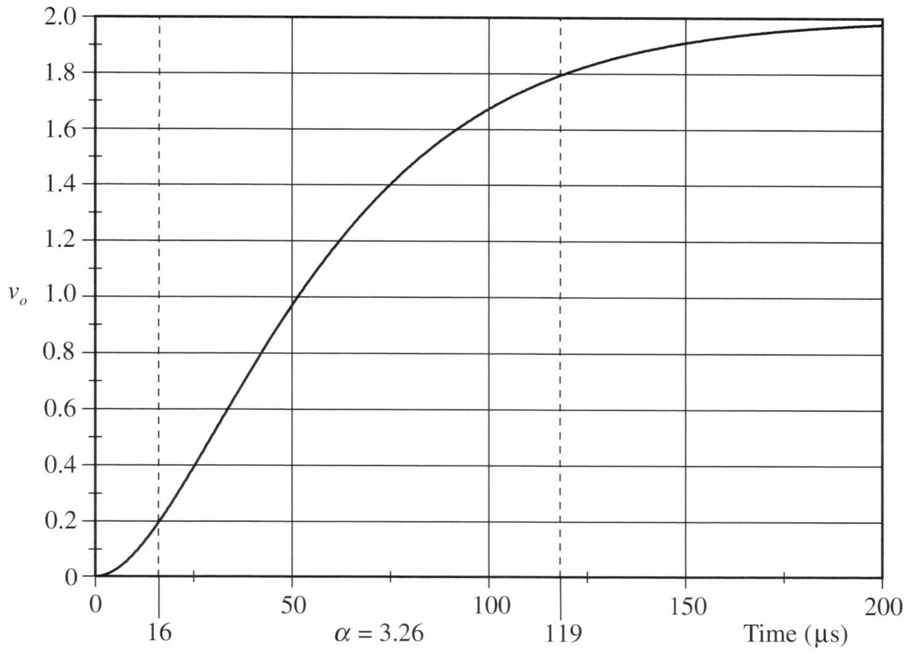


**Fig. 5.12.7** Output $v_o$ given by Eq. (5.12.25).

For a ramp equivalent to one-half of a 1 V peak-to-peak 1 kHz triangular wave $a=2\times10^3$ V s$^{-1}$ (see above Eq. (5.12.20)) and the final output amplitude as $t\to\infty$ is from (5.12.25) and (5.12.24) (the exponential dominates $\alpha t$ after a short time):

$$v_o=\frac{K}{\alpha^2}=aC_4R_3=2\text{ V peak} \tag{5.12.26}$$

and since $R_3=1$ GΩ this corresponds to a value of $i=2\times10^{-9}$ A peak. The calculated waveform of $v_o$ is shown in Fig. 5.12.7 and indicates a 10 to 90% risetime of $\tau_r=103$ μs. The value of $\alpha=3.26\times10^4$ s$^{-1}$ may be determined from Eq. (5.12.24) and since the shape of the response Eq. (5.12.25) is fixed by $\alpha$ we can derive a general relation for the risetime $\tau_r$ from these results:

$$\alpha\tau_r=3.26\times10^4\times103\times10^{-6}=3.36,\quad\text{say}\quad 3.4$$

so that $$\tau_r=3.4\left(\frac{C_UR_3}{\omega_T}\right)^{\frac{1}{2}} \tag{5.12.27}$$

For a triangle wave the *change* of slope is twice that for a simple ramp so the change of current is 4 nA and the output voltage will be 4 V peak-to-peak.

The PSpice simulation of the circuit is shown in Fig. 5.12.8, which shows good agreement with the above analysis indicating that the approximations made are valid. The advantage of PSpice in this application is that one can call for perfect

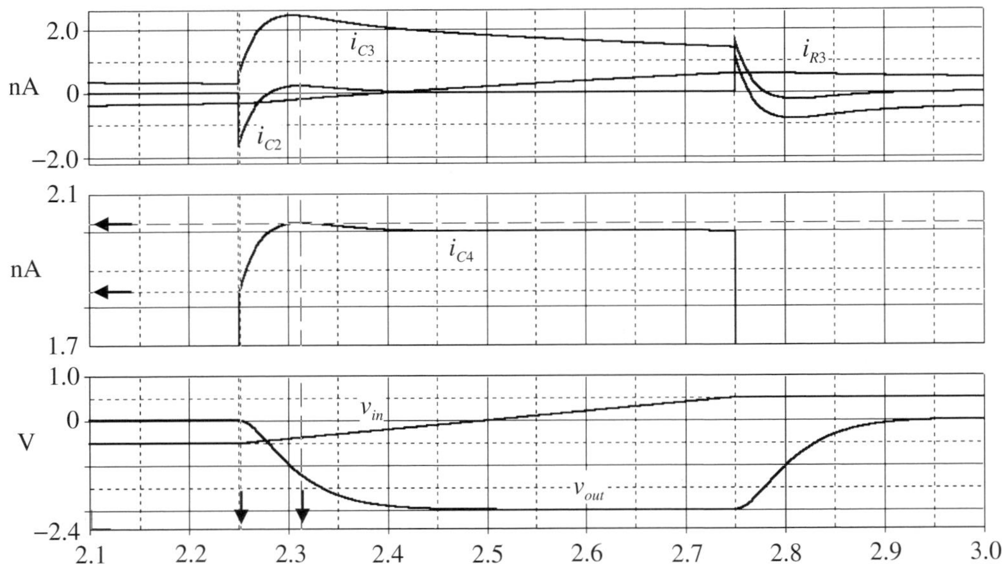


Cursor on $i_{C4}$ at 2.251 ms/1.84 nA Cursor on $i_{C4}$ at 2.312 ms/2.023 nA Time (ms)

**Fig. 5.12.8** PSpice simulation of circuit of Fig. 5.12.5 with values as given in Eq. (5.12.23).

generators, triangular waves with perfect corners and can readily determine currents and voltages that may be difficult to measure in practice.

The input current waveform $i$ under conditions for critical damping and at high frequency may be determined from:

$$\begin{aligned} I &= sC_4(V_{in} - V_1), \quad \text{with} \quad V_o = \frac{-V_{in} s A_0 C_4}{TC_U(s+\alpha)^2} \quad \text{and} \quad V_1 = \frac{-V_o}{A} \\ &= sC_4 V_{in} - \frac{sC_4}{A}\left[\frac{V_{in} s A_0 C_4}{TC_U(s+\alpha)^2}\right] \\ &= sC_4 V_{in}\left[1 - \frac{sA_0C_4}{TC_U(s+\alpha)^2}\frac{sT}{A_0}\right] \\ &= sC_4 V_{in}\left[1 - \frac{s^2 C_4}{C_U(s+\alpha)^2}\right] \end{aligned} \tag{5.12.28}$$

and for a ramp input $V_{in} = a/s^2$ as before:

$$I = \frac{aC_4}{s} - \frac{asC_4^2}{C_U(s+\alpha)^2} \tag{5.12.29}$$

The time function is thus (Table 1.12.1, Nos 3 and 14):

$$i = aC_4 u(t) - \left(\frac{aC_4^2}{C_U}\right)e^{-\alpha t}(1-\alpha t) \tag{5.12.30}$$

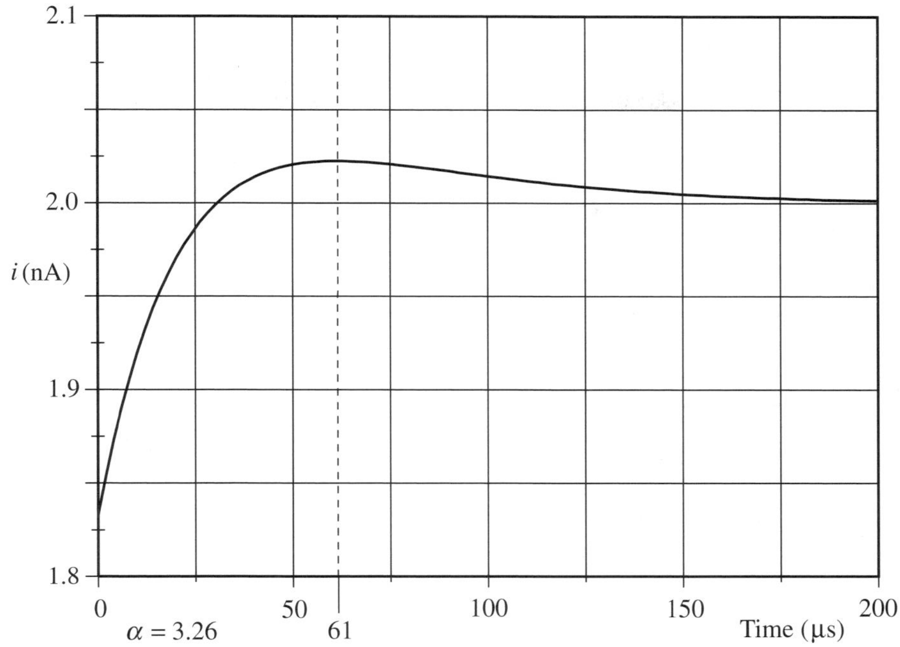


**Fig. 5.12.9** Current input pulse. Most of the initial step to $i=1.833$ nA is omitted for resolution of the overshoot.

where $u(t)$ is the unit step function and for our standard component values:

$$aC_4=2\times10^{-9}\text{ A}\quad\text{and}\quad\frac{aC_4^2}{C_U}=0.167\times10^{-9}\text{ A}\tag{5.12.31}$$

with initial step at $t=0$ of $i_0=2\times10^{-9}-0.167\times10^{-9}=1.833\times10^{-9}$

To match the ramp input $v_{in}=at$ assumed above, the simulation has been carried out using a trapezoidal input rather than a triangular wave. The form of $i$ is shown in Fig. 5.12.9. It is seen that the 'square wave' is slightly distorted and has an overshoot with a peak at a time $t=2/\alpha=61$ μs after the edge. The peak is found by differentiating (5.12.29) and putting the result equal to zero to find the maximum. The form of the $i$ response is just what would be expected for a system with limited high frequency response (Oliver 1961).

It is also easy to examine the effects of deviations from the ideal circuit that would be very difficult to do by normal analysis. Some conclusions are:

(a) Deviation of $\pm10\%$ in the matching of $C_1(R_1+R_2)$ and $C_3R_3$ has minor effect.

(b) Variation in the values chosen for $C_1$ (and hence for $R_1$ to maintain matching) from 1 μF to 10 nF (at least) has negligible effect. Recall, however, that the source impedance of the divider was assumed small relative to the load $Z_3$.

(c) The source resistance of $v_{in}$ over the range 0 to 1 k at least has little effect, i.e. for typical signal generators.

(d) The load on the amplifier should be kept small to avoid excessive power dissipation and hence increased bias current and drift.

(e) Components must be very carefully placed to minimize coupling and capacity, and the feedback resistor must be rigidly fixed in position – minute movement produces changes in the output signal.

## Feedback resistor model

The assumption has implicitly been made that the feedback resistor $R_3$ can be represented as a simple resistor with a parallel capacitor $C_3$. More complex representations have been considered (Kendall and Zabielski 1970) which treat it rather like a transmission line. It would be expected that at higher frequencies the feedback current would flow through $C_3$ rather than $R_3$($Z(C_3) = R_3$ at 167 Hz for our standard values). $C_3$ must of course include the stray amplifier feedback capacity as well as any due to $R_3$. Such small capacities are difficult to measure in practical circumstances. Simulation results demonstrate that at a frequency of 100 Hz, $i(R_3)$ only just reaches the input current. This would suggest that any model of $R_3$ as a sequence of low-pass *RC* sections would be largely immaterial in determining the high frequency response. Simulations have been run with up to 10 *RC* sections which demonstrate that this view is essentially correct. It is difficult to know what values to ascribe to $C_3$ or to the distributed capacity from the bodylength of $R_3$ to common. If we can treat $R_3$ as a conductor above a ground plane, then we may use the formula for microstrip transmission lines (Millman and Taub 1965, p. 85). For a diameter $d = 5$ mm and a height $h = 5$ mm above the plane ($\varepsilon = \varepsilon_0\,\varepsilon_r$, where $\varepsilon_0 = (36\pi \times 10^9)^{-1}$ F m$^{-1}$ and $\varepsilon_r = 1$ for air):

$$C = \frac{2\pi\varepsilon}{\ln\left(\dfrac{4h}{d}\right)} = 40\ \text{pF m}^{-1} \tag{5.12.32}$$

and for a resistor of length 3 cm we get a capacity of about 1 pF.

If $C_3$ *is* the dominant part of the feedback impedance $Z_3$ then we must determine which factor sets the overall bandwidth of the system. The expression for the risetime $\tau_r$ (Eq. (5.12.27)) shows that the ratio $C_U/\omega_T$ is the arbiter. Table 5.12.1 shows the relevant characteristics of a range of amplifiers with the requisite low bias current. The OPA128 has been used for simulation as an excellent representative of the lowest bias current types. If a 1 pA bias current is acceptable then the OPA637 appears to be a good choice. Note that the OPA637 model does not include an input capacitor and the OPA637E model includes both an input capacitor and two common mode capacitors, all of 1 pF (see Biagi et al. 1995).We have used the OPA637 model and included $C_{in}$ in $C_2$.

Table 5.12.1 *Typical parameters for some electrometer type amplifiers*[a]

| Amplifier | $i_{bias}$ (pA) | $A_0$ ($\times 10^6$) | $f_T$ (MHz) | $Z_{in}$ (Ω, pF) | $f_c$ (Hz) | $\tau_r$ (μs) |
|---|---|---|---|---|---|---|
| OPA128 | 0.06 | 2.5 | 2.5 | $10^{13}$, 1 | 1 | 103 |
| OPA637 | 1 | 0.91 | 40 | $10^{13}$, 8 | 40 | 21 |
| OPA671 | 5 | 0.01 | 35 | $10^{12}$, 3.5 | 3500 | 23 |
| AD515A | 0.075 | 0.1 | 1 | $10^{13}$, 1.6 | 10 | 133 |
| AD549 | 0.06 | 1 | 1 | $10^{13}$, 1 | 1 | 133 |
| LMC6001 | 0.025 | 5.6 | 1.3 | $>10^{12}$ | 0.23 | 116 |

[a] The numbers should be taken as approximate and fuller information from the data sheets should be consulted. Data for the OPA128 is from the datasheet rather than the SPICE model as used in Eq. (5.12.23). Values for $\tau_r$ are for $C_U = 12$ p and $R_3 = 1$ GΩ.

It is evident from the open-loop responses of all the amplifiers listed that there is a dominant pole, responsible for the corner frequency $f_C$ which results in a cut-off slope of $-6$ dB/oc to above the transition frequency $f_T$ so easing the achievement of stability against oscillation. Such dominant poles are commonly deliberately introduced in the design of the device by adding capacity at an appropriate place. In the present application the matter of stability is very strongly influenced by the external circuit to an extent that it would seem well worth examining the use of uncompensated versions giving much wider bandwidth even at the cost of much greater amplifier phase shift. (The OPA637 model appears to have a high frequency pole in the region of 12 MHz which makes the transition frequency lower than expected.) A simulation using the OPA637E gives a bandwidth of about 23 kHz and a risetime of 15 μs. The value of $R_2 = 97$ Ω with other values as before. The calculated value $\alpha = 20.53 \times 10^4$ which gives from Eq. (5.12.27) a value of 16.5 μs using the values from Table 5.12.1. Decreasing $C_1$ to 10 nF and hence increasing $R_1 + R_2$ to 100 k, with $R_2 = 974$ Ω, gives the same result.

## The effect of source resistance

The analysis has made the assumption of infinite source resistance, i.e. a perfect current source. In practice the source resistance will vary over a wide range so it is necessary to investigate the limits over which the analysis is valid within acceptable error. As noted before the equations become too complex to solve analytically for the full circuit. We have therefore made use of the power of SPICE to determine the expected response as a function of source resistance relative to the feedback resistance. Using the 'standard compensated circuit' as we have done previously the conclusion is that for source resistances of 1% of the feedback resistance (e.g.

(a)

(b)

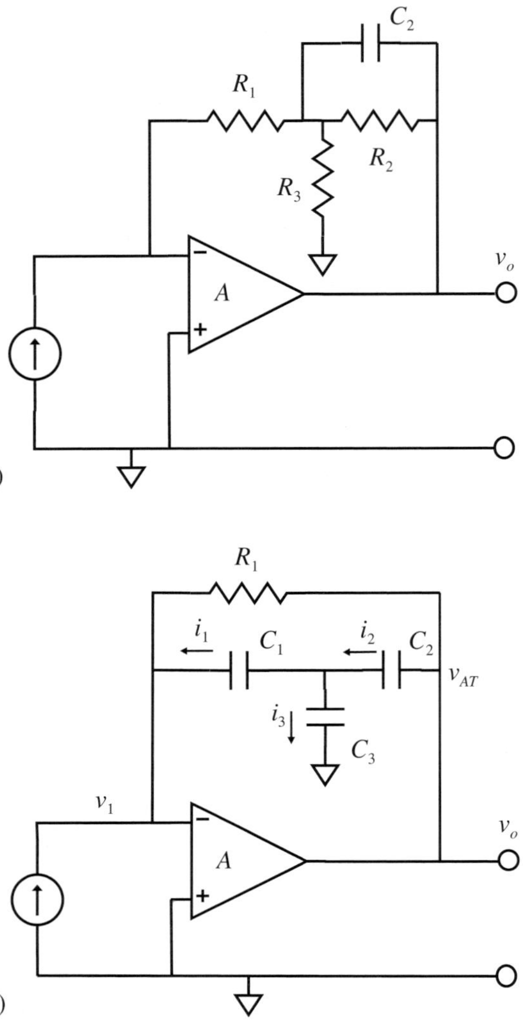


**Fig. 5.12.10** (*a*) Tee feedback resistor arrangement. (*b*) Tee feedback capacitor arrangement.

10 MΩ for our standard value of $10^9$ Ω) an effect is just noticeable. Here the current through the source resistance is about equal to that through $C_2$.

## Tee feedback networks

Difficulties with very high value resistors, particularly with regard to precision and stability, have prompted the use of Tee feedback networks to achieve a high effective resistance using much lower value resistors. The general arrangement is shown in Fig. 5.12.10(*a*) including the normally used compensation capacitor $C_2$.

When using a single feedback resistor it is sometimes necessary to use a variable sub 1 p compensation capacitor, which is difficult to obtain, and again a Tee capacitor network may be used instead as shown in Fig. 5.12.10(*b*) (Burr-Brown 1995; see Figure 4*a*). The requirement now is to determine the relationship between the Tee capacitors and the equivalent single capacitor, say $C_E$. The sums will be found in Section 3.8.

The compensation of the Tee resistor network, as shown in Fig. 5.12.10, is also wanting in terms of the effect on the risetime. Since the Tee acts to attenuate the output signal before feeding this back via $R_1$ then it would seem appropriate to make the $R_2$ and $R_3$ attenuator frequency independent (Section 5.2) by placing a capacitor $C_3$ across $R_3$ as well, with $R_2/R_3 = C_3/C_2$. A simulation test will confirm the effectiveness of this. See also Section 3.8.

Noise characteristics are of increasing importance as the bandwidth is increased. This is considered by Praglin and Nichols (1960), by Cath and Peabody (1971) (some of the noise spectra for the latter are in part in error; Hamilton 1977) and Burr-Brown (1994).

## SPICE simulation circuits

| | |
|---|---|
| Fig. 5.12.3(*a*) | Pioatstb.SCH |
| Fig. 5.12.3(*b*) | Pioatstb.SCH |
| Fig. 5.12.8 | PioatstA.SCH |

## References and additional sources 5.12


Barker R. W. J. (1979): Modern electrometer techniques. *Proc. IEE* **126**, 1053–1068.

Biagi H., Stitt R. M., Baker B., Baier S. (1995): *Burr-Brown SPICE Based Macromodels, Rev.F*, Burr-Brown Application Bulletin AB-020F, (see Figs. A4 and B2).

Briano, Bob (2000): Improved photodiode pre-amp uses current-feedback amplifier. *Electronic Design* 10 July, 131.

Burr-Brown (1994): *Noise Analysis of FET Transimpedance Amplifiers*, Burr-Brown Application Bulletin AB-076 (In Applications Handbook LI459).

Burr-Brown (1995): *Photodiode Monitoring with Op Amps*, Burr-Brown Application Note AB-075, January (In Applications Handbook LI459).

Cath P. G., Peabody A. M. (1971): High-speed current measurements. *Analytical Chem.* **43**, September, 91A–99A.

Graeme J. G. (1996): *Photodiode Amplifiers: Op Amp Solutions*, New York: McGraw-Hill. ISBN 0-07-024247-X.

Hamilton T. D. S. (1977): *Handbook of Linear Integrated Electronics for Research*, London: McGraw-Hill. ISBN 0-07-084483-6.

Holbrook J. G. (1966): *Laplace Transforms for the Electronic Engineer*, 2nd Edn, Oxford: Pergamon Press. Library of Congress Cat. No. 59-12607.

Iida T., Kokubo Y., Sumita K., Wakayama N., Yamagishi H. (1978): A fast response electrometer for pulse reactor experiments. *Nuc. Instrum. Meth.* **152**, 559–564.

Jodogne J. C., Malcorps H. (1982): Ultra-low current measurements using operational amplifiers. *Int. J. Electronics* **52**, 363–368.

Keithley Instruments (1972): *Electrometer Measurements*, Keithley Instruments Handbook.

Keithley Instruments (1984): *Low Current Measurements*, Keithley Application Note No. 100.

Kendall B. R. F., Reiter R. F. (1974): Three-terminal shielded resistors for fast electrometers. *Rev. Sci. Instrum.* **45**, 850–852.

Kendall B. R. F., Zabielski M. F. (1970): Compensated resistors for high frequency electrometer applications. *Electronics Lett.* **6**, 776–778.

Millman J., Taub H. (1965): *Pulse, Digital, and Switching Waveforms*, New York; McGraw-Hill. Library of Congress Cat. Card No. 64-66293.

Oliver B. M. (1961): *Square Wave Testing of Linear Systems*, Hewlett-Packard Application Note 17.

Pease R. A. (1982): Picoammeter/calibrator system eases low-current measurements. *EDN* **27** (7), 143–149.

Pease R. A. (1995): What's all this designer stuff, anyhow? *Electronic Design* 9 January. Reprinted with permission of Electronic Design Magazine Penton Media Publishing.

Pelchowitch I., Zaalberg van Zelst J. J. (1952): A wide-band electrometer amplifier. *Rev. Sci. Instrum.* **23**, 73–75.

Praglin J., Nichols W. A. (1960): High-speed electrometers for rocket and satellite experiments. *Proc. IRE* **48**, 771–779.

Welwyn Components (2000): High resistance chips to 50 GΩ (0503 to 1206 sizes). www.welwyn-tt.co.uk

Wing W. H., Sanders T. M. (1967): FET operational amplifiers as fast electrometers. *Rev. Sci. Instrum.* **38**, 1341–1342.

# 5.13 Three-pole, single amplifier filter

A more important set of instruction books will never be found by human beings. When finally interpreted the genetic messages encoded within our DNA will provide the ultimate answers to the chemical underpinnings of human existence.

James Watson

We consider here the design of a three-pole filter using only a single amplifier. This would require us to solve third order equations which makes the calculation somewhat complex to carry out by hand, but the use of mathematical software packages now makes this relatively simple (Brokaw 1970: Rutschow 1998). Most active filters of this form use unity, or very low, overall gain. This circuit is somewhat unusual in that we have found it to be usable at very high gain; the matter of allowable gain will be discussed later. The circuit is shown in Fig. 5.13.1.

The general transfer function for a three-pole low-pass filter is given by:

$$\frac{V_{out}}{V_{in}} = \frac{A_0}{A_3 s^3 + A_2 s^2 + A_1 s + A_0} \tag{5.13.1}$$

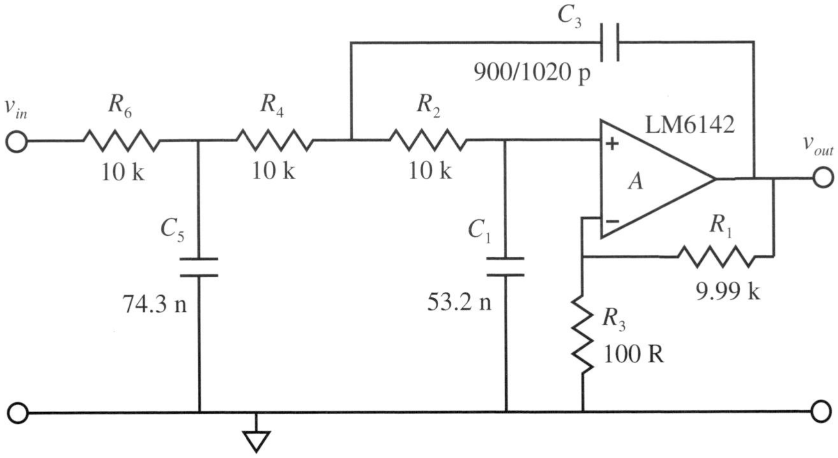


**Fig. 5.13.1** Three-pole single amplifier low-pass filter.

Table 5.13.1 *Three-pole single amplifier filter parameter values for various responses (after Rutschow 1998)*

| Filter type | Features | $A_1$ | $A_2$ | $A_3$ | Corner attenuation (dB) |
|---|---|---|---|---|---|
| Butterworth | Maximally flat passband | 2 | 2 | 1 | 3 |
| Chebyshev −1 dB ripple | Equal 1 dB passband ripples and very rapid cut-off | 2.52071 | 2.01164 | 2.03537 | 1 |
| −3 dB ripple | Equal 3 dB passband ripples and very rapid cut-off | 3.70466 | 2.38334 | 3.99058 | 3 |
| Optimal (Papoulis) | Rapid cut-off and monotonic in passband | 2.35529 | 2.27036 | 1.7331 | 3 |
| Bessel (Thomson) | Approximates Gaussian response. Minimizes phase delay distortion | 1 | 0.4 | 0.06667 | 0.84 |
| Paynter | Excellent time domain response. Minimal overshoot | 3.2 | 4 | 3.2 | 10.4 |

The gain can be normalized to unity by dividing through by $A_0$. A difficulty in determining the relationships between the components arises because there are only three equations for six unknowns. For a corner frequency of $f_0 = \omega_0 / 2\pi$ and gain $K$ these are:

$$A_1 = [(1-K)(R_6+R_4)C_3 + R_6C_5 + (R_2+R_4+R_6)C_1]\omega_0$$
$$A_2 = [(1-K)R_6R_4C_5C_3 + R_6(R_2+R_4)C_1C_5 + R_2(R_4+R_6)C_1C_3]\omega_0^2$$
$$A_3 = R_6R_4R_2C_5C_3C_1\omega_0^3$$
$$K = 1 + \frac{R_1}{R_3} \qquad (5.13.2)$$

Various types of response can be obtained according to the relationships between the $A$'s as shown in Table 5.13.1 (the $A$ values are normalized to $A_0 = 1$; Kuo 1966). To find appropriate values it is necessary to select three and then to seek the best values for the other three. This can be carried out with the aid of Mathcad using the *Given* and *Find* commands, with the capacitors as variables, setting say the three resistor values (set them to the same value to start: as in most circumstances in electronics, if in doubt use 10 k! – as with the three bears, it is not too big, it is not too small and may be just right) and choosing the appropriate values of the $A$'s for the desired filter type. Some nominal capacitor values also need to be entered to provide a starting point for the search. This gives the calculated capacitor values. The nearest standard values for the capacitors are then set and the resistor values determined instead in the same way. The nearest standard values are then used. To show the frequency response the circuit is transferred to PSpice to see if it is acceptable. It should be noted that, as with all circuits in which positive and negative feedback are counterbalanced, the response is somewhat sensitive to gain $K$. It is difficult to provide an expression for the limit since all the components themselves depend on the value of $K$ chosen. Mathematically, the simplest transfer function of the filter is obtained when all the $R$'s and all the $C$'s are equal. Equation (5.13.1) then becomes:

$$\frac{V_{out}}{V_{in}}=\frac{1}{R^3C^3s^3+R^2C^2(5-K)s^2+2RC(3-K)s+1},\quad \text{for}\quad A_0=1 \tag{5.13.3}$$

and we may examine the stability using the Routh rules given in Section 1.12. The values of the three variables are (dividing through by the coefficient of $s^3$):

$$\alpha=\frac{(5-K)}{RC},\qquad \beta=\frac{2(3-K)}{R^2C^2},\qquad \gamma=\frac{1}{R^3C^3} \tag{5.13.4}$$

so $K$ must be less than 3 otherwise $\beta$ will change sign and by Rule 1 the system will be unstable. If $K<3$ so that the first part of Rule 3 is fulfilled, then applying the second part for the *limit of equality*:

$$\beta=\frac{\gamma}{\alpha}\quad \text{gives}\quad \frac{2(3-K)}{R^2C^2}=\frac{1}{R^3C^3}\frac{RC}{(5-K)} \tag{5.13.5}$$

$$\text{or}\quad 2K^2-16K+29=0\quad \text{so}\quad K=5.225\quad \text{or}\quad 2.775$$

using the standard formula for quadratics (Section 1.10). The higher value has already been eliminated and thus the maximum value for $K$ is 2.775 rather than 3. This can now be examined using PSpice. The frequency response will show a large peak around the corner frequency and a transient run, using a short pulse input to nudge the system, will show oscillations. It is found that the $K$ limit is accurate; a gain of 2.8 gives growing oscillations, while a gain of 2.75 gives decaying oscillation. A gain of 2.775 gives effectively constant oscillation. It is instructive to apply the $T$ technique (Section 5.14) by applying a voltage generator at the (+) input of the amplifier (this will mean that the open-loop response will follow $T_v$). If you examine the loop gain at zero phase shift you will find it very close to unity. You should also note the change of phase of the signal at (+).

This limit of $K\approx 3$ is not as restrictive as it appears. The result arises from the arbitrary choice of the components without consideration of the resulting response, but it at least reminds us that the system is capable of oscillation. The Butterworth configuration has been investigated as a function of $K$ up to 1000. It is possible to achieve stability even at this very high gain but the system is now very sensitive to the value of $C_3$ and the frequency response is considerably rounded (suitable values to try for a LM6142 amplifier with 1 kHz cut-off are: $R_2 = R_4 = R_6 =$ 10 k, $C_1 = 118.9$ n, $C_3 = 235$ p, $C_5 = 101.5$ n with the feedback resistors 9.99 k and 10R). Increasing $C_3$ to 240 p results in oscillation while reduction to 200 p gives better transient response. What has not been allowed for in the design equations is the amplifier open-loop response which will become more significant as gain or cut-off frequency increases. It is somewhat astonishing that such high gains are feasible but you need to investigate the performance carefully. In carrying out the calculations it is probably best to start with low gain and increase the gain in limited steps, feeding back the new values as better approximations for the next step to avoid convergence failure.

As an example Fig. 5.13.2(*a*) shows the Butterworth frequency response for a gain of one hundred and a design corner of 1 kHz for two values of $C_3$. The input signal was 1 mV.

The cursors (set two decades apart) indicate a slope of about 60 dB/dec as expected for a third order response. If the frequency response appears satisfactory then the system should also be tested with a pulse input to examine the transient response. Figure 5.13.2(*b*) shows the response for a 1 mV, 10 μs risetime pulse with the same two values of $C_3$; the decrease in overshoot is at the expense of risetime. For better pulse response the Paynter configuration gives a fast risetime with only a small overshoot. Increasing $C_1$ will remove the overshoot at the expense of increased risetime. Even if the gain is low it is as well to use an amplifier with a good gain–bandwidth product so that the amplifier roll-off does not significantly affect the response.

For intermediate response between Butterworth and Bessel (Thomson) reference may be made to Al-Nasser (1972), Melsheimer (1967) or Van Valkenburg (1982). These provide pole locations for variation between Butterworth ($m = 0$) and Thomson ($m = 1$) but some sums are required to derive the equivalent $A$ coefficients. If the real pole is $|a|$ and the complex poles are $|b \pm jc|$ (i.e. ignore the minus signs in the tables), then the denominator polynomial is given by:

$$s^3 + s^2(a + 2b) + s(2ab + b^2 + c^2) + a(b^2 + c^2)$$

$$\text{so} \quad A_3 = \frac{1}{a(b^2 + c^2)}, \quad A_2 = \frac{(a + 2b)}{a(b^2 + c^2)}, \quad A_1 = \frac{(2ab + b^2 + c^2)}{a(b^2 + c^2)}, \quad A_0 = 1 \qquad (5.13.6)$$

The references also give some advice on the *GB* product required for the amplifier. Papoulis (1958) deals with the optimal filter.

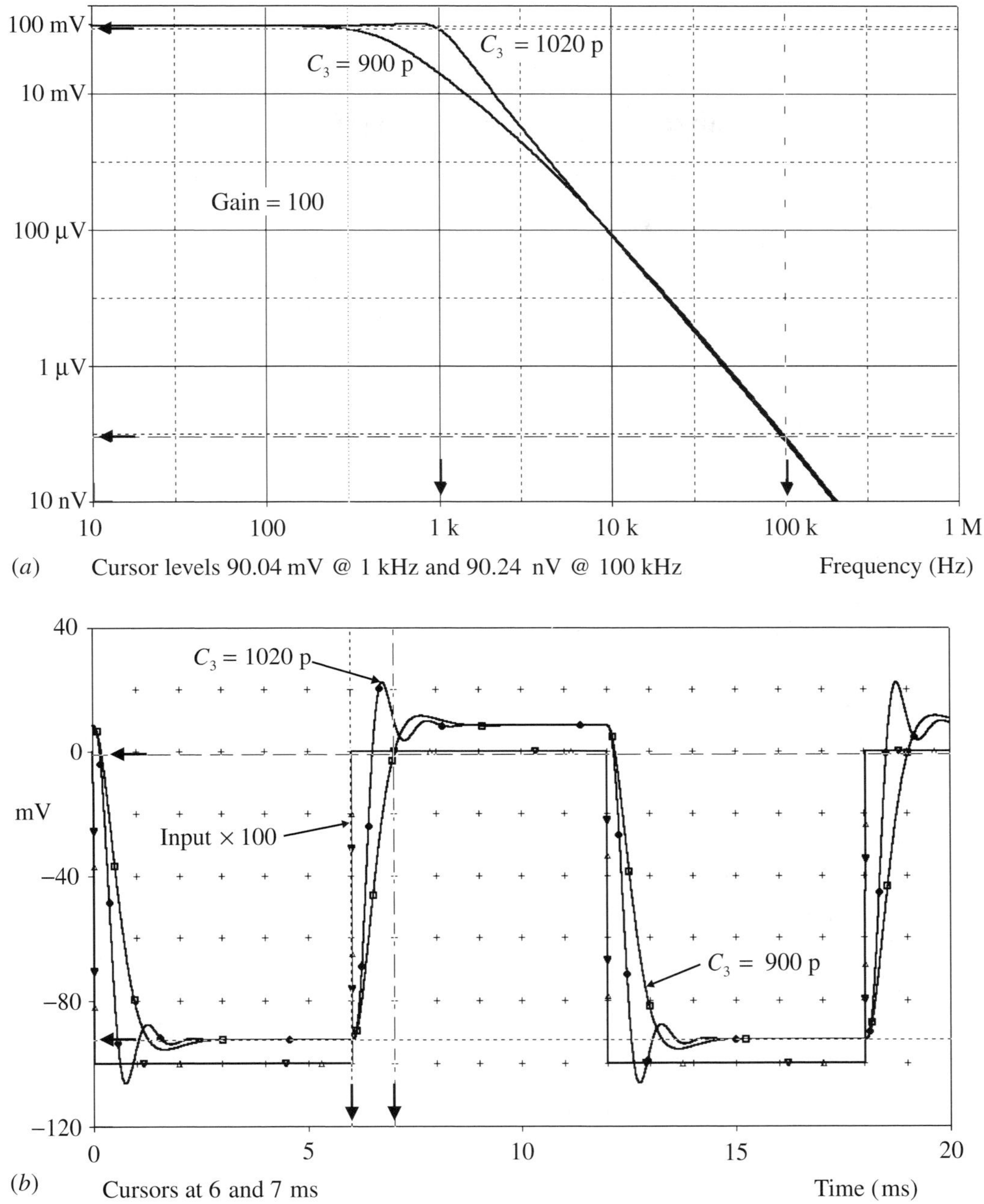


**Fig. 5.13.2** (*a*) Frequency response of circuit of Fig. 5.13.1 with values as shown. (*b*) Transient response for the circuit (the outputs have been offset for clarity).

## SPICE simulation circuits

| | |
|---|---|
| Fig. 5.13.2(*a*) | 3psalpf5.SCH Equal *R*'s and equal *C*'s |
| Fig. 5.13.2(*a*) | 3psalpf7.SCH Gain = 100, values as 5.13.1 |
| Fig. 5.13.2(*b*) | 3psalpf6.SCH Gain = 100, pulse response |
| Paynter response | 3psalpf4.SCH |

## References and additional sources 5.13


Al-Nasser F. (1972): Tables shorten design time for active filters. *Electronics* 23 October, 113–118.

Anderson B. D. O., Moore J. B. (1979): *Optimal Filtering*, Englewood Cliffs: Prentice Hall.

Brokaw A. P. (1970): Simplify 3-pole active filter design. *EDN* 15 December, 23–28.

Budak A. (1965): A maximally flat phase and controllable magnitude approximation. *Trans. IEEE* **CT-12** (2), June, 279.

Hansen P. D. (1963): New approaches to the design of active filters. *The Lightning Empiricist* **13** (1 and 2), January–July. Philbrick Researches Inc. Part II in **13** (3 and 4), July–October 1965. (Discussion of Paynter filters.)

Karatzas T. (1997): Quick and practical design of a high-pass third-order Bessel filter. *Electronic Design* 3 November, 211–212.

Kuo F. F. (1966): *Network Analysis and Synthesis*, New York: John Wiley, ISBN 0-471-51118-8. See p. 385.

Lacanette K. (1991): *A Basic Introduction to Filters – Active, Passive, and Switched-Capacitor*, National Semiconductor Application Note AN-779, April.

Melsheimer R. S. (1967): If you need active filters with flat amplitude and time delay responses, discard the classical approach. A simple method yields the correct circuit quickly. *Electronic Design* **8**, 12 April, 78–82.

Orchard H. J. (1965): The roots of maximally flat delay polynomials. *IEEE Trans.* **CT-12** (3), September, 452–454.

Papoulis A. (1958): Optimum filters with monotonic responses. *Proc. IRE* **46** (3), March, 606–609.

Rutschow C. (1998): Design a 3-pole, single amplifier, LP active filter with gain. *Electronic Design*, 2 November, 154, 156. I am indebted to the author for helpful discussions.

Stout D. F., Kaufman M. (1976): *Handbook of Operational Amplifier Circuit Design*, New York: McGraw-Hill. ISBN 0-07-061797-X. (See p. 10–18; in their expression for $B$ the leading $C_1$ should be $R_3$.)

Thomson W. E. (1952): Network with maximally flat delay. *Wireless Engng* **29**, October, 256–263.

Van Valkenburg M. E. (1982): *Analog Filter Design*, New York: Holt, Rinehart and Winston. ISBN 0-03-059246-1, or 4-8338-0091-3 International Edn.

# 5.14 Open-loop response

I have got the solution, but I do not yet know how I am to arrive at it.

Karl Friedrich Gauss

To understand and examine the response of a feedback system it is necessary to know both the open-loop and the closed-loop response. The basic theory of servo systems has been outlined in Section 3.15. While it is in principle easy to determine the response of the closed-loop system it is usually much more difficult to find that of the open-loop configuration. If the loop is broken at any point then the bias of the system will be affected and the gain will usually be large, both of which make for great difficulty in making any measurements. It would be of great benefit if the open-loop circuit could be determined on the closed-loop circuit, and a way of doing this with PSpice has been presented by Tuinenga (1988). Some further discussion can be found in MicroSim (1993). His approach is somewhat opaque and his result would appear to be slightly wrong, at least as far as PSpice is concerned. The error may appear to be small but it is most significant just where it matters, i.e. the unity-gain crossover point. We will generally follow his approach, filling in many of the intermediate steps, and apply the result to an examination of a very common servo system, a simple voltage regulator in Section 5.10.

The general approach is to inject a current and then a voltage signal at an appropriate point in the closed-loop system and from the two responses the open-loop response can be derived. Recall that the generators we will employ are ideal devices, having infinite and zero impedance, respectively, and the relation between the Thévenin and Norton equivalent circuits (Section 3.1).

To get the general idea of what we are going to do we can first of all look at the application (Fig. 5.14.3(*b*), p. 517) we will later use to demonstrate the operation of the technique. If a current is injected by the current generator *ISIN* then part of this current will flow to the right through $Z_1$ and part to the left through $Z_2$. The current through $Z_1$, which we will call the forward current, feeds a signal around the loop via the buffer $A_2$ and amplifier $A_1$ and the net current through $Z_2$ is the return current. The loop current gain is then the ratio of these two currents. A similar view may be taken with regard to the voltage source as illustrated in Fig. 5.14.3(*a*) to find the loop voltage gain. In both cases the application of the generator does not interfere with the normal operation of the circuit owing to their ideal impedance properties.

When the test circuits of Fig. 5.14.3 were examined using PSpice and the original formula for the loop gain, a good fit was generally found but there were some regions where significant deviations were observed. An initial attempt to derive the result given by Tuinenga led to a relation similar to Eq. (5.14.6) but without the 2 in the denominator. When plotted this gave a closer fit to the directly measured open-loop response but was not quite there. A hunch to change the sign of the 2 gave an excellent fit and prompted a rethink of the analysis. Anything that was tried could always be tested via PSpice to see if things were going the correct way. There is a problem insofar as the interpretation of the signs or senses of the signals, especially since PSpice does not care about the fact that in a negative feedback loop the gain is 'negative' in our normal way of thinking, and just plots its absolute value which is what we will be examining. Taking Tuinenga's result and changing the sign or sense of every '$T$' gave the new relation of Eq. (5.14.6) so it was an indication or warning that we must interpret the signs carefully. It was now a case of 'knowing' the answer and working backwards to see how to get there, a circumstance not unknown to that great mathematician and physicist Gauss as the quotation above suggests. Ultimately, it is PSpice that will be the arbiter and the results presented later will, I hope, prove the matter. I write *prove* here in the sense that Neville Shute used in *No Highway*.

Tuinenga begins his analysis by consideration of the circuits shown in Figs. 5.14.1 and 5.14.2. These present some conceptual difficulties which took some time to elucidate. Considering a particular point in a circuit we can break the loop, inject the test current $i_F$ into the effective impedance $Z$ and represent what may be called the return signal by the controlled current source $i_R$ and add an equivalent impedance $Z$ so that conditions in the open loop are the same as they were in the closed. The position is then as shown in Fig. 5.14.1(*a*).

It is probably instructive to set up the circuits as shown in Fig. 5.14.1 and run simulations to see how they respond. It does give one a better feeling as to what is going on, particularly in the case of current injection. It also serves to check that you have got the sense of the feedback correct so that it is negative rather than positive.

Opening a feedback loop in most circumstances causes considerable difficulty as we may have high gain around the loop and because it may be difficult to maintain the same quiescent conditions. However, it is not necessary to break the loop since we can merge the two sections of (*a*) to get (*b*) which is the same circuit since the two current generators have infinite impedance. In (*b*) the forward current injected is $i_x$ and $i_y$ is the return current, so in this configuration with ideal sources the open-

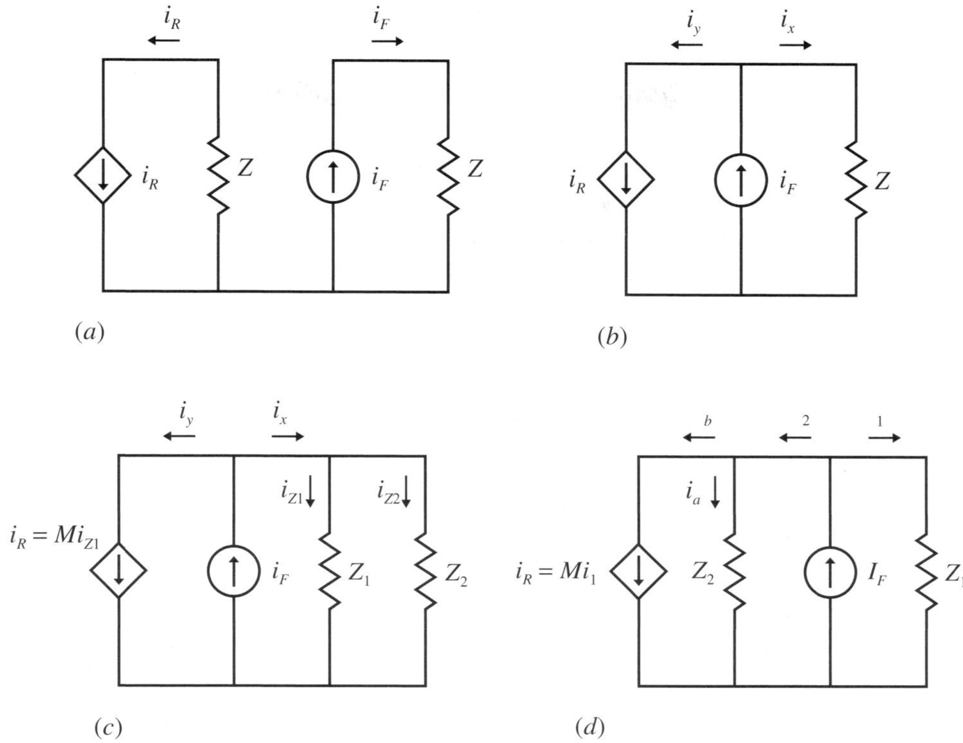


**Fig. 5.14.1** (*a*) Current source signal injection in opened feedback loop. (*b*) Merged circuit with impedance $Z$. (*c*) Impedance $Z$ split into $Z_1$ and $Z_2$, but with the same effective impedance. (*d*) $Z_1$ representing the load impedance and $Z_2$ the source impedance of the controlled current source $i_R$.

loop gain $T$ is given by the current gain $T_i$ (we use the symbol $T$ here to match the use by Tuinenga):

$$T = T_i = \frac{i_R}{i_F} = \frac{i_y}{i_x} \tag{5.14.1}$$

The series of configurations shown in Fig. 5.14.1(*a–d*) represent the change from an injection point where we have a perfect current source to a point where the current source is more realistic with a source impedance $Z_2$. The effective impedance at the injection point is unchanged since $Z$ is equal to $Z_1$ and $Z_2$ in parallel. Since $i_1$ is the forward input, and if $M$ is the rest of the loop gain, then the return current is $Mi_1$, and the two currents $i_1$ and $i_a$ are related by (the return current must be of the opposite sense to the forward current as there is negative feedback):

$$i_a Z_2 = i_1 Z_1 \quad \text{or} \quad i_a = i_1 \frac{Z_1}{Z_2} = i_1 K, \quad \text{where} \quad K \equiv \frac{Z_1}{Z_2}$$

$$\text{so} \quad i_2 = i_a + i_b = i_a + Mi_1 \quad \text{or} \quad i_2 = i_1 K + Mi_1 \tag{5.14.2}$$

$$\text{and}\quad T_i = \frac{-i_2}{i_1} = -(K+M) \quad \text{or} \quad -M = T_i + K \qquad (5.14.2\ \text{cont.})$$

where $T_i$ is what we can actually measure.

We can find the open-loop gain $T$ from Fig. 5.14.1($c$) noting that since the forward impedance is $Z_1$ carrying current $i_{Z1}$, then the controlled current source is $i_y = Mi_{Z1}$:

$$i_{Z1}Z_1 = i_{Z2}Z_2 \quad \text{or} \quad i_{Z2} = i_{Z1}\frac{Z_1}{Z_2} = i_{Z1}K, \quad \text{where} \quad K \equiv \frac{Z_1}{Z_2} \quad \text{as before}$$

$$\text{Now} \quad i_x = i_{Z1} + i_{Z2} = i_{Z1} + i_{Z1}K \quad \text{so} \quad i_{Z1} = \frac{i_x}{1+K} \quad \text{and} \quad i_y = Mi_{Z1} = \frac{Mi_x}{1+K}$$

$$\text{then} \quad T = \frac{-i_y}{i_x} = \frac{-M}{1+K} \quad \text{or} \quad -M = T(1+K) \qquad (5.14.3)$$

and using Eq. (5.14.2) gives

$$T(1+K) = -M = T_i + K \quad \text{or} \quad T = \frac{T_i + K}{(1+K)} \quad \text{or} \quad K = \frac{T_i - T}{T-1}$$

This does not look too promising as we do not know $K = Z_1/Z_2$, but if we follow a similar procedure but inject a voltage, rather than a current, at the same point we can find a second expression for $T$ that will allow us to eliminate the impedance ratio $K = Z_1/Z_2$. We carry out a similar exercise but using a voltage rather than the current source as illustrated in Fig. 5.14.2.

We can now find the open-loop gain $T$ from ($a$) as follows:

$$\frac{v_{Z1}}{v_x} = \frac{Z_1}{Z_1 + Z_2} = \frac{K}{K+1} \quad \text{so} \quad v_{Z1} = \frac{Kv_x}{K+1} \quad \text{and} \quad v_y = -Nv_{Z1} = \frac{-NKv_x}{K+1}$$

$$\text{so} \quad T = \frac{-v_y}{v_x} = \frac{NK}{K+1} \qquad (5.14.4)$$

and from ($b$) we can determine the measured voltage loop gain $T_v$ (remember that the shunt $Z_2$ in Fig. 5.14.1($b$) with a current source is the same impedance $Z_2$ in series with a voltage source in Fig. 5.14.2($b$); see Section 3.1). In this case the controlled voltage source $v_R = -Nv_a$, where we do not have to assume that the loop contribution $N$ is the same as in the current case, so:

$$i = \frac{v_a}{Z_1} \quad \text{and} \quad v_b = -Nv_a + iZ_2 = -Nv_a + v_a(Z_2/Z_1) = -Nv_a + v_a(1/K)$$

$$\text{so} \quad \frac{v_b}{v_a} = -N + (1/K) = -T_v \quad \text{or} \quad N = T_v + (1/K) = \frac{KT_v + 1}{K} \qquad (5.14.5)$$

and from Eq. (5.14.4)

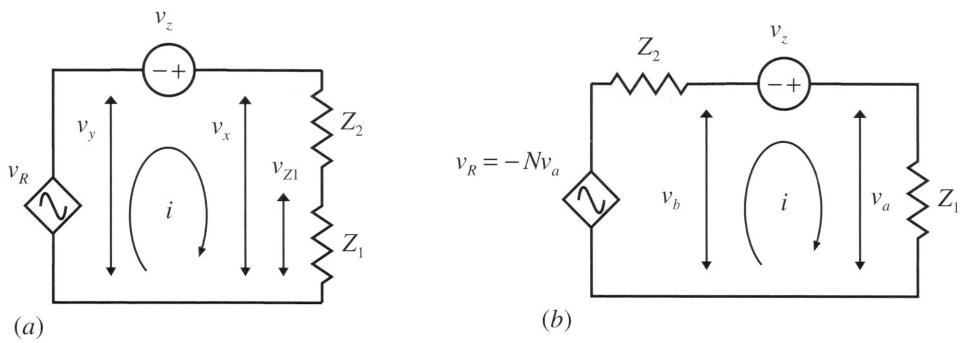


**Fig. 5.14.2** (*a*) Voltage source $v_z$ signal injection inside loop. (*b*) Divided impedances with $Z = Z_1 + Z_2$, with $Z_2$ representing the source impedance of the controlled source $v_R$.

$$T = \frac{NK}{K+1} = \frac{K}{K+1}\left(\frac{KT_v+1}{K}\right) = \left(\frac{KT_v+1}{K+1}\right) \quad \text{or} \quad K = \frac{T-1}{T_v - T} \tag{5.14.5 cont.}$$

From (5.14.3) and (5.14.5) we can eliminate $K$ to get an expression for the loop gain $T$ in terms of the quantities $T_i$ and $T_v$ that we can measure:

$$\left(\frac{T_i - T}{T-1}\right) = K = \left(\frac{T-1}{T_v - T}\right) \quad \text{so} \quad (T-1)(T-1) = (T_i - T)(T_v - T)$$

$$\begin{aligned} &\text{giving} \quad T^2 - 2T + 1 = T_iT_v - TT_v - TT_i + T^2 \\ &\text{so} \quad T(T_v + T_i - 2) = T_iT_v - 1 \\ &\text{and} \quad T = \frac{(T_iT_v - 1)}{(T_v + T_i - 2)} \end{aligned} \tag{5.14.6}$$

so now measurement of $T_i$ and $T_v$ enable determination of $T$. By hand of course this would be a considerable task, but PSpice makes it quite simple. This result differs from that of Tuinenga in that we find –2 in the denominator whereas he finds +2. The simulations we examine below make it evident that the –2 is correct. In fact, if $T \to -T$, $T_i \to -T_i$ and $T_v \to -T_v$ in Tuinenga's equation you will obtain Eq. (5.14.6), so it is really a matter of interpretation of how the equations relate to PSpice. Tuinenga also points out that an equivalent expression to (5.14.6) is given by (we change all his $T$'s to $-T$'s):

$$(T-1) = \frac{1}{\dfrac{1}{(T_i - 1)} + \dfrac{1}{(T_v - 1)}} \tag{5.14.7}$$

as you can show by multiplying this out – it is as if $(T_i - 1)$ and $(T_v - 1)$ were in 'parallel'. This means that if one of these is much smaller than the other then the smaller will dominate in $T$ and give the effective open-loop response. This suggests that you should try to 'break' the loop where the return side is a relatively ideal current or voltage source. In any case, SPICE *PROBE* macros can be used to present Eq. (5.14.6) in the general case (MicroSim 1993). Suitable macros are (we have written $T$ in terms of a general function $X$ in case you wish to use it elsewhere):

$$\begin{aligned} &TV=(V(VY)/V(VX)), \quad TI=(I(Z2)/I(Z1)) \\ &X(A,B)=(((A)^*(B)-1)/((A)+(B)-2)), \quad T=(X(TV,TI)) \end{aligned} \tag{5.14.8}$$

To illustrate the operation of this technique we will first examine the example given by Tuinenga as shown in Fig. 5.14.3: (*a*) shows voltage injection, (*b*) shows current injection, (*c*) shows direct open-loop determination, and (*d*) illustrates how in this case we can determine the closed-loop response and by means of $C_3$ compensate the loop for stability. We will substitute an actual operational model, the LM6142AN, and include a few other manipulations that might be useful elsewhere.

The open-loop gain of the LM6142 (about 110 dB) is slightly higher than that of the simple amplifier in the original circuit (100 dB) so some feedback has been added to adjust the gain; since there is only a small amount of feedback the resistors are not quite as one would expect. Since the amplifier will now have a low input impedance which will severely load the potential divider, a unity-gain high input impedance buffer has been inserted. The bandwidth of this buffer is at least 10 MHz so that there will be negligible interaction. Since we will also measure the open-loop gain directly as a comparison (Fig. 5.14.3(*c*)) there is a considerable problem with such high gain in setting the d.c. levels. This is overcome by introducing the capacitor $C_2$ in series with the 10 Ω resistor which means that the amplifier gain is only unity at z.f. so the bias is no longer a problem. If the capacitor is large enough then the deviation at the frequencies we examine is negligible. The response will of course fall off at some very low frequency but we know that in practice the response will just stay level – in SPICE there is no difficulty in finding a zero leakage capacitor of 1 F! The divider loading in the open-loop circuit (*c*) is simply overcome by making the amplifier non-inverting so it has high input impedance. There is no feedback to worry about and the gain is only changed from $10^5$ to $10^5+1$. The three separate circuits serve to provide (*a*) $T_v$, (*b*) $T_i$ and, in effect, (*c*) $VOL=T$ directly. In (*b*) the resistor $Z_2$ is inserted to allow the measurement of $I_y=I(Z2)$, and $I_x$ is given by $I(Z1)$. Including $Z_2$ in (*a*) as well, allows adjustment of the ratio $K=Z_1/Z_2$ to see how the responses change. The responses are shown in Fig. 5.14.4. The results shown are for the ratios (*a*) 10/9990, (*b*) 9990/10 and (*c*) 5000/5000 so the total $Z=10$ k in all cases.

Figure 5.14.4(*a*) shows the results for the case of $K=Z_1/Z_2 \ll 1$, and so $T_i$ should be a good fit to $VOL$ (we actually plot $VOL/VIN$ to normalize), as it is, though it deviates somewhat at high frequency, and $T_v$ a poor fit. However, $T$ is seen to match $VOL$ very well over the whole range as we hoped. Included in this plot is the curve

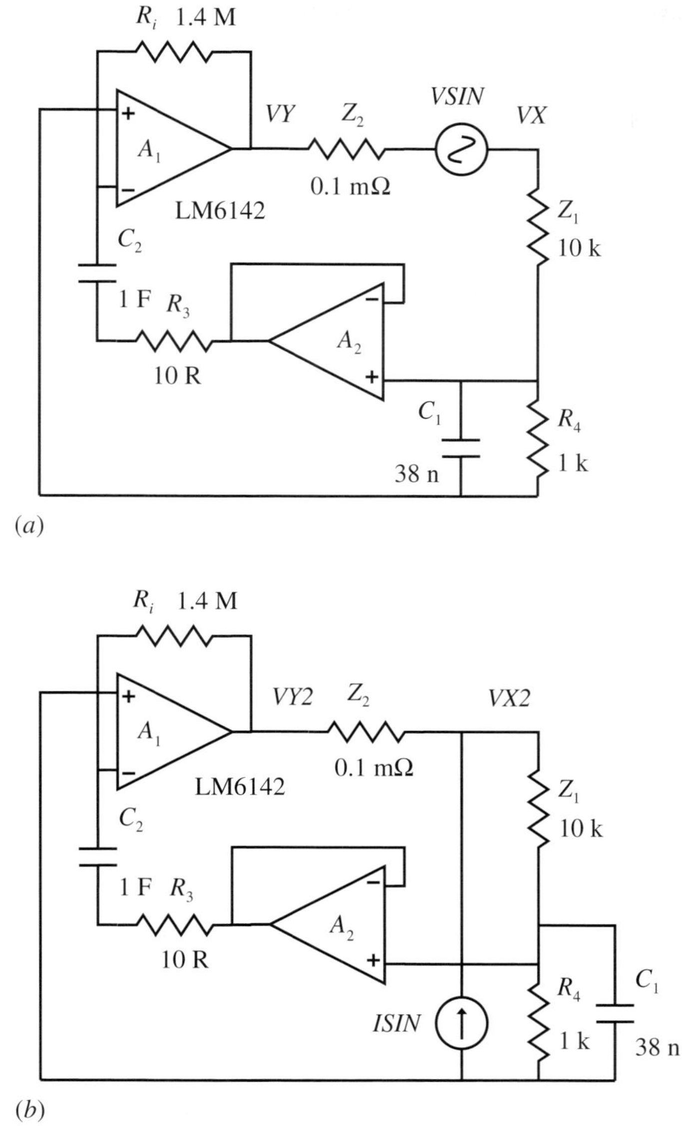


**Fig. 5.14.3** (*a*) Circuit for testing Eq. (5.14.6): voltage injection. (*b*) Current injection. (*c*) Direct open-loop measurement. (*d*) Compensation. (*Fig. cont. overleaf*)

for *VCL*/*VCI* with $C_3 = 3.8$ n, i.e. the compensated response (see below). Figure 5.14.4(*b*) shows the responses for $K = Z_1/Z_2 \gg 1$, so $T_v$ rather than $T_i$ is a good fit except at high frequency but $T$ fits very well as predicted. Also included here is the *VCL*/*VCI* response from Fig. 5.14.3(*d*) with $C_3$ set to a tiny value (1 pF) so that the system is not compensated and the large peak at ≈88.4 kHz appears. Figure 5.14.4(*c*) shows the responses for an intermediate case with $K = Z_1/Z_2 \approx 1$, and neither $T_v$ nor $T_i$ fit *VOL* but $T$ still fits very well. Also shown here is the response *TT* for

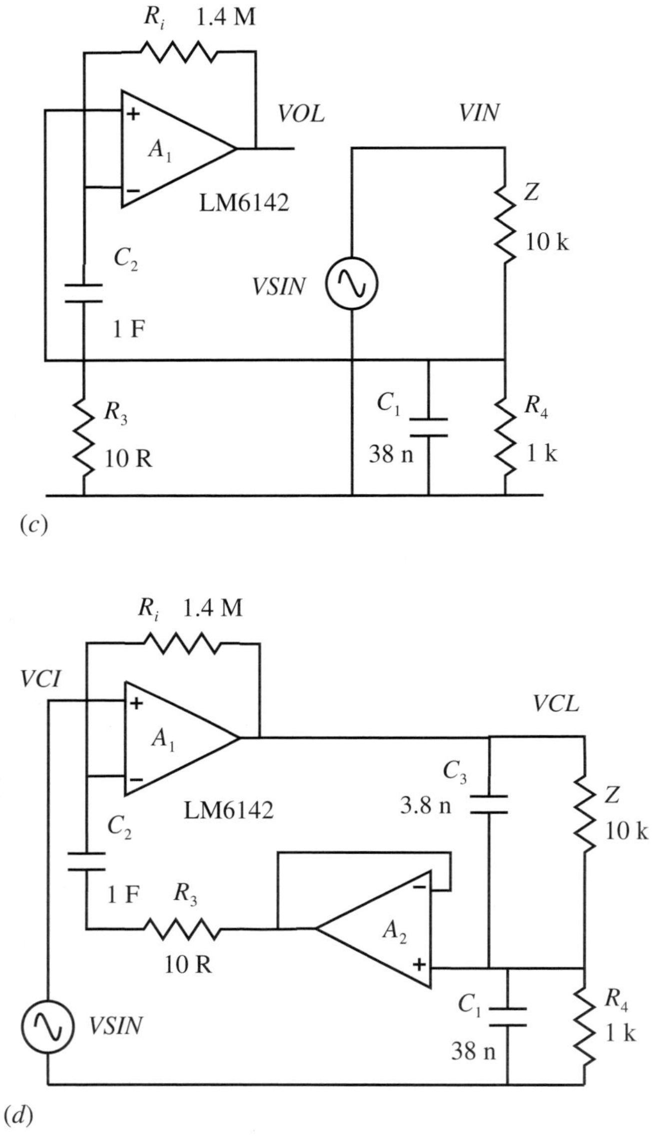


**Fig. 5.14.3** *(cont.)*

Tuinenga's version of *T* and it can be seen that it deviates substantially. Figure 5.14.4(*d*) shows an excerpt from (*b*) around the unity-gain frequency showing again the difference between *VOL*, *T* and *TT*, which was the original cause for concern described at the start of this section (the response without the 2 lies between *T* and *TT*).

In order to determine the stability of the feedback it is necessary to know the phase shift at the unity-gain frequency. The upper plots of Fig. 5.14.4 show the

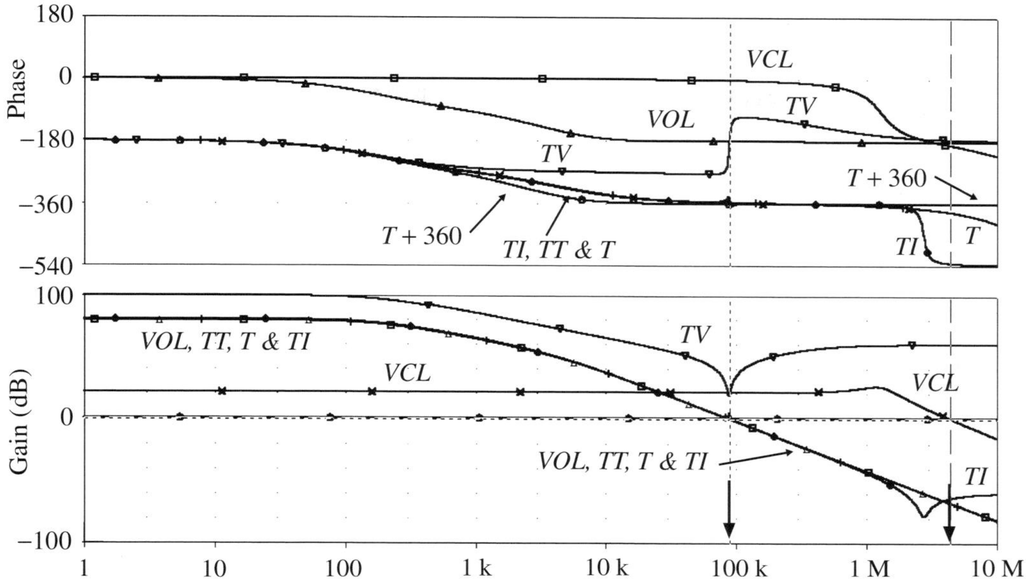


(*a*) Cursors at 89 kHz and 4.4 MHz

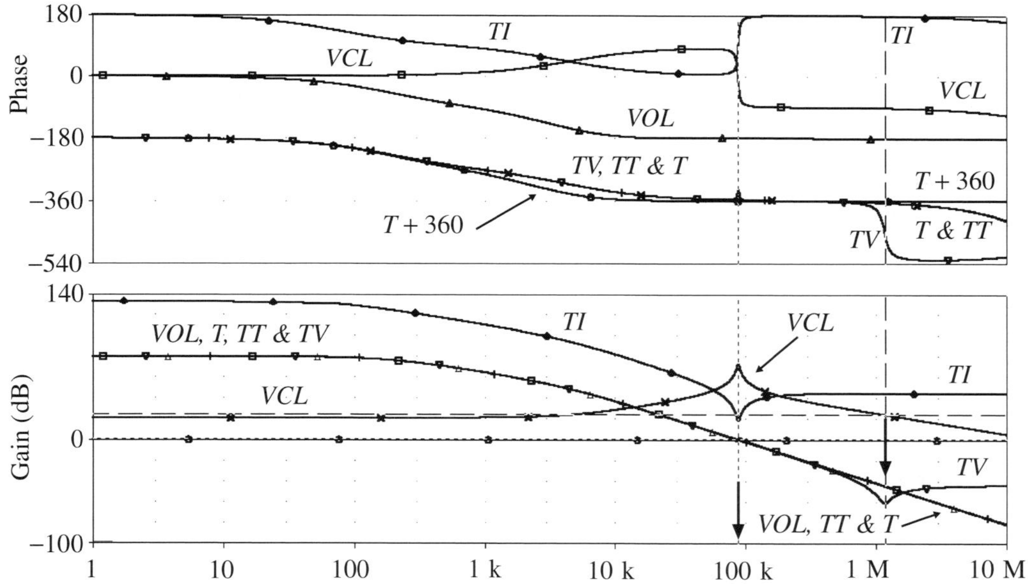


(*b*) Cursors at 89 kHz and 1.17 MHz

**Fig. 5.14.4** (*a*) Voltage mode injection: $Z_1 = 10\ \Omega$, $Z_2 = 9990\ \Omega$. (*b*) Current mode injection: $Z_1 = 9990\ \Omega$, $Z_2 = 10\ \Omega$. (*c*) Mixed mode injection: $Z_1 = 5000\ \Omega$, $Z_2 = 5000\ \Omega$. (*d*) Expanded region around unity gain from case (*b*). (*e*) Transient responses. (*Fig. cont. overleaf*)

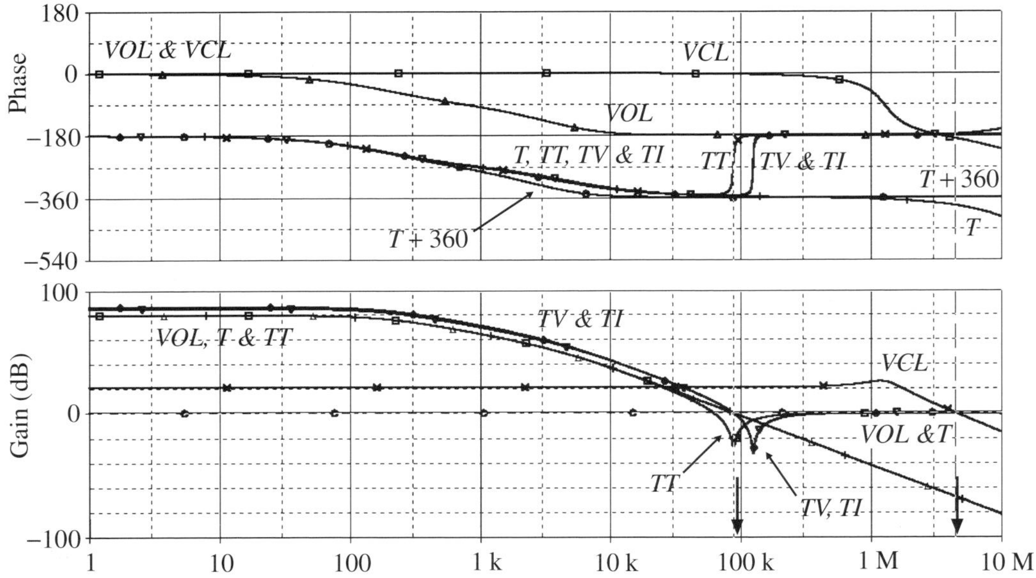


(*c*) Cursors at 89 kHz and 4.4 MHz Frequency (Hz)

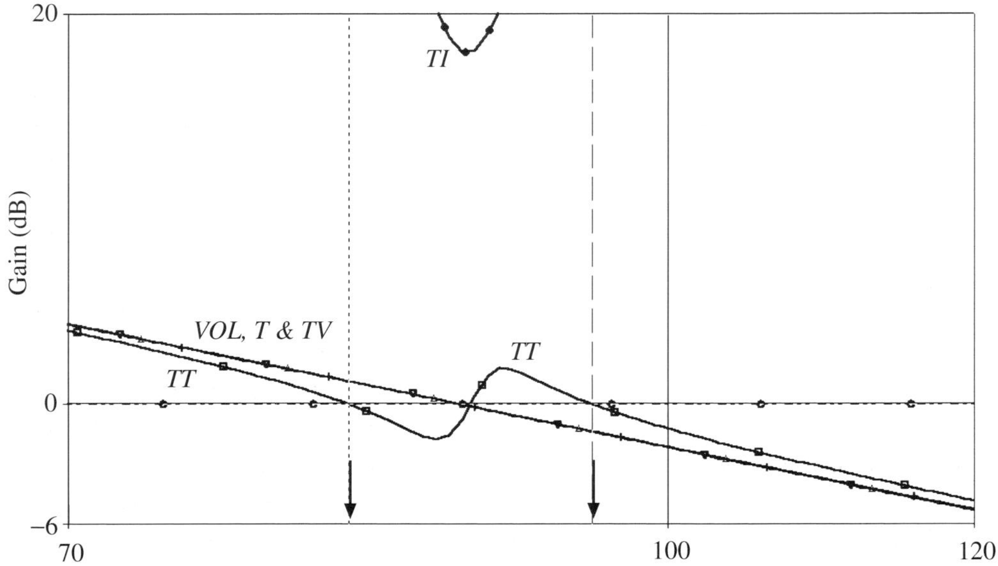


(*d*) Cursors at 88.6 and 95.6 kHz Frequency (kHz)

**Fig. 5.14.4** (*cont.*)

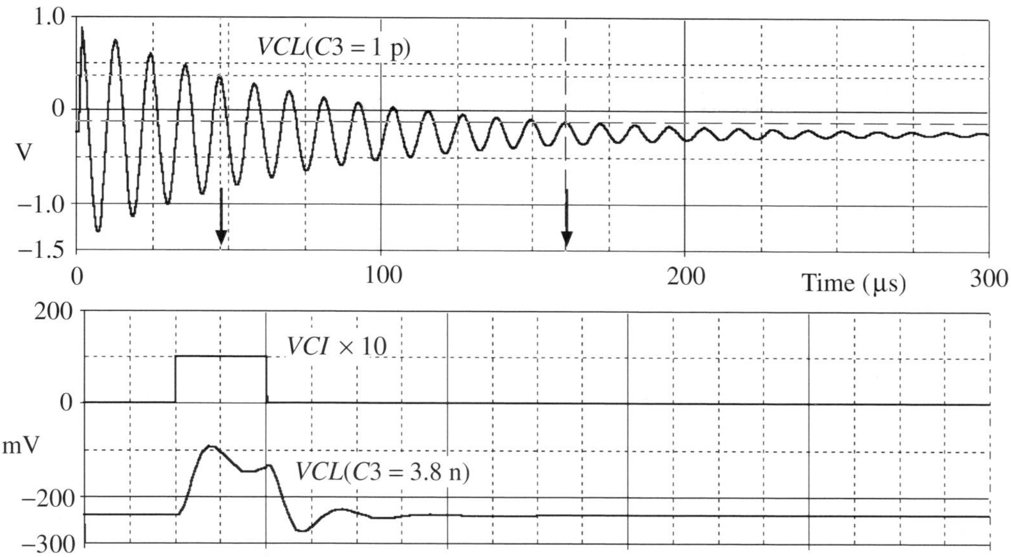


(*e*) Cursor difference 113.6 μs on upper trace Time (μs)

**Fig. 5.14.4** (*cont.*)

phase variation for the various signals. The phase of *VOL* has had 180° subtracted to make up for the non-inverting configuration used here. The loop phase margin is given by $P(T)+360°$ at unity-gain frequency; here we see it is just a few degrees so the system is close to the point of oscillation as one may expect from the amplifier and the divider time constants. Replacement of the *ISIN* generator in Fig. 5.14.3(*b*) with an *IPULSE* generator (1 mA,1 μs) and running a transient response will show a damped oscillation at ≈88.4 kHz which fits with the uncompensated closed-loop response as found from Fig. 5.14.3(*d*).

The performance of the closed-loop system is determined in circuit Fig. 5.14.3(*d*) with $C_3$ set to some tiny value. There is a peak at ≈88 kHz in the closed-loop response (Fig. 5.14.4(*b*)) together with a sharp change of phase. Removal of the pole is achieved by compensating the attenuator by setting $C_3$ to 3.8 nF (based on the inverse ratio of the resistors). The response is now flat out to about 4 MHz and the sharp phase change is removed (see Fig. 5.14.4(*a*) and (*b*), curves *VCL*). Transient response may also be examined by using *VPULSE* (10 mV, 1 μs) source with the results shown in Fig. 5.14.4(*e*) for $C_3=1$ p and 3.8 n. The time for 10 cycles is 113.6 μs giving an oscillation frequency of 88 kHz. $C_3=2$ n gives somewhat better transient response.

The results presented generally confirm the theory.

## SPICE simulation circuits

Consult the SimCmnt.doc file on the CD before running

| | |
|---|---|
| Fig. 5.14.4(*a–c*) | Openlop 5.SCH |
| Fig. 5.14.4(*d*) | Openlop 7.SCH |
| Fig. 5.14.4(*e*) | Openlop 6.SCH |

## References and additional sources 5.14


Mancini R. (1996): *PSPICE Performs Op Amp Open Loop Stability Analysis (HA5112)*, Harris Semiconductor App Note AN9536.1, November.

MicroSim Corporation (1993): *Application Notes Manual*, July, pp. 47–50.

Spohn P. (1963): A quick, convenient method for measuring loop gain. *Hewlett-Packard Journal* **14**, 5–8, January–February. (Uses a clip-on current probe.)

Tuinenga P.W. (1988): *SPICE: A Guide to Circuit Simulation and Analysis Using PSPICE*, Englewood Cliffs: Prentice Hall. ISBN 0-13-834607-0. (3rd Edn, 1995, ISBN 0-13-158775-7, has a fuller description.)

# 5.15 Lumped or distributed?

Begin at the beginning and go on till you come to the end, then stop.

King of Hearts to Alice. Lewis Carroll/Charles Ludwige Dodson

In interconnecting fast systems with short coaxial cables the question arises as to the conditions under which one may consider the cable as a lumped element or whether it must be treated as a transmission line. There is no clear dividing line as it depends in particular on how much distortion you are willing to accept. The significant relationship is between the risetime $\tau_R$ (or bandwidth) of the signal and the time delay $t_d$ of the cable (e.g. Delaney 1969, p. 188). If $t_d$ is short so that the round-trip time $2t_d$ is significantly less than $\tau_R$ then a lumped view is acceptable. It is suggested (Meta Software 1992) that if $t_d > \tau_R/5$ then you should consider it as a transmission line. As an example let us examine the circuit shown in Fig. 5.15.1, which shows the output of a photomultiplier which can produce fast pulses from,

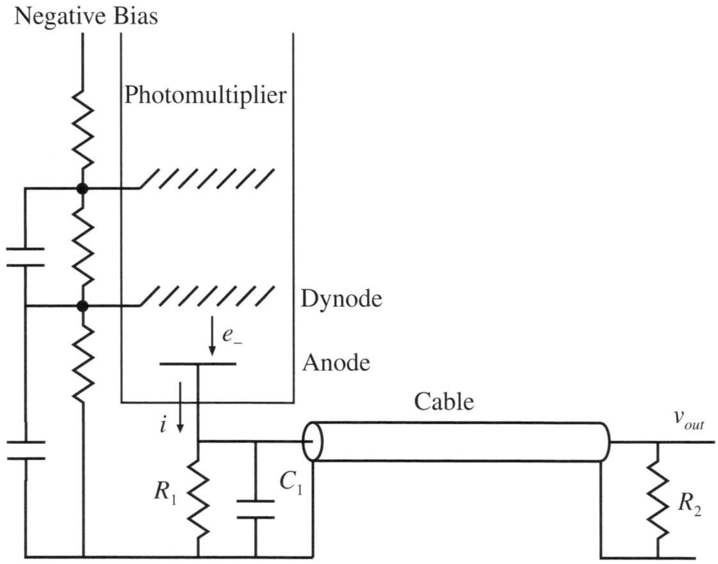


**Fig. 5.15.1** Cable connection example.

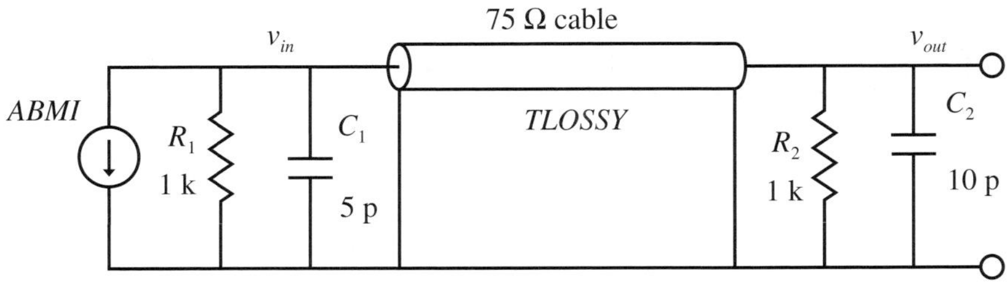


**Fig. 5.15.2** Equivalent circuit model for source, cable and load.

say, a single photon. The output is essentially a current source, so the current is passed through an anode load resistor to generate a voltage. There will of course be some capacity associated with the anode and from any surroundings and the signal is passed through a coaxial cable to the rest of the system.

The nature of the electron multiplication process in the many stages of the device results in an output pulse with a sharper rise than fall which we can represent as a function of time $t$ by a relation of the form (Hamilton and Wright 1956):

$$i = at^n \exp(-bt) \tag{5.15.1}$$

where $a$, $b$ and $n$ are constants. This can be implemented by means of a PSpice analog behavioural model with current output (an *ABMI* device) with the defining expression:

$$PWR(10{,}14)*PWR(TIME{,}2)*EXP(-2*PWR(10{,}8)*TIME) \tag{5.15.2}$$

with $a = 10^{14}$, $n = 2$ and $b = 10^8$ to give a risetime of about 10 ns and an amplitude of about 1.4 mA. We have chosen an anode load of 1 k in parallel with 5 pF, i.e. not matched to the 75 Ω cable. Typical values per metre for such a cable are (MicroSim 1993, p. 163, for RG11 cable; the $R$ value is appropriate for a frequency of a few MHz):

$$R = 0.311\ \Omega\ \mathrm{m}^{-1},\quad L = 0.378\ \mu\mathrm{H}\ \mathrm{m}^{-1},\quad C = 67.3\ \mathrm{pF}\ \mathrm{m}^{-1},\quad G = 6.27\ \mu\mathrm{S}\ \mathrm{m}^{-1} \tag{5.15.3}$$

We will use the *TLOSSY* model for the transmission line even though for the short lengths involved the effect of $R$ and $G$ will be negligible. Figure 5.15.2 shows the circuit for simulation.

In SPICE current sources, the current flows internally from (+) to (−) so the voltage output here will be a negative going pulse as would be obtained from a photomultiplier. It may be convenient to run a second *ABMI* to provide a reference signal to compare with the one driving the line but the load for this will need adjusting to give a comparable signal. You can now run the simulation and see the effects of changing the length of the line. As shown in Section 3.17 the delay $t_d$ is about 5 ns m$^{-1}$ so for half a metre we have $5t_d \approx \tau_R$ so there should not be too much

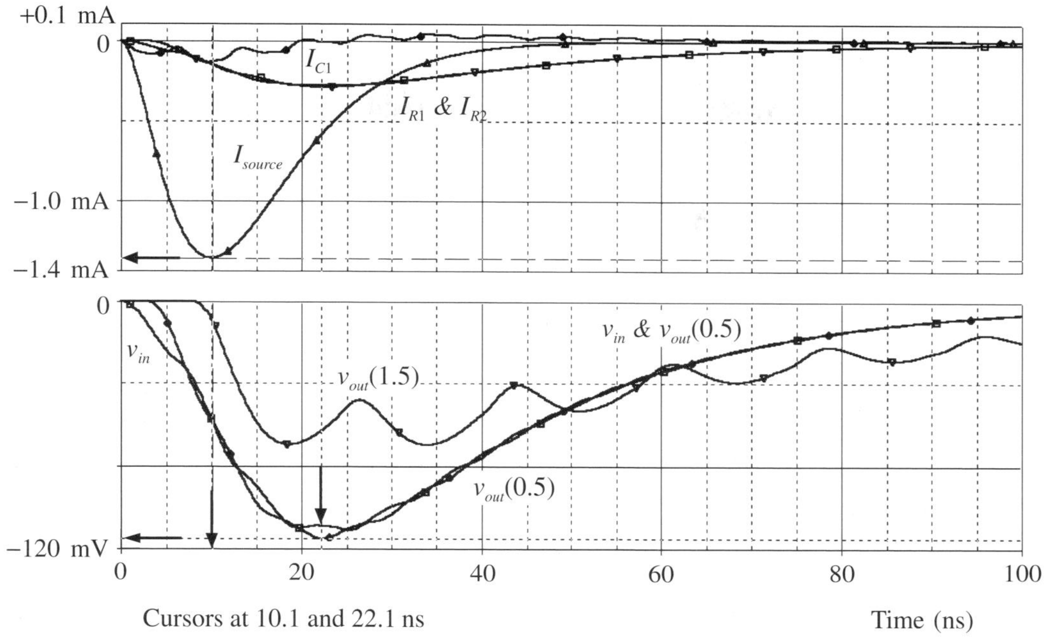


**Fig. 5.15.3** Pulse responses for different cable lengths.

distortion. It is evident from the signal at the sending end that what the current generator sees initially is not so much the 1 k but the characteristic impedance of the line, i.e. 75 Ω, since there has been no time for the signal to propagate down the line and, as it were, to find out what else is there. Thus the choice of a 1 k anode load, in the hope of obtaining a larger signal within the constraints of the time constant set by the local capacity, was misguided. It might be better, if the far end was not properly terminated, to have made the anode load 75 Ω to avoid reflections, but then the actual load seen would only be 75/2 = 37.5 Ω. There is always a price to be paid. Changing the length of the cable to 1.5 m gives a rather undesirable output as shown in Fig. 5.15.3. The currents shown are only for the 0.5 m case.

The effects of reflection in this case are somewhat muted due to the rounded form of the input pulse. If the experiment is repeated with rather sharper pulses the response will be much worse.

## SPICE simulation circuits

| | |
|---|---|
| Fig. 5.15.2 | Tlneld2.SCH |

## References and additional sources 5.15

Delaney C. F. G. (1969): *Electronics for the Physicist*, Harmondsworth: Penguin Books.

Hamilton T. D. S., Wright G. T. (1956): Transit time spread in electron multipliers. *J. Sci. Instrum.* **33**, 36.

Meta Software (1992): *HSPICE User Manual, Vol. 2 Elements and Models*, Ver. H92, Chapter 2.

MicroSim Corporation (1993): *Circuit Analysis Reference Manual*, July, p. 163.

# 5.16 Immittance *Through the Looking Glass*: gyrators, negative immittance converters and frequency dependent negative resistors

No problem can stand the assault of sustained thinking.

Voltaire (1694–1778)

The gyrator is a type of circuit element additional to the simple $L$, $C$ and $R$. It has the property that it can invert impedances, i.e. it can make a capacitor look like an inductor (or vice versa) and hence provides a means of realizing inductorless filters (Roddam 1957; Riordan 1967). As a rather unusual circuit capable of high performance it is worth investigating and the design is considerably eased by PSpice simulation. There are many ways of realizing such a 'device' but we will consider one which has become commonly used (Van Valkenburg 1982). The circuit, proposed by Antoniou (1969), is shown in Fig. 5.16.1.

Taking generalized impedances $Z_x$ as shown and assuming ideal operational amplifiers so that the voltage difference between the inputs of the amplifiers tends to zero, we may derive the input impedance $Z_{11}$ as follows:

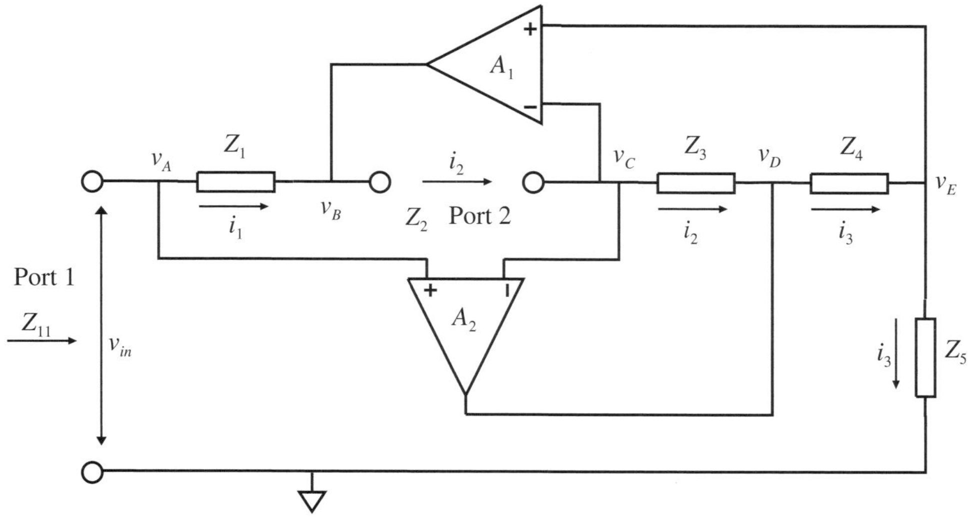


**Fig. 5.16.1** A gyrator circuit.

Table 5.16.1 *Impedance choices for gyrator or for frequency dependent negative resistor (FDNR)*

| | $Z_1$ | $Z_2$ | $Z_3$ | $Z_4$ | $Z_5$ | $Z_{11}$ |
|---|---|---|---|---|---|---|
| Gyrator A | $R_1$ | $1/sC_2$ | $R_3$ | $R_4$ | $R_5$ | $\left(\frac{R_1R_3R_5}{R_4}\right)sC_2 = sL_{11}$ |
| Gyrator B | $R_1$ | $R_2$ | $R_3$ | $1/sC_4$ | $R_5$ | $\left(\frac{R_1R_3R_5}{R_2}\right)sC_4 = sL_{11}$ |
| FDNR A | $1/sC_1$ | $R_2$ | $R_3$ | $R_4$ | $1/sC_5$ | $\left(\frac{R_3}{C_1C_5R_2R_4s^2}\right) = \frac{1}{Ks^2} = \frac{-1}{K\omega^2}$ |
| FDNR B | $R_1$ | $R_2$ | $1/sC_3$ | $R_4$ | $1/sC_5$ | $\left(\frac{R_1}{R_2R_4C_3C_5s^2}\right) = \frac{1}{Ks^2} = \frac{-1}{K\omega^2}$ |

$$v_A = v_C = v_E \quad \text{so that} \quad i_1Z_1 + i_2Z_2 = 0, \quad i_2Z_3 + i_3Z_4 = 0, \quad i_3Z_5 = v_{in}$$

$$\text{so} \quad i_1Z_1 = -i_2Z_2 = \frac{v_{in}}{Z_5}\frac{Z_2Z_4}{Z_3} \quad \text{since} \quad i_2 = \frac{-i_3Z_4}{Z_3} \quad \text{and} \quad i_3 = \frac{v_{in}}{Z_5} \tag{5.16.1}$$

$$Z_{11} = \frac{v_{in}}{i_1} = \frac{Z_1Z_3Z_5}{Z_2Z_4}$$

There are two choices for the $Z_x$ (forms A and B) that give the desired inversion (Chen 1995, p. 353), together with two for the realization of the frequency dependent negative resistor (FDNR) (Burr-Brown 1994), Table 5.16.1.

The entries for the gyrators show how the capacitor ($C_2$ or $C_4$) is inverted to an equivalent inductor $L_{11}$. The entries for the FDNRs show how the configuration is equivalent to a negative resistance $R_{11} = -1/K\omega^2$ (Bruton 1969). This latter form is the basis for a filter topology referred to as a generalized impedance converter or GIC. Sometimes the term immittance is used to indicate impedance or admittance. We will examine an example of such a third order low-pass filter that has particularly good linear-phase response (Burr-Brown 1994, p. 218). We will first consider the design of the FDNR and then examine its performance using SPICE. Filter design theory is often based on passive configurations which can then be transformed into active form. The form given has been 'optimized for phase linearity and stopband attenuation by means of exhaustive computer simulation and empirical analysis'. The response is in-between Butterworth and Bessel, the former being appropriate for smooth in-band response and the latter for phase linearity. The significance of phase linearity is discussed in Section 3.6. The passive base-circuit is shown in Fig. 5.16.2 and the component values are (as is usual) for a cut-off frequency of $\omega_c = 1$ rad s$^{-1}$.

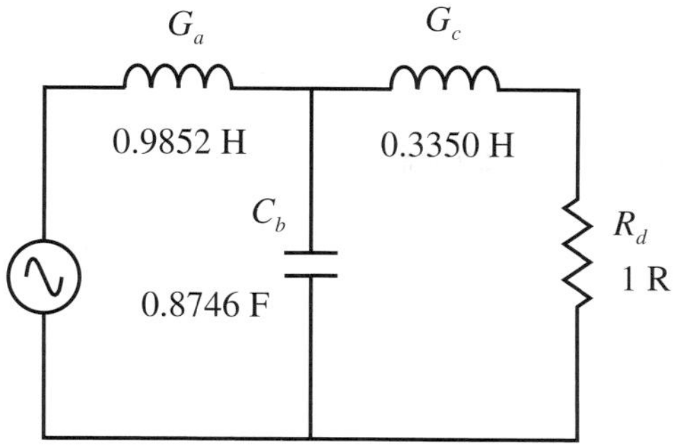


**Fig. 5.16.2** Initial base circuit for a cut-off frequency of 1 rad $s^{-1}$.

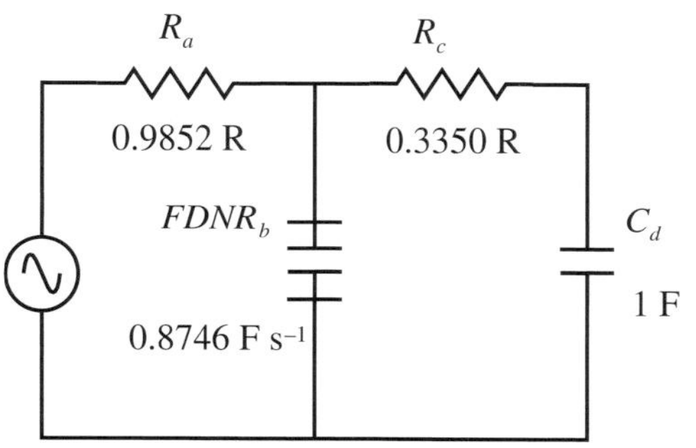


**Fig. 5.16.3** Transformed circuit with all components of the base circuit multiplied by $1/s$.

The components are inverted by multiplication by $1/s$ since we do not want to use inductors. This leads to the configuration shown in Fig. 5.16.3 where the inductors have become resistors, the resistor a capacitor and the capacitor a FDNR.

The transformations can be understood by considering the impedance of the original component multiplied by $1/s$, and the form of the result will tell us what type the new component is:

$$sL \times 1/s = L = R, \quad \text{i.e. a resistor } R \text{ of value } L \text{ ohm}$$
$$R \times 1/s = R/s = 1/sC, \quad \text{i.e. a capacitor } C \text{ of value } 1/R \qquad (5.16.2)$$
$$\frac{1}{sC} \times \frac{1}{s} = \frac{1}{s^2C}, \quad \text{i.e. a FDNR}$$

The FDNR will be realized using version B in the table with the relative values $R_1 = R_2 = 1$ and $C_3 = C_5 = 1$ so now $K$ is set by $R_4$ alone – in this case $Z_{11} = 0.8746$. The units for a FDNR are deduced from the original farad for the capacity divided by $s$ (or effectively $\omega$, which has units $s^{-1}$) so we get F $s^{-1}$. It is now necessary to scale the components to match the cut-off frequency we want, which we will take to be $f_c = 10$ kHz or $\omega_c = 2\pi \times 10^4$, so we divide $C_d = 1$ F (which was, as indicated above, for $\omega_c = 1$) by the new $\omega_c$ which gives the scaled value (in Fig. 5.16.4) of

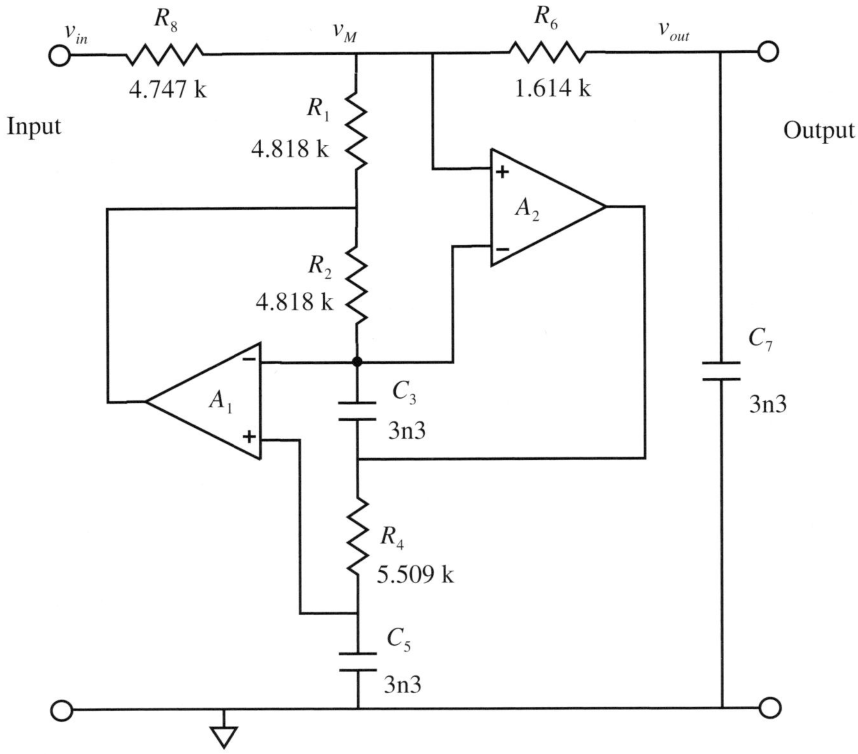


**Fig. 5.16.4** Final transformed and scaled circuit.

$C_7 = 15.9$ μF. To use more manageable capacitors we can scale down their magnitude and correspondingly scale up the resistors. If we select say $C_7 = 3.3$ n (and hence also $C_3$ and $C_5$) then the scaling factor $N$ is:

$N = \dfrac{15900}{3.3} = 4818.2$ and the various components become

$$\begin{aligned} R_8 &= R_a N = 0.9852 \times 4818.2 = 4.747 \text{ k} \\ R_6 &= R_c N = 0.3350 \times 4818.2 = 1.614 \text{ k} \\ R_1 &= R_2 = 1 \times 4818.2 = 4.818 \text{ k} \\ C_3 &= C_5 = C_7 = 3.3 \text{ n} \end{aligned} \tag{5.16.3}$$

$$R_4 = \frac{R_1}{R_2 C_3 C_5 Z_{11}} \times N = \frac{1}{1 \times 1 \times 1 \times 0.8746} \times 4818.2 = 5.509 \text{ k}$$

to give us the final circuit shown in Fig. 5.16.4.

The circuit may now be simulated and results shown in Fig. 5.16.5 were for a LM6142 amplifier with 17 MHz gain-bandwidth. The in-band gain is unity with a –3 dB frequency of 13.7 kHz. This type of filter is somewhat influenced by the

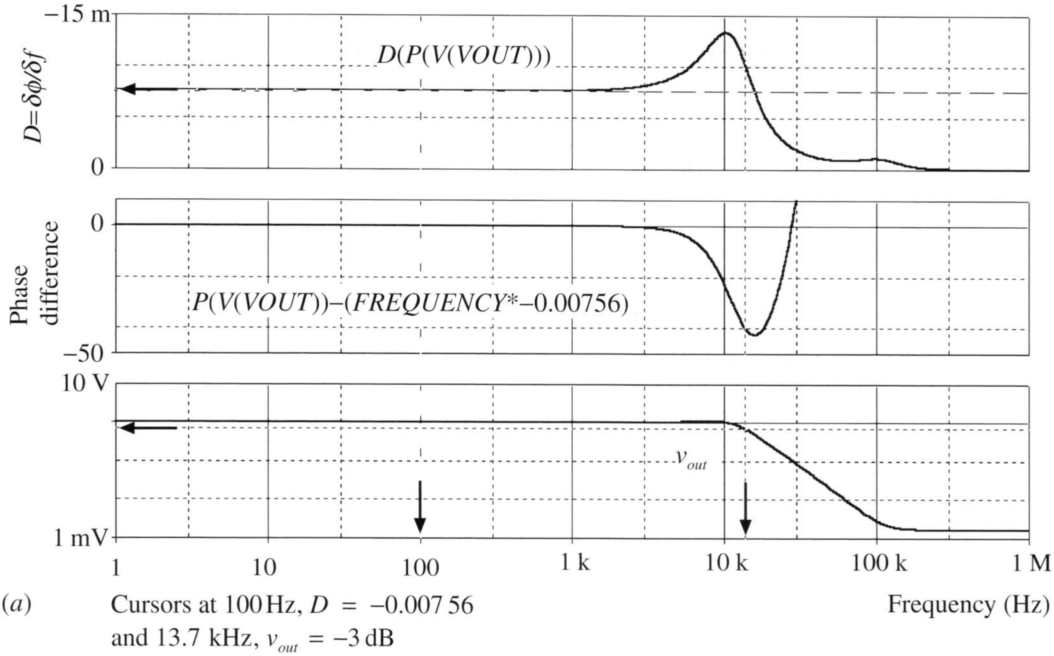


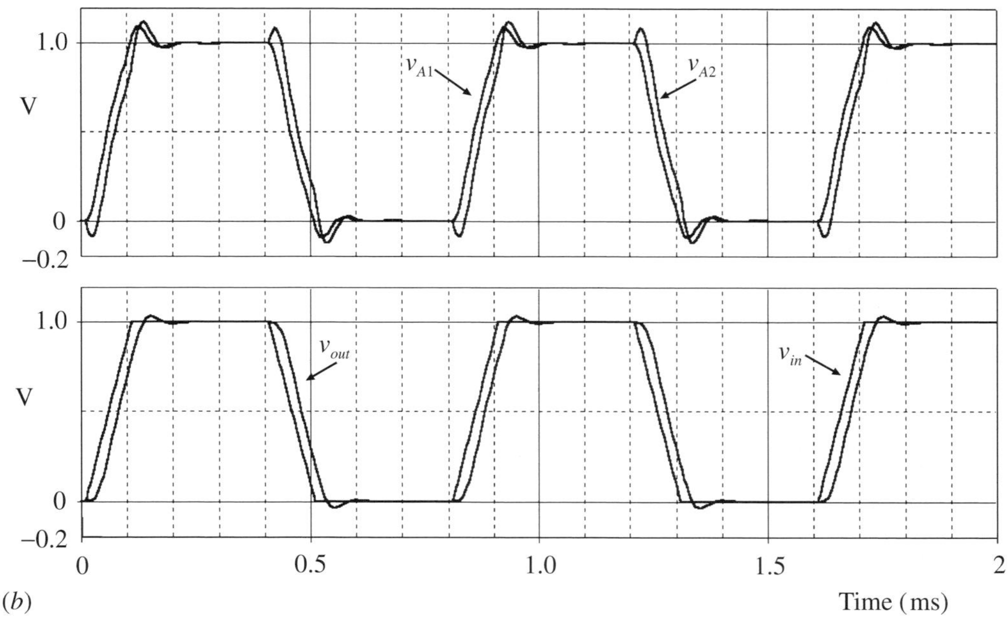


**Fig. 5.16.5** (*a*) Gain and phase response of the low-pass filter circuit of Fig. 5.16.4. (*b*) Pulse response of circuit of Fig. 5.16.4. (*c*) Frequency characteristic of the FDNR. (*Fig. cont. overleaf*)

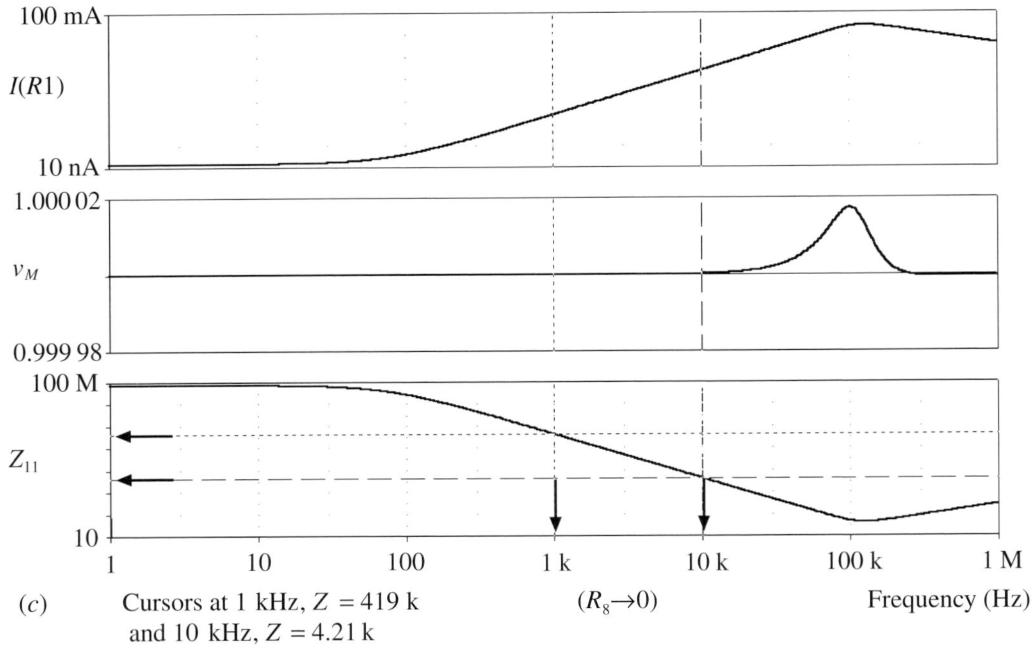


**Fig. 5.16.5** (*cont.*)

gain-bandwidth of the amplifiers used, so suitable types should be used. The performance of the FDNR can be examined by making $R_8$ very small and plotting the ratio of the input voltage to the current through $R_1$. The relation between phase $\phi$ and frequency $f$ is:

$$f=\frac{d\phi}{dt} \quad \text{or in terms of a delay } \delta t=\frac{\delta\phi}{f} \tag{5.16.4}$$

so that for a delay independent of frequency $\delta\phi$ must be proportional to frequency. To check the relation you can plot the differential of the output signal phase (*D*(*P*(*V*(*VOUT*))) in PSpice) which should give a fixed value, depending on the delay, independent of frequency. The phase is very close to the ideal over much of the range and over a more limited range can be very good indeed (Fig. 5.16.5(*a*)). At 100 Hz the differential of the phase is $D=-0.00756$ so to enable comparison the phase difference *P*(*V*(*VOUT*))−(*Frequency***D*) has been plotted (remember that *D* is negative).

The transient response may be examined using a pulse input and it will be found that there is some overshoot and ringing. Changing $R_4$ to 3.3 k largely removed this, but in any case the output from the two amplifiers should be examined as there are large transients for fast input signals which will cause overload well before the filter output would indicate. Substitution of a faster amplifier, a LM6152, with 45 MHz GB, had little effect so it is not a bandwidth limiting effect.

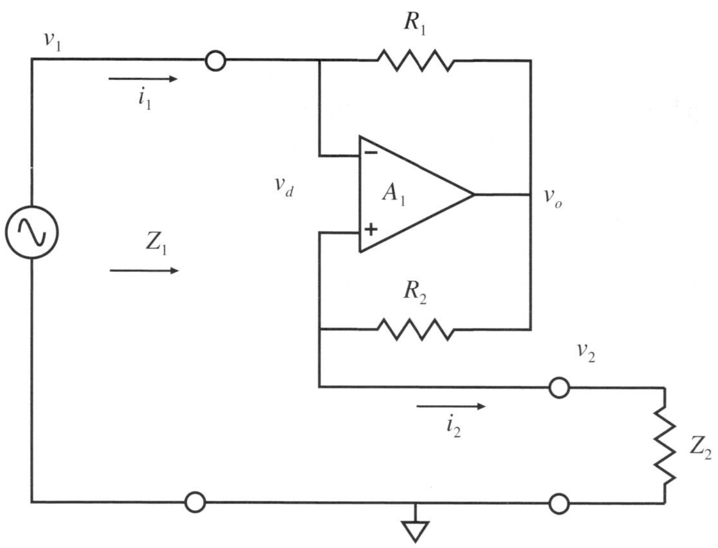


**Fig. 5.16.6** A negative immittance converter circuit (NIC).

If $R_8$ is made negligible then dividing the voltage $v_M$ by the current through $R_1$ will show the impedance characteristic $Z_{11}$ of the FDNR and this is shown in Fig. 5.16.5(*c*). Taking two frequencies a factor 10 apart as shown by the cursors then the impedance ratio is found to be 419/4.21≈100, so the impedance is inversely proportional to the square of the frequency as indicated in Eq. (5.16.2).

A related device is the negative impedance (or immittance) converter (NIC) which has the property that the driving point impedance is the negative of the terminating impedance. A circuit showing one way of achieving this is given in Fig. 5.16.6. We assume an ideal amplifier so that the voltage between the two inputs $v_d \to 0$ and that no current flows into the inputs. We can write the equations:

$$v_d = 0, \quad v_1 = v_2, \quad \text{and} \quad v_2 = v_o\left(\frac{Z_2}{R_2 + Z_2}\right) \quad \text{or} \quad v_o = v_2\left(\frac{R_2 + Z_2}{Z_2}\right) = v_1\left(\frac{R_2 + Z_2}{Z_2}\right)$$

$$v_1 - v_o = i_1 R_1, \quad \text{or} \quad v_1 - v_1\left(\frac{R_2 + Z_2}{Z_2}\right) = i_1 R_1, \quad \text{so} \quad Z_1 = \frac{v_1}{i_1} = -\left(\frac{R_1}{R_2}\right) Z_2 \qquad (5.16.5)$$

so that the input impedance $Z_1$ is the negative of the terminating impedance $Z_2$, times a gain factor $R_1/R_2$ which is commonly made unity. Since there is positive feedback as well as negative there are serious considerations of stability to be addressed which depend on the external impedances as well as those included above (Burr-Brown 1966; Philbrick 1968). The circuit is reciprocal in that the input can be at $v_2$ with the output at $v_1$. The relationship between $Z_1$ and $Z_2$ will be the same but there is one important difference. For the circuit as shown in Fig. 5.16.6

if the source impedance is high then the negative feedback will be greater than the positive and the circuit will be stable; it is said to be open-circuit stable (OCS). For the reciprocal configuration, if the source impedance is low the system will be stable; it is said to be short-circuit stable (SCS). The examples referred to below will illustrate this.

An interesting example of the use of a NIC is given in Section 5.26 where it is used to construct a chaotic circuit, and another in Section 5.17. It may also be noted that the Howland 'current pump' (or current source) also hides within it a NIC (Philbrick 1968, p. 66; Hamilton 1977).

### SPICE simulation circuits

| | |
|---|---|
| Fig. 5.16.5(*a*) | Gic_fil3.SCH |
| Fig. 5.16.5(*b*) | Gic_fil2.SCH |
| Fig. 5.16.5(*c*) | Gic_fil4.SCH |

## References and additional sources 5.16


Antoniou A. (1969): Realization of gyrators using operational amplifiers and their use in RC-active network synthesis. *Proc. IEE* **116**, 1838–1850.

Bruton L. T. (1969): Network transfer functions using the concept of frequency-dependent negative resistance. *IEEE Trans.* **CT-16**, 406–408.

Burr-Brown (1966): *Handbook of Operational Amplifier Active RC Networks*, 1st Edn, Burr-Brown. See p. 48.

Burr-Brown (1994): *A Low Noise, Low Distortion Design for Antialiasing and Anti-imaging Filters*, Burr-Brown Application Bulletin AB-026. (Also in Applications Handbook LI-459.)

Chen, Wai-Kai (Ed.) (1995): *The Circuits and Filters Handbook*, Boca Raton: CRC Press and IEEE Press. ISBN 0-8493-8341-2.

Hamilton T. D. S. (1977): *Handbook of Linear Integrated Electronics for Research*, London: McGraw-Hill. ISBN 0-07-084483-6.

Philbrick/Nexus (1968): *Applications Manual for Operational Amplifiers*, Philbrick/Nexus Research.

Riordan R. H. S. (1967): Simulated inductors using differential amplifiers. *Electron. Lett.* **3**, 50–51.

Roddam T. (1957): The gyrator. *Wireless World* **63**, 423–426.

Skritek P. (1983): Comparison of overload and noise characteristics of FDNR circuits. *IEEE Trans.* **CAS-30**, 500–503.

Van Valkenburg M. E. (1982): *Analog Filter Design*, New York: Holt, Rinehart and Winston. ISBN 0-03-059246-1, or 4-8338-0091-3 International Edn. See Chapter 15.

# 5.17 Maser gain simulation

The birth of the maser required a combination of instincts and knowledge from both engineering and physics.

Charles H. Townes

An example of a circuit that can illustrate the variation of the active medium in a maser is given by Siegman (1971, p. 203). This uses a NIC (Section 5.16) and a tee-pad as a form of directional coupler (Section 5.21). A parallel tuned circuit is used to represent the frequency dependence of the maser and the NIC allows the gain to be varied. The circuit is shown in Fig. 5.17.1.

The tee-pad, consisting of $R_2$–$R_5$, allows the measurement of the reflected signal from the resonator as we shall show (there is an error in Siegman's Fig. 5-14(b) – the conductances $G_0/3$ should be $3G_0$). The series resistance of $L_1$ is deduced from its $Q = 60$ and the resonant frequency of about 3.6 kHz to be 31 Ω. The equivalent parallel resistance is, from Eq. (3.5.17), 120 kΩ, which determines the values of the tee and of the generator source resistance $R_1$. The output signal is the difference $v_B - v_A$ which must be measured by high impedance oscilloscope. We analyse the circuit shown in Fig. 5.17.2 where the 'maser' system is represented by $Y_L$ (or $Z_L = 1/Y_L$) and the various resistors are scaled as shown.

The impedance of the $v_B$ and $v_A$ branches in parallel is:

$$\frac{1}{Z} = \frac{1}{4R} + \frac{1}{R + Z_L}, \quad \text{so} \quad Z = \frac{4R(R + Z_L)}{5R + Z_L} \quad \text{or} \quad \frac{Z}{4R + Z} = \frac{(RY_L + 1)}{2(3RY_L + 1)} \tag{5.17.1}$$

and the voltages can be determined from:

$$\frac{v_2}{v_1} = \frac{Z}{4R + Z}, \qquad \frac{v_B}{v_2} = \frac{3R}{4R}, \qquad \frac{v_A}{v_2} = \frac{Z_L}{R + Z_L} = \frac{1}{1 + RY_L} \tag{5.17.2}$$

$$\text{so} \quad v_B = \left(\frac{v_1 Z}{4R + Z}\right)\left(\frac{3}{4}\right) \quad \text{and} \quad v_A = \left(\frac{v_1 Z}{4R + Z}\right)\left(\frac{1}{1 + RY_L}\right)$$

$$\text{then} \quad \frac{v_B - v_A}{v_1} = \left(\frac{Z}{4R + Z}\right)\left(\frac{3}{4} - \frac{1}{1 + RY_L}\right)$$

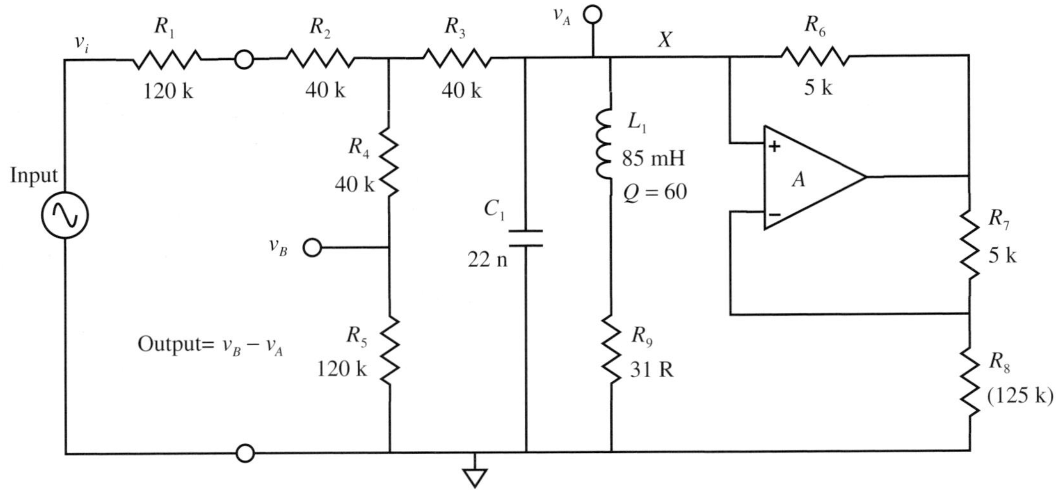


**Fig. 5.17.1** Maser equivalent circuit.

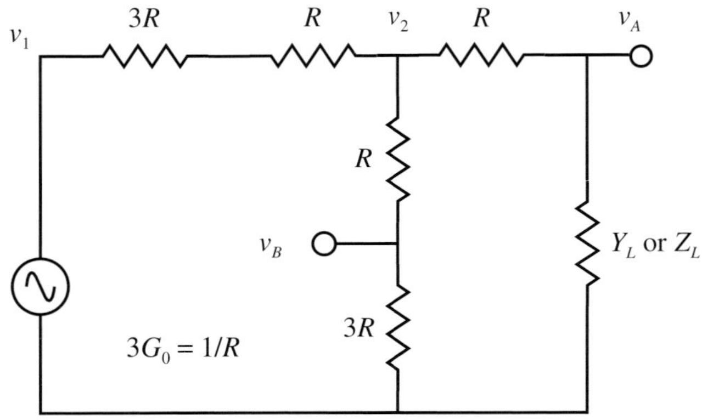


**Fig. 5.17.2** Directional coupler tee-pad.

$$= \left(\frac{Z}{4R+Z}\right)\left[\frac{(3RY_L - 1)}{4(1+RY_L)}\right]$$

$$= \frac{(RY_L + 1)}{2(3RY_L + 1)}\left[\frac{(3RY_L - 1)}{4(1+RY_L)}\right] \qquad (5.17.2\text{ cont.})$$

$$= \frac{1}{8}\left[\frac{3Y_L - \dfrac{1}{R}}{3Y_L + \dfrac{1}{R}}\right]$$

$$= \frac{1}{8}\left[\frac{Y_L - G_0}{Y_L + G_0}\right] \quad \text{since} \quad \frac{1}{R} = 3G_0 \qquad (5.17.2 \text{ cont.})$$

which is in the form of a reflection coefficient $\rho/8$ as given by Siegman (we differ by an overall negative sign but that only means a phase inversion). Since the source impedance for the NIC is high we should use the OCS configuration (Section 5.16) rather than the SCS used by Siegman. To obtain a negative resistance just equal to the positive resistance of the resonator and with $R_6 = R_7$ we require $R_8 = 120\ \text{k}\Omega$. If the circuit is simulated the results shown in Fig. 5.17.3 are obtained.

It is evident that the changeover from loss to gain occurs at about the value of $R_8$ expected. If the SCS circuit is used you will find instability in the results as predicted. If the circuit is broken at point $X$ the response of the resonator by itself will be displayed as shown in Fig. 5.17.3($b$).

## SPICE simulation circuits

| | |
|---|---|
| Fig. 5.17.3($a$) | Siegman1.SCH |
| Fig. 5.17.3($b$) | Siegman2.SCH |

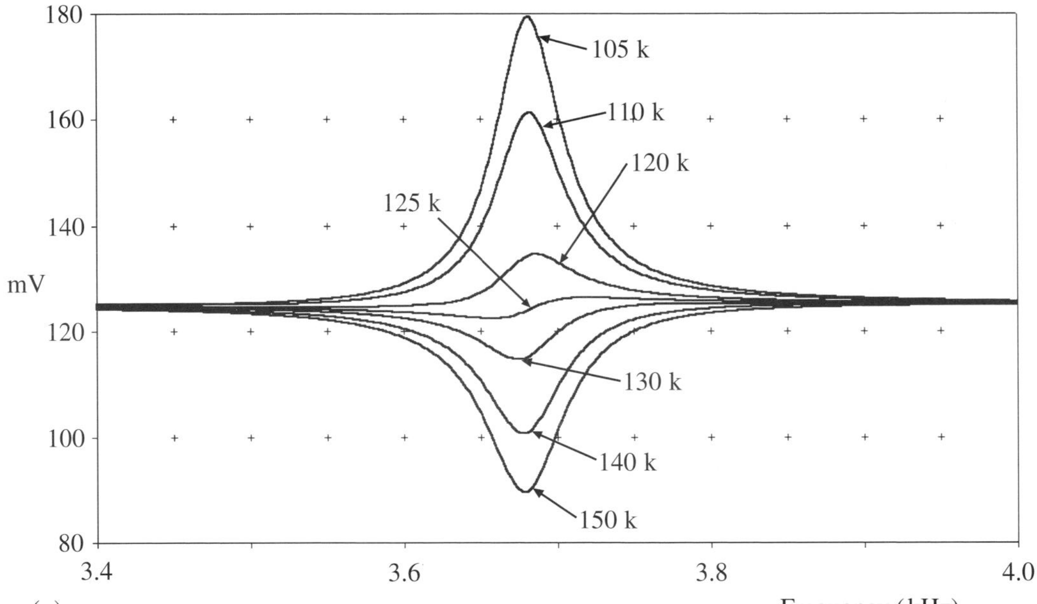


($a$)

**Fig. 5.17.3** Simulation responses for circuit of Fig. 5.17.1. ($a$) Gain or loss as a function of $R_8$. ($b$) Response of resonator alone. (*Fig. cont. overleaf*)

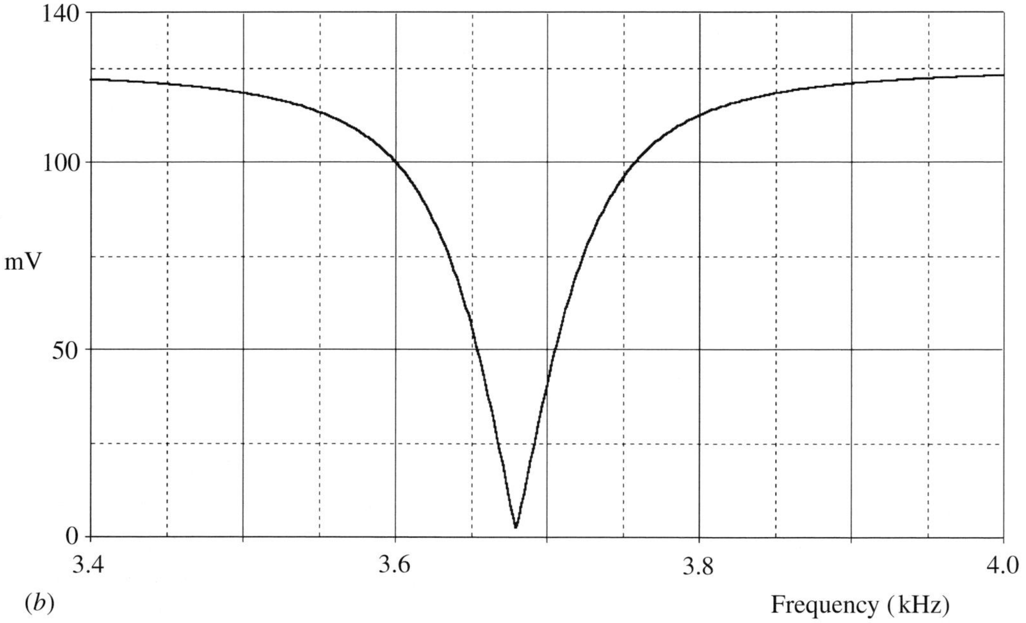


**Fig. 5.17.3** (*cont.*)

## References

Siegman A. E. (1971): *An Introduction to Lasers and Masers*. New York: McGraw-Hill. Library of Congress Cat. No. 79–123189.

# 5.18 Frequency-independent phase shifter

An ounce of action is worth a ton of theory.

Friederich Engels

Phase shifters, by their nature, are generally frequency dependent. If a frequency-independent circuit is required one is faced with a difficult problem. An ingenious network has been proposed (Horowitz and Hill 1989, p. 295; Gingell 1973) which is not the sort of thing that springs to mind. It has the form of a distributed passive *RC* filter as shown in Fig. 5.18.1. The low frequency limit is given by $f_L = 1/2\pi RC$ with $R$ and $C$ the largest (input) components, and for 1 k and 8 n this gives $f_L = 20$ kHz. The high frequency limit is given by the same expression but now with the smallest (output) capacitor of 1 n and we find $f_H = 160$ kHz. Here we have used a factor 2 for the spacing of the poles. Individual phases vary with frequency but the differences remain close to 90°. Though it uses a large number of components (and could have more than shown) it could be fairly compactly implemented using surface mounted or array components. To make the figure less cluttered, corresponding points have been marked with labels (or *GLOBALS* in PSpice).

The circuit does not encourage analysis, but SPICE comes to the rescue. The circuit is fed with a sinusoidal signal $v_{in}$ and its inverse which is readily available. The four outputs are labelled with their relative phases. The relation of these to the input phase is variable over the operational frequency range. For the values of $R = 1$ k and $C = 8$ n to 1 n shown, the range of independence is from about 20 to 200 kHz as can be seen from the SPICE phase curves, about as predicted (Fig. 5.18.2). The limits depend of course on the magnitude of the deviation acceptable.

In the reference above it is stated that for a six section circuit (rather than the four used here) the phases are within $\pm 0.5°$ over a 100:1 frequency range. The outputs should be buffered to avoid additional effects. If the four outputs are connected to symmetrical taps on a continuous potentiometer then the wiper can tap off any phase from 0 to 360°, a technique that has been used for lower frequency generators used for servo testing, though these often used two-phase oscillators to generate the four signals required (Section 5.7). An example of the use of the

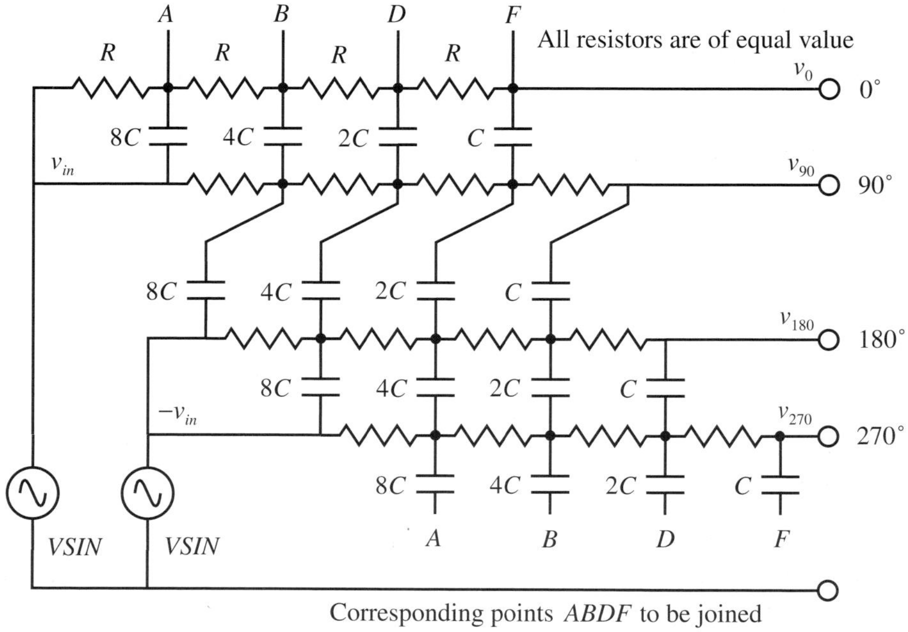


**Fig. 5.18.1** Frequency-independent phase shifter.

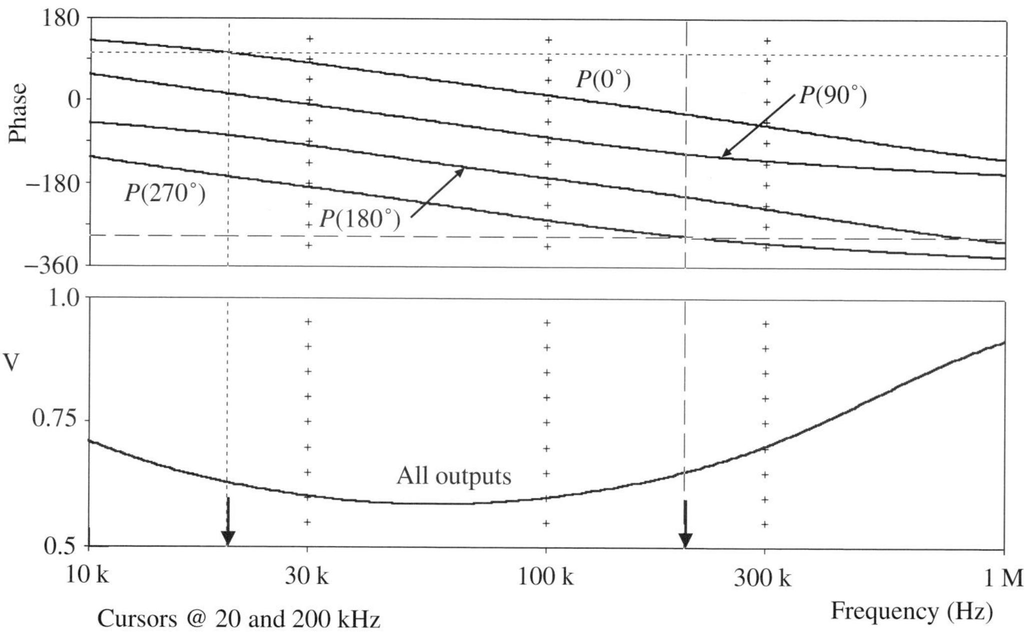


**Fig. 5.18.2** Amplitude and phase response of the circuit of Fig. 5.18.1.

Gingell circuit is given in Thomson (1979), who uses 16 stages with a $\sqrt{2}$ factor between the attenuation poles.

## SPICE simulation circuits

| | |
|---|---|
| Fig. 5.18.2 | Phasqud1.SCH |

## References and additional sources 5.18

Burn C. (1998): Voiceband 90° phaseshifts. *Electronic Engng* January 14, 16.

Gingell M.J. (1973): Single-sideband modulation using sequence asymmetric polyphase networks. *Electr. Commun.* **48** (1, 2), 21–25.

Gingell M. J. (1975): *The Synthesis and Application of Polyphase Filters with Sequence Asymmetric Properties*, PhD Thesis, University of London.

Horowitz P., Hill W. (1989): *The Art of Electronics*, 2nd Edn. Cambridge: Cambridge University Press. ISBN 0-521-37095-7.

Jones B. K., Sharma B. K., Biddle B. H. (1980): A frequency independent phase shifter; *J. Phys. E* **13**, 1346.

Kalinski J. (1970): Variable FET resistance gives 90° phase shift. *Electronics* **43**, 88.

Kasal M., Halámek J., Husek V., Villa M., Ruffina U., Cofrancesco P. (1994): Signal processing in transceivers for nuclear magnetic resonance and imaging. *Rev. Sci. Instrum.* **65,** 1897–1902.

Midihian M., Watanabe K., Yamamoto T. (1979): A frequency independent phase shifter. *J. Phys. E* **12**, 1031.

Thomson F. J. (1979): Frequency multiplexer for ultrasonic Doppler blood flow signals. *Rev. Sci. Instrum.* **50**, 882–887.

# 5.19 Ratemeter

There are defeats more triumphant than victories.

Montaigne

Voltage-to-frequency conversion is a commonly used technique in many applications. The inverse process of frequency-to-voltage conversion, especially when the input is in the form of randomly spaced pulses, is less commonly encountered. This determination of an averaged rate can be achieved using the circuit shown in Fig. 5.19.1.

The 'frequency' is first converted to a pulse waveform of fixed amplitude $v_{in}$. Assuming for the moment ideal diodes, then on the positive going edge of $v_{in}$ diode $D_1$ will conduct so $C_i$ will be charged to $v_{in}$. On the negative going edge $C_i$ will be discharged through $D_2$, and since we have a virtual common all the charge is passed to $C_f$. Since the charge $Q$ on $C_i$ is given by $Q_i = C_i v_{in}$ the step in $v_{out}$ will be:

$$\delta v_{out} = \frac{Q_f}{C_f} = \frac{Q_i}{C_f} = v_{in}\frac{C_i}{C_f} \tag{5.19.1}$$

This charge will leak away via $R_f$ between input pulses. If the average output is $\langle v_{out} \rangle$ for an input repetition rate $f_{in}$, then the average current $\langle i_f \rangle$ through $R_f$ is equal to the input charge per unit time $v_{in} C_i f_{in}$, and:

$$\langle v_{out} \rangle = \langle i_f \rangle R_f = v_{in} C_i f_{in} R_f \tag{5.19.2}$$

so for fixed $v_{in}$ the output voltage is proportional to $f_{in}$. So we have an analog frequency meter or pulse ratemeter. Note that the output does not depend on $C_f$ but this does control the fluctuations in $v_{out}$ (Eq. (5.19.1)). A long time constant $R_f C_f$ gives a smooth output but a slow response to changes in $f_{in}$ and vice versa. If $f_{in}$ is constant then the system can be used as a capacitance meter (commonly used in hand-held test multimeters). Addition of a moving coil meter (which has a very restricted frequency response and so helps smooth fluctuations) to measure the output voltage, together with an audible sounder gives in effect the well known 'Geiger counter' beloved of film makers. With random counts, as from a radioactive source, the capacity $C_f$ must be adjusted to suit the fluctuations. The circuit has

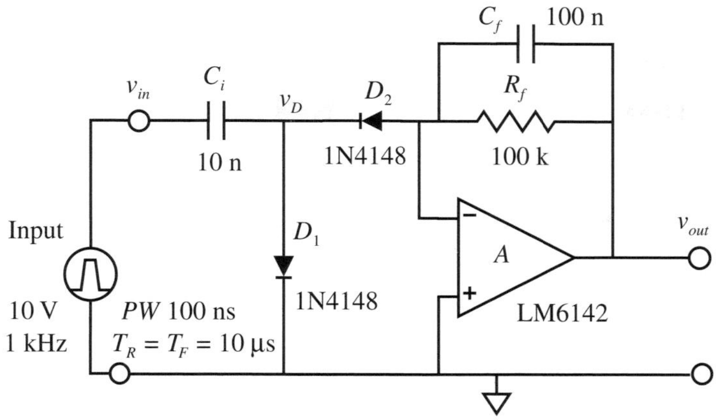


**Fig. 5.19.1** Ratemeter circuit.

been simulated using the components shown in the figure, with some results shown in Fig. 5.19.2.

As the smoothing time constant is about 10 ms it is necessary to set the run time to be at least, say, seven times this for the output to settle (100 ms was used). Allowing for the cut-in voltage of the diodes the measured amplitudes agree with (5.19.1) and (5.19.2).

If $R_f$ is omitted then the output will be a staircase waveform which can be used to generate a raster display for example, but some means must be included to reset before the output reaches the amplifier limit.

## SPICE simulation circuits

| | |
|---|---|
| Fig. 5.19.2(*a*) | F2vconv1.SCH |
| Fig. 5.19.2(*b*) | F2vconv1.SCH (Expanded time region) |

# Additional source 5.19

Cooke-Yarborough E. H., Pulsford E. W. (1951): A Counting-Rate Meter of High Accuracy; *Proc. IEE* **98 Part** (II), 191–195.

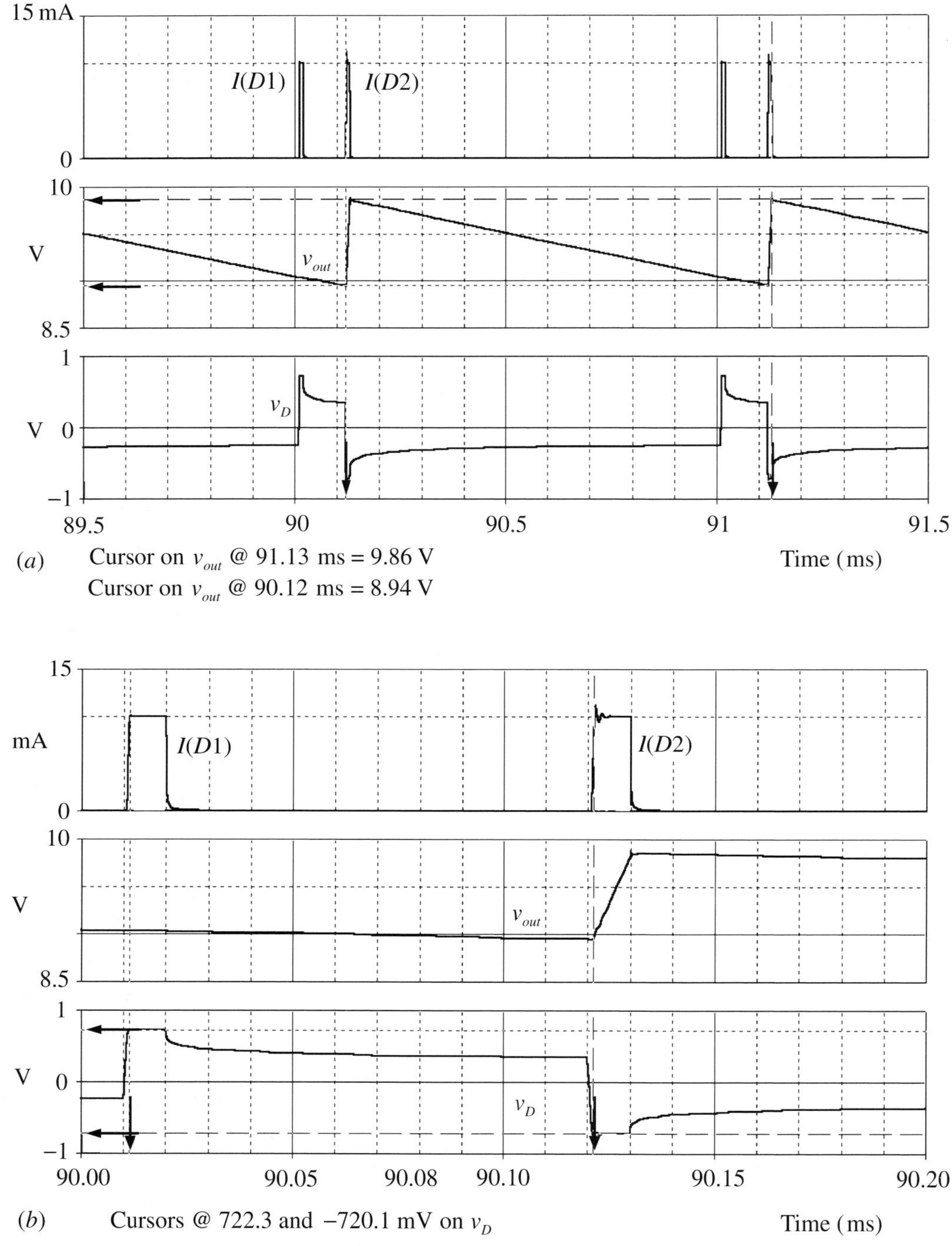


**Fig. 5.19.2** (*a*) Responses of ratemeter. (*b*) Expanded time region.

# 5.20 Baluns and high frequency transformers

The sea squirt, after an active life, settles on the sea floor, and, like a professor given tenure, absorbs its brain.

Steve Jones, in *Almost Like a Whale*

The performance of transformers at high frequencies is considerably compromised by parasitic capacity and inductance. A solution to the problems has been found by constructing the windings in the form of transmission lines where the capacity and inductance of the windings is subsumed into the transmission line characteristic. With such forms very wide band performance can be achieved and they can be used for a number of very useful purposes, though the physical form may look somewhat unusual at first sight. Similar performance may be obtained at very high frequency where it is feasible to use lines of lengths determined by the wavelength but this is unattractive at lower frequency where the physical extent would be prohibitive and in any case they are not broad band (e.g. Terman 1950, pp. 186 and 855; Balanis 1982, Section 8.8).

Discussion of the operation of transmission line transformers provides a useful example of the interaction between the circuit and the electromagnetic field views. We will follow the treatment given by Millman and Taub (1965, Section 3–20). Consider first a simple transmission line as shown in Fig. 5.20.1(*a*). A wave from the generator travels along the line to the load $R_L$ which we will take to be equal to the characteristic impedance $R_0$ of the line. The line is now folded back on itself so that $R_L$ is adjacent to the generator and point $Q$ is connected to point $S$ as shown in (*b*) (ignore the toroid for the moment). Though it may appear that the line is being shorted it must be remembered that the line is not just a zero resistance conductor but that it is a transmission line with inductance and capacity. Thus a wave of polarity as shown on the generator will appear inverted at the output at point $R$, i.e. the circuit constitutes a 1 : 1 inverting transformer. The line used may be either a wire pair (usually twisted) or a coaxial cable. For small transformers a pair of enamel insulated wires is flexible, compact and easy to use but there is no difference in principle and coax can nowadays be obtained in very small diameters.

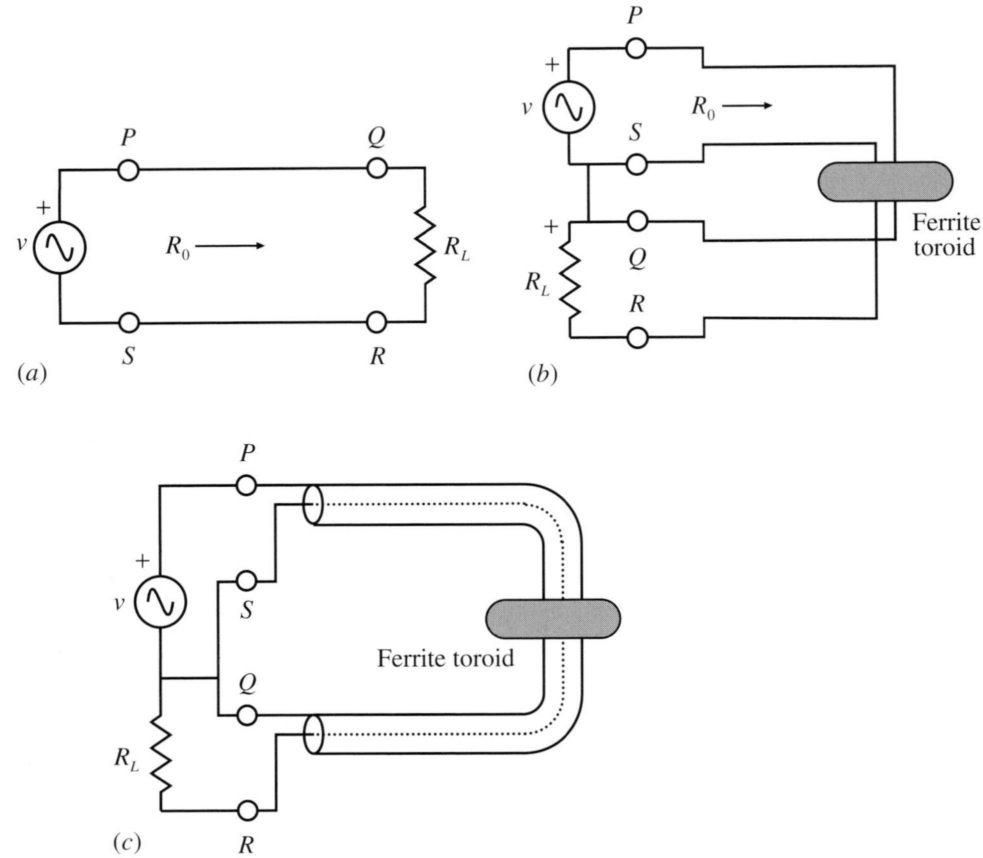


**Fig. 5.20.1** (*a*) Transmission line. (*b*) Inverting transformer with toroid. (*c*) Inverting transformer using coax.

At low frequency the line tends towards a short circuit around the loop *P*–*Q*–*S*–*v*–*P* and this will raise the low frequency cut-off and hence reduce the bandwidth. It is therefore desirable to increase the inductance, and hence impedance, of this loop, which can be achieved by means of a ferrite element (usually a toroid) as shown in (*b*). The increased inductance at low frequency will lower the low frequency cut-off (Section 4.4) so high permeability ferrite is desirable. An apparent drawback of such high permeability ferrites is that their high frequency performance falls off rapidly and they become very lossy. From the inductance point of view this does not matter too much since the line inductance will itself have increasing impedance with frequency, and it turns out that the increase in losses in the ferrite also has a beneficial effect. This is where we must look a little more closely at the system from a field point of view.

It may perhaps be easier to follow the argument with reference to the configuration shown in Fig. 5.20.1(*c*) using a coaxial cable. At high frequency we

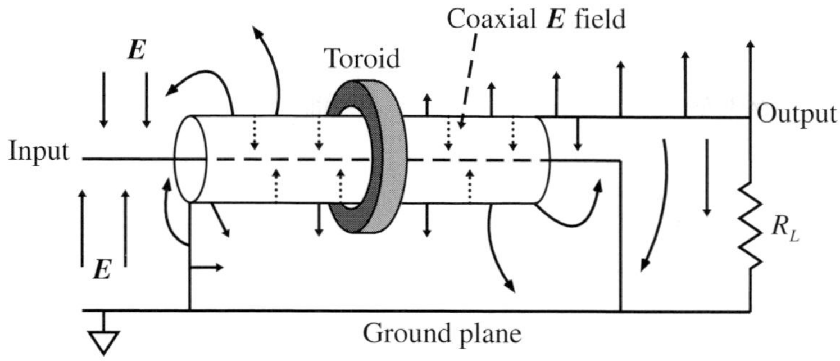


**Fig. 5.20.2** Electric fields for transmission line transformer.

should consider the outer sheath conductor as, in effect, two separate conductors, since the skin effect (Section 2.8) means that current can flow on the outer surface quite independently of that on the inner. Thus when we launch the wave from *PS* there will be an inner component – the usual one we associate with a coaxial cable and shown in Fig. 4.9.1 – and an outer one with a field between the outer sheath and the rest of the world. It is this latter wave that is involved in the 'short circuit' at low frequency and which we wish to inhibit at high frequency. The toroid thus performs both jobs simultaneously! In the case of a twisted pair line, the 'line wave' has its energy concentrated largely in the space between the wires while the common mode wave is around the outside, so we obtain a similar effect. A discussion of this view is given by Talkin and Cuneo (1957) who refer to the two as the transmission line mode with the currents in opposite directions, and as the coil mode, with currents in the wires in the same direction. The coil mode is further considered by Yamazaki et al. (1984). An illustration of the fields in the case of Fig. 5.20.1 is given by Koontz and Miller (1975) and is sketched in Fig. 5.20.2.

The incident wave from the left will see the two channels, one the coaxial cable and the other between the outside the cable and primarily the ground plane. The arrows represent the ***E*** field with the standard TEM field inside the cable. At the output the wave will propagate towards the load $R_L$ and some will be reflected back towards the input. At low frequency the permeability will make the outer channel a high impedance and at high frequency the ferrite will attenuate both external waves. Thus the transfer of energy will be primarily via the proper channel and the coupling between the elements of the cable. It is also of relevance to note that the ferrite will substantially decrease the velocity of propagation of the external wave since this is inversely proportional to $\sqrt{\mu}$ (and $\mu$ can be large) so that any wave through this will arrive too late to have any effect (Rochelle 1952).

This type of wideband transformer is often used to obtain a balanced signal from an unbalanced one and hence are known as baluns. Earlier designs were somewhat complex and bulky (Rochelle 1952; Talkin and Cuneo 1957; Lewis and

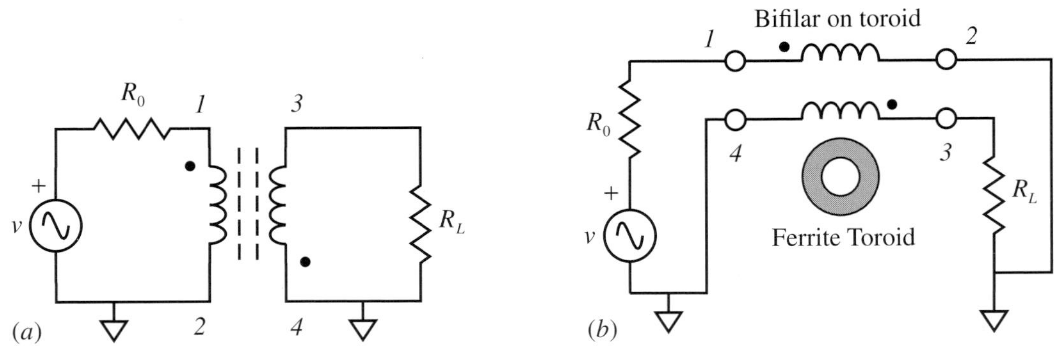


**Fig. 5.20.3** (*a*) Standard inverting transformer. (*b*) Balun inverter.

Wells 1959) but later developments have shown that simple miniature constructions work well (Ruthroff 1959; Winningstad 1959). The simple inverting transformer of Fig. 5.20.1 can be represented in conventional (*a*) and transmission line (*b*) form as shown in Fig. 5.20.3 (Ruthroff 1959), the significant difference being that in (*b*) there is no Galvanic isolation.

If an extra winding 5–6, essentially an extension of the input winding 1–2 is added (Fig. 5.20.4) then with the sense of the windings as shown in (*b*) the signal across the load will be balanced, i.e. equal positive and negative amplitudes as indicated.

For multiwinding transformers the rules for the form and the sense of the windings are that (Ruthroff 1959):

(i) without the load, there must be a closed circuit to allow magnetization of the magnetic core;

(ii) the sense of the windings should be such that the core fields generated aid one another.

If Galvanic isolation is required between input and output then a somewhat different approach is required. Consider two lines connected at input and output as shown in Fig. 5.20.5 (Millman and Taub 1965, Section 3-20).

With equal amplitude generators phased as shown and the symmetry of the circuit it is evident that the potential at $A$–$B$ is the same as that at $C$–$D$. The configuration of each line is that of an inverting transformer as depicted in Fig. 5.20.1 so the voltages across the loads are $v/4$ with the polarities shown. The currents in the lines will be in the sense indicated and be equal. It is then evident that the currents in the connection $A$–$B$, and those in $C$–$D$, will cancel, so we can remove both connections without any consequence to give Fig. 5.20.5(*b*). The directions of the currents in the two loops are shown. Again, to limit the common-mode currents the two lines are wound on a common ferrite core.

So now the primary is isolated from the secondary as in a more conventional

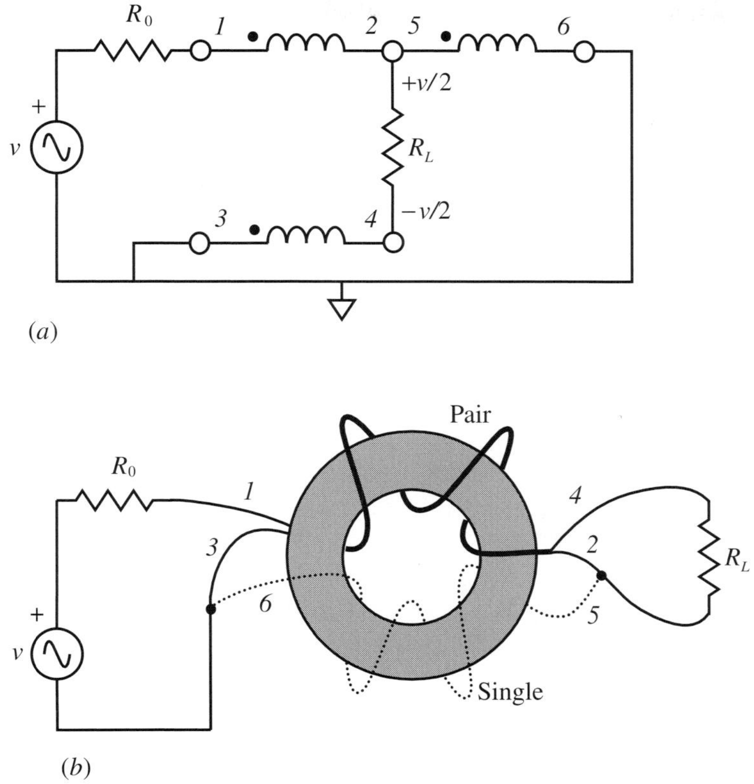


**Fig. 5.20.4** (*a*) Ruthroff transformer schematic. (*b*) Ruthroff transformer windings.

transformer. There is one important difference from the earlier configurations in that the matching resistance is $2R_0$ in this case; there are in effect two transmission lines in series so the matching resistance is doubled. Figure 5.20.5(*c*) redraws (*b*) to demonstrate the equivalence with the usual isolating transformer of Fig. 5.20.3(*a*), and (*d*) is the equivalent circuit using coaxial lines. Koontz and Miller (1975) illustrate such a transformer with a risetime of 1 ns and isolation up to 100 kV.

The performance of the inverting transformer of Fig. 5.20.1 may be examined using SPICE, though there is a difficulty in how to allow for the effect of the toroid. The toroid serves to inhibit the coil-mode current which is in effect the same as a common-mode signal, so that insertion of a common-mode choke (Section 4.4) between the generator and the line will attenuate this while allowing the required differential line currents to pass. To represent the line we need to construct a balanced circuit since both conductors necessarily have impedance and the models provided in PSpice are unsymmetrical. A circuit as shown in Fig. 5.20.6 may be created where we have used only eight segments as an approximation.

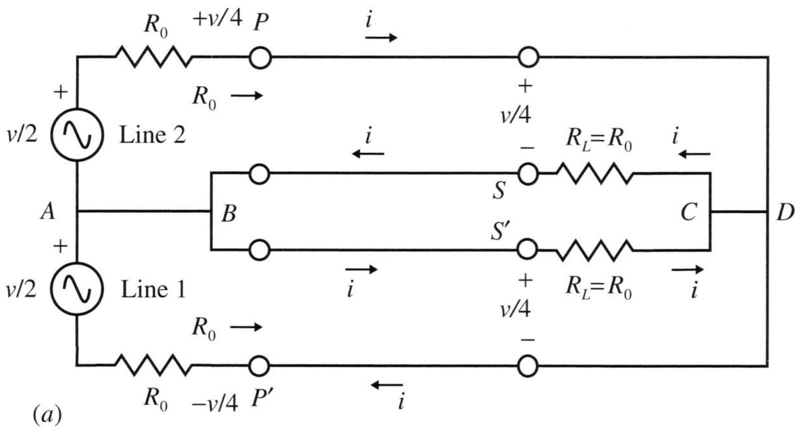


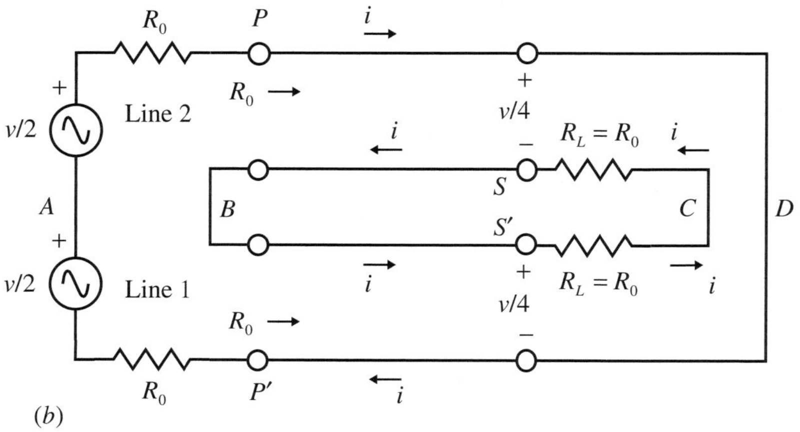


**Fig. 5.20.5** (*a*) Parallel transmission lines. (*b*) Links carrying zero current removed. (*c*) Equivalent standard transformer winding. (*d*) Isolating transformer using coax.

For the component values shown the characteristic impedance $Z_0$ and delay $t_D$ are given by:

$$Z_0 = \left(\frac{L}{C}\right)^{\frac{1}{2}} = \left(\frac{8 \times 16 \times 10^{-9}}{3 \times 8 \times 10^{-12}}\right)^{\frac{1}{2}} = 73\ \Omega \tag{5.20.1}$$

and $t_D = (LC)^{\frac{1}{2}} = (8 \times 16 \times 10^{-9} \times 3 \times 8 \times 10^{-12})^{\frac{1}{2}} = 1.75$ ns

The common-mode choke is formed by $L_1$ and $L_2$ with a coupling factor (*K_Linear*) $k = 1$. If $R_3$ is made large then we have a simple non-inverting configuration which may be tested with output at $Q$ to see if all is well. If $R_3$ is made small then we have the inverting configuration with output at $U$. $R_3$ and $R_4$ allow measurement of current if desired. $R_2$ may be varied from say 50 to 100 Ω to

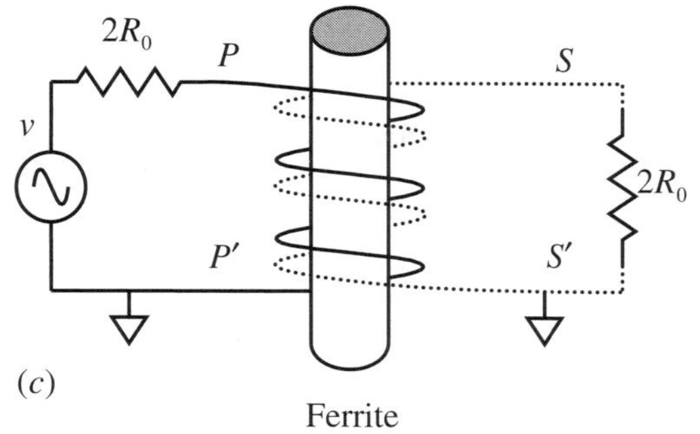


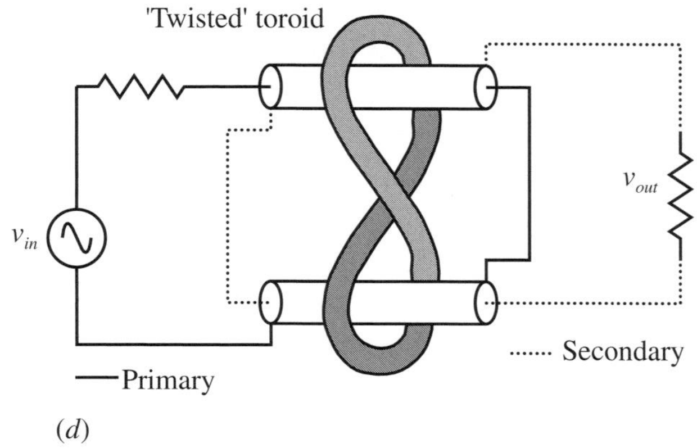


**Fig. 5.20.5** *(cont.)*

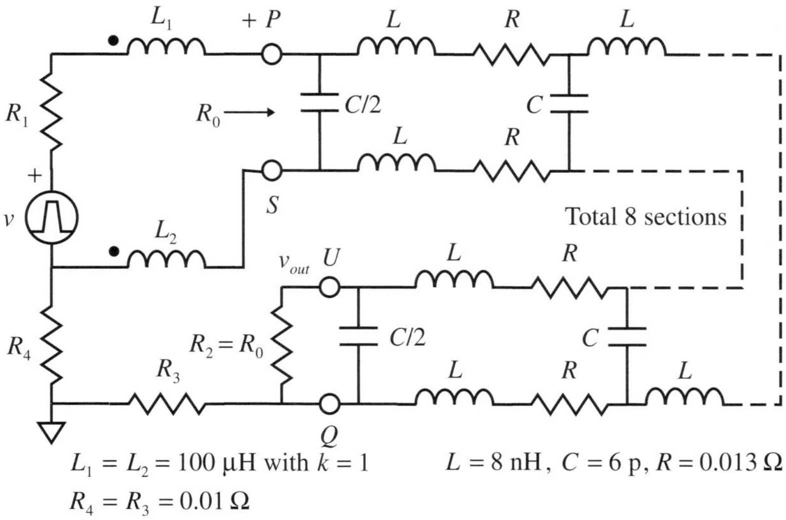


**Fig. 5.20.6** Inverting transformer for simulation.

see the effect of mismatch. $R_1$ may be varied from small (not zero) to, say, 73 Ω to vary the output from $v$ to $v/2$. Currents may be examined to determine magnitude and direction at any point: displaying the currents through the capacitors will show the progressing wave with currents only during the rise and fall of the pulse. The values for $Z_0$ and $t_D$ are found to agree with Eq. (5.20.1).

The performance of the isolating transformer of Fig. 5.20.5 may be examined in a similar manner by constructing a circuit as shown in Fig. 5.20.7 (though here we only show four sections for each line for compactness).

You may examine the operation progressively by starting with two isolated lines as in Fig. 5.20.6 and adding in or deleting the interconnections. As before the two chokes $L_1L_2$ and $L_3L_4$ are used to represent the effect of the ferrite. The signals at $P'$ and $S$ will be found to be the same so these two points may be connected if a 1:1 non-inverting transformer is required (this disagrees in part with the conclusion of Millman and Taub). Other complex arrangements are described by Bosshard et al. (1967) and for very low impedance loads by Tansal and Sobol (1963).

### SPICE simulation circuits

Consult the SimCmnt.doc file on the CD before running

| | |
|---|---|
| Fig. 5.20.6 | BalunT5.SCH |
| Fig. 5.20.7 | BalunT4.SCH |

## References and additional sources 5.20


Balanis C. A. (1982): *Antenna Theory*, New York: John Wiley. ISBN 0-471-60352-X.

Berg R. S., Howland B. (1961): Design of wide-band shielded toroidal RF transformers. *Rev. Sci. Instrum.* **32**, 864–865.

Blocker W. (1978): The behavior of the wideband transmission line transformer for non-optimum line impedance. *Proc. IEE* **65**, 518–519.

Bosshard R., Zajde C., Zyngier H. (1967): Nanosecond bifilar wound transformers. *Rev. Sci. Instrum.* **38**, 942–945.

Gorodetzky S., Muser A., Zen J., Armbruster R. (1962): Application des transformateurs d'impulsions en electronique rapide impulsionelle. *Nucl. Instrum. Meth.* **17**, 353–360.

Granberg H. O. (1980): Broadband transformers. *Electronic Design* 19 July, 181–187.

Granberg H. (1995): *Broadband Transformers and Power Combining Techniques for RF*, Motorola Application Note AN749. Also RF Application Reports Handbook HB215/D.

Irish R. T. (1979): Method of bandwidth extension for the Ruthroff transformer. *Electron Lett.* **15**, 790–791.

Koontz R. F., Miller R. H. (1975): Nanosecond electron beam generation and instrumentation at SLAC. *IEEE Trans.* **NS-22**, 1350–1353.

Kraus H. L., Allen C. W. (1973): Designing toroidal transformers to optimize wideband performance. *Electronics* 16 August, 113–116.

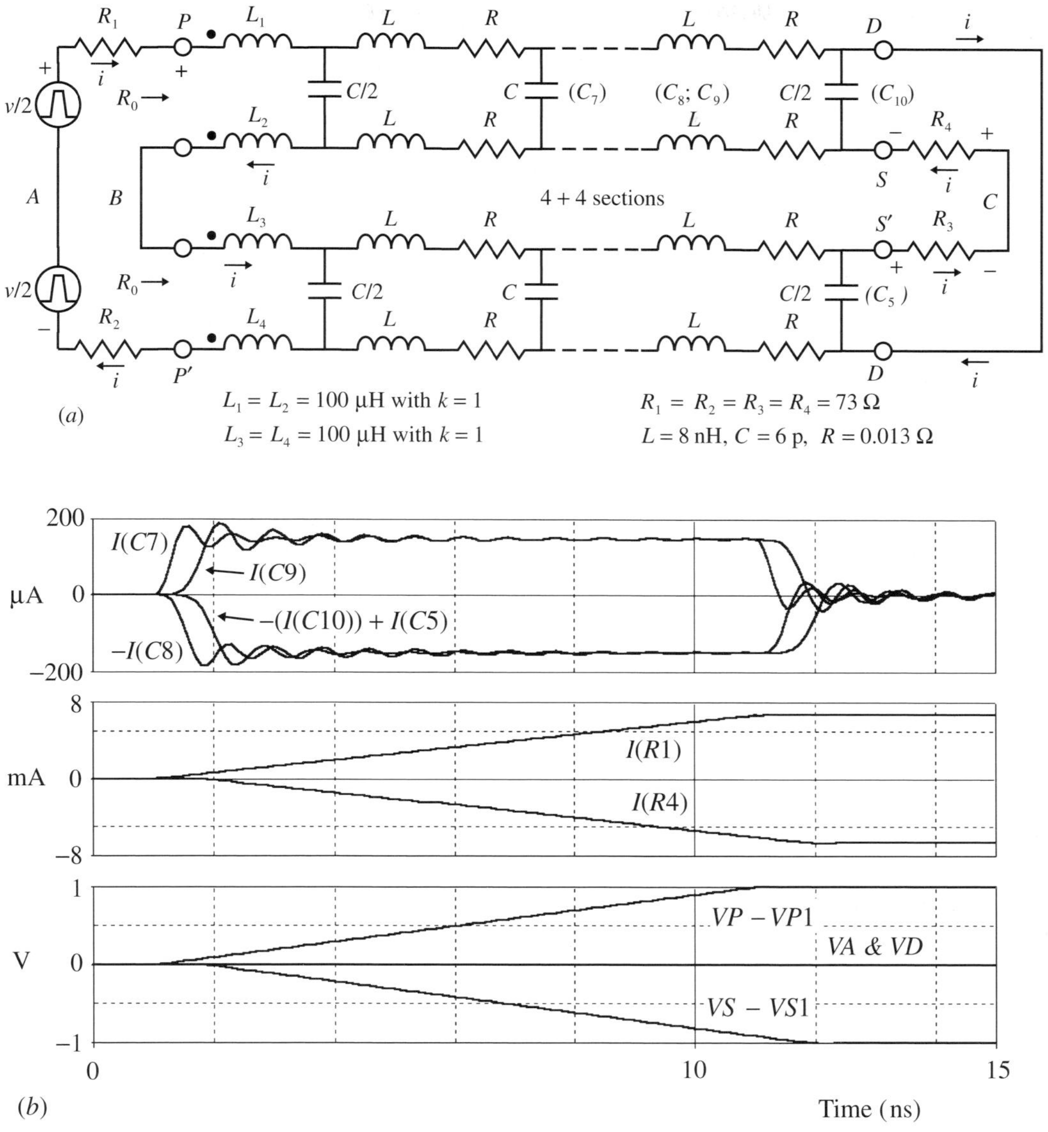


**Fig. 5.20.7** (*a*) Isolating transformer for simulation. (*b*) Simulation waveforms for a pulse input showing the rising edge only: capacitors $C_5$ and $C_{8-10}$ are indicated in (*a*). The currents through these illustrate the propagation along the line and only occur during the pulse edge transitions (alternate currents are shown inverted for clarity).


Lewis I. A. D., Wells F. H. (1959): *Millimicrosecond Pulse Techniques*, Oxford: Pergamon Press. Library of Congress Cat. No. 59-12608.

Matick R. E. (1968): Transmission line pulse transformers – Theory and applications. *Proc. IEEE* **56**, 47–62.

Millman J., Taub H. (1965): *Pulse, Digital, and Switching Waveforms*, New York: McGraw-Hill. Library of Congress Cat. No. 64-66293.

O'Dell T. H. (1991): *Circuits for Electronic Instrumentation*, Cambridge: Cambridge University Press. ISBN 0-521-40428-2.

Oltman G. (1966): The compensated Balun. *IEEE Trans.* **MTT-14**, 112–119.

O'Meara T. R. (1961): A distributed-parameter approach to the high frequency network representation of wide-band transformers. *IRE Trans.* **CP-8**, 23–30.

Pitzalis O., Couse T. P. (1968): Practical design information for broadband transmission line transformers. *Proc. IEEE* **56**, 738–739.

Rochelle R. W. (1952): A transmission-line pulse inverter. *Rev. Sci. Instrum.* 23, 298–300.

Ruthroff C. L. (1959): Some broad-band transformers. *Proc. IRE* **47**, 1337–1342.

Sevick J. (1987): *Transmission Line Transformers*, American Radio Relay League Inc., 225 Main Street, Newington, Connecticut 06111. ISBN 0-87259-046-1.

Talkin A. I., Cuneo J. V. (1957): Wide-band balun transformer. *Rev. Sci. Instrum.* **28**, 808–815.

Tansal S., Sobol H. (1963): Wide-band pulse transformers for matching low impedance loads. *Rev. Sci. Instrum.* **34**, 1075–1081.

Terman F. E. (1950): *Radio Engineers' Handbook*, New York: McGraw-Hill.

Winningstad C. N. (1959): Nanosecond pulse transformers. *IRE Trans.* **NS-6**, 26–31.

Yamazaki H., Homma A., Yamaki S. (1984): Isolation and inversion transformers for nanosecond pulses. *Rev. Sci. Instrum.* **55**, 796–800.

# 5.21 Directional coupler

Euler faced non-mathematical sceptic Diderot with the challenge 'Sir, $a+b^n/n=x$, hence God exists; reply!' Diderot did not reply and Euler's case prevailed.

A directional coupler is used to monitor the power flowing in a particular direction in a conductor. For example, a signal source may be feeding an antenna but if the matching is not correct there will be power reflected from the antenna that travels back towards the source. A circuit that achieves this can be constructed using two transformers as shown in Fig. 5.21.1 (McWhorter 1991). The idea is that power in at port $A$ is transmitted with small loss to port $B$, a fraction is transmitted to port $D$ but nothing passes to port $C$. Reflected power at port $B$ will be partially coupled to port $C$ but not to port $D$. This allows transmitted power to be monitored at $D$ and reflected power at $C$. The degree of attenuation from the main path to the monitor path is determined by the ratio of the primary inductance $L_1$ to the secondary $L_2$. For a turns ratio of, say, 10 so that $L_2/L_1=100$, the power out at $D$ will be –20 dB ('a 20 dB coupler') and that at $B$ will be –0.04 dB, i.e. very little loss.

Since we are demonstrating the principle of operation we use ideal transformers with unity coupling, so that in the relation $M^2=k^2L_1L_2$ the coupling factor $k=1$ (Section 4.3). The labelling of the currents and voltages of the above reference have been preserved to make comparison easier. The circuit can be analysed by writing down the loop equations for the loops indicated in the figure by the four currents $I_1$ to $I_4$:

$$\begin{aligned} -V_1+I_1(R_3+R_1+sL_3)+I_2R_1+I_3sM&=0\\ -V_1+I_1R_1+I_2(R_1+sL_2)-I_4sM&=0\\ I_1sM+I_3(R_2+sL_4)+I_4R_2&=0\\ -I_2sM+I_3R_2+I_4(R_4+R_2+sL_1)&=0 \end{aligned} \tag{5.21.1}$$

and note that since the transformers are identical then $L_3\equiv L_1$ and $L_4\equiv L_2$. We also make the condition that all the resistances are equal, i.e. $R_1=R_2=R_3=R_4=R$, and that in the frequency region over which the coupler will be used $sL_2\gg R$. The solution of Eq. (5.21.1) for the separate currents is a straightforward if rather lengthy

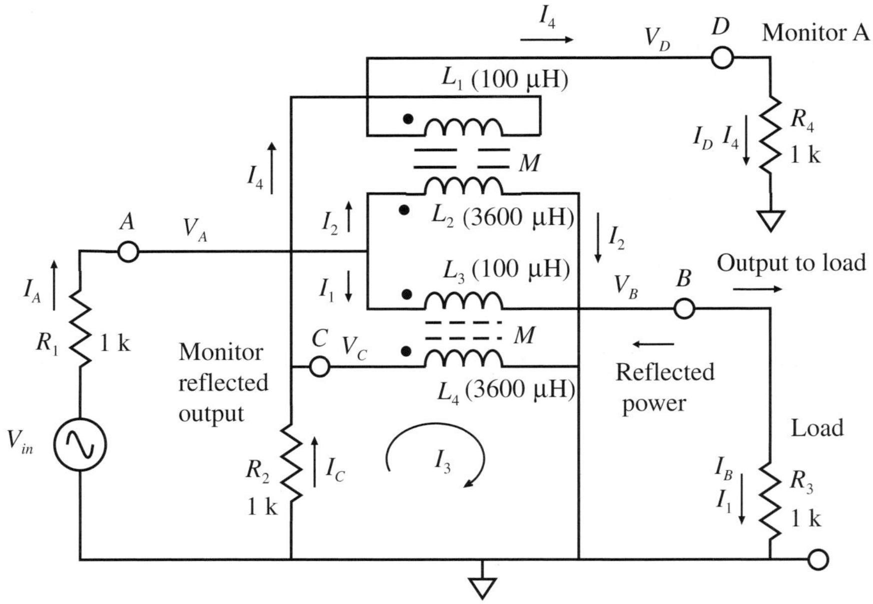


**Fig. 5.21.1** Directional coupler.

process which we will not reproduce here. The solution in terms of determinants (Section 1.10) is illustrated for $I_1$ by the equation:

$$GI_1 = \begin{vmatrix} V_1 & R & sM & 0 \\ V_1 & sL_2 & 0 & -sM \\ 0 & 0 & sL_2 & R \\ 0 & -sM & R & (2R+sL_1) \end{vmatrix} \tag{5.21.2}$$

where the determinant of the coefficients, $G$, is given by:

$$G = \begin{vmatrix} (2R+sL_1) & R & sM & 0 \\ R & sL_2 & 0 & -sM \\ sM & 0 & sL_2 & R \\ 0 & -sM & R & (2R+sL_1) \end{vmatrix} \tag{5.21.3}$$

Equations (5.21.2) and (5.21.3), after using our approximation $sL_2 \gg R$, reduce after much algebra to:

$$GI_1 = V_1(2s^2L_2^2R) \quad \text{and} \quad G = 2s^2R^2L_2(2L_2 + L_1)$$

$$\text{so} \quad I_1 = \frac{V_1}{R}\frac{L_2}{(2L_2 + L_1)} \tag{5.21.4}$$

Solving for the other currents we eventually find the port currents to be:

$$I_A = I_1 + I_2 = \frac{V_1}{R}\left(\frac{L_1 + L_2}{L_1 + 2L_2}\right)$$

$$I_B = I_1 = \frac{V_1}{R}\left(\frac{L_2}{L_1 + 2L_2}\right) \tag{5.21.5}$$

$$I_C = -(I_3 + I_4) = 0$$

$$I_D = I_4 = \frac{V_1}{R}\left(\frac{M}{L_1 + 2L_2}\right)$$

The power at each port $x$ is given by $P_x = I_x^2 R$ and the ratio to the input power $P_A$ by $P_{xA}$:

$$P_A = \frac{V_1^2}{R_1}\frac{(L_1 + L_2)^2}{(L_1 + 2L_2)^2} \approx \frac{V_1^2}{R_1}\frac{L_2(2L_1 + L_2)}{(L_1 + 2L_2)^2}, \quad \text{if} \quad L_1^2 \ll L_2^2$$

$$P_B = \frac{V_1^2}{R_3}\frac{L_2^2}{(L_1 + 2L_2)^2} \quad \text{so} \quad P_{BA} = \frac{L_2}{(2L_1 + L_2)}\left(\frac{R_1}{R_3}\right) \tag{5.21.6}$$

$$P_C = 0 \quad \text{so} \quad P_{CA} = 0$$

$$P_D = \frac{V_1^2}{R_4}\frac{M^2}{(L_1 + 2L_2)^2} \quad \text{so} \quad P_{DA} = \frac{L_1 L_2}{(L_1 + L_2)^2}\left(\frac{R_1}{R_4}\right) \approx \frac{L_1}{(2L_1 + L_2)}\left(\frac{R_1}{R_4}\right)$$

The analysis is clearly rather lengthy and not one that is readily used to investigate variations and the effects of, say, parasitic elements on performance, but at least we now understand what to expect in general. The resistive terminations would normally be equal, which would then cancel out in Eq. (5.21.6). They have been left in place so that the effect of changes may be assessed. It is now with some relief that we can turn to PSpice to see if our computations are correct at least as far as the approximations allow. To discriminate between output and reflected signals $R_3$ can be in part replaced by a *ZX* component with a reference resistor and driven by, say, a *VPULSE* source (see Section 5.27 part (f)). Some simulation results are shown in Fig. 5.21.2.

It can be seen that the output at port *D* is effectively independent of the load at port *B*, and that there is an output at port *C* only when port *B* is not matched ($R_3 = 700\ \Omega$). Some illustrative results are given in Table 5.21.1.

The agreement for the matched case is satisfactory, but for the unmatched case the differences are greater. However, it should be recalled that the analysis was carried out for all the resistors equal so it is not too surprising that there are differences.

A circuit with somewhat similar properties but dependent also on frequency is shown in Fig. 5.21.3 (Ponsonby 1990).

The analysis is again rather extended so we will simply state its properties and

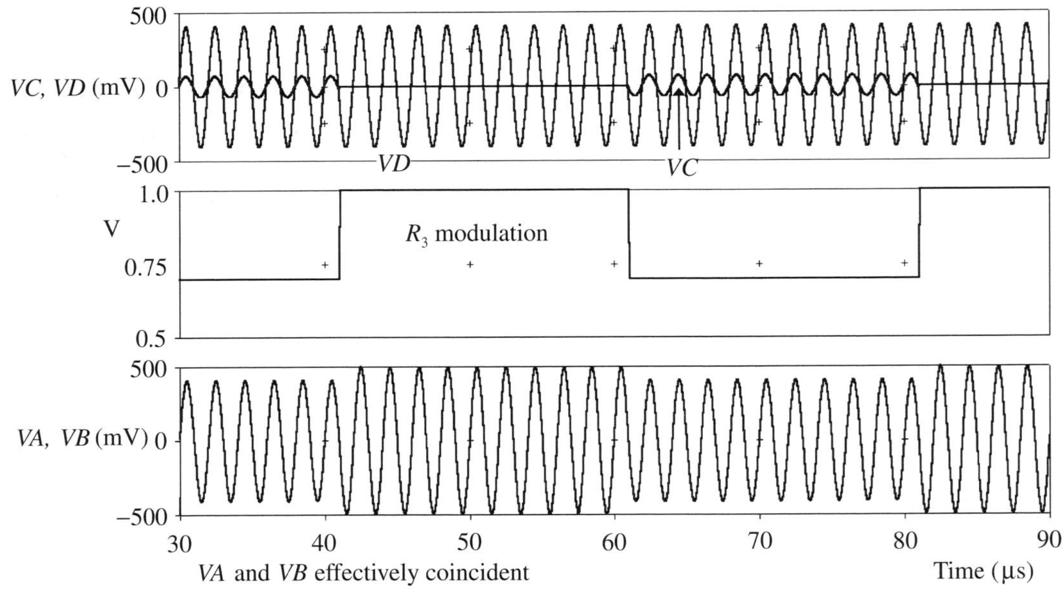


**Fig. 5.21.2** Simulation waveforms for Fig. 5.21.1. Resistors of 5 Ω were inserted in series with each $L_1$ (100 μH) and 30 Ω for each $L_2$ (3600 μH). $R_3$ was modulated from 1 k to 700 Ω. The source frequency was 500 kHz and amplitude 1 V.

Table 5.21.1 *Power ratios at the various directional coupler ports*

| | $R_3=1$ k | | | $R_3=700\ \Omega$ | | |
|---|---|---|---|---|---|---|
| Power ratio | $P_B/P_A$ | $P_C/P_A$ | $P_D/P_A$ | $P_B/P_A$ | $P_C/P_A$ | $P_D/P_A$ |
| Calculated | 0.95 | 0 | 0.026 | 1.35 | 0 | 0.026 |
| Measured | 0.99 | 0 | 0.028 | 1.39 | 0.001 17 | 0.040 |

then examine the performance using PSpice. The resistors in series with the inductors represent their inherent resistance. The 1 : 1 transformer is assumed to have complete coupling so that $M=L$ and the four port resistors are equal. Consider a signal input at $A$. For a low frequency the output will be at port $B$ and for a high frequency at port $D$. There will be no output at port $C$. The crossover will be at the frequency:

$$f_c=\frac{1}{2\pi(LC)^{\frac{1}{2}}},\quad \text{where} \quad C=C_1+C_2$$

$$\text{with} \quad R=\left(\frac{L}{C}\right)^{\frac{1}{2}} \tag{5.21.7}$$

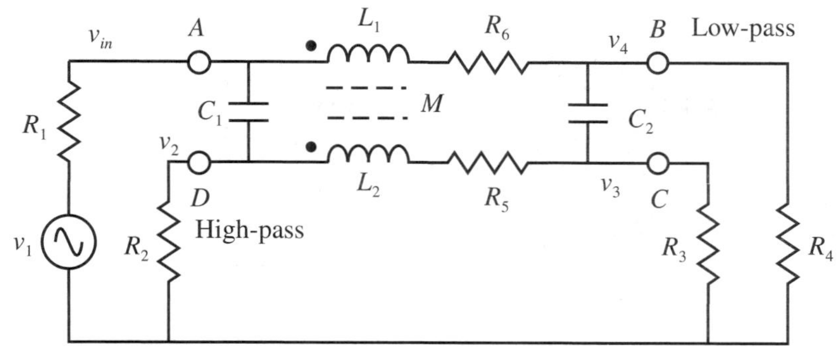


**Fig. 5.21.3** Frequency dependent coupler.

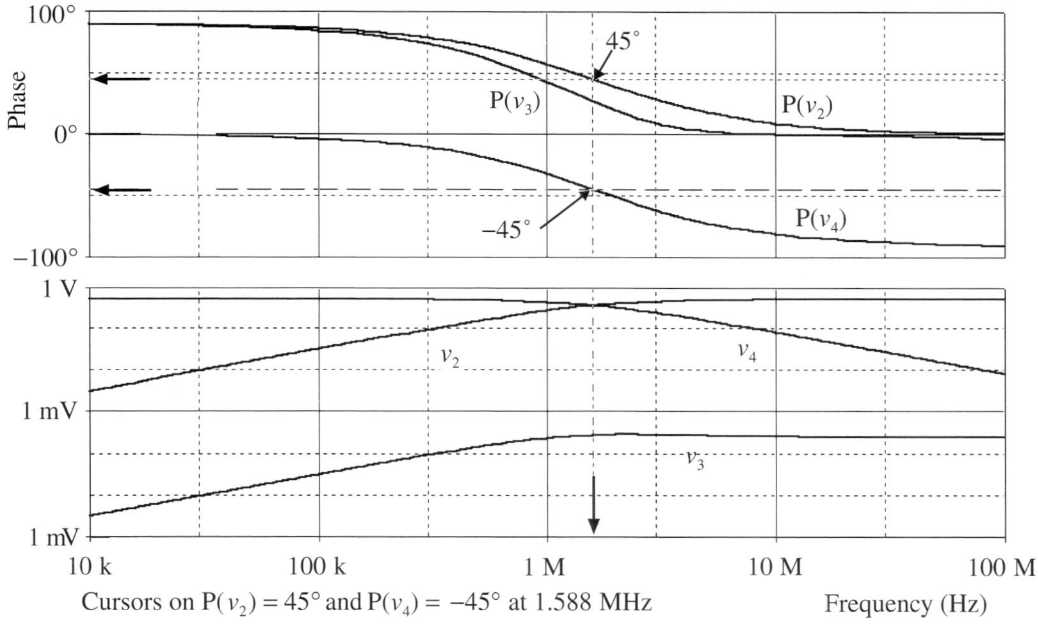


**Fig. 5.21.4** Responses of frequency dependent coupler.

and the phase difference between port *B* and port *D* will be 90°, with each at 45° to port *A*, at *all* frequencies. The circuit is symmetric, and since, for example, in the case of input at *A* there is no output at *C*, then a separate signal may be input at *C* with no output at *A* but with superposed outputs at *B* and *D*. Responses are illustrated in Fig. 5.21.4.

## SPICE simulation circuits

| | |
|---|---|
| Fig. 5.21.2 | Dircpl2.SCH |
| Fig. 5.21.4 | Xfmrcplr.SCH |

## References and additional sources 5.21

Dunbar S. (2000): Build a vector network analyzer. *Electronic Design* 29 May, 107. See also Wiebach W. (2000): Crafting a noncomplex circuit. *Electronic Design* 21 August, 56.

McWhorter M. (1991): Broadband RF transformer directional couplers. *RF Design* **14** (7), 53–58.

Ponsonby J. E. B. (1990): Private communication.

# 5.22 Power control or hotswitch

A cartoonist is someone who does the same thing every day without repeating himself.

Charles Schulz

Controlling power to subcircuits, or hot insertion of devices, is a common requirement. For the more common positive power supply a P-channel FET is most convenient as the switch since the gate can be pulled to zero to switch on and no extra bias voltage is required. N-channel FETs are often used to control large currents as they have lower on resistance but their use necessitates an additional bias voltage. Some IC hotswitch versions incorporate the required bias supply but the r.f. switching frequency may be undesirable. A simple control switch is shown in Fig. 5.22.1. A positive control signal switches $Q_2$ on which pulls the gate of $Q_1$ close to common so that $Q_1$ connects the power supply to the subcircuit and has resistance $R_{DS(on)}$. A choice of transistor to suit the current load should be made. Even tiny SOT-23 FETs can have resistances of a few tenths of an ohm, e.g. the IRLML6302. If some delay before switch-on is required a *RC* time constant on the input gate is sometimes used as shown.

A difficulty with this simple circuit is that the switch closes rather abruptly so that the initial transient supply current arising from the rapid charging of the

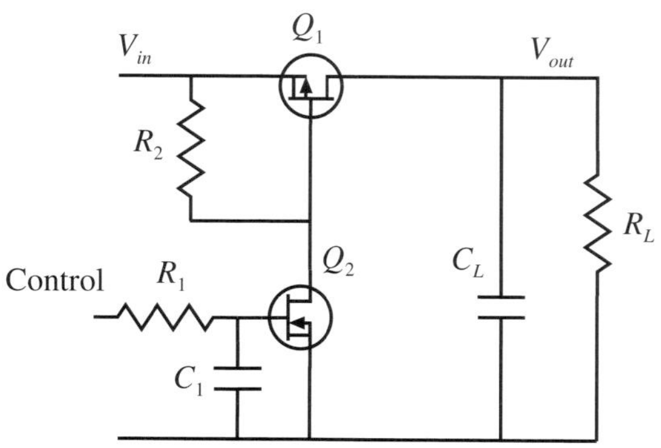


**Fig. 5.22.1** Simple FET power control switch.

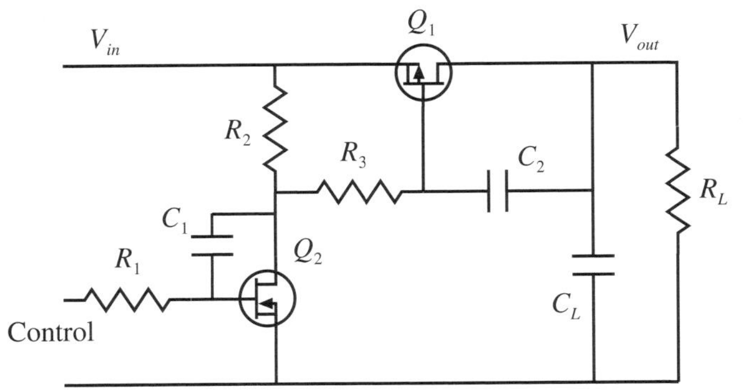


**Fig. 5.22.2** Power control switch using Miller integrator control.

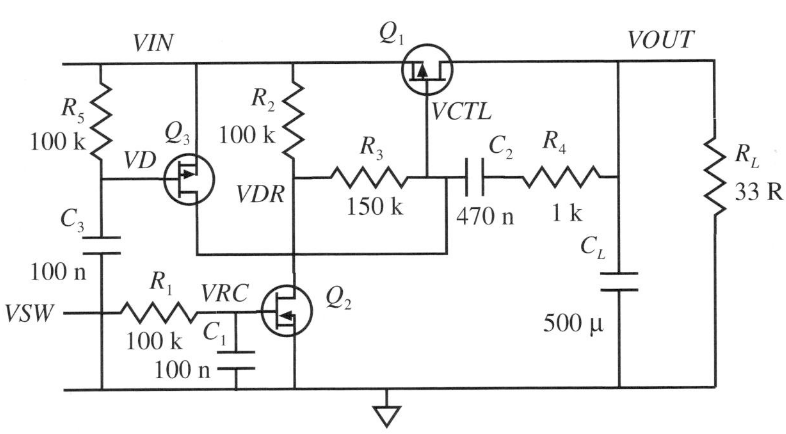


**Fig. 5.22.3** Modified control for initial supply transient.

subcircuit smoothing capacity $C_L$ can be very large resulting in severe enough disturbance to the main circuit to cause it to malfunction, e.g. microprocessor reset or power supply protective shutdown. These currents, depending on the incidental resistance in the system and load capacity $C_L$, can be amperes. What is required is a more gradual power-on sequence that limits the transient inrush current to a tolerable value. FETs turn well on over a small range of gate voltage the values of which are rather variable between FETs even of the same type, so it is difficult to adjust directly for a more gradual turn-on. A very simple variation of the circuit of Fig. 5.22.1 can provide excellent control of both switch-on and of delay. The answer is the use of what is often considered a considerable nuisance, the Miller effect, i.e. the consequence of the feedback capacity between drain and gate and the gain of the FET (Section 3.10). Originally described for vacuum tubes (capacity from plate to grid) the negative feedback through this capacity effectively multiplies its value by the gain. (It is this that makes the operational integrator so

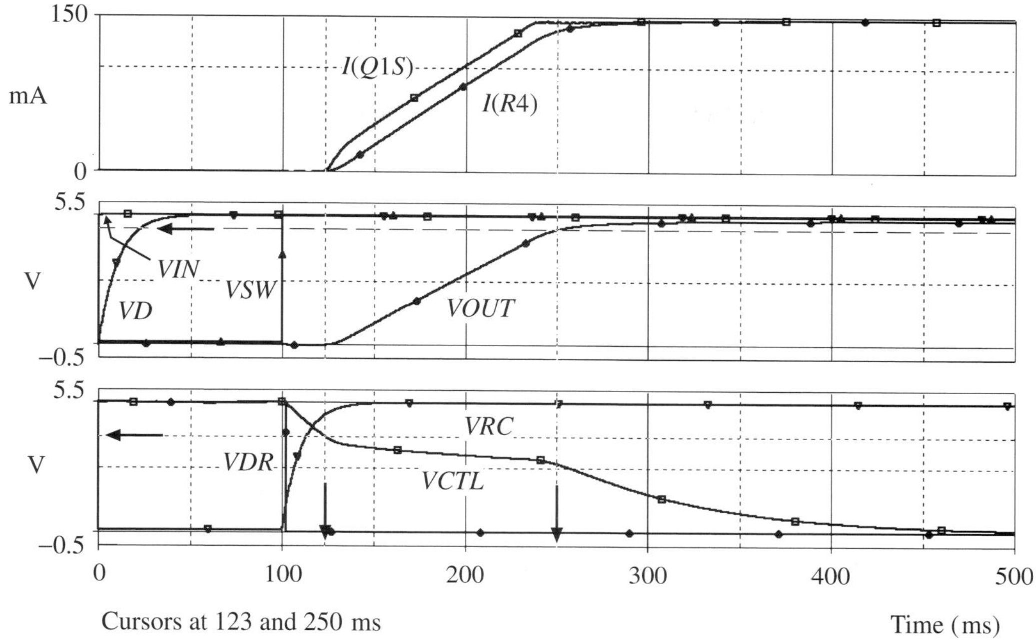


**Fig. 5.22.4** Responses of circuit of Fig. 5.22.3 with application of *VIN* and subsequent switch-on by *VSW* input.

effective and these are sometimes referred to as Miller integrators.) Thus if a capacitor is added as shown by $C_2$ (Fig. 5.22.2) to increase the inherent FET $C_{gd}$ substantially then with $R_3$ we have a Miller integrator with a selectable time constant and the negative feedback dominates the response rather than the particular characteristics of the FET (as in any other application using negative feedback). Variations in $V_{th}$ between FETs are now unimportant. The result is a linear rise in output voltage in a time set by the integrator time constant of $R_3$ times the effective value of $C_2$.

The same Miller technique can be employed to provide the delay-to-on time using $R_1$ and $C_1$ to give the linear ramp down at $Q_2$ drain, but the configuration shown in Fig. 5.22.3 has less interaction with the rest of the circuit. The circuit of Fig. 5.22.2 operates as described if the input supply voltage is permanently applied. However, when $V_{in}$ is initially applied then the gate of $Q_1$ will initially be held at zero volts by $C_L$ and $C_2$ so there will be a very large current transient. To avoid this $Q_3$ is added to hold the gate of $Q_1$ at $V_{in}$, and charge $C_2$, until $C_3$ charges up and $Q_3$ goes off (Fig. 5.22.3).

As shown, with $R_1 = 100$ k, a value of $C_1 = 100$ nF gives a delay of about 30 ms to the *start* of the output ramp. Though the responses shown in Fig. 5.22.4 are those from the simulation, the results in practice are closely similar even using quite different transistors.

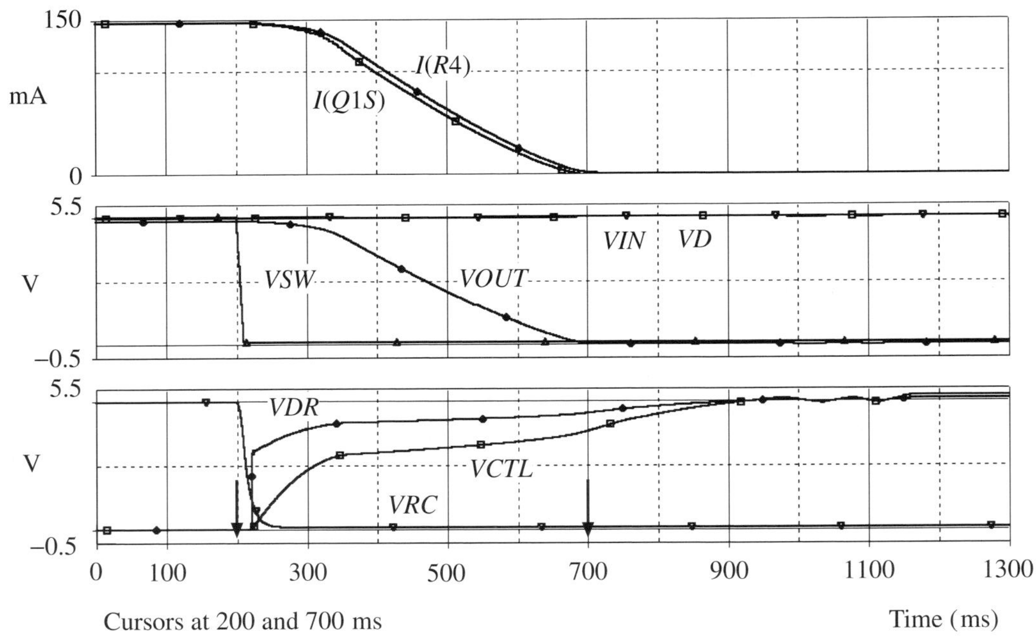


**Fig. 5.22.5** Responses for Fig. 5.22.3 with *VSW* control.

When switching off the circuit behaves in roughly the inverse way to switch-on, but the decay of output voltage and current drawn is rather longer since $Q_2$ is now inactive. The decay can be speeded up by additional means but a slow decay is beneficial as regards the effect on the main circuit voltage regulator. Since voltage regulators are unipolar in the sense that they can actively respond to increases in current load, i.e. the series element is turned on, but they are unable to compensate actively for a decrease in output current demand, i.e. the series element turns off. Thus a sudden decrease in output current arising from switching off the subcircuit is less desirable than a gradual decrease. The responses to switching off by *VSW* are shown in Fig. 5.22.5.

A sudden demand for extra current by the subcircuit causing $V_{out}$ to fall will, via $C_2$, simply turns $Q_1$ further on.

Many manufacturers make a wide variety of hot swap devices to suit differing applications and with desirable protection mechanisms. None that I know of operate in the manner described above or allow ready adjustment of the ramping characteristics.

## SPICE simulation circuits

Consult the SimCmnt.doc file on the CD before running

| | |
|---|---|
| Fig. 5.22.3 | Hotsw3.SCH |
| Fig. 5.22.4 | Hotsw3.SCH |
| Fig. 5.22.5 | Hotsw3.SCH |

## References and additional sources 5.22

Daniels D. G. (1998): *TPS202x/3x and TPS204x/5x USB Power Distribution Application Report*, Texas Instruments Literature Number SLVA049.

Lenz M. (1999): *Power Management for Control Equipment*, PCIM Europe Issue 12.

Reay R., Herr J. (1996): Safe hot board swapping. *Electronic Product Design* December 41–42. (LTC 1421.)

# 5.23 Modulation control of a resonant circuit

Scientists study what is. Engineers create what has never been.

Theodore von Karman

Radiofrequency identification tags are small unpowered devices that can carry information and are used in a similar way to bar-code labels. They differ from the latter in that they can often carry much more information and can be written to. A radiofrequency (r.f.) field is used to provide power to the tag via a pickup coil embedded in the tag and modulation of this field by the tag allows extraction of the information stored in the tag. The only way of writing to the tag is via control of the applied r.f. field by switching it on and off for (short) defined periods. This seems a straightforward operation but since resonant circuits are generally used to develop the required intensity of r.f. magnetic field ($B$) and the modulation times can be as short as, say, three cycles, this would require very low $Q$ resonators. It should be pointed out that the coupling between coil and tag is near-field, i.e. like an air-cored transformer rather than the far-field (radiative) regime.

To obtain high r.f. fields a high $Q$ is required, though this is limited by the bandwidth necessary to receive the data stream from the tag. What is needed is a means of switching the r.f. on and off rapidly while maintaining a high $Q$. Switching off the field is quite straightforward but to get it back up to full amplitude takes a time proportional to $Q$ (see Fig. 5.23.1). Even if the off-time can be adjusted it is evident that the time to reach a given intensity of field that the tag recognizes as *ON* will depend on the distance between the coil and the tag. This is where a little lateral thinking is required or (as I like to think) how a physicist may approach the problem.

A resonator requires two forms of energy storage with the energy flowing back and forth between the two, as for example in a simple pendulum where the energy at the extremes of its motion is completely potential (velocity = 0) and in the centre where it is kinetic (velocity = maximum). For an $LC$ resonator the energy at one point is all stored in the magnetic field of $L$ (energy $=\frac{1}{2}LI^2$) and a half cycle later it is all stored in the electric field of $C$ (energy $=\frac{1}{2}CV^2$). When the field is to be switched off the energy must be dissipated, which can be achieved by shorting the

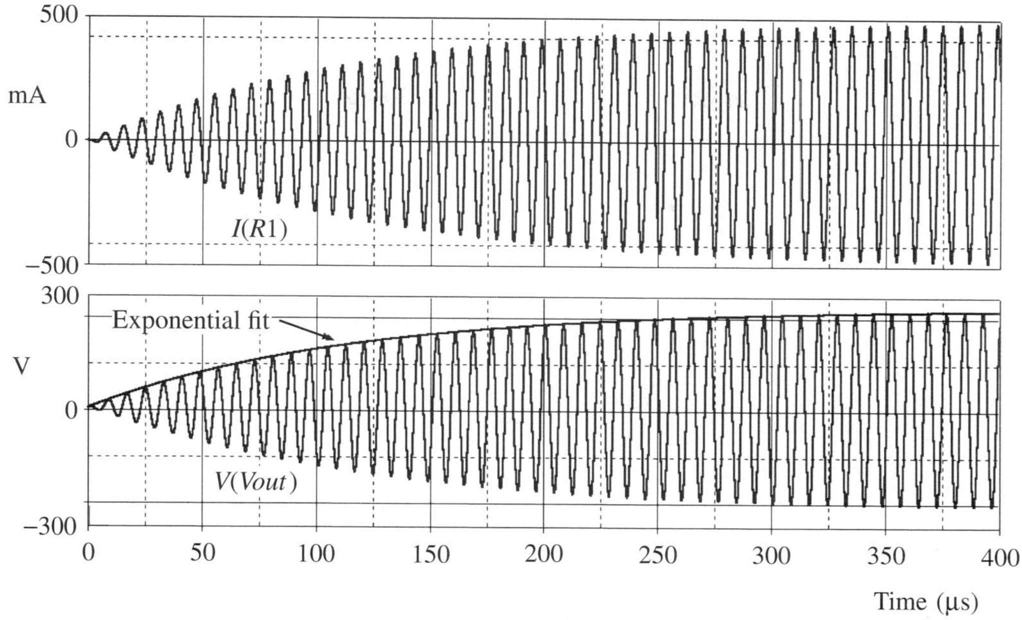


**Fig. 5.23.1** Growth of amplitude in a driven resonant circuit. This waveform is actually the part preceding that shown in Fig. 5.23.3.

resonant circuit so the field drops to zero in the order of a half cycle. The difficulty arises when the field is switched on again as the energy must be returned to the resonator, which as can be seen from Fig. 5.23.1 takes many cycles. The answer to this dilemma lies in the realization that the tag can only detect varying fields (the tag coil voltage is proportional to $\partial B/\partial t$) and does not respond to z.f. fields. Thus it is not necessary to dissipate the resonator energy to give zero effective field, only to stop it varying. If we can clamp the current to its maximum value then, when the clamp is released all the energy will be retained in the field and the oscillation can begin again immediately at full amplitude. It is as if you held the pendulum at one extreme of its motion, thus storing the energy, and then let it go sometime later. The analogy is not quite equivalent in that this would be like storing the energy in the resonator capacitor $C$ rather than that in $L$. Storage in $L$ is akin to storing the kinetic energy of the pendulum, which is a little more difficult to achieve. Thus the inductive 'kick' that can be a nuisance in inductive circuits is here made use of.

A simple demonstration circuit for achieving the desired control is shown in Fig. 5.23.2. A series resonant circuit $L_1C_1$ is driven at the resonant frequency by the push–pull transistors $Q_1$ and $Q_2$. At resonance the impedance is a minimum and is resistive. The $Q$ of the circuit is determined by the effective resistance $R_1$ of $L_1$ (both z.f. ohmic and that arising from the skin effect, which can be considerable) and the external resistance $R_2$ together with any losses in the driver transistors and the clamp. The output clamp is applied by means of $Q_5$ together with $R_5$, $D_1$ and $D_2$,

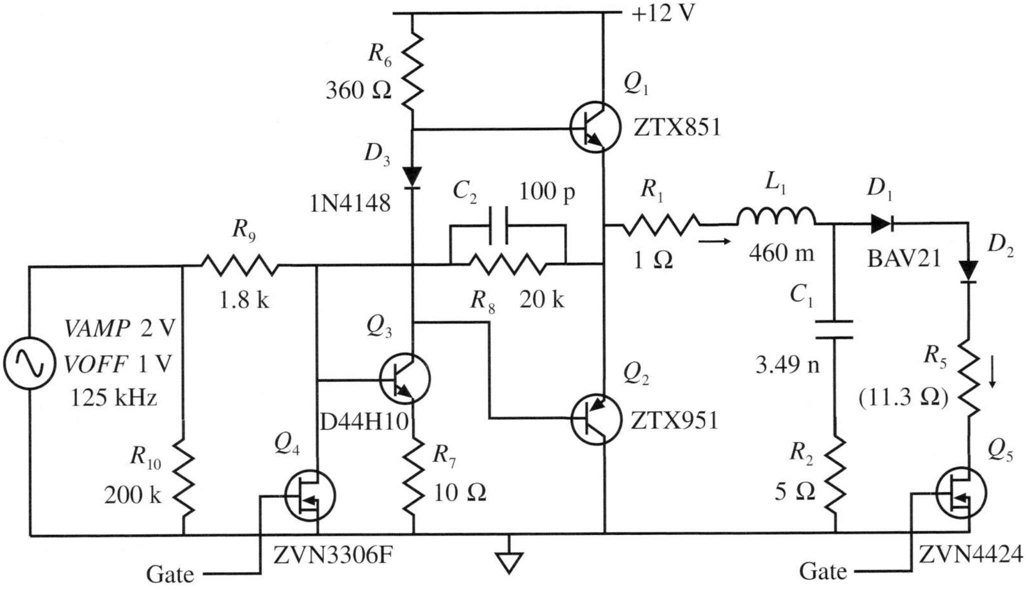


**Fig. 5.23.2** Series resonant circuit driver and gated amplitude modulator.

and the input signal may be clamped by $Q_4$. The resonant current in $L_1$ will be 90° out of phase with the resonant voltage as shown in Fig. 5.23.3. Thus if the *OFF* control is applied late in the negative excursion of $V_{out}$, then as $V_{out}$ subsequently passes through zero the clamp $Q_5$ will conduct, and $Q_4$ will fix the drive signal to keep $Q_1$ on and $Q_2$ off. This point of (near) zero voltage will also be that of maximum inductor current which can now continue to flow through $Q_1$, $L_1$, $D_1$, $D_2$ $R_5$ and $Q_5$. The diodes $D_1$ and $D_2$ are necessary since $Q_5$ would conduct through its internal diode during the reverse polarity of $V_{out}$. Two diodes are used simply for voltage rating reasons. Proper choice of $R_5$, with some allowance for the on resistance of $Q_5$, will maintain the current constant: too high a value results in a decay and too low a value in an increase of current during the *OFF* period with consequent under- or overshoot on resumption. The simulation waveforms illustrated were obtained using *TOPEN* and *TCLOSE* devices rather than the transistors shown, but the latter in actual circuits gave essentially the same result. The simulation shows a peak voltage of about 218 V and peak current of nearly 600 mA from a supply voltage of 12 V. The simple $Q$ of the resonator, using $R_1$ and $R_2$, is about 60. The effective $Q$ derived from the rise of $V_{out}$ using Eq. (3.5.13) is about 44–47 (it is difficult to measure precisely). Fitting an exponential rise to the output as shown in the figure gives a time constant $\tau_a = 115\text{E}-6$. From just above Eq. (3.5.13) we have:

$$\tau_a = \frac{2Q}{\omega_0} \quad \text{or} \quad Q = \frac{\omega_0 \tau_a}{2} = \frac{2\pi \times 125 \times 10^3 \times 115 \times 10^{-6}}{2} = 45.2 \qquad (5.23.1)$$

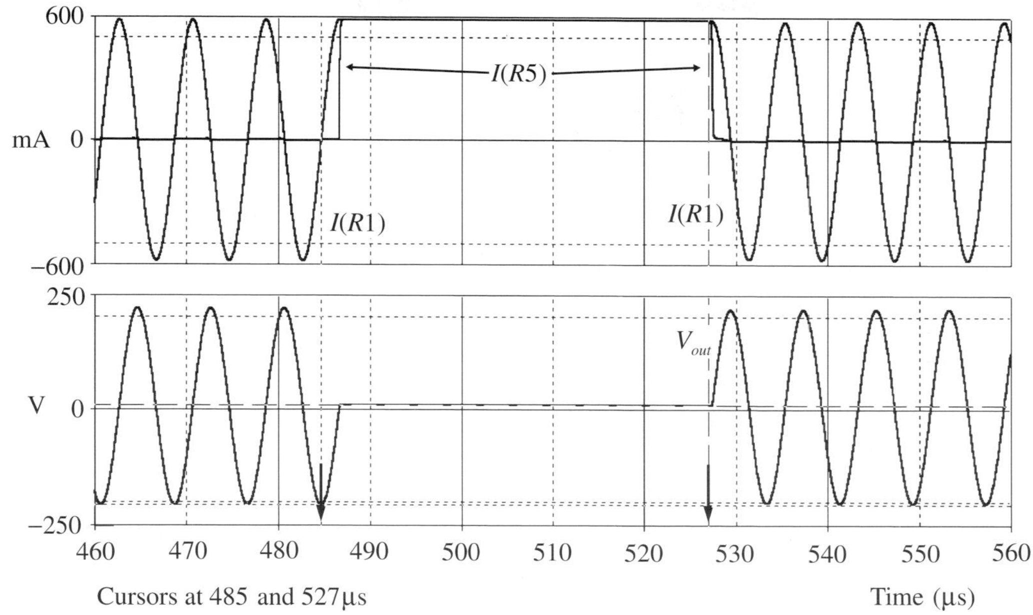


**Fig. 5.23.3** Voltage and current waveforms for circuit of Fig. 5.23.2 in the region of the clamp.

The fit shown is the equation (*TIME* is the SPICE time variable):

$$V_{out} = 218.3[1 - \exp(-TIME/115E - 6)] + 6 \tag{5.23.2}$$

where the +6 arises from the offset at the start, being the bias point of the driver.

When it is required to start the oscillation again the clamp is released but this should be done relative to the drive-signal phase otherwise there will be a hiatus, a period of beats between the drive signal and the voltage induced by the collapsing magnetic field. Figure 5.23.3 also shows the hold current $I(R_5)$ which, by adjustment of $R_5$, remains constant during the *OFF* interval. It is evident that it is now possible to have an *OFF* period of even a single cycle and that the field will rise to the tag trigger level in the same time whatever the distance (provided of course that the field could ever reach the required level). Figure 5.23.4 shows the cases where $R_5$ is too high or low, and the case when the phasing of the *ON* time is not correct relative to the input drive signal. The gate switching times indicated in the figure (485 and 527 μs) do not quite match the zero point times of the signals since allowance must be made for the response times of the switches which were set to 1 μs.

This circuit has been used with resonant voltages of several hundred volts peak-to-peak. The voltage rating of both $Q_5$ and $D_1$, $D_2$ must be appropriate. Since the *OFF* intervals are usually only a few cycles the power dissipation in $Q_5$ is low and small FETs suffice, e.g. the ZVN4424 shown here. This is also desirable since it is beneficial to keep the capacity at the drain low. Though the figures shown are the results of simulation, actual circuits produced almost identical results.

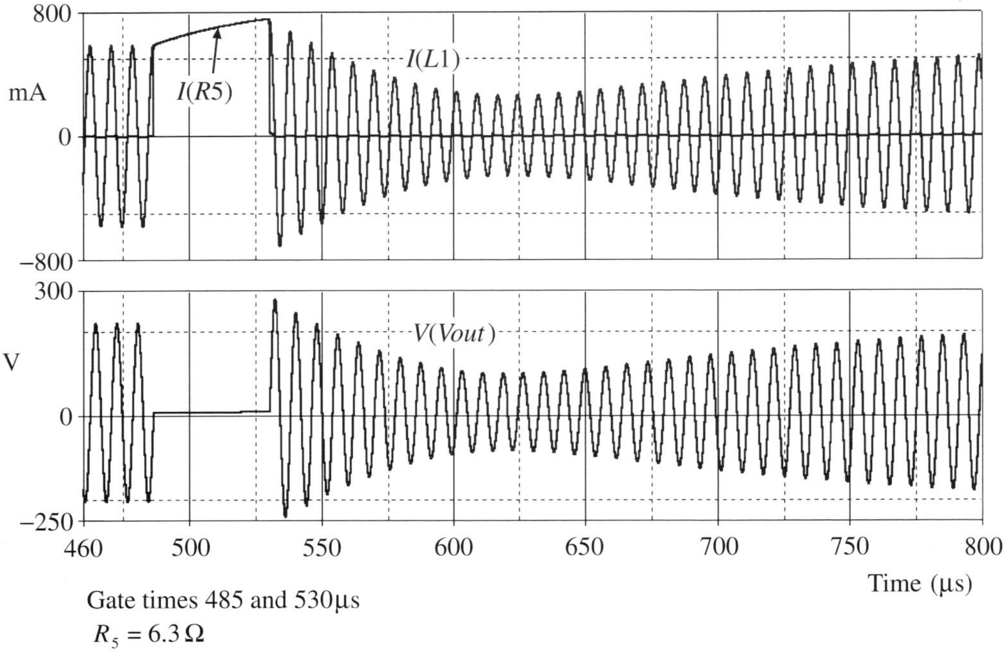


**Fig. 5.23.4** Effect of timing mismatch on resumption of oscillation.

Extending this technique to higher frequencies presents an interesting challenge as there are possibilities for applying it in such areas as, for example, nuclear magnetic resonance and spin echoes.

## SPICE simulation circuits

| | |
|---|---|
| Fig. 5.23.1 | Lcmult17.SCH |
| Fig. 5.23.3 | Lcmult17.SCH |
| Fig. 5.23.4 | Lcmult17.SCH |

# 5.24 Photomultiplier gating circuit

Beginning a story is like making a pass at a total stranger.

Amos Oz in *The Story Begins*

Pulsed gain control of photomultipliers is used in several forms of time resolved spectroscopy (Hamilton 1971; Hamilton and Naqvi 1973) and, for example, in LIDAR. There are several ways of achieving this depending on the time scales involved. The circuit described in the above references operated from 3 μs to infinity and could be readily inserted into an existing system. It is now required to decrease the minimum time of operation to 100 ns so we will use SPICE to examine whether the original circuit can be improved. The original circuit is shown in Fig. 5.24.1 together with the general circuit configuration of a photomultiplier.

The gating circuit is essentially a high voltage bistable which can be triggered into either state by appropriate pulses. The constraints on the circuit are that it must operate on the low current of the dynode bleeder chain and that substantial voltages are involved of about 100 V between dynodes. To switch the photomultiplier off it is necessary to reverse the voltage between a pair of dynodes by at least 30 to 50 V so that the electrons are repelled by the reversed electric field and hence the gain reduced. It has been found in the original circuit that the gain could be reduced by a factor of ≈1000. Changing a potential of 100 V in 100 ns implies a slewing rate of 1000 V $\mu s^{-1}$, which presents some difficulties with only a few mA of bleeder current available. However, the duty cycle is low so we can use large transient currents to obtain the fast transitions but still have an acceptable average current. The switching circuit is also at a fairly high (negative) d.c. voltage so there is also the question of triggering the circuit from one state to the other.

To keep the quiescent current low it is necessary to use the high value resistors shown. This has a significant effect on the time response and some additional measures will be required to give a faster response. The common emitter connection of $Q_3$ and $Q_4$ provides an effective negative bias to allow either of these transistors to be cut off. To obtain faster operation we can first of all seek faster transistors with the requisite voltage rating. A scan through the handbooks (there are not many high voltage high frequency transistors) produced the NPN 2N5550

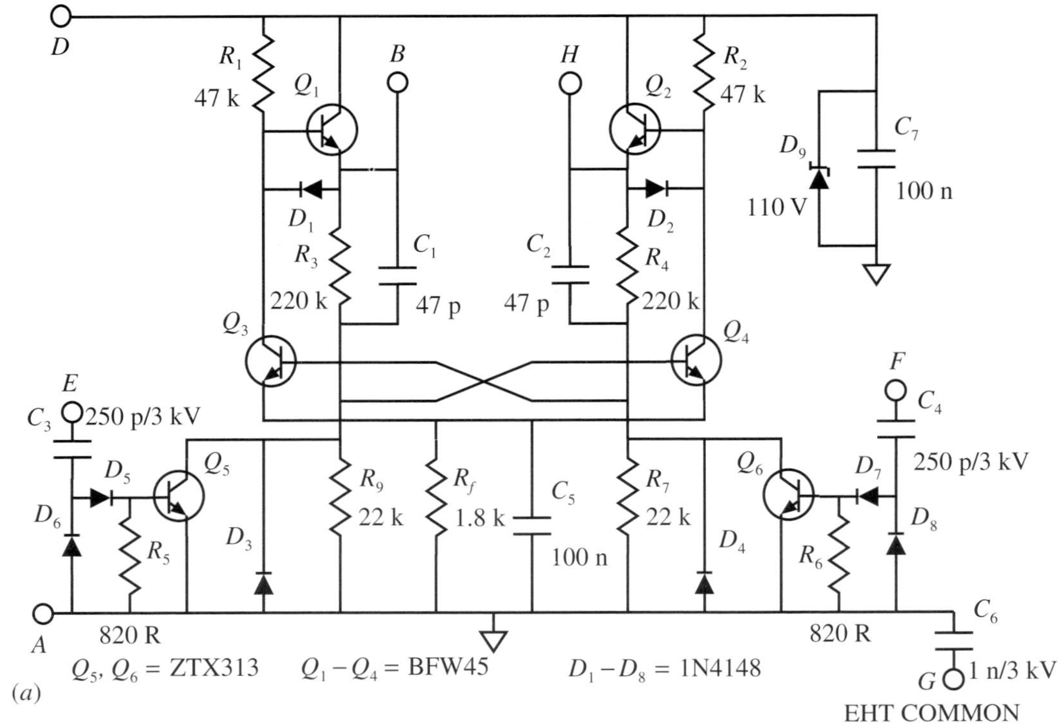


**Fig. 5.24.1** (*a*) Original photomultiplier gating circuit. (*b*) General photomultiplier circuit. Resistors $R_1$–$R_{10}$ are typically ≈100 k. The capacitors $C_1 \approx 5$ p represent the inherent capacities of the multiplier. For high output currents, capacitors as at $C_3$ and $C_4$ are usually included. Maximum output current $i_A$ should be <1% of the bleeder current $i_{10}$ for linear operation. For a secondary electron multiplication coefficient of, say, $\delta \approx 3$ the capacitors $C_3$ and $C_4$ should be at least in the ratio of $\delta$ as illustrated. The charge $q$ sequence between the last few dynodes illustrates the multiplication process, and the dynode currents $i_7$, $i_8$, $i_9$ must supply the difference between the incoming and outgoing charges.

and its PNP complement the 2N5401 (the reason for the latter will be apparent shortly). The operation of the circuit may be considered in three regimes; quiescent, turning *OFF* and turning *ON*. For the quiescent state the power dissipation must be low and the current consumption must match that of the bleeder chain. In the transitions between states, the currents can be large to achieve short times since for low duty cycles there will be little effect on the average current. The final circuit is shown in Fig. 5.24.2 which also shows the quiescent currents (in μA) as found by SPICE, adding up to a total of 2.5 mA.

Consider the interval when $Q_3$ is turning *ON* and the voltage at $B$ is falling. As the voltage at the collector falls $Q_1$ will be turned *OFF* and the voltage at $B$ will fall slowly depending on $R_1$ and the capacity at that point, which must include the dynode capacities. The presence of $D_1$, however, means that $Q_3$ can pull $B$ down

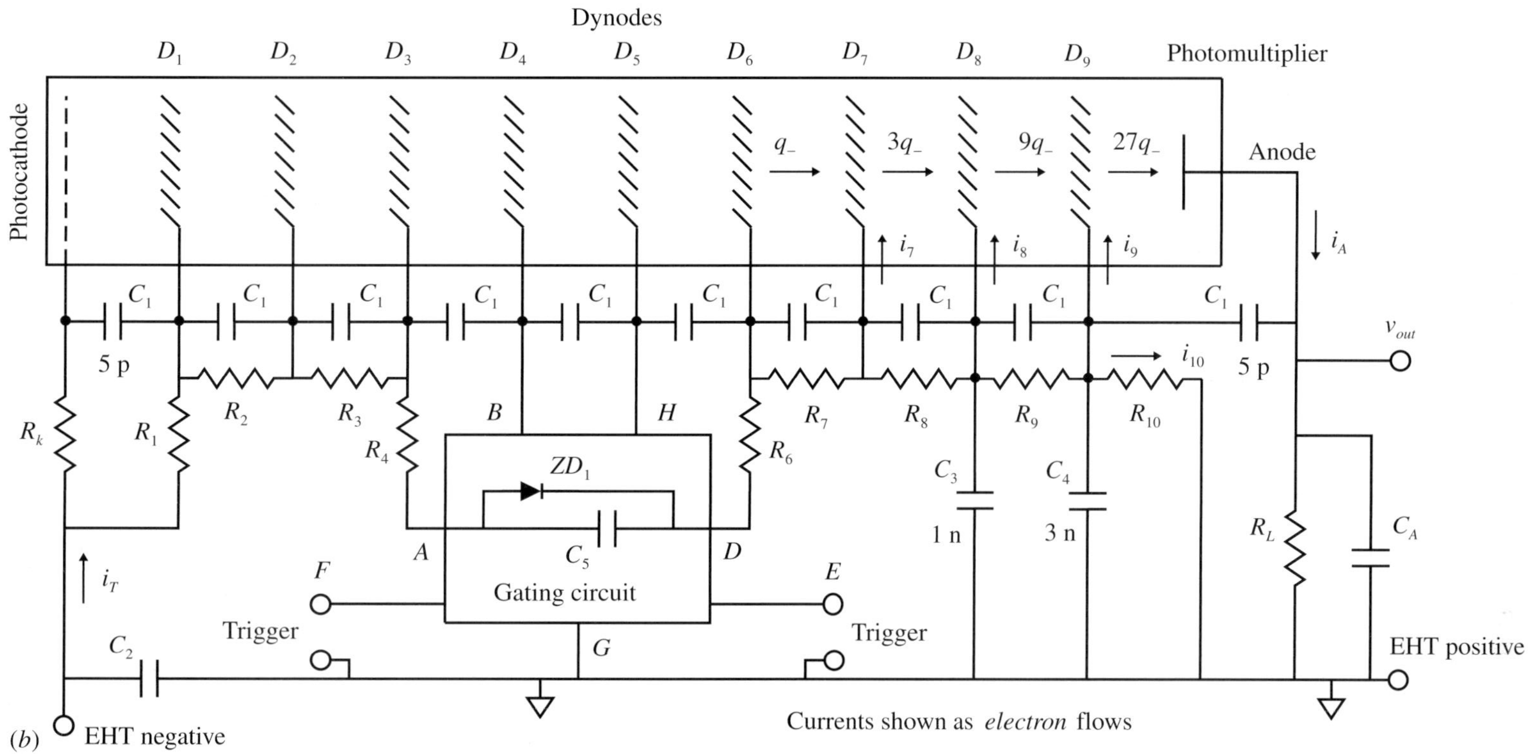

(*b*)

**Fig. 5.24.1** (*cont.*)

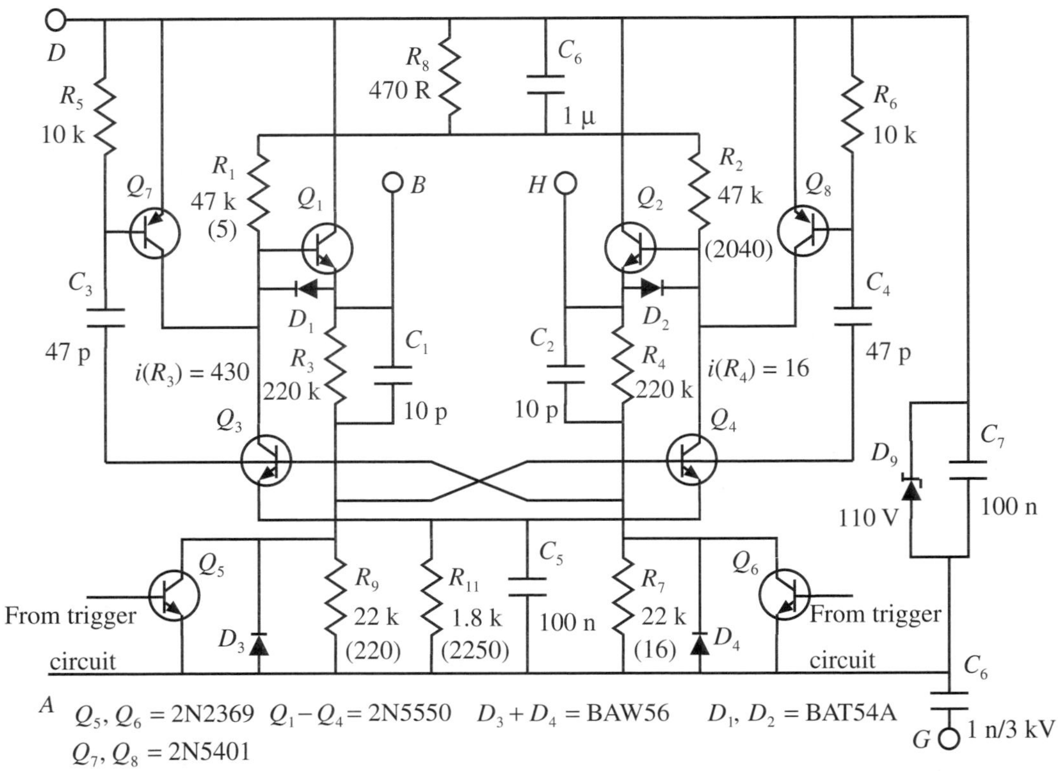


**Fig. 5.24.2** Improved photomultiplier gating circuit (currents shown in µA).

directly though a large transient current will flow. The simulation shows a time interval of about 10 ns with a peak current of about 500 mA, and hence the need for a large bypass capacitor $C_7$ across the circuit. When $Q_3$ is turned *OFF* then *B* will rise according to the time constant $R_1$ and the load capacity which would again be rather long. To speed this up the negative going signal on the base of $Q_3$ is capacitively coupled via $C_3$ to the base of the PNP transistor $Q_7$ to switch it on and hence pull *B* up much more rapidly. $R_8$ with $C_6$ provides a small bias to allow $Q_7$ and $Q_8$ to pull up at the end of the rise more effectively. The simulation shows a time of about 16 ns, and thus the desired minimum time interval of 100 ns is readily achieved (it looks as if it could be shorter if required). The optimum value for $C_{3,4}$ is dependent on the capacity associated with the dynodes being driven so some adjustment may be necessary. The values shown were for interdynode capacities of 5 pF. Figure 5.24.3 shows the simulated waveforms.

A necessary requirement is a fast trigger circuit that can also provide the high voltage isolation to allow common referred timing sources. Many previous designs have used optocouplers to provide the isolation but they are too slow or require significant power supplies for our purposes (see for example the Hewlett-Packard

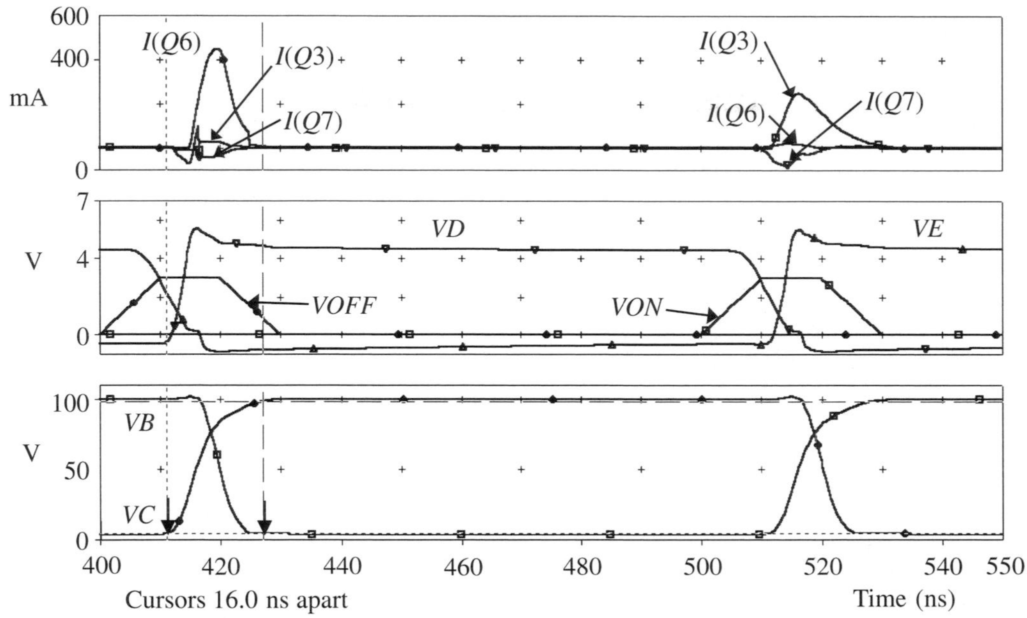


**Fig. 5.24.3** Simulated waveforms for the improved gating circuit.

HCPL7101). A fast (wideband) transformer has some attractions (see Section 5.20) and a suitable circuit has been designed around this as shown in Fig. 5.24.4.

In most cases the photomultiplier unit does not have additional power supplies so the common side of the trigger unit has been designed to run off the trigger source itself and thus only the signal coax cable is required. The trigger source is set to be quiescently at +5 V, i.e. we desire CMOS levels rather than the more loosely defined TTL levels. As noted above, the duty cycle is usually low so there is plenty of time for $C_7$ to be charged via $D_5$ to provide the local power supply at *VS* (rather like many 232-type communication interfaces are operated). When the negative going trigger pulse occurs then the fast CMOS gate operates as normal and $D_5$ isolates *VS*. The quiescent current for the 74AC08 gate is only a negligible 4 μA even though it can actively source and sink at least 24 mA even at logic levels. $D_6$ limits any positive overshoot when the trigger pulse ends. The secondary of the 1 : 1 transformer produces a positive pulse to switch $Q_{5,6}$ *ON* to trigger the bistable circuit. Triggering multivibrators is most efficiently done by turning the *ON* transistor *OFF* as here (Millman and Taub 1965, p. 381). Schottky diode $D_7$ prevents saturation and $D_8$ speeds up the off time of $Q_{5,6}$. Simulation waveforms are shown in Fig. 5.24.5. It should be noted that the power supply attribute for the 74AC08 gate should be changed to $v_C$ (as shown on the schematic) and the current taken by it can be seen from the current through $R_{14}$.

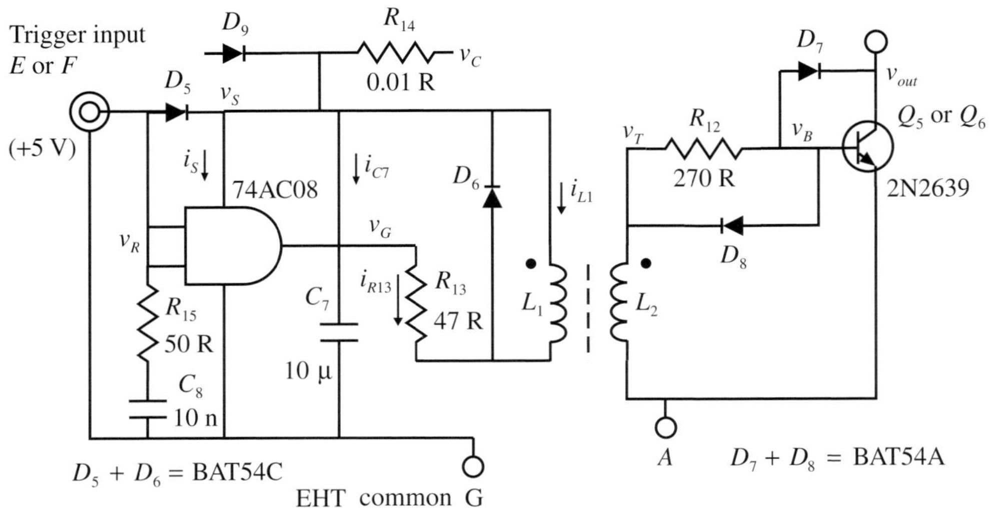


**Fig. 5.24.4** Trigger circuit for the photomultiplier gate.

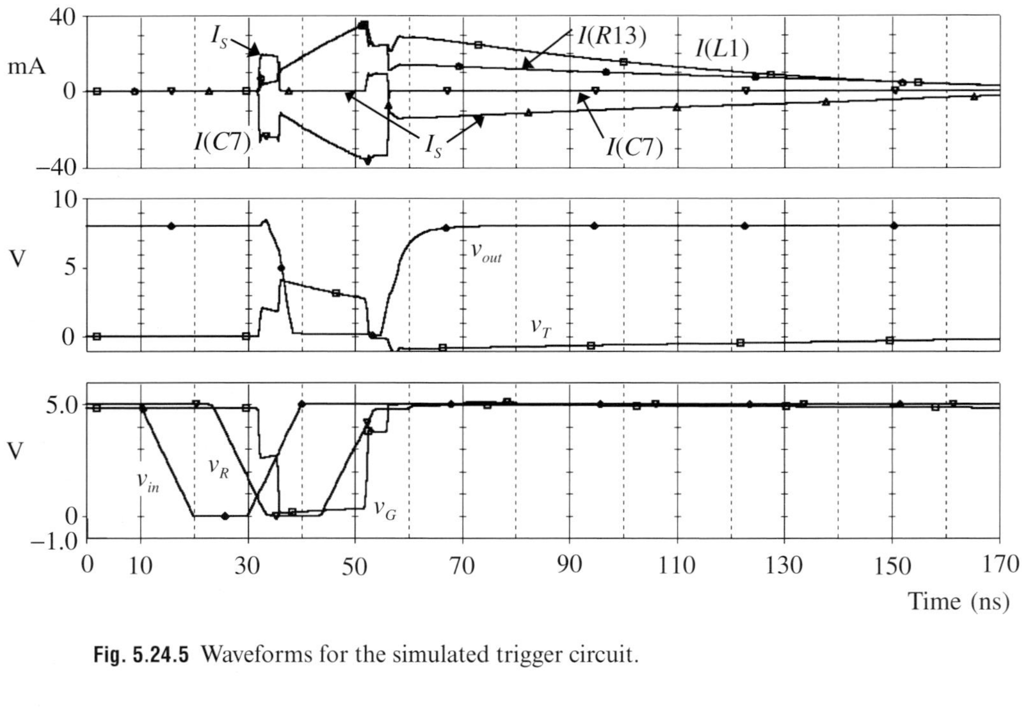


**Fig. 5.24.5** Waveforms for the simulated trigger circuit.

The time delay through the trigger is about 10 ns but there will be some additional delay due to the response of the bistable. As there are separate triggers for *OFF* and *ON*, the second input can be connected as illustrated via $D_9$ and the $v_S$ power supply can be efficiently shared. If, as would be likely, there is some distance between the trigger source and the photomultiplier so that a significant length of

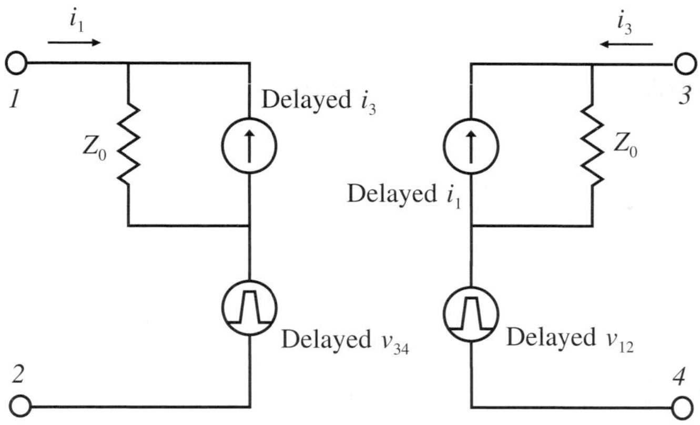


**Fig. 5.24.6** Equivalent circuit of PSpice *T* model transmission line.

coaxial cable is used, it will be necessary to terminate the cable with the characteristic impedance to avoid reflections with fast pulses. This is shown by $R_{15}$, which must also be d.c. isolated by $C_8$ to avoid the large current that would arise from the quiescent +5 V level.

The simulation of the photomultiplier itself in SPICE presents something of a problem. It has been emphasized in earlier parts that the current in a circuit must be continuous to satisfy Kirchhoff's current law. In a photomultiplier we have a case where current leaves the circuit, albeit for a short interval, when the emitted electrons from the cathode or from a dynode pass to the next element. Though the time for the electrons to pass to the next element is short, say 1 ns, it is just these times that are of interest. Starting off with a delta function pulse of current the final output at the anode will have a broadened distribution since all the electrons do not follow the same geometrical path and so arrive at successive stages at different times – a phenomenon known as transit-time spread (e.g. Hamilton and Wright 1956). We will not attempt to simulate the spread but will consider an approximate equivalent circuit that does show the time delays so that we can examine the effects of the growing current intensity on the voltages of the dynodes. Again we will only do this approximately since the secondary emission factor $\delta$ is a function of the interdynode voltage. To provide the delay between stages an ideal transmission line (PSpice *T* device, Fig. 5.24.6) is used. The configuration of this model is not revealed in the SPICE library part but the *Circuit Analysis Reference* manual (MicroSim 1993, 1996) shows it to be two separated generators without a common connection.

A proposed equivalent circuit for the photomultiplier and the resistive bleeder is shown in Fig. 5.24.7. The 'screen' connection of the *T*-line input is connected to circuit common so the input current from the current generator $I_1$ does not upset dynode $D_2$. The benefit of a current source is that it has infinite impedance and

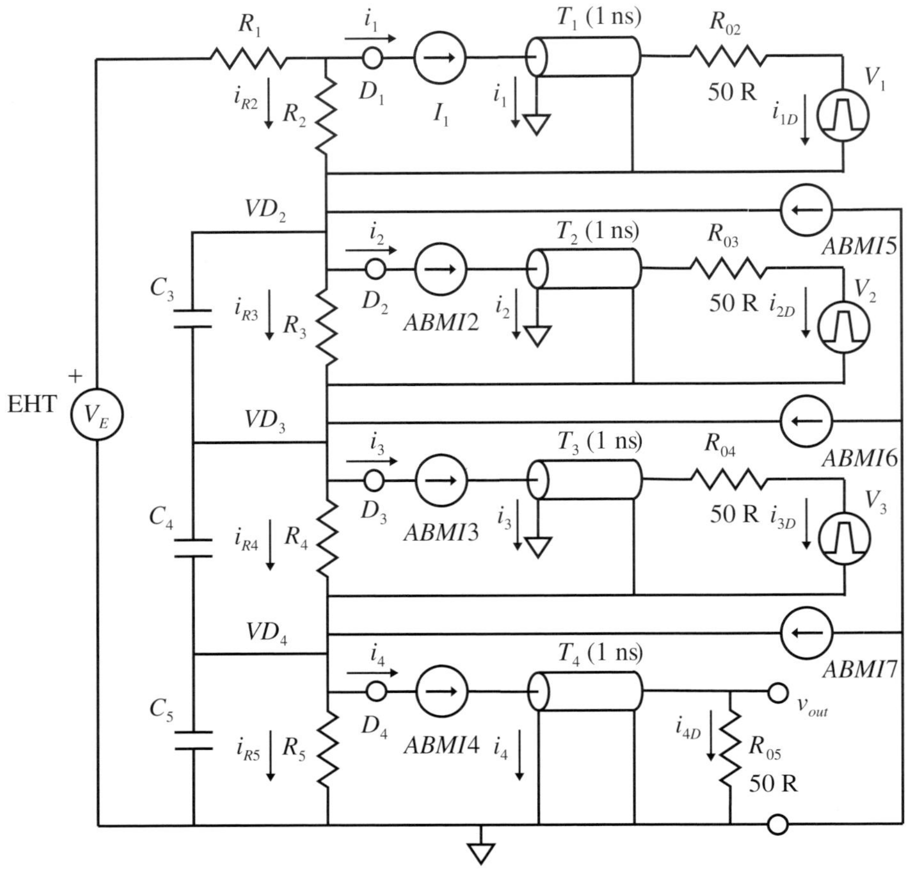


**Fig. 5.24.7** Simulation circuit for a photomultiplier.

hence does not otherwise affect the circuit. The delayed output current $I_{1D}$ from $T_1$, with $Z_0 = 50\ \Omega$ and set for a delay of 1 ns (the sort of time for the charge to pass between dynodes), passes through a null voltage source $V_1$ and the matching resistor $R_{02}$ of 50 Ω to avoid any reflections. This current of course returns via the connection to the screen of the *T*-line and does not pass to the rest of the circuit. The connection of this point to the bleeder is to mollify PSpice which would otherwise flag a floating node. Choosing a value $\delta = 3$, say, the *ABMI2* current source is set to a value of $I_{1D} \times 3 \equiv I(V_1)*3$ and a compensating current of $I_{1D}$ is injected by *ABMI5* to represent the current from $D_1$ arriving at $D_2$. Care should be taken with the sense of the currents; for example, examine the directions of the arrows on the *ABMI* symbols as they are a bit squalid and difficult to make out. To make the arrangement simpler to follow, all the polarities have been inverted as if we were dealing with positrons rather than electrons, so that the conventional current directions are now appropriate. The procedure is continued down the dynodes till we arrive at the anode. If a short pulse is now set for $I_1$, say 0.1 ns rise and fall and 1 ns width, then you can see the propagation of the currents via the dynodes and the final voltage (or current) output. You may find it instructive to insert a number of low value resistors at appropriate points to enable the display of the various currents. In practice the bleeder resistors $R_1$ to $R_5$ have high values of the order of 100 k and the supply voltage is such as to give typically a standing current $I_{R5}$ of, say, 1 mA. The maximum dynode currents are generally kept below 1% of the bleeder current to achieve acceptable linearity, and generally the final few dynodes are decoupled with capacitors to diminish transient voltage changes on the dynodes as the electron bunch grows and progresses. For fast pulses the output current can considerably exceed the bleeder current, in which case dynode decoupling needs very careful consideration. This equivalent circuit allows one to examine such changes and to determine the appropriate capacitors to use. It will be necessary to simulate for considerably longer times than the input pulse since the dynode voltage recovery times are long owing to the high value of the resistors with the decoupling capacitors.

## SPICE simulation circuits

Consult the SimCmnt.doc file on the CD before running

| | |
|---|---|
| Fig. 5.24.1(b) | Pmtgat.SCH |
| Fig. 5.24.3 | Pmtgat3.SCH |
| Fig. 5.24.5 | Pmgttrg 1.SCH |
| (Fig. 5.24.7) | Pmbldr 4.SCH |

## References and additional sources 5.24


Barisas B. G., Leuther M. D. (1980): Grid-gated photomultiplier with subnanosecond time response. *Rev. Sci. Instrum.* **51**, 74–78.

Hamilton T. D. S. (1971): Variable duration photomultiplier gating circuit. *J. Phys. E Sci. Instrum.* **4**, 326–327.

Hamilton T. D. S., Naqvi K. R. (1973): Instrument for time-resolved phosphorimetry using an electronically gated photomultiplier. *Analytical Chem.* **45**, 1581–1584.

Hamilton T. D. S., Wright G. T. (1956): Transit time spread in electron multipliers. *J. Sci. Instrum.* **33**, 36.

Herman J. R., Londo T. R., Rahman N. A., Barisas B.G. (1992): Normally on photomultiplier gating circuit with reduced post-gate artifacts for use in transient luminescent measurements. *Rev. Sci. Instrum.* **63**, 5454–5458.

MicroSim (1993, 1996): *Circuit Analysis Reference Manual*, Version 5.4, p. 161, or *MicroSim PSpice A/D Reference Manual*, pp. 2–99; Microsim Corp. Handbooks.

Millman J., Taub H. (1965): *Pulse, Digital, and Switching Waveforms*, New York: McGraw-Hill. Library of Congress Cat. No. 64-66293.

Wardle R. (1982): *Gating of Photomultipliers*, Electron Tubes Application Report.

Yoshida T. M., Jovan T. M., Barisas B. G. (1989): A high-speed photomultiplier gating circuit for luminescence measurements. *Rev. Sci. Instrum.* **60**, 2924–2928.

# 5.25 Transatlantic telegraph cable

A thousand mile journey starts with a single step.

Lao-tse

This may seem a rather unexpected topic, particularly when we will examine the performance of the original undersea cables laid in the 1850s. The development of the overland telegraph from the early work in the 1830s of Samuel Morse and Charles Wheatstone had resulted in a rash of telegraph connections in many parts of the world. The impact of this means of speedy communication over long distances has been likened to the present impact of the huge growth of the internet in our own time and has been chronicled in *The Victorian Internet* (Standage 1998). The first undersea cable was laid in 1851 from Calais to Dover and a much longer one in 1856 of 740 km across the Black Sea from Balaclava to Varna (in what is now Bulgaria). The success of this encouraged thoughts of a transatlantic cable though the distance of some 4450 km presented great technical and engineering problems. It is, however, the ideas as to how the signals were thought to be transmitted that is our immediate interest. It should be recalled that Maxwell's proposals regarding the existence of electromagnetic waves were still some years away. The role of inductance was little understood and the connection was viewed simply as a distributed *RC* network. In this configuration the equation describing the propagation is the diffusion rather than a wave equation, and is equivalent, for example, to the flow of heat. It was not for many years that the benefit of inductance was appreciated, primarily due to the lonely efforts of Oliver Heaviside who fought a long and controversial battle particularly with Sir William Preece, chief engineer of the Post Office, which controlled telegraphs in Britain. A good account is given by Jordan (1982).

The first proper investigation of the circuit aspects of long cables was the classic paper by William Thomson (1855), the results of which controlled the design for some 40 years but was also the origin of the long battle fought by Heaviside against great opposition from the establishment. This controversy revolved around the role of inductance in the performance of the cables, a factor which Thomson, as he admitted many years after his seminal paper, had decided was insignificant and had

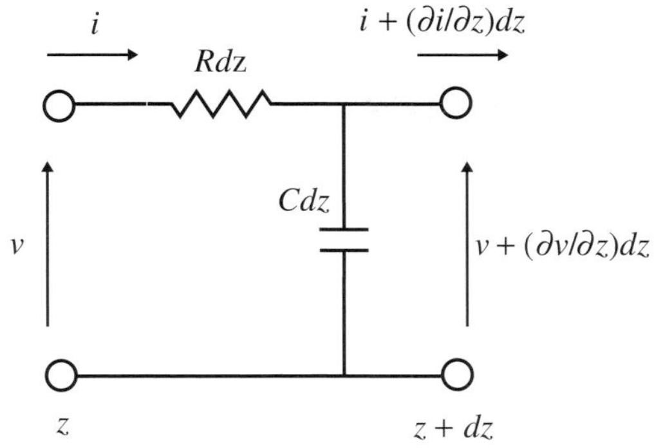


**Fig. 5.25.1** General parameters of a segment of a very low frequency cable.

therefore omitted in his investigation. Heaviside realized that there was great advantage in the presence of inductance, which others had tried to minimize, but Thomson's theory was accepted as the proper one by almost all, including Preece who had the position to control much of what was done, and against whom there are several stirring diatribes in Heaviside's works (1922). We will first outline the approach of Thomson and then see if we can simulate the various effects. It was known that the rate of signalling on long distance telegraphs was rather slow and for the earliest undersea cable, from Britain to France, was even slower. The cable was viewed as a sequence series $R$ (per unit length) and parallel $C$ (per unit length) segments with effectively no inductance or leakage conductance. Consider a length $dz$ of the cable as shown in Fig. 5.25.1. Then we can write the charge $q$ on the length $dz$ as $Cv \cdot dz$ and the resistance of the length as $R \cdot dz$. With the voltages and currents as shown, then the net current will be equal to the rate of change of the charge $q$ and the net voltage around the loop $dv$ will be equal to the voltage drop across $R \cdot dz$:

$$q = Cv\,dz \quad \text{so} \quad \frac{\partial i}{\partial z}dz = \frac{-dq}{dt} = \frac{-\partial}{\partial t}(Cv\,dz) \quad \text{or} \quad \frac{\partial i}{\partial z} = -\,C\frac{\partial v}{\partial t}$$

$$\frac{\partial v}{\partial z}dz = -Ri\,dz \quad \text{or} \quad \frac{\partial v}{\partial z} = -Ri \quad \text{and differentiating with respect to } z \qquad (5.25.1)$$

$$\frac{\partial^2 v}{\partial z^2} = -R\frac{\partial i}{\partial z} = -R\left(-C\frac{\partial v}{\partial t}\right) = RC\frac{\partial v}{\partial t}$$

and this is of the form of a diffusion equation (Pain 1976) and describes phenomena such as the flow of heat or the diffusion of dopants into silicon. As Thomson remarks:

> This equation agrees with the well-known equation of the linear motion of heat in a solid conductor; and various forms of solution which Fourier has given are perfectly adapted for answering practical questions regarding the use of the telegraph-wire.

The particular solution of Eq. (5.25.1) as given in the paper is that the time required for the output current (detecting devices then were current rather than voltage sensitive) to reach the detection level was proportional to $RCz^2$. This apparently simple relation was the origin for most of the difficulties referred to above. The general consideration of heat flow problems and the analogy to electrical diffusion is, for example, discussed by Pipes (1958). If we assume an effectively infinite line then the solution is much simplified, and we will outline the approach of Grivet (1970, Vol. 1, p. 401). The identical heat flow case is solved by means of Laplace transforms by Boas (1966). The conditions applying to the long cables are that $R \gg \omega L$ and $\omega C \gg G$ so that the propagation constant $\gamma$ and the characteristic impedance $Z_0$ are given by (Section 3.17):

$$\gamma = [(R+sL)(G+sC)]^{\frac{1}{2}} = (sCR)^{\frac{1}{2}}$$

$$Z_0\left(\frac{R+sL}{G+sC}\right)^{\frac{1}{2}} = \left(\frac{R}{sC}\right)^{\frac{1}{2}} \tag{5.25.2}$$

and so the general expressions for the voltage $V(z, s)$ and current $I(z, s)$ as functions of position $z$ and frequency $s$ on an infinite line are (Grivet 1970: note 1):

$$\begin{aligned} V(z, s) &= V(0, s)\exp\left[-(sCR)^{\frac{1}{2}} z\right], \quad \text{where } V(0, s) \text{ is the value at } z=0 \\ &= V_0(s)\exp\left(-\alpha\sqrt{s}\right), \quad \text{where} \quad \alpha^2 \equiv CRz^2 \quad \text{and } V_0(s) \equiv V(0, s) \\ I(z, s) &= I(0, s)\exp\left[-(sCR)^{\frac{1}{2}} z\right], \quad \text{where } I(0, s) \text{ is the value at } z=0 \\ &= \frac{V_0(s)}{Z_0}\exp\left(-\alpha\sqrt{s}\right) \\ &= V_0(s)\left(\frac{sC}{R}\right)^{\frac{1}{2}}\exp\left(-\alpha\sqrt{s}\right) \end{aligned} \tag{5.25.3}$$

We now choose an input signal $V_0(s)$ in the form of a step input and find the corresponding time response using Laplace transforms. The transform of the step is $1/s$ so that we have:

$$\begin{aligned} V(z, s) &= \frac{V_0}{s}\exp\left(-\alpha\sqrt{s}\right) \\ \text{or} \quad v(z, t) &= v_0\left[1 - \operatorname{erf}\left(\frac{RCz^2}{4t}\right)^{\frac{1}{2}}\right] \end{aligned} \tag{5.25.4}$$

where we have used transform No. 22 in Table 1.12.1 and the error function erf is described in Section 4.9. For the current we find:

$$I(z, s) = \frac{V_0}{s}\left(\frac{sC}{R}\right)^{\frac{1}{2}}\exp\left(-\alpha\sqrt{s}\right) \tag{5.25.5}$$

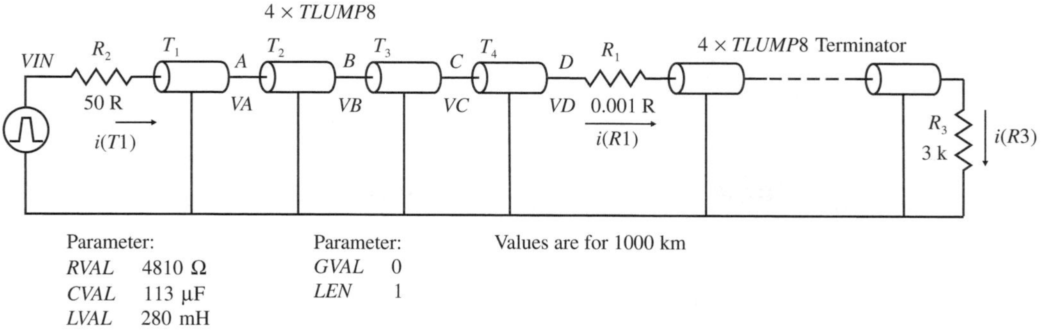


**Fig. 5.25.2** Simulation circuit for the transatlantic cable. $R_1$ is introduced to provide a convenient device to determine the current. Parameter symbols are used to enable global editing of the component values.

$$\text{or}\quad i(z,t) = v_0\left(\frac{C}{\pi R t}\right)\exp(-RCz^2/4t) \qquad (5.25.5\ \text{cont.})$$

where we have used transform No. 23 in Table 1.12.1.

To compare these results with those from PSpice we use the circuit shown in Fig. 5.25.2. The active part of the cable is made up of four *TLUMP*8 devices $T_1$ to $T_4$. A transmission line terminated by its characteristic impedance should appear to the transmitter as an infinite line. The characteristic impedance is, however, a complex quantity as shown by Eq. (5.25.2) so a simple approximation was used in the form of another four sections of *TLUMP*8 so *D* represents the receiving point. *TLUMP*8 is a model of eight *RC* segments giving a total of 32. Increasing the resolution by replacing the *TLUMP*8's with say *TLUMP*32's produced little change. The reason for using four sections rather than one *TLUMP*32 is to allow the ready identification of the intermediate points to see how the waveform changes. All the internal nodes of the transmission lines are 'named' by PSpice but it is more unwieldy than the method actually used and the number of 'available traces' are reduced dramatically (try it and see for yourself, and then try asking Cadence/OrCAD to explain the configuration!). Each *TLUMP*8 will represent 1000 km of cable for a total of 4000 km (the actual transatlantic distance is about 4400 km, but the difference is inconsequential). Values for the resistance and capacitance are taken from Grivet (p. 405) who gives figures per nautical mile which may be converted to km as indicated (1 nautical mile = 1.852 km):

$$\begin{aligned} R &= 8.9\ \Omega\ \text{per nautical mile} = 4.81\ \Omega\ \text{km}^{-1} \\ C &= 0.21\ \mu\text{F per nautical mile} = 0.113\ \mu\text{F km}^{-1} \qquad (5.25.6) \\ L &= 280\ \mu\text{H km}^{-1} \end{aligned}$$

where the value for $L$ has been approximated from ordinary coaxial cable, but as will be found from the simulations its effect is insignificant in practice. The conductance $G$ has been taken as zero though this must be far from the truth with present day good dielectrics let alone the gutta percha used originally. If you make $G$ more significant you will see the decrease in output current. Heaviside (1922, Vol. 1, p. 417 etc.) considered the possibility of using controlled leakage to shorten the received pulses but the consequential loss of amplitude meant that such schemes were unacceptable. However, it then became more evident that inductance would be of benefit and the idea of substantially increasing this either in the cable or by periodic loading was investigated. It was such techniques, primarily introduced by Pupin, that made longer distance telephone communication, with its much higher frequencies, possible. Before this the variation of delay with frequency rapidly made speech unintelligible (e.g. Everitt and Anner 1956, p. 312; Grivet 1970, p. 319).

As Grivet points out, the solution found in (5.25.4) and (5.25.5) is for a rather simplified model for which the output starts immediately with the input and gives no indication of delay or velocity, both of which depend on the inclusion of inductance. However, the simulation results show that the normal magnitude of inductance still has minimal effect. It is as if we were operating with an incompressible fluid in a hydrodynamic application, a point made by Stokes in the original paper (Stokes is perhaps much better known for his work in fluid flow, e.g. the relation giving the terminal velocity of a body falling through a viscous fluid, which was essential in Millikan's determination of the quantization and magnitude of the charge on an electron). In a finite length we would expect the current to settle eventually to the z.f. value which is independent of capacitance. For an infinite line it will of course tend to zero, as indicated by the plot of $i(t)$ (see Eq. (5.25.7)) in Figure 5.25.4. Fig. 5.25.3 shows the frequency response of the circuit.

Figure 5.25.4 shows the pulse response for an input of 1 V, a rise and fall time of 100 μs, and period 4 seconds. It is evident that the voltage modulation ratio is far smaller than that for the current, which is consistent with the frequency responses. This was just as well since the receivers used were current sensitive.

The calculated step response from Eq. (5.25.5) is also shown, where the parameters at the receiving end are:

$$\left(\frac{C}{\pi R}\right)^{\frac{1}{2}} = \left(\frac{113\times10^{-9}}{\pi\times4.81}\right)^{\frac{1}{2}} = 86.5\times10^{-6}$$

$$\left(\frac{CRz^2}{4}\right) = \frac{113\times10^{-9}\times4.81\times16\times10^{6}}{4} = 2.174, \quad \text{for } z = 4000\text{ km} \tag{5.25.7}$$

so $i(t) = [(86.5\mathrm{E}-6)/SQRT(TIME)]^{*}\exp(-2.174/TIME)$

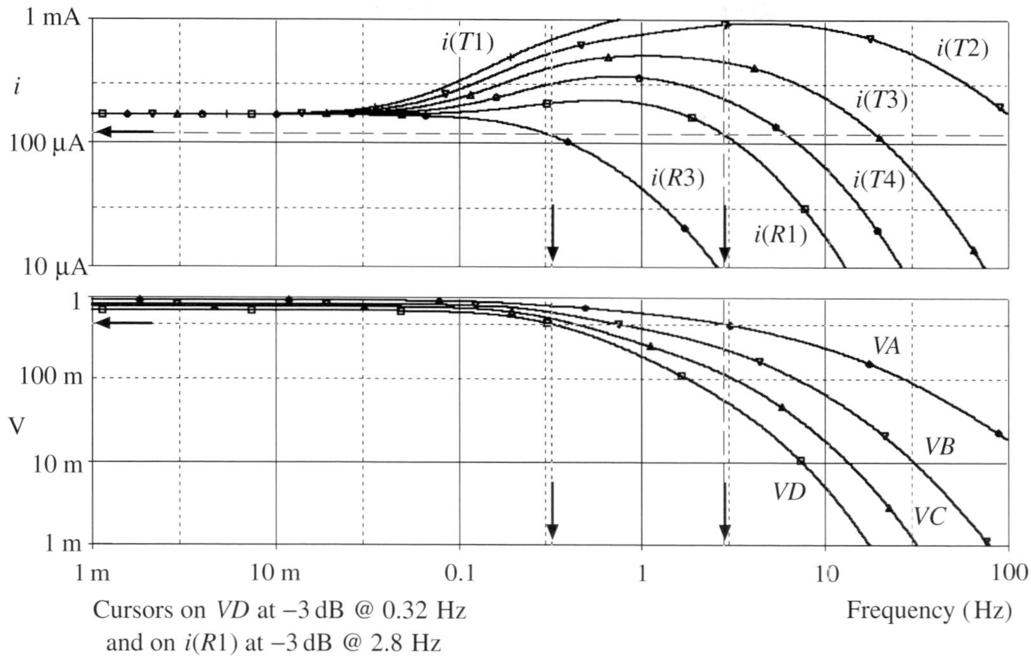


**Fig. 5.25.3** Frequency response of the circuit of Fig. 5.25.2. $i(T1)$ represents the current input to segment $T_1$ etc. The $-3$ dB point for the output at $R_1$ for the current response indicates nearly an order of magnitude improvement relative to that for the voltage response at *VD*.

and the fit is seen to be satisfactory (remember that the calculation is only for a step rather than the series of pulses). The curves of Fig. 5.25.4 illustrate how the initial 'fast' rise of the input is progressively washed out (*VA*, *VB*, *VC* to *VD*) as it passes down the cable. It is evident that signalling at the rate of one dot every four seconds was quite possible, but as the rate is increased the amplitude decreases until the point is reached when the intensity is insufficient to activate the receiver successfully. Thomson and Stokes adopted a limiting point of 55% of the maximum as a criterion, which is why that point is indicated on the curve for $i(t)$. Stokes estimated that to signal half-way round the Earth would allow only one dot per 15 s, which is in reasonable agreement with our simulation allowing for the distance being about three times further.

My interest in this matter dates from my days as a postgraduate in the department of Natural Philosophy of Glasgow University, where in the small museum there was a piece of Kelvin's transatlantic cable. I was intrigued by the problem of how they expected to be able to signal over such a large distance, let alone the mechanical difficulties of manufacturing such a length, and the little they knew about the seabed on which the cable had to lie or the chasms or mountains it had to span. An idea of the form of construction of cables of the time is given by Grivet

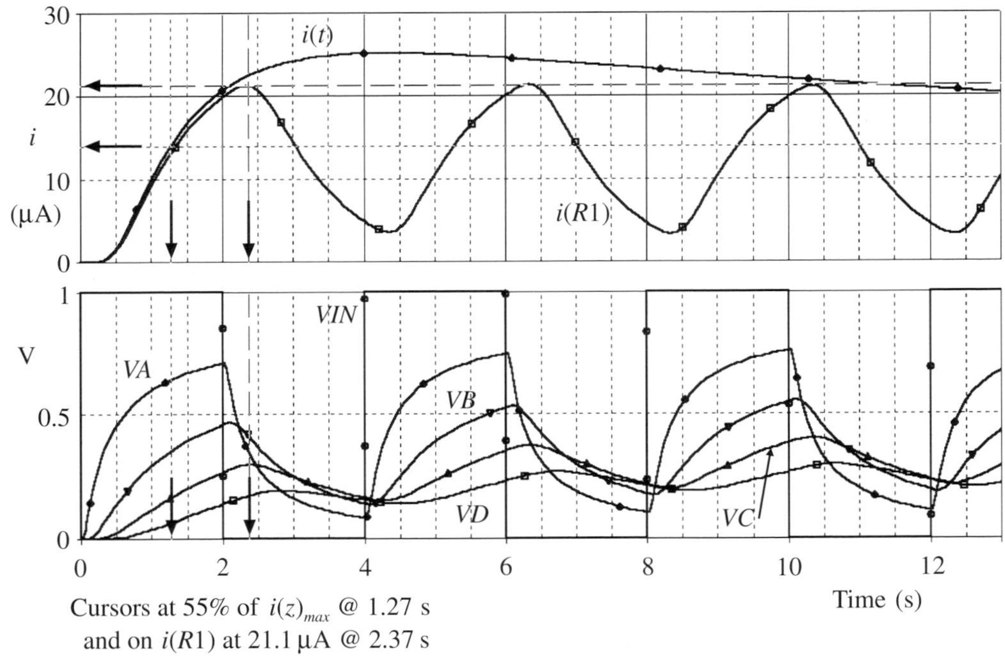


**Fig. 5.25.4** Transient responses of the cable.

(p. 3) and, for example, by Wheatstone (1854/5). The return connection for these cables is the seawater rather than ground currents and Grivet estimates a radius of some 100 m for this 'conductor', the approximate skin depth at a frequency of a few Hz!

## SPICE simulation circuits

Consult the SimCmnt.doc file on the CD before running

| | |
|---|---|
| Fig. 5.25.3 | Tratcbl8.SCH |
| Fig. 5.25.4 | Tratcbl6.SCH |

## References and additional sources 5.25


Boas M. L. (1966): *Mathematical Methods in the Physical Sciences*, New York: John Wiley. Library of Congress Cat. No. 66-17646. See Chapter 14, part 9.

Cohen L. (1928): *Heaviside's Electrical Circuit Theory*, New York: McGraw-Hill. See pp. 66 and 124.

Everitt W. L., Anner G. E. (1956): *Communication Engineering*, 3rd Edn, New York: McGraw-Hill. Library of Congress Cat. No. 55-12099.

Grivet P. (1970): *The Physics of Transmission Lines at High and Very High Frequencies*, Vols 1 and 2, London: Academic Press. ISBN 12-303601-1. Note 1: the equations (47) on p. 402 and (53) on p. 403 are in error.

Heaviside O. (1922): *Electromagnetic Theory*, London: Benn Brothers. Vols I and II.

Jordan D. W. (1982): The adoption of self-induction by telephony 1886–1889. *Annals of Science* **39**, 433–461. See Chapter 18.

Pain H. J. (1976): *The Physics of Vibrations and Waves*, 2nd Edn, New York: John Wiley. ISBN 0-471-99408-1.

Pipes L. A. (1958): *Applied Mathematics for Engineers and Physicists*, New York: McGraw-Hill. Library of Congress Cat. No. 57-9434.

Standage T. (1998): *The Victorian Internet*, London: Walker Publishing. ISBN 08027-13424.

Thomson W. (and Stokes G.) (1855): On the theory of the electric telegraph. *Proc. Roy. Soc.* **7**, 382–399. This is in the form of a letter to G. Stokes and his letter in reply making a very substantial contribution.

Wheatstone C. (1854/5): An account of some experiments made with the submarine cable of the Mediterranean Electric Telegraph. *Proc. Roy. Soc.* **7**, 328–333. (This was the 110 mile cable from (La) Spezia to Corsica.)

# 5.26 Chaos

Civilization is a race between education and chaos.

H. G. Wells

Non-linear systems are very difficult to deal with analytically so for much of the era of electronics recourse has been made to techniques of linearization to allow us to do the sums, but the consequence has been a severe limitation on our understanding and to generalized views that certain things cannot occur. The early experiments by van der Pol (1927, and see Section 1.13) should have been a warning, but the means of doing the difficult sums were not then available. In more recent times the work of Lorenz (1963) investigating the problem of atmospheric weather prediction, together with the availability of powerful computation facilities, awoke the scientific community to the realization that the behaviour of non-linear systems was far more complex than realized and that they were often extremely sensitive to initial conditions as to how they would evolve – the so called butterfly effect.

It is not the intention here to provide an introduction to the theory of chaotic systems but to consider a number of electronic circuits which we can simulate to illustrate some of the effects. The literature is extensive and some appropriate references are given below.

There are a number of introductory papers which are helpful in gaining an insight into this field (e.g. Robinson 1990; Lonngren 1991; Hamill 1993). The original Lorenz equations are derivatives with respect to (time) $t$ in the three dimensions $x$, $y$ and $z$:

$$\frac{dx}{dt}=\sigma(y-x) \qquad \frac{dy}{dt}=rx-y-xz \qquad \frac{dz}{dt}=xy-bz \tag{5.26.1}$$

where the parameters $\sigma$, $r$ and $b$ can vary over various ranges to give solutions of interest. Robinson, for example, describes an analog computer realization for solution of the equations with accessible ranges of $1<\sigma<100$, $0.1<b<10$ and $0<r<625$, though these limits do not represent any absolute magnitudes. Since the solutions can vary dramatically with very small changes in these parameters

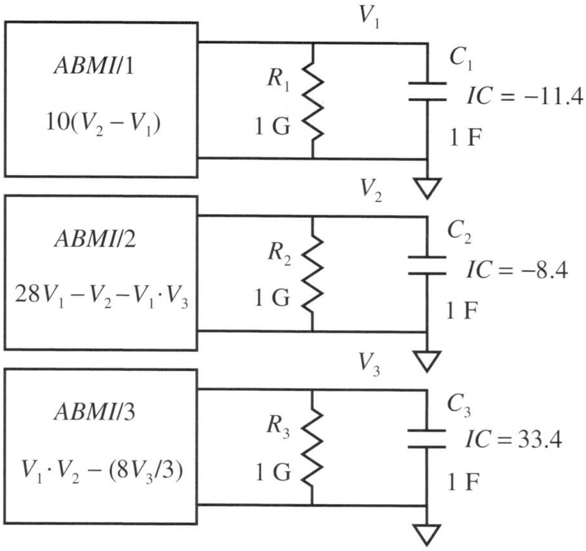


**Fig. 5.26.1** Integrators for the solution of Eq. (5.26.1).

there is much scope for investigation. These equations may readily be solved using SPICE by setting up three integrators as shown by Hamill (1993) using the approach of Prigozy (1989), though we here replace the original *VCCS* with an *ABMI*.

The voltages *V*(*V*1), *V*(*V*2) and *V*(*V*3) represent $x$, $y$ and $z$; the currents provided by the *ABMI*s represent $dx/dt$, $dy/dt$; and $dz/dt$ and the 1 F capacitors serve to integrate the currents to give the voltages (remember that the *ABMI*'s are current sources).

The parameters shown and the initial conditions for the voltages on the capacitors are the 'classic' values illustrated by many authors (Robinson 1990, p. 809), and running the simulation for, say, 40 s (with *Step Ceiling* 0.002 s for smoothness) will give the outputs as functions of time. Change the $x$-axis to *V*(*V*2) and plot *V*(*V*1) and *V*(*V*3) against this to get the usual double scroll portraits. To change the $x$-axis from the normal *Time*, go to *PLOT/UNSYNC PLOT/x-AXIS SETINGS/AXIS VARIABLE* and choose the one you want from the list presented. Figure 5.26.2 shows *V*(*V*1) and *V*(*V*3) as functions of time illustrating the chaotic response, and if you plot *V*(*V*3) against *I*(*C*2) you get Fig. 5.26.2(*b*), which seems appropriate for the equations that led to the idea of the butterfly effect. I will dub this the Lorenz butterfly in his honour (Austin 2000).

As a second example we may examine the apparently simple sinusoidally driven resonant circuit as shown in Fig. 5.26.3 in which the capacitor is non-linear, i.e. the capacity is a function of the applied voltage (Azzouz et al. 1983; Lonngren 1991). For a SPICE simulation we may use the circuit shown in Fig. 5.26.3(*a*) in which a

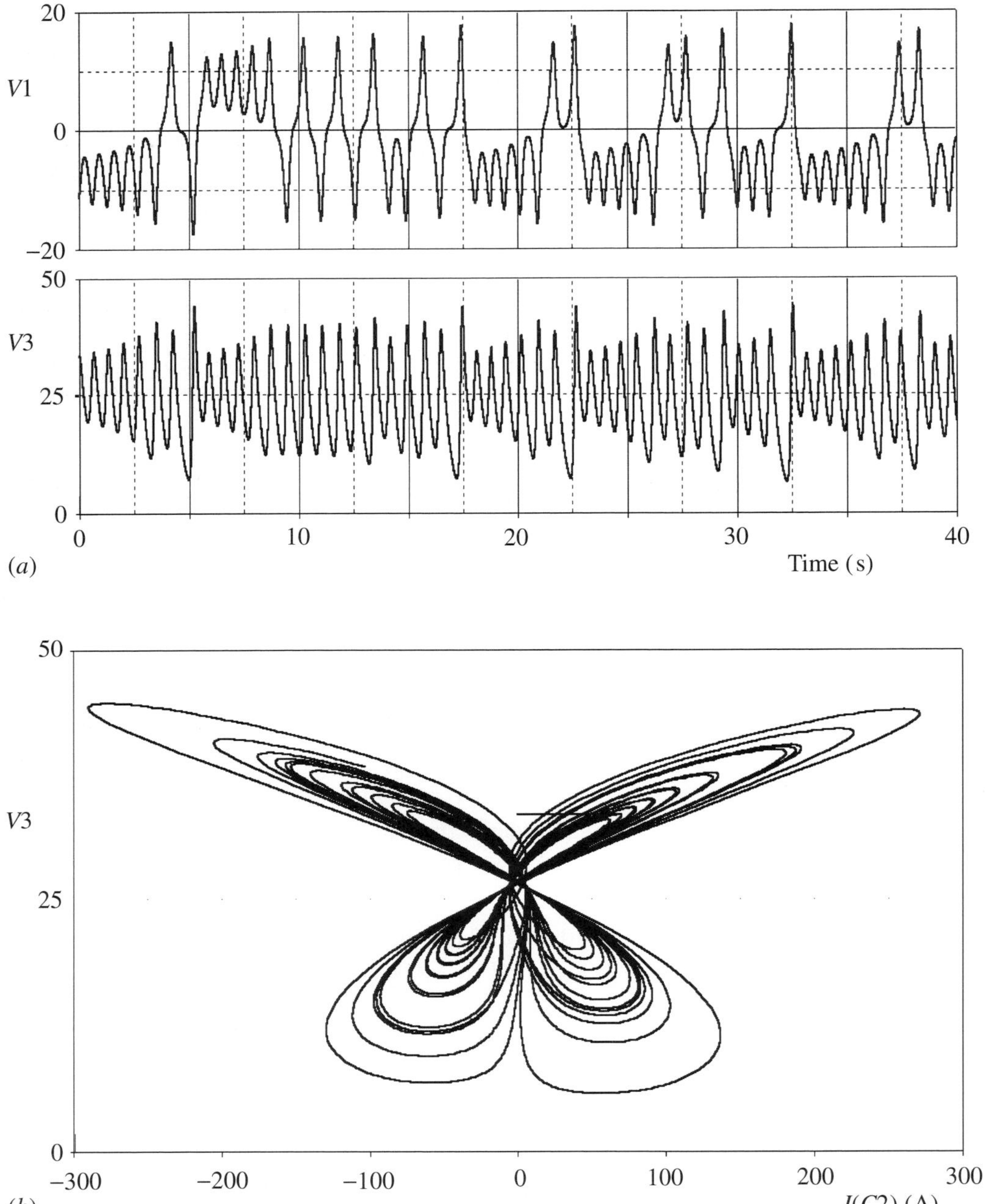


**Fig. 5.26.2** (*a*) Response of the circuit of Fig. 5.26.1. (*b*) The Lorenz butterfly.

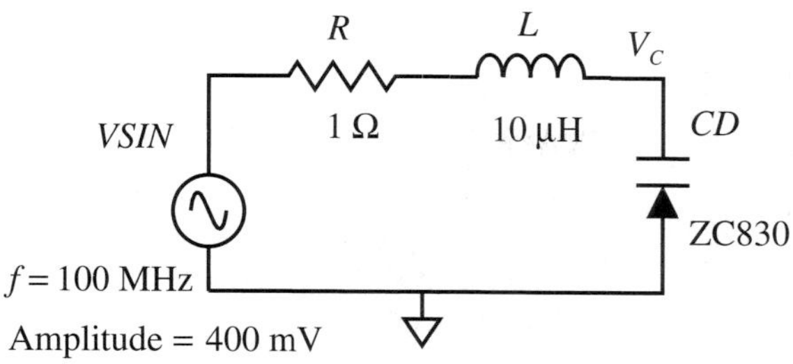


(*a*)

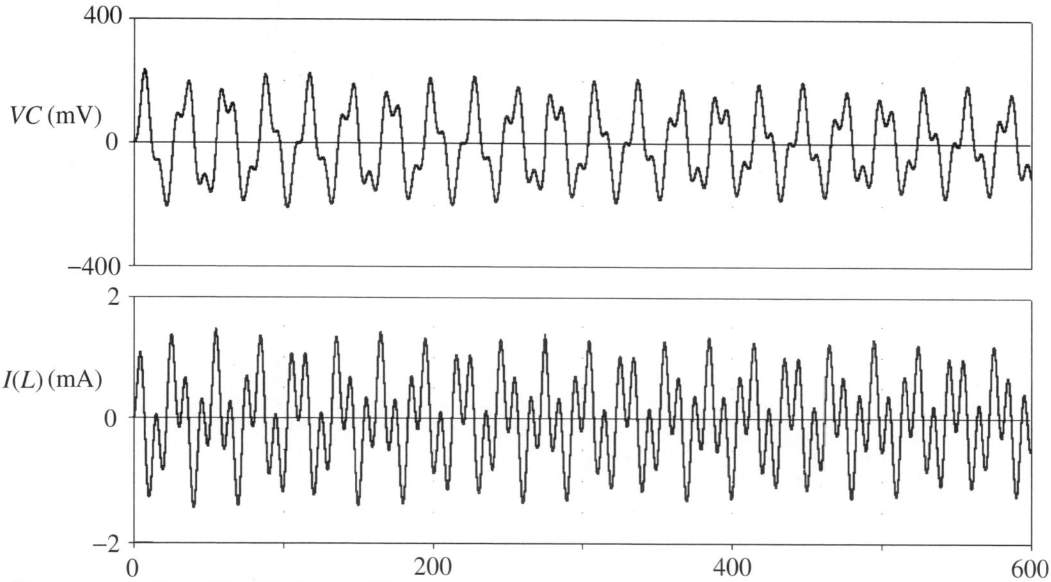


(*b*)

**Fig. 5.26.3** (*a*) Sinusoidally driven resonant circuit with non-linear capacitor. (*b*) Time-dependent output. (*c*) Output for $L = 1$ μH (the reproduction with thicker trace would only allow half the response to be shown – the full response will be shown by the simulation). (*d*) Output for $L = 13$ μH. (*Fig. cont. overleaf*)

high frequency voltage-controlled tuning diode (Zetex ZC830 for which there is a Spice model) has been used. To match the characteristics of the diode the driving frequency has been set to 100 MHz. The characteristics of the diode which are of particular significance are the variation of the capacity as a function of voltage, and the transit time of the carriers. Figure 5.26.3(*b*) shows the time dependent output and Fig. 5.26.3(*c*) shows some results for $L = 1$ μH with the drive amplitude low at 400 mV to minimize the conduction of the diode. Figure 5.26.3(*d*) shows a portrait for $L = 13$ μH to give what looks like a re-entrant barrel or Japanese lantern, but that is an illusion.

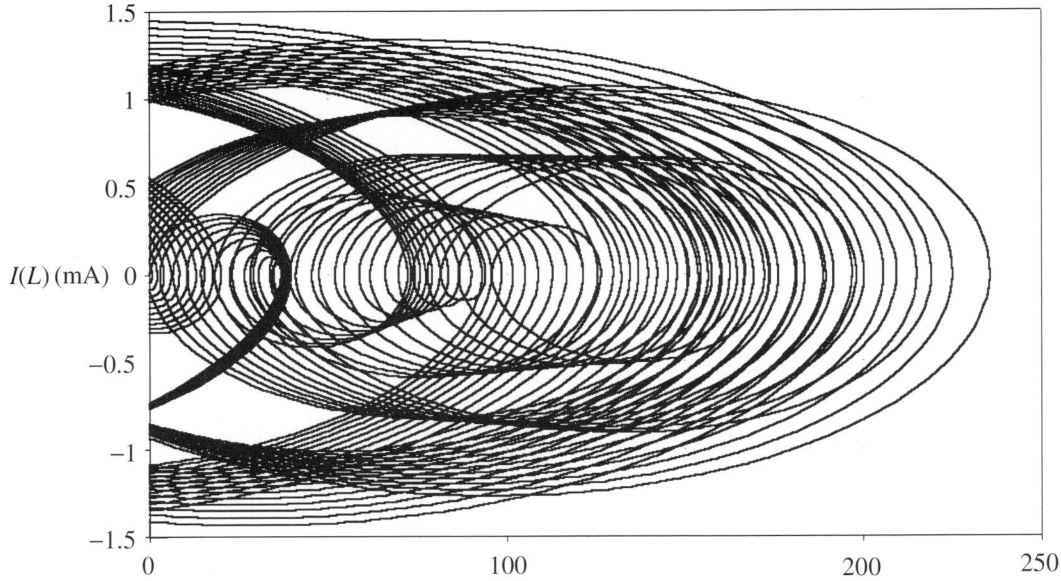


(*c*) Run = 1150 ns, $V = 400$ mV, $L = 1\ \mu$H $V_C$ (mV)

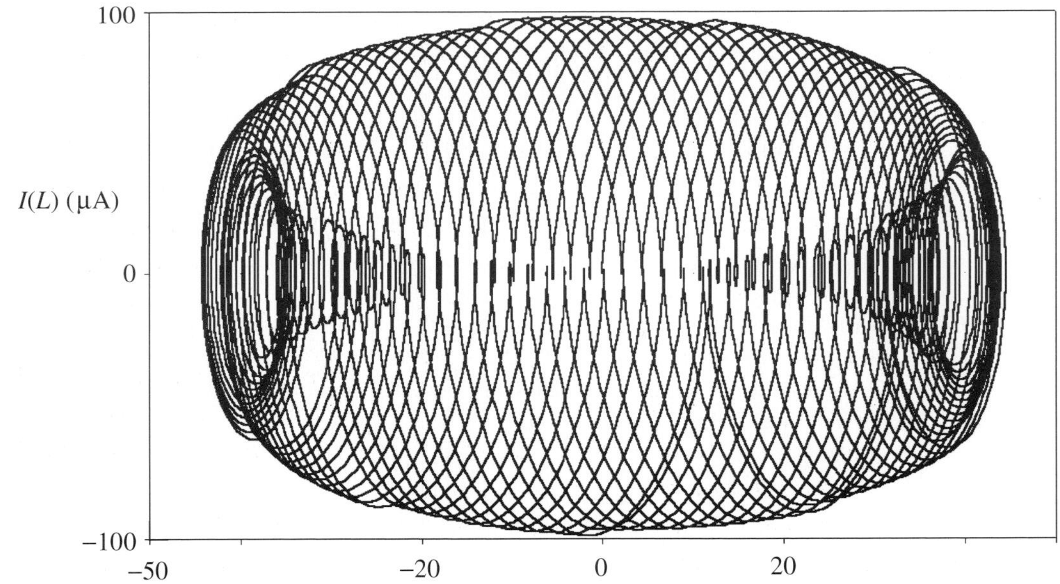


(*d*) Run = 1150 ns, $V = 400$ mV, $L = 13\ \mu$H $V_C$ (mV)

**Fig. 5.26.3** (*cont.*)

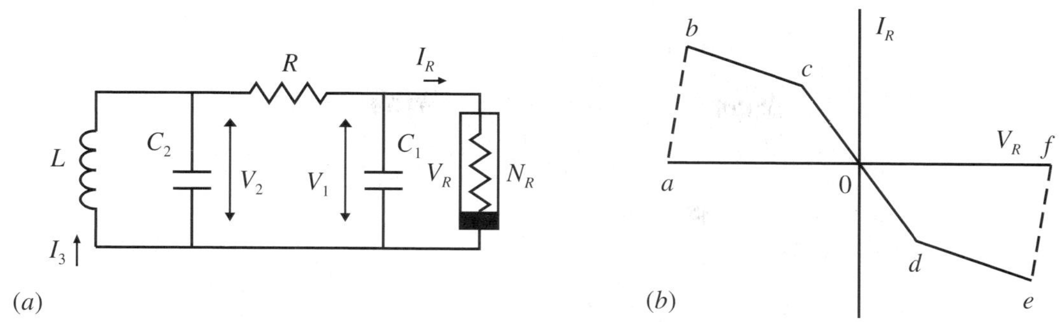


**Fig. 5.26.4** (*a*) Chua's basic circuit. (*b*) Non-linear 'diode' characteristic.

One of the most commonly studied chaotic systems is Chua's circuit or Chua's oscillator. As described by Kennedy (1995a, p. 1116):

> Chaos is characterized by a stretching and folding mechanism; nearby trajectories of a dynamical system are repeatedly pulled apart exponentially and folded back together.
>
> In order to exhibit chaos, an autonomous circuit consisting of resistors, capacitors and inductors must contain (i) at least one locally active resistor, (ii) at least one nonlinear element, and (iii) at least three energy storage elements. The active resistor supplies energy to separate trajectories, the nonlinearity provides folding, and the three-dimensional state space permits persistent stretching and folding in a bounded region without violating the noncrossing property of trajectories.

The simplest circuit that fulfils these requirements is that of Chua and is shown in Fig. 5.26.4. The non-linear element $N_R$, commonly referred to as Chua's diode, has the characteristic illustrated in (*b*). Any physically realizable negative resistance device must for large enough voltages become passive so the power dissipation will be positive, and hence there will be segments of positive slope at each end as shown dashed. If you do a simulation of the driving point impedance of the realization of $N_R$ as shown in Fig. 5.26.5 you will obtain such a form. The usual practical realization of Chua's circuit requires the addition of a resistor in series with the inductor to represent its inherent resistance. The circuit is then referred to as Chua's oscillator and is shown in Fig. 5.26.5 (Kennedy 1995a, b).

In the first paper Kennedy varies $R_0$, and in the second $C_1$, to examine the various responses. In any simulations it is difficult to find the appropriate values to demonstrate the various responses and the time taken in searching can be considerable. If possible, running an actual circuit will be considerably quicker as you can see the consequences immediately. Though it is standard to plot $V_1$ against $V_2$ you should not overlook all the other available variables which can in effect provide differing points of view that allow transitions to be more readily seen. Also, to simplify plots it is helpful to limit the range so that the path can be more readily followed. Figure 5.26.6 shows some examples of what is possible. In these a run of

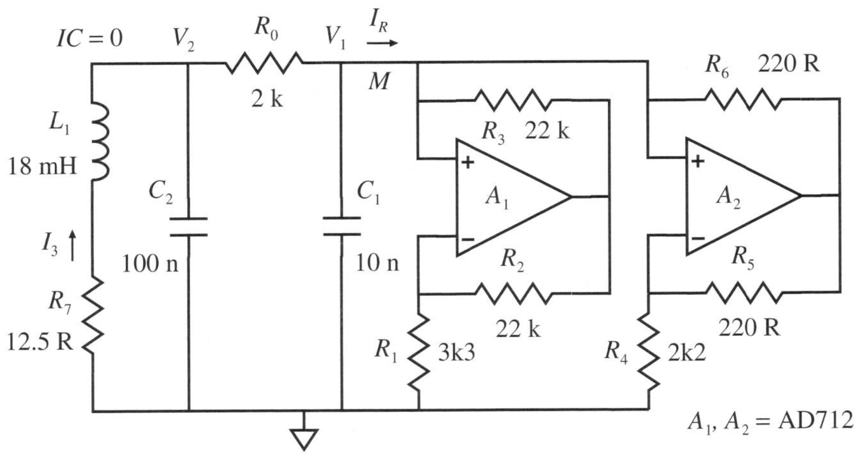


**Fig. 5.26.5** Chua's circuit for simulation.

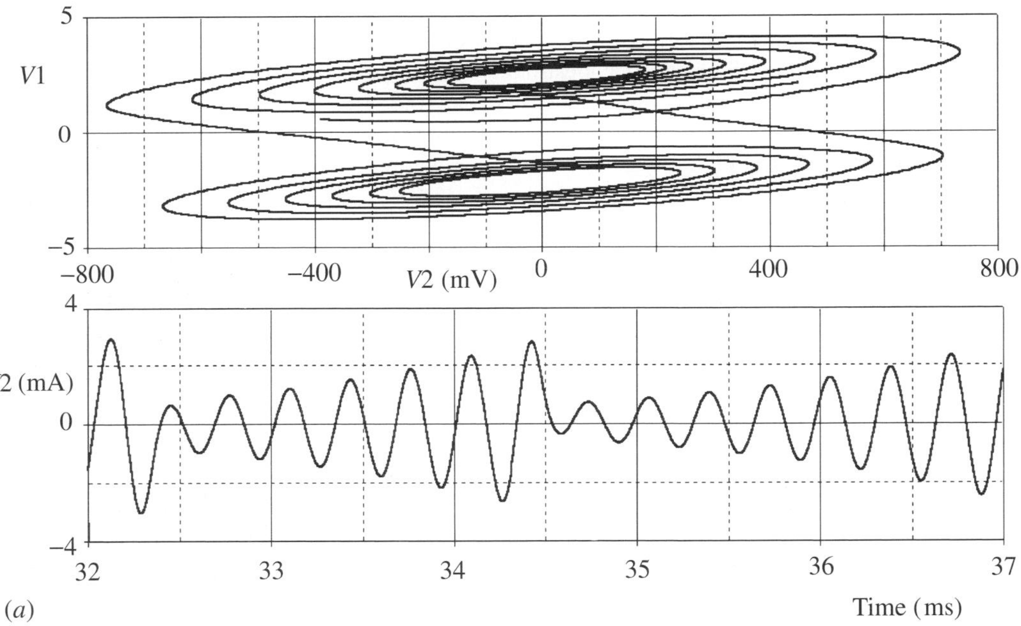


**Fig. 5.26.6** (*a*, *b*, *c*) Assorted illustrative plots for Chua's oscillator. Parameter values were as in Fig. 5.26.5 except for $R_0 = 1k8$ and $C_1 = 9.55$ n. Initial condition on $V_2 = 0$ and step ceiling = 1 μs.

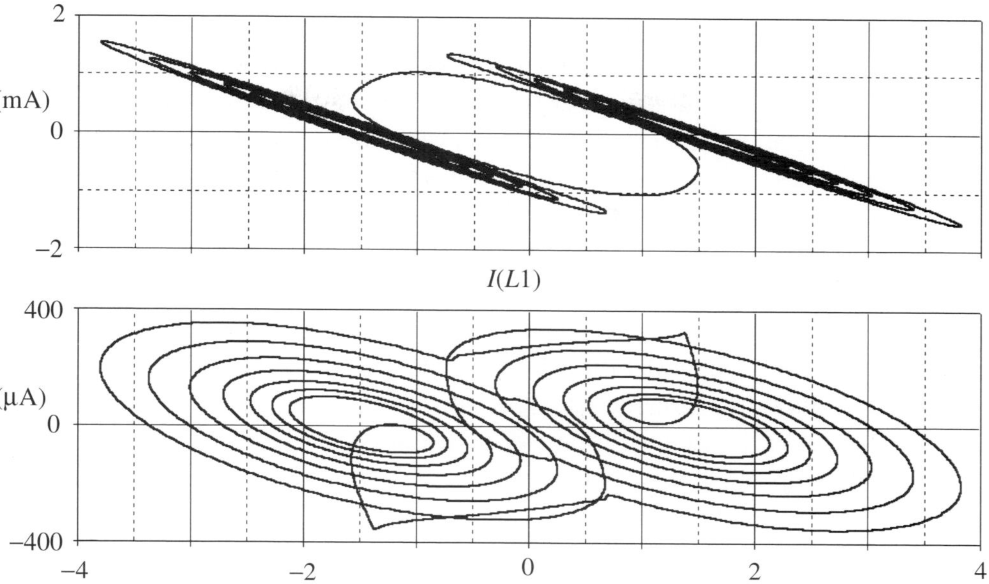


(*b*)

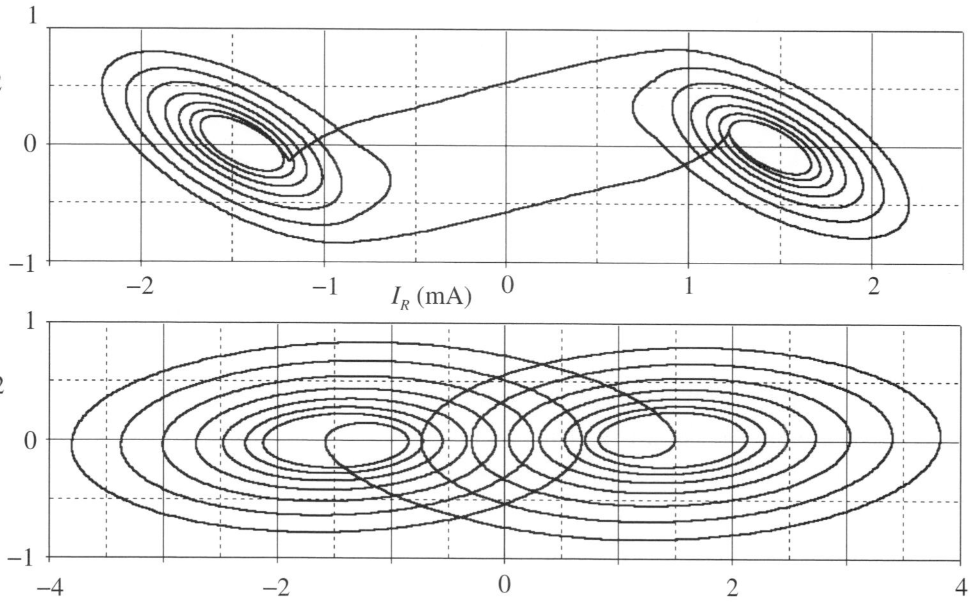


(*c*)

**Fig. 5.26.6** (*cont.*)

50 ms was used and then a suitable segment selected covering an up and a down transition on the time graph of *V*2. Then the simulation was run again with *No Print Delay* set to the earlier time and the *Final Time* to the later time. Note also that the *Step Ceiling* has an effect on what you may see and you should make this as short as you can stand (1 μs was used for the results in the figure). The simulation approximations can lead to deviations in the paths calculated in some cases, but you may want to use a longer step time when searching. The parameter values for the figures are given in the caption.

In part (*c*) the current $I_R$ may be determined by inserting, at point *M*, a *VSRC* set for zero voltage so that you can then ask for the current through it. The portrait shows the 'gravitational' pull of the attractor as a bulge on the penultimate cycle of each scroll, with capture on the next.

Examples of many other aspects of stability and chaotic response are discussed by Pippard (1985). A final warning: fossicking around in chaos circuits can seriously affect your time schedules!

## SPICE simulation circuits

Consult the SimCmnt.doc file on the CD before running

| | |
|---|---|
| Fig. 5.26.2(*a*) | Chaos 2.SCH |
| Fig. 5.26.2(*b*) | Chaos 6.SCH |
| Fig. 5.26.3(*b*) | Chaos 1.SCH |
| Fig. 5.26.3(*c*) | Chaos 5.SCH |
| Fig. 5.26.3(*d*) | Chaos 4.SCH |
| Fig. 5.26.6(*a*) | Chuachs 5.SCH |
| Fig. 5.26.6(*b*) | Chuachs 4.SCH |
| Fig. 5.26.6(*c*) | Chuachs 3.SCH |

Note: In reproducing the chaotic responses generated in the simulations it was necessary to use broader traces than normal in PROBE to allow reduction. PSpice has a limitation on the 'length' of trace that can be broadened so the length shown has had to be restricted. Your simulations will look much better on the screen.

## References and additional sources 5.26

Austin (2000): A cartoon in my newspaper, *The Guardian*, shows two butterflies perched on a twig in the jungle, with one saying to the other 'Let's flap our wings and make it rain over the Easter holiday'.

Azzouz A., Duhr R., Hasler M. (1983): Transition to chaos in a simple nonlinear circuit driven by a sinusoidal voltage source. *IEEE Trans.* **CAS-30**, 913–914.

Chua L. O., Hasler M., Neirynck J., Verburgh P. (1982): Dynamics of a piecewise-linear resonant circuit. *IEEE Trans.* **CAS-29**, 535–547.

Hamill D. C. (1993): Learning about chaotic circuits. *IEEE Trans.* **EDUC-36**, 28–35. Note that the expression for $dx/dt$ in his Eq. 1 should be $10y - 10x$, which then agrees with his Fig. 1.

Jordan D. W., Smith P. (1999): *Nonlinear Ordinary Differential Equations*, 3rd Edn, Oxford: Oxford University Press. ISBN 0-19-856562-3.

Kennedy M. P. (1995a): Bifurcation and chaos. Section 38, in Chen, Wai-Kai (Ed.) *The Circuits and Filters Handbook*; Boca Raton: CRC Press and IEEE Press. ISBN 0-8493-8341-2. The quotation is reprinted with permission of the publishers and is copyright CRC Press, Bocca Raton, Florida. My thanks to the author for a copy of his software package 'Adventures in Bifurcations and Chaos', a simulator for Chua's oscillator.

Kennedy M. P. (1995b): Experimental chaos from autonomous electronic circuits. *Phil. Trans. Roy. Soc. London* **A353**, 13–32.

Kennedy M. P., Chua L. O. (1986): Van der Pol and chaos. *IEEE Trans.* **CAS-33**, 974–980.

Kennedy M. P., Kreig K. R., Chua L. O. (1989): The Devil's staircase: the electrical engineer's fractal. *IEEE.Trans.* **CAS-36**, 1133–1139.

Lonngren K. E. (1991): Notes to accompany a student laboratory experiment on chaos. *IEEE Trans.* **EDUC-34**, 123–128.

Lorenz E. N. (1963): Deterministic nonperiodic flow. *J. Atmospheric Sci.* **20**, 130–141.

Matsumoto T., Chua L.O., Tokumasu K. (1986): Double scroll via a two-transistor circuit. *IEEE Trans.* **CAS-33**, 828–835.

May R. M. (1976): Simple mathematical models with very complicated dynamics. *Nature* **261** (5560), 459–467. A general discussion, particularly of biological applications.

Murali K., Lakshmanan M. (1992): Effect of sinusoidal excitation on the Chua's circuit. 1:– Fundamental theory and applications. *IEEE Trans.* **CAS-39**, 264–270.

Parker T. S., Chua L.O. (1987): Chaos: a tutorial for engineers. *Proc. IEEE* **75**, 982–1008.

Pippard A. B. (1985): *Response and Stability*, Cambridge: Cambridge University Press. ISBN 0-521-31994-3.

Prigozy S. (1989): Novel applications of SPICE in engineering education. *IEEE Trans.* **EDUC-32,** 35–38.

Robinson F. N. H. (1990): Analogue electronic model of the Lorenz equations. *Int. J. Electronics* **68**, 803–819.

Sparrow C. (1983): An introduction to the Lorenz equation. *IEEE Trans.* **CAS-30**, 533–540.

van der Pol B. (1927): Forced oscillations in a circuit with non-linear resistance. *Philosophical Magazine* **3**, 65–80.

# 5.27 Spice notes

There are no facts, only interpretations.

Friedrich Nietzsche

In this section we will outline a number of topics in relation to the use and application of PSpice that have arisen as a consequence of experience over many years. Most of the books at present available that make use of SPICE refer to the pre-Windows versions of the software and are hence much involved in discussion of the construction of netlists and the use of detailed commands. The Windows versions avoid much of this and save enormous effort, though it is desirable that some understanding of the structure is obtained to allow a better appreciation of what goes on behind the scenes, especially when problems arise. This is not intended as an instructional manual for PSpice, which itself comes with the equivalent of several substantial books of instruction. It is intended as a reminder of a number of techniques and of some of the faults you may encounter. I am sure there are many more, but these are matters that I have found relevant to analog simulation. The comments refer to Version 8.

(a) A small but vital component required in every circuit is the common reference point. This is provided by the *AGND* symbol which indicates the zero potential reference point. If it is not present there will be no complaint from SPICE but it will probably wander off and disappear into a state of non-convergence and hence failure.

(b) A necessary stage in any simulation is that the system can find a self-consistent starting configuration from where it can launch itself. If for any reason this cannot be achieved then a fault will be indicated. Problems of this sort are sometimes difficult to overcome as it is not evident what is causing the failure to converge. The system is set to a certain maximum of iterations before it fails, and the number is one of the parameters that can be changed, but it is likely that increasing the number will not help. There are also tolerances on the degree of consistency that is acceptable, and these levels can also be adjusted. It is, for example, often a problem with MOSFETs since the range of currents that are encountered is so large, from picoamps of gate current to amps of drain current. The default values of *VNTOL* (for voltage) and *ABSTOL* (for current) are set to be generally appropriate for low current systems, but if you are interested in high power circuits then it is probably of little significance whether the gate current is known to a picoamp resolution in comparison with a drain current of many amps (*ABSTOL* should be no more than $10^{-9}$ of the highest current). It is sometimes helpful to add highish value resistors around the FET to increase currents associated with the gate but without affecting actual operation. Diodes can also cause problems and often need to be paralleled with high value resistors to enable successful bias point computation. For a full discussion of matters relating to convergence reference may be made to Vladimirescu (1994, Chapter 10). The various settings are to be found under *ANALYSIS/SETUP/OPTIONS* which will provide a list of parameters and their (default) values, and the means to alter them. Do not set *RELTOL* greater than 0.01. Doing so may seriously affect the simulation. Options *ITL*1 and *ITL*2 can often be increased to help find an initial bias point.

(c) *Floating nodes and capacitors*: All nodes must have some d.c. path to common. If not, the initial bias point cannot be determined and an error will be reported. Capacitors are the main culprits and you will need to add, if necessary a very large resistor, e.g. 1 GΩ, to allow simulation to proceed. Diodes can also cause the same problem and can be shunted or otherwise treated if need be.

(d) *Series voltage sources and inductors*: Any voltage source must feed a finite (rather than zero) impedance as otherwise you will have an infinite current and PSpice will object. Inductors are handled like voltage sources and will lead to the same objection. Thus you should always include some resistance in series with an inductor that can represent its inescapable resistance both for the above reason and since with perfect inductors you may get perfect but impossible outputs.

(e) *Analog behavioural model (ABM) devices*: There a number of these devices providing either voltage or current output with either none, one or two input pins. The function of the device is defined by a user-written mathematical algorithm, so they are very flexible. The simplest, for example, appears with only one pin (for the output signal), but there is an implied hidden pin connected to common. The prescription is typed in in the normal form for mathematical expressions and it is wise to use defining brackets to ensure you get what you intended. The text can continue over several lines in the attributes input window and the lines are simply concatenated. An example is given in Section 5.15 where we wished to define a particular form of pulse. An *ABM* definition will accept reference to *V(netname)* or to *I(Vdevice)* but not to say *I(R#)*.

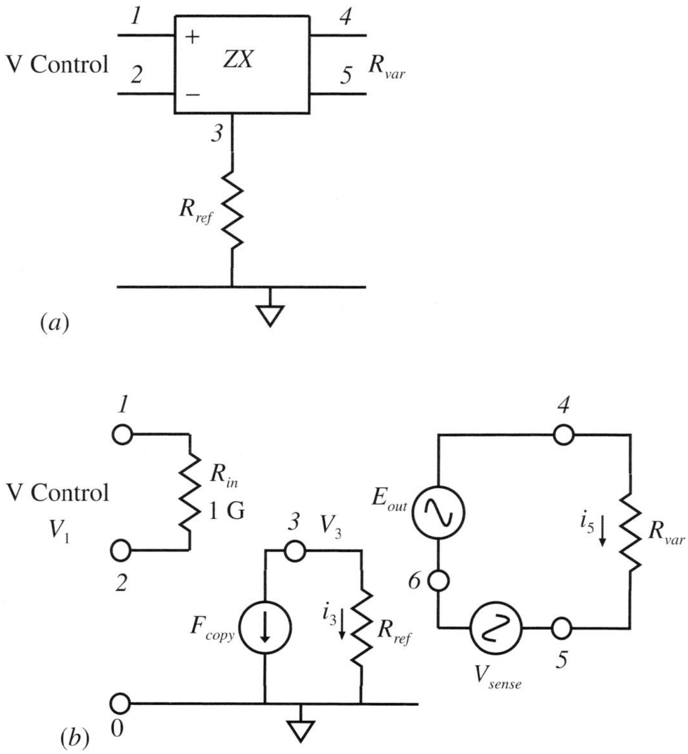


**Fig. 5.27.1** (*a*) Symbol for *ZX* device (*YX* is similar). (*b*) Model of *ZX* device.

(f) *ZX and YX devices*: These two devices are described in Tuinenga (1988) and provide a convenient way of obtaining varying components, i.e. *R*, *L* and *C*. The configuration is shown in Fig. 5.27.1.

The *ZX* version allows a reference *R* or *L*, from terminal 3 to common, to be multiplied by an input voltage across floating terminals 1 and 2, to give the product, in effect a voltage controlled resistor, across floating terminals 4 and 5. The voltage may be one already present in your circuit or you can simply add an appropriate extra generator. You should take care, however, that the control voltage does not go negative as you will otherwise get, say, a negative resistance! The *YX* circuit provides the same function for an admittance or capacitor. These devices are examples of subcircuits and demonstrate how you may make up your own. The description of these subcircuits may be viewed by placing one on your schematic and then going to *EDIT/MODEL/EDIT INSTANCE MODEL* (*TEXT*). It is also discussed in Tuinenga (1988, p. 189) and is:

```
.subcircuit ZX 1 2 3 4 5
Eout 4 6 poly(2) (1,2) (3,0) 0 0 0 0 1
Fcopy 0 3 Vsense 1
Rin 1 2 1G
Vsense 6 5 0
.ends
```

Line 1: the subcircuit name and the internal node numbers.

Line 2: $E$ device ($VCVS$) between nodes #4 and #6, defined by a polynomial in two variables, the voltages between nodes #1 and #2 (say $V_1$) and between #3 and #0 (say $V_3$). The polynomial form is $C_0+C_1V_1+C_2V_3+C_3V_1^2+C_4V_1V_3+\cdots$ so in this case all the coefficients are zero except that for $V_1V_3$.

Line 3: $F$ device ($CCCS$) between nodes #3 and #0 which develops a current proportional to that in $V_{sense}$, the constant of proportionality in this case being 1. Note that there is an implied connection to node #0, which is not shown on the symbol.

Line 4: The input resistance between the control input nodes #1 and #2 and of value 1 GΩ.

Line 5: A voltage source between nodes #6 and #5 of zero voltage. Used to enable measurement of the current in the output circuit.

Line 6: Indicates the end of the definition.

We may understand the operation of the subcircuit as follows, using $V_1$, $V_2$, $i_3$ and $i_5$ as shown in Fig. 5.27.1($b$):

$$E_{out}=V_1\times V_3,\quad V_3=i_3\times R_{ref},\quad i_3=i_5=\frac{E_{out}}{R_{var}}\quad \text{so}$$

$$V_3=\frac{E_{out}\,R_{ref}}{R_{var}}=\frac{V_1\,V_3\,R_{ref}}{R_{var}}\quad \text{or}\quad R_{var}=V_1R_{ref}$$

(g) *Transmission lines*: For simple applications the lossless or ideal $T$ line may be used (see e.g. Section 5.24). If lossy lines are required you need values for the parameters. Though details may differ somewhat between specific coaxial types, an approximate set of values for 50 and 75 Ω cables are:

| Impedance, $Z_0$ | $L$ (μH m$^{-1}$) | $C$ (pF m$^{-1}$) | $R$ (Ω m$^{-1}$) | $G$ (μS m$^{-1}$) |
|---|---|---|---|---|
| 50 Ω | 0.25 | 100 | 0.37 | 7.5 |
| 75 Ω | 0.38 | 67 | 0.31 | 6.3 |

The values for *R* and *G* vary widely depending on cable size, dielectric and frequency so the values are only for guidance, and in any case have little effect unless the cable is long.

(h) *Transformers and dotting*: Indication of the phasing in a transformer is sometimes required and this is not normally provided. You may edit the symbol to make the pin numbers visible and remember which ends are 'dotted' or you can add dots to the symbol to make it obvious. Make up a simple test circuit to confirm the polarities (Hirasuna 1999c).

(i) *Terminal numbers on components*: Pin numbers or designations are not displayed by default. Editing the symbol (click on the symbol and select *EDIT/SYMBOL*) will allow you to enable their display and this is useful in understanding current flow and selection of voltage signals where you have not explicitly named a net. The PSpice library Analog.slb is the source of the standard symbols for *R*, *L* and *C*. There is an alternative version Analog_p.slb which provides visible pin numbers. If you wish to use this instead then this library must be configured before the former so that it is accessed first. This may be done via the *OPTIONS* menu or by editing the *MSIM.INI* file.

(j) *Current directions*: While voltages will be referenced to *AGND* as 0 V, currents provide a more difficult problem in understanding the sense in which they flow. The SPICE manuals will give you some advice but I find that displaying the pin numbers on the schematic is the most useful approach; the currents displayed in *PROBE* will flow from pin 1 to pin 2 in a passive component (*R*, *L* or *C*) and if necessary you can always insert a very low value resistor at any point in the circuit (think of it as the wiring) and then request the current through it. For a voltage source the current is taken to flow from + to − internally so that it appears to flow the 'wrong way' in the external circuit. If you run a bias point analysis then you can display the voltages and currents at the nodes as described in (B). If you click on a current flag you will see on the component to which it refers (shown by a dashed line) a small arrow which indicates the direction of current flow.

(k) *Phase*: In feedback circuits, for example, the phase of signals is of great interest. In the *TRACE/ADD* window you will find on the right-hand side a list of commands or instructions that allow the choice of a wide range of derived traces. For phase click on *P*(), which will appear in the selection box below, and then click on the signal trace of interest which will be inserted into the bracket. You can also ask for sums to be done as, for example, $P(V(VOUT))+180$ where phase is in degrees.

(l) *PROBE trace manipulations*: In displaying traces there are many options in addition to the simple display of traces. You can request any mathematical function of the available traces and involve more than one trace dataset in the function. If the function is extensive, or if it is to be used many times, then it is worth making it into a *MACRO* that can be stored with the *.prb* file and if required transferred to other schematics (just copy the *.prb* file and rename it to match the new schematic). On the *TRACE/ADD* window you will see a selector showing *ANALOG OPERATORS AND FUNCTIONS* with a selector which will alternatively display *MACROS*. Give the macro a unique name and enclose its arguments in () comma separated. In the definition box type the definition and then save the macro as required. An example of this is given in Section 5.14 where macros are used in applying the *T* technique. In general it is undesirable to display current traces and voltage traces on the same plot since currents are commonly numerically much smaller than voltages and the currents simply appear along the *x*-axis. A separate plot is usually more revealing. If in any plot you require a reference trace just ask for one, e.g. add a zero trace 0.

(m) *Markers for signals*: *V* and *I* markers may be placed on the circuit to indicate those that you wish to examine. In *PROBE* setup you can restrict collection of results to these points if you have memory limitations. Otherwise PSpice will retain all the voltages and currents, which can lead to large data files.

(n) *Trace availability listing and reductions*: All the traces that are available are listed upon *TRACE/ADD* and there can be a long list. There is a check box which enables suppression of the display of internal model variables, which will generally not be of interest, and will usually considerably shorten the list.

(o) *Trace editing*: If you wish to edit a trace a double click on the name will bring up the *TRACE/ADD* window and you can adjust the requirements rather than deleting the trace and starting from scratch, e.g. to keep the trace colours the same since you have just learnt which is which in a complex display. The adjustment is not limited to the original signal – you can ask for a completely new one.

(p) *Laplace parts*: The Laplace part, a form of *ABM*, allows you to set up a transfer function by specifying it in the normal transfer function form with a numerator polynomial and a denominator polynomial. Examples of this are shown in Sections 3.4 and 4.3. As this is a frequency domain part, the output will depend on the past inputs as well as the present. In the *MicroSim User's Guide, Cautions and Recommendations for Simulation and Analysis, Laplace Transforms*, there is an error in that it is stated that 'the transform can contain poles in the left-half plane. Such poles will cause an impulse response that increases with time instead of decaying'. It is poles in the right-half plane which will have such an effect. Poles in the left-half plane denote a stable system.

(q) *Mathematical constructions and functions*: Mathematical expressions can use a wide range of operators and functions and the rules of precedence control the basic operators +, −, *, /, but it is usually safer to use brackets to make it quite clear what you want. Check by counting to ensure there are equal numbers of left and right brackets. When typing in an expression it is almost reflex to type things like 2(. . .) rather than 2*(. . .) so read the expression over with a rather pedantic eye. Examine carefully the outcome to ensure you are getting what you expect and test with some numerical values.

(r) *Time*: When using time as a parameter in function definitions, e.g. in a statement for an *ABM*, it must be written explicitly in full as '*time*' and not as say *t*. Similarly '*frequency*' must be used for the variable frequency.

(s) *Net names*: For signals in which you are particularly interested it is convenient to give the net a name rather than relying on the numerical assignments of SPICE. The names should be descriptive so that they are readily identified but kept short especially if you are going to display many waveforms together. Using names beginning with the same letter means that they appear close together in the list and are therefore easier to locate.

(t) *Parametrized variables* (*PARAM*): It is often useful to run a simulation with some component having a number of different values with the results available for plotting on the same graph. This can be carried out with a *PARAM* arrangement. Place a *PARAM* symbol on the schematic and double click to see its attributes. Select *NAME*1 and type *RVAL* (e.g. for a resistor *R*) and for *VALUE*1 type in a value that you wish it to have if you were not varying the value, i.e. as a record of what it was. For the component in question alter its value attribute to {*RVAL*} with curly brackets. Under *ANALYSIS/SETUP* choose *PARAMETRIC*, select *GLOBAL PARAMETER* and *LINEAR*, type in *RVAL* in the *NAME* box and the start, end and increment values if a fixed increment is appropriate. An alternative is to select *VALUE LIST* and type in a series of values separated by commas or spaces.

(u) *Power supplies and amplifier symbols*: Power supplies for amplifiers or other devices may be placed away from the main part of the circuit and the connections indicated by means of a *GLOBAL* symbol attached to the voltage source and the appropriate device pin and given an appropriate name. It is helpful to enable the polarity and value of the voltage supply to be displayed on the schematic, and it is safer to connect the negative terminals of both positive and negative supplies to common and then to set the negative supply explicitly to $-V$ rather than connecting the positive terminal to common. In naming global power supplies you should be careful to avoid the system names reserved for digital devices such as *VCC*, *VEE*, *VDD*, *VSS*. Be warned that, for example, op-amp symbols from libraries are not universally oriented in the same sense, and you may flip a symbol to get the input terminals a more convenient way round. Always check carefully which is the positive and the negative supply pin and make sure they are connected to the correct supply. SPICE will not flag an error if you get it wrong and you can spend a lot of time getting peculiar results.

(v) *Log and linear displays and range selection*: When the traces are displayed by *PROBE* the *Y-axis* scale will be linear so that, for example, frequency responses of an amplifier will look rather different from what is usually presented in the literature. You can go to *PLOT/Y AXIS SETTINGS* and select *LOG* to get the more usual display. The data range for both *X* and *Y* may also be set here if you want a more precise range than you get from the *VIEW* subcommands. The *TOOLS/OPTIONS* menu allows you to select the placing of symbols (small circles, squares, triangles, etc.) on the traces if it is difficult to follow when you have many traces. You cannot obtain a log display if the *y*-axis covers zero or negative values.

(w) *Probe active display*: From *PSPICE A_D* it is possible to set *PROBE* to display the traces as they are computed rather than waiting till the end of the simulation. The traces will be updated at intervals rather than point-by-point. This is particularly useful if the simulation run will be long and you are not sure that the system is responding properly. If it appears that something is wrong the run can be terminated prematurely to allow a fix to be tried.

(x) *Probe setup*: Before starting a run it is usually desirable to go to *ANALYSIS/PROBE SETUP* to choose the way in which *PROBE* will operate. '*AUTOMATICALLY RUN PROBE AFTER SIMULATION*' and '*RESTORE LAST PROBE SESSION*' are probably the most convenient. If you do not choose the latter you will have to type in the traces you want displayed again if you do another run. *PROBE* settings are stored in a file ⟨*Filename*⟩*.prb* so they will be preserved and if you transfer your schematic file to another system it is useful to take a copy of the *.prb* file as well. All other files will be automatically recreated. However, if you change from say a *TRANSIENT* run to an *AC SWEEP* run, the probe file will be wiped and a new version created.

(y) *Time step limitation*: In setting up the parameters of the run (*ANALYSIS/SETUP*) you set the length of the run but you can also set a maximum step size which cannot then be exceeded. PSpice normally chooses the time steps in a dynamic way to suit its own requirements, and if it thinks it can get away with it can take quite long steps to speed up the run in real time. This is often quite satisfactory but, for example, when generating a *Sin* wave it can lead to rather 'distorted' waves made up of straight-line approximations. If you require smooth waveforms you must set the maximum time step to an appropriate value which you can relate to the period of the wave. The penalty is of course that the run will take a rather longer time as more points must be computed. If you watch the left-hand box in the *PSPICE A_D* window during the simulation you can see how the time steps vary in all cases (Version 9 uses a separate window). *PRINT STEP* controls how often the optional text format data are written to the output file and so is usually not of interest. However, when performing a Fourier analysis the print step is used as the sampling interval.

(z) *Averaging*: There are a number of averaging functions available in *PROBE*. Under *ADD/TRACE* you will be presented with a list among which you will find:

*AVG*($x$) Running average over range of $x$ variable
*AVG*($x,d$) Running average over range of $x$ variable from $(x-d)$ to $(x)$
*RMS*($x$) Running average over range of $x$ variable

Note that you will not find these items directly in the *Help Contents*, but under '*Arithmetic in Trace Expressions*'. It is curious how people differ in how they approach such matters.

(A) INI *file setup, colours, libraries and sequence*: The *msim.ini* file, located in the Windows directory, contains all the initialization information for PSpice. You may read and edit it using any text editor but you should be careful not to change anything you do not understand. There are, however, a number of changes you can make to suit your preferences. *PROBE* is initialized for six trace colours but can have twelve. The entries are in the *.ini* file but are disabled by the semicolon at the start of the line. Delete the semicolon to enable the additional colours, which are all used in the sequence in which they are listed. These may be rearranged if you wish but make sure the format and names are exactly as you see them. The colours used for *SCHEMATIC* can be changed from the screen via *OPTIONS/DISPLAY PREFERENCES*. When installed the colours appear to have been selected to suit a white background. If you prefer a black background, for example (as I do for less eyestrain from a bright white screen), then you will want to change many of the colours. The allowed colour names for *PROBE*, and the objects that are particularly significant for *SCHEMATIC* are shown below.

| Probe Colours | Schematic Objects | My Scheme |
|---|---|---|
| Black (background) | Attributes | Green |
| White (foreground) | Background | Black |
| Brightgreen | Foreground | Red |
| BrightRed | Grid | Yellow |
| BrightBlue | Pinname | Magenta |
| BrightYellow | Pinno | Magenta |
| BrightMagenta | Refdes | Cyan |
| BrightCyan | Selection | Brightwhite |
| Mustard | Symbols | Yellow |
| Pink | Wire | Cyan |
| LightBlue | | |
| DarkPink | | |
| Purple | (for others see manual) | |

Colours may also be set as *rgb* values, e.g. *object* = *255 0 0* gives red. The model libraries are also listed in order in the *msim.ini* file so you can see what you have. If you wish to suppress some of them, or add any, for any reason, then the remaining libraries must be in sequentially numbered order without any gaps.

(B) *View bias levels* V *and* I *on schematic*: There are two buttons on the top ribbon that allow you to display directly on the schematic the bias values at each point. This can be useful for seeing whether you are operating at a sensible quiescent condition, or when there is a problem in finding an acceptable starting configuration.

(C) DC, AC *and* TRANSIENT *runs*: You must choose the type of simulation you require from the *ANALYSIS* menu. An *AC* run will find a quiescent operating point and linearize the circuit around that point. It will then determine the frequency (and phase) response over the range specified to be plotted by *PROBE*. The magnitude of the gain response is not limited by the circuit since it is now considered as linear, so you may get some unexpected results. A *DC* run requires the selection of some source and a range over which it is to be varied. Then you can get *PROBE* to plot any of the other variables against this to see, for example, the proper value to set the source to bias your circuit appropriately. A *TRANSIENT* run will plot proper waveforms allowing for all the characteristics of the models and of the circuit. The basic *x*-axis parameter will be *TIME* but this can be subsequently manipulated in many ways only limited by your ingenuity. For an example of this see Figs. 3.12.4 or 5.26.2.

(D) *Attribute display, e.g. volt sources, etc*: It is usually helpful to display on the schematic the attributes of signal sources as a reminder, since with much adjusting when investigating a system you may easily forget. When setting the value of an attribute there is a choice as to whether you wish to display it. Select *CHANGE DISPLAY* and then *NAME AND VALUE*. The text lands up in a heap on the schematic and you have to separate them out to be legible. Setting *CAPS LOCK* on your keyboard makes reading small text easier without enlarging the schematic so far that only a small portion is visible.

(E) *Fourier display*: A spectrum of a waveform may be obtained by running a Fourier analysis. This can, for example, indicate distortion on a sine wave or show the bandwidth required to pass a given signal without too much distortion. This option is available if you do a transient analysis since you must have a wave to analyse.

(F) *Transfer function*: See Laplace (p).

(G) *Model/subcircuit*: When PSpice accesses the library to read the model for a device then it expects the template definition string type to match the type in the library. A problem sometimes arises as when the template gives it as a

*.subckt* and the library shows it as a *.model*. Templates beginning with an *X* refer to a subcircuit.

(H) GLOBAL: A global (which has a symbol just like any other object) allows a given value to be distributed about the schematic without having to use wires. All points with the same *GLOBAL* name will be considered as joined. This is particularly useful for power supplies.

(I) TOPEN, TCLOSE: These switch components are useful to allow something to change at a given time. They are in effect variable resistors (and can be used as such) and are defined with a minimum and a maximum resistance value together with a time at which they begin to operate and a time interval to complete the operation. You should allow an appropriate time for the changeover otherwise the simulator may have difficulty in following too rapid a change satisfactorily.

(J) *Voltage-controlled (and current-controlled) switch*: These devices are in effect voltage- (or current-) controlled resistors. They are listed in the *Get Part* display as *Sbreak* (*Wbreak*). The parameters are:

| *Sbreak* | | *Wbreak* | |
|---|---|---|---|
| *RON* | On resistance | *RON* | On resistance |
| *ROFF* | Off resistance | *ROFF* | Off resistance |
| *VON* | Voltage for *RON* | *ION* | Current for *RON* |
| *VOFF* | Voltage for *ROFF* | *IOFF* | Current for *ROFF* |

To set the parameters you need to place a component and then go to *EDIT/MODEL* and *EDIT INSTANCE MODEL (TEXT)*. The ratio of resistance should be kept below $10^{12}$ and the values must be between zero and 1/*GMIN* (see under *ANALYSIS/SETUP/OPTIONS*). The time for the transition between the limits should not be too short to avoid numerical problems. Time is not directly set but is determined by the rate of change of the control signal.

(K) *Sources*: There is an array of both voltage and current sources available and you should remember that they are perfect devices, i.e. voltage sources have zero internal impedance and current sources have infinite impedance. Thus, for example, a voltage source of zero voltage could be used as a current monitor, or if it generates only a short transient then at other times it is in effect not present. Similarly, a current source may be used to provide a short prod to start an oscillator but after that it also is not present as it has infinite impedance. For more complex or mathematically defined signals you may use one of the *ABM* devices or for graphically or program-controlled sources there is the *STIMULUS EDITOR*. For the latter remember to enter the name and location of the stimulus file in *ANALYSIS/LIBRARY AND INCLUDE FILES*.

(L) *Stimulus generation*: See sources (K).

(M) *Signal generating circuits; multiplying and summing*: Two parts that are useful for generating more complex signals are the *MULT* and *SUM*. Each has two inputs and carries out the indicated process on these. Together, or with others, they can be used to construct or process many types of signal. These are control system parts (there are many more) and can be connected together with no need for dummy load or input resistors.

(N) *Libraries*: PSpice comes with an extensive set of libraries but if you are only using the educational version then the libraries are severely limited. Device manufacturers provide libraries for their own products and you may be able to find what you need from such sources, but the reduced version of PSpice allows only a limited number of libraries to be loaded and a limited number of components in any library. It is then necessary to create your own User Library into which you can put models for those particular devices you need. Libraries are straightforward text files and are reasonably easy to read and to see the extent of a given model. An asterisk indicates a comment line (which is disregarded) and the model ends with a *.END* and usually a *.$*. Cutting and pasting is direct, though PSpice may object when you start as there is an index file (*nom.lib*) which tells PSpice where to find a model, and if the libraries have been altered it will have to make a new index. If the changes are to your User Library then PSpice will only have to make a new index file for this and not for the whole library set which could take a long time. If you do have an additional library file remember again to make this evident under *ANALYSIS/ LIBRARY AND INCLUDE FILES*.

(O) *Significance of parameters in models, e.g. FETs* (Vladimirescu 1994, p. 384) *and BJTs* (Vladimirescu p. 87): The meaning of symbols used for the various parameters in a device model are not usually obvious. It is of course rather dangerous to alter any of these unless you do know what you are doing, but it is sometimes useful to know the value of some parameter being used for a particular device. For example, you may wish to know the junction capacity (*CJO*) or the reverse recovery time for a diode (*TT* or transition time is specified from which $t_{rr}$ may be obtained: see Hambley 1994, p.673), or the breakdown voltage of a Zener diode. Tables of the parameters are given in many books on SPICE, e.g. Vladimirescu (1994), Hambley (1994), Schubert and Kim (1996).

(P) *Indexes*: The compilation of an index for books and manuals is a laborious and unexciting task and many do not seem to have been done carefully or by a considerate mind (there is my hostage to fortune, but I hope you will let me know of my omissions). Just because a word does not appear in the index does not necessarily mean that there is no treatment of that topic so rootle around using a bit of lateral thinking in case it is referred to by another name. For example, the PSpice *Circuit Analysis User's Guide* does not have an entry for phase or for transfer function, both of which do appear in the text.

(Q) *Magnitudes*: There is one magnitude multiplier that can cause an error: remember that both m and M refer to milli ($\times 10^{-3}$) and that $\times 10^{6}$ is indicated by *meg* or *MEG.* It may be easier to stick with k ($\times 10^{3}$) to make up meg(ohms say) as in 1000 k. However, *PROBE* uses m as milli and M as Meg!

(R) There is unfortunately no provision of a parking lot in which to place temporarily unrequired components and where they would be ignored. All pins on any such items must be connected somewhere otherwise PSpice will object, so strap them all to common. This may cause trouble if you do this with an amplifier for example.

(S) The version of PSpice provided with this book is the demonstration (or student) version and is in effect supplied free by Cadence. For evident commercial reasons it is therefore restricted in the complexity of circuit which is permitted. The limitation imposed is a maximum of 64 nodes in the circuit and this includes those internal to models e.g. an op amp. There is also a limit on the number of libraries that may be open at any one time, and in the number of models in a library. Many additional models may be obtained via the Internet to add to those supplied. Though these are supplied free by device manufacturers it should by understood that they are copyright and must not be reproduced for commercial gain without permission. If you try running a simulation and receive an error message to say that there are too many nodes, it may be possible to avoid this by using models with fewer internal nodes, e.g. try replacing the LM6142 with the LF411, though the bandwidth will be reduced.

## References and additional sources 5.27

During the writing of this book MicroSim were taken over by OrCAD and they in turn by Cadence. Any reference to any of these companies should be taken as also referring to the others, particularly in terms of access to information. Presently, the websites are:

| | |
|---|---|
| www.OrCAD.com | www.OrCAD.co.uk |
| www.MicroSim.com | www.PSpice.com |
| www.Cadence.com | |

Agnew J. (1991): Simulating audio transducers with SPICE. *Electronic Design* 7 November, 45, 46, 48, 54, 56, 59.

Albean D. L. (2000): Simulate a neon lamp in SPICE by using a hysteretic resistor. *Electronic Design* 10 January, 136, 138.

Alexander M., Bowers D. F. (1990): *SPICE-Compatible Op Amp Macromodels*, Precision Monolithics Application Note AN-138. See also *EDN* 15 February, and 1 March, 1990.

Al-Hashimi B. (1995): *The Art of Simulation Using PSPICE: Analog and Digital*, Boca Raton: CRC Press.

Antognetti P., Massobrio G. (1993): *Semiconductor Device Modeling with SPICE*, 2nd Edn, New York: McGraw-Hill. ISBN 0-07-002107-4.

Baker B. (1993): *Operational Amplifier Macromodels: A Comparison*, Burr-Brown Application Bulletin AB-046.

Banzhaf W. (1989): *Computer-Aided Circuit Analysis Using SPICE*, Englewood Cliffs: Prentice Hall. ISBN 0-13-162579-9.

Biagi H., Stitt R. M., Baker B., Baier S. (1995): *Burr-Brown Spice Based Macromodels, Rev. F*, Burr-Brown Application Bulletin AB-020F, January.

Boorom K. (1998): *Modeling of a High-Speed Digital Bus with OrCAD PSpice*, OrCAD Technical Note No. 80.

Chadderton N. (1996): *Zetex SPICE Models*, Zetex Application Note 23, March. www.zetex.com/spice1

Christensen T. (1999): *Analog Behavioral Modeling*, Cadence/MicroSim Application Note PSPA025. May.

Coelho J. (1999): A SPICE model for the ideal transformer. *Electronic Design* 28 June, 85.

Com Pro Hard- & Software Vertreibs Gmbh. (2000): *EDALIB Spice Models*, www.edalib.com (40 000 models covering many categories of device from many manufacturers and suitable for most SPICE simulators. A new version with 50 k models is imminent.)

Defalco J. A. (1993): Put a damper on ground bounce. *Electronic Design* 5 August, 71, 72, 74, 75.

Edelmon D. (1994): Simulate laser-trimmed resistors with SPICE. *Electronic Design* 2 May, 105, 106, 108, 109.

Foty D. (1997): *MOSFET Modeling with SPICE. Principles and Practice*, Englewood Cliffs: Prentice Hall. ISBN 0-13-227935-5.

Goody R. W. (1995): *PSPICE for Windows*, Englewood Cliffs: Prentice Hall. ISBN 0-02-345022-3.

Gottling J. G. (1995): *Hands on PSPICE*, New York: John Wiley. ISBN 0-471-12488-5.

Hageman S. C. (1995): PSpice models nickel-metal-hydride cells. *EDN* 2 February, 99.

Hageman S. C. (1996): *Create Analog Random Noise Generators for PSpice Simulation*, MicroSim Application Notes Ver. 6.3, p. 61, MicroSim Corp., April.

Hambley A. R. (1994): *Electronics. A Top-Down Approach to Computer-Aided Circuit Design*, New York: Macmillan. ISBN 0-02-349335-6.

Henn C. (1993): *Macromodels for RF Op Amps are a Powerful Design Tool*, Burr-Brown Application Bulletin AN-189, 1993.

Hirasuna B. (1999a): *Use Constrained Optimization to Improve Circuit Performance*, Cadence/MicroSim Application Note PSPA027, May.

Hirasuna B. (1999b): *Analog Behavioral Modeling using PSpice*, Cadence/MicroSim Application Note PSPA022, April.

Hirasuna B. (1999c): *Using Coupled Inductors and Inductor Cores*, Cadence/MicroSim Application Note PSPA021. April.

Holzer M., Zapsky W. (1999): *Modeling Varistors with PSpice: Simulation Beats Trial and Error*, Siemens + Matsushita (Epcos) Components Application Note, www.epcos.com/pr/inf/70/so/e0000000.htm

Intusoft: www.intusoft.com makes available a range of SPICE models and access to their Newsletters containing useful simulation examples and tips.

Jung W. (1990): *Questions and Answers on the SPICE Macromodel Library*, Linear Technology Application Note 41. *Linear Technology Applications Handbook*, Vol. II, 1993.

Jung W. (1991): *Using the LTC Op Amp Macromodels*, Linear Technology Application Note 48. *Linear Technology Applications Handbook*, Vol. II, 1993.

Jung W. (1998a): Add some SPICE to your analog designs. *Electronic Design*, 9 February, **46** (3), 147–148.

Jung W. (1998b): SPICE programs: computerized circuit analysis for analog EEs. *Electronic Design*, 1 May, **46** (10), 130.

Karki J. (1999a): *THS3001 SPICE Model Performance*, Texas Instruments Application Report SLOA038, October 1999.

Karki J. (1999b): *Building a Simple SPICE Model for the THS3001*, Texas Instruments Application Report SLOA018, April.

Kavalov V. (1993): 'Fooling' SPICE to handle ideal circuits. *Electronic Design* 10 June, 68, 71–72.

Kielkowski R. (1993): *Inside SPICE – Overcoming the Obstacles of Circuit Simulation*, New York: McGraw-Hill. ISBN 0-07-911525-X.

Kraft L. A. (1991): Modelling lightning performance of transmission systems using PSpice. *IEEE Trans.* **PWRS-6**, 543–549.

Meares, L. G., Hymowitz C. E.(1988): *Simulating With SPICE*, Intusoft.

Meares, L. G., Hymowitz C. E. (1990): *SPICE Applications Handbook*, Intusoft.

Muller, K. H. (1990): *A SPICE Cookbook*, Intusoft.

Nance P., März M. (2000): *Thermal Modelling of Power Electronic Systems*, PCIM Europe-Power Electronic Systems 2/2000, 20, 22, 26, 27.

National Semiconductor (2000): Webench™ Online Power Supply Simulation Tools; http://power.national.com (free access).

National Semiconductor (2000): *Simulation SPICE Models for Comlinear's Op Amps*, National Semiconductor Application Note OA-18, May.

Nehmadi M., Ifrah Y., Druckmann I. (1990): Experimental result study and design enhancement of a magnetic pulse compression circuit by using the PSPICE simulation program. *Rev. Sci. Instrum.* **61**, 3807–3811.

O'Hara M. (1998): Modelling board-level DC–DC converters in SPICE. *Electronic Product Design* July, 25, 26, 28.

Oxner E. (1991): Parameter extraction and estimation produce accurate JFET models. *EDN* 19 August, 137–142, 144.

Pease R. A. (1991): *Troubleshooting Analog Circuits*, London: Butterworth-Heinemann. ISBN 0-7506-9499-8.

Prigozy S. (1994): Simplified PSPICE models for power switching devices. *Computers in Education J.* **IV** (3), July–September 42–50.

Rashid M. H. (1990): *SPICE for Circuits and Electronics Using Pspice*, Englewood Cliffs: Prentice-Hall.

Sandler S. M. (1997): *SMPS Simulation with SPICE 3*, New York: McGraw Hill. ISBN 0-07-913227-8.

Schmid R., Blake K. (1994): *Simulation SPICE Models for Comlinear's Op Amps*, Comlinear Application Note OA-18, July.

Schubert T., Kim E. (1996): *Active and Non-linear Electronics*, New York: John Wiley. ISBN 0-471-57942-4.

Tuinenga P. W. (1988): *SPICE: A Guide to Circuit Simulation and Analysis Using PSPICE*, Englewood Cliffs: Prentice Hall. ISBN 0-13-834607-0. (3rd Edn, 1995, ISBN 0-13-158775-7.)

Vladimirescu A. (1994): *The SPICE Book*, New York: John Wiley. ISBN 0-471-60926-9.

Watson D. J. (1997): Voltage controlled attenuator in PSpice. *Electronic Engineering* April, 28.
www.ednmag.com/reg/download/SPICE.asp An EDN magazine site for Spice packages, models and application information. See:
MS735.TXT: Solving Convergence Problems by Hymowitz C.E. (Intusoft).
MS697Z.ZIP: Simulation of common battery types by Hageman S. C.
MS436Z.ZIP: Battery types and chemistry, bibliography of design ideas on chargers, etc. by Hageman S. C.

Some sources of information and Spice models.

| | |
|---|---|
| www.analog.com | Analog Devices |
| www.apexmicrotech.com | |
| www.avxcorp.com | Capacitors, inductors |
| www.beigebag.com | Beige Bag Software for $B^2$ Spice A/D 2000 |
| www.burr-brown.com | Now part of Texas Instruments |
| www.cadence-europe.com | |
| www.coilcraft.com | Inductors |
| www.edalib.com | Com Pro Hard- & Software Vertreibs Gmbh. (2000): EDALIB Spice Models (40,000 models covering many categories of device from many manufacturers and suitable for most SPICE simulators. A new version with 50 k models is imminent they say. Some are downloadable, the bulk are for purchase.) |
| www.elantec.com | |
| www.electronicsworkbench.com and www.ewbeurope.com | |
| www.epcos.com | Select 'Tools for developers' for useful simulation software and models; capacitors and thermistors |
| www.fairchildsemi.com | models/discretes/ . . . Models are unfortunately individual rather than combined in one file. Also http://ensignia.fairchildsemi.com Interface design resource simulator. |
| www.intusoft.com | Listing of Sites with Models for many manufacturers as well as other information (www.softsim.com in Europe) |
| www.irf.com | International Rectifier |
| www.johanson-caps.com | Capacitors |
| www.kemet.com | Capacitors |
| www.linear-tech.com | Linear Technology. See also downloadable SwitcherCAD, FilterCAD and Noise simulators. |
| www.national.com | National Semiconductor. See also WEBENCH free on-line simulator at http://power.national.com for WebSIM, WebTHERM and EasyPLL. |
| www.maxim-ic.com | |
| www.micrel.com | |
| www.motorola.com/rf/models | |
| www.onsemi.com | (Ex Motorola) |
| www.penzar.com | Penzar Development for TopSPICE |

| | |
|---|---|
| www.polyfet.com | |
| www.semiconductors.philips.com | |
| www.semi.harris.com | |
| www.simplorer.de | Simec GmbH |
| www.spectrum-soft.com | |
| www.telcom-semi.com | |
| www.ti.com | Texas Instruments. Now includes Burr-Brown |
| www.zetex.com | |

There is no difficulty in beginning; the trouble is to leave off.

Henry James

# Bibliography

I wanted all my books to be buried in perpetual oblivion, that thus there might be room for better books.

Martin Luther 1545

Abramowitz M., Stegun I. A. (Eds) (1970): *Handbook of Mathematical Functions with Formulas, Graphs and Mathematical Tables*, Washington: National Bureau of Standards, Applied Mathematics Series.

Al-Hashimi B. (1995): *The Art of Simulation Using PSPICE: Analog and Digital*, Boca Raton: CRC Press.

Anderson B. D. O., Moore J. B. (1979): *Optimal Filtering*, Englewood Cliffs: Prentice Hall.

Banes J. R. (1987): *Electronic System Design: Interference and Noise-Control Techniques*, Englewood Cliffs: Prentice-Hall.

Bell D. A. (1960): *Electrical Noise*, New York: Van Nostrand. Library of Congress Cat. No. 59-11055.

Bennett S. (1993): *A History of Control Engineering 1930–1955*, London: Peter Peregrinus and IEE. ISBN 0-86341-299-8.

Bennett W. R. (1960): *Electrical Noise*, New York: McGraw-Hill.

Best R. F. (1984): *Phase-Locked Loops: Theory, Design and Applications*, New York: McGraw-Hill.

Bleaney B. I., Bleaney B. (1957): *Electricity and Magnetism*, Oxford: Oxford University Press. Also 1963.

Boas M. L. (1966): *Mathematical Methods in the Physical Sciences*, New York: John Wiley. Library of Congress Cat. No. 66-17646.

Bode H. W. (1947): *Network Analysis and Feedback Amplifier Design*, New York: Van Nostrand.

Brumgnach E. (1995): *PSpice for Windows*: a Primer, Albany, NY: Delmar Publishers. ISBN 0-8273-6821-6. Library of Congress Cat. No. 94003102.

Buchwald J. Z. (1988): *From Maxwell to Microphysics*, Chicago: University of Chicago Press. ISBN 0-226-07883-3.

Burke H. E. (1986): *Handbook of Magnetic Phenomena*, New York: Van Nostrand Reinhold. ISBN 0-442-21184-8.

Carter G. W., Richardson A. (1972): *Techniques of Circuit Analysis*, Cambridge: Cambridge University Press.

Champeney D. C. (1987): *A Handbook of Fourier Theorems*, Cambridge: Cambridge University Press. ISBN 0-21-26503-7.

Chen Wai-Kai (Ed.) (1995): *The Circuits and Filters Handbook*, Boca Raton: CRC Press and IEEE Press. ISBN 0-8493-8341-2.

Corson D. R., Lorrain P. (1962): *Introduction to Electromagnetic Fields and Waves*, San Francisco: W. H. Freeman. Library of Congress Cat. No. 62-14193.

Courant R. (1937): *Differential and Integral Calculus*, 2nd Edn, Vols 1 and 2, London: Blackie and Son.

de Sa A. (1990): *Principles of Electronic Instrumentation*, 2nd Edn, London: Edward Arnold. ISBN 0-7131-3635-9.

Dirac P. A. M. (1930): *The Principles of Quantum Mechanics*, Oxford: Oxford University Press.

Dobbs E. R. (1993): *Basic Electromagnetism*, London: Chapman and Hall. ISBN 0-412-55570-0.

Dorf R. C., Svoboda J. A. (1996): *Introduction to Electric Circuits*, New York: John Wiley. ISBN 0-471-12702-7.

Dostal J. (1993): *Operational Amplifiers*, 2nd Edn, London: Butterworth-Heinemann.

Dummer G. W. (1978): *Electronic Inventions and Discoveries*, 2nd Edn, Oxford: Pergamon.

Egan W. F. (1981): *Frequency Synthesis by Phase Lock*, New York: John Wiley.

Eisberg R. M. (1961): *Fundamentals of Modern Physics*, New York: John Wiley. ISBN 0-471-23463-X.

Fano R. M., Chu L. J., Adler R. B. (1960): *Electromagnetic Fields, Energy, and Forces*, New York: John Wiley.

Feucht D. L. (1990): *Handbook of Analog Circuit Design*, New York: Academic Press. ISBN 0-12-2542240-1.

Feynman R. P. (1998): *The Meaning of It All*, London: Penguin Press. ISBN 0-713-99251-4.

Feynman R. P., Leighton R. B., Sands M. (1964): *The Feynman Lectures on Physics*, Vols I, II, III, Reading Mass: Addison-Wesley. Library of Congress Cat. No. 63-20717.

Fink D., Christansen D. (1989): *Electronics Engineer's Handbook*, 3rd Edn, New York: McGraw-Hill. ISBN 0-07-020982-0.

Fourier J. B. J. (1822): *Theorie Analytique de la Chaleur*, (English version) New York: Dover.

Franco S. (1988): *Design with Operational Amplifiers and Analog Integrated Circuits*, New York: McGraw-Hill.

Fredericksen T. M. (1984): *Intuitive Op Amps*, National Semiconductor.

Garbuny M. (1965): *Optical Physics*, New York: Academic Press. Library of Congress Cat. No. 65-19999.

Gardner F. M. (1979): *Phaselock Techniques*, 2nd Edn, New York: John Wiley.

Gardner J. W. (1994): *Microsensors*, New York: John Wiley. ISBN 0-471-94136-0.

Gibson W. M. (1969): *Basic Electricity*, Harmondsworth: Penguin Books.

Gillispie C. C. (1997): *Pierre-Simon Laplace, 1749-1827: A Life in Exact Science*, Princeton University Press. ISBN 0-691-01185-0.

Good R. H. (1999): *Classical Electromagnetism*, Fort Worth: Saunders College Publishing. ISBN 0-03-022353-9.

Goody R. W. (1995): *PSPICE for Windows*, Englewood Cliffs: Prentice Hall. ISBN 0-02-345022-3.

Gottling J. G. (1995): *Hands on PSPICE*, New York: John Wiley. ISBN 0-471-12488-5.

Graeme. J. (1995): *Photodiode Amplifiers: Op Amp Solutions*, New York: McGraw-Hill. ISBN 0-07-024247-X.

Graeme J. (1997): *Optimizing Op Amp Performance*, New York: McGraw-Hill. ISBN 0-07-024522-3.

Grant I., Philips W. R. (1975): *Electromagnetism*, London: John Wiley. ISBN 0-471-32246-6. (2nd Edn, 1990, ISBN 0-471-92712-0.)

Gray P. R., Meyer R. G. (1977): *Analysis and Design of Analog Integrated Circuits*, John Wiley. ISBN 0-471-01367-6. (3rd Edn, 1993, ISBN 0-471-57495-3.)

Grivet P. (1970): *The Physics of Transmission Lines at High and Very High Frequencies*, Vols 1 and 2, New York: Academic Press. ISBN 0-12-303601-1.

Grover F. W. (1946): *Inductance Calculations*, New York: Van Nostrand.

Gupta M. S. (1977): *Electrical Noise: Fundamentals and Sources*, New York: IEEE Press. Reprints of important papers on noise.

Hambley A. R. (1994): *Electronics. A Top-Down Approach to Computer-Aided Circuit Design*, New York: Macmillan. ISBN 0-02-349335-6.

Hamilton T. D. S. (1977) *Handbook of Linear Integrated Electronics for Research*, London: McGraw-Hill. ISBN 0-07-084483-6.

Haus H. A., Melcher T. R. (1989) *Electromagnetic Fields and Energy*, Englewood Cliffs: Prentice-Hall. ISBN 0-13-249277-6.

Heald M. A., Marion J. B. (1995): *Classical Electromagnetic Radiation*, Fort Worth: Saunders College Publishing. ISBN 0-03-097277-9.

Heaviside O. (1922): *Electromagnetic Theory*, Vol. I and II. London: Benn Brothers.

Heaviside O. (1950) *Electromagnetic Theory*, New York: Dover Publications.

Hofman J. R. (1995): *André-Marie Ampère*, Cambridge: Cambridge University Press. ISBN 0-521-56220-1.

Holbrook J. G. (1966): *Laplace Transforms for the Electronic Engineer*, 2nd Edn, Oxford: Pergamon Press. Library of Congress Cat. No. 59-12607.

Horowitz P., Hill W. (1989): *The Art of Electronics*, 2nd Edn, Cambridge: Cambridge University Press. ISBN 0-521-37095-7.

Hunt B. (1991): *The Maxwellians*, Ithaca, NY: Cornell University Press.

James G., Burley D., Clements D., Dyke P., Searl J., Wright J. (1996): *Modern Engineering Mathematics*, 2nd Edn, Wokingham: Addison-Wesley. ISBN 0-201-87761-9.

Jiles D. (1991): *Introduction to Magnetism and Magnetic Materials*, London: Chapman and Hall. ISBN 0-412-38630-2.

Johnson D. E., Johnson J. R., Hilburn J. L. (1992): *Electric Circuit Analysis*, Englewood Cliffs: Prentice Hall. ISBN 0-13-249509-0.

Jones B. K. (1986): *Electronics for Experimentation and Research*, Englewood Cliffs: Prentice-Hall. ISBN 0-13-250747-1.

Jordan D. W., Smith P. (1999): *Nonlinear Ordinary Differential Equations*, 3rd Edn, Oxford: Oxford University Press. ISBN 0-19-856562-3.

Kielkowski R. (1994): *Inside Spice – Overcoming the Obstacles of Circuit Simulation*, New York: McGraw-Hill. ISBN 0-07-911525-X.

Kittel C. (1958, 1961): *Elementary Statistical Physics*, New York: John Wiley. Library of Congress Cat. No. 58-12495.

Klein M. V. (1970): *Optics*, New York: John Wiley. ISBN 0-471-49080-6.

Korn G. A., Korn T. M. (1961): *Mathematical Handbook for Scientists and Engineers*, New York: McGraw-Hill. Library of Congress Cat. No. 59-14456.

Lambourne R., Tinker M. (2000): *Basic Mathematics for the Physical Sciences*, New York: John Wiley. ISBN 0-471-185207-4.

Lamey R. (1995): *The Illustrated Guide to PSpice for Windows*, Albany, NY: Delmar Publishers. ISBN 0-827-37068-7.

Langford-Smith F. (1954): *Radio Designer's Handbook*, London: Illife and Sons.

Laplace P. S. (1779): Sur les suites, oeuvres complètes. *Mémoires de l'académie des sciences*, **10**, 1–89.

Levine W. S. (Ed.) (1995): *The Control Handbook*, Boca Raton: CRC Press and IEE Press. ISBN 0-8493-8570-9.

Lighthill M. J. (1955): *Fourier Analysis and Generalized Functions*, Cambridge: Cambridge University Press.

Lipson S. G., Lipson H. (1969): *Optical Physics*, Cambridge: Cambridge University Press. ISBN 0-521-06926-2.

Loudon R. (1973): *The Quantum Theory of Light*, Oxford: Clarendon Press. ISBN 0-19-851130-2.

Lynn P. A. (1986): *Electronic Signals and Systems*, London: Macmillan. ISBN 0-333-39164-0.

McCollum P. A., Brown B. F. (1965): *Laplace Tables and Theorems*, New York: Holt, Reinhart, Winston.

Maxwell J. C. (1891): *A Treatise on Electricity and Magnetism*, New York: Dover Publications 1954. A copy of the 3rd Edition of 1891 in two volumes. ISBN 0-486-60637-6 and 0-486-60636-8.

Millman J., Taub H. (1965): *Pulse, Digital and Switching Waveforms*, New York: McGraw-Hill. Library of Congress Cat. No. 64-66293.

Morrison R. (1977): *Grounding and Shielding Techniques in Instrumentation*, 2nd Edn, New York: John Wiley.

Motchenbacher C. D., Connelly J. A. (1993): *Low Noise Electronic System Design*, New York: John Wiley. ISBN 0-471-57742-1.

Motchenbacher C. D., Fitchen F. C. (1973): *Low-Noise Electronic Design*, New York: John Wiley.

Nahin P. J. (1987): *Oliver Heaviside: Sage in Solitude*, New York: IEEE Press #PC02279.

Nixon F. E. (1965): *Handbook of Laplace Transforms*, 2nd Edn, Englewood Cliffs: Prentice Hall. Library of Congress Cat. No. 65-14937.

Oatley C. (1977): *Electric and Magnetic Fields*, Cambridge: Cambridge University Press. ISBN 0-521-29076-7.

O'Dell T. H. (1988): *Electronic Circuit Design*, Cambridge: Cambridge University Press. ISBN 0-521-35858-2.

O'Dell T. H. (1991): *Circuits for Electronic Instrumentation*, Cambridge: Cambridge University Press. ISBN 0-521-40428-2.

O'Hara J. G., Pricha W. (1987): *Hertz and the Maxwellians*, London: Peter Peregrinus. ISBN 0-86341-101-0.

Ott H. (1988): *Noise Reduction Techniques in Electronic Systems*, 2nd Edn, New York: John Wiley. ISBN 0-471-85068-3.

Pain H. J. (1976): *The Physics of Vibrations and Waves*, 2nd Edn, New York: John Wiley. ISBN 0-471-99408-1.

Pallás-Areny R., Webster J. G. (1999): *Analog Signal Processing*, New York: John Wiley. ISBN 0-471-12528-8.

Panofsky W. K. H., Phillips M. (1962): *Classical Electricity and Magnetism*, 2nd Edn, Reading, Mass: Addison-Wesley. Library of Congress Cat. No. 61-10973.

Pease R. A. (1991): *Troubleshooting Analog Circuits*, London: Butterworth-Heinemann. ISBN 0-7506-9499-8.

Peyton A., Walsh V. (1993): *Analog Electronics with Op-Amps*, Cambridge: Cambridge University Press. ISBN 0-521-33604-X.

Pipes L. A. (1958): *Applied Mathematics for Engineers and Physicists*, New York: McGraw-Hill. Library of Congress Cat. No. 57-9434.

Pippard A. B. (1985): *Response and Stability*, Cambridge: Cambridge University Press. ISBN 0-521-31994-3.

Ramo S., Whinnery J. R. (1944): *Fields and Waves in Modern Radio*, New York: John Wiley. (2nd Edn, 1953.)

Ramo S., Whinnery J. R., van Duzer T. (1965): *Fields and Waves in Communication Electronics*, New York: John Wiley.

Rashid M. H. (1990): *SPICE for Circuits and Electronics Using Pspice*, Englewood Cliffs: Prentice-Hall.

Roberge J. K. (1975): *Operational Amplifiers: Theory and Practice*, New York: John Wiley. ISBN 0-471-72585-4.

Robinson F. N. H. (1962): *Noise in Electrical Circuits*, Oxford: Oxford University Press.

Robinson F. N. H. (1974): *Noise and Fluctuations in Electronic Devices and Circuits*, Oxford: Oxford University Press. ISBN 0-19-859319-8.

Sandler S. (1997): *SMPS Simulation with SPICE 3*, New York: McGraw-Hill. Library of Congress Cat. No. 96037203.

Savant C. J. (1962): *Fundamentals of the Laplace Transformation*, New York: McGraw-Hill.

Schubert T., Kim E. (1996): *Active and Non-linear Electronics*, New York: John Wiley. ISBN 0-471-57942-4.

Schwartz L. (1950, 1951): *Theorie des Distributions*, Vols I and II, Paris: Hermann et Cie.

Senior T. B. A. (1986): *Mathematical Methods in Electrical Engineering*, Cambridge: Cambridge University Press. ISBN 0-521-30661-1.

Siebert W. McC. (1986): *Circuits, Signals, and Systems*, Cambridge, Mass: MIT Press/New York: McGraw-Hill. ISBN 0-07-057290-9.

Simpson T. K. (1997): *Maxwell on the Electromagnetic Field*, New Brunswick: Rutgers University Press. ISBN 0-8135-2363-1.

Singh J. (1999): *Modern Physics for Engineers*, New York: John Wiley. ISBN 0-471-33044-1.

Slichter C. P. (1964): *Principles of Magnetic Resonance*, New York: Harper and Row.

Smith J. I. (1971): *Modern Operational Circuit Design*, New York: John Wiley. ISBN 0-471-80194-1.

Snelling E. C. (1969): *Soft Ferrites. Properties and Applications*, London: Iliffe Books. (2nd edition, Butterworths, 1988.)

Solymar L. (1976): Lectures on Electromagnetic Theory, London: Oxford University Press. ISBN 0-19-856126-1.

Spiegel M. R. (1965): *Schaum's Outline Series, Theory and Problems of Laplace Transforms*, New York: McGraw-Hill.

Springford M. (1997): *Electron: A Centenary Volume*, Cambridge: Cambridge University Press. ISBN 0-521-56103-2.

Standage T. (1998): *The Victorian Internet*, London: Walker Publishing. ISBN 08027-13424.

Stigler S. (1999): *Statistics on the Table: the History of Statistical Concepts and Methods*, Cambridge, Mass: Harvard University Press. ISBN 0-674-83601-4.

Stout D. F., Kaufman M. (1976): *Handbook of Operational Amplifier Circuit Design*, New York: McGraw-Hill. ISBN 0-07-061797-X.

Stuart R. D. (1961): *An Introduction to Fourier Analysis*, London: Chapman and Hall, Science Paperback SP21. No ISBN.

Susskind C. (1995): *Heinrich Hertz: A Short Life*, San Francisco: San Francisco Press. ISBN 0-911302-74-3.

Terman F. E. (1950): *Radio Engineers' Handbook*, New York: McGraw-Hill.

Thomas R. E., Rosa A. J. (1994): *The Analysis and Design of Linear Circuits*, Englewood Cliffs: Prentice Hall. ISBN 0-13-147125-2.

Tinker M, Lambourne R. (2000): *Further Mathematics for the Physical Sciences*, New York: John Wiley. ISBN 0-471-86273-3.

Tobey G. E., Graeme J. G., Huelsman L. P. (1971): *Operational Amplifiers. Design and Applications*, New York: McGraw-Hill. Library of Congress Cat. No. 74-163297.

Tricker R. A. R. (1966): *The Contributions of Faraday and Maxwell to Electrical Science*, Oxford: Pergamon Press. Library of Congress Cat. No. 66-23859.

Tuinenga P. W. (1988): *SPICE: A Guide to Circuit Simulation and Analysis Using PSPICE*, Englewood Cliffs: Prentice Hall. ISBN 0-13-834607-0. (3rd Edn, 1995, ISBN 0-13-158775-7.)

Van der Ziel A. (1954): *Noise*, Englewood Cliffs: Prentice-Hall.

Van Valkenburg M. E. (1960): *Introduction to Modern Network Synthesis*, New York: John Wiley.

Van Valkenburg M. E. (1982): *Analog Filter Design*, New York: Holt, Rinehart and Winston. ISBN 0-03-059246-1, or 4-8338-0091-3 International Edn.

Vladimirescu A. (1994): *The Spice Book*, New York: John Wiley. ISBN 0-471-60926-9.

Walton A. K. (1987): *Network Analysis and Practice*, Cambridge: Cambridge University Press. ISBN 0-521-31903-X.

Webster J. G. (Ed.) (1997): *Medical Instrumentation Application and Design*, 3rd Edn, New York: John Wiley. ISBN 0-471-15368-0.

Williams J. (Ed.) (1995): *Analog Circuit Design: Art, Science, and Personalities*, London: Butterworth-Heinemann. ISBN 0-7506-9640-0.

Williams J. M. (Ed.) (1997): *Linear Applications Handbook, Vol. III, Linear Technology* (also previous volumes in 1987 and 1990).

Williams J. (Ed.) (1998): *The Art and Science of Analog Circuit Design*, Boston: Butterworth-Heinemann. ISBN 0-7506-7062-2.

Wilmshurst T. H. (1985): *Signal Recovery from Noise in Electronic Instrumentation*, Bristol: Adam Hilger. ISBN 0-85274-783-7.

# Name index

*Atque inter silvas Academi quarere verum.*
And seek the truth in the groves of Academia.
Horace, 65–8 BC, Epistles II

All references have been listed including those by companies.

# Subject index

All tragedies are finish'd by a death,
All comedies are ended by a marriage.
The future states of both are left to faith.
Byron in Don Juan

Note: Spice topics in particular are collected under the entry PSpice.

# Part index

I am in a sense something intermediate between God and nought.
René Descartes, *Discourse on Method*

Components referred to in the text or schematics.

## Companion CD-ROM

This CD-ROM contains MicroSim evaluation software and circuit examples from the book provided by the author.

Author-provided circuit examples © Scott Hamilton

MicroSim evaluation software © 1997–2002 Microsim Corporation

Published by Cambridge University Press

Manufactured in the United Kingdom

Not for sale separately

This CD-ROM is provided 'as is' and the publisher accepts no liability for loss or damage of any kind resulting from the use of this product.

### MicroSim evaluation software installation instructions

**Windows 95, 98, 2000**

- Insert the MicroSim CD into your CD-ROM drive. Wait for the MicroSim screen to appear and follow the installation instructions.
- If the MicroSim screen does not appear, select Run from the Start Menu and enter *drive:*\SETUP.EXE (where *drive* is your CD-ROM drive). Wait for the MicroSim screen to appear and follow the installation instructions.

**Windows NT**

- Insert the MicroSIM CD into your CD-ROM drive.
- Select Run from the Start Menu of the Program Manager
- Enter *drive:*\SETUP.EXE (where *drive* is your CD-ROM drive.) Wait for the MicroSim screen to appear and follow the installation instructions.

See the Readme.wri file for last-minute information on this release. To find out what's new in this release and how to get started with MicroSim's programs, see the QuickStart tutorial located in the MicroSim program group.

### Circuit examples from the book

The author has provided circuit examples from the book in the Examples folder on the CD-ROM. It is recommended that you copy these to your hard disk before using them.

### System Requirements

**Hardware**

Any 486 or Pentium PC
16 Mb RAM
CD-ROM drive
Floating point copressor
Mouse

**Operating System**

Microsoft® Windows® 95, 98, 2000 or Windows NT™